Accelerated Chemistry

A Mastery-Oriented Curriculum

Second Edition

CENTRIPETAL PRESS
CLASSICAL ACADEMIC PRESS

Camp Hill, Pennsylvania
2020

Accelerated Chemistry

A Mastery-Oriented Curriculum

Second Edition

John D. Mays

CENTRIPETAL PRESS
CLASSICAL ACADEMIC PRESS

Camp Hill, Pennsylvania
2020

CENTRIPETAL PRESS
CLASSICAL ACADEMIC PRESS

Accelerated Chemistry

Edition 2.0

ISBN: 978-0-9972845-6-0

Classical Academic Press
515 S. 32nd Street
Camp Hill, PA 17011
www.ClassicalAcademicPress.com/Novare/

Cover design by Nada Orlic, http://nadaorlic.info/

PGP/IS.03.24

Acknowledgements

I wish to express my deep appreciation to those who reviewed this text during its development: Chris Mack, PhD; Christina Swan, PhD; and Christian J. Corley. Many thanks also to our excellent copy editor Gerald Tilma. Each of these friends is an expert at what he or she does, and their contributions have been invaluable.

Reviewers

Chris Mack	Adjunct Faculty, University of Texas at Austin PhD, University of Texas at Austin, Chemical Engineering MS, Electrical Engineering, University of Maryland BS degrees in Physics, Electrical Engineering, Chemistry, Chemical Engineering, Rose-Hulman Institute of Technology
Christina Swan	Biology/Chemistry Teacher, Regents School of Austin, Austin Texas PhD, Molecular Pathology, University of California at San Diego BS, Biology, Westmont College
Christian J. Corley	Math-Science Department Chair, Regents School of Austin, Austin, Texas MS, Computer Science, Duke University BS, Geophysics, Texas A& M University

CENTRIPETAL PRESS

CLASSICAL ACADEMIC PRESS

Contents

CENTRIPETAL PRESS
CLASSICAL ACADEMIC PRESS

Preface

For Teachers

Thank you for taking the time to begin here at the beginning. This chemistry text is quite different from other high school texts covering similar material. So to ensure that your experience with this text is successful, it is important to be aware of its unique features and the logic behind its structure.

Several important resources are described in the pages that follow. For your convenience, they are all listed here. Please visit our website (centripetalpress.com) or contact us for more information on all these except the last one. The *Illustrated Guide to Home Chemistry Experiments* is not published or distributed by Centripetal Press, but is readily available from booksellers.

- *Centripetal Chemistry Supplement* (John D. Mays, 2016)
- *Principles of Chemistry* (John D. Mays, 2016)
- *Accelerated Studies in Physics and Chemistry* (John D. Mays, 2015)
- *From Wonder to Mastery* (John D. Mays, planned for 2020)
- *Chemistry Experiments for High School* (Christina Swan and John D. Mays, 2014)
- *Chemistry Experiments for High School at Home* (Christina Swan and John D. Mays, 2014)
- *Science for Every Teacher* (John D. Mays, 2013)
- *The Student Lab Report Handbook* (John D. Mays, 2009)
- *Illustrated Guide to Home Chemistry Experiments: All Lab, No Lecture* (Robert Bruce Thompson, 2008)

1. Student Audience and Preparedness

Accelerated Chemistry is specifically designed to suit the needs of accelerated high school students who have already mastered one year of introductory physics and chemistry. Although variations are certainly possible regarding when and with whom this text is used, my assumption is that most students using this text will be in 10th grade and concurrently enrolled in Algebra 2. Further, in preparing the text I have assumed that students using it have a strong aptitude for science and mathematics and have developed solid study habits. For students who are not operating at the advanced level I've described, I recommend our companion text for grade-level students, *Principles of Chemistry*.

Accelerated Chemistry is designed for students to use the year following a course based on our text *Accelerated Studies in Physics and Chemistry* (*ASPC*). The curriculum in *ASPC* is designed to give students a thorough grounding in concepts from physical science that are prerequisite for success in chemistry, including types of substances, physical and chemical properties, atomic theory, the electromagnetic spectrum, potential and kinetic energy, and density. *ASPC* also requires students to gain full proficiency in basic mathematical skills, including metric prefixes, unit conversions, scientific notation, and significant digits (except the addition rule).

Virtually all first-year chemistry texts begin with a couple of chapters devoted to the math topics I just mentioned; science topics are usually developed at length elsewhere in the text. However, in *Accelerated Chemistry* some of these topics are not treated at an introductory level, and some are not addressed at all. This is because students who have completed *ASPC* have already mastered this set of knowledge and skills—and they are ready to put them to immediate use in chemistry. To begin an exciting new course with a mundane review of topics students mastered the previous year is both tedious and boring, as I know from first-hand discussions with many students in this very position. Accordingly, *Accelerated Chemistry* jumps right in without the warm-up chapters that are familiar to chemistry teachers. Naturally, even with students who have completed *ASPC*, there will be occasions when students need review or even remediation, particularly in math skills. For such occasions, a tutorial addressing the mathematical skills is available in Appendix A.

I recognize that there are excellent, high-aptitude students who can benefit from the unique features of this text but who did not have *ASPC* the previous year. For such students, I have prepared a small textual supplement entitled *Centripetal Chemistry Supplement*. A basic outline of the content included in the *Supplement* is shown in the table below. For students who did not study *ASPC* prior to the use of the present text, I strongly recommend starting the year by going through the chapters in the supplement (and the tutorial in Appendix A). The supplement is a durable, reusable manual, available at a price that we have kept as low as possible.

Contents of the *Centripetal Chemistry Supplement*
• Mass and Energy: mass, energy, the law of conservation of energy, forms of energy
• Atoms and Atomic Theory: review of atomic facts, ions, history of atomic models prior to Niels Bohr
• Substances and Their Properties: types of substances, pure substances, mixtures, physical and chemical properties, density

2. Our Emphasis on Wonder, Integration, and Mastery

Wonder

The study of science should always begin with wonder. The world is a stunning place, full of surprises and jaw-dropping phenomena. But today there are many barriers standing in between our students and the world they might otherwise become fascinated with. Safety concerns interfere with kids playing and exploring outdoors. Liability concerns make it hard to find a decent chemistry set. And unfortunately, it is common today for young people grow up spending most of their time indoors with digital media. The natural draw of nature for developing youths is now commonly missed. Many people have never seen the night sky in an area that is completely dark and have no idea of the stunning beauty of the heavens at night. Most kids have not hiked or camped in the forests and have not learned to listen to the sounds made by animals, insects, and trees.

Additionally, the environmental challenges we face today from pollution, resource exploitation, and, especially, climate change require a new generation of people who care about the earth. But people usually do not care about what they do not love, and they do not love what they do not know about. Helping students to know the natural world has never been more important than it is today. Only if they know the world will students begin to love it, and only then will they be motivated to take care of it. To nurture this love, we begin with the natural wonder we feel when we encounter the natural world.

Integration

A second major aspect to the needed paradigm shift is that instruction must be integrative. The habit of compartmentalizing disciplines of learning must be eliminated. This habit currently pervades everything from problem assignments to lesson presentations to test design. Instead of isolating science content from everything else, critical points of effective integration must be developed. Some critical integration points include:

- *frequent use of mathematical skills in science classes, and frequent science applications in math classes*
- *maximizing opportunities to develop good written expression on exams, lab reports, and papers*
- *developing key historical connections that serve to enhance understanding of science as a process; and*
- *treating, in addition to basic skills, the nature of scientific and mathematical knowledge, and the roles these play in leading us toward truth, goodness, and beauty.*

Naturally, for integration to be effective, specific learning objectives must be developed, explained to students, and incorporated into assessments. Centripetal Press texts include clear learning objectives in every chapter.

Mastery

Mastery essentially means proficiency and long-term retention of course content. The first step toward mastery-learning is to change how we define success. The broken default pattern is what I call the *Cram-Pass-Forget cycle*: students cram for tests, pass them, and forget most of what they crammed in just a few weeks. Success in such an environment is a matter of jumping through assessment hoops. Students are not only cheated by this regimen, they are bored with it. And teachers are demoralized by the results.

By contrast, Centripetal Press advocates methods and curriculum designed to promote proficiency and long-term retention using a *Learn-Master-Retain cycle*. This first involves culling the content scope to an amount that truly can be mastered in the course of a school year. Many educators unthinkingly prioritize quantity over quality. But we believe students should be presented with a right amount of material they can learn deeply rather than a bloated scope of content they will neither comprehend nor remember. Even with a reduced scope, students who study for mastery typically outperform their peers as they move to higher level classes.

Second, leading students to mastery and retention requires teaching methods designed to produce these results. The standard approach used today involves teaching a chapter and giving a test on the chapter. By contrast, pedagogy designed for mastery and retention involves continuous review, ongoing accountability for previously studied material, and embedding of basic skills into new material. Of course, an effective method includes innovative strategies to enable students to master course content.

My ideas about all three parts of this core philosophy are described in detail in my book *From Wonder to Mastery*. This book, along with a more detailed description of our textbook philosophy, may be found on our website.

3. Recommendations for Teaching With This Text

In this section, I make a few brief points about teaching for Mastery and Integration with this text, beginning with Mastery.

First, students need continually to be working with previously learned material to keep old skills fresh. Included in the Chapter Exercises at the end of each chapter is a set of General Review Exercises covering material from previous chapters. Students should always work through these exercises.

Second, the teacher's assessment regimen should support the goal of students retaining previously learned material. For students at this level, I recommend the following four-part assessment regimen: (a) Do not award credit in students' grades for homework. (Do, however, require students to complete their assignments and hold them accountable for doing so.) The logic behind this principle is explained in detail in *From Wonder to Mastery* and in our white paper "Putting Homework in its Place" available on our website. (b) Between each chapter test, administer one or two quizzes covering material presented in class during the previous week. This motivates students to stay current with their studies instead cramming the night before the test. (It also gives them direct information about how prepared they are for that part of the chapter test.) Count the average of all the quizzes for a single semester together as equivalent to one chapter test in the semester grade. (c) Use the "Standard Problems List" technique described in *From Wonder to Mastery*. Inform students at the beginning of the year that each chapter test includes problems from the Standard Problems List on material from previously covered chapters. (d) On each chapter test, allocate about 20% of the exam to material on the Standard Problems List from previously covered chapters. I have placed a Standard Problems List for Chapters 1–12 in the Preface for Students. Individual teachers may wish to modify that list to suit what the teacher feels are the most important topics to emphasize.

Third, consider Chapter 13 (Organic) to be optional. There is a lot of content in any chemistry course, and mastery requires distilling of content to the most important essentials. Loading in content for the sake of "coverage" sends students back into the futility of the *Cram–Pass–Forget cycle*. So if your school year is too short, or if your students are working at a pace that will only allow for the first 12 chapters to be included in the course, consider Chapter 13 optional. Additionally, feel free to cut some of the more advanced material from any chapter if the students in a particular class simply can't take it in fast enough. Classes are different year to year, and some groups of students can handle more than others. Teaching according to mastery principles is quite different from conventional *Cram–Pass–Forget* methods, and it is more demanding for teacher and students alike. However, the rewards are huge, which is why I constantly promote mastery methods. But teachers need to administer course content with wisdom.

Now I will make a few points about how the principle of Integration should work with the course.

First, in addition to computations, I always promote questions on tests and quizzes requiring responses written in complete sentences. In their responses, students should be required to demonstrate competence with standard English. I discourage so-called objective items requiring true/false, matching, and multiple-choice responses.

Second, the incorporation of language skills into your course will be enhanced even further if you require students to write their lab reports from scratch. I address this more in the next section.

Third, a healthy epistemology of science should pervade all science courses. The *Cycle of Scientific Enterprise* model briefly described in the Introduction is the place to start. This topic is developed at length in *From Wonder to Mastery* and in the first chapter of *Introductory Principles in Physics*. I strongly encourage all teachers to draw upon these resources and become fluent in the concepts and terminology about the nature of science and scientific knowledge. Then bring the subject up as often as possible in class. Science teachers are very busy, and there are constant pressures threatening to cause courses to lag behind in the curriculum. It is tempting to stick to the technical content and neglect teaching our students what science *is* and how science *works*. Specifically, only a small percentage of our adult population understands what the statement "science is modeling" means, or has any idea of the distinction between scientific claims and truth claims. One of the results of this massive lack of understanding is fuel thrown on the fires of the "science-faith debate," as if one has to choose between one's faith tradition and robust science. I argue that one does not have to make such a choice. A critical component of the science teacher's role is helping students learn to participate in a healthy, responsible dialog concerning what we know about the world from faith traditions and what we know about the world from scientific inquiry.

Finally, as I state in Appendix C, Wolfgang Goethe once wrote that "the history of science *is* science." As I mention in the previous paragraph, science is about modeling nature. Our models—theories—are never perfect and never complete, and as a result they change over time. One of the best ways to help students grasp this is to get into the history of the subject—the scientists with their theories and discoveries. In my texts for younger students, I place learning objectives pertaining to the history prominently alongside those pertaining to the technical content. In this text, I do not do so. The reason is that there are quite a few historical references in this text, and I do not think it reasonable to require students to memorize a paragraph of information pertaining to every one of them. So instead of stipulating historical learning objectives, I leave it up to the individual teacher to decide how much history to include and which scientists to focus on. However, it is very important that some history is included in student learning objectives and assessments, and that the historical circumstances are related to the nature of scientific knowledge and the *Cycle of Scientific Enterprise* wherever possible.

To assist both teacher and student in managing historical content, all the scientists discussed in the text are listed in Appendix C. I know that some teachers and home school families using this text may wish for more prescriptive requirements when it comes to the history. Accordingly, I offer this proposal: have the students write a paper on the history of chemistry that summarizes the contributions to chemistry from at least six or eight scientists. Require that the contributions discussed be tied together in a unified narrative, rather than simply quoted one after the other from material in the text. Alternatively, have the students write papers in which they describe how several of the experiments or other contributions described in this text relate to the *Cycle of Scientific Enterprise* model discussed in the Introduction.

However you choose to go about it, the history of science is important. The historical material should be an important component of your course, not something students simply skip over.

4. Laboratory Work and Lab Reports

A laboratory component is essential for every high school chemistry class. Not only does a lab *practicum* give students direct knowledge and experience that are virtually impossible to obtain from a text, but the report writing component of the lab work provides a rich enhancement to the overall learning objectives for the course. As I state above, development of English language skills should be deeply integrated into our science courses. Requiring students to write lab reports from scratch—rather than by filling out blank spots in a workbook—provides a premier opportunity to do this.

High school science teachers should require that students write full-length lab reports from scratch, on a computer, five or six times per year. *From Wonder to Mastery* includes assessment guidelines and learning objectives for lab reports at different grade levels throughout high school.

There is a lot that goes into writing a quality lab report, and teaching students how to do it without a guide to help is a difficult and time-consuming task. Thus, I commend to you and your students my manual, *The Student Lab Report Handbook*. I recommend that schools distribute copies at the beginning of the school year to each high school freshman. Let each student keep the book for use at home for the next four years. Students should begin learning how to write lab reports in their freshman science class and continue writing reports in all science classes throughout their high school years. Students trained this way astonish their lab instructors when they get to college, and are prepared for college in a way few students are.

Students in high school chemistry courses typically conduct at least 10 or 12 lab exercises during the year. But requiring students to write 10 or 12 lab reports from scratch in a year would be unduly burdensome. This is why I recommend that you require students to write reports from scratch five or six times during the year—three times in the fall and three times in the spring. For other experiments, use a short-form lab report in which students present data and interact with a few key questions.

To accompany this text in schools, Dr. Christina Swan and I produced *Chemistry Experiments for High School*. The experiments in the book are designed for use in a fully-equipped chemistry lab. (An alternative resource for those without a lab is discussed below.) At the end of each experiment in *Chemistry Experiments*, there are a few pages that may be used as a short-form lab report. The book is printed on perforated sheets that can be written on, removed from the book, and submitted. This provision is *not* intended to suggest that all the experiments should be documented this way. But for those that are, the book makes short-form reports easy.

For those that do not have access to a laboratory—and this includes most home-school students—we offer *Chemistry Experiments for High School at Home*. The experiments in this manual are adapted from *Chemistry Experiments for High School*, substituting alcohol burners for Bunsen burners, gravity filtration for vacuum filtration, etc. Some of the experiments do require special chemicals or apparatus—for students to have a legitimate chemistry laboratory experience there is just no way around it. But we have sought to keep these special items to a minimum.

5. Revisions in the Second Edition

In addition to error corrections and minor editorial revisions, the text has been updated to be consistent with the revisions to the SI unit system that went into effect in 2019. The affected sections are Sections 1.1.1, 1.5.4, 1.5.5, 7.2.2, 3.1.1, A.1.2 in Appendix A, and Table B.2 in Appendix B. The changes to the mole and Avogadro's number are the most significant of these, requiring substantive changes in Sections 1.5.4 and 1.5.5. Section 1.5.2 was also revised to include mention of the dalton (Da).

Preface

For Students

You probably don't normally read the Preface to Teachers in your textbooks. (You may not normally read the one to students either, but I am glad you are reading this one.) However, the books I write are quite different from other textbooks you may have used in the past. So in this case, I recommend that you read the Preface to Teachers so you learn about how this book is structured and what I recommend to teachers about how to use it. In particular, it is my view that students should master and *remember* what they are taught, instead of cramming for tests and forgetting everything a few weeks later.

For this to happen, your teacher needs to teach and test in ways you may not be accustomed to. My recommendation to your teacher is that all your chapter exams include problems and questions from prior chapters. Now, that doesn't mean you need to remember every detail from every chapter. But it does mean that there are certain questions and types of problems that are considered very foundational in any chemistry course. I call these Standard Problems. My advice to your teacher is that you use a list of standard problems in your course, and that your exams always include problems from the standard problems list. At the end of this Preface, you will find the Standard Problems List I recommend to your teacher for this chemistry course.

Naturally, your teacher is the person who decides whether to use the Standard Problems List. But if you do use it, then you need to study in a way that enables you to stay current with the material on the list. Simply doing your homework each night and cramming for tests won't cut it. You must have specific study strategies that help you remember definitions and concepts from previous chapters and how to solve previously learned types problems.

Here are my recommendations for how to do that.

1. Study the Objectives List at the beginning of each chapter carefully. Make it your goal to be able to do everything on the list (that is, for the objectives that have been covered so far in class) before your quizzes and tests occur.
2. Look over the Standard Problems List regularly. Identify any item that you cannot do or cannot remember how to do (assuming the topic has already been covered in class) and follow up on it.
3. Develop, maintain, and practice flash cards for each new chapter Objectives List and each item on the Standard Problems List.
4. Read every chapter in this text at least once, preferably twice. Ideally, every time your instructor covers new material you should read the sections in this book corresponding to that material within 24 hours.
5. Go back and read the chapters in this book again when you are a month or two down the road. You will be amazed at how much easier it is to remember things when you have reread a chapter. (Besides, reading is more fun than rehearsing flash cards.)
6. When you are working on exercises involving computations, check your answers against the answers in the back of the book. Every time you get an incorrect answer, dig in and stay with the problem until you identify your mistake and obtain the correct answer. If you can't figure out a problem after 10 or 15 minutes, raise the question in class.
7. Every time you lose significant points on a quiz or test, follow up and fill in the gaps in your learning. If you didn't understand something, raise the question with your instructor. If you

forgot something, rehearse it more thoroughly until you have it down. If you failed to commit something to memory, or didn't have it in your flash cards, then add it to the cards and commit it to memory. If you were not proficient enough at one or more of the computations, look up some similar problems from the exercises or from previous quizzes and tests and practice them thoroughly.

Another important point to make in this preface is in regards to students' mathematical readiness. In preparing this text, I have assumed that students are:

- familiar with the SI System of units and unit prefixes
- proficient at performing unit conversions
- proficient at applying the basic rules for the use of significant digits in measurements and computations (except for the addition rule, addressed in Chapter 1).

If you are lacking skills in any of these areas, please refer to Appendix A for a tutorial. A complete list of constants, prefixes, and coversion factors students are expected to know from memory is found in Appendix D.

Finally, there is a fair bit of historical material scattered around in this text. It is my view that the history of science *matters* and that students should be held accountable for learning and remembering important historical information. In Appendix C you will find a list of all the scientists discussed in the text. This appendix is there to help you as you study. Your instructor will let you know which scientists to know about and what to know about them. I mention it here so that as you study and review you will know to use Appendix C as review a tool in addition to the Standard Problems List.

Standard Problems List

Given appropriate reference materials, students should be prepared to work the following problems on any chapter exam throughout the course, assuming the material has already been covered in class.

FROM CHAPTER 1

1. Describe the Bohr model and explain how it is able to explain emission spectra.
2. Write electron configurations for non-exceptional elements.
3. Determine molar mass or formula mass and perform unit conversions between moles and grams.
4. Determine numbers of atoms in a given quantity of substance.
5. Determine percent composition, determine an empirical formula from the percent composition, and determine the molecular formula from the empirical formula and the molar mass.

FROM CHAPTER 2

6. Describe the general structure and arrangement of the periodic table.
7. Describe general trends in the periodic table pertaining to atomic size, ionization energy, and electronegativity.
8. Identify the names of the major groups of elements and other significant regions in the periodic table.

FROM CHAPTER 3

9. Name binary ionic compounds and acids and write their formulas, including those incorporating polyatomic ions.
10. Describe ionic and covalent bonding.

11. Draw Lewis structures.

FROM CHAPTER 4

12. Apply VSEPR theory to predict molecular shapes and bond angles.
13. Describe the different intermolecular forces, when they occur, and their relative strengths.

FROM CHAPTER 5

14. Balance chemical equations.
15. Perform stoichiometric calculations.
16. Determine oxidations states of pure elements and elements in compounds.
17. Use the activity series of metals to predict whether a single-replacement reaction will occur.

FROM CHAPTER 6

18. Describe the kinetic-molecular theory of gases and the Maxwell-Boltzmann velocity distribution in gases.
19. Explain the causes of surface tension and capillary action.
20. Describe the four basic states of matter.
21. Calculate the heat that must be added to or removed from a substance to change its temperature or effect a phase transition.
22. Explain the causes of evaporation and vapor pressure.

FROM CHAPTER 7

23. Describe Boyle's law, Charles' law, and Avogadro's law.
24. Solve problems using Boyle's law, Charles' law, the ideal gas law, and the density equation.
25. Distinguish between ideal gases and real gases.
26. Use Dalton's law of partial pressures to compute partial pressures, mole fractions, and the total pressure in a gas mixture.
27. Solve stoichiometric problems involving gas volumes.

FROM CHAPTER 8

28. Describe dissolution, electrolytes, and enthalpy of solution.
29. Use solubility guidelines to classify compounds as soluble or insoluble in aqueous solution.
30. Use notions of ionization and polarity to explain the phrase "like dissolves like."
31. State Le Châtelier's principle and Henry's law and relate both to the equilibrium state of a gas dissolved in liquid solution.
32. Calculate molarity and molality.
33. Use solubility guidelines to predict when precipitates will form.
34. Write ionic equations and net ionic equations describing precipitation reactions.
35. Describe four colligative properties of solutions.
36. Calculate boiling points and freezing points of solutions.

FROM CHAPTER 9

37. Describe the three acid-base theories, including definitions for acids and bases in each.
38. Identify conjugate acid-base pairs.
39. Write ionic and net ionic equations for neutralization reactions.
40. Identify the direction of the equilibrium in a Brønsted-Lowry acid-base reaction.

41. Explain the concept of the self-ionization of water.

42. Describe the pH scale.

43. Compute pH, pOH, $[H_3O^+]$, and $[OH^-]$ from each other and from concentration data.

44. Explain the purpose of titration.

45. Determine $[H_3O^+]$ and $[OH^-]$ from titration data.

FROM CHAPTER 10

46. Define enthalpy, enthalpy of combustion, enthalpy of reaction, and enthalpy of formation.

47. State Hess's law.

48. Calculate enthalpy of reaction and enthalpy of formation.

49. Define and calculate the Gibbs free energy, and use it to predict if a reaction occurs spontaneously.

50. Distinguish between enthalpy, entropy, and free energy.

51. Describe four factors that affect the rate of a chemical reaction.

52. Define reaction mechanism, intermediate, activation energy, and activated complex.

53. Determine rate-limiting steps and rate laws.

FROM CHAPTER 11

54. Use Le Châtelier's principle qualitatively to describe how an equilibrium system responds (that is, in which direction the equilibrium shifts) to changes in temperature, pressure, or concentration.

55. Determine equilibrium constants from concentration data.

56. Determine acid dissociation constants from concentration and pH data.

57. Explain how buffering works.

58. Explain the processes of anion and cation hydrolysis.

59. Calculate the solubility product from concentration data.

60. Determine the solubility of a compound from its solubility product.

61. Use solubility product and concentration data to predict if precipitates form when low-concentration ionic solutions are mixed.

FROM CHAPTER 12

62. Identify oxidation states.

63. Write oxidation and reduction half-reactions.

64. Balance redox equations using the half-reaction method.

65. Describe how an electrochemical cell works.

66. Distinguish between voltaic and electrolytic cells.

67. Define anode and cathode.

68. On a diagram of an electrochemical cell, identify the anode, cathode, direction of electron flow, and directions of ion flow in the salt bridge.

Accelerated Chemistry

A Mastery-Oriented Curriculum

Second Edition

CENTRIPETAL PRESS
CLASSICAL ACADEMIC PRESS

Introduction

What is Chemistry All About?

This computer model depicts three of the *orbitals* available in atoms for holding electrons. Shown are one of the three *p* orbitals in each of three different subshells. The inner pair of orbitals can hold two of the highest energy electrons for elements 5 through 10 in the Periodic Table of the Elements. The middle pair is available to hold two of the highest energy electrons for elements 13 through 18 in the periodic table, and the outer pair can hold two of the highest energy electrons belonging to elements 31 through 36 in the periodic table.

In this Introduction, we touch briefly on electron orbitals. We treat the subject in more depth in Chapter 1.

Objectives for the Introduction

After studying this chapter and completing the exercises, you should be able to do each of the following tasks, using supporting terms and principles as necessary.

1. Briefly explain how electrons, electrical forces, minimizing energy, whole number ratios, and modeling can each be thought of as central to understanding what chemistry is all about.
2. State and explain examples illustrating a system moving to a lower energy state and a system experiencing an increase in entropy.
3. Briefly explain hydrogen bonding and why it plays such a large role in mixtures containing water.
4. Explain the relationship between energy and atomic orbitals.

I.1 A Few Major Themes

Chemistry is the study of the elements, how they combine to form mixtures and compounds, the properties of these substances, and the processes involved. One of the astonishing things about the physical world is that as complex as the details are, we can understand a lot about how it works in terms of just a few basic principles from physics. This striking situation is a direct result of the fact that nature is governed by an orderly, mathematical set of physical laws.

Because of the existence of the physical, chemical, and biological laws of nature, nature functions in an orderly, predictable way. These laws have been present since the beginning of the universe, and the more we study them, the more wonderful they seem. Equally wonderful is the fact that humans can *understand* these laws. Of course, we don't understand everything, and there may be things about the way nature works that we will never be able to understand. Nevertheless, what we do understand points to a deep structure in the way nature works.

Studying chemistry involves learning a great deal of terminology, and exploring quite a few different types of processes. The amount of information involved can be daunting! But one way to help organize all this information is to be alert to a few fundamental principles that just keep showing up over and over. In this introductory chapter, we take a brief look at a few of these principles. As you read through the chapters ahead, you will see again and again that we can understand a lot about topics such as molecular structure, solubility, and chemical reactions in terms of a few basic concepts.

I.1.1 Chemistry Is All About Electrons

You recall that atoms consist of a tiny nucleus containing particles called protons and neutrons, and that the nucleus is surrounded by cloud-like regions containing the atoms' electrons. As it turns out, a lot of chemistry can be understood in terms of the atoms' electrons—where they are, how many there are, whether an atom has ionized by gaining or losing electrons, whether an atom is sharing electrons with another atom, and so on.

The cloud-like regions containing an atom's electrons are called *orbitals*, and electrons reside in different orbitals according to how much energy they have. The arrangement of the orbitals is the same for all atoms, although the specific energies associated with each orbital vary from atom to atom, depending on the size of the nucleus and how many electrons an atom has. The orbitals in atoms are grouped into different energy groupings called *shells*. There are seven main shells containing the orbitals with the electrons of all the elements discovered so far. There are additional shells above these that high-energy electrons can move into when they absorb more energy.

Figure I.1. Representations of the H_2O water molecule (top), the NH_3 ammonia molecule (middle), and the CH_4 methane molecule (bottom).

In each shell, there is a specific number of orbitals, and each orbital and set of orbitals holds a specific number of electrons. One of the essential facts about atomic behavior is that atoms seek to gain, lose, or share electrons until they have just the right number of electrons so that they have only full shells, without any extra electrons and without electrons missing from any orbitals in the full shells. If only the first shell is full, an atom has two electrons. If the first two shells are full, 10 electrons. If the first three are full, 28 electrons, and so on. Significantly, these numbers relate to the numbers of elements in the rows of the Periodic Table of the Elements. With this one fact, we can understand a great deal about how atoms of one element bond with atoms of other elements to form compounds.

The position of the electrons within an atom also has a lot to do with how an atom behaves. One aspect of atoms that affects the position of electrons is the shapes of the different orbitals. Some orbitals are spherically shaped, some are shaped in pairs of protruding lobes often described as "dumbbells," and some are shaped as rings. There are other more complex shapes as well. Since all electrons repel each other due to their negative electrical charge, electrons located in the lobes of dumbbell-shaped orbitals push away from each other, resulting in molecules with very particular shapes. Examples are the water, ammonia, and methane molecules illustrated in Figure I.1.

Electron position is also affected by the fact that within molecules some atoms attract electrons more strongly than others, an effect denoted by a value called the *electronegativity* of the atom. We discuss this in more detail later, but I will mention an important example here to illustrate this point. The electronegativity values for oxygen and hydrogen are 3.44 and 2.20, respectively. This means the oxygen atoms in water molecules attract electrons more strongly than the hydrogen atoms do. As a result, the four bonding electrons in the molecule crowd over toward the oxygen atom, making the oxygen region of the water molecule more electrically negative and the hydrogen regions more electrically positive. These differences make the water molecule electrically imbalanced—or *polar*, as we say—negative on one side and positive at the ends on the other side, as illustrated in Figure I.2. In this diagram, the arrows point from the positive regions of the molecule toward the negative region of the molecule.

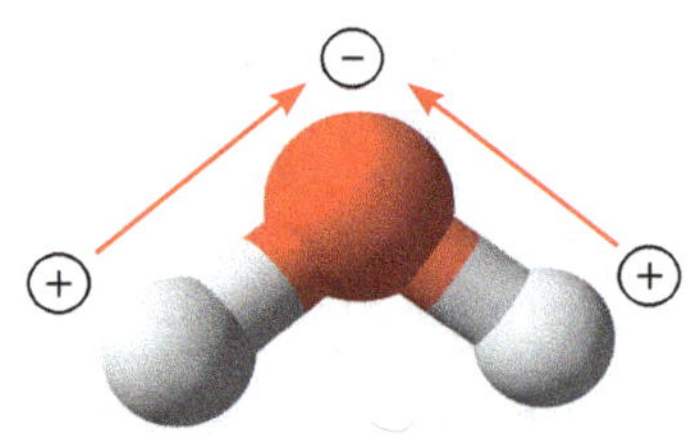

Figure I.2. The higher electronegativity of oxygen atoms compared to hydrogen atoms results in the polar water molecule.

The shapes and polarizations affect atomic behavior because of electrical attractions and repulsions, the basic theme we discuss in the next section.

I.1.2 Chemistry Is All About Electrical Forces

You know that there are two types of electrical charge: protons are positively charged, and electrons are negatively charged by exactly the same amount. Like charges repel each other (such as two positive charges) and opposite charges attract (positive and negative).

The reason these electrical attractions and repulsions are so important for chemistry is that atoms and molecules are as prickly as porcupines with charges that repel or attract other charges.

Some of these attractions and repulsions are stable and long-lasting, like the attraction between positive sodium ions and negative chlorine ions that holds together the atoms in the crystal lattice of sodium chloride (table salt). All ionic compounds are held together in rigid crystal structures by the strong forces of electrical attraction and repulsion between the ions in the crystal lattice.

Other electrical interactions are sort of semi-stable, you might say, given the fact that molecules are moving around all the time. The world-class example of this is *hydrogen bonding*, which we examine in detail later. The most common example of hydrogen bonding takes us back to the polar water molecule described in the previous section. Since water molecules are polar, the positive regions of one water molecule are attracted toward the negative regions of other water molecules, as illustrated in Figure I.3.

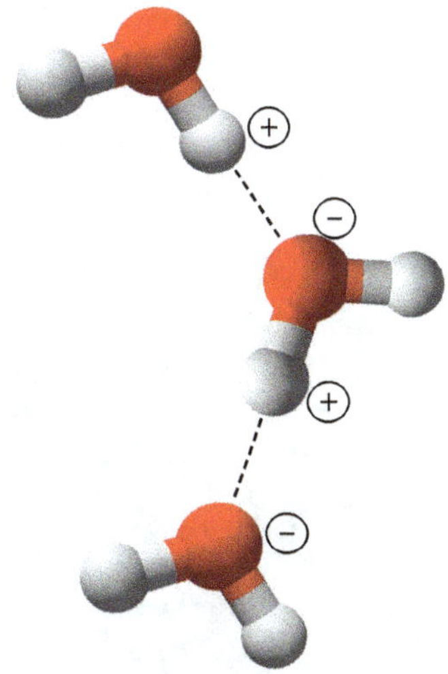

Figure I.3. Hydrogen bonding in water molecules. Dashed lines indicate hydrogen bonds between water molecules.

The importance of the hydrogen bonding phenomenon cannot be overstated. Water is everywhere, and thus so is hydrogen bonding. Hydrogen bonding explains why so many things dissolve in water, it explains why water travels upwards against the force of gravity when soaking into the fibers of a towel, and it explains why water gets less dense right before it freezes (which in turn explains why ice floats). Figure I.4 is a model of how the water molecules are arranged in ice. The dashed lines in the figure indicate the hydrogen bonds between water molecules. The result of these bonds is the three-dimensional, hexagonal structure of ice.

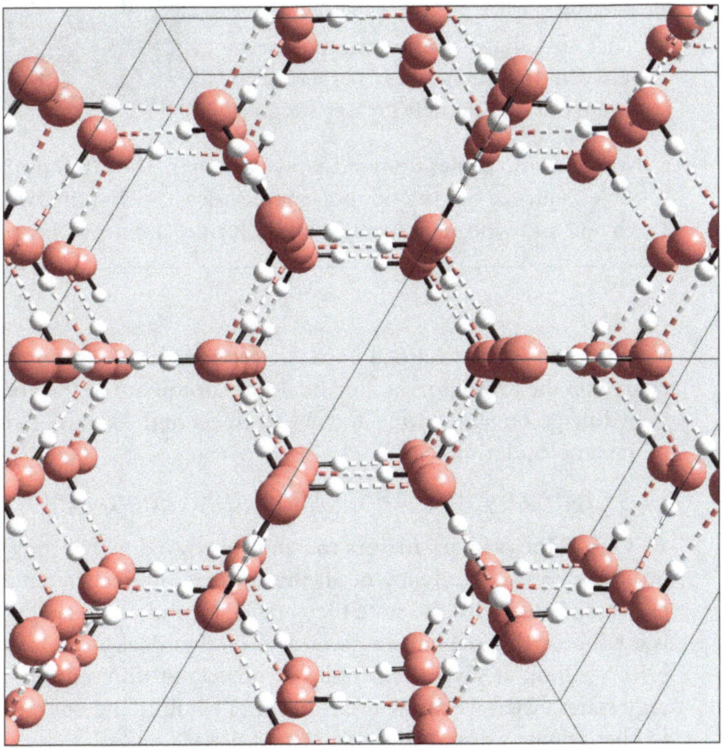

Figure I.4. The crystal structure in ordinary water ice. In this model, oxygen atoms are red and hydrogen atoms are white. Hydrogen bonds are shown as dashed lines. Thick black lines indicate the bonding of hydrogen to oxygen inside individual water molecules.

There are several other ways electrical forces between atoms and molecules are made manifest. In general, these different attractions and repulsions are called *intermolecular forces*. There is an electron cloud around every atom (except in the case of a hydrogen atom that has lost its only electron due to ionization). There is also an electron cloud around and between the atoms of every molecule. As the electrons swarm around in these clouds, there are moments when some regions in the molecule are more negatively charged because of electrons crowding together. There are other moments when regions

Hmm... Interesting. Why water forms beads

As described in this chapter, the polarity of water molecules makes them cling to one another. When nonpolar molecules are in contact with water, the water molecules are attracted to each other but not to the nonpolar molecules. The molecules in waxy leaves and oil-based wood finishing products are nonpolar. When water molecules rest on a surface of nonpolar molecules, they cling to each other but not to the surface, and the result is the formation of water drops. Small drops are nearly spherical because this shape minimizes the energy between the molecules. Larger drops flatten out due to their greater weight.

Nonpolar molecules do not dissolve in water. The attractions between the polar water molecules squeeze out the nonpolar molecules, causing the two substances to separate. This is why oil and vinegar separate—oil molecules are nonpolar and vinegar is mostly water.

are more positively charged because electrons have temporarily moved away and the positive charge on the protons in atomic nuclei are dominant in the area. These electron movements and crowding go on all the time and at extremely high speeds, giving rise to ever-changing patterns of intermolecular forces.

I.1.3 Chemistry Is All About Minimizing Energy

One of the primary drivers causing atoms to do what they do is the natural tendency of all things to minimize the energy associated with the state they are in. Minimizing energy is a concept that explains a great deal of chemical behavior. Here we look at several examples of objects in different energy states. Then we apply the concept of minimizing energy to phenomena we see occurring in chemistry.

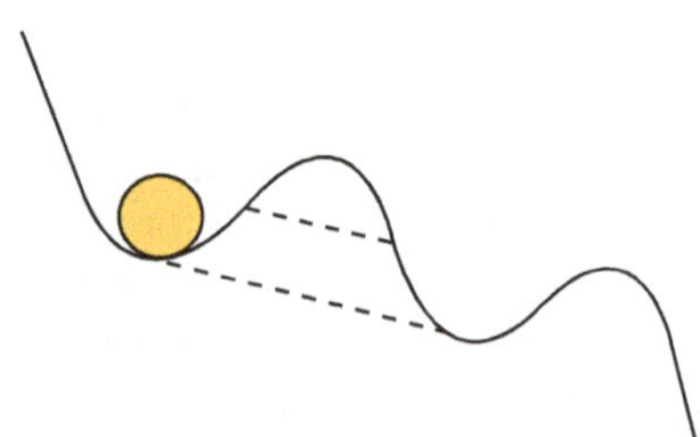

Figure I.5. The ball is trapped in the lowest-energy region in its vicinity. However, if a tunnel to a lower-energy region opens up, the ball goes there.

To begin, in a previous science course you may have studied different forms of potential energy. For example, gravitational potential energy is the energy an object has after being lifted up in a gravitational field. Figure I.5 shows a ball up on the side of the hill. The ball is trapped in a small valley or depression. The ball is located up above the ground, so it has gravitational potential energy. The ball always acts to reduce its potential energy if given a chance. If a tunnel opens up to a lower energy state, the ball goes there, releasing potential energy into some other form of energy (such as kinetic energy) as it goes. Another way for the ball to release potential energy and move to a lower state is for someone to hit it or kick it so that it has enough kinetic energy to get over the small hill where it is trapped. The point is that given the chance, the ball releases potential energy and moves to a lower energy state.

Another example of this idea is shown in Figure I.6. A cone held on its point has potential energy that is released if the cone is released and allowed to fall. In this case, the cone doesn't even need any kind of push or kick; it spontaneously moves to the lower energy state (laying down on its side) if released.

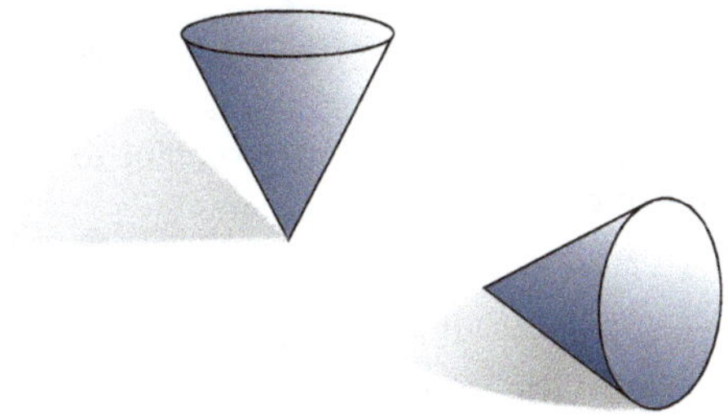

Figure I.6. The cone on the left is in a higher energy state. When released, it falls to the lower energy state shown on the right.

As a third example, consider the act of stretching a rubber band. To stretch out a rubber band, you have to supply energy. That is, you have to do mechanical work on the rubber band. If you release the stretched rubber band, it spontaneously contracts back to its lower energy (unstretched) state.

Let's now apply the idea of minimizing energy to a familiar chemical reaction: the combustion of hydrogen to produce water. The reactants are molecules of hydrogen and oxygen. Each of these gases exists as diatomic molecules, meaning that each hydrogen molecule is a pair of hydrogen atoms bonded together, and each oxygen molecule is a pair of oxygen atoms bonded together, as illustrated in Figure I.7. At room temperature, these gas molecules zoom around inside their container, colliding with one another several billion times per second, but otherwise nothing else happens.

In terms of the energy of these molecules, they are in a situation similar to the ball in Figure I.5: there are lower energy states the molecules can go to, releasing energy in the process, but they can't get there without a boost of energy to get the process started. Now, if a spark or flame is introduced to this gas mixture, the heat from the spark or flame excites the nearby molecules, causing them to move much faster and slam into each other with enough energy to break the bonds holding the molecules together. The result—which only lasts for a tiny fraction of a second—is a soup of unbonded gas atoms.

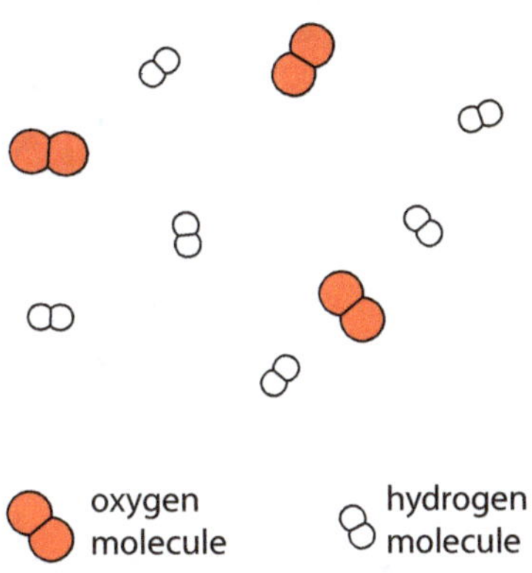

Figure I.7. Diatomic oxygen and hydrogen gas molecules.

At this point, we have a situation similar to the stretched rubber band the instant after being released, before it has had a chance to shrink. Electrical attractions between the protons and electrons in the isolated atoms of oxygen and hydrogen draw the atoms toward each other at an extremely high rate. Consider the collapse of the rubber band after it is released. It collapses to its unstretched state—a lower energy state—releasing energy in the process. The energy released might result in a snap (kinetic energy) that stings your hand and a sound wave (kinetic energy in moving air molecules) producing a snapping sound. Just as the relaxed rubber band is at a lower energy state and releases energy to get there, the hydrogen and oxygen atoms collapse together to the

Figure I.8. Energy released as heat and light as hydrogen and oxygen atoms combine to form water molecules.

lowest energy state they can find, which is to form water molecules (H_2O). As they do so, they release a lot of energy in the form of light and heat and all this happens in an instant. This is the explosion of hydrogen, captured in the photograph of Figure I.8.

This release of heat indicates the reaction is *exothermic*—the reaction releases energy. When considering the way energy relates to various chemical processes, I have found it very helpful to remember the rubber band and to compare it in my mind to the way positive and negative ions are attracted to each other. If separated positive and negative ions are released and allowed to fly together, energy is released—the light and heat of the exothermic reaction—as the ions move to a lower energy state. To separate them, one has to pull them apart by putting in energy (doing work on them) and thus moving them to a higher energy state—just like stretching the rubber band. This is an *endothermic* process, where energy is being absorbed by the ions. The most well-known example of an endothermic chemical reaction is the photosynthesis reaction that occurs in plants, depicted in Figure I.9.

And I can't help pointing out in passing the exquisite elegance of the photosynthesis reaction, a process both simple and incredibly complex that happens automatically and continuously all over the world. Consider how wonderful it is that oxygen-breathing creatures like ourselves live on a planet covered with oxygen-producing vegetation. Of course, every school kid learns about photosynthesis, but do we also learn about the delicate balance displayed everywhere in the environment around us in nature? I encourage you, as a young student, to study the complex relationships in nature and the responsibility we have to live in a way such the delicate balance is maintained.

$$\underset{\text{carbon dioxide}}{6CO_2} + \underset{\text{water}}{6H_2O} \xrightarrow{\text{light}} \underset{\text{sugar}}{C_6H_{12}O_6} + \underset{\text{oxygen}}{6O_2}$$

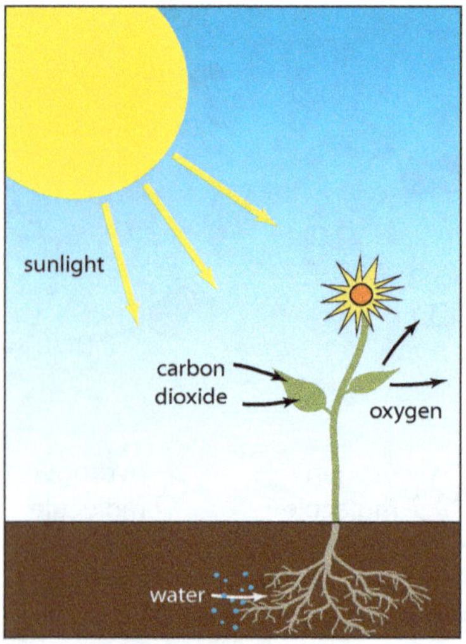

Figure I.9. The photosynthesis reaction is endothermic, as light from the sun is used by plants to convert water and carbon dioxide into sugar and oxygen.

There are two more important concepts about the role of energy in chemistry to note here. The first involves a quantity called *entropy*. Entropy is a term that originated in the field of thermodynamics. Entropy is a measure of the *disorder* present in a system, and the second law of thermodynamics states that left to themselves, physical processes go in a direction that increases the entropy (disorder) in the system. As an example, consider a glass of water you may be holding in your hand versus a broken glass on the floor with water splashed everywhere. While the glass is intact with the water contained in it, the system of glass and water is in an orderly state. When you release the glass, disorder increases—the glass breaks and the water goes everywhere on the floor. If you leave the mess like this, the disorder continues to increase: the water evaporates and the water molecules are not even together any more at all. Instead, they are randomly distributed around in the atmosphere. And with time, the chunks of glass get trampled and broken more and more until the remnants of the glass are completely gone. You will never see this process occur in reverse!

For some chemical processes, the minimizing of energy and the increase of entropy both pull in the same direction. In other cases, they try to pull the system in opposite directions. This sets up a sort of tug of war, and the process goes in the most favorable direction. We discuss this in more detail later.

Finally, some detail is in order regarding the boundaries surrounding energy minimization. Try this little thought experiment: imagine a hydrogen ion, which is simply a proton with its positive charge. Nearby is an negatively charged electron, as illustrated in Figure I.10. Since these particles have opposite charges, they are strongly attracted to each other, and since the proton's mass is 1,836 times greater than the electron's mass, the electron dashes toward the proton while the proton essentially stays put, waiting for the electron to arrive. You might expect that the electron would crash right into the proton, bringing the potential energy between them right down to zero. But this is not what happens.

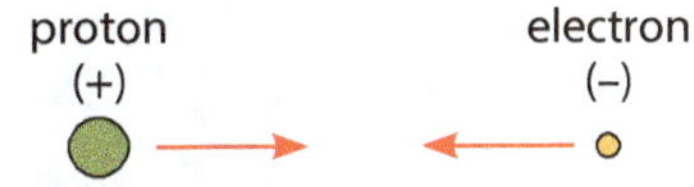

Figure I.10. Oppositely charged particles strongly attracted to each other.

In 1905, Albert Einstein theorized that energy is *quantized*—it comes in discrete chunks or packets. Since 1905, a host of scientists have explored the quantization of energy, confirming Einstein's proposal over and over and giving birth to the now well-developed theory of quantum mechanics. What quantum mechanics suggests for our proton-electron scenario is that an electron in an orbital of an atom cannot possess just any old amount of energy; it can only possess particular values of energy. In the context of dropping into one of the orbitals surrounding the proton, the electron can only possess an amount of energy corresponding to the one of the energies of the proton's orbitals. The bottom line is that instead of crashing into the proton and sticking to it like cat hair sticking to your pants, the electron instead pops into the lowest energy orbital available around the proton and stays there, captive, buzzing around furiously like a bee in a bottle. (But though this analogy may be suggestive, it is strictly metaphorical. Electrons are not at all like bees. For one thing, they don't have wings. And they don't make honey, either.)

I.1.4 Chemistry Is All About Whole Number Ratios of Atoms

It is strange to think that even as recently as the beginning of the 20th century there was no consensus among scientists as to whether atoms even existed. In 1803, English scientist John Dalton put forward the first detailed, scientific atomic theory. Dalton proposed that all material substances are composed of atoms, and that the way different compounds are formed is by atoms combining together. Since various substances are composed of discrete, individual particles and not just a continuum of matter, there is always a whole number of each type of atom in the substance.

Although many scientists throughout the 19th century refused to accept the existence of atoms, we now agree that Dalton was correct. Compounds do form with whole-number ratios of the atoms involved. (Back then, those who accepted the existence of atoms were called "atomists." Today, everyone is an atomist, so we don't need a name for this view any more.) As an example, sulfuric acid, H_2SO_4, has two hydrogen atoms, one sulfur atom, and four oxygen atoms in every molecule, so the ratio of oxygen atoms to hydrogen atoms in the molecule is 2 to 1. The ratio of oxygen atoms to sulfur atoms is 4 to 1. Of course, there are strange exceptions to every rule, including this one. Nevertheless, it is correct to say that just about every compound, regardless of how the atoms are structured, consists of atoms of different elements joined together in predictable whole-number ratios. Figure I.11 contains a photo of the mineral fluorite along with a computer model of the crystal structure of fluorite, or calcium fluoride, which has the formula CaF_2. In this crystal structure, the ratio of fluorine atoms to calcium atoms is 2 to 1. The caption in the figure explains this, using the computer model of the crystal structure as an aid.

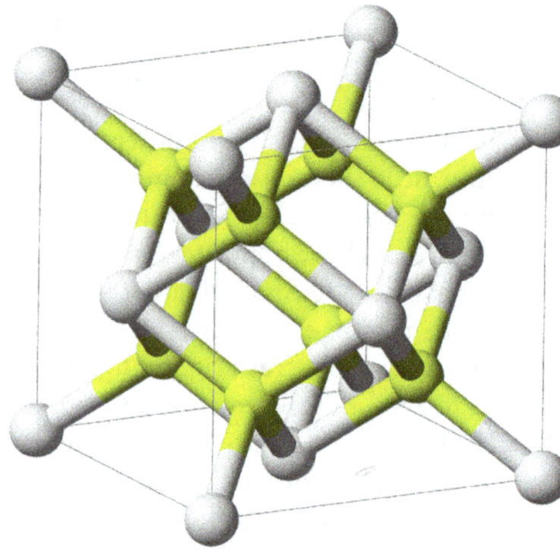

Figure I.11. The ratio of fluorine atoms (yellow) to calcium atoms (white) in fluorite is 2 to 1. To see this, note that there are 8 fluorine atoms completely within this crystalline cell. Each of the 8 calcium atoms at the corners is shared by 8 cells—the one shown and 7 other surrounding cells. Eight calcium atoms each shared 8 ways contributes a net of 1 calcium atom to the cell. Then there are 6 calcium atoms on the faces of the cell, each shared by the cell shown and the adjacent cell. Six atoms each shared 2 ways contributes a net of 3 atoms to the cell. In total then, this cell claims 8 fluorine atoms and 4 calcium atoms, a ratio of 2 to 1.

The fact that atoms combine in whole-number ratios is a powerful computational tool. When we get into the math behind chemical reactions (*stoichiometry*, as it is called), we will appeal often to the whole-number ratios of atoms involved in order to compute how much of one compound reacts with a given quantity of another compound.

I.1.5 Chemistry Is All About Modeling

Chemical reactions are happening around us all the time. Just pour a can of soft drink into a glass and watch the carbonic acid (H_2CO_3) in the can convert into the carbon dioxide bubbles and water (CO_2 and H_2O). Light up the gas grill and watch methane (CH_4) reacting with the oxygen (O_2) in the air to produce carbon dioxide (CO_2) and water (H_2O). Heat up a pan of cake batter in an oven and the rather complicated molecules in the batter reacts and change into different complicated molecules in a cake.

In these examples, even though we see bulk materials going into a chemical reaction (the *reactants*) and resulting from the chemical reaction (the *products*), we are not able to see the actual atoms and molecules as they zoom around, combining with and separating from each other. Understanding the behavior of things we cannot see is tricky business.

As mentioned in the previous section, even in recent scientific history the existence of atoms was debated for a hundred years. The issue was finally resolved with experiments in the early 20th century that gave more and more support to the theory that material substances were composed of atoms. We certainly know a lot more today about atoms and their internal structure than we did just a few decades ago. We can even put this knowledge to use in designing amazing new engineering materials, specialty drugs, and even chemical delivery systems to get the drugs into our bodies. But there remains much we do not understand about atoms.

It is helpful to think of science as the process of building "mental models" of the natural world. These mental models are called *theories*. The information we use to build our mental models—scientific facts—comes from experiments, observations, and inferences from these.

Since chemistry deals so much with atoms and molecules, which we can't see, we are almost completely dependent on inferences to develop atomic models describing how the atomic world works. Knowing that the gunpowder in a firecracker explodes when ignited doesn't require a model. It is obvious to all of us that gunpowder is explosive. But *why* is it explosive? What are the rules governing how the atoms in those compounds behave? Understanding why gunpowder explodes does require a model. And the models we work with in chemistry come at us from two different directions.

First, there is the information we gather from experiments. Chemical experimentation has been going on for hundreds of years. In the early days of the scientific revolution, scientists were amazed to discover quantitative laws such as Dalton's whole number ratios and the inverse relationship between the pressure and volume of a gas, a relationship known as Boyle's Law. Second, there is the theoretical modeling that occurs when scientists attempt to apply physical principles from quantum mechanics, thermodynamics, and statistical mechanics to the solution of chemical problems. The shapes and sizes of the atomic orbitals, which we address in Chapter 1, are an example of this type of theoretical modeling.

The theoretical models developed by scientists are the basis for our entire understanding of how the natural world functions. Successful theories are those that account for the facts we know and lead to new hypotheses (predictions) that can be put to the test. It is helpful to think about the relationship between facts, theories, hypotheses, and experiments as illustrated in Figure I.12. This diagram illustrates what I call the *Cycle of Scientific Enterprise*. It is important for every student to develop a correct understanding of the kind of knowledge scientific study provides

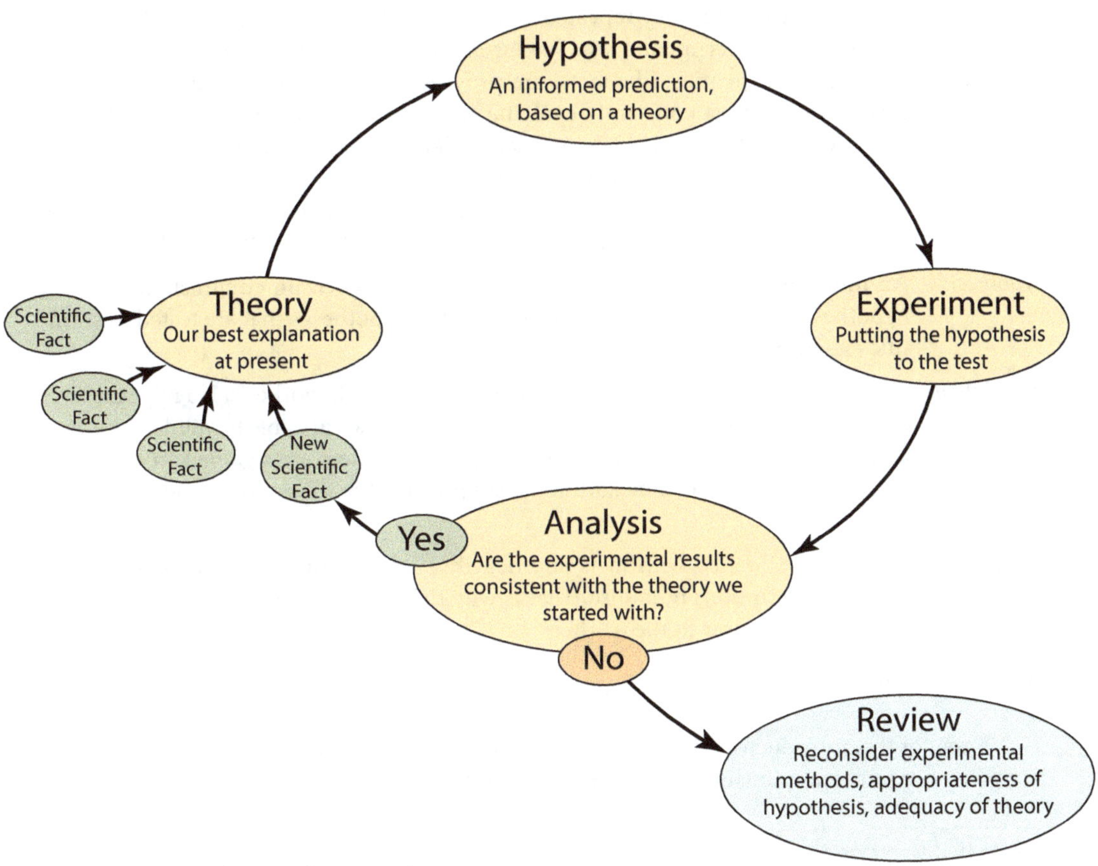

Figure I.12. The Cycle of Scientific Enterprise.

for us. The *goal* of science is to uncover the truth about how nature works, but scientific theories are always works in progress. Even our best theories are provisional and subject to change. For this reason, science is not in the business of making truth claims about scientific knowledge. Science is in the business of modeling how nature works with theories based on research.

As our theories develop over time, our hope is that they get closer and closer to the truth—the amazing and profound truth about mysteries such as what protons and electrons are, why they have the properties they have, and how the two most successful theories of the 20th century—quantum mechanics and general relativity—can be reconciled with each other. But the truth about nature is always out in front of us somewhere, always outside our grasp. To know the truth about nature, we would have to possess total knowledge about everything in nature. We are nowhere close to that.

Here are some definitions to keep in mind as you consider the models we discuss in future chapters.

Fact A proposition based on a large amount of scientific data that is correct so far as we know. Facts are discovered by experiment, observation, and inferences from experiments and observations. Facts can and do change as new scientific knowledge—new data—is acquired. Since facts are always subject to change, we generally avoid terms like *true* or *proven*. Instead, we say a fact is correct so far as we know.

Theory A mental model that accounts for the data (facts) in a certain field of research, and attempts to relate them together, interpret them, and explain them. Scientific theories are successful if they repeatedly allow scientists to form new hypotheses that can be put to experimental test. *Successful theories are the glory and goal of science.* Nevertheless, theories, like facts, are provisional and subject to change. Indeed, theories are almost constantly evolving as research continues. And as with facts, when referring to theories we avoid terms like *true* or *proven*. Instead, we speak in terms of how successful theories have been in generating hypotheses that are confirmed by experiments. A widely accepted scientific theory should be understood as our best explanation at present—our best *model* of how nature works.[1]

Hypothesis An informed prediction about what will happen in certain circumstances. Every hypothesis is based on a particular theory. It is hypotheses that are tested in scientific experiments.

Experiment A test designed to confirm or disconfirm a particular hypothesis. If a hypothesis is confirmed through experiment, and if other scientists are able to validate the confirmation by replicating the experiment, then the new facts gained from the experimental results become additional support for the theory the hypothesis came from.

Chemistry is a subject loaded with facts and heavily based on theories—*models*—that we know are incomplete descriptions of nature. That is why the research continues, as our models (hopefully) get nearer and nearer to the truth.

I.2 Conclusion

The goal of this introductory chapter is simply to alert you to some of the key concepts undergirding our understanding of chemical processes. Over and over in the coming chapters, you

1 Note that the term *law* is simply an obsolete term for what we call a theory. For historical reasons, the term is still in use.

will find that thinking about the content in terms of one or more of these central ideas will help you develop a better grasp of the material.

Introduction Study Questions

1. Write five brief paragraphs summarizing the main ideas behind the titles of Sections I.1.1 through I.1.5.
2. Describe two examples, other than those in this Introduction, of a system of some kind spontaneously (without help) moving from a higher energy state to a lower energy state. (Hint: If energy is being released, it means the entities involved in the process are moving to a lower energy state.)
3. Describe two examples, other than those in this Introduction, of a system that will move to a lower energy state if allowed to, but which needs an initial boost of energy to get started (like the ball in Figure I.5 being kicked and then having enough energy to get out of the valley).
4. Describe two examples of processes in which entropy *decreases*. In each case, describe what source of energy and/or intelligence must be present for the decrease in entropy to occur. Here is an example to assist your thinking: an oxygen tank contains pressurized oxygen gas. The oxygen in this tank is more ordered than the oxygen in air because it has been separated from the air; there is a boundary (the tank) between the oxygen and the air. And if the valve on the tank is opened, the oxygen flows out into the air to increase the entropy (disorder). What we will never see: opening the tank valve and oxygen atoms from the atmosphere spontaneously flow into the tank. But the oxygen is put into the tank somehow, and the process that put it there decreases the entropy of that oxygen.
5. What is the ratio of nitrogen atoms to hydrogen atoms in ammonia molecules? What is the ratio of hydrogen atoms to carbon atoms in propane molecules?
6. Why are water molecules polar and what is the significance of this fact?
7. If oppositely charged objects attract, why can't a free electron and a free proton collide into one another and stick together because of their opposite charges?
8. A hydronium ion is a water molecule that has gained an extra proton. (A proton is identical to a hydrogen ion.) Hydronium ions form spontaneously in water, and are formed in greater quantities any time an acid is poured into water. What is the ratio of hydrogen atoms to oxygen atoms in hydronium ions?
9. What is hydrogen bonding?
10. Distinguish between endothermic and exothermic processes.
11. In a previous course, you may have learned about the "gold foil experiment" conducted by Ernest Rutherford in 1909. This experiment led Rutherford to propose that the positive charge in atoms is concentrated in a tiny nucleus in the center of the atom. Think about this experiment and explain why Rutherford had to depend on inference as he interpreted his experimental data.
12. Why doesn't oil dissolve in water?
13. Distinguish between facts, theories, and hypotheses.
14. Explain why it is scientifically inappropriate to say, "no theory is true until it is proven."

Chapter 1

Atomic Structure

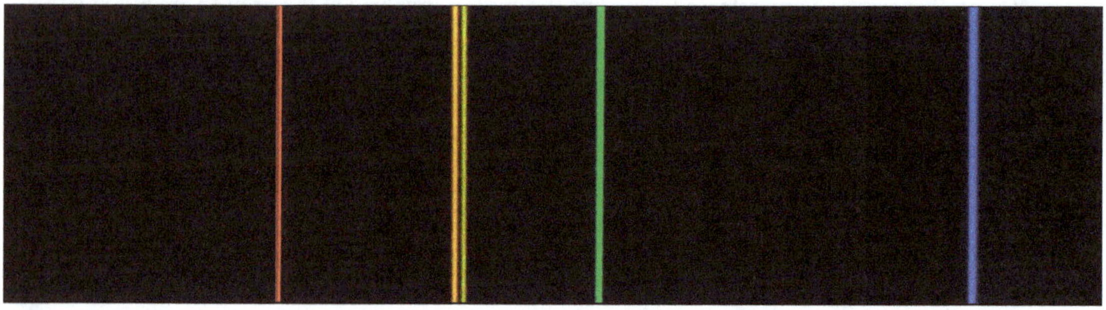

Hg

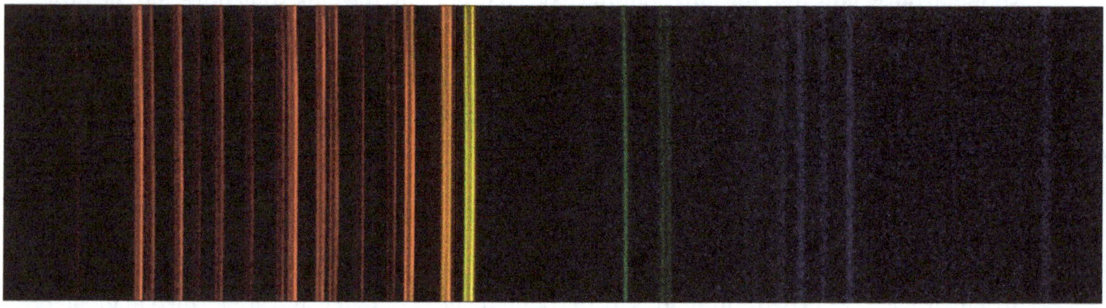

Ne

When excited by heat or electricity, the atoms of each element emit a specific, unique set of wavelengths of light—the atomic spectrum for that element. The visible spectra emitted by mercury and neon are shown above. Atomic spectra were known and studied in the 19th century, but there was no theory at that time that could explain the source of the colors different elements emit. Then in 1913, Niels Bohr published his new model of the atom, locating the electrons in atoms in specific energy levels. Bohr theorized that when excited, electrons jump to higher energy levels, and that to drop back down to a lower energy level an electron emits a packet of electromagnetic energy—what we now call a photon. In 1901, Max Planck had published the equation relating specific amounts of energy to specific wavelengths (colors) of light. Bohr's successful explanation for atomic spectra opened the door for detailed study of the internal structure of atoms.

The two spectra shown above were imaged in the Laser Optics Lab at Regents School of Austin in Austin, Texas.

Objectives for Chapter 1

After studying this chapter and completing the exercises, you should be able to do each of the following tasks, using supporting terms and principles as necessary.

SECTION 1.1

1. Describe the electromagnetic spectrum and state approximate wavelengths for the ends of the visible spectrum.
2. Define *quantum* and explain what it means for energy to be quantized.
3. Given the Planck relation and Planck's constant, determine the energy of a photon of a given wavelength and vice versa.

SECTION 1.2

4. Describe the Bohr model of the atom and relate how the model explained the phenomenon of atomic spectra.

SECTION 1.3

5. Distinguish qualitatively between the orbital energies in the hydrogen atom and those of other atoms.
6. Describe the two main ways that atoms can possess energy.
7. State and describe the four quantum numbers required to describe the quantum state of an electron.
8. For the first three principle quantum numbers, describe the orbitals available for electrons.
9. State the Aufbau principle, the Madelung rule, and Hund's rule, and relate them to the way electrons are located in atoms.
10. State the Pauli exclusion principle and explain its relationship to the placement of electrons in atoms.

SECTION 1.4

11. Given the periodic table, write electron configurations (full and condensed) and draw orbital diagrams for all elements in the first five periods (including the nine *d*-block elements with anomalous configurations).

SECTION 1.5

12. Given the periodic table, determine the number of protons, electrons, and neutrons in the atoms of a given nuclide.
13. Given isotope mass and abundance data, calculate the atomic mass of an element.
14. Define the unified atomic mass unit, u, and state two definitions for the mole.
15. Calculate the molar mass of a compound or molecule, and calculate the mass in grams of a given mole quantity of a compound or molecule, or vice versa.
16. Calculate the number of atoms or molecules in a given quantity of substance.
17. Given mass data for an unknown compound, determine the percent composition and empirical formula of the compound.
18. Use the percent composition along with the molar mass or molecular mass of an unknown compound to determine the molecular formula for the compound.
19. Correctly use the rules for significant digits in computations, including the addition rule.

1.1 Atomic Spectra

1.1.1 The Electromagnetic Spectrum

Understanding our present theory of atomic structure and the story of how it unfolded requires a basic understanding of the electromagnetic spectrum. We thus begin this chapter with a brief review of this topic.

The spectrum of visible light is shown in Figure 1.1. The visible spectrum runs through the colors of the rainbow—red, orange, yellow, green, blue, violet—and includes wavelengths from about 750 nm (red) down to about 400 nm (violet).

Figure 1.1. Colors and approximate wavelength ranges in the visible portion of the electromagnetic spectrum.

Visible light is just a small portion of a vast spectrum of electromagnetic radiation that occurs in nature. Figure 1.2 shows the most important regions of the electromagnetic spectrum, from radiation with wavelengths in the range of 1 km, the region of AM Radio waves, down to the high-energy Gamma Rays, with wavelengths in the range of 1 picometer (pm). (The metric prefix *pico–*, which may be new to you, means 10^{-12}. One picometer is one thousandth of a nanometer.) As you see from the figure, the solar emission spectrum runs from wavelengths of about 1 mm down to wavelengths of about 0.1 μm. The solar spectrum includes the infrared, visible, and ultraviolet regions, and is strongest in the middle of the visible spectrum.

The contemporary theory of light (a shorter term for electromagnetic radiation in general) holds that light exhibits both wave-like properties and particle-like properties. Since light behaves like waves, we can refer to the wavelengths of particular colors. But light also behaves like particles. We call these particles *photons*, and each photon represents a single packet of energy. The packet of energy in a photon is also called a *quantum* of energy. (The plural is *quanta*.)

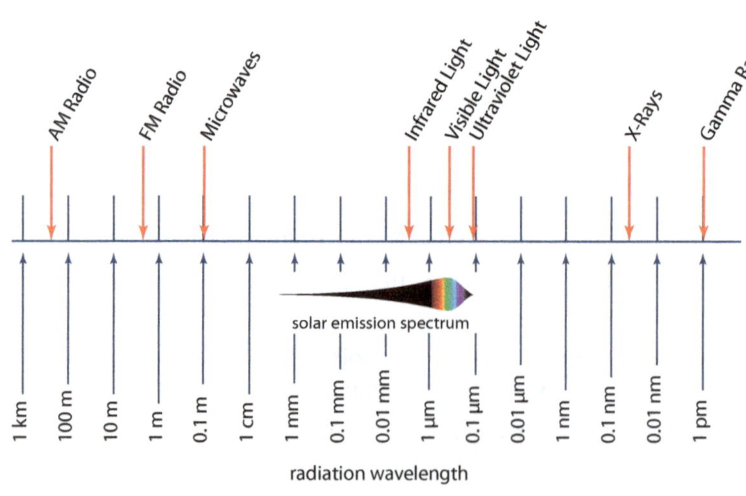

Figure 1.2. The electromagnetic spectrum.

When thinking of light as waves, we characterize those waves by the wavelength. When considering light as discrete packets of energy, we tend to think of the amount of energy in each packet (each photon). It turns out there is a simple equation relating these together.

It was in 1901 that German physicist Max Planck (Figure 1.3) conceived of treating energy as if it were quantized. He was working on a different problem at the time (the so-called blackbody radiation problem), and he did not imagine that energy really *is* quantized. However, he introduced what he thought was a mathematical trick—the quantization of energy—and in the process, quantum theory was born. Planck won the Nobel Prize in Physics for this work in 1918. (It's ironic, isn't it, to win the Nobel Prize for a major discovery that the scientist thinks is just a mathematical trick?) Four years later in 1905, German physicist Albert Einstein (Figure 1.4) proposed that energy really is quantized and used this idea to solve another problem (explaining the photoelectric effect). For this, Einstein won the Nobel Prize in Physics in 1921.

Figure 1.3. German physicist Max Planck (1858–1947).

Figure 1.4. German physicist Albert Einstein (1879–1955).

The equation Planck introduced is called the *Planck relation*. This important equation is:

$$E = hf \qquad (1.1)$$

In this equation, E is the energy in the photon in joules (J). The next term in the equation, h, is a constant known as the *Planck constant*. The value of h is defined as exactly

$$h = 6.62607015 \times 10^{-34}\ \text{J} \cdot \text{s} \qquad (1.2)$$

The last term in the Planck relation, f, is the frequency of the wave. The frequency and wavelength of a wave are related by the equation

$$v = \lambda f \qquad (1.3)$$

In this equation, v is the velocity of the wave, which is the speed of light in this case (2.9979×10^{8} m/s). The wavelength in meters (m) is represented by the Greek letter λ (lambda, the Greek lower-case letter l). If we solve Equation (1.3) for the frequency and insert it into the Planck relation, we have

$$E = \frac{hv}{\lambda} \qquad (1.4)$$

With this equation, we can compute the energy in a single photon of light at any wavelength, or vice versa. To illustrate such a calculation, we have our first example problem. As you see below, example problems in this text are set off by two red triangles. In presenting this example, I am assuming you are:

- familiar with the SI System of units and unit prefixes
- proficient at performing unit conversions
- proficient at applying the basic rules for the use of significant digits in measurements and computations (except for the addition rule, covered later in the chapter).

If you are lacking skills in any of these areas, please refer to Appendix A for a tutorial.

Example 1.1

The bright blue line in the mercury vapor spectrum (see the upper image on the opening page of this chapter) has a wavelength of 435.8 nm. Determine the energy contained in a single photon of this blue light.

We begin by writing down the given information and the unknown we seek to find.

$$\lambda = 435.8 \text{ nm}$$
$$E = ?$$

Next, we convert the given wavelength into the MKS[1] length unit, meters.

$$\lambda = 435.8 \text{ nm} \cdot \frac{1 \text{ m}}{10^9 \text{ nm}} = 4.358 \times 10^{-7} \text{ m}$$

From this value, along with the Planck constant and the speed of light, we calculate the energy of a photon with this wavelength.

$$E = \frac{hv}{\lambda} = \frac{\left(6.626 \times 10^{-34} \text{ J} \cdot \text{s}\right)\left(2.9979 \times 10^8 \ \frac{\text{m}}{\text{s}}\right)}{4.358 \times 10^{-7} \text{ m}} = 4.558 \times 10^{-19} \text{ J}$$

This is an extremely small amount of energy, less than a billionth of a billionth of a joule. The given wavelength and the value for the Planck constant each have four significant digits. Thus, the result is stated with a precision of four significant digits.

1.1.2 Energy in Atoms

As mentioned in the caption on the opening page of this chapter, the atoms of every element emit a specific set of colors when excited. In the context of atomic theory, the term *excitation* refers to the absorption of energy by atoms, either from electromagnetic radiation (light) or from collisions with other particles.

Let's spend a moment considering the ways an individual atom can possess energy. There are two basic mechanisms by which atoms can possess energy. First, all atoms possess *kinetic energy*, the energy associated with motion. Kinetic energy in atoms is illustrated in Figure 1.5. In solids, the atoms are fixed in place and are not free to move around, so the kinetic energy is manifest in the atoms' vibrations. In liquids and gases (fluids), atoms are free to move around, so the energy possessed by atoms in fluids is in their translational kinetic energy. Also, when atoms in fluids are bound together in molecules, the molecules can tumble and rotate, so some of their kinetic energy is in the energy of rotation. Atoms in molecules also vibrate, just as balls attached to one another by springs can wiggle back and forth. In all these cases, the kinetic energy in atoms and

1 MKS stands for meter-kilogram-second. The MKS system is a subset of the metric or SI System of units. Using MKS units for computations is always wise practice because the units of measure in the computation will all be consistent with each other, and the result of the computation always comes out in MKS units. For more on this, see Appendix A.

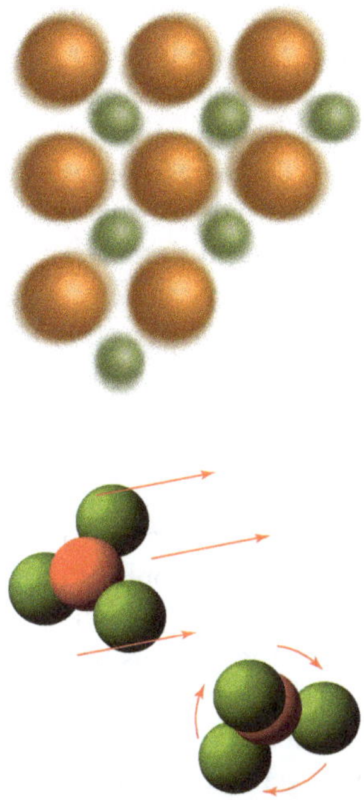

Figure 1.5. The atoms in the solid crystal at the top are blurred to illustrate their vibrations. The gas molecules at the bottom are translating and tumbling.

molecules correlates directly to their temperature. The hotter they are, the more vigorously they vibrate and the faster they move.

The second basic way an atom can possess energy is in the energies of the atom's electrons. As mentioned in the Introduction and Chapter 1 (and discussed in detail later in this chapter), the electrons in atoms are located in various orbitals, and different orbitals are associated with different amounts of electron energy. Atoms can absorb quanta of energy from the photons of electromagnetic radiation and from collisions with other particles, such as ions and free electrons. When an atom absorbs energy in this way, the quantum of energy absorbed by the atom is manifest in one or more of the atom's electrons moving into higher-energy orbitals. This is atomic excitation.

When an atom's electrons are all in their lowest-energy orbitals, the atom is said to be in the *ground state*. Excitation occurs when an atom absorbs a quantum of energy causing an electron to move to a higher-energy orbital. When this happens, the atom is said to be in an *excited state*. Atoms tend not to remain in excited states. Instead, after becoming excited an atom typically heads straight back to the ground state, generally by emitting the energy it absorbed in the form of one or more new photons.[2]

A newly emitted photon may not have the same amount of energy as the original quantum of energy the atom absorbed. To explain this, we need to introduce a commonly used graphical representation of the different energies electrons can possess. For now, let's call these *energy levels*. We will relate these more carefully to the energies of electrons in atoms in the next few sections.

Figure 1.6 is a diagram representing four different energy levels actually available to an electron in a hydrogen atom, labeled in the figure $n = 1$, $n = 2$, and so on. I explain this diagram carefully below, but first we need to pause here to revisit one of the points made in the Introduction. In Section I.3, I note that when an electron is

2 In some substances, electrons in atoms can remain in excited states for an extended period of time. As the atoms in such a substance return to the ground state over time the substance gradually radiates the energy away. This is the way *phosphorescence* (glowing in the dark) works.

Hmm... Interesting. **Neon signs and phonons**

When excited atoms in gases return to the ground state they do so by emitting photons. Neon signs are tubes of gas excited by high-voltage electricity. Their glowing colors are caused by the atoms returning to the ground state. When excited atoms in solids (and some liquids) return to the ground state, they can do so by emitting photons as gases do, but they can also emit *phonons*, packets of vibrational energy. Phonons can travel as waves through the crystal lattice in a solid, displacing the atoms from their equilibrium positions. In the image on the right, the wavelength of the emitted energy is shown in red (and the displacement of the atoms is greatly exaggerated).

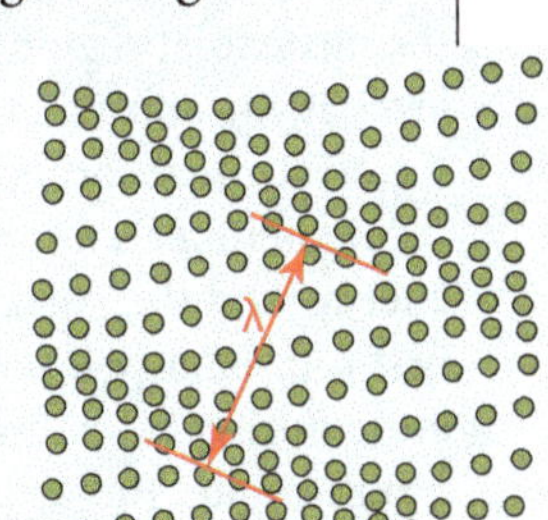

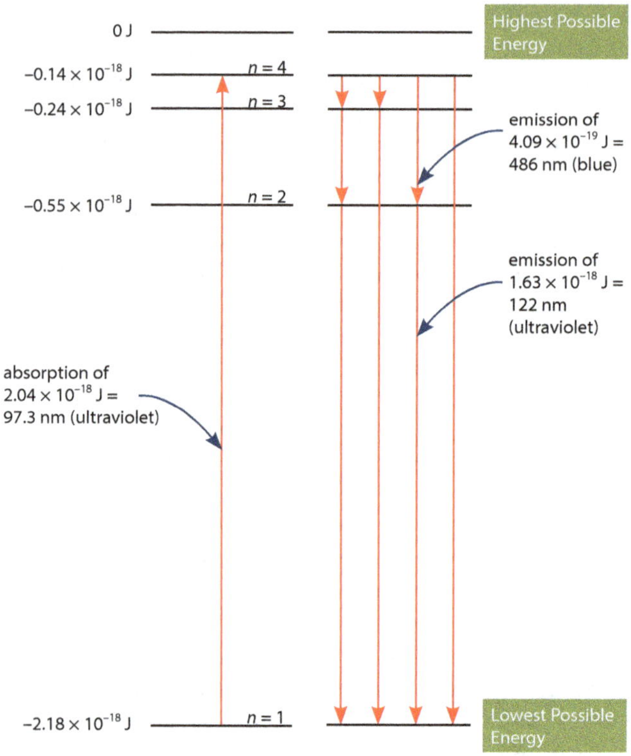

Figure 1.6. An energy-level diagram illustrating how quanta of energy are absorbed and released by electrons in hydrogen atoms.

held at a certain distance away from the positive nucleus, it has a high potential energy. As the electron is allowed to move closer to the nucleus, its potential energy decreases just as the gravitational potential energy of an object above the ground decreases with decreasing height. Since the electron's potential energy decreases as the electron moves closer to the nucleus, the electron releases energy as it gets closer and closer to the nucleus.

In discussions of the energies of electrons in atoms, it is customary to assign a reference value of 0 joules to the energy an electron has when it is completely free from the nucleus—in other words, when it is very far away. Then, since the electron's potential energy decreases as it gets closer to the nucleus, the energy an electron has is expressed as a negative value. This happens because instead of setting the potential energy to be zero at the nucleus, we set the zero energy reference to be when the electron is far away from the nucleus. So don't let the negative energy values bother you. It makes sense that the electron's energy is set to zero when it is far away from the nucleus because at that point the electron really has nothing to do with that nucleus. But to be consistent with what we know about potential energy, the electron's energy must decrease as it enters the area near the nucleus, so its energy takes on increasingly negative values relative to the zero-energy reference.

Going back now to Figure 1.6, let's assume that an electron in a hydrogen atom is in the ground state, which means it has the lowest possible energy. Since hydrogen atoms only have one electron, this places the electron at the bottom of the figure at the first energy level, $n = 1$. An energy of -2.18×10^{-18} J is the energy an electron has when it is in this first energy level. Assume now that this electron absorbs a quantum of energy equal to 2.04×10^{-18} J from an incoming photon, indicated by the arrow pointing upward on the left side of the figure. From the Planck relation, you can verify that this corresponds to a wavelength of 97.3 nm, placing this photon in the ultraviolet region of the electromagnetic spectrum. The electron is now in the fourth energy level. If you subtract the energy of $n = 1$ from that of $n = 4$, the difference is the amount of energy the electron absorbed, 2.04×10^{-18} J.

Remember, energy in atoms is quantized. Electrons can only have certain specific values of energy, and the permissible values of energy an electron in a hydrogen atom can have (for the first four energy levels) are the energies listed down the left side of Figure 1.6. Very quickly the atom emits this energy in the form of new photons and the electron drops back down to the ground state. But as you can see from the right side of the figure, the electron in the hydrogen atom has four different ways of doing this.

First, the electron can release the smallest permissible amount of energy each time it emits a photon. This causes it to drop down one energy level at a time, emitting three separate photons on its way back to the ground state, as shown by the sequence of three downward pointing arrows leading from $n = 4$ to $n = 1$. Second, the electron can first drop from $n = 4$ to $n = 3$ and then drop to $n = 1$. Third, the electron can first drop from $n = 4$ to $n = 2$ and then drop to $n = 1$. For this possibility, the emitted amounts of energy are shown in the figure. The energy emitted when dropping from $n = 4$ to $n = 2$ is 4.09×10^{-19} J. Using the Planck relation, you can calculate the wavelength of a photon with this energy. Doing so gives a wavelength of 486 nm, which is in the visible portion of the electromagnetic spectrum. A check of Figure 1.1 indicates that this wavelength corresponds to blue light. The drop from $n = 2$ to $n = 1$ is a much larger energy drop, 1.63×10^{-18} J. This energy corresponds to a wavelength of 122 nm, which is in the ultraviolet region and is not visible.

Finally, the electron can drop back to $n = 1$ by emitting 2.04×10^{-18} J, the same amount of energy it absorbed in the first place. This energy produces a new ultraviolet photon with the same wavelength as the photon originally absorbed by the atom.

Note from Equation (1.4) that energy and wavelength are inversely proportional. Longer wavelengths represent lower energies; shorter wavelengths represent higher energies. This relationship is illustrated in Figure 1.7.

The spectrum of wavelengths emitted by each element is unique, which means that light spectra can be used to identify the element's presence in a gas or solution. The science of such identifications is called *spectroscopy*.

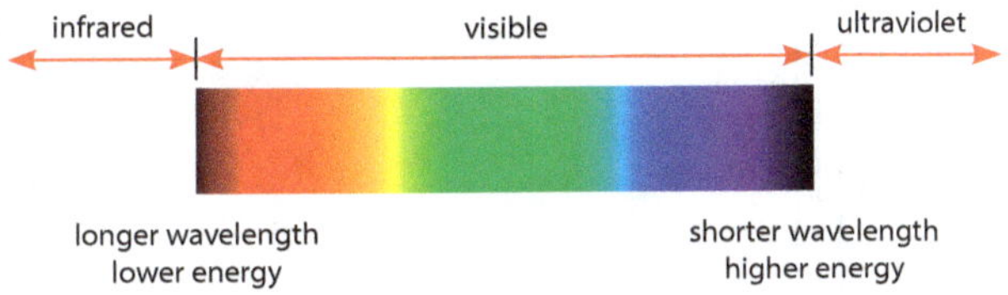

Figure 1.7. Wavelengths and energies in and near the visible spectrum.

1.1.3 The Hydrogen Atom

The hydrogen atom is the simplest atom, with only one electron, and thus it has been studied extensively. The wavelengths for the possible electron energy transitions in the first six energy levels of the hydrogen atom are shown in Figure 1.8. All the arrows in this diagram are shown pointing in both directions because the wavelengths shown can represent either the absorption or emission of energy. These are the energies hydrogen atoms can absorb and emit.

In 1885, Swiss mathematician and physicist Johann Balmer discovered the formula that predicts the lines in the visible hydrogen spectrum. This series of lines is now called the Balmer series. In 1888, Swedish physicist Johannes Rydberg worked out the more general formula for all the hydrogen wavelengths. The ultraviolet and infrared lines in the hydrogen spectrum were not known initially (because they are invisible). But in 1906, American physicist Theodore Lyman observed the ultraviolet series that bears his name, and in 1908, German physicist Friedrich Paschen observed the infrared series of lines in the hydrogen spectrum.

The Rydberg formula that predicts all these wavelengths has an interesting mathematical structure, and is worth showing here. Here it is:

$$\frac{1}{\lambda} = R\left(\frac{1}{n_1^2} - \frac{1}{n_2^2}\right) \tag{1.5}$$

The R in this equation is the so-called Rydberg constant (1.097×10^7 m^{-1}), and n_1 and n_2 represent the numbers for the two energy levels in question.

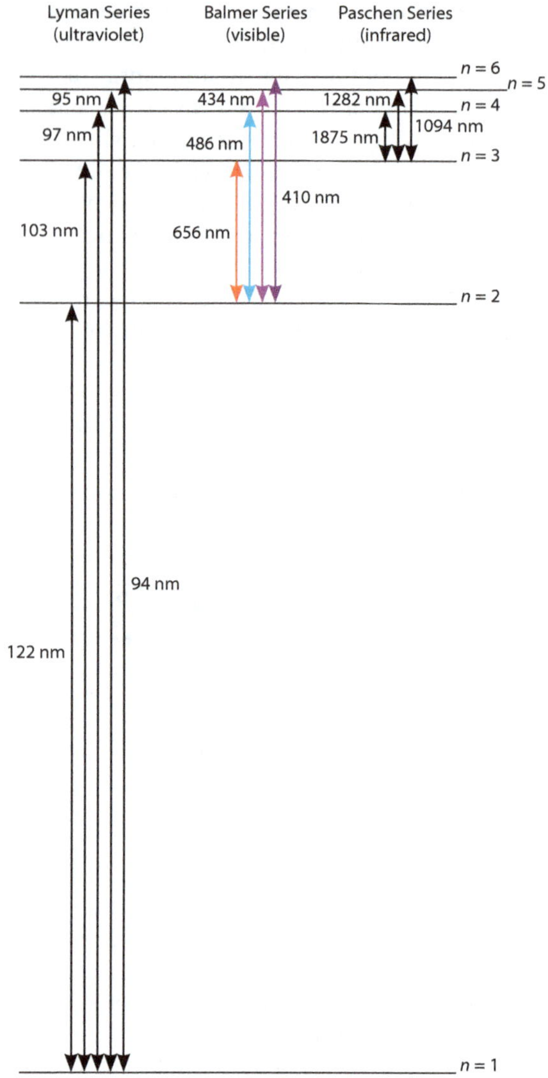

Figure 1.8. Wavelengths for electron transitions in the first six energy levels of the hydrogen atom.

I love pointing out to students that mathematical relationships are all over the place in nature—from the orbits of the planets, to the structure of gravity, to the ways atoms emit wavelengths of light. This is one of the reasons the natural world is so full of wonder. It is also wonderful that humans are able to identify and understand these mathematical relationships.

1.2 The Bohr Model of the Atom

As mentioned on the opening page of this chapter, Danish physicist Niels Bohr introduced his new model of the atom in 1913. This new atomic model was of tremendous importance in the development of atomic theory. Rydberg's formula predicting the lines in the hydrogen spectrum had been known since 1888, but until Bohr's model there was no theoretical basis for the observed spectrum.

In Bohr's atomic model, the electrons orbit the nucleus like planets orbiting the sun. In the model, the electrons have fixed energies, the same energies as those shown in Figures 1.6 and 1.8. These different energy levels correspond to different orbits around the nucleus.

Bohr correctly described the cause of the specific lines in the emission spectra of atoms—electrons absorbing energy and moving to higher energy levels and then releasing photons at specific energies as they move back to lower energy levels.

Another significant feature of the Bohr model is the number of electrons that he permitted at each energy level. These numbers are shown in Figure 1.9. If you compare these numbers to the Periodic Table of the Elements shown on the inside back cover of this text, you see that the number of electrons in each energy level corresponds to the number of elements in each period (row) of the table: two in the first, eight in the second, eight in the third, and 18 in the fourth, etc.

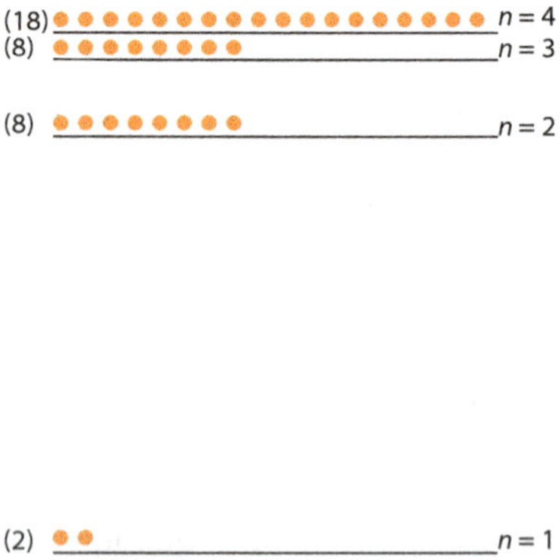

Figure 1.9. Number of permissible electrons in the first four energy levels of the Bohr model of the atom.

As powerful as it is, the Bohr model was known to have weaknesses from the start. For one thing, there is no explanation for why electrons are able to stay in their orbits. Electrons moving in circles radiate energy, so one would think electrons in orbits would gradually lose energy, slow down, and spiral in to the nucleus. Another issue is that for atoms other than hydrogen, the energies between the energy levels do not match up precisely with the observed wavelengths in the emission spectra of elements.

1.3 The Quantum Model of the Atom

1.3.1 Schrödinger and Pauli

In 1926, Austrian physicist Erwin Schrödinger (Figure 1.10) published what is now called the *Schrödinger equation.* This was a landmark achievement and one of the hallmarks of 20th-century physics. For this work, Schrödinger received the Nobel Prize in Physics in 1933.

Figure 1.10. Austrian physicist Erwin Schrödinger (1887–1961).

Solutions to the Schrödinger equation are now understood to provide us with the details of the internal structure of energy levels in atoms. With the arrival of the Schrödinger equation, the *quantum model* of the atom began to unfold. The history of quantum physics is still being written. There are many mysteries associated with the behavior of electrons as described by the quantum model. But quantum theory has a colossally impressive string of achievements, and its success in predicting atomic behavior is undeniable. We know the quantum model will continue to evolve, and may some day even be replaced. But the details we consider in this section are now generally accepted as correct. Remember, chemistry is all about modeling—developing theories. Theories are explanations, and the quantum model is widely accepted as our best explanation of how atoms are structured.

In addition to Schrödinger's equation, there is one other theoretical milestone that we need to have in hand to understand the details to follow. In 1925, while Schrödinger was working on his equation, another Austrian physicist, Wolfgang Pauli (Figure 1.11) formulated what is now known as the *Pauli exclusion principle.* In short, *the Pauli exclusion principle holds that no two electrons in the same atom can occupy the same quantum state.* We will unpack this further as we go along. For this important contribution to quantum theory, Pauli won the Nobel Prize in Physics in 1945.

Figure 1.11. Austrian physicist Wolfgang Pauli (1900–1958).

1.3.2 Shells, Subshells, and Orbitals

The quantum state of an electron in an atom—its unique address, we might say, within the atomic quantum realm—is specified by four different *quantum numbers.* According to the Pauli exclusion principle, every electron in an atom has a unique quantum state. This is one of the laws of nature governing the way atoms are structured. This situation in atoms is analogous to postal addresses. Every postal customer in the U.S. has a unique address. For a house, this unique address requires four pieces of information—the street number, street name, city, and state. For an apartment complex, an apartment number is also required. (The zip code doesn't have any additional location information in it; it just helps speed things up.)

The physics behind these quantum numbers is quite complicated, and as an introductory chemistry student you would not normally be required to get much into that. However, introductory chemistry classes do generally now require students to learn the arrangement of shells, subshells, and orbitals in atoms for "energy levels" $n = 1$ through $n = 4$ because knowing this structure allows us to specify where the electrons are in an atom. And as you recall from the Introduction, chemistry is all about electrons! So here we go. There is a lot of detail in this section, and it is all important.

The phrase *energy levels* is in quotes just above for an important reason. We are transitioning now from the energy levels in Bohr's atomic model to those of the far more accurate quantum model. In Bohr's model, and in the hydrogen atom as we still understand it, there is only one energy level for each value of n. The quantum model is quite different, as we will see.

Recall from Figure 1.9 that the numbers of electrons permitted in the first four levels of Bohr's model are 2, 8, 8, and 18. These numbers correspond to the number of elements in the first four periods of the periodic table. Bohr was on the right track, but did not initially perceive the correct pattern. We now refer to n as the *principle quantum number*, and in every atom except hydrogen there are multiple energy levels associated with each value of n. As explained in detail below, the number of electrons allowed for each value of n is actually $2n^2$. This gives us 2, 8, 18, and 32 electrons in the various energy levels associated with $n = 1$ through $n = 4$.

The clusters of energy levels associated with each value of n are commonly called *shells*. As I state just above, the quantum state of an electron in an atom, including its energy, is specified by four quantum numbers; the principle quantum number—the shell number—is the first of them. So beginning with the principle quantum number you are already familiar with, here is a list of the names and other details for the four quantum numbers:

1. Principle Quantum Number, n Values for n are the integers 1, 2, 3, 4, 5, ... These are the main clusters of energy levels in the atom, also called shells. So far as we know, there is no highest value for n.

2. Azimuthal Quantum Number, l Within each shell except the first one ($n = 1$), there are *subshells*. The number of subshells in a shell is equal to the principle quantum number. For example, for $n = 3$ there are three subshells. Values for l are integers ranging from 0 to $(n - 1)$. Typically, these subshells are referred to by the letters *s*, *p*, *d*, *f*, and *g* rather than by the values of l. These common letter designations are shown Table 1.1. The azimuthal quantum numbers describe specific types of subshell configurations. So for example, within any shell the *s* subshell is always structured the same way. Likewise, the *p* subshell has the same general structure in every shell except $n = 1$ (since $n = 1$ doesn't have a *p* subshell). Again, the number of subshells in a given shell is equal to the principle quantum number. So, in the first shell there is one subshell, denoted as 1*s*. In the $n = 2$ shell, there are two subshells, denoted as 2*s* and 2*p*, and so on. (Note: The azimuthal quantum number is also sometimes called the *angular momentum quantum number*.)

l value	Common Letter Designation
0	*s*
1	*p*
2	*d*
3	*f*
4	*g*

Table 1.1. Letters used to designate values of the azimuthal quantum number, l.

3. Magnetic Quantum Number, m_l Within each subshell (numbered l), the possible values for m_l are the integers ranging from $-l$ to l. So, in a subshell with $l = 2$, the values for m_l are −2, −1, 0, 1, and 2. The magnetic quantum number is associated with specific shapes and orientations of orbitals within a subshell. A important point to note is that *any orbital in an atom can hold at most two electrons.*

4. Spin Projection Quantum Number, m_s As you recall, the Pauli exclusion principle requires every electron in an atom to be in a unique quantum state. That is, each electron has a unique set of quantum numbers. And since each orbital can hold two electrons, we need one more piece of information to distinguish from one another the quantum states of the two electrons. This characteristic is called *spin*. Unfortunately, it's a very misleading term because electrons aren't really spinning. In fact, it's pretty hard to say exactly *what* they are doing. But anyway, accepting spin as a real property analogous in some way to spinning, any two electrons in the same orbital have opposite spins. The two possible values for electron spin are $m_s = +1/2$ and $m_s = -1/2$, and we call these "spin up" and "spin down." At this point in your career you really don't need to worry about what these strange names and numbers mean. The fact is, if there are two electrons in the same orbital (and there can be at most two) one has spin up and one has spin down. This final quantum specification allows each electron in every atom to inhabit a unique quantum state.

All this information pertaining to the first three quantum numbers is summarized in Table 1.2. Hopefully your understanding of all these shells, subshells and orbitals will be enhanced by looking at images of computer models of the orbitals. Let's be a bit clearer about what these orbitals are: they represent the solutions to the Schrödinger equation for electrons in atoms with different energies. Table 1.3 depicts the orbitals in the various subshells associated with the first three shells, $n = 1$ through $n = 3$. Note first that in each shell there is an *s* orbital. These are spherical in shape. The models shown depict the sphere cut in half so you can see the relative sizes. If you look carefully at the 2*s* orbital, you can see the tiny 1*s* orbital inside it. And inside the 3*s* orbital you can see both the 1*s* and 2*s* orbitals inside it. We are coming back to electron energy soon, but for now note that within any shell, the *s* orbital is the lowest energy orbital in that shell. Note also that for all these orbital arrangements, the atomic nucleus is at the center. All orbitals are symmetric about the nucleus.

Beginning with $n = 2$, there is a *p* subshell in each shell, and beginning with $n = 3$ there is also a *d* subshell in each shell. The orbitals in the *p* subshell are usually described as resembling "dumbbells" because of their twin lobes. There are three of these twin-lobed orbitals in each *p* subshell, each oriented at right angles to the other two. For this reason, they are designated the p_x, p_y, and p_z orbitals—they can be thought of as lined up along the *x*, *y*, and *z* axes in a three-dimensional Cartesian coordinate system as depicted in Figure 1.12. (I explain the elongated appearance of the orbitals shown in Figure 1.12 shortly.) In Table 1.3, in order to make the *s* orbitals large enough to see and still have room to fit the *p* and *d* orbitals on the page, the *p* and *d* orbitals are shown much smaller than their actual size relative to the *s* orbitals.

Looking now at the $n = 3$ orbitals in Table 1.3, note that the 3*p* orbitals are shown surrounding the 2*p* orbitals. The three 3*p* orbitals are superimposed on each other just as the 2*p* ones are

n	Possible Values of l	Subshell Name	Possible Values of m_l (Each value corresponds to one orbital.)	Number of Orbitals in the Subshell	Total Number of Orbitals in the Shell ($= n^2$)
1	0	1*s*	0	1	1
2	0	2*s*	0	1	4
	1	2*p*	−1, 0, 1	3	
3	0	3*s*	0	1	9
	1	3*p*	−1, 0, 1	3	
	2	3*d*	−2, −1, 0, 1, 2	5	
4	0	4*s*	0	1	16
	1	4*p*	−1, 0, 1	3	
	2	4*d*	−2, −1, 0, 1, 2	5	
	3	4*f*	−3, −2, −1, 0, 1, 2, 3	7	

Table 1.2. Subshells and orbitals for the $n = 1$ through $n = 4$ shells.

(Figure 1.12). Finally, you can see that shapes of the five 3*d* orbitals are pretty bizarre. These orbitals are also superimposed on each other, and the whole bunch of them is superimposed on top of all the other orbitals in the table. Then, of course, there are all the orbitals for higher principle quantum numbers superimposed on top of them. Beginning with the $n = 4$ shell (not shown in the table) there is an *f* subshell in each shell. There are seven orbitals in each *f* subshell,

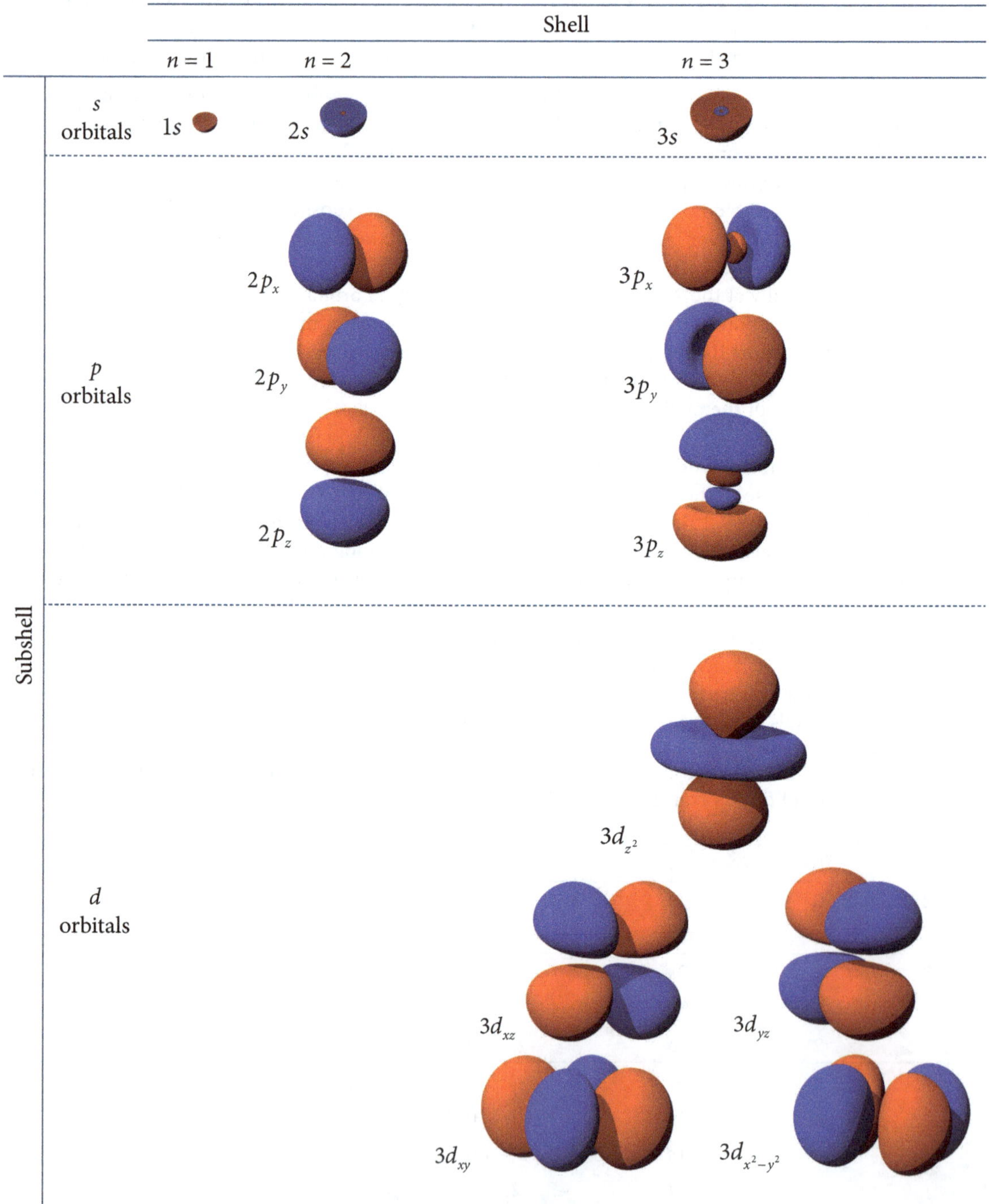

Table 1.3. Shapes of *s*, *p*, and *d* orbitals for the $n = 1$, $n = 2$, and $n = 3$ shells.

and they sort of resemble the *d* orbitals, only with six or eight lobes instead of four.

Recall that each orbital can house a maximum of two electrons (with opposite spins). For example, just to be clear, the $2p_x$ orbital with its two lobes is a single orbital (even though Table 1.3 shows the two lobes in different colors). Likewise, the $3d_{z^2}$ orbital with its two lobes and doughnut around the middle is also a single orbital. With a maximum of two electrons in each orbital, you can see that the first shell, $n = 1$, can hold at most two electrons, both in the 1*s* orbital. The $n = 2$ shell can hold a maximum of eight electrons: two in the 2*s* orbital, and two in each of the three 2*p* orbitals. The $n = 3$ shell can hold up to 18 electrons: two in the 3*s* orbital, six in the 3*p* orbitals, and a total of 10 in the 3*d* orbitals.

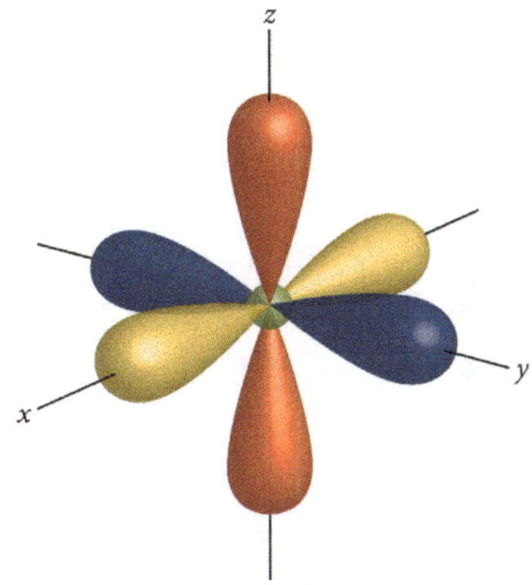

Figure 1.12. The elongated "probability distributions" of the 2*s* and 2*p* orbitals shown together. (2*s* = green; $2p_x$ = yellow, $2p_y$ = blue, $2p_z$ = red)

Before we move on and get back to talking about energy, one more important point should be made about orbitals. As noted above, the orbitals shown in Table 1.3 are the solutions to the Schrödinger equation. However, it is not correct to think of these shapes as locating where the electrons *are*. (Remember, the world of quantum mechanics is weird.) But it turns out that if we *square* the solutions to the Schrödinger equation we get shapes indicating *probabilities* of where the electrons are. This is what is depicted in Figure 1.12. Squaring the 2*p* solutions elongates the shapes of the orbitals. These orbitals that come from squaring solutions to the Schrödinger equation are called *probability distributions*. They should be envisioned as fuzzy at the edges and denser in the middle, indicating a lower probability that an electron is at the edge of the orbital and a higher probability that an electron is in the center part of the orbital.

Note just one more feature of the orbital arrangements: just because an orbital has more than one part—like the two lobes of a *p* orbital—does not mean that one electron is in one lobe and the other electron is in the other lobe. Instead, both electrons inhabit both lobes. Even stranger, to pass from one lobe to the other the electron somehow passes right through the atomic nucleus. (Don't hurt your brain by trying too hard to understand this. No one else understands it either! Electrons are very strange.)

1.3.3 The Aufbau Principle, the Madelung Rule, and Hund's Rule

Now that you know how the orbitals are arranged, we return to the topic of electron energies. Let's begin by recalling how one knows how many electrons an atom has. Unless it has ionized, an atom has the same number of electrons as protons, and the number of protons is given by the *atomic number* (*Z*). For example, if you check the periodic table inside the back cover of the book, you see that iron is element 26. This means an atom of iron has 26 protons and 26 electrons. The protons are all in the nucleus with the neutrons. The electrons are distributed around in various orbitals.

Figure 1.13 is another type of energy level diagram and illustrates the energies associated with the different orbitals. In this diagram, each little square represents an orbital, and each string of connected squares represents a subshell. On the left are the orbital energies for the hydrogen atom. As you see, all orbitals associated with a given principle quantum number have the same energy. These are the energies shown in Figures 1.6 and 1.8.

On the right side of Figure 1.13 is a general arrangement depicting the energies for atoms other than hydrogen. Here the energies go up with each subshell. For example, subshell 4*f* has

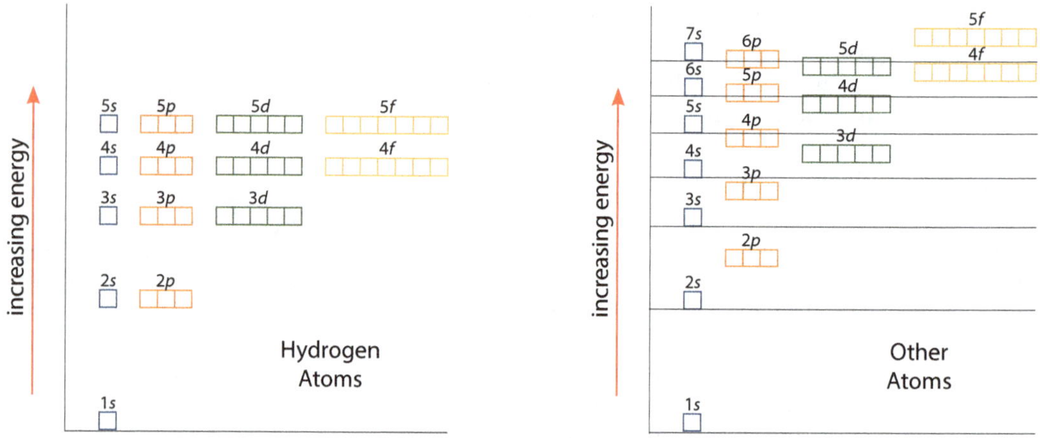

Figure 1.13. In hydrogen atoms, all orbitals within a given shell are at the same energy level. In atoms of other elements, orbital energy increases with increasing azimuthal quantum number, and the sequence of energies follows the Madelung rule.

a higher energy than 4*d*, which has a higher energy than 4*p*, which has a higher energy than 4*s*. Also, note especially that the energies associated with different principle quantum numbers (shells) overlap. Thus, subshell 4*s* has a lower energy than subshell 3*d*.

An important point to note about the right side of Figure 1.13 is that the exact energies associated with various subshells are different for every atom. With only one electron, the orbitals in an energy level of a hydrogen atom are basically all the same. But with multiple electrons repelling each other in an atom, the subshells begin spreading upward and each subshell is at a different energy. The amount of spread—and thus the exact energy associated with each subshell—is different for every atom. The important consequence of this for what we have covered so far in this chapter pertains to atomic spectra. The energy released by an electron transition from, say, a 5*d* orbital to a 4*p* orbital depends on the atom—that is, the element—involved. As you know, the energy in an emitted photon determines its wavelength and color (the Planck relation). The fact that the energies for the different orbitals depend on the element is the reason why spectroscopy can be used to identify the presence of elements in a sample. Each atom emits its own spectrum of wavelengths corresponding to the unique energy differences between the orbitals in that particular kind of atom.

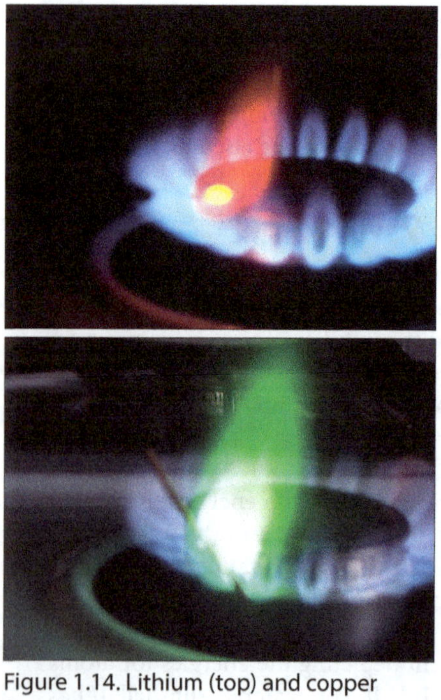

Figure 1.14. Lithium (top) and copper flame tests.

The colors emitted by excited atoms in two metals are illustrated in Figure 1.14. The images show lithium and copper wires heated in a stove-top gas flame, causing electrons in the metal atoms to absorb photons of heat energy (electromagnetic radiation in the infrared region). As the electrons return to the ground state, they emit photons of visible light, and the colors produced depend on the energies of the subshells in the atoms of the respective metals. If the flames are observed through a prism, the colors in the flames are separated into a line spectrum and the individual color wavelengths can be identified. A test like this that uses flame as the energy source for exciting the metal is called a *flame test*.

With the energy picture under our belts, we are finally ready to describe how electrons are arranged in atoms. We are describing here the electron positions when atoms are in the ground state. You already know that when atoms are excited, electrons jump from ground state energies up to higher energies.

There are three principles involved in determining electron arrangement in ground state atoms. The first is the *Aufbau principle*, named after a German word meaning "building up." The Aufbau principle states that *electrons fill places in orbitals in order of increasing energy, starting from the lowest energy orbital and going up from there*. Remember: chemistry is all about minimizing energy. Electrons in a ground-state atom go into the lowest energy orbitals available.

The second principle is the *Madelung rule*. This principle specifies the order of the shells and orbitals for increasing energy. On the right side of Figure 1.13, the sequence the orbitals are in as energy increases follows the Madelung rule. Another common way of depicting the sequence of energies according to the Madelung rule is shown in Figure 1.15. If you start at the top and follow the arrows in descending order, you get the same sequence of orbitals as shown on the right side of Figure 1.13. The mathematical principle involved is that each arrow in Figure 1.15 represents a particular value of the sum of the principle quantum number and the azimuthal quantum number, $n + l$.

For example, look at the arrow starting at the $3d$ orbital. For $3d$, $n = 3$ and $l = 2$, so $n + l = 5$. (The values of l are shown in Table 1.1.) For $4p$, $n = 4$ and $l = 1$, and $n + l = 5$. For $5s$, $n = 5$ and $l = 0$, so $n + l = 5$. So the subshells fill up in the sequence shown in Figure 1.15.

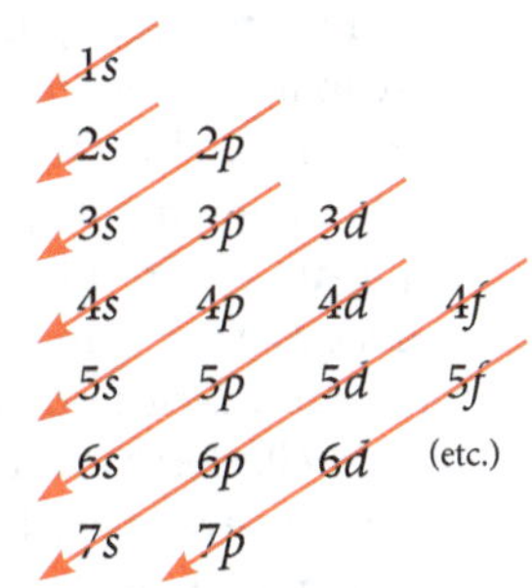

Figure 1.15. Since each red arrow represents a particular value of $n + l$, this diagram shows the order in which the subshells fill with electrons, according to the Madelung rule.

The third principle involved in electron arrangements is *Hund's rule*, which applies to the case of subshells that are only partially filled. Hund's rule states that *if orbitals of equal energy are available within a subshell, electrons fill them all up singly before they begin doubling up in orbitals*. For example, as you can see from the right side of Figure 1.13, the $3d$ subshell contains five orbitals of equal energy. According to Hund's rule, if there are electrons in this subshell, but not enough electrons to fill the subshell, the electrons go into the orbitals as one electron per orbital until each of the five orbitals has one electron in it. After that, any remaining electrons go in as the second electron in each orbital until each electron has a place. And again, remember that all orbitals can hold at most two electrons.

The principle at work behind Hund's rule is again energy minimization. Spreading single electrons in the orbitals of unfilled subshells is a lower energy configuration than putting pairs of electrons together when other orbitals remain empty. Minimizing the energy this way also makes the atom more stable, just as the cone on its side in Figure I.6 is more stable than the cone on its point.

1.4 Electron Configurations

1.4.1 Electron Configurations and Orbital Diagrams

You may be pleased to know that the ocean of information described in the previous section will be a lot easier to remember after you have had a bit of practice writing *electron configurations* to indicate where all the electrons are in an atom of a given element. The electron configuration for a given element is a list, written in a particular format, of all the subshells in use in an atom and how many electrons are in each one. As an example, consider iron, atomic number 26

($Z = 26$). There are 26 electrons in an atom of iron. The subshells required to hold them all, in order of increasing energy according to the Madelung rule, are as follows:

1*s*	2*s*	2*p*	3*s*	3*p*	4*s*	3*d*
holds 2 electrons	holds 2 electrons	holds 6 electrons	holds 2 electrons	holds 6 electrons	holds 2 electrons	holds 6 electrons in 10 places

The electron configuration is formed simply by chaining these together, placing the numbers of electrons as superscripts on the subshells they go with, without any punctuation. In front of the electron configuration, it is customary to place the element's chemical symbol followed by a colon. So, the electron configuration for iron is written as follows:

Fe: $1s^2 2s^2 2p^6 3s^2 3p^6 4s^2 3d^6$

Electron configurations only indicate subshells; they do not indicate which orbitals electrons are in inside the subshells. But we can use an *orbital diagram* similar to Figure 1.13 to show more precisely where the electrons are. Remember, Hund's rule comes into play, requiring that orbitals of equal energy each receive one spin-up electron before any of them take a second spin-down electron. A great metaphor for this was first used by Wolfgang Pauli, who formulated the Pauli exclusion principle. Pauli said that when filling up the orbitals in a subshell, electrons are like passengers filling a bus. Each takes a seat by himself until every seat has one person in it. After that, people start doubling up.

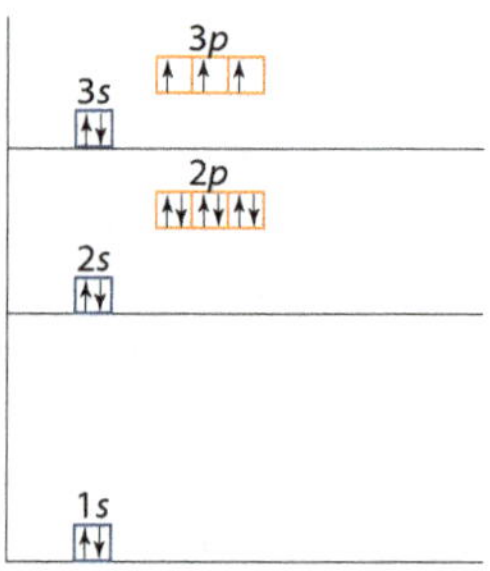

Figure 1.16. The electron arrangement for phosphorus, with 15 electrons.

Figure 1.16 shows the electron arrangement for phosphorus, $Z = 15$. Each of the little arrows represents one electron, with upward arrows representing spin up and downward arrows representing spin down. Notice that the three electrons in the 3*p* subshell are placed so that each orbital contains one spin-up electron, as Hund's rule requires.

To make an orbital diagram, you simply show the orbitals in order, side by side, and put in the arrows representing the electrons. Thus, the orbital diagram and electron configuration for phosphorus are as follows:

Orbital Diagram	Electron Configuration
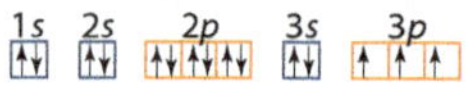	P: $1s^2 2s^2 2p^6 3s^2 3p^3$

Here are three more examples: sodium (Na) with 11 electrons, chlorine (Cl) with 17 electrons, and nickel (Ni) with 28 electrons:

Z	Orbital Diagram	Electron Configuration
11		Na: $1s^2 2s^2 2p^6 3s^1$
17		Cl: $1s^2 2s^2 2p^6 3s^2 3p^5$
28		Ni: $1s^2 2s^2 2p^6 3s^2 3p^6 4s^2 3d^8$

Note in each case that the superscripts add up to the number of electrons being represented in the notation, 11, 17, and 28 in the three examples above.

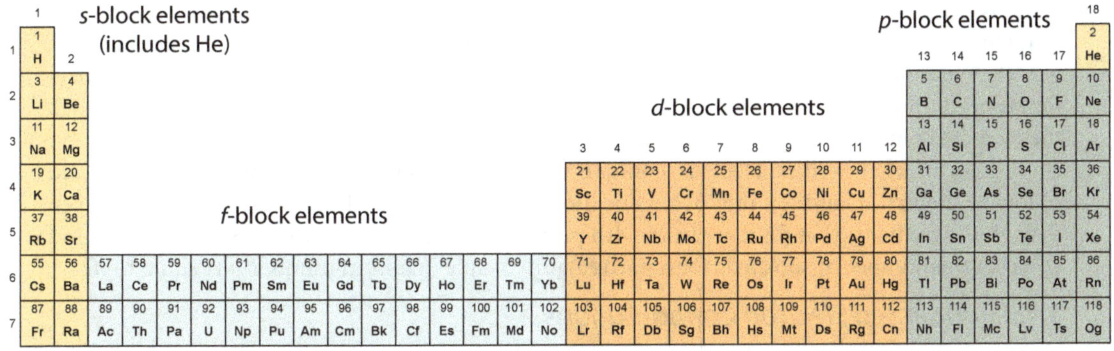

Figure 1.17. Common terms applied to groups of elements based on the type of suborbital that is being filled as we move through the block in a given period (row). Group numbers shown are the numbers typically shown on the standard table with the *f*-block elements taken out.

To write an electron configuration, you need to know the atomic number (*Z*) to get the number of electrons, the number of electrons that can reside in each of the types of subshells (*s*, *p*, *d*, and *f*; we won't deal with *g* subshells), and the energy sequence according to the Madelung rule. The periodic table itself is the best aid to writing an electron configuration. Figure 1.17 is a depiction of the periodic table in its full, long form. The rare-earth elements, normally shown separated out beneath the main table, are shown in their rightful place.[3] Notice the captions on the different blocks of elements in the table. There are two groups (columns) in the *s* block, because *s* subshells can only hold two electrons. The *p* subshells can hold six electrons because there are three orbitals in each *p* subshell, and the *p* block is six groups wide. In the same way, the *d* block is 10 groups wide, and the *f* block is 14 groups wide. (See Table 1.2 to confirm the number of orbitals in a subshell, keeping in mind that each orbital can hold up to two electrons.)

Now look what happens as we move in order through elements in the table:

- Hydrogen (H) has one electron, helium (He) has two. In the element at the end of the 1st period (row)—helium—the 1*s* subshell is full.
- Lithium (Li) has three electrons, beryllium (Be) has four. With Be, the 1*s* and 2*s* subshells are full. The 5th electron possessed by boron (B) goes into the 2*p* subshell. So do all the additional electrons added in elements 6 through 10. Neon (Ne) has 10 electrons, so in the element at the end of the 2nd period (neon), subshells 1*s*, 2*s*, and 2*p* are all full.
- Sodium (Na) has 11 electrons, and the 11th one goes in the 3*s* subshell. Magnesium (Mg) has 12, and the 12th one fills the 3*s* subshell. Aluminum (Al) has 13, and the 13th one goes in the 3*p* subshell. In the element at the end of the third period—argon (Ar)—the 1*s*, 2*s*, 2*p*, 3*s*, and 3*p* subshells are full.
- The 4th period begins with potassium (K) and calcium (Ca), in which new electrons are placed in the 4*s* subshell. Then look what happens next: the next element, scandium (Sc), has 21 electrons and the 21st one goes into a *d* subshell, the 3*d* subshell, in fact. Each new electron for elements 21 through 30 goes into the 3*d* subshell. Then the 31st electron in gallium (Ga) is placed in the 4*p* subshell. In the element at the end of the 4th period—krypton (Kr)—the 1*s*, 2*s*, 2*p*, 3*s*, 3*p*, 4*s*, 3*d*, and 4*p* subshells are full.

---

3 The *f*-block elements shown are also called the *inner transition metals* or *rare-earth elements*. They are usually removed from the table and shown beneath it for the simple reason that with them in place, the table is inconveniently wide. Since most of our work in chemistry is with *s*-, *p*-, and *d*-block elements, the removal of the *f*-block elements in the standard representation of the periodic table doesn't cause much trouble.

This pattern continues, and the order of new shells coming into use continues to follow the Madelung rule. Now you can see why the blocks of elements in Figure 1.16 are identified the way they are. Moving from left to right across any period in the table, the additional electron for atoms of the next element goes into a subshell of the type indicated by the name of the block the element is in.

Be careful when writing electron configurations for elements at the beginning of the *d* and *f* blocks. The first *d* subshell that occurs in the sequence of shell filling is the 3*d* subshell, even though the elements that fill it are in the 4th period. Table 1.2 and Figure 1.15 will help remind you that the number of subshells for a particular principle quantum number *n* is equal to *n*.

1.4.2 Condensed Electron Configurations

You may already know that the elements in Group 18, the noble gases, are very unreactive. In fact, these elements eluded discovery by researchers for a long time. Since they don't form compounds, scientists didn't even know they existed!

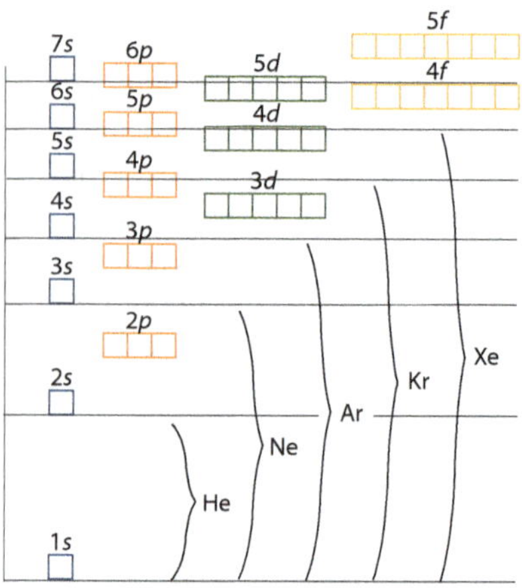

Figure 1.18. The orbital filling pattern of the noble gases.

The reason the noble gases are so nonreactive is that their electron arrangements are very stable, low-energy configurations. Obviously, each of the noble gases is at the end of a period in the periodic table (Group 18). Figure 1.18 illustrates the pattern that occurs in the orbital filling of these elements. In each case, all orbitals are filled up to but not including the *s* orbital of the next principle quantum number.

Since the noble gases are so stable, occurring as they do at the end of each period, the chemical symbols of the noble gases are used to form the so-called *condensed electron configurations*. The condensed electron configuration is a shorter, more convenient form.

Here's an example to show how this works. A glance at the periodic table shows that the only difference in the electron configurations of, say, titanium (Ti, Z = 22) and argon (Ar, Z = 18), is that titanium has four extra electrons. The electron configurations for argon and titanium are:

Ar: $1s^2 2s^2 2p^6 3s^2 3p^6$

Ti: $1s^2 2s^2 2p^6 3s^2 3p^6 4s^2 3d^2$

The condensed electron configuration for any element is written by using the chemical symbol of the noble gas in the previous period to represent all the filled orbitals up to that point, and then just adding on the orbitals in the period where the element is. The noble gas chemical symbol is always written in square brackets. The condensed electron configuration for titanium is written as

Ti: $[\mathrm{Ar}]4s^2 3d^2$

As one more example, the condensed electron configuration for phosphorus (Z = 15) is written as

P: $[\mathrm{Ne}]3s^2 3p^3$

In the condensed electron configuration, the inner-shell electrons lumped together under the noble gas symbol are called the *core electrons*.

1.4.3 Anomalous Electron Configurations

There are a few elements with electron configurations that are notable exceptions to the ordinary rules. All these exceptions are *d*-block or *f*-block elements, as shown in Figure 1.19.

We would expect the condensed electron configuration for chromium ($Z = 24$) to be

Cr: $[\text{Ar}]4s^2 3d^4$

However, it is not! One of the 4*s* electrons goes into the 3*d* subshell instead, giving

Cr: $[\text{Ar}]4s^1 3d^5$

Copper ($Z = 29$), does the same thing. We would expect the condensed electron configuration to be

Cu: $[\text{Ar}]4s^2 3d^9$

Instead, it is

Cu: $[\text{Ar}]4s^1 3d^{10}$

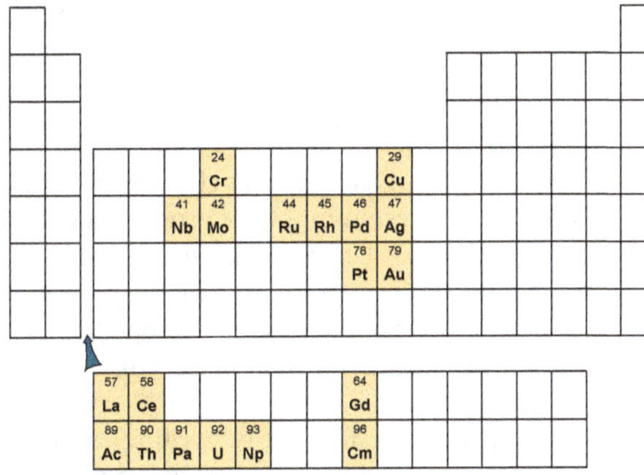

Figure 1.19. Elements with anomalous electron configurations.

The electron configurations for all the anomalous elements are shown in Table 1.4. Notice that palladium ($Z = 46$) is so anxious to get electrons into the 4*d* subshell that it steals both the electrons that ordinarily would be in the 5*s* subshell and places them in the 4*d* subshell to fill it up.

The electron configurations for these elements are not difficult to remember. For the *d*-block exceptions, except for palladium, they move one *s*-subshell electron into a *d* subshell at the highest energy in the atom. Palladium is the exception among exceptions: it moves two.

The *f*-block exceptions move one *f* subshell electron into a *d* subshell at the highest energy in the atom. As with the *d*-block exceptions, there is one exception among the *f*-block anomalies: thorium moves two *f*-subshell electrons into the *d* subshell.

Although the details are complex, the bottom line for these exceptions to the normal rules is that

Element	Atomic Number, Z	Symbol	Condensed Configuration
chromium	24	Cr	Cr: $[\text{Ar}]4s^1 3d^5$
copper	29	Cu	Cu: $[\text{Ar}]4s^1 3d^{10}$
niobium	41	Nb	Nb: $[\text{Kr}]5s^1 4d^4$
molybdenum	42	Mo	Mo: $[\text{Kr}]5s^1 4d^5$
ruthenium	44	Ru	Ru: $[\text{Kr}]5s^1 4d^7$
rhodium	45	Rh	Rh: $[\text{Kr}]5s^1 4d^8$
palladium	46	Pd	Pd: $[\text{Kr}]4d^{10}$
silver	47	Ag	Ag: $[\text{Kr}]5s^1 4d^{10}$
lanthanum	57	La	La: $[\text{Xe}]6s^2 5d^1$
cerium	58	Ce	Ce: $[\text{Xe}]6s^2 4f^1 5d^1$
gadolinium	64	Gd	Gd: $[\text{Xe}]6s^2 4f^7 5d^1$
platinum	78	Pt	Pt: $[\text{Xe}]6s^1 4f^{14} 5d^9$
gold	79	Au	Au: $[\text{Xe}]6s^1 4f^{14} 5d^{10}$
actinium	89	Ac	Ac: $[\text{Rn}]7s^2 6d^1$
thorium	90	Th	Th: $[\text{Rn}]7s^2 6d^2$
protactinium	91	Pa	Pa: $[\text{Rn}]7s^2 5f^2 6d^1$
uranium	92	U	U: $[\text{Rn}]7s^2 5f^3 6d^1$
neptunium	93	Np	Np: $[\text{Rn}]7s^2 5f^4 6d^1$
curium	96	Cm	Np: $[\text{Rn}]7s^2 5f^7 6d^1$

Table 1.4. Elements with anomalous electron configurations.

there are other factors coming into play in their atomic structure. At high energy levels, the orbital energies are very close together. The anomalous configurations represent the arrangement of electrons that minimizes the energy in the atom.

1.5 Isotopes and Atomic Masses

1.5.1 Isotopes

As you know, the atomic number (Z) of an element designates the number of protons in the nucleus of an atom of that element. For a given element, the atomic number is fixed: if an atom has a different number of protons, it is an atom of a different element. But the number of neutrons that may be present in the nucleus is not fixed. For most elements, there are variations in the number of neutrons that can be present in the nucleus. The varieties are called *isotopes*. For most elements, there is one isotope that is the most abundant in nature and several other isotopes that are also present but in smaller quantities. The general term for any isotope of any element is *nuclide*.

Isotopes are designated by writing the name of the element followed by the number of nucleons (protons and neutrons) in the isotope. The number of nucleons in a nucleus is called the *mass number*. For example, the most common isotope of carbon is carbon-12, accounting for about 98.9% of all the naturally occurring carbon. In the nucleus of an atom of carbon-12 there are six protons and six neutrons. There are two other naturally occurring carbon isotopes. Carbon-13 with seven neutrons accounts for about 1.1% of natural carbon. Atoms of carbon-14, of which only a trace exists in nature, have eight neutrons in the nucleus.

1.5.2 The Unified Atomic Mass Unit

The mass of a single atom is an extremely small number. But so much of our work in chemistry depends on atomic masses that scientists having been using units of *relative* atomic mass for a long time—all the way back to John Dalton, before actual masses of atoms were even known. Prior to the discovery of isotopes in 1912, the so-called *atomic mass unit* (amu) was defined as 1/16 the mass of an oxygen atom. After the discovery of isotopes, physicists defined the amu as 1/16 the mass of an atom of oxygen-16, but the definition used by chemists was 1/16 the average mass of naturally occurring oxygen, which is composed of several isotopes. To eliminate the confusion resulting from these conflicting definitions, the new *unified atomic mass unit* (u) was adopted in 1961 to replace them. Many texts continue to use the amu as a unit, but they define it as the u is defined. Strictly speaking, the amu is an obsolete unit that has been replaced by the u, now also called the dalton (Da). The u and the Da are alternative names (and symbols) for the same unit. The use of the dalton has increased in recent years, particularly in molecular biology.

The unified atomic mass unit, u, *is defined as exactly 1/12 the mass of an atom of carbon-12.* Table 1.5 lists a few nuclides and their atomic masses using the u as a unit of mass. All the elements listed exist as other isotopes in addition to those shown, but as you see from the percentage abundances, the ones shown are the major ones for the elements represented in the table.

1.5.3 Atomic Masses

In addition to the atomic number, the Periodic Table of the Elements lists the atomic mass in unified atomic mass units (u) for each element. But since there are multiple isotopes for just about every element, *the atomic mass values in the periodic table represent the weighted average of the masses of naturally occurring isotopes.*

An example of a weighted average is the average age of the students in the sophomore class at your school. Let's say there are 47 sophomores, 40 of whom are 15 years old and 7 of whom are 16 years old at the beginning of the school year. To determine the average age of these students, let's first determine the proportion of the students at each age.

$$\frac{40}{47} = 0.851 \ (85.1\%)$$

$$\frac{7}{47} = 0.149 \ (14.9\%)$$

To calculate the average age, we first multiply each student age by the proportion of students of that age to find the contribution to the average from each age group. Then we add the contributions together to find the weighted average age for the sophomore class.

$$\begin{array}{r} 15 \text{ years} \cdot 0.851 = 12.8 \text{ years} \\ +16 \text{ years} \cdot 0.149 = 2.38 \text{ years} \\ \hline = 15.2 \text{ years} \end{array}$$

We perform a similar calculation when computing the average atomic mass of an element from the masses of its isotopes, as shown in the following example.

Z	Nuclide	Mass (u)	Abundance (%)
1	hydrogen-1	1.0078	99.9885
1	hydrogen-2	2.0141	0.0115
6	carbon-12	12.0000	98.93
6	carbon-13	13.0034	1.078
14	silicon-28	27.9769	92.223
14	silicon-29	28.9765	4.685
14	silicon-30	29.9738	3.092
17	chlorine-35	34.9689	75.76
17	chlorine-37	36.9659	24.24
20	calcium-40	39.9626	96.941
20	calcium-42	41.9586	0.647
26	iron-54	53.9396	5.845
26	iron-56	55.9349	91.754
26	iron-57	56.9354	2.119
26	iron-58	57.9333	0.282
29	copper-63	62.9296	69.15
29	copper-65	64.9278	30.85
92	uranium-235	235.0439	0.7204
92	uranium-238	238.0508	99.2742

Table 1.5. Major isotopes for a few elements.

▼ Example 1.2

Given the isotope masses and abundances for copper-63 and copper-65 in Table 1.5, determine the atomic mass for naturally occurring copper.

Multiply each isotope's mass by its abundance to get the isotope's contribution to the average atomic mass of the element. Then add together the contributions from each isotope. The data from the table are:

copper-63: mass = 62.9296 u, abundance = 69.15%

copper-65: mass = 64.9278 u, abundance = 30.85%

$$\begin{array}{r} 62.9296 \text{ u} \cdot 0.6915 = 43.52 \text{ u} \\ + \ 64.9278 \text{ u} \cdot 0.3085 = 20.03 \text{ u} \\ \hline = 63.55 \text{ u} \end{array}$$

Compare this value to the value shown in the periodic table inside the back cover of the text.

▲

The unified atomic mass unit, u, is defined as 1/12 the mass of an atom of carbon-12. Although the value of this mass is quite close to the masses of the proton and neutron, it is not exact because of the mass of the electrons in atoms of carbon-12, and also because of the mass-en-

Particle	Mass
proton	1.007277 u
neutron	1.008665 u
electron	0.0005486 u

Table 1.6. Masses in u of the three basic subatomic particles.

ergy involved in binding the nucleus of the atom together. (The mass of nucleons bound together in a nucleus does not equal the sum of their individual masses.) Table 1.6 shows the masses of the three basic subatomic particles in unified atomic mass units.

Still, the proton and neutron masses are very close to unity (one) and the electron mass is extremely small. This means that for elements with a very large abundance of one isotope we can use the atomic mass and atomic number in the periodic table to determine the numbers of protons and neutrons in the nucleus of the most common isotope. For example, from Table 1.5, the mass of uranium-238 is very close to 238 u. Since an atom of uranium-238 has 92 protons, the balance of the mass is essentially all neutrons. Thus, there are 238 – 92 = 146 neutrons in uranium-238.

1.5.4 The Mole and the Avogadro Constant

When solving problems in chemistry, we are generally working with chemical reactions in which huge numbers of atoms are involved, including all the naturally occurring isotopes, so performing reaction calculations with the masses of individual atoms is not practical. However, the average mass of a given multiple of some kind of atom is simply that multiple times the atomic mass. The mass of one million atoms of aluminum is 1,000,000 times the atomic mass of aluminum.

In chemistry, the standard bulk quantity of substance used in calculations is the *mole* (mol). The mole is a particular number of particles of a substance, just as the terms *dozen*, *score*, and *gross* refer to specific numbers of things (12, 20, and 144, respectively). A mole is exactly $6.02214076 \times 10^{23}$ particles of a substance.

This value is known today as the *Avogadro constant*, N_A. More formally, the Avogadro constant is defined as exactly:

$$N_A = 6.02214076 \times 10^{23}\ \text{mol}^{-1} \tag{1.6}$$

Usually, we just round this value to 6.022×10^{23} mol^{-1}. In the next section, I'll describe why this value is what it is, instead of being a more convenient round number. For the moment, let's focus on what it means. Now, don't freak out over the unit of measure. Allow me to explain. Raising a unit of measure to the power –1 is mathematically equivalent to placing the unit in a denominator because $x^{-1} = \frac{1}{x}$. In other words, Equation (1.6) is the same thing as saying "$6.02214076 \times 10^{23}$ per mole." To make things even clearer, it's okay to say it this way: "N_A is about 6.022×10^{23} particles per mole." This is the way I like to think of it when performing unit conversions, as we do quite a lot in coming chapters. Without the units of measure, the value $6.02214076 \times 10^{23}$ is called *Avogadro's number*. With the units, it is called the Avogadro constant. Using this terminology, the mole can be defined this way:

A mole is the amount of a pure substance (element or compound) that contains Avogadro's number of particles of the substance.

Let's now consider what we mean when we refer to *particles* of a substance. For substances that exist as molecules, the particles are the molecules. For substances that exist as individual atoms, the particles are the individual atoms. Metals are like this, since a pure metal is composed of individual atoms of the same element joined together in a crystal lattice. The noble gases are also like this. The noble gases are located in the far right-hand column of the Periodic Table of the Elements. As I discuss more in coming chapters, atoms of noble gases are almost completely unreactive—they don't bond with other atoms at all. At ordinary temperature and pressure, the noble gases are gases composed of individual atoms.

For crystalline compounds, the "particles" in a mole of the substance are the *formula units* in the crystal lattice. A formula unit is one set of the atoms represented by the chemical formula of the compound. For example, the chemical formula for calcium carbonate is $CaCO_3$. One formula unit of calcium carbonate includes one calcium atom, one carbon atom, and three oxygen atoms.

The value of the Avogadro constant was determined approximately by French Physicist Jean Perrin (Figure 1.20) in the early 20th century. Perrin determined the value of the constant through several different experimental methods. In the 19th century, many scientists did not yet accept the existence of atoms as a scientific fact and Perrin's research put the atomic nature of matter beyond dispute. For this work, he received the Nobel Prize in Physics in 1926. Perrin proposed naming the constant after Amedeo Avogadro, a 19th-century Italian scientist who was the first to propose that the volume of a gas at a given temperature and pressure is proportional to the number of particles of the gas (atoms or molecules), regardless of the identity of the gas. In fact, at 0°C and atmospheric pressure, one mole of any gas occupies a volume of 22.4 L.

Figure 1.20. French physicist Jean Perrin (1870–1942).

1.5.5 Molar Mass and Formula Mass

Since 2019, the value of the Avogadro constant in Equation (2.2) is exact by definition. But the number has the value it does because it was originally chosen so that the average atomic mass in u of a molecule of a compound, as computed from the mass values in the periodic table, would be numerically equivalent to the mass of one mole of the compound in grams per mole. (We address these calculations below.) Now, recall that the definition of the unified atomic mass unit (or dalton) is such that an atom of carbon-12 has a mass of exactly 12 u. According to the original definition of Avogadro's number, there were also exactly 12 grams of carbon-12 in one mole of carbon-12. So according to these definitions, an atom of carbon-12 has a mass of exactly 12 u, and a mole of carbon-12 had a mass of exactly 12 grams. This quantity, the mass of one mole of a substance, is called the *molar mass*. Because of the way the molar and atomic masses were defined, the molar mass for an atom was numerically equivalent to the atomic mass. As a result of the 2019 redefinition of Avogadro's number, the atomic mass in u and the molar mass in g/mol are no longer exactly equivalent. However, they are extremely close and may still be treated as equal for practical purposes. (The difference is a factor of only about 4×10^{-10}.)

Even though the exact equivalence ended in 2019, these are still very handy definitions! For example, from the periodic table we find that the average mass of one atom of silicon ($Z = 14$) is 28.0855 u. This also tells us that the mass of one mole of silicon is 28.0855 g, so the molar mass of silicon is 28.0855 g/mol. Likewise, from the periodic table we find that the average mass of one atom of copper ($Z = 29$) is 63.546 u. This also tells us that the mass of one mole of copper is 63.546 g, so the molar mass of copper is 63.546 g/mol. For the elements that exist as single atoms, the molar mass in g/mol and the atomic mass in u are numerically equivalent (almost).

From the periodic table, we can also determine the molar mass of compounds—the mass of a mole of the compound. We simply add up the molar masses for the elements in the chemical formula, taking into account any subscripts present in the formula, and we have the molar mass for the compound in g/mol. If we add up the element atomic masses in unified atomic mass

Quantity	Units	Definition
molar mass	g/mol	The mass of one mole of a substance, approximately equal to the sum of the atomic masses of the elements in a chemical formula, taking into account the subscripts indicating atomic ratios in the compound.
formula mass	u	The mass of one formula unit of a substance. Numerically nearly equivalent to the molar mass.
molecular mass	u	The average mass of a single molecule of a molecular substance. Numerically equivalent to the formula mass. (May also be converted to grams and expressed in grams, see Section 2.4.4.)

Table 1.7. Definitions and units for molar mass, formula mass, and molecular mass.

units we obtain what is called the *formula mass* of the compound in u. If the compound is molecular, then the formula mass may also be referred to as the *molecular mass*, the average mass of a single molecule of the substance.

The details of these three different mass terms are summarized in Table 1.7.

▼ Example 1.3

Determine the formula mass and molar mass for water, H_2O. We note that since water is composed of molecules, the formula mass may also be called the molecular mass.

From the periodic table, the atomic masses of hydrogen (H) and oxygen (O) are:

H: 1.0079 u

O: 15.9994 u

There are two hydrogen atoms and one oxygen atom in each water molecule, so we multiply these numbers by the element atomic masses and add them up to get the formula mass of H_2O.

$$(2 \times 1.0079\ \text{u}) + (1 \times 15.9994\ \text{u}) = 18.0152\ \text{u}$$

Thus, the formula mass for water is 18.0152 u. This is also the molecular mass. The calculation of the molar mass is identical, except we use units of g/mol instead of u.

From the periodic table, the molar masses of hydrogen (H) and oxygen (O) are:

$$\text{H: } 1.0079\ \frac{\text{g}}{\text{mol}}$$

$$\text{O: } 15.9994\ \frac{\text{g}}{\text{mol}}$$

There are two hydrogen atoms and one oxygen atom in each water molecule, so we multiply these numbers by the element masses and add them up to get the molar mass of H_2O.

$$\left(2 \times 1.0079\ \frac{\text{g}}{\text{mol}}\right) + \left(1 \times 15.9994\ \frac{\text{g}}{\text{mol}}\right) = 18.0152\ \frac{\text{g}}{\text{mol}}$$

▲

▼ Example 1.4

Determine the molar mass for nitrogen gas, N_2.

From the periodic table, the atomic mass of nitrogen (N) is:

$$\text{N: } 14.0067\ \frac{\text{g}}{\text{mol}}$$

There are two nitrogen atoms in each molecule, so we multiply the atomic mass by two to get the molar mass of N_2.

$$\left(2\times 14.0067\ \frac{\text{g}}{\text{mol}}\right)=28.0134\ \frac{\text{g}}{\text{mol}}$$

▲

▼ Example 1.5

Determine the mass in grams of 2.5 mol sodium bicarbonate, $NaHCO_3$ (baking soda).

In any problem like this, we first find the molar mass of the given compound. Then we simply use that molar mass to compute the mass of the given quantity. From the periodic table, the atomic masses of the elements in the compound are:

$$\text{Na: } 22.9898\ \frac{\text{g}}{\text{mol}}$$

$$\text{H: } 1.0079\ \frac{\text{g}}{\text{mol}}$$

$$\text{C: } 12.011\ \frac{\text{g}}{\text{mol}}$$

$$\text{O: } 15.9994\ \frac{\text{g}}{\text{mol}}$$

The oxygen appears three times in the formula, so its mass must be multiplied by three and added to the others.

$$22.9898\ \frac{\text{g}}{\text{mol}}+1.0079\ \frac{\text{g}}{\text{mol}}+12.011\ \frac{\text{g}}{\text{mol}}+\left(3\times 15.9994\ \frac{\text{g}}{\text{mol}}\right)=84.007\ \frac{\text{g}}{\text{mol}}$$

This value is the molar mass for $NaHCO_3$. To find the mass of 2.5 mol we multiply:

$$2.5\ \text{mol}\cdot 84.007\ \frac{\text{g}}{\text{mol}}=210\ \text{g}$$

▲

▼ Example 1.6

A scientist measures out 125 g of potassium chloride (KCl). How many moles of KCl does this quantity represent?

First, determine the molar mass of KCl. From the periodic table:

$$K: 39.098 \frac{g}{mol}$$

$$Cl: 35.4527 \frac{g}{mol}$$

The formula includes one atom of each, so we add them to obtain the molar mass:

$$39.098 \frac{g}{mol} + 35.4527 \frac{g}{mol} = 74.551 \frac{g}{mol}$$

Beginning now, always think of the molar mass of any substance as a conversion factor that can be written right side up or upside down to convert grams to moles or vice versa. For KCl, 74.551 g is equivalent to 1 mol, so these quantities can be written as conversion factors, like this:

$$\frac{74.551 \text{ g}}{1 \text{ mol}} = \frac{1 \text{ mol}}{74.551 \text{ g}}$$

This makes the last step of this problem easy. Just select the way of writing the molar mass conversion factor that cancels out the given units (g) and gives the units required (mol). This is nothing but a unit conversion.

$$125 \text{ g} \cdot \frac{1 \text{ mol}}{74.551 \text{ g}} = 1.68 \text{ mol}$$

▲

The photograph in Figure 1.21 shows one mole of each of four substances. The first is one mole of copper, equal to 63.5 g. The second is a 250-mL beaker containing one mole of water. As you can see, this is not much water—only 18 mL. In the upper right is a weigh tray containing one mole of sodium chloride, 40.0 g. (This is just under 1/4 cup.) Finally, one mole of baking soda, 84.1 g. (This is right at 1/3 cup.)

Figure 1.21. Clockwise from left are shown 1 mole of copper, 1 mole of water, 1 mole of table salt, and 1 mole of baking soda.

▼ Example 1.7

Calculate the number of water molecules in a 1.00-liter bottle of water.

The logic of this problem, in reverse, is as follows: To calculate a number of molecules, we must use the Avogadro constant. To use the Avogadro constant, we need to know the number of moles of water we have. To determine the number of moles, we need to know both the molar mass and the mass of the water. To determine the mass from a volume, we use the density equation.

So we begin with the given information and the density equation to determine the mass of water we have. The given information and unit conversions are as follows:

$$V = 1.00\ \text{L} \cdot \frac{1000\ \text{cm}^3}{1\ \text{L}} = 1.00 \times 10^3\ \text{cm}^3$$

$$\rho = 0.998\ \frac{\text{g}}{\text{cm}^3}$$

$$m = ?$$

Now we write down the density equation and solve for the mass:

$$\rho = \frac{m}{V}$$

$$m = \rho \cdot V = 0.998\ \frac{\text{g}}{\text{cm}^3} \cdot 1.00 \times 10^3\ \text{cm}^3 = 998\ \text{g}$$

Next we need the molar mass of water. We calculated this in Example 1.3 and obtained 18.0152 g/mol. We use this molar mass as a conversion factor to convert the mass of water into a number of moles of water:

$$998\ \text{g} \cdot \frac{1\ \text{mol}}{18.0152\ \text{g}} = 55.40\ \text{mol}$$

This intermediate result has four significant digits—one more than we need in the final result. Finally, with the number of moles in hand we use the Avogadro constant to determine how many particles of water this is, which is identical to the number of water molecules.

$$55.40\ \text{mol} \cdot \frac{6.022 \times 10^{23}\ \text{particles}}{\text{mol}} = 3.34 \times 10^{25}\ \text{particles}$$

▲

1.5.6 Gram Masses of Atoms and Molecules

The molar mass from the periodic table and the Avogadro constant can be used to calculate the mass in grams of an individual atom. Recall that the atomic mass value in the periodic table gives both the average atomic mass in u, and the molar mass in g/mol. Knowing the molar mass in g/mol we can simply divide by the number of atoms there are in one mole to find the mass of one atom in grams. Although this kind of calculation is quite simple, I have found that it is *very* easy for students to get confused and not be able to determine whether one should multiply or divide or what. So here's a problem solving tip: let the units of measure help you figure out what to do. If you include the units of measure in your work and pay attention to how the units cancel out or don't cancel out, these calculations are pretty straightforward. *Keep this principle firmly in mind throughout your study of chemistry!* Units of measure are not an annoying burden; they are the student's friend.

▼ Example 1.8

Determine the average mass in grams of an atom of boron.

From the periodic table we find that the molar mass of boron is 10.811 g/mol. One mole consists of Avogadro's number of atoms of boron, so if we divide the molar mass by the Avogadro constant, we will have the mass of a single atom of boron. Let's begin by setting up the division I just described, and then use the old invert-and-multiply trick for fraction division to help with the unit cancellations.

$$\frac{10.811\ \frac{\text{g}}{\text{mol}}}{6.0221\times10^{23}\ \frac{\text{particles}}{\text{mol}}}=10.811\ \frac{\text{g}}{\text{mol}}\cdot\frac{1}{6.0221\times10^{23}}\ \frac{\text{mol}}{\text{particles}}$$

$$=\frac{10.811}{6.0221\times10^{23}}\ \frac{\text{g}}{\text{particle}}=1.7952\times10^{-23}\ \frac{\text{g}}{\text{particle}}$$

So the average mass of one boron atom is 1.7952×10^{-23} g. Note that I use five digits in the value of the Avogadro constant to preserve the precision we have in the molar mass.

▲

For molecular substances, the molar mass can be used to compute the molecular mass in grams, the average mass of one molecule. This is done by first computing the molar mass of the compound, just as we did before. Then we simply divide by the Avogadro constant to obtain the mass of one molecule.

Like the atomic mass, the molecular mass is an average mass, since the atomic masses used in calculating the molar mass are all based on the average mass of different isotopes with their abundances taken into account. The molecular mass for a *specific* molecule would have to be calculated based on the specific masses of the nuclides in the molecule.

▼ Example 1.9

Determine the mass in grams of one molecule of carbon tetrachloride, CCl_4.

From the periodic table we find that the molar masses of carbon and chlorine are 12.011 g/mol and 35.4527 g/mol, respectively. From this we calculate the molar mass of CCl_4:

$$\left(1\times12.011\ \frac{\text{g}}{\text{mol}}\right)+\left(4\times35.4527\ \frac{\text{g}}{\text{mol}}\right)=153.822\ \frac{\text{g}}{\text{mol}}$$

With this molar mass we use the Avogadro constant to get the molecular mass in grams. This time, instead of writing the Avogadro constant in the denominator of a big fraction, I simply treat it as a conversion factor and write it in the equation such that the mole units cancel out. (This is the way I always perform such calculations.) I also use six digits in the Avogadro constant to preserve the precision we have in the molar mass.

$$153.822\ \frac{\text{g}}{\text{mol}}\cdot\frac{1\ \text{mol}}{6.02214\times10^{23}\ \text{particles}}=2.55427\times10^{-22}\ \frac{\text{g}}{\text{particle}}$$

▲

1.5.7 Percent Composition and Empirical Formulas

Laboratory chemical analysis of a substance enables a chemist to determine the *percent composition* of the substance. When a new compound is discovered, chemists place a high priority on determining the percentages, by mass, of each element in the substance. This is the percent composition. From the percent composition a so-called *empirical formula* for the substance can be worked out—a formula that represents the ratios of the elements in the substance.

For example, suppose laboratory analysis of a 221.6-g sample of ascorbic acid (vitamin C) results in the following mass data:

H: 10.15 g
C: 90.68 g
O: 120.8 g

If we divide each of the mass values by the total mass of the sample we have the percent composition of the sample:

$$\text{H: } \frac{10.15 \text{ g}}{221.6 \text{ g}} = 0.04580$$

$$\text{C: } \frac{90.68 \text{ g}}{221.6 \text{ g}} = 0.4092$$

$$\text{O: } \frac{120.8 \text{ g}}{221.6 \text{ g}} = 0.5451$$

Thus, the percent composition is 4.58% hydrogen, 40.92% carbon, and 54.51% oxygen. Note that we expect these percentages to add up to 100%, but due to limits on the precision of the data they add to 100.01%.

Given either the percent composition or the actual masses from a sample we can determine the empirical formula for a substance. The empirical formula may differ from the actual *molecular formula* of the substance. An empirical formula represents the smallest whole number ratios of the elements in the substance, while the molecular formula represents the actual numbers of each element in the molecule.

For example, hydrogen peroxide, H_2O_2, is a common household disinfectant. Molecules of hydrogen peroxide contain two atoms of hydrogen and two atoms of oxygen, so the molecular formula for this substance is H_2O_2. But the empirical formula is HO, because the empirical formula contains the smallest whole number values that can represent the ratios in the compound. Since each molecule of H_2O_2 contains two atoms of H and two atoms of O, the ratio of H to O in the molecule is 1 : 1, giving an empirical formula of HO.

In many cases, the empirical and molecular formulas are identical. The molecular formula for methane, for example, is CH_4. This formula indicates a ratio of carbon to hydrogen atoms in the molecule of 1 : 4. This same formula is the empirical formula, because 1 and 4 are the smallest whole numbers that can represent this ratio.

To determine the empirical formula from percent composition, assume you have a sample of the substance with a mass of exactly 100 g. Use the percent composition to determine the masses of each element in the 100-g sample, then use the mass data from the periodic table to convert each of these masses to numbers of moles. Finally, divide each of the mole values by the smallest number of moles to determine the whole number ratios in the formula. An example illustrates the calculation.

▼ Example 1.10

Given percent composition data for ascorbic acid (see above), determine the empirical formula for this substance.

We assume a sample with a mass of exactly 100 g. We begin by using the percent composition to obtain masses in grams for each element in the substance. Assuming a 100-g sample just makes this easy. Since hydrogen is 4.58% of the 100-g sample, the mass of the hydrogen in the sample is 4.58 g. Similarly, the masses of the carbon and oxygen are 40.92 g and 54.51 g, respectively.

Next, we use the molar masses for each element to convert each of these masses to number of moles. We use the molar mass as a conversion factor, just as we have before.

$$4.58\text{ g H}\cdot\frac{1\text{ mol}}{1.0079\text{ g}}=4.54\text{ mol H}$$

$$40.92\text{ g C}\cdot\frac{1\text{ mol}}{12.011\text{ g}}=3.407\text{ mol C}$$

$$54.51\text{ g O}\cdot\frac{1\text{ mol}}{15.9994\text{ g}}=3.407\text{ mol O}$$

Next, to determine the ratios of elements in the substance, divide each of these mole amounts by the smallest of them.

$$\frac{4.54\text{ mol}}{3.407\text{ mol}}=1.33$$

$$\frac{3.407\text{ mol}}{3.407\text{ mol}}=1.00$$

$$\frac{3.407\text{ mol}}{3.407\text{ mol}}=1.00$$

These values tell us that the ratio of hydrogen to carbon to oxygen in ascorbic acid is 1.33 : 1.00 : 1.00. Now, we need the smallest whole numbers that preserve this same ratio. Noting that the value 1.33 is very close to 4/3, we multiply all the values by 3 to get whole number ratios of 4 : 3 : 3 for hydrogen : carbon : oxygen. Finally, we use these ratios to write the empirical formula. In formulas containing these three elements it is traditional to write the elements in the formula in the order C—H—O. Doing so gives us

$C_3H_4O_3$.

▲

1.5.8 Determining a Molecular Formula from an Empirical Formula

The empirical formula determined in the previous example relates to the molecular formula by some simple multiple. Recall that the subscripts in the molecular formula of hydrogen peroxide, H_2O_2, are simply double the subscripts in the empirical formula, HO. We can determine the molecular formula for a compound from the empirical formula if we have access to the molecular mass of the compound. We do this by computing the formula mass for the empirical formula and comparing this to the molecular mass to see what the multiple is between the empirical formula mass and the molecular mass. Then we can multiply the subscripts in the empirical formula by the same multiple to get the molecular formula. In other words,

$$\text{whole number multiple} = \frac{\text{molecular mass}}{\text{empirical formula mass}}$$

This calculation is illustrated in the following example. Note that although this example uses atomic masses and molecular mass in u, the same computation can be performed using molar masses in g/mol.

▼ Example 1.11

The experimentally determined molecular mass for ascorbic acid is 176.1 u. Use this value and the empirical formula from Example 1.9 to determine the molecular formula for ascorbic acid.

We begin by determining the formula mass for the empirical formula, $C_3H_4O_3$.

C: 12.011 u

H: 1.0079 u

O: 15.9994 u

$$(3 \times 12.011\ \text{u}) + (4 \times 1.0079\ \text{u}) + (3 \times 15.9994\ \text{u}) = 88.063\ \text{u}$$

Next we calculate the whole number ratio by dividing the molecular mass by the empirical formula mass:

$$\text{whole number multiple} = \frac{176.1\ \text{u}}{88.063\ \text{u}} = 2.000$$

Finally, we multiply all the subscripts in the empirical formula by this multiple to obtain the molecular formula:

$C_6H_8O_6$.

▲

1.5.9 Significant Digit Rules for Addition

As noted just before Example 1.1, I have been assuming in this chapter that students using this text are already familiar with the use of significant digits in scientific measurements and computations. If you are not, then now is the time to study the tutorial on the subject in Appendix A.

If you have used the significant digits rules prior to this course, your experience with the use of significant digits may be limited to computations involving multiplication and division. The rule for these kinds of computations is based on the number of significant digits in the values used in the computation: the result must have the same number of significant digits as the least precise value in the computation. With this rule, the limitation on the result is the *number of significant digits* in the least precise value used in the computation.

For addition, a completely different rule applies. If you pay attention to the significant digits in Example 1.3, you notice that our result contains six significant digits, even though one of the values used in the computation has only five significant digits. This is a result of the addition rule.

When performing addition, it is not the number of significant digits that governs the precision of the result. Instead, it is the *place value of the last digit that is farthest to the left in the numbers being added* that governs the precision of the result. To illustrate, consider the following addition example:

$$\begin{array}{r} 13.65 \\ 1.9017 \\ \underline{+\ 1{,}387.069} \\ 1{,}402.62 \end{array}$$

Of the three values being added, 13.65 has digits out to the hundredths place, the second number goes out to the ten thousandths place, and the last number goes out to the thousandths place. Looking at the final digits of these three, you can see that the final digit farthest to the left is the 5 in 13.65, which is in the hundredths place. This is the digit that governs the final digit of the result. There can be no digits to the right of the hundredths place in the result. The justification for this rule is that one of our measurements is precise only to the nearest hundredth, even though the other two are precise to the nearest thousandth or ten thousandth. Since one of our values is precise only to the nearest hundredth, it makes no sense to have a result that is precise to a place more precise than that, so hundredths are the limit.

Correctly performing addition problems in science (where nearly everything is a measurement) requires that you determine the place value governing the precision of your result, perform the addition, then round the result. In the above example, the sum is 1,4602.6207. Rounding this value to the hundredths place gives 1,4602.62.

Going back to Example 1.3, performing the multiplications gives the following addition problem:

$$\begin{array}{r} 2.0158 \\ \underline{+\ 15.9994} \\ 18.0152 \end{array}$$

Both values are precise to the nearest ten thousandth, and so is the result. In this case, we gain precision because now we have a value with six significant digits. The same thing occurs in the illustration above. One of the values in the addition has only four significant digits, but the result has six.

Chapter 1 Exercises

SECTION 1.1

1. Determine the energy in a photon of light from a green laser with a wavelength of 543 nm.
2. An atom absorbs a photon, causing one of its electrons to move to an orbital associated with 2.2718×10^{-19} J higher energy. Determine the wavelength of the absorbed photon and state what region of the electromagnetic spectrum it is in.
3. For a single photon to ionize a ground-state hydrogen atom, its energy has to raise the energy of the atom's electron to 0 J. What wavelength of light does this and what part of the electromagnetic spectrum is it in?
4. Calculate the energies for the four lines in the visible spectrum of the hydrogen atom.

SECTION 1.2

5. What are two of the limitations of the Bohr model of the atom?
6. In the Bohr model, how many electrons would you expect the 5th energy level to be able to hold? Explain your response.

SECTION 1.3

7. A certain atom is in the ground state. The 3*p* subshell of this atom is 2/3 full.
   a. Identify the element this atom represents.
   b. How many unpaired electrons are there in the atom? (A paired electron is one in an orbital with another one possessing opposite spin.)
8. In a certain ground-state atom, the 4*d* subshell has two electrons in it.
   a. Identify the element this atom represents.
   b. How many unpaired electrons are there in the atom?
9. How do the values of the azimuthal quantum number and magnetic quantum number relate to the principle quantum number?
10. Demonstrate mathematically that the 4*f* subshell can accommodate 14 electrons.
11. Generally speaking, what is the explanation for an atom's electron configuration not following the sequence described by the Madelung rule?

SECTION 1.4

12. For each of the following elements, draw the orbital diagram and write the full-length electron configuration.

a. chlorine	b. oxygen	c. ruthenium
d. potassium	e. vanadium	f. bromine

13. For each of the following elements, write the condensed electron configuration.

a. chlorine	b. nitrogen	c. aluminum
d. yttrium	e. strontium	f. tungsten
g. cesium	h. iodine	i. neodymium

14. Compare the electron configurations for beryllium, magnesium, and calcium. Formulate a general rule for the condensed electron configuration of a Group 2 element.
15. For which group of elements does the electron configuration always end with np^2? Explain how you know.
16. Write the condensed electron configurations for ytterbium, einsteinium, and nobelium.

SECTION 1.5

17. Which two nuclides in Table 1.5 have 20 neutrons?
18. In Table 1.5, how many neutrons are there in the heaviest nuclide listed? How many neutrons are there in the lightest nuclide listed?
19. Determine the number of atoms in each of the following.

a. 73.2 g Cu	b. 1.35 mol Na	c. 1.5000 kg W

20. Determine the mass in grams for each of the following.

a. 6.022×10^{23} atoms K	b. 100 atoms Au	c. 0.00100 mol Xe
d. 2.0 mol Li	e. 4.2120 mol Br	f. 7.422×10^{22} atoms Pt

21. Determine the number of moles present in each of the following.

a. 25 g $Ca(OH)_2$ b. 286.25 g $Al_2(CrO_4)_3$ c. 2.111 kg KCl
d. 47.50 g $LiClO_3$ e. 10.0 g O_2 f. 1.00 mg $C_{14}H_{18}N_2O_5$

22. As mentioned in the text, the sum of the masses of the particles in an atom does not equal the mass of the atom. Some of the mass of the individual particles is converted to energy, and the atom weighs less than the sum of the weights of its parts. How much mass is converted into energy when the individual protons, neutrons, and electrons are assembled to form an atom of uranium-238?

23. Referring to Table 1.5, calculate the atomic mass for silicon, calcium, iron, and uranium. Compare your results to the values shown in the periodic table.

24. Calculate the molar mass for each of the following compounds or molecules.

a. ammonia, NH_3 b. carbon dioxide, CO_2 c. chlorine gas, Cl_2
d. copper(II) sulfate, $CuSO_4$ e. calcium nitrite, $Ca(NO_2)_2$ f. sucrose, $C_{12}H_{22}O_{11}$
g. ethanol, C_2H_5OH h. propane, C_3H_8 i. glass, SiO_2

25. Determine the formula masses for these compounds:

a. $MgCl_2$ b. $Ca(NO_3)_2$ c. $(SO_4)^{2-}$ (The 2– indicates this is an ion with an electrical charge of –2. The charge does not affect your calculation.)
d. $CuSO_4$ e. BF_3 f. CCl_4

26. Determine the mass in grams of 2.25 mol silver nitrate, $AgNO_3$.

27. Given 2.25 kg CCl_4, answer these questions:

a. How many moles CCl_4 are present?
b. How many carbon atoms are present?
c. Approximately how many carbon-13 atoms are present?

28. Given 1.00 gal H_2O at 4°C, answer the questions below. (Hint: You must use the appropriate volume conversion and the density of water to determine the mass of 1.00 gal H_2O. See the information in Tables B.3 and B.5 in Appendix B.)

a. How many moles H_2O are present?
b. How many hydrogen atoms are present?
c. Approximately how many deuterium (hydrogen-2) atoms are present?

29. Automobile antifreeze is composed of ethylene glycol. This green liquid is 38.7% C, 9.7% H, and 51.6% O by mass. The molecular mass is 62.1 u. Determine the empirical formula and the molecular formula for ethylene glycol.

30. A scientist isolates 47.593 g of a new, unidentified substance. The scientist also determines the following masses for the elements in the substance: carbon: 43.910 g; hydrogen: 3.683 g. Finally, the scientist is also able to determine the molecular mass of the substance to be 78.11 u. From these data, determine:

a. the percent composition
b. the empirical formula
c. the molecular formula

31. Hydrogen chlorate, $HClO_3$, is a molecular substance that becomes chloric acid when dissolved in water. Determine the number molecules present in 125.0 g $HClO_3$.

32. Determine the percentage composition of these compounds:

a. sodium bicarbonate, $NaHCO_3$

b. sodium oxide, Na_2O

c. iron(III) oxide, Fe_2O_3

d. silver nitrate, $AgNO_3$

e. calcium acetate, $Ca(CH_3COO)_2$

f. aspirin, $C_9H_8O_4$

33. A *hydrate* is a compound with water molecules trapped in the crystal lattice. Determine the mass percentage of water in zinc sulfate septahydrate, $ZnSO_4{\cdot}7H_2O$. (The coefficient on the H_2O indicates the number of water molecules present for each unit of $ZnSO_4$.)

34. The results of quantitative analysis show that a compound contains 22.65% sulfur, 32.38% sodium, and 44.99% oxygen. Determine the empirical formula for this compound.

35. A compound has an empirical formula of CH_2O and a molar mass of 120.12 g/mol. Determine the molecular formula for this compound.

36. Determine the empirical and molecular formulas for each of the following:

 a. caffeine, which contains 49.5% C, 5.15% H, 28.9% N, and 16.5% O by mass, and has a molecular mass of 195 u.

 b. ibuprofen, which contains 75.69% C, 8.80% H, and 15.51% O by mass, and has a molar mass of 206 g/mol.

 c. propane, which contains 81.71% C and 18.29% H by mass, and has a molar mass of 44.096 g/mol.

 d. aspartame, a sugar-free sweetener, which contains 57.14% C, 6.16% H, 9.52% N, and 27.18% O, and has a molecular mass of 294.302 u.

 e. acetylene, a gas used in cutting torches, which contains 92.26% C and 7.74% H, and has a molar mass of 26.038 g/mol.

37. Toluene is a solvent commonly found in chemistry labs. An analysis of a 10.5-g sample shows that the sample contains 9.581 g carbon and 0.919 g hydrogen. If the molar mass of toluene is 92.140 g/mol, determine the percent composition, empirical formula, and molecular formula.

Chapter 2

The Periodic Law

Reihen	Gruppe I. — R^2O	Gruppe II. — RO	Gruppe III. — R^2O^3	Gruppe IV. RH^4 RO^2	Gruppe V. RH^3 R^2O^5	Gruppe VI. RH^2 RO^3	Gruppe VII. RH R^2O^7	Gruppe VIII. — RO^4
1	H=1							
2	Li=7	Be=9,4	B=11	C=12	N=14	O=16	F=19	
3	Na=23	Mg=24	Al=27,3	Si=28	P=31	S=32	Cl=35,5	
4	K=39	Ca=40	—=44	Ti=48	V=51	Cr=52	Mn=55	Fe=56, Co=59, Ni=59, Cu=63.
5	(Cu=63)	Zn=65	—=68	—=72	As=75	Se=78	Br=80	
6	Rb=85	Sr=87	?Yt=88	Zr=90	Nb=94	Mo=96	—=100	Ru=104, Rh=104, Pd=106, Ag=108.
7	(Ag=108)	Cd=112	In=113	Sn=118	Sb=122	Te=125	J=127	
8	Cs=133	Ba=137	?Di=138	?Ce=140	—	—	—	— — — —
9	(—)	—	—	—	—	—	—	
10	—	—	?Er=178	?La=180	Ta=182	W=184	—	Os=195, Ir=197, Pt=198, Au=199.
11	(Au=199)	Hg=200	Tl=204	Pb=207	Bi=208	—	—	
12	—	—	—	Th=231	—	U=240	—	— — — —

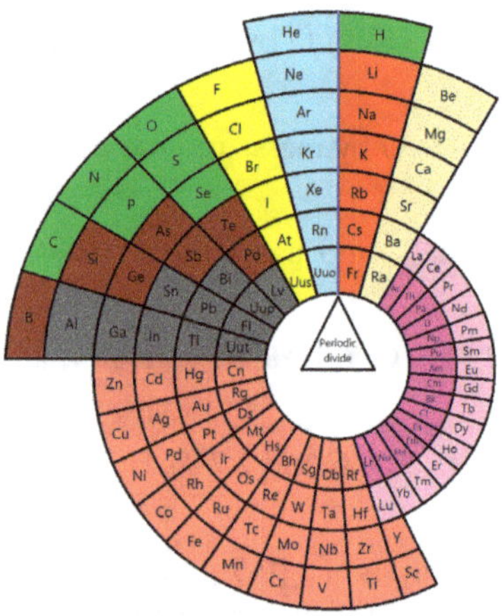

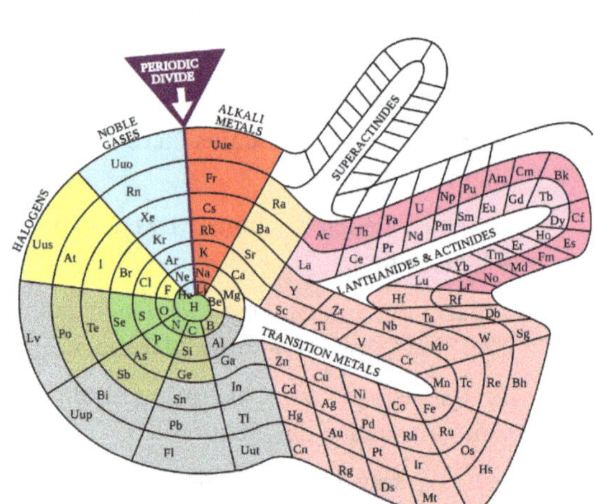

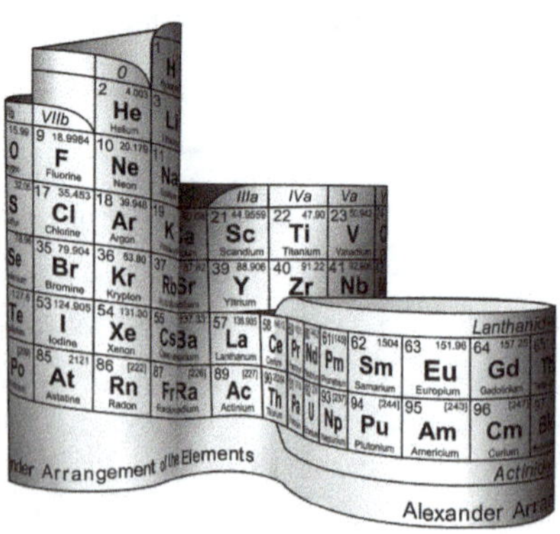

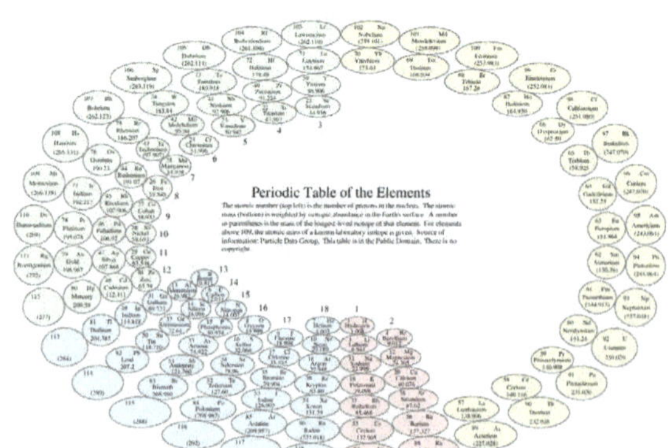

The patterns among the elements have tempted many scientists to try their hands at developing different forms of the Periodic Table of the Elements. At the top is the original periodic table, published by Russian scientist Dmitri Mendeleev in 1869.

Objectives for Chapter 2

After studying this chapter and completing the exercises, you should be able to do each of the following tasks, using supporting terms and principles as necessary.

SECTION 2.1

1. Describe the general structure and arrangement of the Periodic Table of the Elements.

SECTION 2.2

2. Identify the names and locations of the major regions in the periodic table.
3. Identify the names and locations of the elements in Groups 1, 2, 3–12, 16, 17, and 18. Also identify the groups known as the rare-earth elements and the names of the two rows of these elements.
4. State the two collective names for Groups 1–2 and 13–18.
5. State the chief chemical property that distinguishes the metals from the nonmetals.

SECTION 2.3

6. Define the length unit known as the *angstrom* (Å).
7. Use diatomic bond length and atomic radius data to estimate bond lengths in molecules.
8. Describe the trends in atomic radii in the periodic table across periods and down groups.
9. Explain three factors influencing the trends in atomic radii in the periodic table.
10. Compare ionic radii to atomic radii for metals and nonmetals.
11. Use your knowledge of atomic and ionic size trends in the periodic table to arrange lists of elements and ions in order by size.

SECTION 2.4

12. Distinguish between core electrons and valence electrons.
13. Estimate effective nuclear charge, Z_{eff}.
14. Define *ionization energy*, describe the trends for ionization energy across periods and down groups in the periodic table, and use Z_{eff} and other factors to account for these trends.
15. Write regular and condensed electron configurations for an ion in a given oxidation state.
16. Use Z_{eff} to explain the large difference in ionization energy between an atom's core electrons and its valence electrons.
17. Predict oxidation states for metals in Groups 1–4 and nonmetals in Groups 15–17.
18. Define *electron affinity*, use Z_{eff} to account for the high electron affinity of elements in Groups 16 and 17, and explain why electron affinity values are not available for the noble gases.
19. Define *electronegativity*.
20. Describe the trends for electronegativity across periods and down groups in the periodic table. Name the elements with the lowest and highest electronegativity values.
21. Use Z_{eff} to account for the trend in electronegativity across periods.

SECTION 2.5

22. Explain why hydrogen is located in Group 1 and why it acts like the Group 17 elements.

2.1 The Periodic Table of the Elements

The contemporary Periodic Table of the Elements is shown again in Figure 2.1 and in the inside the rear cover of the text. Discovering new elements and figuring how they relate to one another was one of the hottest issues in science in the 19th century. By the 1860s, several scientists had noticed *periodicities* in the properties of the known elements. A periodicity is a regular, cyclic variation of some kind. These scientists noticed, for example, that when elements were listed in order according to atomic weight, the physical property of density increased and decreased in a cyclic fashion. In 1864, German chemist Lothar Meyer published a paper describing cyclic variation in the chemical property known as *valence*, an important property related to the number of bonds an atom makes with other atoms to form compounds.

Credit for the discovery of the *periodic law* and the development of the first Periodic Table of the Elements is generally given to Russian scientist Dmitri Mendeleev, who published his table of the elements in 1869. Mendeleev (Figure 2.2) had not only noticed the periodicities, but he also arranged some 67 elements into a table and predicted the existence of several unknown elements based on gaps in the table as he had organized it. The elements Mendeleev predicted included those now known as gallium, germanium, technetium, and others. The properties Mendeleev predicted for these elements included valence, density, atomic weight, and color. His predictions of not only the existence of these elements but also their properties is the reason for the general credit Mendeleev gets for discovering the periodic law.

Figure 2.2. Russian scientist Dmitri Mendeleev (1834–1907).

As Mendeleev organized the elements, he ordered them by atomic weight and aligned them into rows and columns based on their chemical and physical properties. Today, just as back then, one of the most important things to know about the periodic table is that *elements in the same group (column) exhibit very similar chemical properties.*

Mendeleev's original table is shown on the opening page of this chapter. In that image, just above the center and a bit to the left you can see gaps at the atomic mass values 68 and 72. These are the positions now occupied by gallium and germanium, discovered in 1875 and 1886, respectively. Mendeleev was a brilliant scientist. His fields of expertise included physics, chemistry, and a host of areas of technology. Mendeleev taught in St. Petersburg at several different institutions, and because of his work there St. Petersburg became internationally known for prominence in chemical research.

There were some debates in the 19th century regarding proper placement for four of the elements. If you look at tellurium ($Z = 52$) and iodine ($Z = 53$) in the periodic table to the right, you see that the atomic mass of tellurium is the larger of the two. Because of this, many scientists felt that iodine should come before tellurium in the table. But iodine exhibits all the properties of the Group 17 elements and Mendeleev argued that it should come after tellurium. This problem was resolved when scientists realized that the atomic number was the correct parameter to use for ordering the elements in the table. Mendeleev had ordered them by atomic mass (known then as atomic weight); today the atomic number governs the order. A similar debate over placement surrounded the elements cobalt ($Z = 27$) and nickel ($Z = 28$).

The noble gases were not known in Mendeleev's time and he provided no place for them in his table. Interestingly, when they were eventually discovered Mendeleev was resistant to accept the discovery because the new elements didn't fit into his table. The solution to this little problem was simply to add another column for them, now known as Group 18.

Figure 2.1. The Periodic Table of the Elements.

	1 1A	2 2A	3 3B	4 4B	5 5B	6 6B	7 7B	8 8B	9 8B	10 8B	11 1B	12 2B	13 3A	14 4A	15 5A	16 6A	17 7A	18 8A
1	1 H Hydrogen 1.0079																	2 He Helium 4.0026
2	3 Li Lithium 6.941	4 Be Beryllium 9.0122											5 B Boron 10.811	6 C Carbon 12.011	7 N Nitrogen 14.0067	8 O Oxygen 15.9994	9 F Fluorine 18.9984	10 Ne Neon 20.1797
3	11 Na Sodium 22.9898	12 Mg Magnesium 24.3050											13 Al Aluminum 26.9815	14 Si Silicon 28.0855	15 P Phosphorus 30.9738	16 S Sulfur 32.066	17 Cl Chlorine 35.4527	18 Ar Argon 39.948
4	19 K Potassium 39.098	20 Ca Calcium 40.078	21 Sc Scandium 44.9559	22 Ti Titanium 47.88	23 V Vanadium 50.9415	24 Cr Chromium 51.9961	25 Mn Manganese 54.9380	26 Fe Iron 55.847	27 Co Cobalt 58.9332	28 Ni Nickel 58.6934	29 Cu Copper 63.546	30 Zn Zinc 65.39	31 Ga Gallium 69.723	32 Ge Germanium 72.61	33 As Arsenic 74.9216	34 Se Selenium 78.96	35 Br Bromine 79.904	36 Kr Krypton 83.80
5	37 Rb Rubidium 85.468	38 Sr Strontium 87.62	39 Y Yttrium 88.9059	40 Zr Zirconium 91.224	41 Nb Niobium 92.9064	42 Mo Molybdenum 95.94	43 Tc Technetium 98.9072	44 Ru Ruthenium 101.07	45 Rh Rhodium 102.9055	46 Pd Palladium 106.42	47 Ag Silver 107.8682	48 Cd Cadmium 112.411	49 In Indium 114.82	50 Sn Tin 118.710	51 Sb Antimony 121.76	52 Te Tellurium 127.60	53 I Iodine 126.9045	54 Xe Xenon 131.29
6	55 Cs Cesium 132.905	56 Ba Barium 137.327	71 Lu Lutetium 174.967	72 Hf Hafnium 178.49	73 Ta Tantalum 180.9479	74 W Tungsten 183.85	75 Re Rhenium 186.207	76 Os Osmium 190.2	77 Ir Iridium 192.22	78 Pt Platinum 195.08	79 Au Gold 196.9665	80 Hg Mercury 200.59	81 Tl Thallium 204.3833	82 Pb Lead 207.2	83 Bi Bismuth 208.9804	84 Po Polonium 208.9824	85 At Astatine 209.9871	86 Rn Radon 222.0176
7	87 Fr Francium 223.0197	88 Ra Radium 226.0254	103 Lr Lawrencium 262.11	104 Rf Rutherfordium 261.11	105 Db Dubnium 262.114	106 Sg Seaborgium 263.118	107 Bh Bohrium 262.12	108 Hs Hassium (265)	109 Mt Meitnerium (266)	110 Ds Darmstadtium (281)	111 Rg Roentgenium (281)	112 Cn Copernicium (285)	113 Nh Nihonium (284)	114 Fl Flerovium (289)	115 Mc Moscovium (288)	116 Lv Livermorium (293)	117 Ts Tennessine (294)	118 Og Oganesson (294)

■ liquid at room temperature

■ radioactive

57 La Lanthanum 138.9055	58 Ce Cerium 140.115	59 Pr Praseodymium 140.9077	60 Nd Neodymium 144.24	61 Pm Promethium 144.9127	62 Sm Samarium 150.36	63 Eu Europium 151.965	64 Gd Gadolinium 157.25	65 Tb Terbium 158.9253	66 Dy Dysprosium 162.50	67 Ho Holmium 164.9303	68 Er Erbium 167.26	69 Tm Thulium 168.9342	70 Yb Ytterbium 173.04
89 Ac Actinium 227.0278	90 Th Thorium 232.0381	91 Pa Protactinium 231.0359	92 U Uranium 238.0289	93 Np Neptunium 237.0482	94 Pu Plutonium 244.0642	95 Am Americium 243.0614	96 Cm Curium 247.0703	97 Bk Berkelium 247.0703	98 Cf Californium 251.0796	99 Es Einsteinium 252.083	100 Fm Fermium 257.0951	101 Md Mendelevium 258.10	102 No Nobelium 259.1009

Note: The upper set of group numbers has been recommended by the International Union of Pure and Applied Chemistry (IUPAC) and is now in wide use. The lower set of numbers is still in common use in America.

The first 92 elements in the periodic table are found in nature; elements 93–118 have been synthesized in laboratories. The "discovery" (by synthesis) of elements 114 and 116 was confirmed in 2011, and in January 2016 official confirmation of elements 113, 115, 117, and 118 was announced. The reason confirmations for the last few elements took so long is that once the nucleus of one of these heavy elements is assembled it doesn't stay around very long—far less than one second.

2.2 Periodic Table Nomenclature

The columns in the periodic table are called *groups* and the rows are called *periods*. The images in Figures 2.3 and 2.4 identify several different specific regions of elements in the periodic table. In the long form of the table shown in Figure 2.3, the elements are classified as *metals*, *nonmetals*, and *metalloids*. Note that hydrogen (H) is classified as a nonmetal, even though it is positioned with the metals in Group 1. Hydrogen's location in Group 1 is due to the fact that hydrogen has one valence electron, which I address in more detail later in the chapter.

As you probably know, metals possess a number of properties in common. Common physical properties include high electrical and thermal conductivity, malleability, ductility, and shininess or *luster*. Chemically, the metals are known for ionizing by losing electrons to form positive ions, known as *cations* (pronounced cat-ion). As positive ions, they bond with negative ions to form ionic compounds. (Again, we address compounds in detail later.) People commonly think of metals as shiny conductors of electricity. Chemists think of them as elements that form positive ions.

The metalloids possess properties that are neither clearly metallic nor clearly nonmetallic. For example, under some conditions they conduct electricity and under other conditions they don't. This property is the reason why some of the metalloids are the elements used to manufacture computer "semiconductors."

The nonmetals have their own distinguishing properties, such as ionization by gaining electrons to form negative ions, called *anions*. They also bond with each other—something metals almost never do.

In the standard form periodic table of Figure 2.4 are shown the common names for several specific groups (columns) of elements. All these names are used frequently in scientific discourse and you need to commit them to memory. Groups 1–2 and 13–18 are also collectively referred to as the *main group elements* or *representative elements*.

You probably noticed that in both the figures elements 113–118 are not included with the rest of the elements in the different classes or groups. This is due to the fact that they have only existed for extremely short periods of time in laboratories and little is known about their properties. By the way, you don't need to feel sorry for Groups 13–15 not having nicknames. They do,

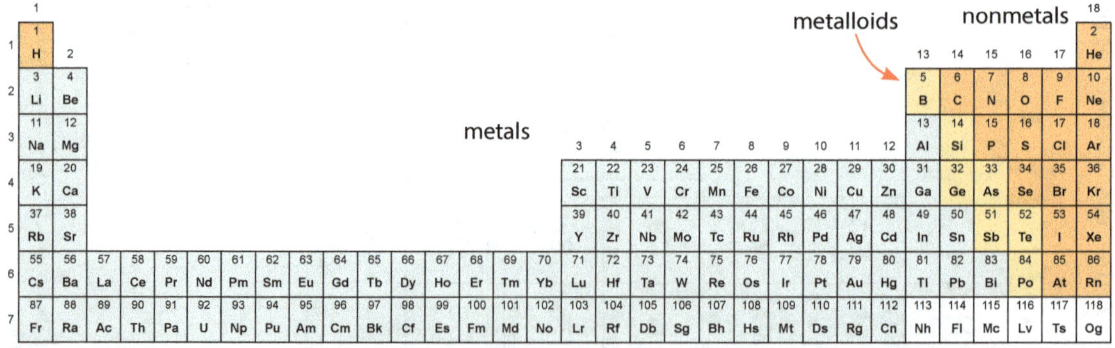

Figure 2.3. The long form of the periodic table indicating the three major classes of elements.

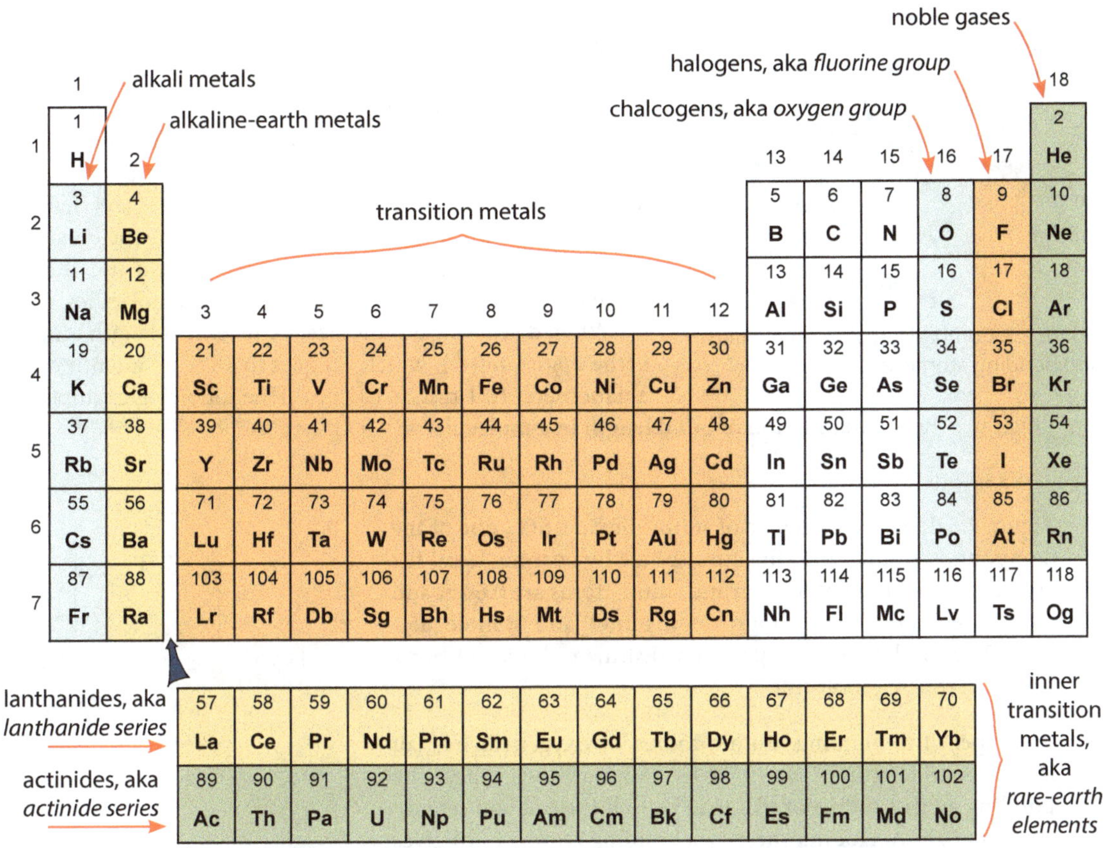

Figure 2.4. The standard form table indicating many of the common names for particular groups of elements.

but the names are seldom used. (Okay, Group 13 is called the *icosagens*, Group 14 the *crystallogens*, and Group 15 the *pnictogens*. See? Well, now you know.)

2.3 Periodic Physical Properties

Physical properties of substances are properties such as color, density, boiling point, electrical conductivity, malleability, and many others. Many of these properties exhibit strong periodicity—cyclic rises and falls with cycles that correspond to the periods in the periodic table. In this section, we look at one of the most important of these—atomic size. As we look at periodic trends in this section and the next, you should closely compare the charts presented with the periodic table in Figure 2.1.

2.3.1 Atomic Radius and Bonding Atomic Radius

Defining the size of atoms is a bit tricky. The nucleus of the atom is extremely small, and virtually all the space an atom takes up is defined by the electrons in their orbitals surrounding the nucleus. The size of the orbitals themselves is defined by the probability distribution of where electrons may be found, and this probability does not drop cleanly to zero at the edge of the orbital. Instead, it fades toward zero, making the atomic radius fuzzy. Still, the electrons in the orbitals create a negatively charged shell around the atom that strongly resists penetration by the shells around other atoms. Thus scientists can measure atomic radius by firing two atoms at one another and examining how closely the two atoms come together before bouncing apart.

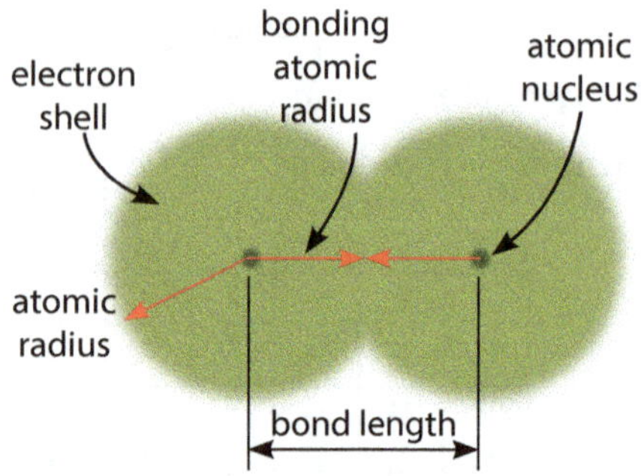

Figure 2.5. Comparison of atomic radius and bonding atomic radius.

The electron clouds repel each other so stiffly that atoms bounce apart as if they were steel spheres.

Figure 2.5 shows two atoms bonded together in a molecule, such as the diatomic oxygen molecule, O_2. As you see, when atoms are bonded together this way the electron shells do interpenetrate one another. The *bond length*, defined as the distance between nuclei, is less than twice the atomic radius. Accordingly, we define the *bonding atomic radius* as half the bond length and use this value to estimate the bond length in various molecules.

When discussing atomic size, a convenient length unit is the *angstrom* (Å), which is equal to 10^{-10} m, a tenth of a nanometer (0.1 nm or 100 pm). The following example illustrates the approximation of bond length in a molecule.

Example 2.1

Figure 2.6 depicts a molecule of sulfur dioxide, SO_2, one of the compounds emitted into the atmosphere by burning fossil fuels. The atomic radii of oxygen and sulfur atoms are 0.64 Å and 1.04 Å, respectively. The bond lengths in O_2 and S_2 molecules are 1.21 Å and 1.89 Å, respectively. Estimate the S—O bond length in the SO_2 molecule.

We expect the bonding atomic radius of oxygen to be half the O_2 bond length, or about 0.61 Å. Similarly, we expect the bonding atomic radius of sulfur to be half the S_2 bond length, or about 0.95 Å. Adding these two bonding atomic radii gives 1.56 Å. This figure compares pretty well with the actual bond length in the SO_2 molecule of 1.43 Å (about 9% difference).

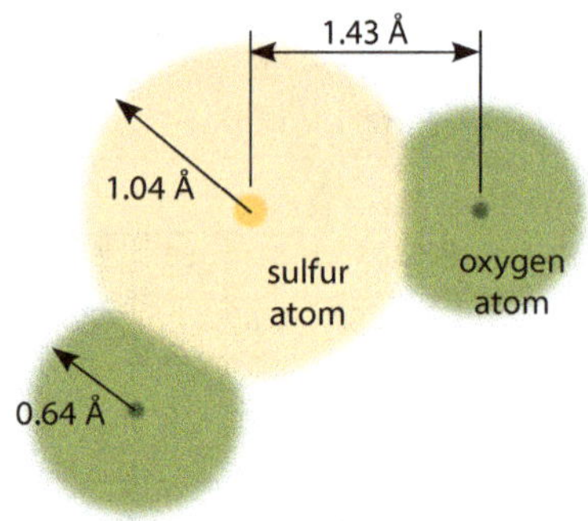

Figure 2.6. Comparison of atomic radius and bonding atomic radius in an SO_2 molecule.

Figure 2.7 shows the presently accepted values for the atomic radii of the first 102 elements. As you see, typical radii are in the range of 1–2 Å, or 0.1–0.2 nm. This means that typical atomic diameters are in the range of 0.2–0.4 nm.

The most striking things about the graph are the strong increase that occurs at the beginning of each new period and the gradual decrease (usually) in atomic size from one element to the next within the period. There are at least three different effects present governing the size of atoms and leading to these patterns.

The first is the number of subshells in use holding the atoms' electrons. At the beginning of each period, an electron appears with a new, higher principle quantum number, *n*, and a new *s* subshell associated with that principle quantum number. This new shell enables electrons to be much farther from the nucleus and is the major cause of the increase in atomic radius at the beginning of each period, as well as the overall upward trend in atomic size.

The second effect is the increasing attraction between the nucleus and the electrons as we move from left to right across the periodic table in any given period, or from top to bottom in a given group. Moving left to right across the periodic table, with each new element comes a new proton in the nucleus. The increasing positive electrical charge at the center of the atom tends to pull the atom's electrons in tighter and tighter. Thus, with a few exceptions, atomic size decreases

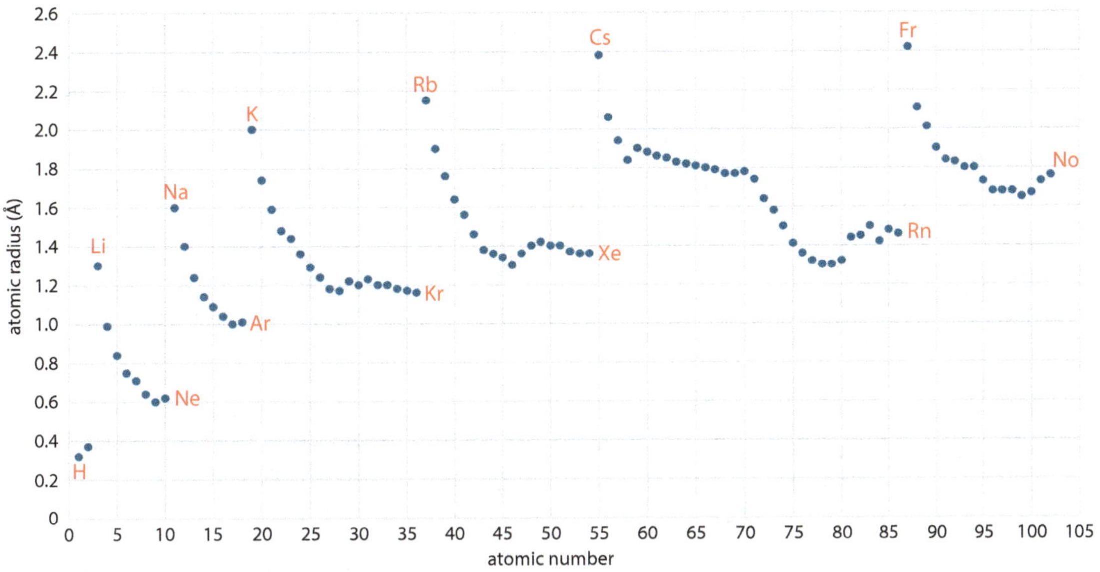

Figure 2.7. Atomic radius values. The six peaks are the alkali metals in Periods 2–7.

from left to right in a period. In the chart of Figure 2.7, this decrease in size after the start of a new period is quite pronounced.

The third effect is called the *shield effect* or *atomic shielding*. The electrons in shells with lower values of *n* effectively form an electrical screen around the nucleus and to some extent shield off the attraction of the positive nucleus for electrons in higher shells. To quantify this screening effect, chemists use a parameter called the *effective nuclear charge* (Z_{eff}). The details of the effective nuclear charge are easier to explain a bit later, so we return to it in Section 2.4.2.

A second feature to note from Figure 2.7 is that the radius of each of the alkali metals is larger than the one just above it in the periodic table. The same holds for each of the noble gases. In fact, this same trend is present in every group in the table. Down every group, the atomic radius of each element is greater than that of the element above it, and this results in the overall upward trend in atomic size shown in the chart. In summary, atomic size generally decreases from left to right in a period and increases from top to bottom in a group.

2.3.2 Ionic Radius

Figure 2.8 depicts the sizes of atoms and their ions in four of the groups of the representative elements. The blue disks are the atomic sizes and the yellow disks are the ionic sizes, both in angstroms.

Metals ionize by losing electrons to form cations (positive ions). The loss of an electron leaves an atom with more protons than electrons, and thus the atom possesses a net positive charge. As we will discuss more in Section 2.4, Group 1 metals always ionize by losing one electron to become ions with a charge of +1. With sodium, for example, we write this ion as Na^{+}. Group 2 metals always ionize by losing two electrons to become ions with a charge of +2. Thus, for the calcium ion we write Ca^{2+}.[1] Nonmetals ionize by gaining electrons to become anions (negative ions). Group 17 elements ionize by gaining one electron to become ions with a net charge of −1. Group 16 elements ionize by gaining two electrons to become ions with a net charge of −2.

The blue circles in Figure 2.8 represent the atomic sizes, and yellow circles represent ionic sizes. Since the metals always ionize by losing the electrons in their highest *s* orbital, their di-

1 The most common convention is to write the sign of the charge after the value of the charge on the chemical symbol for an ion.

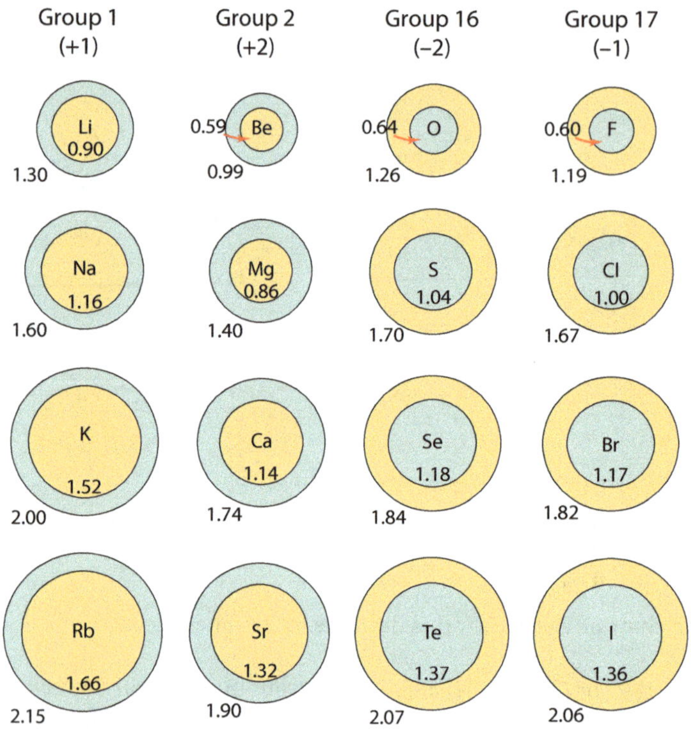

Figure 2.8. Atomic radii (blue) and ionic radii (yellow) for the Group 1 and 2 positive ions and Group 16 and 17 negative ions. All values are in angstroms.

ameters decrease considerably when they ionize. The opposite happens to the nonmetals. Gaining electrons adds to the mutual electron repulsion in the highest-energy orbitals, increasing atomic size.

Just as with neutral atoms, ionic sizes decrease in a period from left to right in the periodic table and increase going down a group. Further, we can summarize the paragraphs above by saying that cations are smaller than their neutral atoms and anions are larger than their neutral atoms.

One final ionic trend needs to be mentioned: among neutral atoms or ions with the same number of electrons, atomic size always decreases as the number of protons increases. Based on what we have seen so far, this should make perfect sense. Among atoms with the same number of electrons—a so-called *isoelectronic series*—there are no new shells to consider and there are no variations in the amount of shielding produced by changing electron configurations. Thus, the only variation is in the positive charge in the nucleus, and as this goes up, the attraction of the nucleus for the electrons goes up, pulling the electrons in tighter and making the atom smaller.

▼ Example 2.2

Based on your knowledge of trends in the periodic table, place the following atoms and ions in order of decreasing size: S, Se^{2-}, O, and S^{2-}.

Anions are larger than their neutral atoms, so $S^{2-} > S$. Elements and ions farther down in a group are larger than those above, so S > O. Also, Se^{2-} is larger than S^{2-}, so

$Se^{2-} > S^{2-} > S > O$.

▲

▼ Example 2.3

Arrange the following ions in order of size from largest to smallest: O^{2-}, Na^{+}, F^{-}, and Mg^{2+}.

We can address this question without any published data on atomic size. Referring to the periodic table, you see that each of the ions listed has 10 electrons. For example, magnesium (Mg) is element 12 and has 12 electrons before ionizing. But the charge on the ion is +2, which means the magnesium atom has lost two electrons and now has 10. Therefore, the order from largest to smallest follows the atomic number order from smallest to greatest, giving the sequence O^{2-}, F^-, Na^+, Mg^{2+}.

▲

2.4 Periodic Chemical Properties

2.4.1 Core and Valence Electrons

Chemistry is all about electrons, and when it comes to the chemical bonding that occurs in chemical reactions, the *valence electrons* of the elements involved determine what kinds of compounds form. To illustrate, consider the elements sulfur (Z = 16) and cobalt (Z = 27) located in the partial periodic tables of Figure 2.9.

The electron configurations for these elements are as follows:

Z	Electron Configuration	Condensed Electron Configuration
16	S: $1s^22s^22p^63s^23p^4$	S: $[Ne]3s^23p^4$
27	Co: $1s^22s^22p^63s^23p^64s^23d^7$	Co: $[Ar]4s^23d^7$

As we saw in the previous chapter, when moving from one element to the next in the periodic table, we add one electron and one proton for each new element. As we look at an element's position in the periodic table, we can think of each of the elements before it as representing the position of one of the element's electrons, arranged according to the Madelung rule and Hund's rule.

Now consider Figure 2.9 and the electron configuration for sulfur. The first 10 electrons in a sulfur atom go to completely filling the n = 1 and n = 2 shells. The next six electrons go into the

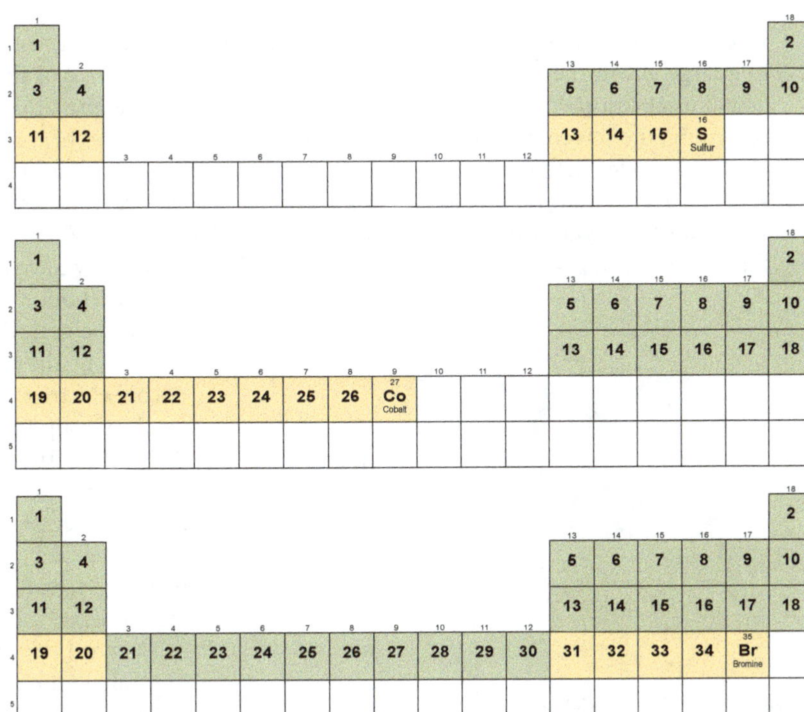

Figure 2.9. Core electrons (green) and valence electrons (yellow) for sulfur, cobalt, and bromine.

$n = 3$ shell. This shell can hold eight electrons, so it is only partially full. The electrons in the full shells are called *core electrons*. The electrons in the partially filled shell are called *valence electrons*, and the shell they are in is called the *valence shell*. The core electrons are the ones involved in screening the nucleus from the outer electrons in the valence shell. These core electrons have more negative energies than the valence electrons, which means they are more tightly bound to the nucleus. As a result, the core electrons are not involved in the electron swapping and sharing that takes place in chemical reactions. That involvement is limited to the valence electrons.

When you look at sulfur's position in the periodic table, just a quick glance indicates that sulfur has six valence electrons, all of them in the third shell, $n = 3$. It's as simple as observing that sulfur is in the third period and counting columns from the left over to where sulfur is. Notice that in the condensed electron configuration notation, all of sulfur's core electrons are represented by [Ne]. The valence electrons are still shown explicitly as $3s^2 3p^4$.

Looking now at cobalt, the element positions colored in green represent cobalt's core electrons. The first 18 electrons are the core electrons, represented by [Ar] in the condensed electron configuration. The next nine electrons are the valence electrons, easily seen from the fact that cobalt is in the fourth period and the ninth column from the left side of the periodic table.

Finally, note that for *p*-block elements in Periods 4–7, the electrons in filled, lower *d* subshells do not act as valence electrons. The only valence electrons in these atoms are those in the *s* and *p* subshells of the unfilled shell. This is illustrated by the diagram for bromine at the bottom of Figure 2.9.

2.4.2 Effective Nuclear Charge

Now we return to the idea of *effective nuclear charge*, Z_{eff}, mentioned a few pages back in Section 2.3.1. Effective nuclear charge is a useful concept when it comes to explaining some of the atomic size variations we see in the periodic table. Z_{eff} is calculated as

$$Z_{eff} = Z - S$$

where Z is the atomic number and S is a value called the *screening constant*. The idea here is that to a large extent the core electrons screen off the positive charge of the nucleus, reducing the actual electrical pull on the valence electrons. For a rough approximation of the effect of this screening, the screening constant can be taken to be the number of core electrons in the atom.

As examples, let's consider again sulfur and cobalt. For sulfur (Figure 2.10), $Z_{eff} = Z - S \approx 16 - 10 \approx +6$, and for Cobalt, $Z_{eff} = Z - S \approx 27 - 18 \approx +9$. Compare these two values to the values for the Group 1 metals in the same periods. For sodium, $Z_{eff} = Z - S \approx 11 - 10 \approx +1$ and for potassium, $Z_{eff} = Z - S \approx 19 - 18 \approx +1$. The alkali metals only have one valence electron, so Z_{eff} comes out to +1. Moving to the right across the period, additional protons are accrued in the nucleus but the number of core electrons remains the same as it is for the Group 1 metal at the beginning of the period. The increase in Z_{eff} from left to right in a period explains the fact that atomic radii decrease from left to right. The value of Z_{eff} can also be used to explain the trends in ionization energy we look at next.

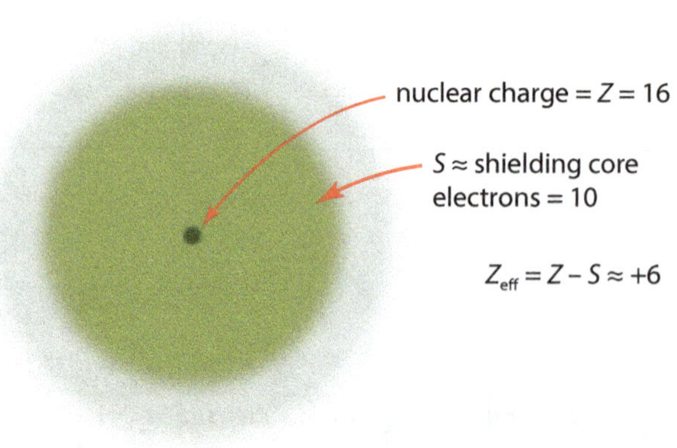

Figure 2.10. Effective nuclear charge for sulfur.

A word of caution is in order here. The calculation outlined above for Z_{eff} is quite simplified, and the results it gives are only approximate. Simply using the number of core electrons for S gives $Z_{eff} = +1$ for each of the alkali metals. In fact, the values of Z_{eff} obtained from more sophisticated calculation methods for the first three alkali metals, lithium ($Z = 3$), sodium ($Z = 11$), and potassium ($Z = 19$), are +1.3, +2.5, and +3.5, respectively.

2.4.3 Ionization Energy

Ionization energy is defined as the amount of energy required to remove a ground-state electron from an isolated, gaseous atom. Recall from our discussion in the previous chapters that adding an electron to an atom releases energy because the electron is going into a lower energy state. Conversely, removing an electron from an atom requires an input of energy, which is the work required to pull the electron from its negative energy state up to zero energy, where it is free from the nuclear attraction of the atom.

Using hydrogen as an example, the removal of the electron from an atom of hydrogen is modeled by the following *ionization equation*:

$$H(g) \rightarrow H^+ + e^- \qquad \Delta E = 2.18 \times 10^{-18}\ \text{J} \qquad (2.1)$$

This expression shows the neutral hydrogen atom on the left, with (g) indicating that the atom is in the gaseous state. On the right, H^+ indicates a hydrogen ion with a charge of +1, and e^- indicates a free electron. In an equation like this, it is customary to write the energy change (ΔE) that occurred in the atomic system (the atom and its electron) during the process. Notice that ΔE is positive, meaning that a certain amount of energy has to be added in order to accomplish the ionization. This amount of energy is the ionization energy.

Because the amounts of energy involved are so small, it is more convenient in discussions of this sort to use an energy unit called the *electron volt* (eV). The electron volt is defined as

$$1\ \text{eV} = 1.60218 \times 10^{-19}\ \text{J} \qquad (2.2)$$

For the ionization of hydrogen, converting the energy above into eV gives

$$\Delta E = 2.18 \times 10^{-18}\ \text{J} \cdot \frac{1\ \text{eV}}{1.602 \times 10^{-19}\ \text{J}} = 13.6\ \text{eV}$$

The energy required to remove one electron from a neutral atom is called the *first ionization energy*. The additional energy required to remove a second electron is called the *second ionization energy*, and so on.

Figure 2.11 charts the first ionization energies of the elements. The strong upward trend in each period is easily accounted for by the increasing value of Z_{eff} moving from left to right in the period. The higher Z_{eff} is, the stronger the nucleus attracts the outermost electrons and the greater the energy required to remove one of them. Notice from the Group 1 and Group 18 elements labeled in the figure that the trend down a group is for the ionization energy to decrease, even though Z_{eff} actually increases down the group, as we saw previously. This is accounted for by the fact that the elements in each new period have their valence electrons in a shell with a higher principle quantum number. These electrons are thus farther from the nucleus, have a less negative energy, and are easier to remove.

Writing electron configurations for ions requires you to keep a somewhat surprising rule in mind: *the electrons an element loses during ionization come from the orbital with the highest principle quantum number, n*. Further, *among the orbitals associated with the highest principle quantum number, the electrons come from the orbitals with the highest value of the azimuthal quantum number, l*. This means, for example, that sodium ionizes (as we would expect) by losing

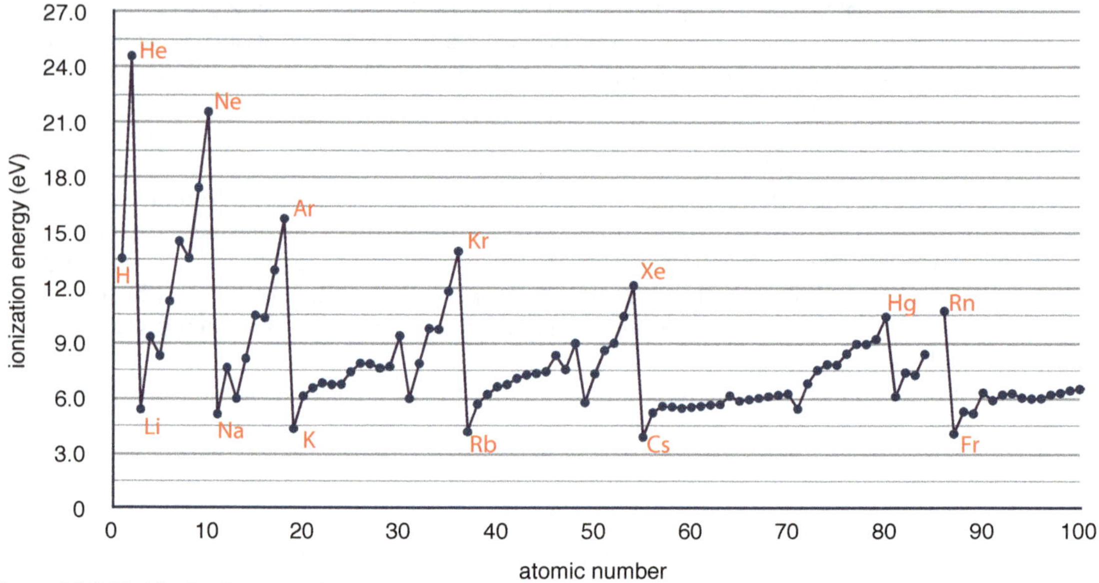

Figure 2.11. First ionization energies.

its 3*s* electron. However, scandium ionizes by losing a 4*s* electron, not a 3*d* electron. The following example illustrates further.

Example 2.4

Write the electron configurations for Sn^{4+} and Cr^{3+}.

Sn is a *p*-block metal in Period 5 and Group 14 with configuration

Sn: $[Kr]5s^2 4d^{10} 5p^2$

Its valence electrons are therefore in the 5*s* and 5*p* subshells. It first loses its two 5*p* electrons, because these have $n = 5$ and $l = 1$. The next two to go are the 5*s* electrons which have $n = 5$ and $l = 0$. The electron configuration of the ion is:

Sn^{4+}: $[Kr]4d^{10}$

Cr is a transition metal with the anomalous electron configuration:

Element	1st	2nd	3rd	4th	5th	6th	7th
Li	5.38	75.64	122.45				
Be	9.32	18.21	153.90	217.72			
B	8.30	25.15	37.93	259.38	340.23		
C	11.26	24.38	47.89	64.49	392.09	489.99	
N	14.53	29.60	47.45	77.47	97.89	552.07	667.05
O	13.62	35.12	54.94	77.41	113.90	138.12	739.29
F	17.42	34.97	62.71	87.14	114.24	157.17	185.19
Ne	21.56	40.96	63.45	97.12	126.21	157.93	207.28

Table 2.1. Ionization energies (eV) for Period 2 elements.

Cr: $[Ar]4s^1 3d^5$

To ionize to Cr^{3+}, the atom first loses the 4*s* electron, which has $n = 4$ and $l = 0$. Then it loses two of its 3*d* electrons, which have $n = 3$ and $l = 2$. Thus, the configuration of the ion is:

Cr^{3+}: $[Ar]3d^3$

▲

Table 2.1 lists the ionization energies up through the 7th ionization energy for the Period 2 elements. Values without shading represent the energies required to remove valence electrons. The yellow shading indicates where core electrons are being removed. Notice the whopping increase in the ionization energy once we start getting into an atom's core electrons. This is explained by the shield effect we discussed previously. Core electrons shield the nucleus from exerting its full attraction on the valence electrons, an effect quantified by the value of Z_{eff}. But inside the core, the full attraction of the nucleus is felt and the energy required to remove an electron is dramatically greater.

This discussion about ionization is an appropriate time to show representative values for ionizations exhibited by the main group elements and transition metals. These values, displayed in Figure 2.12, are called *oxidation states*. I address the origin and use of the term "oxidation" in a later chapter. In this context, it just means that when atoms of a given element ionize, these are the values of charge they typically acquire. The first and most important thing to notice is that on both ends of the periodic table, elements ionize so as to end up with either an empty valence shell or a full one, depending on whether the valence shell was closer to being full or empty to start with. All Group 1 metals and hydrogen ionize by losing their only valence electron and retaining their core electrons. After this ionization, the Group 1 metals are left with one more proton than electron, so they are cations with a charge of +1. Similarly, Group 2 metals lose two electrons and Group 3 metals lose three electrons to end up as cations with charges of +2 or +3, respectively.

On the other end of the table, the noble gases don't readily ionize, so no oxidation states are shown. The nonmetals ionize by taking on electrons to become anions. The halogens all have *p* subshells containing five electrons, so they ionize by gaining one electron to fill up the *p* sub-

	1	2	3	4	5	6	7	8	9	10	11	12	13	14	15	16	17	18
1	H^+																	He
2	Li^+	Be^{2+}											B^{3+}	C^{4+} C^{4-}	N^{3-}	O^{2-}	F^-	Ne
3	Na^+	Mg^{2+}											Al^{3+}	Si^{4+}	P^{3-}	S^{2-}	Cl^-	Ar
4	K^+	Ca^{2+}	Sc^{3+}	Ti^{4+}	V^{5+} V^{4+}	Cr^{6+} Cr^{3+}	Mn^{4+} Mn^{2+}	Fe^{3+} Fe^{2+}	Co^{3+} Co^{2+}	Ni^{2+}	Cu^{2+} Cu^+	Zn^{2+}	Ga^{3+}	Ge^{4+} Ge^{2+}	As^{3+} As^{5+}	Se^{2-}	Br^-	Kr
5	Rb^+	Sr^{2+}	Y^{3+}	Zr^{4+}	Nb^{5+}	Mo^{6+} Mo^{4+}	Tc^{7+} Tc^{4+}	Ru^{4+} Ru^{3+}	Rh^{3+}	Pd^{4+} Pd^{2+}	Ag^+	Cd^{2+}	In^{3+}	Sn^{4+} Sn^{2+}	Sb^{5+} Sb^{3+}	Te^{2-}	I^-	Xe
6	Cs^+	Ba^{2+}	Lu	Hf^{4+}	Ta^{5+}	W^{6+}	Re^{7+} Re^{6+}	Os^{4+}	Ir^{4+} Ir^{3+}	Pt^{4+} Pt^{2+}	Au^{3+} Au^+	Hg^{4+} Hg^{2+}	Tl^+	Pb^{4+} Pb^{2+}	Bi^{5+} Bi^{3+}	Po^{4+}	At	Rn

Figure 2.12. Representative oxidation states.

shell, becoming anions with a charge of −1 in the process. Similarly, Group 16 nonmetals gain two electrons and Group 15 nonmetals gain three electrons to become anions with charges of −2 or −3, respectively.

All the transition metals have multiple oxidation states, and the oxidation state they assume depends on what other elements are around to swap electrons with. The values shown are the most common states. This chart makes it easy to visualize the metalloids as the boundary between the cations and the anions. In the cases of lutetium (Lu) and astatine (At), these elements have multiple oxidation states with none preferred, and for this reason none are shown.

It should be clear now that although ionization energy is defined as the energy required to *remove* an electron from an atom, which is what happens to metals when they ionize, the nonmetals ionize by *gaining* electrons to form anions. The energy involved in a neutral atom gaining an electron is called the *electron affinity*, and is our next topic.

2.4.4 Electron Affinity

Electron affinity is defined as the amount of energy *released* when adding an electron to a ground-state, isolated, gaseous atom. The term "affinity" indicates that electron affinity is a measure of how eager the atoms of a given element are to take on another electron, an important measure when trying to understand or predict chemical reactions.

Here is yet another instance in which energy relationships are crucial for understanding chemical behavior. Most atoms release energy when an electron is added, and accept the additional electron into an orbital to become an anion with a charge of −1. The major exceptions to know about are the noble gases. These elements do not release energy when an electron is added. Instead, energy is required to attach the electron and as a result the atom is not stable. Such an ion immediately rejects the electron and returns to its neutral state. This is because the valence shells in the noble gas atoms are full—the most stable electron configuration there is.

Using sulfur as an example, the gaining of an electron is modeled with an ionization equation as follows:

$$S(g)+e^{-}\rightarrow S^{-}(g) \qquad \Delta E=2.077\ \text{eV} \qquad (2.3)$$

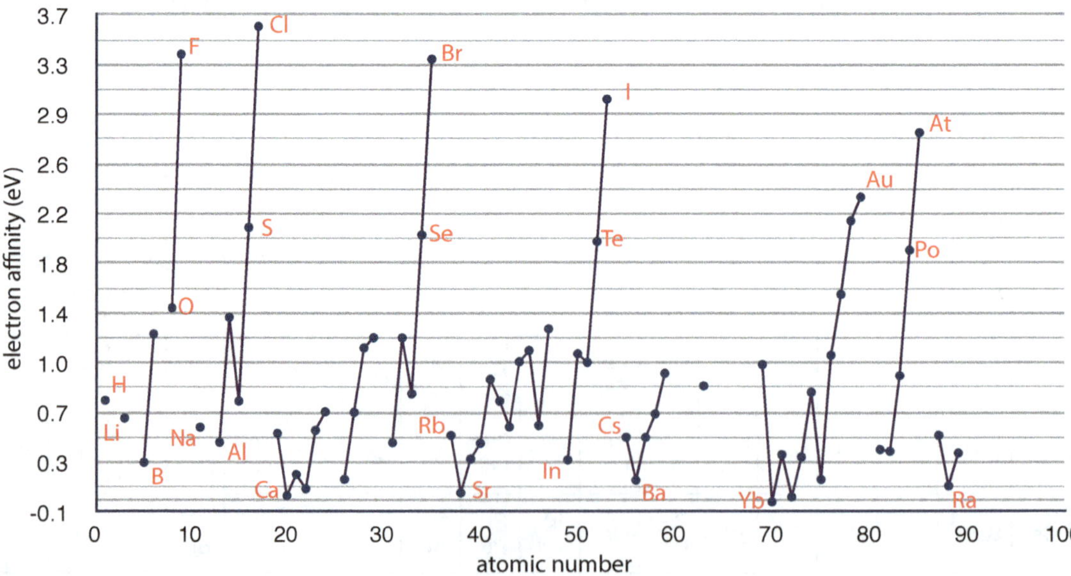

Figure 2.13. Electron affinities.

As with ionization Equation (2.1), the (*g*) indicates that the sulfur is in the gaseous state, and the energy change is stated to the right of the equation.

It is important for you to distinguish clearly in your mind between ionization energy and electron affinity. Ionization energy is the energy required to *remove* an electron from a neutral atom (creating a cation). This energy is positive for every element. Electron affinity is the energy released when an electron is *added* to a neutral atom (creating an anion). Defined this way, this energy is also positive for every element except for the noble gases and a few other elements.

Figure 2.13 shows the electron affinities for the stable atoms. As with other properties we have addressed, there is an obvious periodicity in the affinity values corresponding to the periods in the periodic table. We can again explain the large values associated with the Group 16 and Group 17 elements in terms of the effective nuclear charge, Z_{eff}. In these two groups, the value of Z_{eff} is as high as it gets, so the nuclei of these elements exert the highest possible attraction on a free electron, causing the atom/electron system to release as much energy as possible when the electron drops down into the lower energy state in one of the atom's orbitals.

The breaks between connected dots in Figure 2.13 indicate that several other elements are missing besides the noble gases. These are Be, N, Mg, Mn, and Zn. All these elements exhibit the same unstable behavior the noble gases do when an electron is added to them. Most of these can be at least partially explained by referring to the periodic table. Beryllium has two electrons in the 2*s* subshell. An added electron would go into the 2*p* subshell, where its energy is too high for the small nucleus to hold it. The same is the case for magnesium. Nitrogen has three electrons in the 2*p* subshell—a stable, low-energy state. Adding a fourth electron results in repulsion between two electrons in the p_x orbital, and once again the small nucleus does not have a strong enough attraction to wrangle those electrons and keep them both there. Similar considerations apply for manganese and zinc. Manganese has five electrons in the 3*d* subshell, so the 3*d* subshell is exactly half full with one electron in each orbital. This puts manganese in a situation similar to that of nitrogen. Zinc has 10 electrons in the 3*d* subshell and none in the 4*p* subshell, so zinc is in a situation similar to that of beryllium and magnesium.

There is one element with a very slightly negative value for electron affinity that will still accept an electron. This element is ytterbium (Yb, $Z = 70$). Ytterbium is at the far right of the

Term	Definition Used In This Text	If the energy quantity ΔE is positive, it means:
ionization energy	energy *required* to *remove* an electron	energy flows *into* the atomic system from outside
electron affinity	energy *released* when *adding* an electron	energy flows *out of* the atomic system

Term	Alternative Definition	If the energy quantity ΔE is positive, it means:
ionization energy	energy *involved* when removing an electron	energy flows *into* the atomic system from outside
electron affinity	energy *involved* when adding an electron	energy flows *into* the atomic system from outside (thus, ΔE and the electron affinity values are nearly always negative)

Table 2.2. Definitions used here and elsewhere for ionization energy and electron affinity.

rare-earth elements, in Period 6. Its 4*f* subshell is full and an additional electron goes into the 5*d* subshell.

There is one final point to make before we move on. I have used care in defining ionization energy and electron affinity so that it is clear which way the energy is going when energy values are positive or negative. But you should be aware that some texts and other sources use definitions that may not be as clear, possibly leading to confusion. Instead of using the terms *required* or *released* to define the direction of positive energy flow, some sources use the term *involved*, and define positive energy flow in terms of energy going into or out of the atomic system. Both definitions are valid, as long as one specifies the direction of energy flow that constitutes a positive value of energy.

The definitions we have seen in the last two sections are fairly standard. But to help you avoid confusion, Table 2.2 spells out my definitions and the alternative definitions side by side.

2.4.5 Electronegativity

Figure 2.14. American chemist Linus Pauling (1901–1994).

American chemist Linus Pauling (Figure 2.14) was one of the most important chemists of the 20th century. His work in the 1930s, 1940s, and 1950s on the nature of chemical bonds remains foundational for our understanding of chemistry to this day. For his work, Pauling won the Nobel Prize in Chemistry in 1954. Beginning in the mid-1940s, Pauling showed a deep concern for the negative health effects due to nuclear fallout from nuclear weapons testing, and he presented a petition signed by 11,000 scientists to the United Nations in 1958 to urge a ban on nuclear weapons testing. A scientific study that came out in 1961 showed that radioactivity contamination was indeed widespread in the population, leading to nuclear test ban treaties with the Soviet Union. All this led to Pauling winning the Nobel Peace Prize in 1962. To this day, Pauling remains the only person ever to win two unshared Nobel Prizes.

In 1932, as part of his efforts to understand chemical bonding, Pauling introduced what is now called the *Pauling electronegativity scale*. The electronegativity scale uses a dimensionless quantity called *electronegativity*, with values running from 0.7 (francium) to 3.98 (fluorine). Electronegativity is a measure of how strongly atoms attract the electrons shared between atoms inside molecules. The higher an element's electronegativity relative to the other elements in a molecule, the more an atom attracts the shared electrons in the molecule toward itself.

Once again, the importance of electrical attraction in the atomic world is crucial. Recall from the Introduction that many of the important properties of water are due to the difference between the electronegativities of oxygen (3.44) and hydrogen (2.20) in the water molecule. The oxygen atom attracts shared electrons more strongly than the hydrogen atoms do, and as a result the four bonding electrons in the molecule crowd over toward the oxygen atom. The result is that the oxygen region of the water molecule is more electrically negative and the hydrogen regions are more electrically positive, so water molecules are *polar*—negative at the elbow and positive at the ends, as shown in Figure 2.15. In this diagram, the arrows point from the positive region of the molecule toward the negative region of the molecule. We consider electronegativity further in the next chapter in the context of covalent bonding.

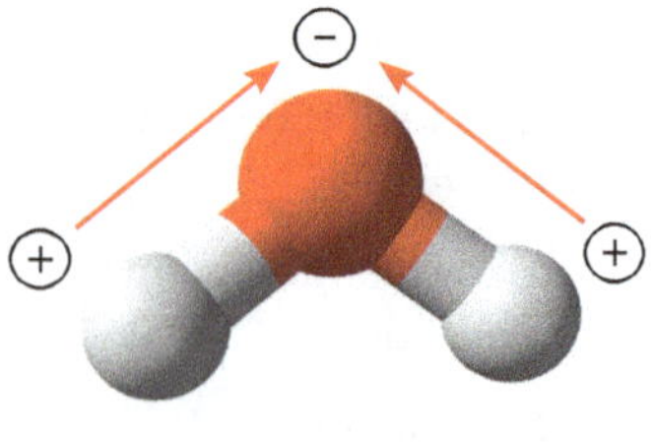

Figure 2.15. The higher electronegativity of oxygen atoms compared to hydrogen atoms results in the polar water molecule.

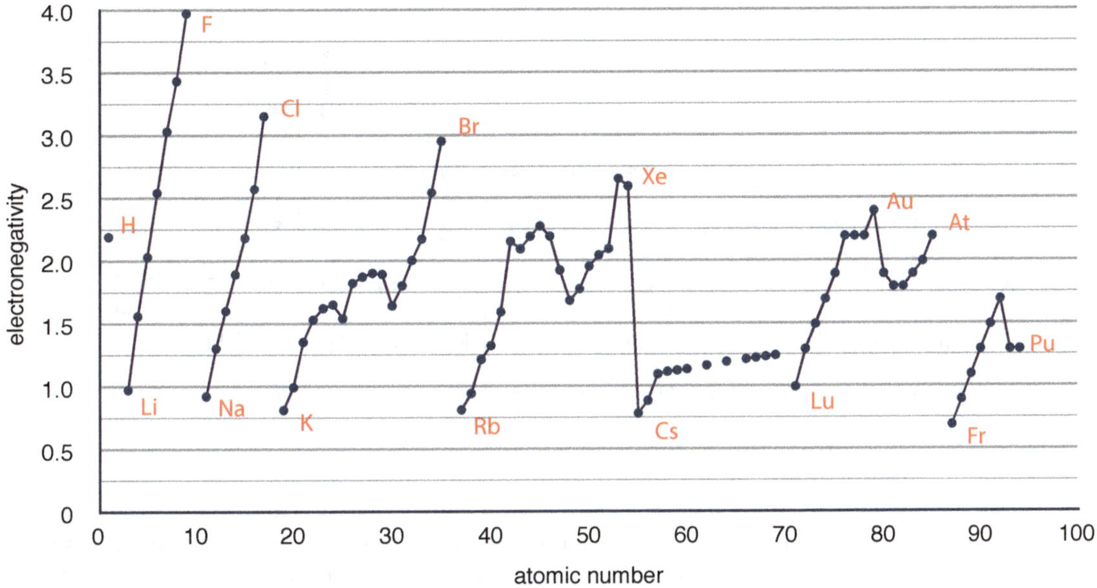

Figure 2.16. Electronegativity values, according to the Pauling electronegativity scale.

Figure 2.16 shows the periodicity of electronegativity and the trend of increasing electronegativity from left to right in the periodic table. There are gaps between the halogens and the alkali metals because the noble gases are not included. This is because electronegativity measures atomic attraction within molecules, and the noble gases don't form bonds with other elements, molecular or otherwise, except under the extreme and unusual conditions that can be produced in a specialized laboratory.

As with electron affinity and ionization energy, the trends in electronegativity can be understood in terms of the effective nuclear charge, Z_{eff}, which increases from left to right as the number of protons in the nucleus increases while the number of core electrons screening the nucleus remains constant. A larger value of Z_{eff} means not only that an atom attracts its own valence electrons more tightly. It also means the atom attracts the valence electrons shared with neighboring atoms more strongly relative to the attraction of neighboring atoms.

The diagram in Figure 2.17 summarizes the general trend of electronegativity values over the periodic table from the lowest value held by francium, to the highest value held by fluorine. You may need various electronegativity values for exercises in this and later chapters, so they are all shown in Figure 2.18 and inside the back cover. If you look at the values toward the right end of the transition metals, you see that around Groups 10–13 there is a small decline in the electronegativities, going against the general upward trend in values from lower left to upper right.

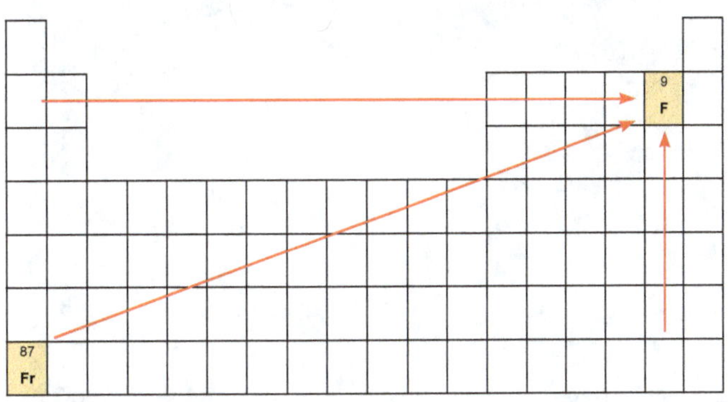

Figure 2.17. The arrows show the trends of increasing electronegativity from the lowest value at francium to the highest value at fluorine.

Period	1	2	3	4	5	6	7	8	9	10	11	12	13	14	15	16	17	18
1	1 H 2.20																	2 He
2	3 Li 0.98	4 Be 1.57											5 B 2.04	6 C 2.55	7 N 3.04	8 O 3.44	9 F 3.98	10 Ne
3	11 Na 0.93	12 Mg 1.31											13 Al 1.61	14 Si 1.90	15 P 2.19	16 S 2.58	17 Cl 3.16	18 Ar
4	19 K 0.82	20 Ca 1.00	21 Sc 1.36	22 Ti 1.54	23 V 1.63	24 Cr 1.66	25 Mn 1.55	26 Fe 1.83	27 Co 1.88	28 Ni 1.91	29 Cu 1.90	30 Zn 1.65	31 Ga 1.81	32 Ge 2.01	33 As 2.18	34 Se 2.55	35 Br 2.96	36 Kr
5	37 Rb 0.82	38 Sr 0.95	39 Y 1.22	40 Zr 1.33	41 Nb 1.6	42 Mo 2.16	43 Tc 2.10	44 Ru 2.2	45 Rh 2.28	46 Pd 2.20	47 Ag 1.93	48 Cd 1.69	49 In 1.78	50 Sn 1.96	51 Sb 2.05	52 Te 2.1	53 I 2.66	54 Xe
6	55 Cs 0.79	56 Ba 0.89	71 Lu 1.0	72 Hf 1.3	73 Ta 1.5	74 W 1.7	75 Re 1.9	76 Os 2.2	77 Ir 2.2	78 Pt 2.2	79 Au 2.4	80 Hg 1.9	81 Tl 1.8	82 Pb 1.8	83 Bi 1.9	84 Po 2.0	85 At 2.2	86 Rn
7	87 Fr 0.7	88 Ra 0.9	103 Lr	104 Rf	105 Db	106 Sg	107 Bh	108 Hs	109 Mt	110 Ds	111 Rg	112 Cn	113 Nh	114 Fl	115 Mc	116 Lv	117 Ts	118 Og

Figure 2.18. Electronegativity values, according to the Pauling electronegativity scale.

2.5 A Few Notes About Hydrogen

Hydrogen is located in Group 1 for a couple of reasons. First, it is an *s*-block element. Its lone electron is in a partially filled *s* subshell, just like the alkali metals. Second, hydrogen usually ionizes by losing an electron to form a cation, like all metals. In fact hydrogen forms the most basic of all cations—a lone proton. Third, in an aqueous solution (a solution with water as the solvent), acids, which are formed with hydrogen, dissociate (come apart) just as ionic compounds do. And just as the metal in a soluble ionic compound is a cation, so is the hydrogen

Hmm... Interesting. Hydrogen in space

Hydrogen is the most abundant element in the universe, making up about 75% of the mass of all matter. The image below, taken by the Hubble Space Telescope, is of the NGC 604 nebula, an enormous region of ionized hydrogen gas (a plasma) in the constellation *Triangulum*. The gas cloud is about 1,500 light years across, and has been called a "nursery of new stars" because of all the new stars formed within it. At the center are over 200 hot stars, each 10–15 times the size of our sun. These hot stars excite the hydrogen atoms, causing them to fluoresce, and heat the nebula to 10,000 kelvins—about twice the temperature at the surface of our sun.

from a dissolved acid. Hydrogen's atomic structure and its normal ionization as a cation with a charge of +1 indicate that hydrogen belongs at the top of Group 1.

But though hydrogen's *structure* assures that its place in Group 1 is probably not going to change, hydrogen's chemical *behavior* is more like a halogen than an alkali metal. For one thing, hydrogen shares electrons in covalent bonds with other elements to form molecules, just as nonmetals do. In contrast, when metals bond they form crystals. A second factor is that hydrogen atoms bond to *themselves* to form molecules of H_2, something no metal does. (An exception is the ion Hg_2^{2+}.) Third, hydrogen can ionize by *gaining* an electron to fill the 1*s* subshell and become H^-, an anion known as *hydride*.

With its only subshell half full with one electron, easily emptied by losing one electron and easily filled by gaining one electron or sharing one pair of electrons (details next chapter), hydrogen is unique among the elements in the periodic table.

Chapter 2 Exercises

SECTION 2.1

1. What entire group of elements did not appear in Mendeleev's original periodic table? Why were they left out and how were they put in?
2. Write a paragraph explaining the general structure and arrangement of the periodic table.

SECTION 2.2

3. State the chief chemical property that distinguishes the metals from the nonmetals.
4. Distinguish between cations and anions.
5. What is the "long form" of the periodic table, and why are there two forms?

SECTION 2.3

6. Use the data in the table below to estimate the following bond lengths. In each case, determine the percent difference between the accepted value given and your estimate (see Section A.4.2 in Appendix A).
   a. the N—H bond length in a molecule of ammonia, NH_3 (accepted value: 1.012 Å).
   b. the C—O bond length in carbon monoxide, CO (accepted value: 1.128 Å).
   c. the C—O bond length in carbon dioxide, CO_2 (accepted value: 1.160 Å).
   d. the P—F bond length in phosphorus trifluoride, PF_3 (accepted value: 1.570 Å).
7. Describe the trend in atomic radius going down a group and across a period.
8. Using the concept of effective nuclear charge, Z_{eff}, write a description accounting for the trends in atomic size in the periodic table.
9. From your knowledge of the periodic table, put the elements rubidium (Rb), silver (Ag), xenon (Xe),

Element	*Z*	Atomic Radius (Å)	Bonding Atomic Distance of Diatomic Molecule (Å)
hydrogen	1	0.32	0.741
carbon	6	0.75	1.242
nitrogen	7	0.71	1.098
oxygen	8	0.64	1.208
fluorine	9	0.60	1.413
phosphorus	15	1.09	1.893
sulfur	16	1.04	1.889
chlorine	17	1.00	1.987
selenium	34	1.18	2.166
bromine	35	1.17	2.281
iodine	53	1.36	2.666

and yttrium (Y) in order of increasing atomic radius. Explain your order by referring to trends in the periodic table.

10. From your knowledge of the periodic table, put the elements sodium (Na), barium (Ba), cesium (Cs), and magnesium (Mg) in order of increasing atomic radius. Explain your order by referring to trends in the periodic table.

11. Based on your knowledge of trends in the periodic table, place the following atoms and ions in order of decreasing size: Be^{2+}, Mg, Ca, and Mg^{2+}.

12. Based on your knowledge of trends in the periodic table, arrange the following atoms and ions in order of size from largest to smallest: S^{2-}, Ar, K^+, Cl^-, and Ca^{2+}.

SECTION 2.4

13. Aluminum and scandium both ionize to +3, even though scandium is in Group 3 and aluminum is in Group 13. Explain why this is.

14. Estimate the effective nuclear charge, Z_{eff}, for vanadium (V), magnesium (Mg), chlorine (Cl), and arsenic (As).

15. Define ionization energy and describe the trends for ionization energy in the periodic table across periods and down groups.

16. Referring again to Table 2.1, explain the large increase in ionization energy that occurs in the yellow shaded region of that table.

17. Write a description accounting for the trends in ionization energy in terms of effective nuclear charge, Z_{eff}, and other factors.

18. Write the condensed electron configurations for Cu^{2+}, As^{5+}, Ag^+, and Au^{3+}.

19. Why are ionization energies so much higher when core electrons are involved than they are when only valence electrons are involved?

20. Based on your knowledge of trends in the periodic table, place the following atoms in order of increasing ionization energy: Ar, Sr, P, Mg, and Ba.

21. Distinguish between ionization energy and electron affinity.

22. Based on your knowledge of trends in the periodic table, place the following atoms in order of increasing electron affinity: Br, Rb, and S.

23. Explain what the electronegativity scale is used for and how it arose.

24. Of the following cations, which is least likely to form: Ca^{3+}, Mg^{2+}, K^+? Explain your response.

25. Why do the chalcogens form ions with a charge of −2?

26. Distinguish between electron affinity and electronegativity.

27. Referring to the periodic table, describe the chemical properties of potassium (K), sulfur (S), xenon (Xe), iodine, (I), and manganese (Mn).

28. Which of the following is likely to have the greatest difference between the third and fourth ionization energies: Cl, Sc, Na, C?

29. Using only the periodic table as a reference (without electronegativity data), predict the relative electronegativities of these elements and put them in order from least to greatest: Ni, Ta, Se, F, Cs, Cl.

30. Develop an explanation for why the electron affinity values for the chalcogens are each significantly lower than those of their halogen neighbors.

31. Why don't the noble gases have electronegativity values listed in Figure 2.16?

SECTION 2.5

32. Describe the chemical properties that place hydrogen in Group 1, and the chemical properties hydrogen shares with Group 17 elements.

GENERAL REVIEW EXERCISES

33. Determine the energy released by the largest electron transition in the Lyman series (see Figure 1.8). State your answer in eV.

34. If a beam of laser light consists of photons with energies of 5.09×10^{-19} J, is the light visible? Explain your response.

35. Identify the block, period, and group for the elements represented by each of the following condensed electron configurations:

    a. $[\mathrm{Ne}]3s^2 3p^3$

    b. $[\mathrm{Xe}]6s^2 4f^{14} 5d^{10} 6p^1$

    c. $[\mathrm{Kr}]5s^1 4d^5$

    d. $[\mathrm{Ar}]4s^2 3d^3$

36. How many orbitals are there in the shell associated with $n = 4$? How many electrons can this shell hold?

37. Determine the number of carbon atoms present in 112 g CO_2.

38. Given 35.0 g H_2SO_4 (hydrogen sulfate, which is called sulfuric acid in aqueous solution), determine the percent composition. Then determine numbers of hydrogen, sulfur, and oxygen atoms present.

39. Identify some specific differences between the chemical properties of the alkali metals and those of the transition metals.

40. Naturally occurring bromine consists of the two isotopes bromine-79 and bromine-81. Given bromine's atomic mass of 79.904 u, is the percentage of bromine-79 found in nature likely to be closer to 25%, 52%, 67%, or 80%? Explain your response.

41. How many grams of calcium are there in 3.00 mol $CaBr_2$?

42. Analysis of a certain sample finds that the sample consists of 53.64% chlorine and 46.36% tungsten. Determine the empirical formula for this compound.

43. What is meant by the phrase, "chemistry is all about modeling"?

44. Why is it that scientists, when they are being accurate in their speech, avoid using the term *truth*? What are they likely to say instead?

45. Is there a difference between scientific facts and historical facts? If so, what is it?

46. What is the difference between molar mass and molecular mass?

47. Why must a calculation of molecular mass based on periodic table data necessarily be an average mass and not the mass of a specific molecule?

48. Distinguish between the two definitions for the mole.

Chapter 3

Chemical Bonding

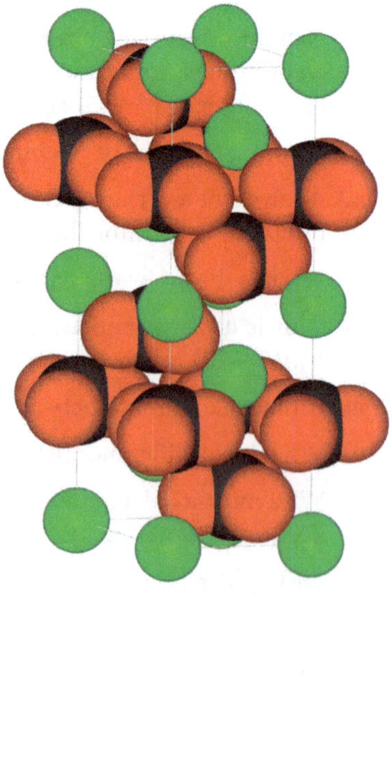

Above are pictured the famous white cliffs of Dover at the southeast corner of England. The cliffs are chalk, one of the many naturally occurring forms of calcium carbonate, $CaCO_3$. Limestone, marble, and sea shells are also made of $CaCO_3$. Marine animals combine the calcium Ca^{2+} ions and carbonate CO_3^{2-} ions dissolved in the water to form their shells. In fact, the limestone and marble were themselves formed by ancient layers of tiny sea creatures with their calcium carbonate shells depositing on the ocean floor as they died.

The space-filling model to the right is of the mineral calcite, also made of $CaCO_3$. The green spheres represent the calcium ions and the black and red spheres are the CO_3^{2-} ions.

As shown later in Figure 3.15, the crystal form of calcium carbonate, calcite, possesses an interesting optical property.

Objectives for Chapter 3

After studying this chapter and completing the exercises, you should be able to do each of the following tasks, using supporting terms and principles as necessary.

SECTION 3.1

1. Define and distinguish between pure substances and mixtures, elements and compounds, and homogeneous mixtures and heterogeneous mixtures.
2. Explain the octet rule.

SECTION 3.2

3. Describe the details of ionic bonding and write formulas for ionic compounds formed from given elements.
4. Given a chart with common oxidation states for transition metals, give names for ionic compounds formed from given elements using the Stock system and the older system based on Latin names.
5. Explain the terms *lattice energy* and *hydrate.*
6. Explain why ionically bonded compounds tend to be brittle and dense and have high melting points.

SECTION 3.3

7. Draw Lewis structures for atoms, molecules, ions, and polyatomic ions.
8. State the names of ionic compounds formed with polyatomic ions.
9. State the names of acids formed from given atoms and polyatomic ions.
10. Describe and explain the three general cases of exceptions to the octet rule.
11. Explain what resonance structures are and depict them in Lewis structures.
12. Give the names for binary molecular compounds.
13. Compute the bond number for a given bond in a given molecule, and relate it to bond strength and bond length.
14. Describe some physical properties of molecular substances and compare the properties of covalently bonded substances to those of ionically bonded substances.

SECTION 3.4

15. Use the electronegativity scale to explain molecular polarity.
16. Given electronegativity values, predict the polarity and the percent ionic character of a given molecule, and characterize bonds as nonpolar covalent, polar covalent, or ionic.
17. Use percent ionic character to give a general account for the colors of aqueous solutions.

3.1 Preliminaries

3.1.1 Types of Substances

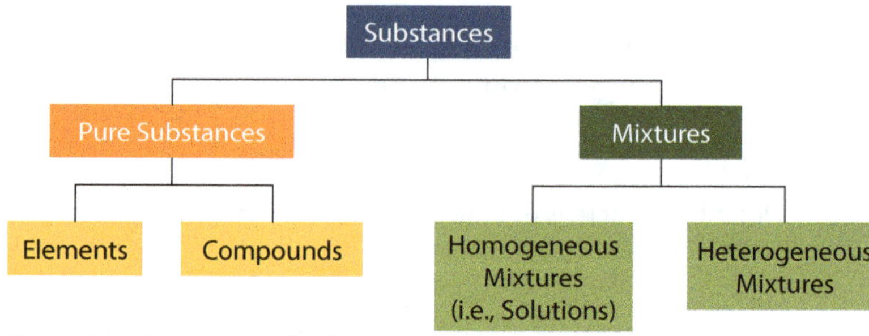

Figure 3.1. Major types of substances.

Before we get into the details of chemical bonding, we should pause for a moment and review the different types of substances, shown in Figure 3.1. The two major classes of substances are *pure substances* and *mixtures*. In pure substances, the chemical composition is uniform throughout the substance, including at the atomic level. If all the atoms in the substance have the same number of protons, the substance is an *element*, such as the elemental gold ingot shown in Figure 3.2. The elemental gold can be purified even further down to a single isotope, so that every atom in the substance is identical, but chemically, all isotopes of gold act the same, so as long as all the atoms are gold atoms the substance is considered chemically pure. The stamp on the ingot indicates it is 99.99% pure gold. If a pure substance contains more than one element but the atoms are bonded together in the same way throughout the substance, the substance is a *compound*. As we have seen, in compounds atoms are joined together in consistent whole number ratios throughout.

Figure 3.2. A 1-kg gold ingot.

The other major class of substance is *mixtures*. When substances are combined, if no chemical reaction occurs a mixture is formed. The components in a mixture retain their individual properties and can be separated by physical means. Again there are two types, referred to as *homogeneous mixtures* and *heterogeneous mixtures*. Homogeneous mixtures have uniform composition down to, but not including, the groups of atoms at the atomic level. The implication of this definition is that homogeneous mixtures are identical with *solutions*, mixtures in which one pure substance (or more than one) is dissolved in another pure substance. One cannot see the different components in a homogeneous mixture, not even with the most powerful microscope. Particles at the atomic level—atoms, molecules, ions—are too small to reflect visible light, and so cannot be seen in the ordinary way regardless of the magnification.

In contrast, in a heterogeneous mixture there are lumps of different materials mixed together. You might be able to see the lumps with the naked eye, as with the mixture of spices in meat seasoning. Or the different substances may be visible only under a microscope, such as microscopic particulates in well water. Either way, if the different substances can be seen, the mixture is heterogeneous.

If no chemical reaction occurs when substances are combined, a mixture is formed. The components in a mixture retain their individual properties and can be separated by physical means such as boiling and filtration.

In previous chapters, our study of atomic structure, electron arrangement, and isotopes has focused on elements. In the balance of this chapter we focus on compounds. In a later chapter, we look closely at solutions and at a group of substances known as colloids that inhabit the boundary between solutions and heterogeneous mixtures.

3.1.2 Chemical Possibilities

As with the atomic orbital structures we have seen in previous chapters, much of our understanding about chemical bonding derives from theoretical calculations. When it comes to chemical bonding, some of those calculations involve the Schrödinger equation; others involve ionization energy, electron affinity, and electronegativity. For all but the last of these, our considerations in this text are primarily qualitative. In other words, we will not go into the calculations themselves; we simply look at what the results tell us about how atomic bonding works. The calculations involving electronegativity are simple, so we will have a look at those.

Much of chemistry is about understanding the mechanisms that bond atoms to other atoms. *And guess what that means*—electrical attractions, electrons, minimizing energy, whole-number ratios, and modeling! Everything from the Introduction is on center stage here.

In some cases, atomic bonds hold atoms together in rigid, extended, regular structures called *crystal lattices*. In other cases, atoms are bonded together in the clusters called *molecules*. And a third type of structure again involves regular arrangement of atoms in a strong crystal lattice, but with much less rigidity and much less brittleness. These three bonding types—named *ionic bonds*, *covalent bonds*, and *metallic bonding*—are each caused by a particular combination of electrical attraction and electron location that minimizes the energy in a given structure. We explore the first two in this chapter and save metallic bonding for the next chapter.

3.1.3 The Octet Rule

For main group elements, the electrons in the *s* and *p* subshells are the valence electrons. Adding the two possible electrons in an *s* subshell to the six possible electrons in a *p* subshell gives eight electrons—an *octet*. The noble gases (except helium) each have an octet without ionizing. This is the most stable of possible electron configurations, where all electrons are in a stable state that minimizes their energies, and for this reason, atoms of noble gas elements have high ionization energies and do not readily bond with other atoms. Other main group elements (and many of the transition metals) tend to ionize or share electrons with other atoms in such a way that they acquire a noble gas electron configuration, the stablest way to be. In other words, *atoms tend to gain or lose electrons to achieve an octet of electrons in the occupied shell with the highest-energy*. This is such a common characteristic that it has been dubbed the *octet rule*. People say that atoms try to "surround themselves" with an octet of electrons. The obvious main group exceptions to the octet rule are hydrogen and helium. These elements are each satisfied with a single pair of electrons in the 1*s* subshell. There are some other noteworthy exceptions (such as boron), which we explore later in the chapter.

The octet rule is an easy way to think about what happens with the electrons in both ionic bonding and covalent bonding. When ionic bonding occurs, atoms ionize, and the oxidation states they take on when they ionize usually involve achieving an octet. In covalent bonding, atoms share electrons in pairs, and each atom involved in the sharing acts as if all shared electrons are its own. In covalent bonds, atoms generally try to share the correct number of electron pairs so that they get their octets. In fact, we even use diagrams that symbolize this, such as the diagram of the hydrogen and oxygen atoms in a water molecule shown in Figure 3.3. We elaborate on this type of diagram when we get to covalent bonding. But for now, you see that each hydrogen is adjacent to a pair of electrons and the oxygen is surrounded by its octet.

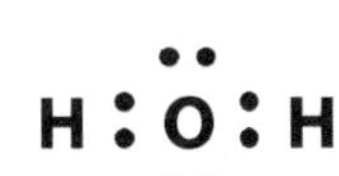

Figure 3.3. Life is good. Each hydrogen has its pair and oxygen has its octet.

3.2 Ionic Bonding

3.2.1 Ionic Bonds and Crystals

As its name implies, ionic bonds are bonds between ions—the symmetrical electrical attraction between cations and anions is what bonds the atoms together. As you know, the minimum energy arrangement for objects with opposite electrical charge is for them to get as close together as possible. Minimum energy for like charges is to be as far away as possible. To minimize the energy of the attractions and repulsions between the ions, the ions arrange themselves into a crystal lattice, as illustrated in Figure 3.4. The figure shows four cells of the cubic crystal lattice formed by +1 cations (alkali metals) and –1 anions (halogens). Sodium chloride is the usual example given. In this crystal structure, the ratio of cations to anions is one to one. Figure I.11 shows the more complicated structure of calcium fluoride, in which there are two fluorine atoms for each calcium atom.

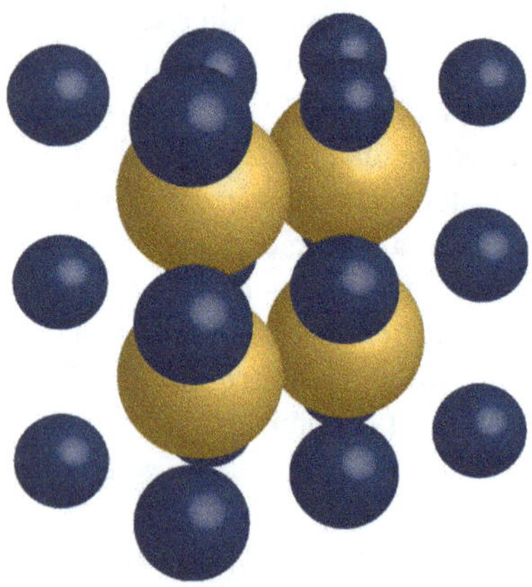

Figure 3.4. The cubic crystal structure formed by ionic bonds between +1 cations and –1 anions. In the lattice, the ratio of cations to anions is 1 : 1.

In the previous chapter, we saw that metals ionize to become cations and nonmetals ionize to become anions. It follows then that ionic bonds form between metals and nonmetals. More generally, ionic bonds form between any type of positive and negative ions, even if the ions are molecules instead of single atoms.

As illustrated in Figure 3.5, we usually think of ionic bonding as a process of electron *transfer*. A metal ionizes by the removal of one or more electrons, and a nonmetal ionizes by acquiring one or more electrons. In the process, electrons are transferred from the metal atoms to the nonmetal atoms. The figure illustrates the electron transfer that occurs between sodium and chlorine. A glance at the periodic table shows that sodium ionizes by the removal of its one valence electron to become a +1 cation. Chlorine, being a Group 17 nonmetal, needs only one additional electron for its valence shell to be full—an attractive, low-energy state—so it becomes a –1 anion. The 3*s* electron from the sodium atom moves into the 3*p* subshell of the chlorine atom. In the process, both sodium and chlorine acquire octets.

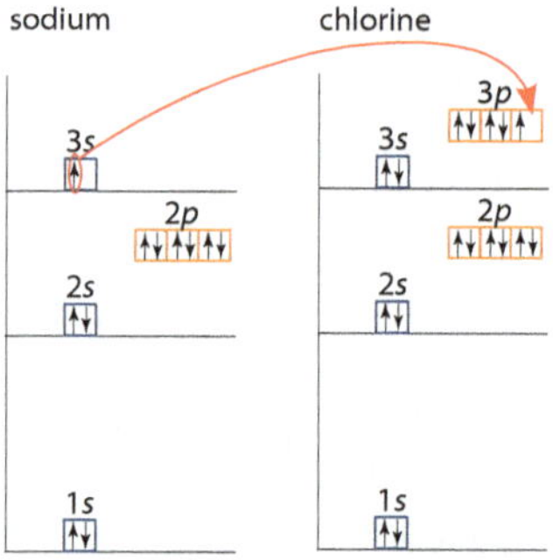

Figure 3.5. Electron transfer between sodium and chlorine atoms. The sodium becomes Na^+, the chlorine becomes Cl^-. In the process, both sodium and chlorine acquire octets.

Figure 3.6 illustrates the electron transfer that occurs between magnesium and chlorine. Magnesium is an alkaline-earth metal. As such, it has two valence electrons and ionizes to become a +2 cation. To accommodate both these electrons, two atoms of chlorine are required, since each takes on only one electron to fill the 3*p* subshell.

When writing the chemical formula for an ionic compound, the convention is to place the chemical symbol for the cation first followed by the chemical symbol for the anion. Thus, the formula for sodium chloride is NaCl. In the formula, subscripts are used to indicate the whole number ratio of atoms in the compound. If the number of atoms in the ratio is one (as with sodium and chlorine being in a one to one ratio), no subscript is used. In magnesium chloride, two atoms of chlorine are needed for each atom of magnesium, so the formula is $MgCl_2$. The subscripts show the whole-number ratios of atoms in the compound.

▼ Example 3.1

Write the chemical formula for calcium sulfide.

As an alkaline-earth metal, calcium ionizes to +2. Sulfur, a chalcogen, ionizes to –2. Since two electrons are removed from each calcium atom and two electrons are added to each sulfur atom, the ratio of calcium atoms to sulfur atoms is one to one. Thus, the formula is CaS.

▲

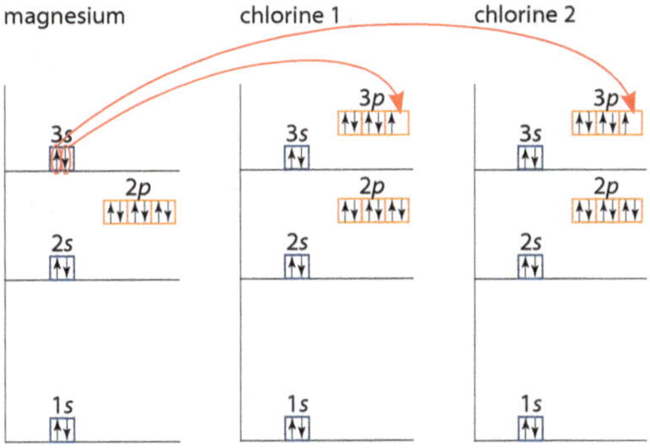

Figure 3.6. Electron transfer between magnesium and chlorine atoms. The magnesium becomes Mg^{2+}, each of the two chlorines becomes Cl^-. All three atoms acquire octets.

▼ Example 3.2

Write the chemical formula for beryllium phosphide.

As an alkaline-earth metal, beryllium ionizes to +2. Phosphorus, a Group-15 nonmetal, ionizes to –3. Two electrons are removed from each beryllium atom, but three electrons are added to each phosphorus atom. The only way for this to work out is for the electrons to be exchanged in quantities based on the least common multiple of two and three, which is six. As shown in Figure 3.7, three beryllium atoms provide a total of six electrons, and these are divided equally among two phosphorus atoms. Since the ratio of beryllium atoms to phosphorus atoms is three to two, the formula is Be_3P_2.

▲

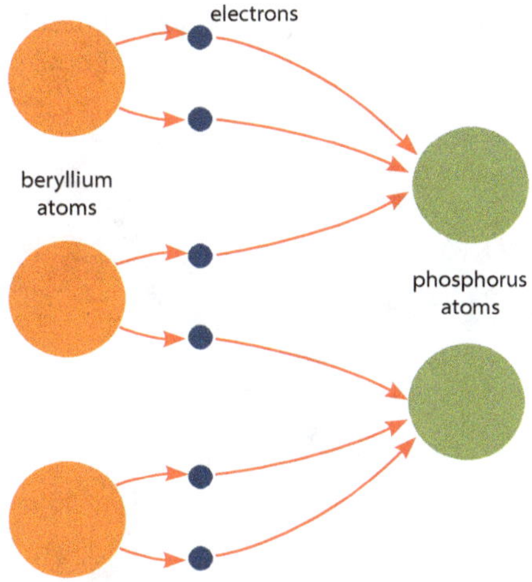

Figure 3.7. Electron transfer between +2 cations and –3 anions requires the ratio of cations to anions to be 3 : 2.

The easiest way to work out formulas for ionic bonds involving metals from Groups 1–4 is to think about the number of valence electrons involved (equal to the group number of the metal) and the numbers of cations and anions that are needed for the electron transfer to work out evenly. But when working out the formulas for ionic compounds with metals in Groups 5 and higher, visualizing numbers of valence electrons from the metal's position in the periodic table isn't so easy. It's easier just to select subscripts for the cation and anion so the product of the subscript and the oxidation state is the same for the cation and anion. This is equivalent to making the amounts of positive and negative charge match, as illustrated in the next example.

▼ Example 3.3

Write the chemical formula for iron oxide, assuming the iron ionizes as Fe^{3+}.

Oxygen ionizes as O^{2-} (Figure 2.12). To make the positive and negative charges match, we need to use the least common multiple of three (from the Fe^{3+}) and two (from the O^{2-}), which is six.

With two +3 cations and three −2 anions the charges are equal. Thus, since the ratio of iron atoms to oxygen atoms is two to three, the formula is Fe_2O_3.

▲

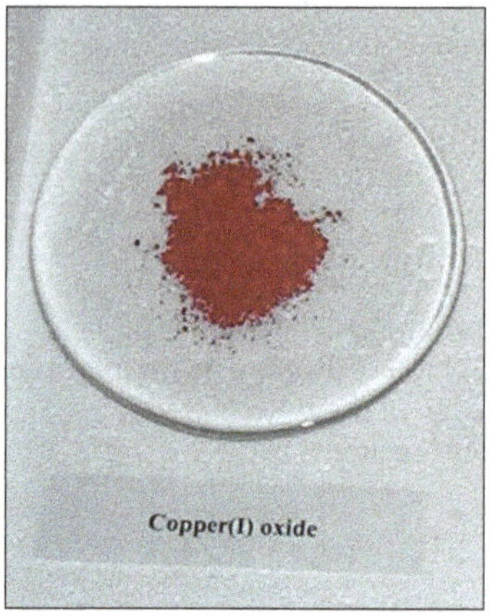

Figure 3.8. Copper(I) oxide, Cu_2O.

Formulas for ionic compounds are called *binary formulas* because there are two parts in the formula: the symbol for the cation and the symbol for the anion. The formulas for compounds formed with cations and anions from main group elements are easy to figure out because main group metals (Groups 1 and 2) and main group nonmetals (from Groups 14–17) ionize very predictably. All you have to do is look at what group an element is in in the periodic table to see what its oxidation state is when the atom gets its octet. But many transition metals have more than one common oxidation state and can thus make more than one compound with a given nonmetal.

For example, as listed in Figure 2.12, copper commonly forms both Cu^+ and Cu^{2+} ions and as a result can form two different compounds when bonding with oxygen. As Cu^+, copper forms Cu_2O, known as cuprous oxide or copper(I) oxide. (We address these naming schemes next.) This substance is a red powder, shown in Figure 3.8. Figure 3.9 shows one cell in the lattice for one of the two possible bonding structures for Cu_2O.

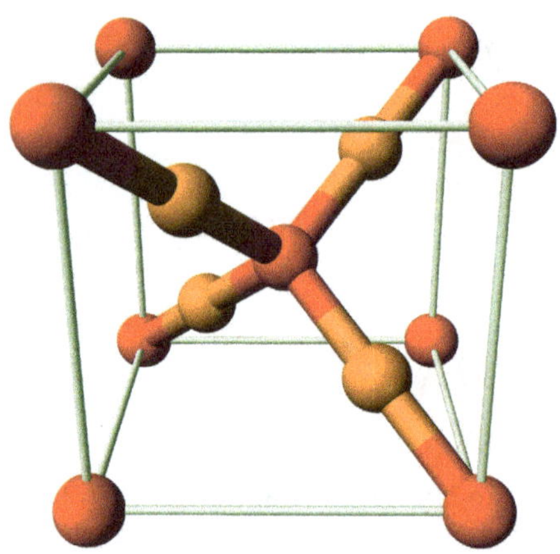

Figure 3.9. A single cell in the lattice structure (one of two possible structures) of copper(I) oxide, Cu_2O. Oxygen atoms are shown in red.

As Cu^{2+}, copper bonds with oxygen in a one-to-one ratio producing CuO, known as cupric oxide or copper(II) oxide. This substance is a black powder, shown in Figure 3.10. One cell of the lattice structure is shown on the left in Figure 3.11. On the right of the figure is an image showing how the crystal lattice looks when several cells are joined together.

3.2.2 Naming Ionic Compounds

Names of ionic compounds always have two parts—the name of the cation placed first, followed by the name of the anion.

The first case to consider is when an ionic compound is formed by a *monovalent* cation and a *monatomic anion*. Monovalent cations have only one oxidation state, and monatomic ions are composed of only one atom, such as all the ions we have discussed so far. The rule for this case is that the name of the cation is simply the name of the metal, and the name of the anion is the name of the nonmetal with the suffix *-ide* attached.

▼ Example 3.4

Write the names and formulas for the compounds formed by the following pairs of ions: Ca^{2+} and F^-; K^+ and O^{2-}; and Al^{3+} and Br^-.

In each case, we state the name of the metal followed by the name of the nonmetal with the *-ide* suffix. The formulas must balance the charges of the ions, just as we have previously seen.

Figure 3.10. Copper(II) oxide, CuO.

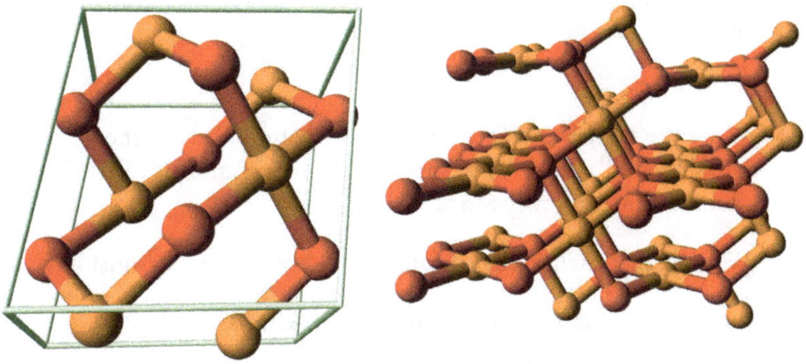

Figure 3.11. One cell of the lattice for copper(II) oxide, CuO (left), and an extended portion of the lattice structure (right).

calcium fluoride, CaF_2

potassium oxide, K_2O

aluminum bromide, $AlBr_3$

▲

The second case is for *multivalent* cations, cations with more than one oxidation state, and monatomic anions. There are two different naming systems in use. In both systems, the cation is written first, just as with monovalent cations, and the anion is written with the suffix *–ide*.

The contemporary system is called the *Stock system*, illustrated in Table 3.1. In the Stock system, the cation is the name of the metal followed by parentheses with upper-case Roman numerals indicating its oxidation state. (No space is placed between the name of the metal and the parentheses.) The multivalent cations are the transition metals and some of the metals in Groups 13–15. The most common oxidation states are shown in Figure 2.12. You may assume all Group 1 and Group 2 metals are monovalent, along with Al^{3+}, Ni^{2+}, Zn^{2+}, and Ag^{+}. (Each of these last four elements has other oxidation states, but they are quite uncommon.) The parentheses are unnecessary in compounds with monovalent cations. The rest of the metals should have the oxidation state written in parentheses as described above.

Oxidation State	Example Cation	Example Compound
lower	Fe^{2+} iron(II) Cu^{+} copper(I)	iron(II) oxide, FeO copper(I) fluoride, CuF
higher	Fe^{3+} iron(III) Cu^{2+} copper(II)	iron(III) sulfide, Fe_2S_3 copper(II) chloride, $CuCl_2$

Table 3.1. Examples of the Stock system for naming ionic compounds.

The older system, illustrated in Table 3.2, is important to know because it still appears often in reference literature and on bottles of chemicals. In this system, the Latin name of the cation is

Oxidation state	Cation Ending	Example Cation	Example Compound
lower	*–ous*	Fe^{2+} ferrous Cu^{+} cuprous	ferrous oxide, FeO cuprous fluoride, CuF
higher	*–ic*	Fe^{3+} ferric Cu^{2+} cupric	ferric sulfide, Fe_2S_3 cupric chloride, $CuCl_2$

Table 3.2. Latin-based cation naming system.

used with the suffix *-ous* to represent the lower oxidation state or the suffix *-ic* to represent the higher oxidation state.

▼ Example 3.5

Write the names and formulas for the compounds formed by the following pairs of ions: Hg^{2+} and Cl^-; Mn^{4+} and S^{2-}; Pb^{2+} and O^{2-}; and Sn^{4+} and Br^-. Include names in both the older system and the contemporary Stock system.

We note that each of the cations is multivalent and so must have the oxidation state specified.

mercury(II) chloride, mercurous chloride, $HgCl_2$

manganese(IV) sulfide, manganic sulfide, MnS_2

lead(II) oxide, plumbous oxide, PbO

tin(IV) bromide, stannic bromide, $SnBr_4$

Note that for some of the metals with a chemical symbol based on a Latin name (e.g., copper, iron, lead, tin, and gold) the old style name is based on the Latin element name. In other cases (e.g., mercury, cobalt, chromium, and manganese), English names are used.

▲

3.2.3 Energy in Ionic Bonds

Typically, the formation of ionic bonds is an exothermic process—the reaction releases energy in the form of heat and light. We can analyze the process in terms of the ionization energy and electron affinity. Remember, chemistry is all about minimizing energy, so we can understand ionic bonding in these terms.

Let's use sodium chloride as an example. The reaction between sodium and chlorine happens spontaneously and a great deal of energy (light and heat) is released. Since the reaction is spontaneous, any energy required to make the reaction occur comes from the reactants themselves.

The ionization energy—energy that must supplied—for a sodium atom is 5.14 eV. The electron affinity—energy released—for a chlorine atom is 3.61 eV. Now, these values may suggest that we have a problem. We have to put more energy in to ionize the sodium than we get back out when the chlorine atom takes on the electron! This seems to indicate that the reaction between sodium and chlorine couldn't release so much heat, or even occur at all. In fact, our energy problem is even worse. For the sodium and chlorine to react, a piece of sodium metal must be vaporized—the atoms have to be in the gaseous state so they can interact with individual chlorine atoms—and this requires 1.12 eV per atom. Also, the diatomic chlorine molecules have to be separated, requiring an additional 1.25 eV per atom. So all together, 5.14 eV + 1.12 eV = 6.26 eV of energy must be supplied per sodium atom, plus 1.25 eV per chlorine atom, for sodium and chlorine to react, a total of 7.51 eV per pair of NaCl atoms is required. However, only 3.61 eV per atom is released when a chlorine atom ac-

Compound	Calculated Lattice Energy, kJ/mol
LiF	1030
NaF	910
KF	808
LiCl	834
NaCl	769
KCl	701
NaBr	732
NaI	682
$MgCl_2$	2477
$CaCl_2$	2268
Na_2O	2481
K_2O	2238
MgO	3795
CaO	3414

Table 3.3. Lattice energies for a few ionic compounds.

cepts an electron. And further, there is energy in the sodium chloride crystal lattice after the reaction is complete. Where does all the extra energy come from? Obviously, there's more to this story.

The ionization energy is the energy required to remove an electron far away from the cation so the electron becomes completely free. Likewise, the electron affinity is the energy released when an electron comes from far away to be united to a neutral atom. This means that in energy terms, we imagine that the electron transfer from the sodium atoms to the chlorine atoms happens with all these atoms far apart. Then after the electron transfer is over, we allow all the ions to collapse together, releasing energy in the process, to form the crystal lattice.

Of course, if we actually combine sodium metal and chlorine gas, the atoms do not spread out, react, and then attract back together from electrical forces. But to perform the energy accounting using the definitions and energy values we have, we imagine they do. Imagining it in energy terms, there is a whopping amount of energy released when all the cations and anions come crashing together. This is where the extra energy comes from, even though it is not apparent simply by looking at the ionization energy and electron affinity values.

Part of the energy released goes to binding the ions together in the crystal lattice. This energy is called the *lattice energy*. Considered this way, the lattice energy is a negative value because the ions go into lower energy states when they come together. Another way to think about lattice energy is as the energy required to separate completely all the ions in the crystal lattice from each other—to pull the crystal apart, ion by ion. Considered this way, the lattice energy is positive. Both definitions are acceptable and common.

Lattice energy values for a few common ionic compounds are given in Table 3.3. To help you get a feel for the quantities of energy involved, let's run through an energy calculation *just for fun*. You may find the results surprising.

▼ Example 3.6

Compare the ionization energy of sodium and the electron affinity of chlorine to the lattice energy of sodium chloride.

It makes intuitive sense to specify lattice energies in terms of a bulk quantity such as a mole. So instead of working on an atom-by-atom basis, we work on a per-mole basis. Let's begin by converting the energy values given in the text above from eV for a single particle to kJ/mol:

$$\frac{7.51\text{ eV}}{\text{particle}} \cdot \frac{1.602 \times 10^{-19}\text{ J}}{1\text{ eV}} \cdot \frac{6.022 \times 10^{23}\text{ particle}}{\text{mol}} = 725{,}000\ \frac{\text{J}}{\text{mol}} = 725\ \frac{\text{kJ}}{\text{mol}}$$

So for one mole of sodium and one mole of chlorine, 725 kJ are required to get the Na and Cl atoms apart and ionize the sodium atoms. Similarly, 349 kJ are released when the chlorine atoms receive their electrons. This is an energy deficit of 376 kJ. The energy released as the lattice forms supplies this, and the lattice energy as well. From Table 3.3, the lattice energy of NaCl is 769 kJ/mol. So bringing the crystal together releases enough energy to cover the energy deficit and still have 769 kJ to form the crystal lattice. This is the lattice energy in one mole of NaCl, which is only 58.44 grams (see Figure 1.18).

How much energy is 769 kJ? Well, one watt of power is an energy rate of one joule per second. Let's pick an electrical device we are familiar with and calculate how long it would run with this much energy to power it. Since we are all familiar with good old fashioned 60-W light bulbs (at least for one more generation or so of students) we will use that:

$$\frac{769{,}000\ \text{J}}{60\ \frac{\text{J}}{\text{s}}} = 12{,}817\ \text{s} \cdot \frac{1\ \text{hr}}{3600\ \text{s}} = 3.6\ \text{hr}$$

Have you ever felt the heat from a 60-W light bulb? If you could trap all this heat and light for 3.6 hours you would have an amount of energy equal to the lattice energy holding together the crystals in 1/4 cup of salt!

When 1/4 cup of salt is dissolved in a beaker of water, 769 kJ of energy must be supplied to the salt to pull apart the crystal lattice and allow the ions to go into solution. Now, all the energy needed for this to happen comes from the water. So why doesn't the water's temperature plummet when salt is dissolved in it? The answer relates to the polarity of the water molecule and the ever-present tendency to minimize energy. Since the water molecules are polar, they rush toward and surround the dissolved ions, releasing enough energy in the process to keep the water temperature roughly constant. It is amazing that there is so much energy involved in the process at the atomic level, and that when we stir salt into a glass of water we are completely unaware of it. We return to this topic when we get to the chapter on solutions.

3.2.4 Hydrates

Figure 3.12. Crystals of copper sulfate pentahydrate, $CuSO_4 \cdot 5H_2O$.

In some ionic compounds, it is common for water molecules to integrate into the crystal lattice if the crystals are grown in an aqueous (water) solution. These compounds are call *hydrates*. The presence of the water molecules can allow compounds to form continuous crystals, when without the water molecules the compound would decompose into a powder. A nice example is the compound copper sulfate pentahydrate, $CuSO_4 \cdot 5H_2O$, shown in Figure 3.12. The coefficient in front of the H_2O, 5 in this case, indicates how many water molecules are in the crystal lattice for each formula unit of atoms in the ionic formula.

The complex structures of $CuSO_4$ (without the water molecules) and $CuSO_4 \cdot 5H_2O$ are shown in Figure 3.13. If you study these images closely you can see that in the $CuSO_4$ lattice,

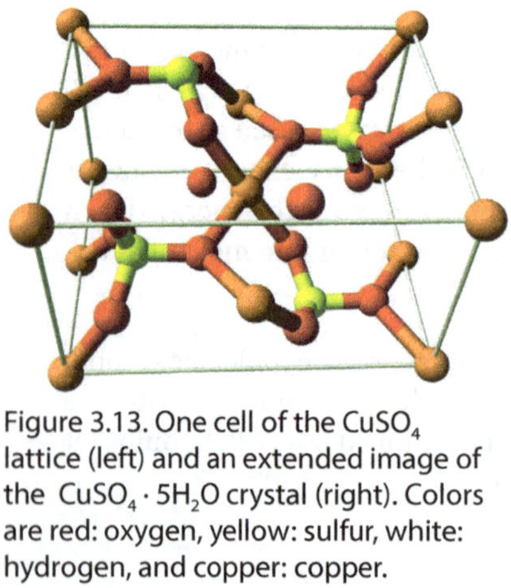

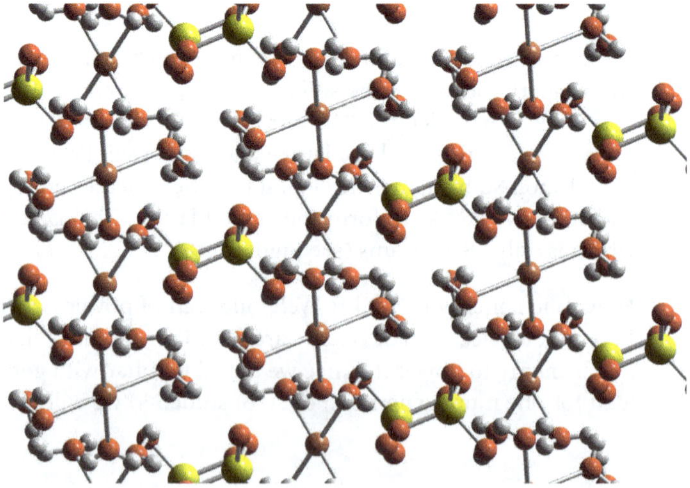

Figure 3.13. One cell of the $CuSO_4$ lattice (left) and an extended image of the $CuSO_4 \cdot 5H_2O$ crystal (right). Colors are red: oxygen, yellow: sulfur, white: hydrogen, and copper: copper.

the copper-colored copper atoms are attached to the red oxygen atoms in the yellow and red sulfate ions (SO_4^{2-}). But interestingly, the water molecules in the hydrate crystal (shown in red and white) are connected to the copper atoms and the sulfate ions are left floating in the lattice.

3.2.5 Intensive and Extensive Properties

We often have occasion to discuss the physical properties of substances. In that regard, chemists tend to classify the properties of substances into two types: *intensive* and *extensive*. An *intensive property* is a bulk property, meaning the property does not depend on how much of the material is present. Examples of intensive properties include brittleness, electrical conductivity, boiling point, and density. *Extensive properties* depend on how much material is present. Examples of these include mass and volume.

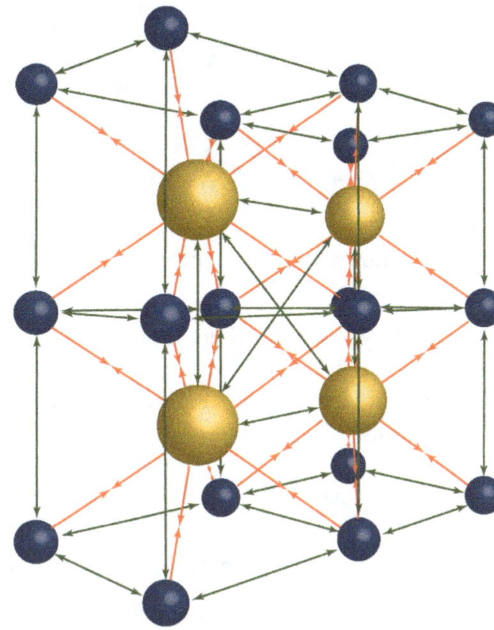

Figure 3.14. The symmetry of attractive and repulsive electrical forces in the ionic bond make the crystal lattice very rigid—if stretched too far it will break rather than bend.

3.2.6 Physical Properties of Ionically Bonded Substances

The high lattice energies in ionic compounds are due to the strong, symmetrical network of electrical attractions and repulsions in the crystal lattice, illustrated in Figure 3.14. The green arrows in this figure indicate electrical repulsions, and the red arrows indicate electrical attractions. (For clarity, the diagonal repulsions between the blue ions are not shown.) As you see, each ion is bound in place by the symmetrical forces all around it.

Many of the intensive physical properties of ionic compounds are due to the strong electrical forces in the crystal lattice. One of these is brittleness, the tendency of a substance to shatter instead of bend when stressed. The classic shattering of a crystal glass from the sound waves from an opera singer's voice (a legend confirmed by MythBusters, although a bit of amplification was required) is a great example of this. Ionic crystals can flex to some extent. This is why vibrating pieces of quartz are used now to regulate the speed of just about every cheap battery-powered clock in the world. But when subjected to a high enough stress, a crystal lattice breaks apart instead of flexing. Whack a lump of iron with a hammer and you may put a dent in it. Whack any ionic compound with a hammer (including all the rock minerals) and watch it shatter. Although the iron is crystalline, like all pure metals, it is not held together by ionic bonds. We consider the bonding in metals in the next chapter.

Compound	Melting Point, °C	Density, g/cm^3
LiF	848.2	2.64
NaF	996	2.78
KF	858	2.48
LiCl	610	2.07
NaCl	800.7	2.17
KCl	771	1.99
NaBr	747	3.20
NaI	661	3.67
$MgCl_2$	714	2.33
$CaCl_2$	775	2.15
Na_2O	1134	2.27
K_2O	740	2.35
MgO	2825	3.6
CaO	2613	3.34

Table 3.4. Two physical properties for a few ionic compounds.

High melting point and high density are two other physical properties of ionic compounds. A few of these values are shown in Table 3.4. The contrast with covalently bonded

Figure 3.15. Calcite exhibits the double refraction known as birefringence.

compounds should be obvious. Water and ammonia, for example, melt at 0°C and –78°C, respectively. And water, with a density of 1 g/cm^3, is fairly dense for a covalent compound, but garden variety ionic compounds are two to three times as dense.

An interesting (although somewhat esoteric) property some ionic compounds exhibit is *birefringence*, or double refraction. Figure 3.15 shows this happening as light passes through a piece of calcite, which is calcium carbonate, $CaCO_3$. If you look closely, you can see there is are two refracted images. This is caused by light reflecting from the letters on the paper and refracting twice as it passes upward through the crystal. Birefringence is caused by the polarization of the light waves interacting with the planes of atoms in the crystal structure.

3.3 Covalent Bonding

3.3.1 Covalent Bonds and Molecules

Covalent bonds are formed when atoms of nonmetals bond with each other. Instead of transferring electrons to achieve octets, atoms *share electrons in pairs* to accomplish the same thing. Since there are no ions involved, we don't have thousands of electrical forces binding the atoms into a lattice. Instead, it is the actual sharing of electrons in orbitals that binds the atoms together. And when atoms bond this way, they bond in groups or clusters—that is, molecules.

Later, we get into the details of theoretically modeling the orbitals of atoms. For now, we focus on the numbers that make the bonds work. The goal is for atoms to share valence electrons in pairs so that all atoms achieve their octets (or for hydrogen, its complete pair). There are a few important exceptions to the ways atoms follow the octet rule, which we will address in a bit.

To begin, let's consider water, H_2O. To visualize the bonding, we draw the symbol for each of the atoms in the molecule along with dots to represent their valence electrons, a representation called a *Lewis symbol.* Each hydrogen atom has one valence electron and the oxygen atom has six valence electrons. This makes a total of eight valence electrons that are involved in the bonding. For a given atom, the first four valence electrons are drawn with one on each side of the element symbol. After each side of the element symbol has one electron, the remaining electrons are placed to form pairs, with a maximum of two electrons on each side of the element symbol. The Lewis symbols for the atoms involved in the water molecule are:

$$\mathrm{H}\cdot \quad \mathrm{H}\cdot \quad \cdot\underset{\cdot\cdot}{\ddot{\mathrm{O}}}\cdot$$

Covalent bonding involves atoms sharing electrons in pairs to achieve their octets (or pair for hydrogen). Each atom contributes one electron to make up a shared pair. The reason they share in pairs is that each atomic orbital holds two electrons, and in covalent bonding the orbitals of two atoms meld together into a single *molecular orbital* that holds the shared pair of electrons.

In the case of the water example, each of the hydrogen atoms shares one pair of electrons with the oxygen atom, represented this way:

$$\mathrm{H}:\underset{\cdot\cdot}{\ddot{\mathrm{O}}}:\mathrm{H}$$

As you see from this representation, we still have eight dots, representing the eight valence electrons. Each hydrogen atom has a pair of electrons next to it, representing its filled 1*s* subshell. The oxygen atom is surrounded by eight electrons, representing its complete octet.

Our next example is carbon dioxide, CO_2. Here are the atoms, each with their valence electrons:

The carbon atom is in the center with an oxygen bonded to either side of it, sharing pairs of electrons. Let's start there.

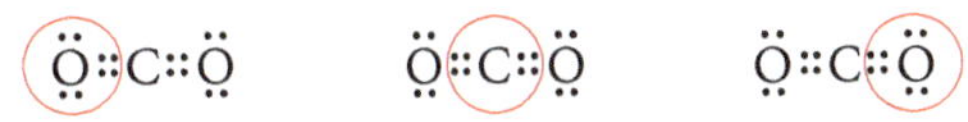

Figure 3.16. Circles indicate the octets for each of the three atoms in the CO_2 molecule.

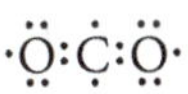

Now we see there are still four unpaired electrons, and none of the atoms has an octet yet. This problem is solved by the carbon atom sharing not one but *two pairs* of electrons with each of the oxygen atoms, symbolized like this:

The double pairs of electrons symbolize two pairs of electrons that are being shared. Both pairs are counted in the octet for each atom in the bond. Now, every atom in the molecule has an octet, as illustrated in Figure 3.16. In this figure, the electrons associated with each atom are circled to show the atom's octet of eight electrons.

Sharing two pairs of electrons is called a *double bond*. There are also triple bonds in which three pairs of electrons are shared. This is illustrated best by the *diatomic molecules* formed by nitrogen atoms. We should pause here for a bit of historical background.

In 1811, Italian scientist Amedeo Avogadro introduced a new molecular theory of gases. Part of Avogadro's theory was the idea that atoms of some elements form molecules this way—by atoms bonding in groups to make molecules, including diatomic molecules. At the time, Avogadro's theory was widely rejected; even John Dalton, the father of modern atomic theory, rejected it. But in 1860, almost 50 years later and just a few years after Avogadro's death, an Italian chemist named Stanislao Cannizzaro gave a presentation in Germany at the first international conference for chemists showing that Avogadro had been correct all along. Acceptance of Avogadro's theory began gaining ground after that.

At the end of Chapter 1, I mention the main substance of Avogadro's new theory—that the volume of a gas at a given temperature and pressure is proportional to the number of particles of the gas (atoms or molecules), regardless of the identity of the gas. Because of his work establishing molecular theory, Avogadro is now considered the father of atomic-molecular theory. And as you have already learned, Jean Perrin honored Avogadro by naming the Avogadro constant after him.

There are seven elements that appear in nature as diatomic molecules. These are identified in Figure 3.17. Atoms of hydrogen and all the halogens are lacking only one electron to have a full octet (or pair, for H). Atoms of each of these elements bond together in pairs and share a single pair of electrons, as shown in Table 3.5. As the table shows, the diatomic oxygen molecule is

Figure 3.17. The seven elements that form diatomic molecules.

Molecule	Lewis Symbol Representation
H_2	H:H
N_2	:N:::N:
O_2	Ö::Ö
F_2	:F̤̈:F̤̈:
Cl_2	:C̤̈l:C̤̈l:
Br_2	:B̤̈r:B̤̈r:
I_2	:Ï̤:Ï̤:

Table 3.5. Lewis symbol representations for the diatomic molecules.

held together by a double bond. Since oxygen atoms have six valence electrons, they need to share two pairs to achieve an octet. Nitrogen atoms have only five valence electrons. Filling up the octet means sharing three pairs of electrons, and thus the *triple bond*. There are many more molecules to explore. We take these up shortly with an enhanced symbolism after an excursion into the topic of polyatomic ions.

3.3.2 Polyatomic Ions

For many covalent molecules, when the atoms all get their octets there are more electrons in the molecule than protons, giving the molecule an overall negative charge. These molecules are called *polyatomic ions*, and these ions will form ionic bonds with cations. In two common cases, the ammonium and hydronium ions, the polyatomic ion has an excess of positive charge.

Sulfate, SO_4^{2-}, carbonate, CO_3^{2-}, and nitrate, NO_3^-, are common polyatomic ions. The bonding in the sulfate ion is as follows:

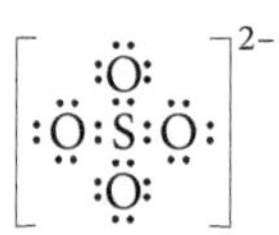

There are 32 valence electrons shown in this molecule. But if we add up the valence electrons from sulfur (6) and four oxygen atoms (6 each) we get 30. Since there are two extra electrons in the molecule than there would be if only the five neutral atoms bonded together, the molecule has a net charge of −2. When drawing the bonding arrangements for molecules that have a net charge, we place the molecule inside square brackets and show the net charge on the upper right corner.

Similarly, the nitrate ion bonds as follows:

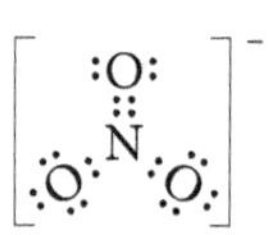

Each atom has an octet, and there are a total of 24 electrons. But adding up the valence electrons for three oxygen atoms and one nitrogen atom gives a total of 23. For this reason, the molecule has a net charge of −1. We discuss molecular shape in much more detail later. But for now you should take note of the general principle that atoms bonded to a central atom distribute themselves evenly around the central atom, so the structures we draw typically simulate this.

Most of the common polyatomic ions have a negative charge, but there are two common positively charged polyatomic ions—ammonium and hydronium. The ammonium molecule is just like the ammonia molecule except with an extra hydrogen atom in the molecule. Ammonia is NH_3, and the three hydrogen atoms make three single bonds with the central nitrogen atom. In ammonia, there is a spare pair of electrons on the central nitrogen atom, and it is easy for a hydrogen ion to attach to this pair to form NH_4^+. In a very similar way, the hydronium ion is just like the water molecule with an extra hydrogen atom attached to the central oxygen atom.

Some of the most common polyatomic ions are listed in Table 3.6. The hydronium ion is included because of its importance later when we discuss acids. But it is highly unlikely that you will find the hydronium ion attached to an anion in an ionic compound.

3.3.3 Ionic Compounds with Polyatomic Ions

Writing the binary formulas for ionic compounds incorporating polyatomic ions is similar to the process we saw earlier for formulas involving only monatomic ions, ions composed of a single atom. When writing formulas involving polyatomic ions, you should determine the subscripts by making the charges in the formula balance. In other words, the ion subscript times the ion's charge must yield the same product (with opposite signs) for both cation and anion.

Ion	Formula
ammonium	NH_4^+
hydronium	H_3O^+
acetate	CH_3COO^-
bicarbonate	HCO_3^-
chlorate	ClO_3^-
hydroxide	OH^-
nitrite	NO_2^-
nitrate	NO_3^-
permanganate	MnO_4^-
carbonate	CO_3^{2-}
chromate	CrO_4^{2-}
sulfite	SO_3^{2-}
sulfate	SO_4^{2-}
phosphate	PO_4^{3-}

Table 3.6. Common polyatomic ions.

▼ Example 3.7

Write the chemical formulas for calcium sulfate, potassium chromate, ammonium sulfite, and magnesium phosphate.

Since calcium ionizes to Ca^{2+}, its charge already balances with sulfate, so the formula is $CaSO_4$.

For K^+ and CrO_4^{2-} to balance, the K^+ needs a subscript of 2, giving the formula K_2CrO_4.

NH_4^+ needs a subscript of 2 to make its charge balance with SO_3^{2-}, giving $(NH_4)_2SO_3$. Note here that placing a subscript on a polyatomic ion requires the ion's formula to be inside parentheses.

To balance Mg^{2+} with PO_4^{3-} requires subscripts on both that multiply their charges to the least common multiple (6), resulting in the formula $Mg_3(PO_4)_2$.

▲

3.3.4 Polyatomic Ion Names

Most of the time you will write names or formulas of compounds with polyatomic ions by using the information in Table 3.6 from memory. However, there are quite a few more polyatomic ions than those listed there, and some involve prefixes. Lest you spend the rest of your student days in bewilderment as to the logic behind these names, let's take a moment here to consider the conventions. (Some day you'll thank me.)

First, there are a few polyatomic ions with *-ide* suffixes, such as:

hydroxide, OH^- cyanide, CN^- peroxide, O_2^{2-}

Second, many polyatomic ions contain oxygen and are accordingly called *oxyanions*. In the common oxyanions formed from elements in Groups 14–17, the number of oxygen atoms bonded in the molecule with the other atoms can vary. The convention for these oxyanions is to assign the suffix *-ate* to the most common one, and the suffix *-ite* to the oxyanion with the same charge but with one oxygen atom fewer. Common examples are:

Most Common	sulfate	SO_4^{2-}	nitrate	NO_3^-
Same Charge, One Less Oxygen	sulfite	SO_3^{2-}	nitrite	NO_2^-

Third, when a series of oxyanions extends to four members, as with those involving halogens, the prefixes *per-* and *hypo-* are used to form the names of the ions with the most and the least oxygen atoms. The most common example is the chlorine series of oxyanions:

One More Oxygen than the *-ate* Ion	perchlorate	ClO_4^-
Most Common	chlorate	ClO_3^-
One Less Oxygen	chlorite	ClO_2^-
One Less Oxygen than the *-ite* Ion	hypochlorite	ClO^-

It can definitely be challenging to remember all the different charges and oxygen numbers for the different oxyanions. A section of the periodic table is shown in Figure 3.18 to help you see and remember the pattern. Each oxyanion shown is the most common for the particular element except perchlorate. The ions in Period 2 have three oxygens and those in Period 3 have four. Beginning with the rightmost oxyanion in each period, charges start at −1 and become more negative toward the left. If you remember the oxygen numbers and charges for the oxyanions in this diagram, especially the element in each period that starts the sequence from right to left, you can deduce the charges and oxygens for the others.

	14	15	16	17
2	**C** carbonate, CO_3^{2-}	**N** nitrate, NO_3^-		
3		**P** phosphate, PO_4^{3-}	**S** sulfate, SO_4^{2-}	**Cl** perchlorate, ClO_4^-

Figure 3.18. Patterns in the oxyanions: oxygens increase going down in the periodic table, charges increase going to the left, and all are the common *-ate* ions except perchlorate.

The fourth and final rule for naming the polyatomic ions pertains to the addition of one or more hydrogen atoms to the ion. If one hydrogen atom is added, the word *hydrogen* is placed before the name of the ion. If two hydrogen atoms are added, the term *dihydrogen* is placed before the name of the ion. Notice that adding hydrogen atoms reduces the charge on the ion by one unit per hydrogen atom. This indicates that hydrogen atoms join these molecules as ions, that is, as lone protons. Examples:

HCO_3^-	hydrogen carbonate
$H_2PO_4^-$	dihydrogen phosphate

An older convention, still very common, is to use the prefix *bi-* for the single hydrogen oxyanions. Thus, hydrogen carbonate, HCO_3^-, is still commonly called bicarbonate, and hydrogen sulfate, HSO_4^-, is sometimes called bisulfate.

3.3.5 Naming Acids

Our detailed study of acids comes later in the course. But while we are studying names of polyatomic ions, let's also take a moment to discuss naming of acids.

For our present purposes, you may think of acids as compounds that contribute H^+ ions when they dissociate in water (that is, when the acid molecules come apart and form + and − ions in solution). You can recognize acids by the H in the front of the formula, most often followed by a halogen or a polyatomic ion. (Occasionally other anions such as S^{2-} are involved.)

Examples are hydrochloric acid, HCl, and nitric acid, HNO_3. Strictly speaking, we say that HCl is hydrogen chloride, named according to the covalent molecule naming rules we address later. But when HCl is dissolved in water—an aqueous solution—we have hydrochloric acid. Here are the conventions for acid naming:

First, acids containing anions with an *-ide* suffix have the *-ide* changed to *-ic* with the prefix *hydro-* added. Examples:

Acid Formula	Anion	Acid Name
HF	fluoride, F^-	hydrofluoric acid
HCl	chloride, Cl^-	hydrochloric acid
HCN	cyanide, CN^-	hydrocyanic acid

Second, acids formed from anions with *-ate* or *-ite* suffixes are named by changing *-ate* to *-ic* and *-ite* to *-ous*. Prefixes on the anions are retained in the acid name. Examples:

Acid Formula	Anion	Acid Name
H_2SO_4	sulfate, SO_4^{2-}	sulfuric acid
HNO_2	nitrite, NO_2^-	nitrous acid
$HClO_4$	perchlorate, ClO_4^-	perchloric acid

An important general term to know is *oxyacid*, which is an acid containing oxygen and at least one other element—usually a nonmetal—bonded to the hydrogen.

▼ Example 3.8

Name these acids: HBr, H_2CO_3, and H_2SO_3.

HBr is formed from the bromide ion, Br^-. Thus, the name is hydrobromic acid.

H_2CO_3 is formed from the carbonate ion, CO_3^{2-}. Thus, the *-ate* is changed to *-ic* and the name is carbonic acid (an oxyacid).

H_2SO_3 is formed from the sulfite ion, SO_3^{2-}. Thus, the *-ite* is changed to *-ous* and the name is sulfurous acid (another oxyacid).

▲

3.3.6 Lewis Structures

In this section, we streamline our Lewis notation a bit and present a formal procedure to enable you to construct diagrams of covalent molecules. We are streamlining the Lewis symbols we have been using in two ways. First, we are replacing shared pairs of electrons involved in bonds, the *bonding pairs*, with a short line segment to indicate the bond. Second, we will now show the other pairs of electrons, the *nonbonding pairs*, only if they are on the central atom in the molecule. This is fine to do because nonbonding pairs only affect the shape of

Molecule	Lewis Structure
H_2O	$H-\ddot{\underset{..}{O}}-H$
CO_2	O=C=O
H_2	H—H
N_2	N≡N
O_2	O=O
F_2	F—F
Cl_2	Cl—Cl
Br_2	Br—Br
I_2	I—I

Table 3.7. Lewis structures for some covalent molecules.

the molecule if they are on the central atom. Otherwise, we know they are there but we can leave them off because they don't affect anything else. The diagrams we produce with this notation are called *Lewis structures*.

Lewis structures were introduced by the great American chemist Gilbert Lewis, frequently referred to as G.N. Lewis. Although he was at the forefront of the theory of chemical bonding in the early 20th century, Lewis never won the Nobel Prize. In a 1916 paper, Lewis first proposed the idea that atoms bonded by sharing electrons in pairs, the phenomenon now known as covalent bonding. He is also the one who in 1926 coined the term *photon* for a quantum of light energy (although this is sometimes erroneously attributed to Albert Einstein).

Table 3.7 shows the Lewis structures for some of the molecules we have discussed so far. In the water molecule the nonbonding pairs are shown on the oxygen atom because it is the central atom in the molecule. The twin pairs of electrons in the double bonds in CO_2 and O_2 have been replaced by double line segments, and the triple bond in N_2 is represented by triple line segments. Each line segment in a Lewis structure represents one pair of electrons.

Our next task is to outline a procedure you can use for constructing Lewis structures on your own for any given simple molecule. Then we consider several examples showing how to use the procedure. Drawing Lewis structures is one of the most effective ways to learn about how covalent bonding works.

To build a Lewis structure, follow the steps described in Table 3.8. Read through the steps before reading the examples that follow.

Steps for Constructing Lewis Structures	
1	Write down the chemical symbols for the atoms in the molecule. If there are more than two, then one of them is in the center. If carbon is present, it is the one in the center. If the molecule contains one of one kind of atom and two or more of a different atom, then the single one is in the center. Otherwise, the least electronegative element is in the center. Hydrogen is never in the center. Halogens usually aren't in the center unless there is one of them and more than one of something else.
2	Arrange the atoms around the central atom. If there is more than one carbon, the carbons chain together to make a central spine instead of a single central atom.
3	Add up all the valence electrons. For polyatomic ions, add another electron to account for each negative charge on the molecule. Or, subtract one electron to account for each positive charge on the molecule.
4	Arrange the electrons in pairs to give an octet around each of the atoms bonded to the central atom (except hydrogen, which only gets a pair).
5	Place the remaining electron pairs on the central atom, even if this means the central atom has more than an octet. (There are some exceptions to the octet rule that we address shortly.)
6	If you don't have enough electron pairs to give the central atom an octet, then you need a double bond (or rarely, triple). (The exception is boron, which can bond without making an octet.)
7	If there is a nonbonding pair on the central atom, give it space around the atom as if there were an atom bonded to it.
8	One more tip: almost always, carbon makes four bonds, halogens make one, and hydrogen makes one. Use these guidelines to narrow your options as you try to figure out the bonding arrangement in a molecule.

Table 3.8. Steps for constructing Lewis structures.

Example 3.9

Draw the Lewis structures for methane, CH_4, phosphorus trichloride, PCl_3, sulfur dioxide, SO_2, and boron trifluoride, BF_3.

Beginning with methane, we place the carbon in the center with the hydrogens around it. There are 4 + 4(1) = 8 valence electrons. These we place around the central atom to give each hydrogen a pair, forming an octet on the carbon in the process. Finally, we replace the electron pairs with line segments. The steps look like this:

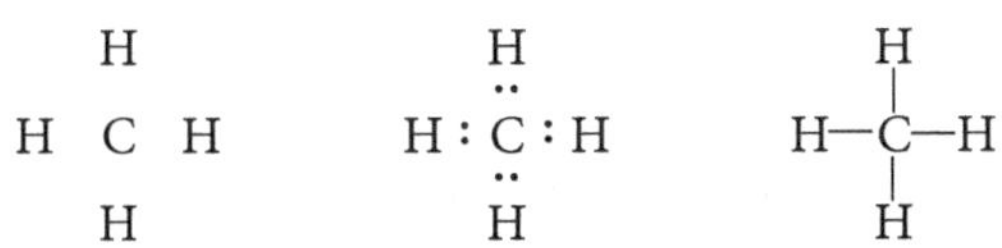

For PCl_3, the phosphorus is in the center with the chlorines around it. There are 5 + 3(7) = 26 valence electrons. Place them to make an octet around each chlorine atom. Place the remaining pair on the central phosphorus atom. The nonbonding pair on the phosphorous atom takes its own place around the central atom. Finally, replace the bonding pairs with line segments. Here are the steps:

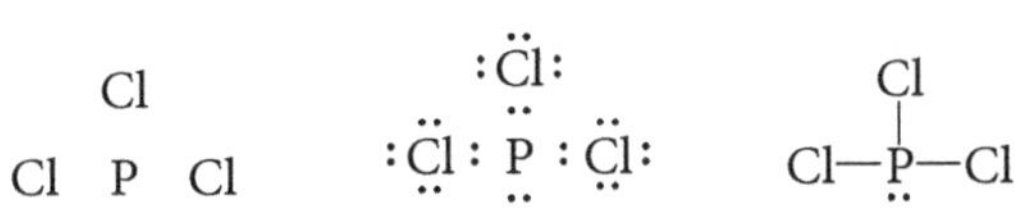

For SO_2, we begin with the sulfur in the center. There are 6 + 2(6) = 18 valence electrons. Begin by forming octets around the oxygen atoms and placing the remaining pair on the central sulfur atom. The sulfur does not have an octet, so a double bond is required. We remove one pair from one of the oxygen atoms and use it to make a double bond between that oxygen and the central sulfur atom. This gives us an octet on each atom.

Obviously, we could have formed the double bond on the other side, O—S̈=O . The two structures differ only in the way the double bonds are formed, and are called *resonance structures*. We discuss resonance structures in more detail soon. Finally, I drew the structures above as linear molecules—all the atoms are in a straight line. But the nonbonding pair arranges symmetrically around the central atom with the bonding pairs, resulting in the two resonance structures looking like this:

We discuss the theory of molecular geometry in the next chapter.

Any time you make a molecule with boron, you should be alert to the fact that boron often breaks the octet rule. For BF_3, begin with boron in the center. We have 3 + 3(7) = 24 valence electrons. Place them around the fluorine atoms to make their octets. This leaves the boron with only six electrons, short of an octet. With any other central atom except boron, we would try forming a double bond to get an octet. But in boron's case, this is not energetically favorable. Now that we know there is no nonbonding pair on the central atom, we need to distribute the three fluorine bonds symmetrically around the central atom. The sequence of steps to work through this structure looks like this:

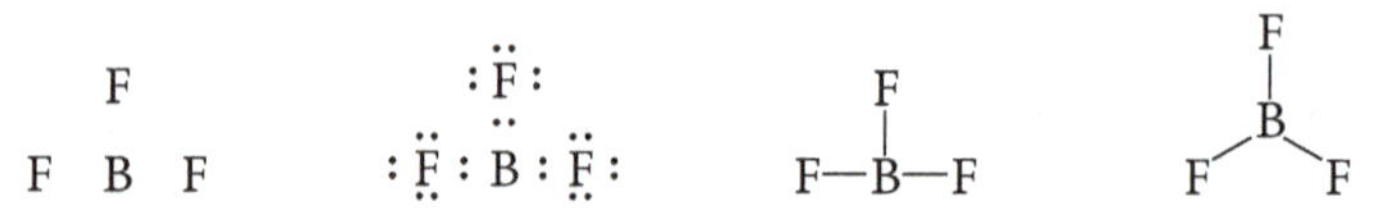

Next, we work through some examples of drawing Lewis structures for polyatomic ions. Notice that in each case, the structure is drawn inside square brackets with the charge of the ion shown at the upper right.

Example 3.10

Draw the Lewis structures for carbonate, CO_3^{2-}, chlorite, ClO_2^-, and acetate, CH_3COO^-.

For carbonate, we begin with the carbon atom in the center. We know in advance that carbon needs to make four bonds. There are 4 + 3(6) = 22 valence electrons. To this, we add two electrons to account for the –2 charge on the molecule, giving a total of 24 electrons. We then place an octet around each oxygen atom. Carbon has now only three bonds, and does not have an octet. So we select one of the oxygen atoms and move one of its nonbonding pairs around to make a double bond with the carbon. With this double bond, carbon now has its four bonds and its octet. Finally, we arrange the three bonded oxygen atoms symmetrically around the central atom. The steps to drawing this structure are as follows:

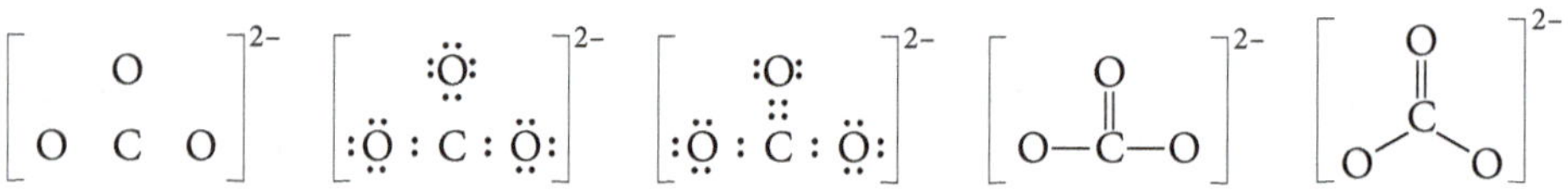

Once again, the choice of the upper oxygen atom for the double bond was arbitrary. We could have chosen either of these as well:

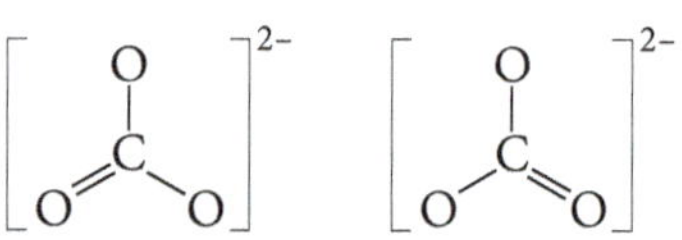

These three structures are identical except for the placement of the double bond, so once again they are resonance structures.

In the chlorite ion, there are 7 + 2(6) = 19 valence electrons, and we add one more electron to account for the –1 charge on the ion, giving us a total of 20. With the chlorine in the center, we place an octet around each oxygen, and the two remaining pairs are placed around the central atom giving it an octet. In the final drawing, the nonbonding pairs on the central atom are shown because although this molecule looks linear on paper it is not, as we will see later.

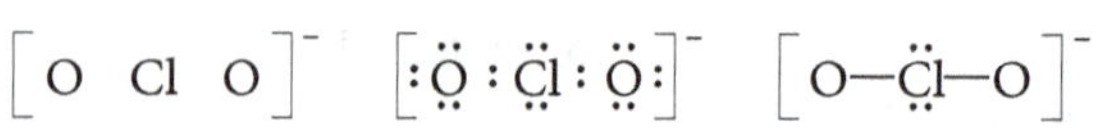

The acetate ion is an example of a molecule with a carbon chain in it, although the chain is a short one with only two atoms. Notice that the way the molecular formula is written gives a clue about the structure. The CH_3 indicates that the three hydrogens are all attached to one of the carbon atoms. Knowing this, and that the carbons chain together, and that since the first carbon now has four bonds the two oxygens must attach to the other carbon, we can begin by laying out the general arrangement of atoms. There are 2(4) + 3(1) + 2(6) = 23 valence electrons. We add one more electron to account for the –1 charge on the ion, giving a total of 24. Place pairs at each hydrogen and an octet around each oxygen. The carbon on the left is complete, but the one on the right has only three bonds. So move one pair from one of the oxygens and use it to form a double bond between the carbon atom and that oxygen atom. The steps are shown below.

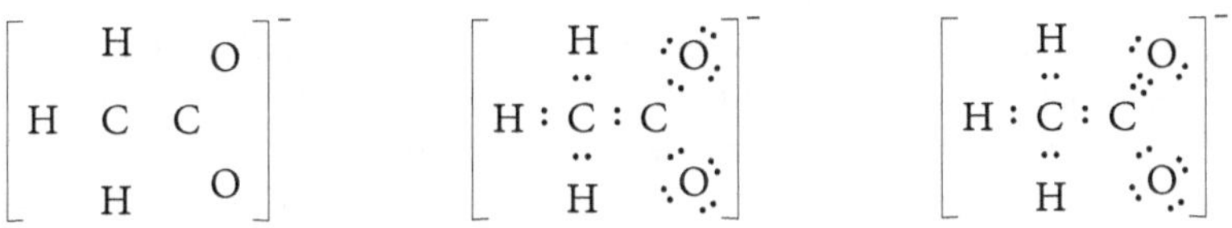

The placement of the double bond could have been on either oxygen atom, so these two arrangements are resonance structures.

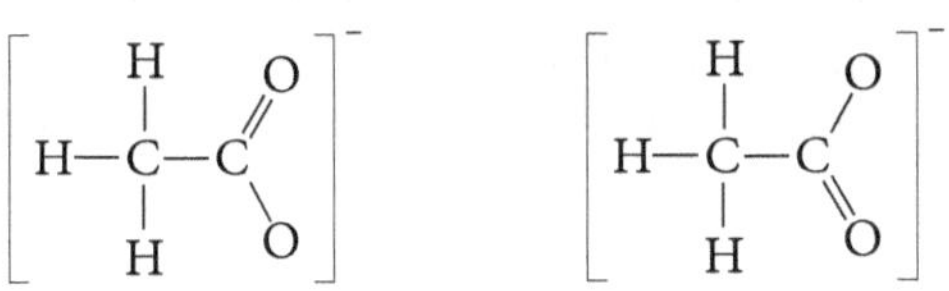

3.3.7 Exceptions to the Octet Rule

We have already seen that boron sometimes acts as an exception to the octet rule by bonding without a full octet. Boron is not the only element that does this, but it is the most common example. You know that elements in the same group in the periodic table exhibit similar chemical properties, so it should come as no surprise that aluminum does it too. There are theoretical ways of explaining why a molecule forms without an octet on the central atom rather than forming a double bond somewhere. As usual, minimizing the energy in the molecular structure is the idea.

There are two other types of exceptions. One is the case when there is an odd number of valence electrons, as with nitric oxide, NO. There are 5 + 6 = 11 valence electrons to share between the nitrogen and oxygen atoms, but the odd number is going to leave one of them short one electron. There are two Lewis structures for this molecule, giving us another pair of resonance structures.

The NO molecule is an example of a chemical species known as a *free radical.* A free radical is any atom, molecule, or ion that has a lone, unpaired electron. Because of the unpaired electron, free radicals are highly reactive, and bond with just about anything—including other entities like itself—to obtain the additional electron and complete the octet.

The third exception to the octet rule is atoms that support *more* than an octet of electrons. An interesting example is the compound antimony pentafluoride, SbF_5. At room temperature, this compound is a colorless oily liquid, and it is famous for its property of reacting with nearly everything. Its Lewis structure is simple to work out if you know that the Sb can have five pairs of electrons around it:

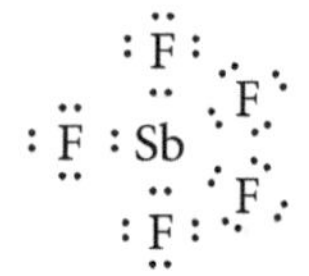

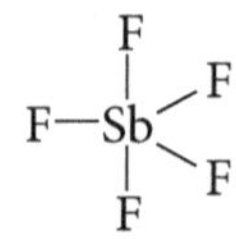

The reason for drawing the Lewis structure the way I did instead of arranging the fluorine atoms evenly around the antimony atom is because as shown, the diagram resembles the way the molecule looks in three dimensions. Molecular structure is one of our topics in the next chapter.

The elements that bond this way are *p*-block elements in periods three and higher. The explanation normally given for why the central atom in these compounds accepts more than an octet of electrons is that since the central atom is a *p*-block element, it has a completely unfilled *d* subshell that can take on the extra pair. Antimony, for example, has valence electrons in the 5*s* and 5*p* subshells, but it also has a 5*d* subshell with nothing in it. (It also has a full 4*d* subshell, of course, but recall that for *p*-block elements the electrons in filled, lower *d* subshells do not act as valence electrons.) Some chemists are now of the opinion that there are different factors behind this type of exception to the octet rule, so you see that our efforts at improving our models continue.

3.3.8 Resonance Structures

We have already seen several instances of resonance structures, such as the two for SO_2:

The name for these structures originated because chemists formerly thought that molecules exhibited a sort of resonance where their structures would switch back and forth from one structure to the other. However, experiments have shown that the molecules don't switch structures this way. Instead, the molecular structure is now understood to be a blend of the separate resonance structures. The electrons in such structures are said to be *delocalized*, being shared in orbitals that extend across several atoms. Resonance structure bonds are typically depicted with a dashed line. For SO_2 this depiction is as follows:

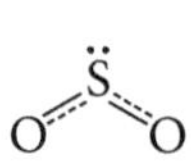

Another example is the carbonate ion, CO_3^{2-}. The Lewis structures we saw previously were:

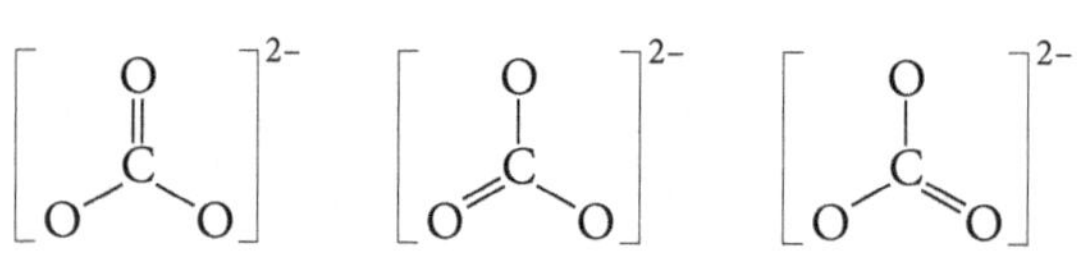

Shown as a single resonance structure this Lewis structure is like this:

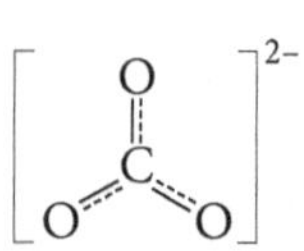

Perhaps the most famous of resonance structures is the benzene ring. Benzene, C_6H_6, is a highly flammable molecule with a sweet or *aromatic* odor found naturally in crude oil. It is an important industrial compound because it is used as a starting point or *precursor* for the manufacture of many other chemicals. English scientist Michael Faraday (famous for Faraday's

Law of Magnetic Induction) was the first to isolate and identify this compound in 1825. The Lewis structures for benzene are:

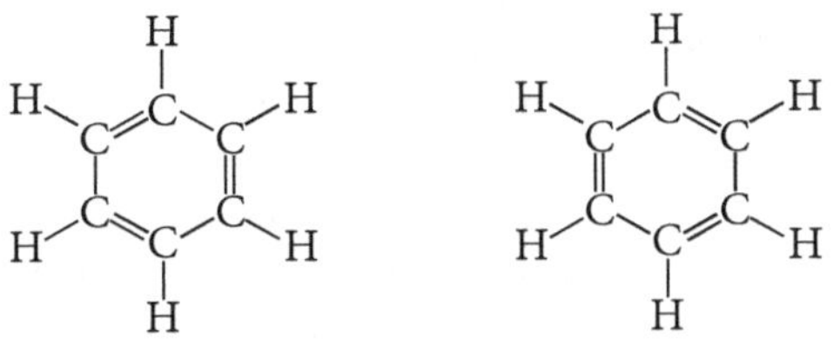

Figure 3.19. German chemist August Kekulé (1829–1896).

The structure of this molecule was so puzzling that it took another forty years before anyone figured it out. The solution finally came to a German chemist named Friedrich August Kekulé (Figure 3.19). Kekulé had long been working on the nature of carbon bonds and had already figured out the four-bond pattern characteristic of carbon. Kekulé recounted that after working on the bonding structure for carbon for many years, the structure for benzene finally came to him in a daydream or vision in which he saw a snake eating its tail. The discovery of the benzene structure has been of enormous significance for the field of organic chemistry—the study of compounds based on carbon. Because of his ground-breaking work, Kekulé is considered to be the father of chemical structure theory.

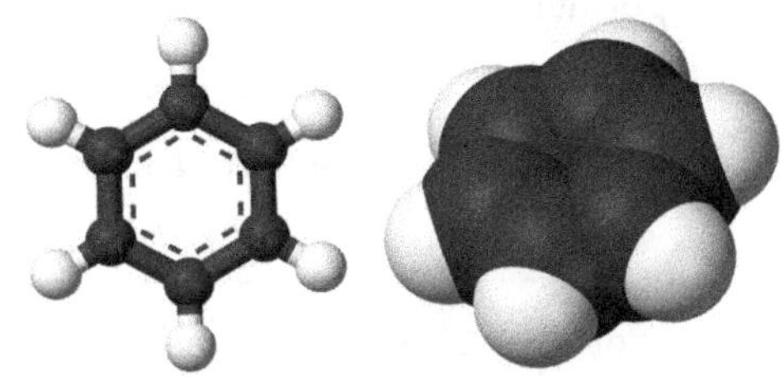

Figure 3.20. Ball-and-stick (left) and space-filling (right) models of the benzene molecule.

Shown as a resonance structure, the Lewis structure for benzene is depicted as shown below.

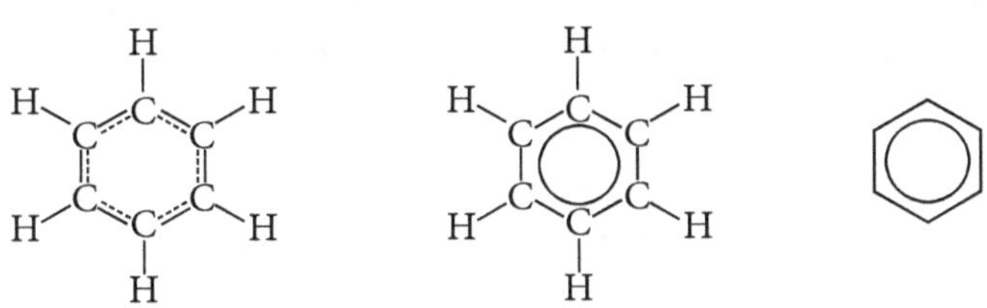

The benzene ring comes up so often in organic chemistry that for simplicity it is often shown with a circle replacing the dashed lines, as in the second image. And for the utmost in simplicity, the ring is simply shown as a hexagon with a circle inside, the carbon atoms, hydrogen atoms, and resonance structure all being assumed. Figure 3.20 shows computer models of the benzene molecule. The model on the left is called a *ball-and-stick model*. This type of model is good for showing the chemical bonds between specific atoms. The model on the right is called a *space-filling* model. The space-filling model gives a more accurate portrayal of the space the molecule takes up.

Number of Atoms	Prefix
1	*mono–*
2	*di–*
3	*tri–*
4	*tetra–*
5	*penta–*
6	*hexa–*
7	*hepta–*
8	*octo–*
9	*nona–*
10	*deca–*

Table 3.9. Prefixes used for naming binary covalent compounds.

3.3.9 Naming Binary Covalent Compounds

When nonmetals form molecules by covalent bonding, the numbers of atoms in the molecules are not predictable the way they are in ionic bonds. Thus, in naming covalent compounds we use prefixes on the atoms involved to denote how many of them are present in the molecule. The Greek prefixes used are listed in Table 3.9.

Many covalent compounds consist of two elements. These are called *binary compounds* and naming them is our topic here. The rules for naming larger organic molecules are addressed in Chapter 13. The conventions for naming covalent binary compounds are as follows:

First, as with ionic compounds, the second element has the suffix *–ide*. Second, both elements in the name have a prefix indicating the number of atoms present in the molecule. However, the *mono–* prefix is not typically used with the first element named. Third, the element farthest to the left in the periodic table is named first unless it is oxygen, which is never written first. If both elements are in the same group in the periodic table, the element with the highest atomic number is written first. Fourth, when a prefix is applied to oxygen, the vowel at the end of the prefix is dropped (except for *di–* and *tri–*).

Examples are as follows:

Formula	Name
SO_2	sulfur dioxide
PCl_3	phosphorus trichloride
N_2O_4	dinitrogen tetroxide

Note that there are other naming conventions in use in addition to the one described here. The compounds NO and N_2O are commonly called nitric oxide and nitrous oxide, respectively. And some sources use the Stock system, as in the name phosphorus(III) bromide, PBr_3.

3.3.10 Energy in Covalent Bonds

From the perspective of quantum mechanics, the covalent bond and the ionic bond are quite different. The ionic bond is held together by the electrostatic attraction between the ions, whereas the covalent bond is held together because of the energy minimization that occurs within the atomic orbitals when the orbitals share electrons to fill subshells. Nevertheless, we can analyze the *bond energy* in a covalent bond in the same way we did for the lattice energy in ionic bonds because in both cases, the atoms seek the arrangement that minimizes the electrostatic potential energy between them.

Consider two hydrogen nuclei far enough apart from each other so that the potential energy between them is essentially zero. As illustrated on the right side of Figure 3.21, as the nuclei approach each other, the potential energy between them decreases. This is due to the attraction between the electron cloud around each atom and the nucleus of the other atom. The minimum energy point is reached when the hydrogen nuclei are 74 pm apart (the bond length). The molecule is now in an energy valley, quite analogous to the ball in a valley of Figure I.5. If the nuclei are slightly pushed together or pulled away from each other, they spring back to the minimum-energy distance of 74 pm. Separating

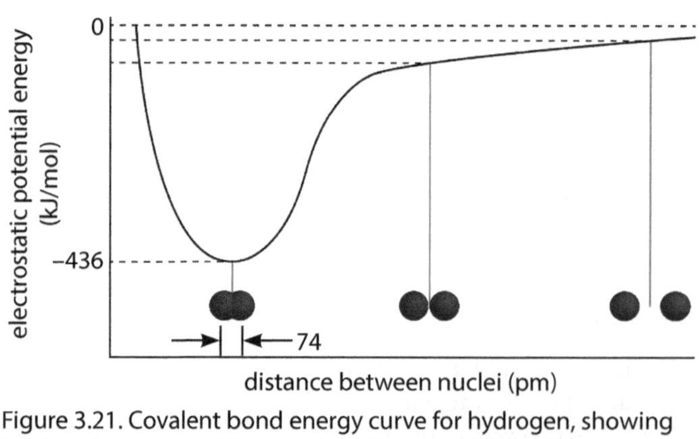

Figure 3.21. Covalent bond energy curve for hydrogen, showing the bond energy and internuclear distance at the stable minimum energy point.

the two hydrogen atoms requires an energy input of 436 kJ/mol.

Table 3.10 shows the bond energies and bond lengths for a number of different covalent bonds. There are two key points to notice from these data. The first is that, generally speaking, double bonds are stronger (higher bond energy) than single bonds, and triple bonds are stronger than double bonds. Stronger bonds mean greater amounts of energy are required to break them. This helps explain why nitrogen gas is so unreactive. The triple-bonded N_2 molecule makes up about 79% of our atmosphere. But while double-bonded oxygen molecules react readily (fires, explosions, rust, organic decay, metabolism) the nitrogen molecules don't.

As you see, HF and S_2 are exceptions to the general trend. The high bond energy of HF is explained by the fact that fluorine is the smallest of the halogens, so the electrons are more tightly bound to the nucleus. The S_2 exception is the opposite: sulfur is below oxygen in the period table, and thus has its valence electrons in a higher shell where they are bound less tightly by the nucleus.

The second point to notice in the table is that as the strength of the bond goes up, the length of the bond shortens. Again, there are exceptions, particularly in molecules incorporating the small hydrogen atom, but in general, tighter bonds pull the atoms closer together.

Bond	Bond Energy (kJ/mol)	Bond Length (pm)
H—H	-436	74
H—F	-570	92
H—Cl	-431	127
C—C	-348	154
F—F	-159	141
Cl—Cl	-436	199
C=C	-618	124
S=S	-425	199
O=O	-498	121
N=O	-632	115
C≡C	-839	121
C≡N	-750	117
C≡O	-1076	113
N≡N	-945	110

Table 3.10. Covalent bond energies and lengths.

3.3.11 Bond Number

One way of describing the strength of covalent bonds is with the notion of *bond number*. For molecules without resonance structures, the bond number is equal to the number of bonds between two atoms in a molecule. Thus, the bond number of the single bond in the Cl_2 molecule is 1, and the bond number of the triple bond in the N_2 molecule is 3.

To determine the bond number for a molecule with resonance structures, the resonance structure bond is divided up across the number of bonds involved, and added to the number of fixed bonds. As an example, consider again the carbonate ion, CO_3^{2-}. The Lewis structure for this molecule is

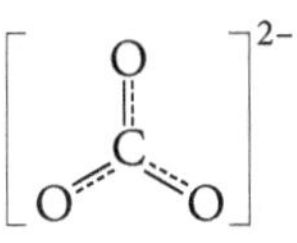

Between the carbon atom and each of the oxygens is one fixed bond and a share in the resonance structure bonds. The resonance structure bonds are single bonds that can be identified with three locations in the molecule, so each place is credited with 1/3 of that bond. (To see this, look again at the Lewis structures shown a few pages back. There is one double bond that could have been drawn in three different places, so half of this double bond contributes 1/3 to the bond number in each place.) Adding this to the fixed bond, the bond number for each of the C—O bonds in the molecule is seen to be

$$1+\frac{1}{3}=\frac{4}{3}$$

As another example, let's consider the benzene ring. Between each of the carbon atoms is one fixed bond and one share of the resonance structure bonds. There are three resonance structure bonds being shared between six places in the molecule, so each place is credited with 3/6 = 1/2 of the bond. Adding this to the fixed bond gives a bond number of

$$1+\frac{1}{2}=\frac{3}{2}$$

The punch line is that the higher the bond number, the stronger and more stable the bond is.

3.3.12 Physical Properties of Covalently Bonded Substances

The lattice energies in ionic compounds run from several hundred to several thousand kJ/mol. Some of the covalent bond energies shown in Table 3.10 are a bit lower, but in general they are in the same league of several hundred kJ/mol. But there are huge differences between the physical properties of ionic compounds and those of covalent compounds. The basic reason for the differences is that the lattice energy in ionic compounds binds the entire crystal structure together. But the bond energy in covalent compounds only holds the individual molecules together. The forces holding one molecule to another, which we study in the next chapter, are much weaker.

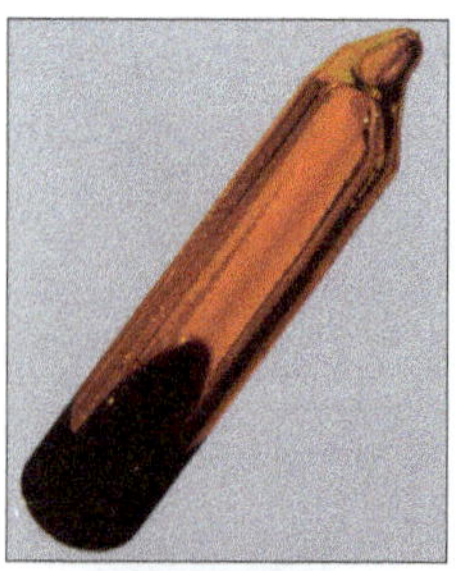

Figure 3.22. Liquid bromine (above) and chlorine gas (below).

One of the results of the weak forces holding molecules together is that many common covalent compounds are gases or liquids at room temperature. Some common covalent compounds have characteristic colors, such as the bromine and chlorine shown in Figure 3.22, but many other common compounds are colorless. The melting and boiling points (intensive properties, as you recall) for a few common covalent compounds are shown in Table 3.11. Compare these values to the melting points for ionic compounds listed in Table 3.4. The ionic compounds are all several hundred or thousand degrees, whereas all the covalent melting points are below or close to zero.

Compounds that are liquids at room temperature include water, bromine (one of only two elements that are liquids at room temperature), acetic acid, ethanol, benzene, and acetone. The rest of the elements and compounds in the table are gases at room temperature.

Compared to ionic compounds, covalent compounds are generally less soluble in water, less electrically conductive, and less thermally conductive.

Compound	Melting Point, °C	Boiling Point, °C
H_2O	0	100
O_2	–219	–183
N_2	–210	–196
Cl_2	–102	–34
Br_2	–7.2	59
CO_2	–79 (sublimes)	
ammonia, NH_3	–78	–33
methane, CH_4	–182	–161
propane, C_3H_8	–188	–42
acetic acid, CH_3COOH	17	118
ethanol, C_2H_5OH	–114	78
benzene, C_6H_6	5.5	80
acetone, $(CH_3)_2CO$	–95	56

Table 3.11. Melting and boiling points for a few covalent compounds.

3.4 Electronegativity, Polarity, and Bond Character

3.4.1 Polarity and Dipoles

At the end of Chapter 2, we discussed the fact that differences in electronegativity between atoms in a molecule cause the molecule to be polar. Electronegativity is a measure of the relative attraction atoms have for the electrons involved in the bond between them. Electronegativity values are lowest in the lower left of the periodic table (Fr) and highest at the upper right (F), as shown in Figures 2.17 and 2.18.

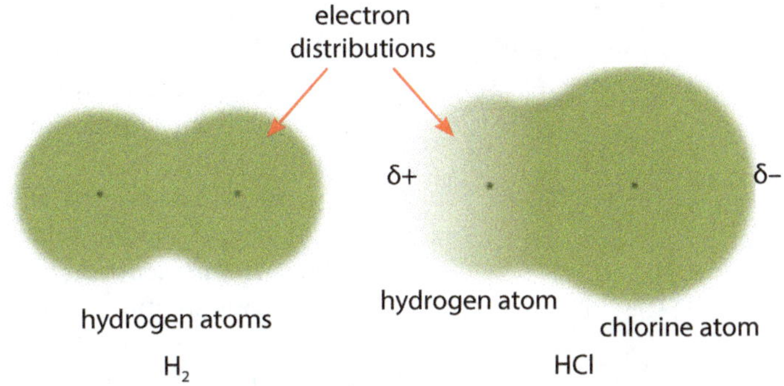

Figure 3.23. Molecules illustrating a nonpolar covalent bond (left) and a polar covalent bond (right).

In a molecule composed of two identical atoms such as H_2, the electronegativity values of the two atoms are the same (2.20 for H). This results in a nonpolar covalent bond, as illustrated by the image on the left side of Figure 3.23. However, in a molecule such as HCl, the electronegativity values differ: 3.16 for Cl and 2.20 for H. The higher electronegativity of Cl results in the shared electrons being pulled over toward the Cl atom, as illustrated on the right side of the figure. The polar object that results from the electron imbalance is called a *dipole*. (Actually, the term *dipole* applies to any object that has one end more positive than the other, regardless of how that happens.) For polar molecules, we symbolize the positive and negative ends of the dipole with the delta-plus (δ+) and delta-minus (δ−) symbols, respectively, as shown in the HCl molecule in the figure. In print, the polarity is shown like this: $\overset{\longleftrightarrow}{\text{H—Cl}}$. In this notation, an arrow is drawn above the element symbols, pointing toward the most electronegative element. A crossbar above the least electronegative element suggests the positive end of the molecule.

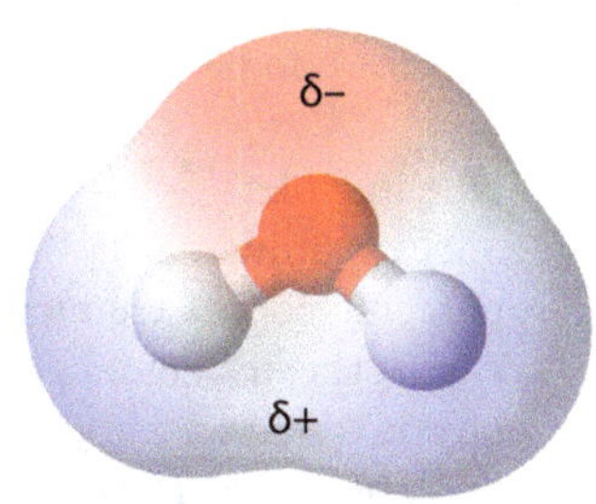

Figure 3.24. The shading shows how the electrical potential around the water molecule changes from positive to negative in the region around the molecule.

The polarity of the water molecule that we discussed previously is so important that we consider it again here just for a moment. The shading of the region around the water molecule in Figure 3.24 illustrates how the electrical potential changes from positive on the hydrogen side to negative on the oxygen side of the molecule.

3.4.2 The Nature of the Bond

Let's briefly recall again that chemistry is all about modeling complex entities we cannot see. Our descriptions of the electrostatic forces in ionic bonds and the electron sharing in covalent bonds are simplistic and both fall short of being complete descriptions. In fact, we now understand most bonds to include both ionic and covalent characteristics, depending on the relative electronegativities of the atoms. We can improve our models of atomic bonding by taking this into account.

The scale in Figure 3.25 is used to characterize atomic bonds as nonpolar covalent, polar covalent, or ionic. If the electronegativities are the same or nearly the same, the bond is considered a nonpolar covalent bond. If the difference is significant but is below 1.7, the bond is considered to be polar covalent. When the difference is 1.7, the bond is considered to be 50% ionic, and for

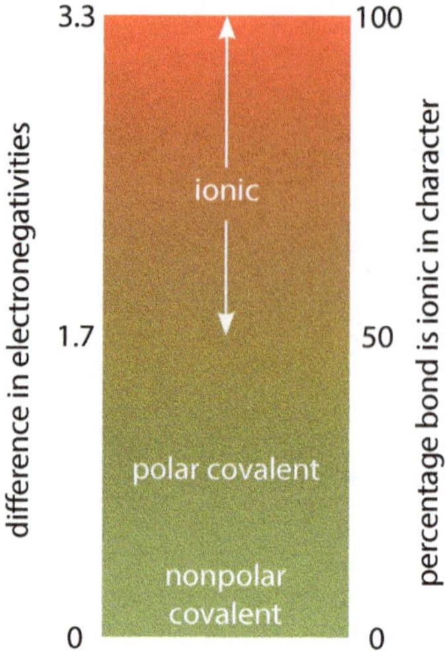

Figure 3.25. Scale for describing the nature of the bond.

differences greater than 1.7, the bond is considered to be essentially ionic in character.

The difference between oxygen (3.44) and hydrogen (2.20) is 1.24, indicating a distinctly polar covalent bond in water molecules. The same is the case for the HCl bond pictured above, with a difference of 0.96. By contrast, the electronegativity difference for NaCl is 2.23, indicating a distinctly ionic bond, though not without a little bit of covalent character as well.

There is a close correlation between the percent ionic character of a bond and the color of a compound in aqueous solution. When the percent ionic character is above 20%, compounds in aqueous solution are colorless. As the percent ionic character decreases from 20% down to 10%, solutions take on colors beginning with lemon yellow and going through the sequence of colors shown in Figure 3.26. For example, when bonded to cations of increasing electronegativity, bromine forms solutions that go from yellow to red to brown. Copper forms solutions that are blue and head toward indigo as they become more covalent. Nickel forms compounds that go from yellow to green to blue to violet.

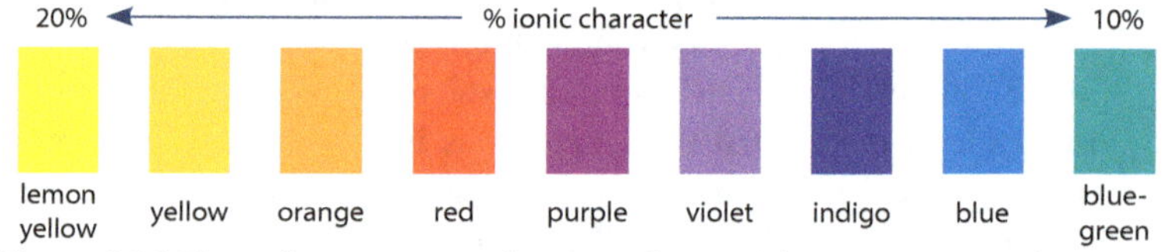

Figure 3.26. The color sequence that correlates with percent ionic character.

Explaining *why* the percent ionic character correlates to color goes beyond what we can cover here, but it has to do with the way electrons in the orbitals of various compounds absorb light. A more complete explanation will have to wait for future study,[1] but at least now you have some idea of where the beautiful colors in aqueous solutions originate.

Chapter 3 Exercises

SECTION 3.1

1. Classify each of the following as element, compound, homogeneous mixture, or heterogeneous mixture.

a. water	b. cesium chloride	c. pond water	d. methane
e. a soft drink	f. nitric acid	g. black coffee	h. argon
i. air	j. hydrogen nitrate	k. exhaust fumes	l. quartz
m. brass	n. hydrogen gas	o. hydrogen cyanide	p. mouthwash
q. platinum	r. dirt	s. radon	t. a smoothie

2. Explain why salt water and sugar water are homogeneous mixtures while automotive paint, which contains invisible particulates, is not.

1 A short discussion that is not too technical for an interested high school student can be found in Chapter 3 of Linus Pauling's text, *The Nature of the Chemical Bond*. If this interests you, you should be able to find this text at a college library.

Hmm... Interesting. The molecular structure of glass and quartz

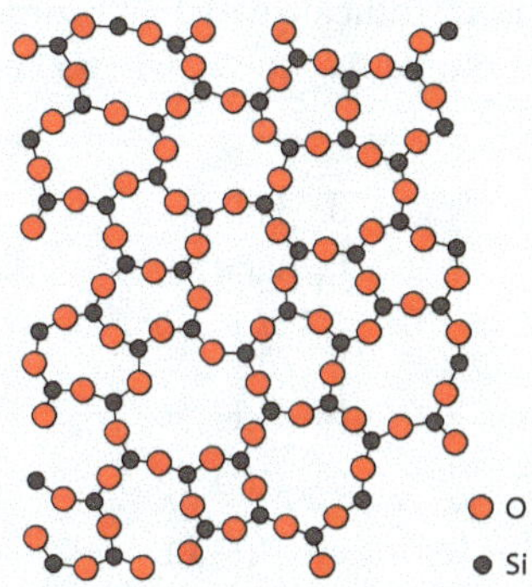

In this chapter we have looked at ionically bonded substances, which have regular crystal lattices. We have also looked at molecular substances, which bond in clusters of atoms. But there is a common substance that bonds in an *irregular* lattice—glass, known in the chemistry world as vitreous silicon dioxide, SiO_2. Because of its completely irregular bonding structure, glass is said to be *amorphous*, a term from Greek that means "no form." Molecular solids, in general, are amorphous, and glass can be thought of as one giant molecule.

There are also forms of SiO_2 with a regular crystal lattice—quartz. The varieties of quartz are endless because of different metallic impurities that are found in the crystal lattice. These impurities cause the quartz crystals to have different variations in color. The color variations make them desirable as gem stones, especially since the regular crystal lattice means they can be cut into the beautiful symmetrical shapes desired in gems.

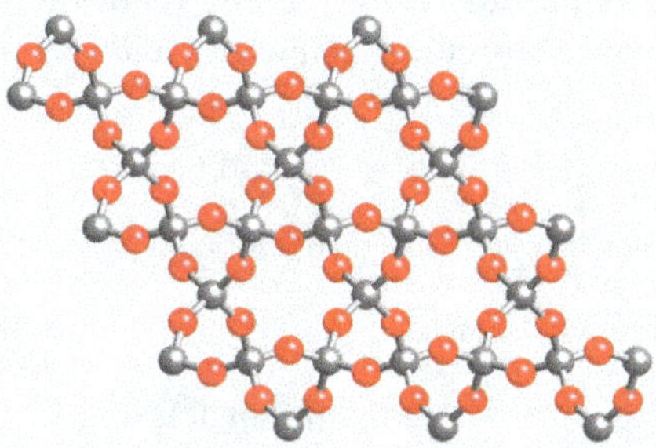

3. Explain the octet rule.

SECTION 3.2

4. Explain what lattice energy is and why the lattice energy of ionic compounds is so large.
5. Distinguish between ionic and covalent bonds.
6. Write the formula for the ionic compound formed by each of the following cation-anion pairs:

a. Sr and Br	b. Mg and O	c. Cu(II) and Cl	d. Li and Se
e. Ni^{2+} and P	f. K and N	g. Be and Cl	h. Na and I
i. Cr^{3+} and O	j. Ag and Se	k. Mn(IV) and F	l. Sb^{3+} and Br
m. Rb and S	n. Sc and O	o. Al and F	p. Mg and F
q. Zn and Cl	r. Li and S	s. Co(III) and O	t. V^{5+} and I

7. Write the names for each compound in the previous question. For multivalent cations, write the name in both the Latin-based system and the Stock system.
8. Explain what a hydrate is and the conditions under which a hydrate forms.
9. Explain why ionic compounds cannot be bent into shapes the way metals can.
10. Explain why ionic compounds generally have high melting points.

11. Since energy is required to separate the ions in a crystal (the lattice energy), why doesn't the temperature of the water drop when salt is dissolved in water?

SECTION 3.3

12. Draw the Lewis structures for the following compounds and ions. Indicate which of the molecules have resonance structures and show them. (In each of the items shown, when H and O are both present, the H bonds to O.)

a. C_2HCl	b. NH_4^+	c. CCl_4	d. NO_3^-
e. PCl_5	f. NO_2^-	g. CO	h. $SiCl_4$
i. O_3	j. CrO_4^{2-}	k. AlF_3	l. NOCl
m. PO_4^{3-}	n. H_2S	o. N_2O_4	p. C_2H_6
q. H_3O^+	r. BCl_3	s. $HClO_4$	t. ClO_2^-
u. H_2CO_3	v. CCl_2S	w. C_2H_4ClF	x. $C_2H_6O_2$

13. State the names for each of the binary compounds listed in the previous exercise. (Several of the compounds are not binary compounds.)

14. Draw the Lewis structures for the following compounds and ions. Identify (i) those that do not obey the octet rule (explain why), (ii) resonance structures (show all), and (iii) free radicals (explain why).

a. AlH_3	b. CS_2	c. $SbCl_5$	d. N_3
e. CH_2Cl_2	f. H_2SO_3	g. OH	h. NO_2

15. State the names of the following acids.

a. H_2SO_4	b. HF	c. H_2CO_3	d. $HClO_4$
e. HNO_2	f. $HBrO_2$	g. HBr	h. H_3PO_4

16. Explain why covalent compounds tend to be gases or liquids at room temperature, while ionic compounds tend to be solids.

17. Describe the three general cases of exceptions to the octet rule, and give an example for each. Show the Lewis structures of your examples.

18. Write the formulas for the ionic compounds formed from each of the ion pairs given below, and state the name of each using the Stock system. (Of course, naming these is a snap.)

a. magnesium and carbonate	b. sodium and chromate
c. calcium and phosphate	d. ammonium and phosphate
e. copper(II) and sulfate	f. sodium and bicarbonate
g. iron(III) and chlorate	h. aluminum and hydroxide
i. chromium(VI) and sulfite	j. calcium and acetate

19. Explain what resonance structures are.

20. Compute bond numbers for the following, and use the results to arrange these three bonds in order of increasing strength. Also predict which bond is the longest and which is the shortest.

    a. the C—S bond in CS_2.

b. the P—Cl bond in PCl_5.

c. the N—O bond in NO_2^-.

SECTION 3.4

21. Explain why water molecules are polar and chlorine molecules are not.

22. For each of the following sets of bonds, compute the difference in electronegativity and arrange the bonds in order of increasing percent ionic character. Identify the most polar and least polar in each set. Identify each bond as nonpolar covalent, polar covalent, or ionic in character.

    a. Be—F, O—F, and C—F

    b. F—F, B—F, and S—O

    c. O—Cl, S—Cl, and C—P

23. When chlorine bonds to the three ions below, it forms compounds that are yellow, orange, or red in aqueous solution. Predict the color each compound displays.

    a. As(III)

    b. Cd(II)

    c. Sb(III)

GENERAL REVIEW EXERCISES

24. Determine the number of oxygen atoms present in 100.00 g limestone, which is composed of calcium carbonate.

25. Determine the percent composition for any sample of sulfur hexafluoride.

26. In Bohr's model of the atom, the first four energy levels could hold 2, 8, 8, and 18 electrons, respectively. The values 2, 8, 18, and 32 were later found to be the correct values for n = 1, 2, 3, and 4. Explain where Bohr's numbers came from, and what determines the numbers of electrons in the shells. Hint: compute the difference between each of the electron quantities and relate that difference to the quantum model of the atom.

27. Write the electron configurations and draw the orbital diagrams for Cl, Sc, Ag, and Gd.

28. Using the data in Table 1.5, determine the atomic mass of carbon. Compare your result to the value given in the periodic table by calculating the percent difference (see Section A.4.2).

29. Determine the number of chlorine atoms present in 1.000 cm^3 magnesium chloride. The density of this compound is listed in Table 3.4.

30. Why are the noble gases unreactive?

31. Without referring to any actual radius data, determine the correct order by size, from smallest to largest, for the following ions and atoms: Mg^{2+}, Rb^+, Ca^{2+}, K^+, Rb.

32. Determine the empirical formula for a sample with a percent composition of 12.84% S, 25.45% Cu, 25.63% O, and 36.07% H_2O by mass. Knowing that the empirical formula is the same as the molecular formula, name this compound.

33. In a certain laser, the electron transition generating the laser beam has an energy of 3.73×10^{-19} J. Determine the wavelength of the beam generated by this laser.

Chapter 4

Molecular Theory and Metallic Bonding

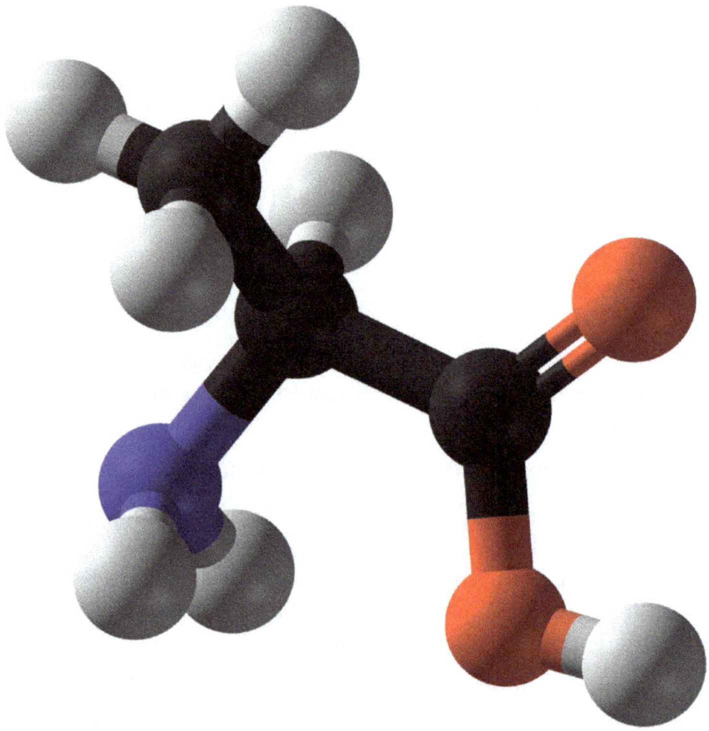

There are 22 "standard" amino acids, of which 21 are found in humans. These molecules are joined together in chains to produce the all-important proteins that are at the center of nearly every cellular process.

The model above is a representation of an amino acid called *L-alanine*. Carbon, hydrogen, oxygen, and nitrogen atoms are shown in black, white, red, and blue. In this chapter, we discuss theoretical models that account for the orientations of the bonds that join atoms together. The examples we consider in the chapter entail simpler molecules than this one, but the same ideas apply to L-alanine, the other amino acids, and much larger molecules as well.

Objectives for Chapter 4

After studying this chapter and completing the exercises, you should be able to do each of the following tasks, using supporting terms and principles as necessary.

SECTION 4.1

1. Describe the central theoretical problem addressed by VSEPR theory.
2. Distinguish between electron domain geometry and molecular geometry.
3. Use VSEPR theory to predict the electron domain geometries and molecular geometries of given molecules.
4. State the names of the five basic electron domain geometries and the seven basic molecular geometries and draw diagrams illustrating each.
5. Explain the conditions under which the bent, trigonal bipyramidal, and octahedral geometries form.
6. Explain how nonbonding electron pairs affect the bond angles in molecules and use this idea to predict when bond angles will depart from the standard angles of pure geometric figures.
7. Apply the VSEPR theory and the principles of electron domain geometry to describe the geometries and estimate bond angles of larger molecules.
8. Describe orbital hybridization theory and use it to explain why the shapes of molecules differ from what would be expected based on the shells and orbitals of unbonded atoms.
9. Describe sp, sp^2, and sp^3 hybridization, and predict which is involved in a given covalent bonding scenario.
10. Briefly articulate the goal and purpose of valence-bond theory.

SECTION 4.2

11. Give a detailed description of metallic bonding.
12. Compare and contrast covalent bonds and metallic bonding.
13. Explain why the lattice structures of atoms in metals are referred to as "close packing."
14. Explain why metals are good conductors of electricity and heat, exhibit malleability and ductility, have a high luster, and are typically silver in color.

SECTION 4.3

15. Define hydrogen bonding and give several examples of phenomena caused by it.
16. Describe the three different types of Van der Waals forces and give specific examples of each.

4.1 Molecular Structure

4.1.1 Covalent Bond Theory

In the previous chapter, we used the shared-pair bonding theory developed by G.N. Lewis as a first step in understanding molecular bonding. In the present chapter, we add another layer to our theoretical understanding about molecules and covalent bonds. I like to think of these theoretical components as addressing specific questions. So we begin by articulating the question addressed in what we have seen so far, and then we move on to the second question addressed in this new chapter.

Here is the question addressed by the 1916 covalent bonding theory of Gilbert Lewis:

1. What drives atoms to cluster together in particular ways to form molecules?

Lewis' covalent bonding theory addresses this question. As we saw, atoms seek to fill up their octets and form different clusters depending on how many valence electrons each atom brings to the party. The bonds occur by atoms sharing electrons in pairs, and the atomic clusters formed when the atoms do this are called molecules.

4.1.2 Valence Shell Electron Pair Repulsion (VSEPR) Theory

2. What factors govern the shapes of the molecules that form when atoms are covalently bonded together?

The Lewis structures you learned how to draw in the previous chapter show how many and what kind of atoms bond together in molecules, and whether the bonds involve one, two, or three electron pairs. But Lewis' covalent bond theory doesn't explain the shapes of the resulting molecules. The theory that addresses this question is called *Valence Shell Electron Pair Repulsion (VSEPR) theory*. VSEPR theory is quite successful in explaining molecular geometry.

In VSEPR theory, we consider the regions around atoms in molecules where the electrons are found. These regions are called *electron domains*, and there are two basic types. The first is the *bonding domains* between atoms in the molecule, where shared pairs of electrons reside. These electron domains could be due to single bonds, double bonds, or triple bonds; VSEPR theory treats them equally. The second type of electron domain is the *nonbonding domain*, the region where nonbonding pairs (often called *lone pairs*) on a central atom reside. The bonding domains are located in a region between atoms; the nonbonding domains are located on the side of a central atom where no other atom is attached.

For example, recall the Lewis structure for the water molecule:

$$\mathrm{H{:}\ddot{\underset{\cdot\cdot}{O}}{:}H}$$

There are four electron domains around the oxygen atom in this molecule. Two of them are bonding domains—the two regions where the hydrogen atoms are attached by sharing an electron pair with the central oxygen atom. The other two are the nonbonding domains associated with the nonbonding pairs shown in the Lewis structure.

The basic idea behind VSEPR theory is simple. All the electron domains around an atom repel each other, and thus try to push as far away from each other as possible. The result of this is very predictable molecular shapes as the electron domains spread themselves out in two or three dimensions around the central atom in the molecule.

As we look at these geometries, it is important to keep in mind that nonbonding domains on a central atom repel other electron domains just as bonding domains do. In fact, they push a bit harder than the bonding domains do, a detail we will come back to later. We also need

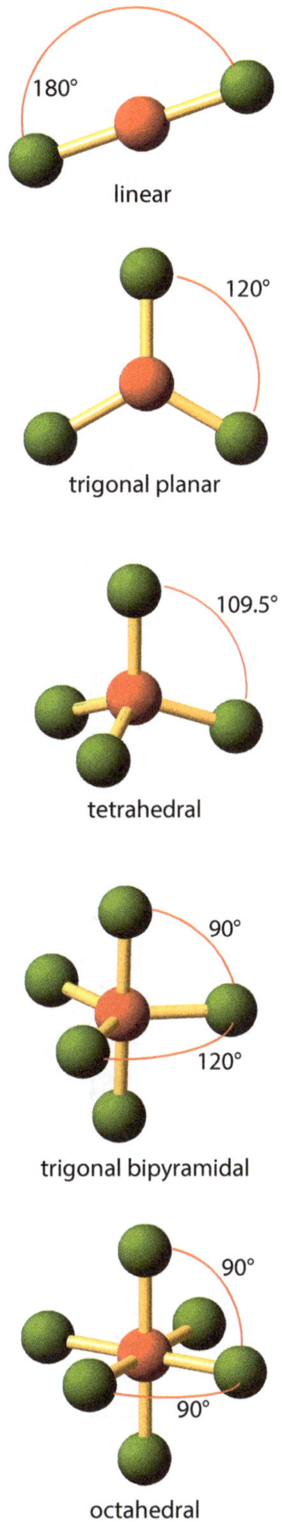

Figure 4.1. The five basic electron domain structures.

to distinguish between the electron domain geometry and the molecular geometry. Sometimes these are the same. However, since nonbonding domains on a central atom influence molecular shape just as bonding domains do, the geometry of the molecule itself can be different from the geometry of the electron domains.

There are five basic geometries that can form as the electron domains around a central atom push away from each other, shown in Figure 4.1. Outlines of these geometries are related to standard polyhedra in Figure 4.2. More details concerning the five geometries, as well as example molecules, are shown in Table 4.1. Refer to all these figures and tables as we go through the details.

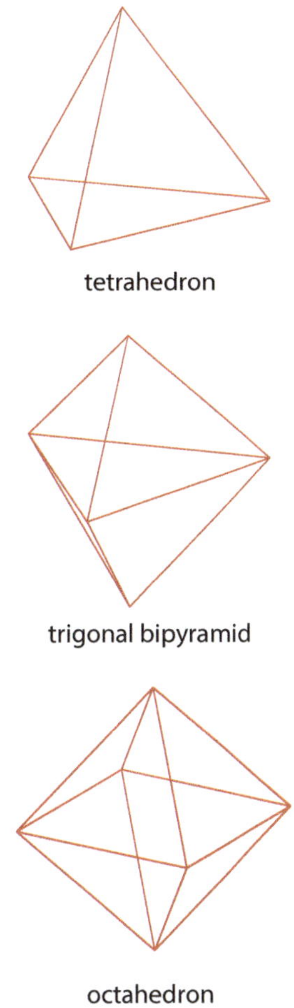

Figure 4.2. Outlines of the three 3-D molecular shapes.

The first geometry is the *linear* arrangement. (The linear arrangement also occurs, of course, with just two atoms, but that arrangement is trivial and not shown.) A linear arrangement occurs when two atoms are bonded to a central atom and there are no nonbonding domains. Generically, we can designate molecules like this as of the type AB_2 (two B atoms attached to one A atom). A classic example is the CO_2 molecule. As the Lewis structure in Table 4.1 shows, there are two oxygen atoms bonded to the carbon and no nonbonding domains.

In the *trigonal planar* geometry, three electron domains spread around the circumference of the central atom, resulting in a *bond angle*—the angle between bonds—of 120°. There are two ways in which electron domains can take on this geometry. The first is with a molecule of the generic type AB_3 such as boron trifluoride, BF_3. The second way is with a molecule that has two bonding domains and one nonbonding domain on the central atom, such as SO_2. We can represent this structure generically as type AB_2E, where the E stands for the nonbonding domain.

Figure 4.3 illustrates the molecular geometry resulting from molecules of the type AB_2E. In the figure, the black line is in the same position as the third bond in a molecule of the type AB_3. But in this case, there is no atom attached there; instead, there is a nonbonding domain there. This molecular geometry is referred to as *bent*.

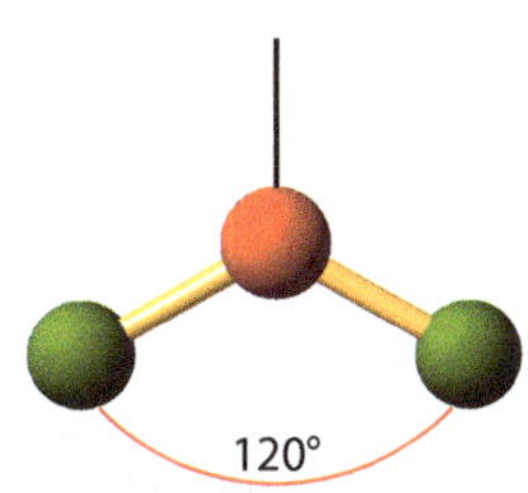

Figure 4.3. The bent geometry resulting (ideally) from molecules of the type AB_2E. (The actual angle is 119°, as discussed in the next section.)

Bond Type	Bonding Pairs	Nonbonding Pairs	Electron Domain Geometry	Molecular Geometry	Example Molecule	Example Lewis Structure
AB_2	2	0	linear	linear	CO_2	O=C=O
AB_3	3	0	trigonal planar	trigonal planar	BF_3	
AB_2E	2	1	trigonal planar	bent	SO_2	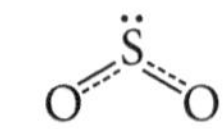
AB_4	4	0	tetrahedral	tetrahedral	CH_4	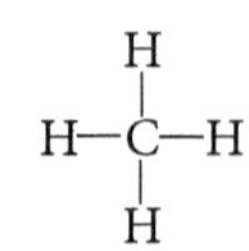
AB_3E	3	1	tetrahedral	trigonal pyramidal	NH_3	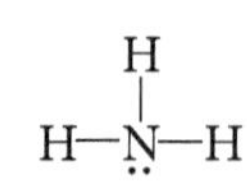
AB_2E_2	2	2	tetrahedral	bent	H_2O	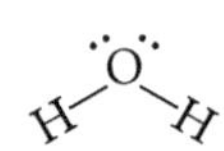
AB_5	5	0	trigonal bipyramidal	trigonal bipyramidal	SbF_5	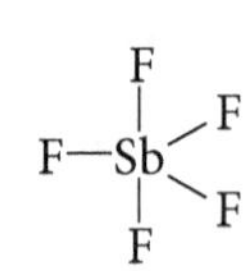
AB_6	6	0	octahedral	octahedral	SF_6	

Table 4.1. Bond types and their geometries.

So for an AB_2E molecule, the electron domain geometry is trigonal planar, but the molecular geometry is bent.

The third basic geometry is the *tetrahedral* geometry, so called because lines drawn connecting the four outer atoms form the outline of a tetrahedron. When there are more than three atoms bonding to the central atom, the electron domains minimize their distance from each other by spreading out in three dimensions.

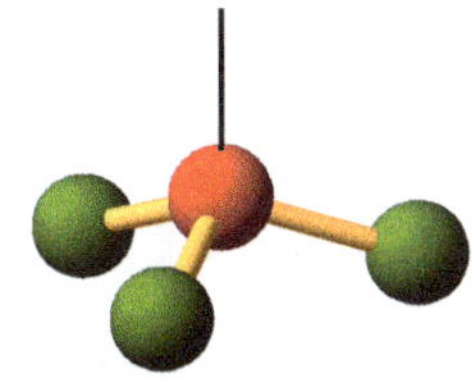

Figure 4.4. The pyramidal molecular geometry resulting from molecules of the type AB_3E.

As the table shows, there are three ways this electron domain geometry can form, represented generically as AB_4, AB_3E, and AB_2E_2. Methane, CH_4, is a perfect example of the AB_4 type. This molecule is shaped just like the tetrahedral shape shown in Figure 4.1. If the molecule has a nonbonding domain in place of one of the bonding domains, as with ammonia, NH_3, the molecular geometry is as shown in Figure 4.4. This is called a *pyramidal* geometry. As before, the black line in this figure represents a nonbonding domain. There are only three atoms attached to the central atom, but since the nonbonding domain pushes just like the three bonding domains do, we have a tetrahedral electron domain geometry and a pyramidal molecular geometry.

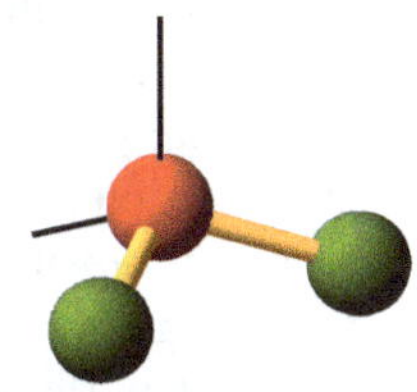

Figure 4.5. The bent geometry resulting from molecules of the type AB_2E_2, of which of H_2O is the most important example.

The water molecule, H_2O, is the perfect example of the AB_2E_2 type. Figure 4.5 shows how a tetrahedral electron domain geometry with only two bonding domains results in the molecular geometry being bent. Note from Figure 4.1 that the bond angle in the tetrahedral geometry is smaller than the bond angle in trigonal planar geometry. As a result, the H_2O and SO_2 molecules are both bent, but the angle in the H_2O molecule is smaller.

As should be clear from our study of covalent bonding in the previous chapter, the last two geometries pertain to molecules that break the octet rule. Antimony pentafluoride, SbF_5, has five bonding domains around the central atom. This results in a *trigonal bipyramidal* geometry. Three of the Sb—F bonds lie in the same plane, 120° apart. The other two are on a line perpendicular to this plane, resulting in 90° bond angles between them and the other three.

Sulfur hexafluoride has six bonding domains surrounding the central atom, resulting in the *octahedral* geometry where every bond angle is 90°.

As with the AB_3 and AB_4 types, there are cases of molecules with AB_5 or AB_6 electron domain geometries, but with fewer than five or six atoms attached to the central atom. This happens when the total number of bonding domains and nonbonding domains on the central atom is greater than four. We do not explore the details of these geometries further here.

Example 4.1

Use VSEPR theory to predict the molecular shapes of ozone, O_3, and the sulfite ion, SO_3^{2-}.

The Lewis structures for O_3 are

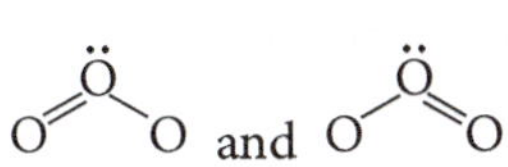

These two resonance structures may be represented as 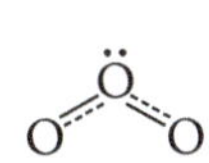 . This molecule is type AB_2E. Thus, the electron domain geometry is trigonal planar and the molecular geometry is bent.

As you can verify, the Lewis structure for the sulfite ion is

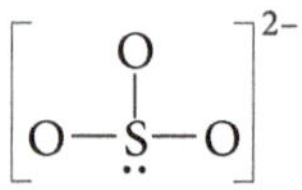

This molecule is of the form AB_3E. With four electron domains, the electron domain geometry is tetrahedral. But since there are only three bonded atoms, the molecular geometry is pyramidal.

▲

4.1.3 The Effect of Nonbonding Domains on Bond Angle

Figure 4.6 illustrates the relative shapes of the two types of electron domains. Nonbonding domains, like the two lone pairs on the oxygen atom in a water molecule, extend out in a wider region from the atom than the bonding domains do, and this has a slight but definite effect on molecular geometry. You can think of it this way: when the electron domains spread themselves out in 3-D space around an atom, repelling the other domains because of the negative charges inside them, the nonbonding domains push a bit harder than the bonding domains do because they are wider. The result is that the bond angles between atomic bonds in the molecule are less than they would be if all were equal. This effect is seen consistently in covalent molecules. Figure 4.7 illustrates how the effect works in the cases of SO_2, an AB_2E type molecule, and H_2O, an AB_2E_2 type. The electron domain geometry for SO_2 is trigonal planar, with angles of 120°. But with a nonbonding domain in there, the angle between the other two domains is squeezed down to 119°. The same thing happens with H_2O. In this case, the 109.5° bond angle of the ideal tetrahedron is reduced to 104.5°.

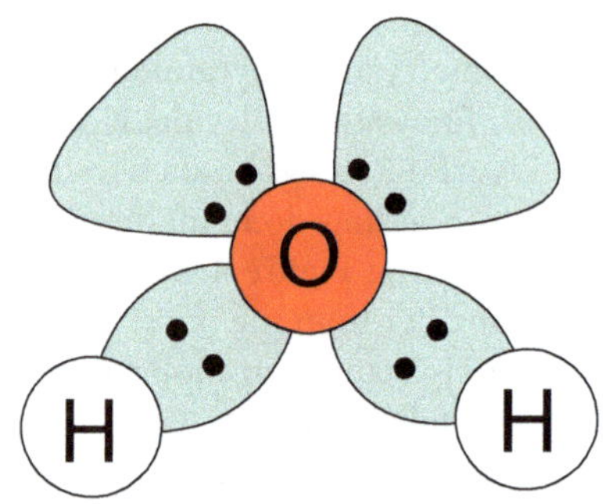

Figure 4.6. The influence of bonding domains does not extend as far to the side as it does for nonbonding domains.

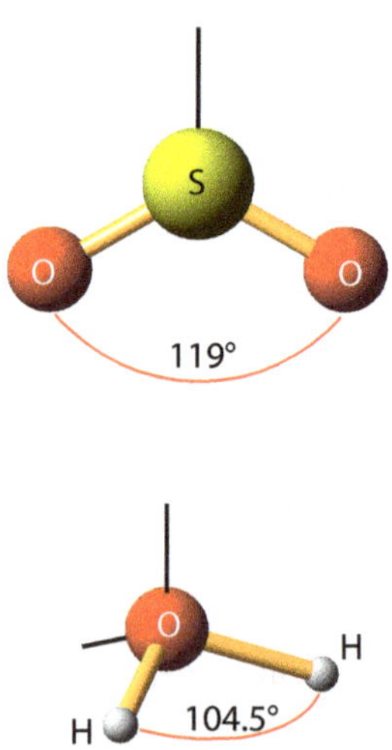

Figure 4.7. In a type AB_2E molecule such as SO_2, the nonbonding domain pushes the bonding domains away slightly, reducing the bond angle from the ideal 120° to 119°. In a type AB_2E_2 molecule like H_2O, the bond angle is reduced from the ideal 109.5° to 104.5°.

Notice in Figure 4.7 that both the molecular geometries are bent, but the bond angles are different. In the case of SO_2, the ideal angle of the trigonal planar geometry is 120°, and the single nonbonding domain reduces that by 1°. In the case of water, the ideal bond angle of the tetrahedron is 109.5°, but in this case, there are two nonbonding domains involved and together they squeeze down the bond angle by 5°.

▼ Example 4.2

Use VSEPR theory to predict the bond angle in the ammonia molecule, NH_3.

The Lewis structure for NH_3 is

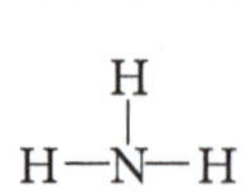

This is a type AB_3E molecule. The electron domain geometry is tetrahedral and there is one nonbonding domain. Since the two nonbonding domains in H_2O (which also has tetrahedral electron domain geometry) reduce the bond angle by 5°, we should expect the bond angle in NH_3 to be reduced by about half as much, putting it at around 107°.

(In fact, the bond angle for NH_3 is 107°.)

▲

4.1.4 Orbital Hybridization Theory

3. How do we understand the shapes and orientations of the electron domains in VSEPR theory in light of what we know about the shapes and orientations of atomic orbitals from the Schrödinger equation?

The VSEPR theory does a nice job of accounting for the molecular shapes that experiments have identified. But these shapes—tetrahedral, trigonal bipyramidal, and so on—don't at all resemble the locations and arrangements of the orbitals described in Chapter 1. We reconcile these through *orbital hybridization theory*, first proposed in 1930 by our friend Linus Pauling. Pauling was working on understanding the shape of the methane molecule in terms of orbital theory.

In the hybridization process, two or more orbitals on the central atom of a molecule blend together to form new *hybrid orbitals*, with different energies and shapes than those of the orbitals that went into making the hybrids. When this happens, the number of hybrid orbitals formed equals the number of individual orbitals that went into the hybridization process.

For example, as illustrated in Figure 4.8, one *s* orbital and one *p* orbital can hybridize to create two so-called *sp* orbitals. The two new orbitals have their main lobes pointing in opposite directions. If an *s* orbital and two *p* orbitals hybridize, the result is three sp^2 (pronounced "*s–p–*two") orbitals. The sp^2 orbitals resemble the *sp* orbitals shown in the figure, but the three of them are oriented at angles of 120° relative to each other. If an *s* orbital and three *p* orbitals hybridize, the result is four sp^3 orbitals. The sp^3 orbitals again resemble *sp* orbitals shown in the figure, but the four of them are oriented to point at the corners of a tetrahedron and are at angles of 109.5° relative to each other.

To illustrate how hybridization works, let's consider the methane molecule Pauling was working on. The orbital diagram for carbon is

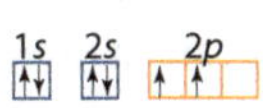

When a carbon atom bonds with four hydrogen atoms to make CH_4, all four of the carbon's valence electrons are involved. Experimental data tell us that after bonding, these four valence electrons are equally distributed around the central atom—each having the same energy, each being in an orbital with a shared electron from a hydrogen atom, and each being in a region pointing out from the central carbon atom

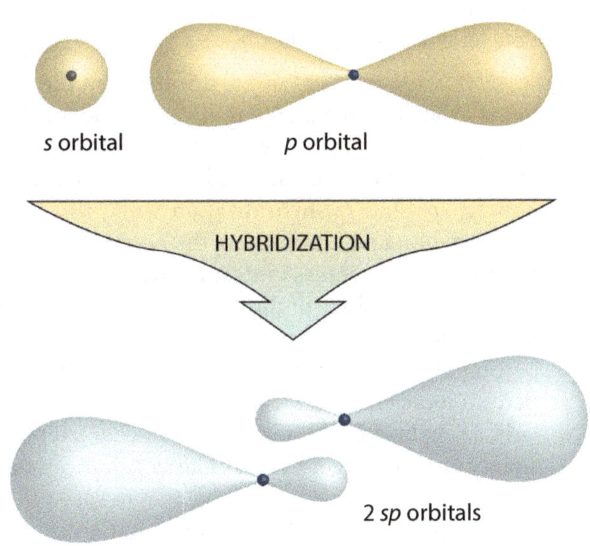

Figure 4.8. Hybridization between one *s* orbital and one *p* orbital produces two *sp* orbitals.

toward a corner of a tetrahedron where a hydrogen atom is located.

But in the unbonded carbon atom, these four electrons are not equally situated. Two of them have the energy of the 2*s* orbital, and two of them have the energy of the orbitals in the 2*p* subshell. Moreover, the orbitals are not shaped alike since the 2*s* orbital is spherical and the 2*p* orbitals each have twin lobes.

Pauling proposed that in the carbon atom, the 2*s* orbital and the three 2*p* orbitals hybridize to form four equal sp^3 orbitals, shown in an orbital diagram this way:

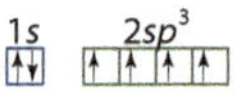

Figure 4.9. The four hybrid sp^3 orbitals point toward the corners of a tetrahedron.

Now that the carbon's 2*s* and 2*p* orbitals have hybridized, they have equal energies and they are spread apart from each other at equal angles in 3-D space, which means they point toward the corners of a tetrahedron, as shown in Figure 4.9. Figure 4.10 shows the carbon hybrid orbitals with the hydrogen atoms attached, each 1*s* hydrogen orbital overlapping the region of one of carbon's sp^3 orbitals. The shared electron pair associated with each bond is in the region where an *s* orbital overlaps with an sp^3 orbital.

Figure 4.10. The *s* orbitals of hydrogen atoms overlap with the sp^3 hybrid orbitals of the carbon atom to form the complete tetrahedral CH_4 molecule.

There are a few more odds and ends about hybridization theory that need to be pointed out. First, hybridization does not happen only to *s* and *p* orbitals. Orbitals from other shells can be involved as well, depending on where the valence electrons are and which ones are involved in the bonding. For example, SF_4 uses one of sulfur's unfilled 3*d* orbitals to make five sp^3d hybrid orbitals.[1]

Second, the notation *sp*, sp^2, and sp^3 tends to be misleading. In cases of *s* and *p* orbital hybridization, there is a match between the number of letters in the notation and the number of orbitals formed. That is, if you consider p^2 to mean "*p* squared," then sp^2 means *s·p·p*, indicating that three orbitals went into the hybridization process (one *s* and two *p*s) and three hybrid orbitals were formed. Now, although reading the notation this way is handy for remembering how *sp*, sp^2, and sp^3 hybridization work, this numerical match does not hold for all cases of hybridization and interpreting the notation this way is not the best approach. It is preferable to think of the notation as indicating the percent character of the hybrid orbitals. Seen this way, sp^2 means that one share of the orbital character is *s*-like, and two shares of the orbital character are *p*-like. The resulting character of the hybrid orbital is 1/3 (33.3%) *s* and 2/3 (66.7%) *p*. (Does this percent character interpretation sound familiar? We used the same language when discussing percent ionic character using electronegativities, another model Linus Pauling invented.) Similarly, sd^3 hybrid orbitals are 25% *s* in character and 75% *d* in character. There is even the possibility of non-integer hybrids like $sp^{2.5}$, indicating a bond character that is 28.6% *s* and 71.4% *p*.

1 The explanation for this is as follows. Of sulfur's six valence electrons, four bond with fluorine atoms, leaving one lone pair. The resulting molecular geometry is trigonal bipyramidal, with the lone pair in one of the middle positions. This molecule has a so-called *see saw* shape.

Example 4.3

Describe the orbital hybridization that occurs on the central atoms in boron trifluoride, BF_3, and the nitrite ion, NH_2^-.

Boron trifluoride is an exception to the octet rule. The Lewis structure for BF_3 is

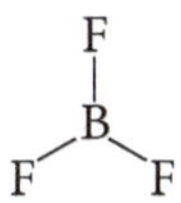

The orbital diagram for boron is

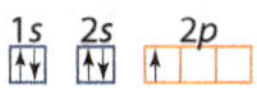

The boron atom shares three pairs of electrons with fluorine atoms. There is no way to do this using only *p* orbitals and still maintain the trigonal planar symmetry in the molecule that we expect from VSEPR theory. This is because the three *p* orbitals are all at right angles to each other, so the fluorine bonds would be as well. Thus, the boron atom uses its 2*s* orbital and two of its 2*p* orbitals to create three *sp* orbitals—sp^2 hybridization. Its new orbital diagram is

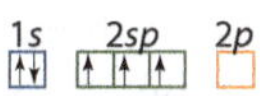

The three new *sp* orbitals have their large lobes lying in the same plane at angles of 120°. The fluorine atoms attach by overlapping their *s* orbitals with the three new *sp* orbitals in the boron atom. In fact, sp^2 hybridization always accompanies trigonal planar molecules composed of four atoms.

The Lewis structure for NH_2^- is

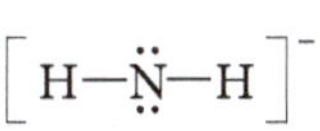

The orbital diagram for nitrogen is

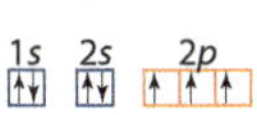

In order to deploy the two equal nonbonding electron pairs and two equal bonding pairs in the Lewis structure, four orbitals are required. Thus the 2*s* orbital and all three 2*p* orbitals hybridize to produce four sp^3 orbitals. In fact, the NH_2^- polyatomic ion has a tetrahedral electron domain structure, and that structure always requires sp^3 hybridization.

The hybridization discussion of Example 4.3 indicates that trigonal planar electron domain geometries always make use of sp^2 hybridization, and tetrahedral electron domain geometries always make use of sp^3 hybridization. If this pattern suggests to you that linear molecules always use *sp* hybridization, you are correct!

4.1.5 Valence-Bond Theory

4. It seems like there's more to covalent bonding than two atoms' nuclei both pulling on the same pair of electrons. That's a classical explanation, not a quantum one. Since atomic bonding occurs down on the atomic level, shouldn't we expect the solutions to the Schrödinger equations for the two atoms to interact? And if so, wouldn't this make explanations of the covalent bond much more complex?

Yes, indeed. Recall that our models of *s*, *p*, *d*, and *f* orbitals in atoms emerge out of the solutions to the Schrödinger equation for the atom. Chemists have long been aware that a correct quantum mechanical model of atomic bonding has to take into account what happens to those solutions when two or more atoms are in close proximity to one another. Solutions to the Schrödinger equation can be interpreted as telling us probabilities of where electrons may be found. (That is, the squares of the solutions can be so interpreted.) But if electrons from one atom can be found in the orbitals of another atom and vice versa, and if the shapes of the orbitals are due to hybridization, then the solutions to the Schrödinger equation should reflect this.

Chemists have been thinking about this since around 1930 when Pauling published his work on orbital hybridization. One approach to the problem is called *valence-bond theory*. Valence-bond theory attempts to construct new equations based on linear combinations of the individual atomic Schrödinger equations. Solutions are then sought to these combination equations.

However, the equations involved are enormously complex, and when valence-bond theory originated, there was no way to solve them. Even the advent of electronic computing in the 1940s wasn't much help for another 50 years or so. Computers simply weren't fast enough to handle the calculations until more recently. But beginning in the late 1990s, scientists have been able to use computers to work out the solutions they are seeking. Needless to say, there remains much work to do. But so far, numerical solutions to the Schrödinger equation now being produced by computers have confirmed that Pauling's hybridization model works quite well.

4.2 Metallic Bonding

4.2.1 Metallic Lattices

We come at last to metallic bonding. The atoms in a lump of metal arrange themselves in a crystal lattice that minimizes the energy between them, just as the atoms in ionic solids do. However, since all the atoms in an elemental metal are the same (except for isotopic variations, which do not affect metallic bonding), there are just two main lattice structures that occur. These structures are described as "close packing" because the atoms are more densely packed than they would be if the atoms were arranged in a rectangular grid or if the layers were directly above each other.

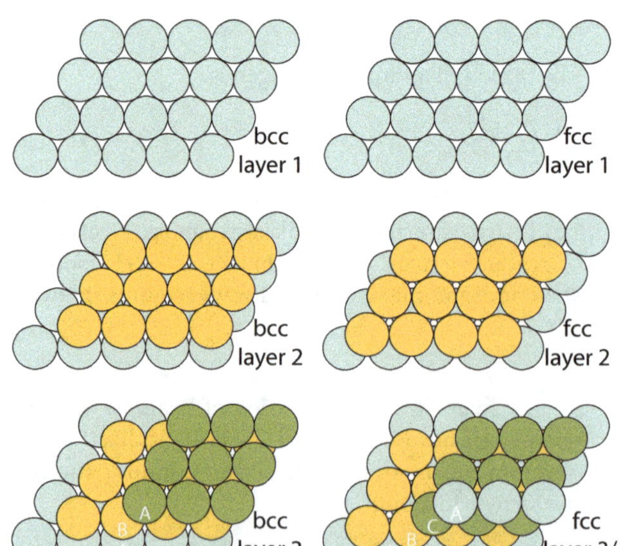

Figure 4.11. Metal atoms can stack in alternating layers (left) of the bcc lattice, or in layers that repeat in sets of three (right), in the fcc lattice.

The first close-packing arrangement is called a *body-centered cubic* (bcc) arrangement, shown down the left side of Figure 4.11. In this arrangement, the third layer is directly above the first layer. The second type of lattice is called a *face-centered cubic* (fcc) arrangement, shown down the right side of Figure 4.11. In the fcc lattice, the positions of the atomic layers repeat in sets of three. To make the geometry of these layers more clear, Figure 4.12 shows outlines of one cell of each.

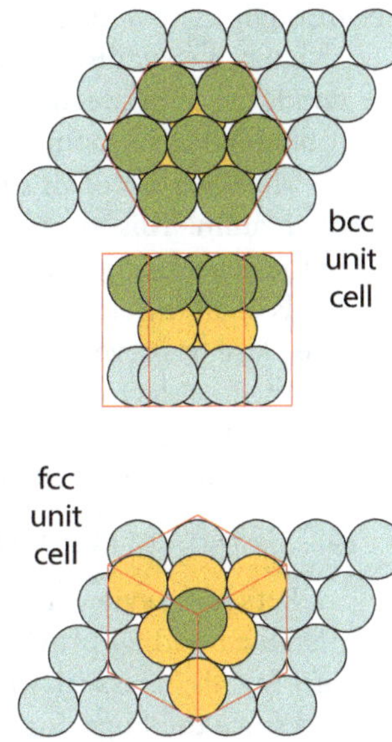

Figure 4.12. One cell of the bcc lattice (top and center), and one cell of the fcc lattice (bottom).

Some metals take on both these structures depending on the temperature. For example, iron, the primary metal in steel, transitions from the bcc structure, called *ferrite*, to the fcc structure, called *austenite*, above 906°C. Stainless steel alloys are made of austenitic iron with 18–20% chromium and 8–10% nickel in the alloy. Once blended in the alloy, the iron preserves its austenitic character when cooled and the result is a family of alloys that possess many superior mechanical qualities. These stainless steel alloys are designated by numbers such as 304 and 316, and are widely used in industry for high-performance machine parts, surgical instruments, and other applications.

The atoms in a metal are held together by the same mechanism that holds together the atoms in a molecule—electrons occupying the orbitals of more than one atom at a time. When an electron simultaneously occupies the orbitals of two or more atoms, each atom has a grip on the electron and the atoms are thus bound to each other.

However, there are major differences between covalent bonds and metallic bonding. This should be obvious, since solid metals possess vastly different properties than the gases and liquids of covalently bonded substances. Metal atoms have a large number of unfilled orbitals and these orbitals overlap one another all over the place in the regions between the metallic nuclei. Consequently, all the valence electrons of the atoms freely move about

Hmm... Interesting. — Tin pest

At cold temperatures, tin transitions from a type of bcc structure called white tin or beta-tin to a semiconductor state called gray tin or alpha-tin. In the alpha state, tin completely loses its metallic properties. This transition is called "tin pest" because once it starts it affects the entire metal. If cold enough for long enough, mechanical objects made of tin simply turn to powder. The lore surrounding tin pest goes way back, and some have blamed Napoleon's defeat by the Russians in 1812 on the deterioration of the soldiers' tin buttons in the Russian winter. Tin pest has also been blamed for the deterioration of food and fuel canisters on Robert Scott's 1910 expedition to the South Pole. Roald Amundsen beat him there, but on Scott's return trip in 1912, the canisters failed, causing the loss of food and fuel for Scott and his team. Scott and his two remaining men died 11 miles from a supply depot. The role of tin in both these incidents has been seriously questioned, but without final verdict. Thus, the legends continue.

through the overlapping orbitals in the lattice, and are delocalized—just like the electrons in resonance structures. It is the very nature of metallic bonding that it cannot be thought of as taking place between just two or three atoms at a time. It is the result of the large numbers of orbitals all overlapping one another in a continuous array of atoms.

The diameter of an atomic nucleus is on the order of 100,000 times smaller than the diameter of an atom. This means that the atomic nuclei occupy very little space in the lattice compared to the space occupied by the overlapping orbitals. So the trillions of electrons from the valence shells of all the atoms have plenty of room to move around and almost nothing to hold them in one place or prevent them from moving in the presence of electrical forces.

The free electrons in the metallic lattice are called *conduction electrons*, and there are so many of them that they are often described as an ocean or sea of electrons, or even as a gas of electrons flowing through the lattice. Commonly, the conduction electrons are called the *electron sea*.

4.2.2 Physical Properties of Metals

Let's now review some of the intensive properties we associate with metals, and consider how metallic bonding provides us with explanations for them. The electron sea obviously explains why metals have such *high electrical conductivity*. Electric current is the flow of electric charge, and since the huge numbers of conduction electrons in metals can freely slide from orbital to orbital in the metallic lattice, all one has to do is establish a difference in voltage (also called *electrical potential*) in one part of the metal relative to another part and the electrons, being negatively charged, all go screaming toward the most positive end of the metal.

The conduction electrons are also one of the reasons metals have *high thermal conductivity*. Heating a metal adds kinetic energy to the electrons, causing them to collide with each other, spreading the thermal energy throughout the lattice. The other reason metals conduct heat so well is the ease with which vibrations are passed from atom to atom through the crystal lattice. But in some metals, the conduction electrons can make a big difference. As examples, Table 4.2 lists the thermal conductivity values for several metals and compares them to the thermal conductivities of ionic compounds made with the same metals. The thermal conductivity of the pure metal is anywhere from two to 80 times as high as that of the ionic compound.

Metal	Thermal Conductivity (W/m·K)	Ionic Compound	Thermal Conductivity (W/m·K)
Na	141	NaCl	≈6
Cu	401	Cu_2O	≈5
Mn	7.8	MnO	≈4
Al	237	Al_2O_3	≈30
K	102	KCl	≈7
Ca	200	$CaCO_3$	≈4

Table 4.2. Thermal conductivities of some metals and their ionic compounds. The unusual units W/m·K work like this: To determine the rate of heat flow through a sample, you multiply the thermal conductivity value by the cross-sectional area of the material, and divide by the length. This gives the rate of heat flow in watts per degree of temperature difference at the two ends.

Malleability and *ductility* are well-known intensive properties of metals. Malleability indicates the ability of a material to deform rather than shatter when compressed, such as when a blacksmith shapes a metal by pounding it. Ductility indicates the ability of a material to deform under tension, such as when copper is drawn into a wire. The explanation of why metals exhibit these

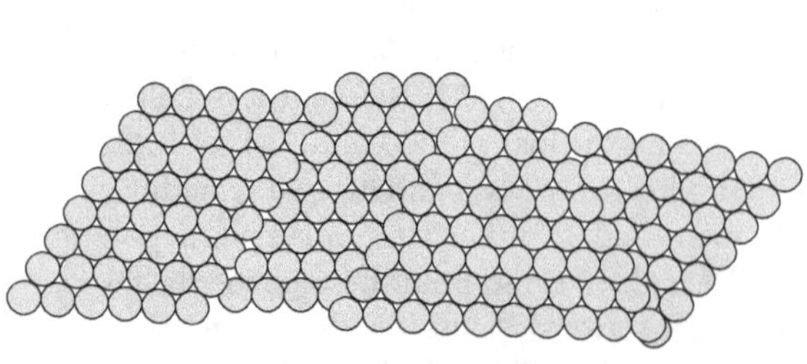

Figure 4.13. Planes of metal atoms can slide past each other without fracturing the metallic lattice.

properties lies in the fact that the atoms are held together in the lattice by the conduction electrons sharing orbitals. Planes of atoms can slide past each other without fracturing the lattice, as illustrated in Figure 4.13. This is because movement of one section of atoms still allows the orbitals of the atoms in that section to overlap with the orbitals of an adjacent section of the metal. By contrast, ionic crystals are held together by balanced forces between ions. We saw in the previous chapter that when an ionic material is stressed enough to change the alignment between positive and negative ions, the layers of atoms push apart instead of pulling together and the material shatters.

Figure 4.14. The silvery-gray color of chromium is typical of most pure metals.

Metals are also known for their silvery *luster*, typified by the sample of chromium shown in Figure 4.14. The shininess of metals is due to the large numbers of unfilled orbitals in the atoms' crystal lattice. There are so many orbitals available for electrons to occupy when they absorb incoming energy that metals absorb and re-emit wavelengths across the visible spectrum. Since the combination of all colors produces white light, this is what makes them shiny and neutral in color (silver/gray).

There are a few metals that are not silver or gray, and that have natural colors instead. These include copper (red-orange), gold (yellow), and osmium (bluish tint). In copper's case, you may recall from Chapter 1 that copper's anomalous electron configuration is $[Ar]4s^1 3d^{10}$. The energy difference between the filled 3*d* subshell and the half-filled 4*s* subshell corresponds to orange light. Electrons hopping back and forth between the orbitals in the 4*s* and 3*d* subshells are responsible for copper's unique color. The explanation for gold's color is essentially the same, gold being in the same group as copper in the periodic table.

4.3 Intermolecular Forces

To conclude our study of chemical bonding, we review the forces we have encountered so far that operate at the atomic level. We also add hydrogen bonding to the list, and several forces known collectively as *Van der Waals forces*.

4.3.1 Bonding Forces

We have seen two basic types of forces that hold atoms together in chemical bonds. The first is the electrostatic force between ions, between polyatomic ions and ions, and between polyatomic ions and other polyatomic ions. These electrostatic forces hold together the ions in the crystal lattice of an ionic compound.

The second bonding force occurs in covalent bonds and metallic bonding. The force is caused by electrons residing in the orbitals of two or more atoms simultaneously, and since each positive nucleus has the electron trapped in the "energy well" of one of its orbitals, the involved atoms become indirectly attached to one another. The force of this kind of bond can occur on a small scale, as when two atoms bond in a molecule, or on a massive scale, as in the vast array of atoms in a metallic lattice.

4.3.2 Intermolecular Forces

The forces *between* molecules—that is, *intermolecular forces*—include the force of hydrogen bonding and *Van der Waals forces*. Intermolecular forces are far weaker than the bonding forces that hold molecules and crystals together, which means that much less energy is required to break the bonds. The lattice energy in a typical ionic crystal might be 2200 kJ/mol (23 eV/particle), and bond energies in common covalent molecules are in the range of 600 kJ/mol (6 eV/particle). By contrast, the total energy involved in hydrogen bonding in a substance might be 60 kJ/mol (0.6 eV/particle). Energies associated with Van der Waals forces are down in the range of 5 kJ/mol (0.05 eV/particle), or even less.

4.3.3 Hydrogen Bonding

We encountered hydrogen bonding back in the Introduction. Now it is time for a more detailed treatment. The term *hydrogen bonding* refers to the attraction between nonbonding electron pairs on the central atom of a molecule and hydrogen atoms that are attached to oxygen, nitrogen, or fluorine atoms. The high electronegativities of O, N, and F ensure that when bonded to hydrogen, the hydrogen atom has a substantially positive character, symbolized by δ^+. These δ^+ hydrogen atoms are attracted to the nonbonding pairs of electrons on surrounding molecules, regardless of whether those molecules are of the same compound.

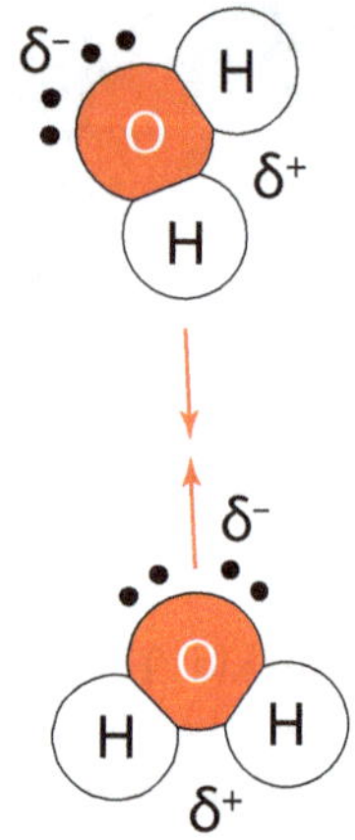

Figure 4.15. The attraction present in hydrogen bonding.

The abundance of water on this planet and in our bodies means that hydrogen bonding involving water molecules is going on everywhere around us and inside us. Figure 4.15 illustrates hydrogen bonding at work between a pair of water molecules. The oxygen atom in water has two nonbonding pairs of electrons. Additionally, the greater electronegativity of oxygen causes the shared electrons in the water molecule to crowd over toward the oxygen atom. The result, as we have seen, is a dipole molecule that is attracted to other similar molecules.

The attraction between water molecules is what causes water molecules to spread out and orient themselves in hexagonal patterns as they cool, causing the density of water to decrease right before it freezes. Thus, ice floats and ponds freeze from the top down, allowing aquatic wildlife to flourish. This extremely helpful behavior of water is very unusual. Almost nothing else expands as it freezes, but water does and it just so happens that our planet is covered with it.

The expansion of water just before freezing is just one of the many unique properties of water that make water the perfect life-support environment. Life isn't just marginally present on earth; it *flourishes* on earth, in part due to the amazing (and wonderful) properties of water.

Another excellent example of hydrogen bonding at work is in the absorbance of cotton towels. Cotton is composed of *polymers*, long chain molecules. The polymers in cotton are called *cellulose*, and the smaller molecular units—or *monomers*—the polymers are made of are called *β-glucose* (beta-glucose), a carbon ring molecule depicted in Figure 4.16. In this figure, carbon atoms are black, oxygens are red, and hydrogens are

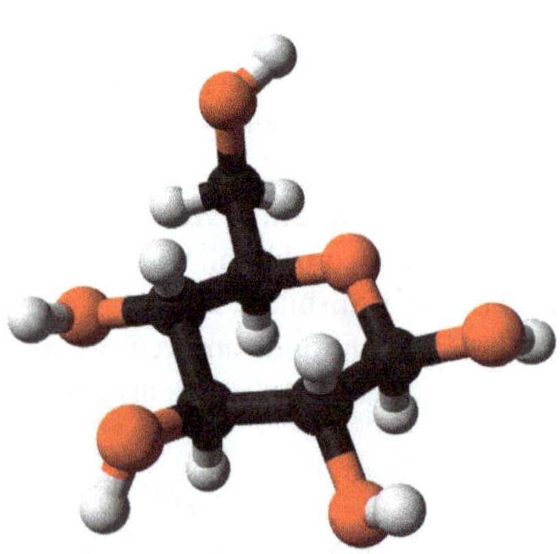

Figure 4.16. The β-glucose molecule.

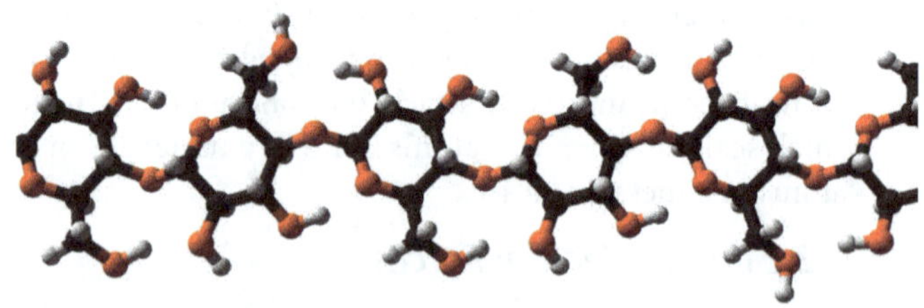

Figure 4.17. A short section of cellulose.

white. In cellulose, hundreds or thousands of β-glucose monomers are attached together in long chains. A short section of such a chain is shown in Figure 4.17. As you can see, the outside of this molecule is covered in —OH pairs or *hydroxyl groups*—hydrogen atoms bonded to oxygen atoms that are in turn bonded into the molecule. The hydrogen in every one of these hydroxyl groups is δ^+ in character because it is bonded to an oxygen atom. As a result, the nonbonding pairs of any water molecules that may be around are attracted to the cellulose chain at one of the hydrogen sites.

Figure 4.18. High-performance athletic socks use hydrogen bonding to keep feet dry.

Cotton absorbs moisture in every direction and holds it there, and this explains why it takes your blue jeans a long time to dry if they get wet. But the manufacturers of high-performance athletic wear have figured out how to engineer the hydrogen bonding process so that moisture is pulled away from the skin and sent to the outside of the garment where it can evaporate. This action is called "wicking," and those of us who have discovered the power of wicking to keep our feet absolutely dry, like the marathon runners in Figure 4.18, will never run in regular cotton socks again. (Please do not reply that you run without socks. That's just nasty.)

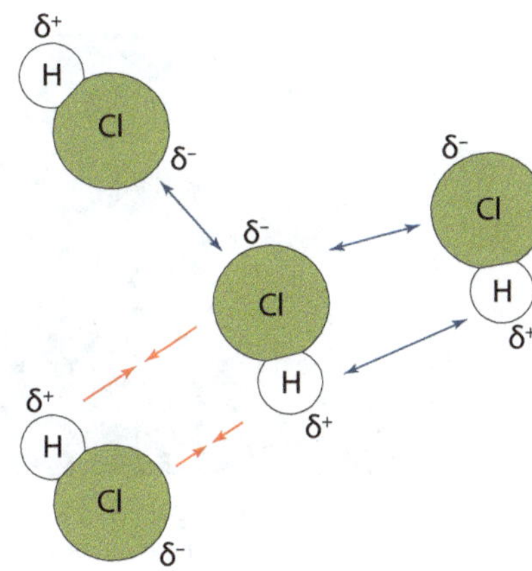

Figure 4.19. Keesom forces are forces between permanent dipoles.

4.3.4 Van der Waals Forces

In addition to hydrogen bonding, there are three other force mechanisms between molecules. These all have to do with attractions between dipoles, and collectively they are referred to as *Van der Waals forces*. Since hydrogen bonding also has to do with forces between dipoles, you should take note that Van der Waals forces are all the forces involving dipoles *other* than hydrogen bonding.

The first is the *Keesom force*, a force between two *permanent dipoles*. The term *permanent dipole* is to be understood as distinguishing between molecules that are dipoles inherently, because of the difference in electronegativities of the atoms in the molecule, and those that possess only momentary and fleeting dipole character, which we get to in a moment. An example of a permanent dipole we have met before is the HCl molecule, illustrated in Figure 4.19. Chlorine is substantially more electronegative than hydrogen, so the molecule is a permanent dipole. The δ^+ end of the HCl dipole attracts the δ^- end and repels the δ^+ end of other dipoles. As you see in Figure 4.19, Keesom forces can be attractive or repulsive, depending on the orientation of the two dipoles involved.

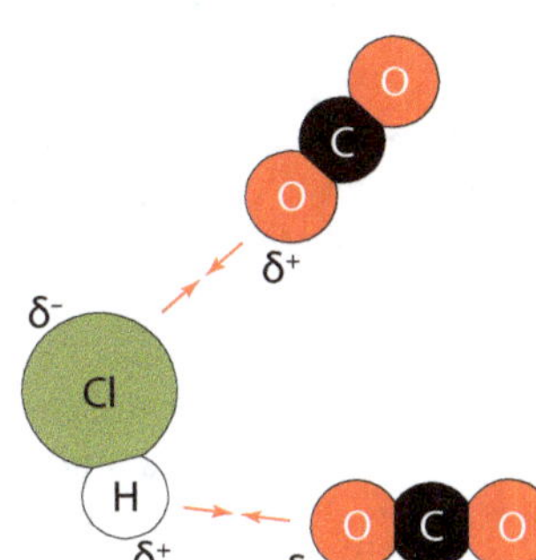

Figure 4.20. The permanent dipole HCl molecule inducing temporary dipoles in CO_2 molecules, resulting in Debye forces between the HCl and CO_2 molecules.

The second Van der Waals force is the *Debye force*. At this point you need to try to imagine the unimaginable—the micro world of molecules. Molecules are always present in unbelievable numbers and moving at unbelievable speeds. So imagine that a permanent dipole like an HCl molecule has a close encounter with a symmetrical, nonpolar molecule, such as CO_2. As illustrated in Figure 4.20, the charged character of the ends of the HCl dipole can temporarily induce dipole character on another molecule. This happens, for example, because the δ+ hydrogen atom attracts the electrons in the nonpolar molecule. Since the charge distribution in the nonpolar molecule is symmetrical to begin with, when the electrons are attracted toward one end of the molecule, that end becomes negative relative to the other end, and the molecule momentarily takes on a temporary dipole character. Debye forces can be induced on any neutral atom or nonpolar, symmetrical molecule.

Note that since Debye forces are induced, the induced charge is always the opposite of the charge inducing it, leading to attraction between the permanent dipole and the induced dipole. Thus, Debye forces are attractive. Note also that since molecules are constantly in motion, the Debye force is fleeting and temporary, and weaker than the Keesom force.

The third of the Van der Waals forces is the *London dispersion force*. London dispersion forces are forces between two temporary dipoles and are the weakest and most fleeting of all intermolecular forces. London forces are a result of the constant motion of the electrons in orbitals. As the electrons move around, there are very brief moments when the electrons surrounding a molecule or an individual atom (as in a noble gas, for example) happen randomly to congregate on one side of the molecule. They don't stay like that, of course, because they repel each other and quickly redistribute themselves. Nevertheless, for an instant, the atom or molecule becomes a dipole: the negative end is where the electrons are gathered and the positive end is where they came from. The force between two such instantaneously created dipoles is the London dispersion force.

Figure 4.21. The ability of a gecko to walk up a plate of glass is probably due to Van der Waals forces.

It should be apparent that since London dispersion forces are caused by the random motions of fast moving electrons, they appear constantly in gases and liquids all around us but only last for an instant.

Figure 4.21 is an image of a gecko walking up a pane of glass. Researchers have published one study after another analyzing the ability of a gecko's feet to stick to a pane of glass, but the issue does not seem to be settled

Bonding Forces	
electrostatic	force between ions and/or polyatomic ions
covalent	force holding atoms together due to shared electrons in overlapping orbitals

Intermolecular Forces		
hydrogen bond		force between nonbonding electron pairs and a hydrogen atom bonded to an oxygen, nitrogen, or fluorine atom
Van der Waals	Keesom force	force between two permanent dipoles
	Debye force	force between a dipole and an induced dipole caused by it
	London dispersion force	force between two instantaneously formed dipoles

Table 4.3. Summary of the various atomic-level forces involved in chemical bonding and molecular attraction.

yet. However, many who have studied the question are convinced that Van der Waals forces are responsible, which means electrical forces. Once again, we can say that chemistry is all about electrical forces!

Table 4.3 summarizes all the forces we have reviewed here.

Chapter 4 Exercises

SECTION 4.1

1. Describe the central theoretical problem addressed by VSEPR theory.
2. Use VSEPR theory to predict the shapes of the following molecules. For each one listed, draw the Lewis structure, state the electron domain geometry, the molecular geometry, and sketch a simple ball-and-stick figure of the molecule.

   a. phosphine, PH_3 b. carbon tetrachloride, CCl_4 c. sulfur dioxide, SO_2
   d. SCl_2 e. NH_2Cl f. NH_4^+
   g. ClO_2^- h. ozone, O_3 i. acetylene, C_2H_2
   j. CS_2 k. SO_3 l. Cl_2O
   m. hydrochloric acid, HCl n. hydronium, H_3O^+ o. hydrogen cyanide (known in the old days as prussic acid), HCN

3. Indicate the hybridization used by the central atom in the following molecules (Note: one of these molecules is an exception to the octet rule):

   a. NH_3 b. $SiCl_4$ c. H_2S
   d. HF e. HNO_2 f. PH_3

4. Explain what is meant by the term *electron domain*.
5. Give approximate bond angles for the bonds indicated in the following molecules:

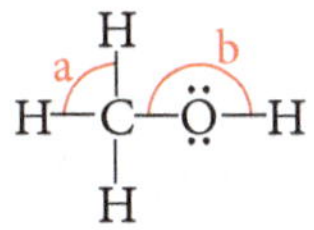

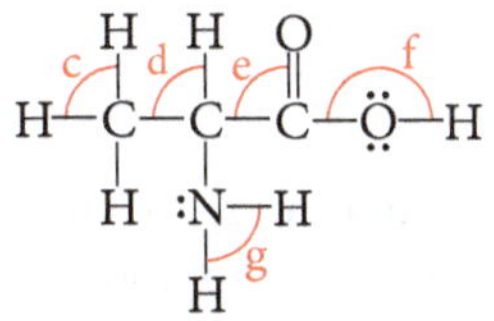

6. Explain why some bond angles are slightly smaller than the angles in pure geometric figures.
7. Consider the molecules described below and state the electron domain geometry and the molecular geometry for each.

   a. two bonding domains and two nonbonding domains.
   b. four bonding domains
   c. three bonding domains and one nonbonding domain
   d. five bonding domains

   e. four bonding domains and two nonbonding domains (Hint: The nonbonding domains are on opposite ends of the molecule, and the resulting molecular shape uses a common geometrical term not introduced in the chapter.)

8. Under what conditions does octahedral molecular geometry occur?
9. Consider a molecule with four bonding domains and one nonbonding domain.
   a. Describe the two options for the molecular geometry of this molecule.
   b. In fact, one of the two options minimizes the energy in the molecule more than the other and is the shape the molecule takes. Identify which one this is and explain why.
10. Look again at the L-alanine molecule on the opening page of this chapter, and shown to the right. Use VSEPR theory to explain the geometry of the molecule at each of the five connection points. These are circled in red in the figure to the right.

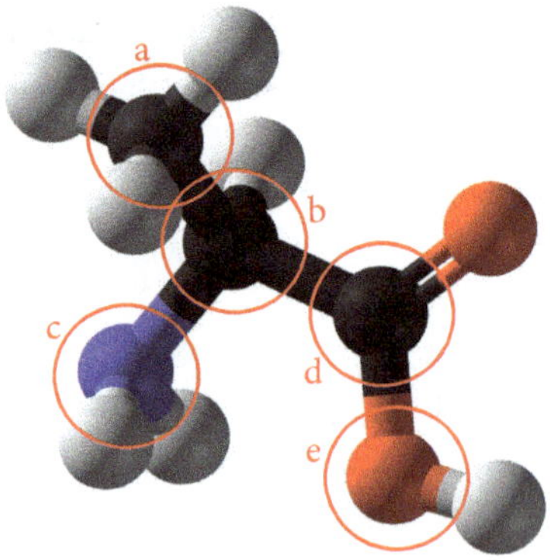

11. Describe the goal and purpose of valence-bond theory.

SECTION 4.2

12. Explain what holds the atoms together in metallic bonding.
13. Explain why it is inaccurate to say that metallic bonding is just like covalent bonding.
14. For each of the following typical properties of metals, use what you know about metallic bonding to explain why metals exhibit the property.
    a. high electrical conductivity
    b. high thermal conductivity
    c. high luster
    d. silver or gray color
    e. malleability
15. Explain why copper has a natural orange-red color.

SECTION 4.3

16. Use what you have learned about intermolecular forces to explain why the boiling point of water is so much higher than that of other molecular compounds of comparable molar mass.
17. Explain what a dipole is.
18. Distinguish between permanent and temporary dipoles.
19. Describe two ways a temporary dipole can form.
20. Explain why cotton towels can absorb so much moisture.
21. Why are London dispersion forces the weakest of the intermolecular forces?
22. What are Van der Waals forces?

GENERAL REVIEW EXERCISES

23. Determine the molar mass and the percent composition of L-alanine, pictured above.

24. The bright orange line in the copper emission spectrum has a wavelength of about 601 nm. Determine the energy associated with each photon of light at this wavelength. State your answer in eV.

25. Identify each of the following bonds as covalent, polar covalent, or ionic in character.

    a. H—N
    b. H—F
    c. Br—Br
    d. C—O
    e. Cl—S
    f. P—O

26. Use VSEPR theory and electronegativities to determine which of the following molecules are polar.

    a. CF_4
    b. Cl_2O
    c. NH_3
    d. SO_2
    e. N_2
    f. SF_2

27. Use your knowledge of atomic and ionic size trends in the periodic table to place the following atoms and ions in order of increasing size: Cl^-, F^-, Te^{2-}, F, and I^-.

28. Write the condensed electron configurations for Hg, Ru, Br, Ba, Cr, and S.

29. Define the unified atomic mass unit.

30. Define the mole.

31. Determine which of the following is likely to have the greatest difference between the second and third ionization energies: Sc, Ar, F, P, and Sr.

32. Distinguish between electronegativity and electron affinity.

33. Determine the number of moles present in each of the following:

    a. 65.5 g water
    b. 1,250 mg vitamin C (ascorbic acid), $C_6H_6O_6$
    c. 400 mg aspirin, $C_9H_8O_4$
    d. 14.0 kg saltpeter, KNO_3
    e. 1,050 g perchloric acid
    f. 953.00 g hydrogen fluoride

34. Determine the number of metal atoms present in each of the following:

    a. 55 g mercury(II) sulfide
    b. 3.00 kg iron(III) oxide
    c. 1.0000 mol calcium carbonate
    d. 45 g strontium nitrite
    e. 2.000 mol sodium chromate
    f. 5.05 kg calcium acetate

35. Draw the orbital diagrams for the following: Sc, Fe, Ne, As, and I.

36. Use data from Table 1.5 to determine the atomic mass of iron.

37. Explain the Aufbau principle, the Madelung rule, and Hund's Rule.

38. State the Pauli exclusion principle.

39. Describe the different ways that atoms can possess energy.

40. Glucose is a carbohydrate that is the major building block for the polymers cellulose and starch. A certain sample of glucose consists of 4.62 g C, 0.776 g H, and 6.154 g O. The molar mass of glucose is 180.157 g/mol. Determine the percent composition, the empirical formula, and the molecular formula of glucose.

Chapter 5

Chemical Reactions and Stoichiometry

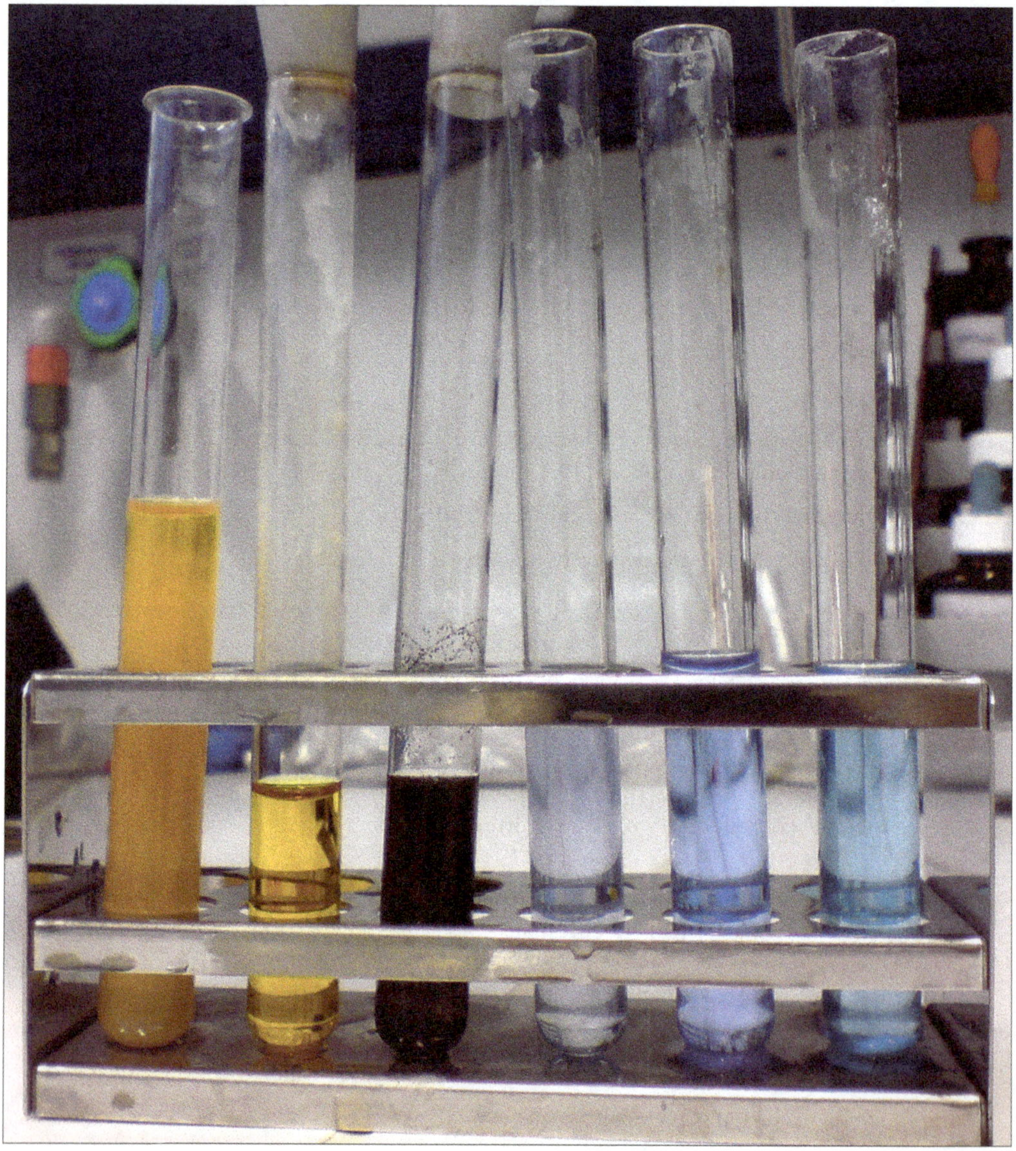

These test tubes contain various solutions and reaction products involving copper(II) nitrate, $Cu(NO_3)_2$. In some cases, precipitates formed; in others, the products remained in solution. We use some of these reactions as examples in this chapter.

Objectives for Chapter 5

After studying this chapter and completing the exercises, you should be able to do each of the following tasks, using supporting terms and principles as necessary.

SECTION 5.1

1. State the law of conservation of mass for chemical reactions.
2. Recognize, interpret, and use the standard symbols involved in chemical equations.
3. Given a verbal description of the reactants and products in a chemical reaction, write the formula equation and balance it.
4. Define the terms *precipitate*, *oxidation*, *reduction*, *oxidizer*, *oxidation state*, and *oxidation number*.
5. Use the standard oxidation state guidelines from memory to determine the oxidation states of atoms in pure elements, in polyatomic ions, and in compounds.

SECTION 5.2

6. Classify chemical reactions as synthesis, decomposition, single replacement, double replacement, combustion, acid-base neutralization, or oxidation-reduction. Also, identify when a reaction may be classified as two or more of these types simultaneously.
7. Use the activity series of metals to predict whether a given single replacement reaction with metals will occur.
8. Recognize and identify common hydrogen compounds as acids and common hydroxide compounds as bases.
9. Explain what happens in an acid-base neutralization reaction and use this information to form a definition for the term *salt*.
10. Explain the term *redox reaction* and describe what happens in such reactions.

SECTION 5.3

11. Describe the origin of the term *stoichiometry*.
12. Given a mole or mass quantity of one of the reactants or products in a specified chemical reaction, perform stoichiometric calculations to determine the quantities required or produced, in moles or mass, of any of the other reactants or products in the reaction.
13. Define the terms *limiting reactant* and *limiting reagent*.
14. Use a balanced chemical equation and available quantities of reactants to identify the limiting reactant in the reaction.
15. Use the available quantity of a limiting reactant to determine the theoretical yield for the reaction.
16. Use the theoretical yield and actual yield in a reaction to determine the percent yield for a given reaction product.

5.1 Introduction to Chemical Equations

5.1.1 Fascinating Chemistry

Chemical reactions take place around us—and in us—all the time, but unless a fire or explosion occurs, we often don't notice. You probably already know that fires and explosions are chemical reactions, but so are the drying of paint, the filling of the air bags in a car during a collision, and the digestion of your food. Chemical reactions are of supreme importance in industry, and without them you and I might have electricity and running water, but we would be without nearly all the products we enjoy. Without the chemical research of the 19th and 20th centuries, we would not have computers, plastic products, batteries, synthetic fibers (such as nylon), or just about anything else composed of nonmetallic man-made materials.

Figure 5.1. Copper carbonate decomposes to copper oxide when heated.

There are several tell-tale signs that a chemical reaction is taking place. The first is the release of heat and light. When you are burning logs at a camp fire, you can be assured a chemical reaction is taking place. The second is the production of a gas, such as the carbon dioxide that fizzes out of a soft drink after the can has been opened. A third is a color change. Figure 5.1 shows the dramatic color change that occurs when beautiful blue-green copper carbonate is heated to above 290°C. The copper carbonate decomposes to dark brown copper oxide, releasing carbon dioxide gas. Finally, a *precipitate* may form. A precipitate is any solid substance that forms when two solutions are combined. Figure 5.2 shows a solid copper(II) hydroxide precipitate coming out of solution because of the reaction that occurs when aqueous solutions of $CuSO_4$ and NaOH are combined.

Figure 5.2. Solid copper(II) hydroxide precipitates out of solution when aqueous solutions of $CuSO_4$ and NaOH are combined.

Now, you may be of the opinion that chemistry is really only enjoyed by science nerds, but I suggest it should be otherwise. Because chemistry is so important in contemporary society, and because so many of the processes and products around us involve chemical reactions that are easily understood, I would expect anyone with a normal, healthy level of curiosity about the world to be interested—if not fascinated—to learn about it. In this chapter, we begin diving in to some of these ubiquitous chemical reactions.

5.1.2 The Law of Conservation of Mass in Chemical Reactions

French chemist Antoine Lavoisier (Figure 5.3) is regarded as the father of modern chemistry. He was the first to understand that combustion was a chemical reaction involving oxygen. He was also responsible for helping to turn chemistry from a descriptive science into a quantitative one. This was because of the famous tin experiments Lavoisier conducted in 1774. In his tin experiments, Lavoisier investigated the formation of tin oxide from tin, and in the process he discovered that when tin turns to tin oxide, it does so by reacting with the oxygen in the air. Before the experiments, no one knew what air was composed of. Through the experiments, Lavoisier demonstrated that air is composed of more than one gas, and that oxygen accounted for about 20% of it.

Figure 5.3. French chemist Antoine Lavoisier (1743–1794).

Also in 1774, English chemist Joseph Priestley (Figure 5.4) was the first to isolate oxygen, which he accomplished by heating mercury oxide. Priestley assumed he had produced a particularly pure form of air. Lavoisier identified the gas that Priestley had produced as being the same as the gas that reacted with the tin and named the gas *oxygen*.

Figure 5.4. English chemist Joseph Priestley (1733–1804).

In the tin experiments, Lavoisier heated tin in sealed vessels so that he could carefully weigh the contents before and after the reaction. He found the weights before and after to be the same. As a result of his measurements, Lavoisier demonstrated conclusively the principle now known as the *law of conservation of mass in chemical reactions: in any chemical reaction, the total mass of the products equals the total mass of the reactants*. This law is the basis for the methods we use to balance chemical equations, a critical step that undergirds all the calculations involving chemical reactions. We address the procedure for balancing chemical equations momentarily. First, we consider the notation used in writing chemical equations.

5.1.3 Reaction Notation

The chemical reaction Joseph Priestley used to isolate oxygen is the following:

$$2HgO(s) \xrightarrow{\Delta} 2Hg(l) + O_2(g)$$

Figure 5.5. Mercury(II) oxide, HgO.

To perform this reaction, solid mercury(II) oxide, shown in Figure 5.5, is heated to produce liquid mercury and gaseous diatomic oxygen. The formulas for each of these compounds and elements are written in the equation with their subscripts. Subscripts in chemical equations are part of the chemical formulas that identify the specific compounds involved in a reaction. The coefficients in front of the compound formulas are determined during the process of balancing the equation.

The symbols used in a chemical equation convey a lot of information about the reaction. The physical states of each of the compounds are shown in parentheses after the chemical symbol or formula. In addition to *s* (solid), *l* (liquid), and *g* (gas) in the equation above, (*aq*) is used to indicate a substance in aqueous solution. The Δ symbol (*delta*, the Greek upper-case D) over the "yields" arrow indicates heating, and if heating to a particular temperature is required, this temperature is written below the arrow. Other symbols may also be placed over the arrow to indicate

Symbol	Meaning
$\rightarrow$	yields
$\rightleftharpoons$	reaction is reversible; proceeds in both directions
s	solid
l	liquid
g	gas
aq	in aqueous solution
$\xrightarrow{\Delta}$ or $\xrightarrow{\text{heat}}$	heating
$\xrightarrow[290°C]{\Delta}$	heat to 290°C
$\xrightarrow{2.5\text{ atm}}$	reaction occurs at a pressure of 2.5 atm
$\xrightarrow{\text{Fe}}$	chemical symbol or formula for a catalyst that must be present

Table 5.1. Symbols used in chemical equations.

reaction conditions. The symbols commonly used are listed in Table 5.1.

5.1.4 Balancing Chemical Equations

As you will see throughout this chapter, the chemical equation is a very important tool for analysis of chemical reactions. To write a chemical equation, first the formulas of the reactants and products are written. This is called the *formula equation*. Using Priestley's mercury(II) oxide reaction again as an example, the formula equation is

$$HgO(s) \xrightarrow{\Delta} Hg(l) + O_2(g) \quad \text{(formula equation)}$$

Once the correct formulas for the reactants and products are present, the equation must be balanced so that it conforms correctly to the law of conservation of mass in chemical reactions. As it stands, the formula above has HgO on the left side, a formula indicating one atom of oxygen. But the O_2 on the right indicates a molecule containing two oxygen atoms.

The *balanced equation* for the mercury(II) oxide reaction is

$$2HgO(s) \xrightarrow{\Delta} 2Hg(l) + O_2(g) \quad \text{(balanced equation)}$$

The coefficients in front of the HgO and Hg apply to the entire compound they precede. Thus, 2HgO means "two units of HgO, and thus two mercury atoms and two oxygen atoms." Note that this is similar in meaning to the mathematical expression $2(a + b)$, but it is completely different in meaning from the expression $2ab$.

While balancing equations, it is most convenient to read a chemical equation as representing numbers of atoms. Read this way, we can simply count atoms of each element on both sides of the equation and manipulate the coefficients until the equation is balanced. But very importantly, the equation can also be read in terms of the bulk quantity moles. And since most reactions involve bulk quantities of atoms, this is the usual way to read an equation once the balancing is complete. Read this way, the balanced mercury(II) oxide equation reads, "two moles of solid mercury(II) oxide yield two moles of liquid mercury and one mole of diatomic oxygen gas."

Figure 5.6. Pouring liquid mercury.

Just in case you've never seen liquid mercury before, it is shown in Figure 5.6. It may look hot, as if it were molten metal, but it is not. It is one of only two elements that are liquids at room temperature. (The other is bromine.) When I was in high school, we were allowed to play with mercury in our hands. It is almost as dense as lead (look at the periodic table), so a blob of mercury feels unbelievably heavy in your hand. Alas, however, playing with mercury is no longer allowed. Once scientists discovered that it is absorbed through the skin, is quite toxic, can enter the body by breathing the vapor, and that its effects are cumulative, the party was over. With prolonged exposure, mercury destroys brain cells and as a result, people lose brain function. Thus the expression, "mad as a hatter." Hat makers in the old days used mercury to shape the felt hats they made. Not good. Oh, and please make sure you and your family don't toss used fluorescent lamps or compact fluorescent lamps in the trash. They all contain mercury. Take them to a recycle center such as one of those giant hardware stores. Not kidding.

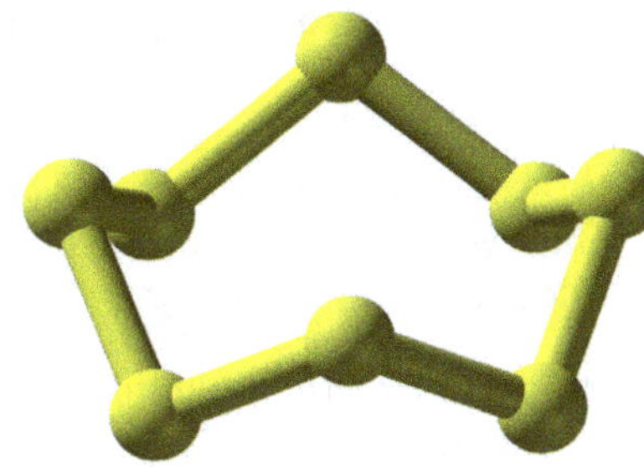

Figure 5.7. A sample of sulfur (above) and a model of cyclooctosulfur, S_8.

Now, back to balancing equations. There are some key points to attend to when writing down the formula equation. Once those are satisfied, then the equation may be balanced.

1. All formulas for compounds must be correct, with the correct subscripts.
2. For pure elements in an equation, there are several cases when a pure element appears in molecular form, and has a subscript. Seven of these are the seven diatomic substances: H_2, N_2 O_2, F_2, Cl_2, Br_2, and I_2. Sulfur and phosphorus form molecules with themselves in the natural state and appear as the molecules S_8 and P_4. The most common form of sulfur, for example, takes the form of the eight-atom ring called *cyclooctosulfur*. A sample of sulfur and the cyclooctosulfur ring are shown in Figure 5.7. And since we are talking about sulfur, I can't resist showing the beautiful images of burning sulfur in Figure 5.8. When it burns, sulfur melts into a blood-red liquid. The flame is bright blue, but cannot be seen in the daytime photo. The second photo, made in the dark, shows it well.
3. The equation must contain all the reactants and products that are part of the reaction that actually occurs.

Figure 5.8. Burning sulfur in the daylight (above) and in the dark (below).

The process of balancing a formula equation is the process of adding coefficients in front of either reactants, products, or both until the equation displays equal numbers of atoms of each element involved on both sides of the equation. For the relatively simple reactions we consider in this chapter, balancing is performed by inspection and trial and error, although there are some guidelines that help expedite the process. Let's consider the guidelines, and then move to examples. Here they are:

1. If an element appears in an odd number on one side of the equation and an even number on the other, you need to make them both even. Start by placing a coefficient (usually 2) on the formula with the odd number.
2. Usually, treat polyatomic ions in formulas as single units. It is easier to count sulfates than it is to count individual

sulfur and oxygen atoms. However, if the polyatomic ion does not appear on both sides of the equation, you must count individual atoms.

3. Tackle the more complicated formulas first. Save placing coefficients on elements that appear alone in the equation (such as the Hg and O_2 terms in the mercury(II) oxide reaction) until last. Placing a coefficient on one element doesn't affect anything else. Placing coefficients on compounds affects two or more elements simultaneously.
4. When you finish balancing, make sure there is no common multiple (other than one) in all the coefficients in an equation. So for example, if all your coefficients are even, divide them all by two.

We now illustrate the process of balancing equations with a number of examples.

Example 5.1

Write the formula equation and balanced equation for the following reaction: zinc metal is combined with hydrochloric acid in aqueous solution. The reaction produces a solution of zinc chloride and hydrogen gas.

We begin by writing the formula equation. Hydrochloric acid is HCl. From Figure 2.12, the oxidation state of zinc is Zn^{2+}, so the formula for zinc chloride is $ZnCl_2$.

$$Zn(s) + HCl(aq) \rightarrow ZnCl_2(aq) + H_2(g) \qquad \text{(formula equation)}$$

Notice that in the reactants there is a single atom of chlorine and a single atom of hydrogen, but in the products there are two of each. We resolve this odd-even problem by placing a coefficient of 2 on the HCl.

$$Zn(s) + 2HCl(aq) \rightarrow ZnCl_2(aq) + H_2(g) \qquad \text{(balanced equation)}$$

This appears to have balanced the equation, but count to make sure. Zn: one on the left, one on the right. H: two on the left, two on the right. Cl: two on the left, two on the right. The equation is balanced.

Example 5.2

One of the forms rust can take is iron(III) oxide (Figure 5.9). Write the formula equation and balanced equation for the combination of iron with oxygen to produce rust.

Figure 5.9. Rust on iron.

Oxygen is a diatomic gas, so it appears as O_2 in the reactants.

$$Fe(s) + O_2(g) \rightarrow Fe_2O_3(s) \qquad \text{(formula equation)}$$

We have the odd-even problem with both Fe and O. Oxygen, with two on the left and three on the right, is more complicated, so we start there. To make the oxygen match, we need the least common multiple of two and three, which is six. To get six on each side, we need a coefficient of 3 on the left and 2 on the right. Note that the coefficient on the right goes in front of the entire compound formula.

$$Fe(s) + 3O_2(g) \rightarrow 2Fe_2O_3(s)$$

The oxygens are balanced now, but the iron is not. There are four iron atoms represented in the product. Since iron is by itself in the reactants, a simple coefficient on the iron finishes the balancing to give us

$$4Fe(s) + 3O_2(g) \rightarrow 2Fe_2O_3(s) \qquad \text{(balanced equation)}$$

We now double check the count of each atom to verify that the equation is balanced.

▲

▼ Example 5.3

Aluminum sulfate (Figure 5.10) is used in water treatment plants to remove impurities in the water. When combined in aqueous solution with calcium hydroxide, the reaction produces aluminum hydroxide and calcium sulfate, neither of which are soluble in water. The two solids that are formed attract particulate impurities in the water, and as they settle, the impurities are taken with them. Write the formula equation for this reaction and balance it.

Figure 5.10. Aluminum sulfate.

The formula equation is straightforward, given the oxidation states of aluminum and calcium (Figure 2.12) and the polyatomic ion list (Table 3.6).

$$Al_2(SO_4)_3(aq) + Ca(OH)_2(aq) \rightarrow Al(OH)_3(s) + CaSO_4(s) \qquad \text{(formula equation)}$$

There are two polyatomic ions that appear on both sides of this equation. We treat these as single objects during the balancing. The aluminum sulfate is as good a place to start as any. There are three sulfates on the left and one on the right, so we place a coefficient of 3 on the calcium sulfate.

$$Al_2(SO_4)_3(aq) + Ca(OH)_2(aq) \rightarrow Al(OH)_3(s) + 3CaSO_4(s)$$

Calcium is now out of balance, so we address that with a coefficient of 3 on the calcium hydroxide.

$$Al_2(SO_4)_3(aq) + 3Ca(OH)_2(aq) \rightarrow Al(OH)_3(s) + 3CaSO_4(s)$$

Now we look at the hydroxides. We have six on the left, so a coefficient of 2 on the aluminum hydroxide gives us six on the right.

$$Al_2(SO_4)_3(aq) + 3Ca(OH)_2(aq) \rightarrow 2Al(OH)_3(s) + 3CaSO_4(s) \qquad \text{(balanced equation)}$$

By adding the coefficient of 2 on the aluminum hydroxide we see that the aluminum becomes balanced at the same time as the hydroxide. We now double check the count of each atom and polyatomic ion to verify that the equation is balanced.

▲

▼ Example 5.4

On the opening page of this chapter, the first test tube contains a solution of white copper(I) iodide precipitate in an aqueous solution of iodine (shown again in Figure 5.11). These products are actually formed in a second reaction that is part of a two-step reaction sequence. The sequence begins with combining aqueous solutions of copper(II) nitrate and potassium iodide to produce an aqueous solution of potassium nitrate and solid copper(II) iodide. The copper(II) iodide immediately decomposes to form white copper(I) iodide precipitate and iodine in aqueous solution. Write the formula equations for each of these reactions and balance them.

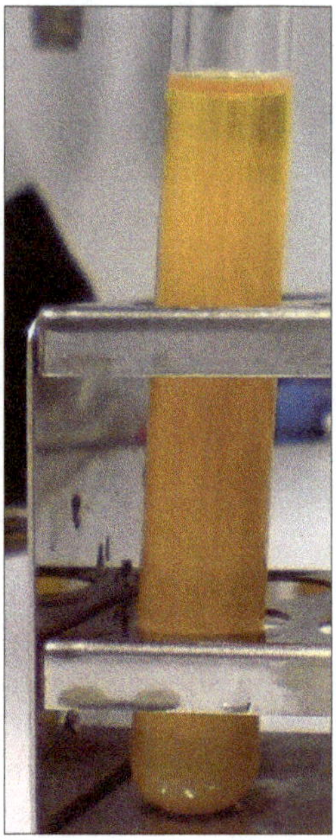

Figure 5.11. White copper(I) iodide in a yellow iodine solution.

The formula equation for the first reaction is as follows:

$$Cu(NO_3)_2(aq) + KI(aq) \rightarrow KNO_3(aq) + CuI_2(s) \qquad \text{(formula equation)}$$

We start with the nitrate. There are two on the left and one on the right, so we add a coefficient of 2 on the potassium nitrate.

$$Cu(NO_3)_2(aq) + KI(aq) \rightarrow 2KNO_3(aq) + CuI_2(s)$$

Now we see that the potassium and iodine are both single on the left and double on the right. Adding a coefficient of 2 on the potassium iodide takes care of both of them and gives us the balanced equation.

$$Cu(NO_3)_2(aq) + 2KI(aq) \rightarrow 2KNO_3(aq) + CuI_2(s) \qquad \text{(balanced equation)}$$

Be sure to double check the count of each atom and polyatomic ion to verify that the equation is balanced.

The formula equation for the second reaction is

$$CuI_2(s) \rightarrow CuI(s) + I_2(aq) \qquad \text{(formula equation)}$$

With three iodines on the right, the thing to do is double the single one so we have an even number on the right. This gives

$$CuI_2(s) \rightarrow 2CuI(s) + I_2(aq)$$

We now have four iodines on the right and two on the left. Placing a coefficient of 2 on the CuI_2 balances the iodine and the copper at the same time.

$$2CuI_2(s) \rightarrow 2CuI(s) + I_2(aq) \qquad \text{(balanced equation)}$$

Finally, double check the count of each atom and polyatomic ion to verify that the equation is balanced.

▲

5.1.5 Oxidation States

We introduced the concept of oxidation states back in Chapter 2 in the context of ionization. It is now time to develop this further so that you are able to determine the oxidation state of any element in any compound. When an atom, such as a metal, loses one or more electrons, we say it has been *oxidized*. When an atom gains one or more electrons we say it has been *reduced*.

The terms *oxidation* and *reduction* go back to the days of Antoine Lavoisier, just a few years before the French Revolution. At the time, it was thought that oxygen was the only element that caused oxidation, hence the name of the process. Lavoisier coined the term reduction, referring to the weight loss of metals during smelting. In this process, ore is converted to pure metal by the removal of oxides and other substances (giving the metal atoms their electrons back). The resulting pure metal weighs less than the ore, so it has been "reduced."

We will use sodium and chlorine to illustrate oxidation and reduction. If an atom of sodium is ionized and loses its one valence electron to a chlorine atom, it is *oxidized* to become Na^+ and its oxidation state is +1. Chlorine is the *oxidizer* or *oxidizing agent* in the process. When the chlorine atom receives that electron, the chlorine atom is reduced to become Cl^- and its oxidation state is −1. Sodium is the *reducer* or *reducing agent*. These two ions join together by electrostatic attraction to become part of the crystal lattice of NaCl.

If this chemistry is driven in the opposite direction (which can be done by running an electric current through molten sodium chloride), the Cl^- ion loses an electron to become Cl; the chlorine ion is now the one that is oxidized. The sodium ion receives that electron to become Na, so the sodium ion is reduced. This is what happens during smelting: the metal atoms in the ionic compounds in the rock get their electrons back, cease to be ions, and become elemental metal. A tried and true mnemonic to help you remember oxidation and reduction is shown in Figure 5.12.

The oxidation state of an element is indicated by an oxidation number such as +2, +1, 0, −2, −1, and so on. The oxidation state of any neutral unbonded atom is 0. As illustrated just above, in ionic compounds the oxidation state of an element is directly related to the number of electrons the element has gained or lost, and the oxidation number is simply the charge on the ion. This charge is usually obvious because of the element's position in the periodic table, or because it is indicated in the name of the compound. In magnesium chloride, $MgCl_2$, the oxidation number of magnesium is +2 because as an alkaline-earth element, magnesium ionizes by losing two electrons. The oxidation number of chlorine is −1 because chlorine, like all the halogens, ionizes by gaining one electron. In copper(I) iodide, CuI, the Roman numeral in the name indicates the oxidation state of the metal (+1).

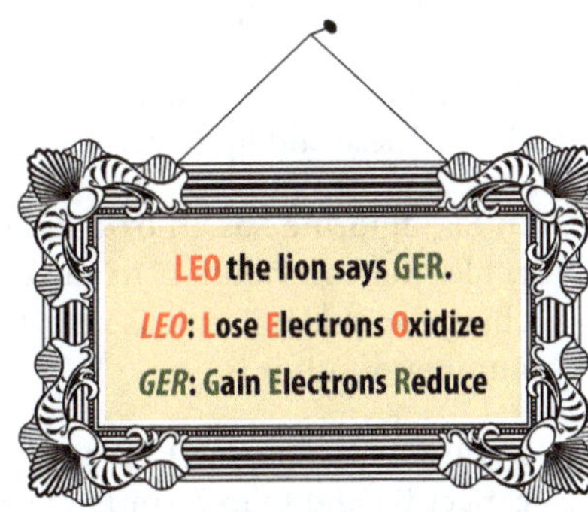

Figure 5.12. A classic memory aid for oxidation and reduction. Say it over and over to yourself. Say it now.

However, in covalent bonds where electrons are being shared, determining the oxidation state of an element is not as simple as keeping track of gaining or losing electrons. Instead, we use electronegativities and a list of rules of thumb to determine oxidation states. In such compounds, we think of the oxidation state as indicating *partial* gain or loss of electrons. Because of this, it is probably more helpful to you if you think of oxidation and reduction this way: *oxidation occurs any time the oxidation state of an element increases; reduction occurs any time the oxidation state of an element decreases*. This definition also aids your memory—you can associate the term *reduction* with a *decrease* in the oxidation state (and the oxidation number).

Here are the rules we follow to determine the oxidation state of elements in covalent compounds:

1. The operating rule is this: *in a molecular compound, the oxidation states of the elements involved must add up to the charge on the molecule*. In a neutral molecule, the oxidation numbers must add up to zero. In a polyatomic ion, they must add up to the charge on the ion.

2. In pure elements in their natural state, the oxidation number is 0. This also applies to diatomic molecules such as H_2 and the other elements that naturally make molecules with themselves, S_8 and P_4.
3. The most electronegative element in the molecule has the oxidation state it would have if it were an anion. Since fluorine has the highest electronegativity, its oxidation state is always −1. Oxygen has the second-greatest electronegativity, so it is usually assigned an oxidation state of −2. One exception is when oxygen forms peroxides as the molecule O_2^{2-}, in which case its oxidation state is −1. A common peroxide is hydrogen peroxide, H_2O_2. Another exception is when oxygen bonds with fluorine to form OF_2, in which case its oxidation state is +2 (since F is always −1).
4. The oxidation number for halogens is −1. The exception is when they bond with oxygen to form polyatomic ions with negative charge (oxyanions). In that case, they have positive oxidation states (except for fluorine).
5. When hydrogen forms a compound with an element that is more electronegative than itself (a nonmetal), its oxidation state is +1. In compounds with metals, the oxidation state of H is −1.

Example 5.5

Determine the oxidation states for each atom in the following compounds or ions: $MgCl_2$, SF_6, BaO_2, $HClO_3$, PO_4^{3-}.

$MgCl_2$ is an ionic compound. The oxidation numbers for Mg and Cl are thus +2 and −1, respectively. These add up to zero: $2 + 2(-1) = 0$.

In SF_6, fluorine has an oxidation number of −1. Since there are six fluorine atoms in the molecule, the contribution to the overall charge of the molecule is $6(-1) = -6$. Since the overall charge on the molecule is zero, the oxidation numbers must add up to zero. Thus, the oxidation number of sulfur is +6.

Barium peroxide, BaO_2, is a nontypical ionic compound—it is a peroxide. We would normally expect Ba and O to form BaO since their typical ionizations are Ba^{2+} and O^{2-}. But in the compound BaO_2, the oxidation state of the metal is the one that runs the show. The oxidation number for Ba is +2 so the oxidation number for O must be −1. Oxygen is not very happy in peroxides, so the compounds are unstable and like to go *boom*. See the accompanying box for more about that exciting subject.

Chlorine and oxygen are both more electronegative than hydrogen, so in $HClO_3$ the oxidation state of H is +1. The chlorate ion, ClO_3^-, is an oxyanion, so we expect the chlorine to have a positive oxidation state. The oxidation state of oxygen is −2 and there are three of them, so the charge contribution from oxygen is $3(-2) = -6$. Adding the hydrogen gives $-6 + 1 = -5$. For the oxidation numbers in the molecule to add to zero, the oxidation state of chlorine must be +5.

Oxygen is more electronegative than phosphorus, so its oxidation state in PO_4^{3-} is −2. The four oxygens make a charge contribution of $4(-2) = -8$. The oxidation numbers must add up to the charge on the ion, which is −3. Thus, the oxidation state of phosphorus is +5.

Hmm... Interesting. Why nitrates and nitros blow up

To a chemist, any substance that causes an element to oxidize is an oxidizing agent, or oxidizer. The oxygen in air causes iron to oxidize and form rust, Fe_2O_3, and is an oxidizing agent in this general chemical sense.

But a more narrow definition of an oxidizing agent is used in classifying substances as dangerous materials. In such classifications, an oxidizing agent is a substance that causes other substances to combust rapidly. In barium peroxide, BaO_2, the oxidation state of oxygen is −1, whereas oxygen's preferred oxidation state, the state in which its valence shell is full, is −2. This means the compound is unstable and the oxygen in BaO_2 aggressively reacts with other substances. BaO_2 is used in fireworks to produce an intense green color. Hydrogen peroxide is also an aggressive oxidizer and is sometimes used as a rocket fuel. It was also the oxidizer used in the 2005 London bombings that killed 52 people. But don't worry about the hydrogen peroxide solution you may have in the medicine chest at home. That's only a 3% aqueous solution of H_2O_2. (Researchers now say that while hydrogen peroxide solution is useful for disinfecting inanimate objects and removing blood stains, it is not the best for disinfecting wounds because it causes tissue damage and thus slows healing.)

Potassium nitrate, KNO_3, is the oxidizer used in gunpowder. Nitrate compounds in general are aggressive oxidizers. Ammonium nitrate, NH_4NO_3, a common fertilizer, was used as an explosive in the 1995 Oklahoma City bombing (which killed 168 people) and the 1993 World Trade Center bombing (which killed 6 people but injured several thousand). (The WTC bombing occurred before the WTC was destroyed by terrorists on September 11, 2001.) Ammonium nitrate is also the substance that exploded and destroyed a fertilizer plant in West, Texas in 2013, killing 15 people. In the 1947 Texas City disaster, a ship containing 2,300 tons of ammonium nitrate detonated, killing 581 people—the deadliest industrial accident in U.S. history.

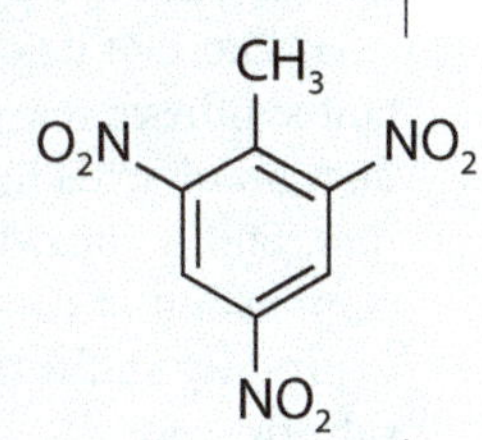

When it comes to explosive compounds, so-called *nitro* compounds are the most explosive of all. Nitro compounds include TNT, nitroglycerine (which can explode just by being jostled), and the stabilized form of nitroglycerine called dynamite. Nitro compounds contain carbon rings with nitronium ions (NO_2^+) attached, as illustrated by the Lewis structure for TNT, trinitrotoluene, shown above. Nitro compounds are not oxidizers, but are explosive for the same reason the nitrates are: because of the nitrogen and oxygen atoms they contain. When the nitrogen atoms are freed from the molecule, they form diatomic molecules of nitrogen gas, N_2. The oxygen atoms also form gases during reactions, typically CO_2, but also CO. The N_2 (and CO) molecules contain triple bonds, which means they are very strong and release a lot of energy while forming (see Table 3.9). The presence of a lot of heat while gases are forming means the gases expand *very* rapidly. Put all this together and you have huge explosions. The reason the nitro compounds explode more vigorously than the nitrate oxidizers is that the nitro compounds have multiple carbons (which get oxidized in the explosion, releasing additional heat), nitrogens, and oxygens *in the same molecule*. You can't get any closer together—thus faster—than that.

One more interesting tidbit: dynamite was invented by Alfred Nobel, and the fortune he made from it was used to create the Nobel Foundation, which awards the annual Nobel Prizes.

5.2 General Types of Chemical Reactions

5.2.1 Synthesis Reactions

In *synthesis reactions*, separate elements or compounds are combined to form a single, new compound. Synthesis reactions are described by the general equation form

$$A + B \rightarrow AB$$

The classic example of a synthesis reaction is the formation of water by the combustion of hydrogen:

$$2H_2(g) + O_2(g) \rightarrow 2H_2O(g)$$

Another example is the production of calcium hydroxide from calcium oxide:

$$CaO(s) + H_2O(l) \rightarrow Ca(OH)_2(s)$$

Figure 5.13. Calcium hydroxide.

Calcium hydroxide, shown in Figure 5.13, is known as "slaked lime" for the same reason people use the phrase "slake your thirst": both are accomplished by adding water. The production of $Ca(OH)_2$ from CaO is an important step in the curing of concrete.

Because of its environmental importance, an important synthesis reaction to know about involves the production of acids in the atmosphere, which leads to *acid rain*. Sulfur dioxide, SO_2, is a common product of the combustion of coal in electric power stations. In the atmosphere, SO_2 combines with moisture to produce sulfur trioxide, SO_3. This compound reacts easily with moisture in the air to produce sulfuric acid:

$$SO_3(g) + H_2O(l) \rightarrow H_2SO_4(aq)$$

Acidic rainfall was first detected back in the 17th century, but after the Industrial Revolution, atmospheric levels began to skyrocket. By the late 1960s, forests like the one in Figure 5.14 were being destroyed, as well as outdoor stone sculptures and architectural works made of marble or stone. New legislation to control emissions of SO_2 and other pollutants from power stations went into effect in the 1980s and 1990s, and as a result acid rain levels have dropped by 65% since 1976 (for a total cost much lower than originally predicted).

Figure 5.14. A forest destroyed by acid rain.

5.2.2 Decomposition Reactions

In synthesis reactions, compounds or elements are brought together; in *decomposition reactions*, compounds are taken apart. The general equation form is as follows:

$$AB \rightarrow A + B$$

The decomposition reaction you see happening most often is probably the decomposition of carbonic acid, H_2CO_3:

$$H_2CO_3(aq) \rightarrow H_2O(l) + CO_2(g)$$

Carbonic acid, an oxyacid, is added to soft drinks to make them fizzy when the can is opened. At atmospheric pressure, carbonic acid spontaneously decomposes to water and carbon dioxide. The CO_2 that *evolves* (i.e., is given off) during the reaction is the gas that makes the drink fizzy.

The heating of mercury oxide by Joseph Priestley, mentioned at the beginning of the chapter, is a decomposition reaction in which oxygen evolves:

$$2HgO(s) \xrightarrow{\Delta} 2Hg(l) + O_2(g)$$

A third example goes back to the calcium oxide synthesis reaction mentioned above. To obtain the CaO in the first place, limestone, which is calcium carbonate, $CaCO_3$, is roasted (that's what they call it) at high temperature, producing the following decomposition reaction:

$$CaCO_3(s) \xrightarrow{\Delta} CaO(s) + CO_2(g)$$

This is the starting point for the manufacture of "Ordinary Portland Cement," the most common cement in the world.

5.2.3 The Activity Series of Metals

To understand single replacement reactions (coming up next), you need to know about the *activity series of metals*, a list of metals in order of their *chemical activity*, that is, how aggressively they react relative to one another. The activity series is shown in Table 5.2. The metals in this list are listed from top to bottom in order of their chemical activity, with the most chemically aggressive metals at the top and the metals that react the least at the bottom. In aqueous solutions of compounds of these metals, any metal on the list replaces any metal below it in a solution. Figure 5.15 illustrates this with a solution of silver nitrate, $AgNO_3$, which is colorless. When a coil of copper wire is placed in the solution, the copper begins going into solution, which turns the solution blue-green. The silver comes out of solution, precipitating in fluffy tufts onto the copper wire.

Figure 5.15. In a solution of silver nitrate, the copper replaces the silver in solution.

The chemical activity of the metals was determined empirically, that is, through experiments. Using the information on the list, you can determine if certain reactions occur. Table 5.2 also indicates the reactivity of the metals with water, oxygen, and acids. Active metals like lithium and sodium react very vigorously in water. All metals except the most unreactive react with oxygen to

Activity of Metals	Activity with Water	Activity with Oxygen	Activity with Acids
Li Rb K Ba Sr Ca Na	reaction	reaction (to form oxides)	reaction (replacing hydrogen)
Mg Al Mn Zn Cr Fe Cd	reaction with steam, but not cold water	reaction	reaction
Co Ni Sn Pb	no reaction	reaction	reaction
Sb Bi Cu Ag	no reaction	reaction	no reaction
Hg Pt Au	no reaction	no reaction	no reaction

Table 5.2. The activity series of metals and other activity information.

form oxides. Most metals also react with acids.

▼ Example 5.6

What do you expect to happen if a steel nail is placed in a container of copper(II) chloride solution?

Steel is mostly iron and iron is above copper in the activity series of metals. When placed in the solution, ions of iron begin going into solution, and copper ions begin precipitating out of solution by attaching to the nail.

▲

The phenomenon described in the previous example is shown in the sequence of images in Figure 5.16. After about 30 minutes, the solution becomes colorless, indicating that all the copper ions have been reduced to copper atoms and have precipitated out of solution. The reaction is a single replacement reaction, our next topic.

5.2.4 Single Replacement Reactions

In *single replacement reactions*, a lone element replaces another in a

Figure 5.16. Iron displaces the copper from solution, and the nail is buried in copper.

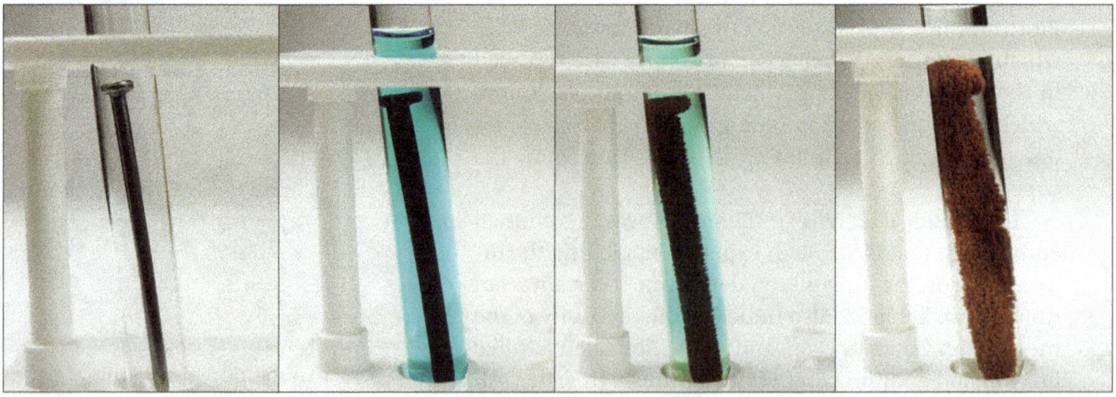

empty test tube with nail | $CuCl_2$ solution added | Cu precipitation underway | Cu precipitation complete

compound. Typically, this happens in aqueous solutions, and may be generically represented by the equation

$$A + BC \rightarrow B + AC$$

Three common single replacement reactions are: one metal replacing another in solution, as in the copper(II) chloride example above; a metal replacing some of the hydrogen atoms in water to form a hydroxide solution; and a metal replacing the hydrogen atoms in an acid to form a salt.

Iron replacing copper in solution in Figure 5.16 is a typical example of the first type. The equation for the reaction is as follows:

$$Fe(s) + CuCl_2(aq) \rightarrow Cu(s) + FeCl_2(aq)$$

In this single replacement reaction, the copper(II) chloride is in aqueous solution. As we explore further in Chapter 8, this means the copper and chlorine ions have *dissociated* in the water—the crystal comes apart and the individual ions are separated from each other. Iron is a more active metal than copper, so when the steel nail is placed in the solution, the Fe atoms are oxidized to become Fe^{2+} ions in solution. The Cu^{2+} ions in the solution are reduced to become Cu atoms, copper metal. The steel nail provides a landing place where atoms of solid Cu can attach to a crystal lattice as they precipitate.

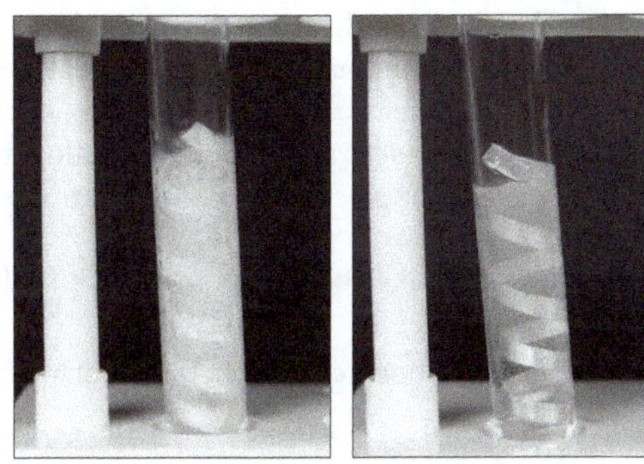

Figure 5.17. Magnesium replaces the hydrogen in the hydrochloric acid molecules, forming a salt and evolving hydrogen gas. The photo on the left is at the beginning of the reaction, when H_2 is rapidly evolving. The photo on the right was taken after about 5 minutes.

The reaction of sodium in water is a well-known example of the second type of single replacement reaction. The equation is:

$$2Na(s) + 2H_2O(l) \rightarrow 2NaOH(aq) + H_2(g)$$

Sodium is high in the activity series, and reacts very vigorously in water. The hydrogen gas evolves quickly and a lot of heat is released—so much, in fact, that the hydrogen easily combusts, causing an explosion.

The third common single replacement reaction is when a metal replaces the hydrogen in an acid. A typical example is the formation of the metal salt magnesium chloride when magnesium is placed in hydrochloric acid:

$$Mg(s) + 2HCl(aq) \rightarrow MgCl_2(aq) + H_2(g)$$

Figure 5.18. *Aqua regia*.

This reaction is shown in Figure 5.17. The bubbles you see are the evolving hydrogen gas. As you see from Table 5.2, every metal in the activity series from lithium down through lead reacts this way with acids. The last seven metals listed generally do not. However, both platinum and gold dissolve in a mixture of acids known from the medieval period as *aqua regia*, Latin for "royal

Hmm... Interesting. **A story about aqua regia**

After 1935, Adolph Hitler prohibited Germans from accepting Nobel Prizes. German physicists Max von Laue and James Franck sent their Nobel Prize gold medals to Niels Bohr at the Bohr Institute in Denmark for safe keeping. In 1940, Germany invaded Denmark. Fearing that the Nobel Prize medals might be confiscated by the Nazis, Bohr directed Hungarian chemist George de Hevesy to dissolve the medals in *aqua regia*, and place the flask of red liquid with the gold in it on a shelf in the lab. The Nazis never knew what was in that flask and after the war it was still sitting there. The Nobel Foundation took back the gold and recast it into two Nobel Prize medals.

water." This extremely corrosive substance, shown in Figure 5.18, is composed of one part nitric acid and two or three parts hydrochloric acid.

Here is one final type of single replacement reaction: the replacement of one halogen by another. The chemical activity of the halogens is highest with fluorine and decreases down Group 17 in the periodic table. Thus, any halogen replaces any of the ones below it. This is illustrated by fluorine gas replacing the chlorine atoms in sodium chloride:

$$F_2(g) + 2NaCl(aq) \rightarrow 2NaF(aq) + Cl_2(g)$$

When the fluorine replaces the chlorine in solution, chlorine gas evolves.

5.2.5 Double Replacement Reactions

Double replacement reactions may be represented by the following generic equation:

$$AB + CD \rightarrow AD + CB$$

In this type of reaction, the four ions in two compounds all "switch partners." Notice that in the product ionic compounds in this generic equation, the cations A and C are still written first.

Double replacement reactions often result in the formation of a precipitate or the evolution of a gas when product compounds are insoluble in water. One example is the reaction discussed in Example 5.4. Another, also involving copper(II) nitrate, is the following reaction:

$$Cu(NO_3)_2(aq) + H_2S(aq) \rightarrow CuS(s) + 2HNO_3(aq)$$

The copper monosulfide[1] product is a black powder that is insoluble in water, so it forms a precipitate that turns the solution black, as shown in the third test tube from the left in the image on the opening page of this chapter, and in Figure 5.19.

The evolution of a gas in a double replacement reaction is illustrated by the reaction of iron(II) sulfide and hydrochloric acid:

$$FeS(s) + 2HCl(aq) \rightarrow H_2S(g) + FeCl_2(aq)$$

1 There are quite a few different copper sulfide compounds. To distinguish CuS from the others, the prefix *mono-* is added to the sulfide in the naming of the compound.

Notice that here, just as in the Mg and HCl single replacement reaction, a metal replaces hydrogen in an acid. In the single replacement reaction, a sample of pure metal is placed in the acid solution. Here, an ionic compound of the metal is placed in the acid solution.

Figure 5.19. The black precipitate copper monosulfide forms when copper nitrate combines with hydrogen sulfide.

5.2.6 Combustion Reactions

Combustion reactions involve the reaction of a fuel with oxygen to produce carbon dioxide and water. Combustion reactions are not actually a separate class of reactions; they are oxidation-reduction reactions (discussed below), and some are also synthesis reactions. But they are so common and important that they warrant their own section here.

The natural gas commonly burned in homes for heating and cooking is primarily composed of methane, CH_4. The methane combustion reaction is as follows:

$$CH_4(g) + 2O_2(g) \rightarrow CO_2(g) + 2H_2O(g)$$

The combustion of most fuels, such as methane, propane, and all other hydrocarbon fuels, produces carbon dioxide and water. One important exception is the combustion of hydrogen, a synthesis reaction:

$$2H_2(g) + O_2(g) \rightarrow 2H_2O(g)$$

Burning hydrogen produces nothing at all but pure water vapor, and this makes hydrogen the ultimate clean fuel. If we could develop efficient and safe ways to prepare, transport, and burn hydrogen, pollution and CO_2 emissions from cars could be eliminated (although the air in cities would be somewhat more humid).

5.2.7 Acid-Base Neutralization Reactions

In Chapter 9, we explore what acids and bases are and look with more detail into acid-base chemistry. However, *acid-base neutralization reactions* are so important that they warrant a brief mention here. Careful definitions for the terms *acid* and *base* must wait for Chapter 9. However, we can say roughly that acids are recognizable as compounds with hydrogen and a halogen, or hydrogen and one of the polyatomic ions. Examples are HCl and H_2SO_4. Equally roughly, bases are recognizable as compounds composed of a metal and hydroxide such as NaOH and $Ca(OH)_2$.

Though it may be tricky to define acids and bases, it is not tricky to define what you get when an acid reacts with a base: you get a *salt*, and often nothing else but water. In fact, that is the definition for a salt—the product of an acid-base reaction. Acid-base chemistry typically takes place in aqueous solution, and is so indicated in the chemical equations. The reaction between an acid and a base is referred to as *neutralization*. Both acids and bases are corrosive. But when an acid and a base are combined in chemically equivalent ratios, their corrosive properties are neutralized as they are converted into a salt and water.

A good example of an acid-base reaction involves the acid in the human stomach, gastric acid, HCl. Gastric acid plays a key role in food digestion by causing the large protein molecules in the foods we eat to unravel so that the digestive enzymes in the stomach can break down the molecules in the food. However, an unbalanced diet sometimes leads to excess gastric acid, a condition known as *acid indigestion* or *heartburn*. For minor symptoms, over-the-counter antacids can be taken to neutralize excess gastric acid. Rolaids, a popular antacid, contains magnesium hydroxide. The gastric acid neutralization reaction from taking Rolaids is as follows:

$$2HCl(aq) + Mg(OH)_2(aq) \rightarrow MgCl_2(aq) + 2H_2O(l)$$

The magnesium chloride salt formed from this reaction is composed of the cation from the base and the anion from the acid. This is always the case in acid-base neutralization reactions.

Tums, another popular antacid, contains calcium carbonate, which, you may recall, is the same compound that composes chalk and limestone. Even though $CaCO_3$ does not contain the hydroxide ion, it still acts as a base to neutralize the HCl in the stomach. The gastric acid neutralization reaction from taking Tums is as follows:

$$2HCl(aq) + CaCO_3(aq) \rightarrow CaCl_2(aq) + H_2O(l) + CO_2(g)$$

The carbonate base in this reaction results in the production of CO_2.

5.2.8 Oxidation-Reduction Reactions

As with acid-base reactions, oxidation-reduction reactions, or *redox* reactions as they are fondly known, are so important that we treat them separately in a later chapter. But since we are talking about different types of chemical reactions here, we need to include a brief look at redox reactions.

Any time an element's oxidation state increases, oxidation occurs. And any time oxidation occurs, reduction occurs as well. These two sentences apply in reverse as well. If an element's oxidation state decreases, reduction occurs, and oxidation also occurs at the same time.

Combustion reactions are always redox reactions because oxidation always occurs. (The oxidation is caused by the oxidizer, or oxidizing agent, which we discussed a few pages back.)For example, here is the reaction for the combustion of propane, C_3H_8:

$$C_3H_8(g) + 5O_2(g) \rightarrow 3CO_2(g) + 4H_2O(g)$$

Let's look at the oxidation states of the carbon in propane and carbon dioxide. The electronegativity of carbon is higher than that of hydrogen, so hydrogen's oxidation state is +1 in propane. There are eight hydrogen atoms in propane, giving a total charge contribution from hydrogen of 8(+1) = +8. The oxidation numbers must all add up to zero, which means carbon's oxidation state in propane, x, must solve the equation: $3x + 8 = 0$, giving an oxidation state of –8/3.

In carbon dioxide, the oxidation state of oxygen is –2, so the oxidation state for carbon must be +4. Thus, the oxidation state of carbon increases from –8/3 to +4. The carbon is oxidized.

The oxidizing agent in this reaction is O_2. In redox reactions, the oxidizing agent is the compound that causes the oxidation of another compound in the reaction. The result for the oxidizer itself is that it is reduced. We can easily see this in the chemical equation because as O_2, the oxidation state of oxygen is 0, but its oxidation state in both CO_2 and H_2O is –2. The oxygen is reduced.

Note that although the carbon is the element oxidized in this reaction, it is common parlance to say that the *propane* is oxidized. I know this sounds confusing, but it is just the way everyone talks, so students must simply get used to it. At first, this kind of language will drive you crazy because you will be wondering about specifically which element is the one that is oxidized in the reaction. To preserve your sanity, just sit down and figure it out like we did above with the carbon in propane. It takes a lot of practice for one's intuitions about oxidation and reduction to become automatic. Just stay with it.

Tossing a metal into an acid solution also produces a redox reaction. Let's look again at the Mg and HCl reaction from a few pages back:

$$Mg(s) + 2HCl(aq) \rightarrow MgCl_2(aq) + H_2(g)$$

Here the pure metal Mg is oxidized to Mg^{2+} to form the ionic compound $MgCl_2$. The oxidizing agent is the hydrochloric acid, but here we should note that more specifically it is the hydrogen in the HCl that is reduced. In HCl, hydrogen's oxidation state is +1; in H_2 it is 0. Thus, the hydrogen is reduced.

The number of different redox reactions is beyond telling. We look at a few more of them in the redox chapter.

5.3 Stoichiometry

Everyone who has ever taken a chemistry class remembers performing stoichiometric calculations. The term *stoichiometry* comes from the Greek words *stoicheion*, meaning "element," and *metron*, meaning "measure." When we do stoichiometry, we are measuring the elements. There are many different basic types of calculations we perform in chemistry, but stoichiometric calculations might just be the most basic of the basic. And here's the good news: even though this section has a long, strange foreign name, the calculations themselves are *easy*.

You really have to know only four things to do stoichiometric calculations. First, you need to be able to set up and solve a proportion, a basic math skill you no doubt learned how to do years ago in Prealgebra. Second, you need to be able to compute the molar mass of a compound. This topic is covered in Chapter 1 and should not be difficult at this point. Third, you need to be able to convert from moles to grams and vice versa. This, too, is covered in Chapter 1 and is not difficult. Finally, you need to be able to set up and balance a chemical equation, which we have just covered and which you have (no doubt) just mastered.

So now, with your confidence meter reading "high," let us proceed.

5.3.1 Stoichiometric Calculations

Stoichiometry is all about calculating the quantities of compounds involved in chemical reactions. For example, just look back at previous page at the reaction of C_3H_8 and O_2. If we have, say, 10.0 kg of propane, how much oxygen is required to burn it all? How much CO_2 and water are formed in the process? These calculations are standard stoichiometry.

Let's begin with RULE NUMBER 1.

RULE NUMBER 1
Perform stoichiometric calculations in moles.

If the problem asks for masses in grams, or supplies you with masses in grams, that's fine. If your given quantities are masses in grams, you simply convert given masses into number of moles, as you already know how to do. If you are required to state your answer as a mass in grams, you just convert to grams at the end. But the stoichiometric calculation is performed in moles.

Now for RULE NUMBER 2.

RULE NUMBER 2
The mole ratios for performing the stoichiometric calculations come from the coefficients in the balanced chemical equation.

That's right. That's why this is easy. Remember that while we are balancing a chemical equation, we think of the coefficients as helping us to figure *numbers of atoms*. But remember also that earlier in this chapter, in Section 5.1.4 on balancing chemical equations, I write that chemical reactions happen with bulk quantities of atoms, and in chemistry our favorite bulk quantity is moles. So think of the coefficients in the balanced equation as telling us the *numbers of moles* of compounds or elements that participate in the reaction.

Now, thinking in moles, look again at that propane equation on the previous page. The coefficients in the equation say this: *one* mole of propane reacts with *five* moles of oxygen to produce *three* moles of carbon dioxide and *four* moles of water. That's how you do it. The rest is just the details, which I demonstrate below in a few examples.

Before we start in on the examples, let's briefly consider how to use *mole ratios* to perform calculations when the quantities you are given do not match the coefficients. (They never do.) Looking again at the propane equation, the equation says that for every five moles of oxygen involved, four moles of water form. If you are given a quantity of oxygen and asked to determine how much water forms, you simply use the 5 : 4 ratio as a conversion factor to figure it out. We can write this conversion factor two ways:

$$\frac{5 \text{ mol } O_2}{4 \text{ mol } H_2O} \quad \text{or} \quad \frac{4 \text{ mol } H_2O}{5 \text{ mol } O_2}$$

Now here's how you use one of these ratios to solve a problem. Suppose the problem is to determine how many moles of water are produced by the reaction of 13.55 mol O_2 with propane, assuming an unlimited supply of propane. Just take your given quantity and multiply it by the mole ratio that appropriately cancels with the given quantity to give you the quantity you need:

$$13.55 \text{ mol } O_2 \cdot \frac{4 \text{ mol } H_2O}{5 \text{ mol } O_2} = \frac{13.55 \cdot 4}{5} \text{ mol } H_2O = 10.84 \text{ mol } H_2O$$

Like I said, easy. Now two more comments before we proceed. First, it should be pretty obvious to you that I wrote the mole ratio in that computation as I do because I am given moles of oxygen and I want that to cancel to give me moles of water. That's why the oxygen is on the bottom in the conversion factor. If I am given moles of water and asked to find moles of oxygen, I flip the conversion factor the other way. Second, the coefficients in the chemical equation are *exact*; they are not approximations. When the equation says 5 mol O_2 react to produce 4 mol H_2O, these figures are exact. Accordingly, they play no part in determining the significant digits to use in your results. In the computation above, the given quantity of oxygen has four significant digits. That's why the result is written with four significant digits.

Now we are ready to dive into some examples. The following examples should illustrate everything you need to know. Keep Rule Number 1 in mind: perform the computations in moles. If you are given a quantity as a mass, use the compound's molar mass to convert it to number of moles for the calculation. If you are asked for a result to be stated as a mass, use the molar mass to convert it at the end after you determine the result in moles.

Example 5.7

In submarines and spacecraft, the breathing air has to be "scrubbed" of excess carbon dioxide to remove the carbon dioxide continuously being exhaled by those on board. One way to purify the breathing air of excess CO_2 is with the compound lithium hydroxide, shown in Figure 5.20. Solid lithium hydroxide reacts with carbon dioxide to produce solid lithium carbonate and water. If the average amount of carbon dioxide exhaled each day is 880 g/person, determine the number of moles of lithium hydroxide consumed per person per day in the reaction.

First, the given quantity is in grams, so we convert this to number of moles. Referring to the periodic table, we determine the molar mass of CO_2, which is 44.010 g/mol. Converting the given quantity to moles, we have

Figure 5.20. Lithium hydroxide.

$$880\text{ g } CO_2 \cdot \frac{1\text{ mol}}{44.010\text{ g}} = 20.0\text{ mol } CO_2$$

We know our final answer must be stated with two significant digits because of the given quantity. I have kept an extra significant digit in the gram/mole conversion because it is an intermediate result. Final rounding is performed at the end.

Next, write the formula equation:

$$LiOH(s) + CO_2(g) \rightarrow Li_2CO_3(s) + H_2O(l)$$

Then, balance the equation.

$$2LiOH(s) + CO_2(g) \rightarrow Li_2CO_3(s) + H_2O(l)$$

Now we are ready for the stoichiometry. In the problem, we are given a quantity of CO_2 and asked to find the corresponding quantity of LiOH in the reaction. From the coefficients in the balanced equation, we see that the mole ratio of LiOH to CO_2 in the reaction is 2 : 1. So we use this mole ratio as a conversion factor to convert the mole quantity of CO_2 into a corresponding mole quantity of LiOH.

$$20.0\text{ mol } CO_2 \cdot \frac{2\text{ mol LiOH}}{1\text{ mol } CO_2} = 40.0\text{ mol LiOH}$$

This is our final result, but it is stated with three significant digits. However, we require two significant digits in our result, and the only way to round 40.0 down to two significant digits is to write it in scientific notation. Thus, our answer is

4.0×10^1 mol LiOH

▲

▼ Example 5.8

One of the body's metabolic processes—from which we get our fuel—is the oxidation of solid glucose by reaction with the oxygen we get from breathing.[2] Glucose, $C_6H_{12}O_6$, is one of the fundamental carbohydrates we take in from eating starch foods such as potatoes and rice. The products of the oxidation reaction are carbon dioxide, which we exhale, and water. (We eliminate more water than we take in; the extra water comes from the oxidation of glucose.) Determine the mass of water produced by the oxidation of 2.00 g glucose.

---

2 Here is another instance of using the term *oxidation* in reference to an entire molecule. If you look at the oxidation states, you see that it is the carbon in the glucose that is being oxidized.

Again, we are given a quantity in grams, so we convert this to moles right at the beginning. Using data from the periodic table, the molar mass of glucose is found to be 180.157 g/mol. Converting the given mass of glucose into moles, we have

$$2.00 \text{ g } C_6H_{12}O_6 \cdot \frac{1 \text{ mol } C_6H_{12}O_6}{180.157 \text{ g } C_6H_{12}O_6} = 0.01110 \text{ mol } C_6H_{12}O_6$$

Again, I have an extra significant digit in this intermediate calculation to avoid the build up of rounding error. I round to the required three significant digits at the end. Next, we write the formula equation for the oxidation of glucose:

$$C_6H_{12}O_6(s) + O_2(g) \rightarrow CO_2(g) + H_2O(l)$$

Now we balance the equation:

$$C_6H_{12}O_6(s) + 6O_2(g) \rightarrow 6CO_2(g) + 6H_2O(l)$$

This problem gives us a quantity of glucose and asks about a quantity of water, so these are the two compounds to look at in the equation. From the coefficients in the balanced equation, we see that the mole ratio of glucose to water in this reaction is 1 : 6. Now we write down the given quantity of glucose (in moles) and use the mole ratio to convert this into a mole quantity of water.

$$0.01110 \text{ mol } C_6H_{12}O_6 \cdot \frac{6 \text{ mol } H_2O}{1 \text{ mol } C_6H_{12}O_6} = 0.06660 \text{ mol } H_2O$$

Now that we have the amount of water produced, we simply need to use the molar mass of water to convert this into grams as the problem requires. From the periodic table, we determine that the molar mass of water is 18.02 g/mol. Using this value as a conversion factor, we determine the mass of water produced by the reaction:

$$0.06660 \text{ mol } H_2O \cdot \frac{18.02 \text{ g } H_2O}{1 \text{ mol } H_2O} = 0.06660 \cdot 18.02 \text{ g } H_2O = 1.20 \text{ g } H_2O$$

Notice, finally, that this result is rounded to three significant digits as the problem requires.

▲

It should be apparent from these two examples that a stoichiometric calculation can be performed using any or all of the products or reactants in a reaction. It doesn't matter which ones are asked about in the problem; you simply set up mole ratios using two compounds (or elements) at a time. We use this principle in our final example.

▼ Example 5.9

The photosynthesis reaction in plants is identical to the oxidation of glucose running in reverse. In the plant, CO_2 from the air and water from the soil are converted (using sunlight as an energy source, of course) into glucose, which becomes fuel and building material for the plant, and oxygen, which is released into the atmosphere. For each 5.00 g H_2O consumed by a plant, determine the masses of the CO_2 consumed, the O_2 released, and the glucose produced.

Since this reaction is identical to the one in the previous problem, only running in reverse, we can write down the balanced equation immediately:

$$6CO_2(g) + 6H_2O(l) \rightarrow C_6H_{12}O_6(s) + 6O_2(g)$$

In this example, we are required to calculate everything, so we may as well calculate all four molar masses right now. Since the mass given in the problem has three significant digits, we write down each of the molar masses with four significant digits.

CO_2:	44.01 g/mol
H_2O:	18.02 g/mol
$C_6H_{12}O_6$:	180.2 g/mol
O_2:	32.00 g/mol

Next, let's convert the given quantity from grams to moles.

$$5.00 \text{ g } H_2O \cdot \frac{1 \text{ mol } H_2O}{18.02 \text{ g } H_2O} = 0.2775 \text{ mol } H_2O$$

(This value has the extra significant digit required for an intermediate calculation.) We are given an amount of H_2O in the problem, and from this quantity we must calculate quantities of three other substances. So we need mole ratios for water and the three other substances. From the coefficients in the balanced equation, these ratios are

$H_2O : CO_2$	6 : 6, which is equal to 1 : 1
$H_2O : C_6H_{12}O_6$	6 : 1
$H_2O : O_2$	6 : 6, which is equal to 1 : 1

Now we just use the given amount of water, in moles, with one of these ratios to compute the quantities of the other three substances.

$$0.2775 \text{ mol } H_2O \cdot \frac{1 \text{ mol } CO_2}{1 \text{ mol } H_2O} = 0.2775 \text{ mol } CO_2$$

$$0.2775 \text{ mol } H_2O \cdot \frac{1 \text{ mol } C_6H_{12}O_6}{6 \text{ mol } H_2O} = 0.04625 \text{ mol } C_6H_{12}O_6$$

$$0.2775 \text{ mol } H_2O \cdot \frac{1 \text{ mol } O_2}{1 \text{ mol } H_2O} = 0.2775 \text{ mol } O_2$$

The last step is to use the molar masses to convert each of these mole quantities into grams.

$$0.2775 \text{ mol } CO_2 \cdot \frac{44.01 \text{ g } CO_2}{1 \text{ mol } CO_2} = 12.2 \text{ g } CO_2$$

$$0.04625 \text{ mol } C_6H_{12}O_6 \cdot \frac{180.2 \text{ g } C_6H_{12}O_6}{1 \text{ mol } C_6H_{12}O_6} = 8.33 \text{ g } C_6H_{12}O_6$$

$$0.2775 \text{ mol } O_2 \cdot \frac{32.00 \text{ g } O_2}{1 \text{ mol } O_2} = 8.88 \text{ g } O_2$$

These quantities are all rounded to three significant digits as required. We have completed the problem, and there are a lot of calculations involved. A great way to perform an overall error check on the calculations is to verify that the law of conservation of mass in chemical reactions is satisfied. From the equation, the masses of the water and carbon dioxide must add up to equal the masses of the glucose and oxygen.

mass CO_2 + mass H_2O = 12.2 g + 5.00 g = 17.2 g

mass $C_6H_{12}O_6$ + mass O_2 = 8.33 g + 8.88 g = 17.21 g

Rounding that second value to three significant digits, we see that the reaction consumes 17.2 g of reactants and produces 17.2 g of products. Mass conservation is verified, so it is highly likely that our calculations are correct.

▲

5.3.2 Limiting Reactant

Let's say you have a portable propane burner and a small canister of propane to fuel it. We know that the combustion of propane is a chemical reaction with oxygen that produces carbon dioxide and water. In the combustion of the propane in this canister, what is the limiting factor that determines how much water and carbon dioxide are produced? By *limiting factor* what I really mean with this question is *limiting reactant*—the reactant that runs out first, once the reaction starts.

Clearly, the oxygen available for this reaction, which is in the atmosphere, is available without limit. Thus, the limiting reactant is the propane, and when the propane is completely consumed, the reaction ceases. Every chemical reaction has a limiting reactant. One of the reactants is in shorter supply than the others and runs out first. When it does, the reaction ceases. The quantities of products produced by the reaction are determined by the quantity of the limiting reactant available for the reaction.

The limiting reactant is also often called the *limiting reagent* (pronounced ree-A-gent), a reagent simply being one of the compounds consumed in the reaction.

If you have known quantities available for a chemical reaction, you can determine which reactant is the limiting reactant simply by looking at the mole quantities available and the mole ratios from the balanced chemical equation. Once you know which compound is the limiting reactant, you can calculate the quantities of the products produced by the reaction based on the quantity of the limiting reactant supplied. As with the stoichiometric calculations we just reviewed, all the calculations associated with determining the limiting reactant are performed in moles. When you are given mass quantities, begin your solution by converting masses into numbers of moles.

Example 5.10

One of the most important industrial chemical processes in the world is the Haber-Bosch process for producing ammonia from nitrogen in the air. Grains need nitrogen as a nutrient, but cannot get it from the air because the nitrogen molecules are held together by the strong nitrogen triple bond. Bacteria in the soil produce the "fixed" nitrogen that plants use, but bacteria cannot supply the large quantities of nitrogen needed for industrial scale agriculture. The nitrogen-based fertilizers used are made from ammonia. Some have estimated that one third of the world's population is sustained by food grown using fertilizer produced by ammonia made with the Haber-Bosch process.

Figure 5.21. This high-pressure reactor from 1921 is on display at the Karlsruhe Institute of Technology in Germany.

The process occurs at extremely high pressures—around 200 atmospheres, or 3,000 psi—in reactors like the historical 1921 reactor shown in Figure 5.21. For years chemists thought the reaction was impossible, but German chemist Fritz Haber solved the problem in 1909, and with engineering assistance from Carl Bosch, had the commercial production of ammonia up and running by the end of 1910. The reaction is as follows:

$$N_2(g) + 3H_2(g) \rightarrow 2NH_3(g)$$

Consider a test run of this process, with 652 kg N_2 and 175 kg H_2 available for the reaction. Determine whether the N_2 or the H_2 is the limiting reactant. Then assuming that the limiting reactant is completely consumed in the reaction, calculate the mass of ammonia produced by the time the reaction ceases.

We begin by converting our mass quantities from kilograms to grams, and from grams to moles. We use the periodic table to determine the molar masses of H_2 and N_2, and then we use these as conversion factors to get mole quantities:

$$175{,}000 \text{ g } H_2 \cdot \frac{1 \text{ mol } H_2}{2.016 \text{ g } H_2} = 86{,}810 \text{ mol } H_2$$

$$652{,}000 \text{ g } N_2 \cdot \frac{1 \text{ mol } N_2}{28.01 \text{ g } N_2} = 23{,}280 \text{ mol } N_2$$

These mole quantities are each written with four significant digits, one more than required in the final result. Looking now at the balanced equation for the reaction, we see that the ratio of hydrogen to nitrogen required is 3 : 1. In order to consume the 23,280 moles of nitrogen we have available, we need 3(23,280) = 69,840 moles of hydrogen. We have a lot more hydrogen than this available, so nitrogen is the limiting reagent.

Now we perform a standard stoichiometric calculation to determine the amount of ammonia produced from 23,280 mol N_2. The mole ratio of N_2 to NH_3 is 1 : 2, so we use this ratio as a conversion factor to determine the amount of NH_3 produced:

$$23{,}280 \text{ mol } N_2 \cdot \frac{2 \text{ mol } NH_3}{1 \text{ mol } N_2} = 46{,}560 \text{ mol } NH_3$$

Finally, we convert this mole quantity to mass as the problem requires. This requires calculating the molar mass of NH_3 to use as a conversion factor.

$$46{,}560 \text{ mol NH}_3 \cdot \frac{17.03 \text{ g NH}_3}{1 \text{ mol NH}_3} = 792{,}900 \text{ g NH}_3 \cdot \frac{1 \text{ kg}}{1000 \text{ g}} = 793 \text{ kg NH}_3$$

This final result is rounded to three significant digits as the given information requires.

▲

▼ Example 5.11

A 5.00-g strip of magnesium metal is placed in an aqueous solution containing 15.0 g $ZnCl_2$, causing the following reaction:

$$\text{Mg}(s) + \text{ZnCl}_2(aq) \rightarrow \text{Zn}(s) + \text{MgCl}_2(aq)$$

Determine (a) the limiting reactant, (b) the mass of $MgCl_2$ produced, and (c) the mass of the excess reactant that remains.

Converting the given quantities to moles, we have:

$$5.00 \text{ g Mg} \cdot \frac{1 \text{ mol Mg}}{24.305 \text{ g Mg}} = 0.2057 \text{ mol Mg}$$

$$15.0 \text{ g ZnCl}_2 \cdot \frac{1 \text{ mol ZnCl}_2}{136.3 \text{ g ZnCl}_2} = 0.1101 \text{ mol ZnCl}_2$$

The mole ratio for Mg and $ZnCl_2$ in the balanced equation is 1 : 1, so the $ZnCl_2$ is the limiting reactant.

The mole ratio of $ZnCl_2$ and $MgCl_2$ is also 1 : 1, so 0.1101 mol $ZnCl_2$ produces 0.1101 mol $MgCl_2$, giving a mass of

$$0.1101 \text{ mol MgCl}_2 \cdot \frac{95.21 \text{ g MgCl}_2}{1 \text{ mol MgCl}_2} = 10.5 \text{ g MgCl}_2$$

Since the mole ratio for Mg and $ZnCl_2$ is 1 : 1, 0.1101 mol $ZnCl_2$ consume 0.1101 mol Mg. The reaction began with 0.2057 mol Mg, so 0.2057 – 0.1101 mol Mg = 0.0956 mol Mg are left over. This converts to a mass of

$$0.0956 \text{ mol Mg} \cdot \frac{24.305 \text{ g Mg}}{1 \text{ mol Mg}} = 2.32 \text{ g Mg}$$

▲

5.3.3 Theoretical Yield and Percent Yield

In the example of the Haber-Bosch process to produce ammonia, the 793 kg of ammonia we calculated that are produced by the reaction is called the *theoretical yield*. Let's now assume we perform the reaction using the quantities of reactants listed in that example and find that we actually end up with 742 kg NH_3 when all the N_2 is consumed and the reaction ceases. This quantity is called the *actual yield*. We use this quantity to calculate the *percent yield* as follows:

$$\text{percent yield} = \frac{\text{actual yield}}{\text{theoretical yield}} \times 100\%$$

Accordingly, the percent yield associated with an actual yield of 742 kg NH_3 is:

$$\text{percent yield} = \frac{742 \text{ kg } NH_3}{793 \text{ kg } NH_3} \times 100\% = 93.6\%$$

When chemists develop procedures to synthesize compounds, the percent yield is a measure of the effectiveness of the procedure. Yields above 80% are considered very good.

Chapter 5 Exercises

SECTION 5.1

1. Butane, C_4H_{10}, is the fuel used in disposable lighters. Butane burns in oxygen to produce carbon dioxide and water. Write the formula equation for this reaction and balance it.
2. Iron(II) sulfide reacts with hydrochloric acid to produce hydrogen sulfide and iron(II) chloride. Write the formula equation for this reaction and balance it.
3. Acetylene, C_2H_2, is a gas that produces an extremely hot flame when combusting with oxygen, so hot that cutting torches use this gas to cut through steel. The products of the combustion are carbon dioxide and water. Write the formula equation for this reaction and balance it.
4. With the aid of a catalyst, liquid methanol, CH_3OH, is produced from a reaction of carbon monoxide gas and hydrogen gas. Write the formula equation for this reaction and balance it.
5. Potassium carbonate is used to neutralize the sulfuric acid in well water caused by acid rain. The products of the neutralization reaction are potassium sulfate, carbon dioxide, and water. Write the formula equation for this reaction and balance it.
6. The tungsten metal used as filaments in light bulbs is produced by reacting tungsten oxide with hydrogen gas to produce tungsten metal and water. Write the formula equation for this reaction and balance it.
7. Sodium azide, NaN_3, is the original compound formerly used in automobile air bags. Other compounds were used to initiate the reaction, but the bags were filled by the conversion of solid sodium azide into sodium metal and nitrogen gas. Write the formula equation for this reaction and balance it.
8. Identify the oxidation state for each element in the following compounds, elements, and ions:

a. PBr_3	b. As_2O_5	c. ClO_4^-	d. F_2
e. UF_6	f. CO_3^{2-}	g. H_3PO_4	h. $Zn(NO_3)_2$
i. C_4H_{10}	j. $Cr_2O_7^{2-}$	k. KH	l. Fe

9. Why must chemical equations be balanced?

SECTION 5.2

10. For the reactions in exercises 1–7 above, identify the reactions that can be classified as synthesis, decomposition, single replacement, or double replacement.

11. In each of the following reactions, determine which element is oxidized and which is reduced. Also, where possible, identify the reaction as synthesis, decomposition, single replacement, or double replacement.

    a. $3Fe(NO_3)_2(aq) + 2Al(s) \rightarrow 3Fe(s) + 2Al(NO_3)_3(aq)$

    b. $PbS(s) + 4H_2O_2(aq) \rightarrow PbSO_4(s) + 4H_2O(l)$

    c. $N_2(g) + 3H_2(g) \rightarrow 2NH_3(g)$

    d. $Cl_2(aq) + 2NaI(aq) \rightarrow I_2(aq) + 2NaCl(aq)$

12. Explain what a redox reaction is.

13. For each of the following reactions, write the formula equation, balance it, and identify the reaction by type.

    a. magnesium hydroxide yields magnesium oxide and water
    b. sodium chloride and sulfuric acid yield sodium sulfate and hydrogen chloride gas
    c. calcium oxide and water yield calcium hydroxide
    d. ammonium bicarbonate and sodium chloride yield sodium bicarbonate and ammonium chloride

14. Use the activity series of metals to determine which of the following reactions occur. For those that do, complete the equation and balance it.

    a. $Ca(s) + O_2(g) \rightarrow$

    b. $Fe(s) + ZnCl_2(aq) \rightarrow$

    c. $Pt(s) + CuSO_4(aq) \rightarrow$

    d. $Al(s) + H_2SO_4(aq) \rightarrow$

    e. $Ni(s) + CuCl_2(aq) \rightarrow$

    f. $Ba(s) + H_2O(l) \rightarrow$

    g. $Cu(s) + FeSO_4(aq) \rightarrow$

    h. $Al(s) + Pb(NO_3)_2(aq) \rightarrow$

    i. $Li(s) + KI(aq) \rightarrow$

    j. $Au(s) + O_2(g) \rightarrow$

15. Assuming the following compounds are the result of decomposition reactions, identify the reactant, write the formula equation, and balance it.

    a. lithium chloride and oxygen

b. copper(II) oxide and water

c. carbon dioxide and water

SECTION 5.3

16. Silver bromide is used on film for producing holograms. If 3.5 mol silver nitrate react in a double replacement reaction with excess sodium bromide to produce silver bromide, what mass of silver bromide is formed?

17. Hydrogen, the perfect clean fuel, combusts with oxygen to produce water vapor. If 1,575 mol H_2 react, determine the mole quantities of oxygen required and water produced.

18. Aluminum sulfide reacts with water to produce dihydrogen sulfide (usually just called hydrogen sulfide) and aluminum hydroxide. If 2.290 kg aluminum sulfide react completely with excess water, what mass of aluminum hydroxide is produced?

19. Aluminum hydroxide is a popular compound for use in antacid tablets. This compound reacts with stomach acid, HCl, to produce the salt aluminum chloride and water. If 750 mg aluminum hydroxide are consumed, determine:

    a. the number of moles of hydrochloric acid neutralized.

    b. the mass of water produced.

20. Iron ore contains iron(III) oxide along with several other substances. The iron is reduced by heating it to about 1,250°C in the presence of carbon monoxide (a so-called *reducing atmosphere*) in a blast furnace. This process, which has been in use since around 1491, is shown to the right. The reaction produces iron metal and carbon dioxide. What is the theoretical yield in kilograms of iron from this reaction if 2.50×10^4 kg iron(III) oxide react with excess carbon monoxide?

21. A common acid-base reaction is between sulfuric acid and sodium hydroxide to produce sodium sulfate and water. If 29.55 g sodium hydroxide react with 44.11 g sulfuric acid, determine:

    a. the limiting reactant.

    b. the mass of sodium sulfate produced.

    c. the mass of water produced.

22. Octane, C_8H_{18}, is the major component of gasoline. The combustion reaction between octane and oxygen is:

    $$2C_8H_{18}(l) + 25O_2(g) \rightarrow 16CO_2(g) + 18H_2O(g)$$

    a. How many moles of oxygen gas are required to burn 350.0 moles of octane?

    b. How many kilograms of water are produced if the combustion consumes 3.5 kg O_2?

    c. The density of octane is 0.692 g/mL. Determine the mass of oxygen required to burn 11.5 gal octane. (Recall, conversion factors are in Appendix B, Table B.3.)

23. Nitric acid is used in the process of *nitration*, in which organic molecules are converted into explosive compounds. The first step in the industrial production of nitric acid involves the following reaction between ammonia and oxygen to produce nitric oxide and water:

$$4NH_3(g) + 5O_2(g) \xrightarrow{\text{cat}} 4NO(g) + 6H_2O(g)$$

Assume that 855 g NH_3 are combined with 1,750 g O_2 in this reaction. Determine:

a. the limiting reagent.

b. the theoretical yield of nitric oxide, assuming all the limiting reagent is consumed.

c. the percentage yield of NO, assuming 1,272 g NO are actually produced.

24. Aluminum hydroxide undergoes a double replacement reaction with sulfuric acid. Assume that 31.8 g sulfuric acid are combined with 25.4 g aluminum hydroxide with 100% yield. Determine the following:

a. the limiting reactant.

b. the mass of excess reactant remaining.

c. the mass of each product formed.

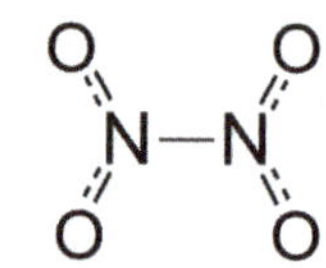

25. Dinitrogen tetroxide, N_2O_4, is an important rocket fuel. Its structure is pictured to the right. On one of the Apollo lunar missions, N_2O_4 powered the lunar landing module as it ascended from the surface of the moon. During the ascent, 1.200×10^3 kg N_2H_4, 1.000×10^3 kg $(CH_3)_2N_2H_2$, and 4.500×10^3 kg N_2O_4 were available for the following reaction:

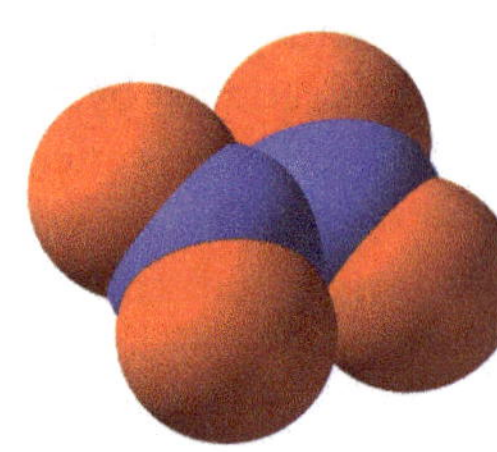

$$2N_2H_4 + (CH_3)_2N_2H_2 + 3N_2O_4 \rightarrow 6N_2 + 2CO_2 + 8H_2O$$

a. Which of the compounds was consumed first in this reaction?

b. How many kilograms of waste water were spewed into space by the reaction?

c. How many nitrogen atoms were left in space by the reaction?

26. For reasons we explore in a later chapter, hydrofluoric acid is classified as a weak acid. Nevertheless, HF is so corrosive it eats right through glass, silicon dioxide, SiO_2, and is used for glass etching. (Needless to say, HF is not stored in glass bottles.) HF has also been used for etching silicon wafers in the process of making semiconductors, although the use of HF has waned in recent years. The reaction between HF and SiO_2 is

$$SiO_2(s) + 6HF(aq) \rightarrow H_2SiF_6(aq) + 2H_2O(l)$$

a. How many moles of HF are needed to react with 542 g SiO_2?

b. How many grams of H_2SiF_6 form when 4.25 mol HF react completely?

c. How many moles of HF are required to produce 2.0 gallons of water? (Remember, the density of water is 0.998 g/mL at room temperature.)

d. If 2.50 kg HF are available to react with 1.205 kg SiO_2, what is the limiting reactant?

e. What is the theoretical yield of H_2SiF_6 if the limiting reagent is completely consumed, assuming the reactant quantities listed in part (d)?

f. Using the theoretical yield from the previous question, what is the percentage yield of H_2SiF_6 if 2.657 kg H_2SiF_6 are produced?

27. When benzene, C_6H_6, reacts with bromine (a type of reaction called *halogenation*), the reaction produces bromobenzene, C_6H_5Br, and hydrogen bromide.
    a. If 45.0 g of benzene reacts with 97.5 g of bromine, what is the theoretical yield of bromobenzene?
    b. If the actual yield of bromobenzene is 63.25 g, what is the percent yield of bromobenzene?
28. To monitor levels of ozone, O_3, in the air, an air sample is processed by an instrument that causes ozone to react with sodium iodide in the following reaction:

$$O_3(g) + 2NaI(aq) + H_2O(l) \rightarrow O_2(g) + I_2(s) + 2NaOH(aq)$$

    a. Determine the number of moles of sodium iodide required to process 2.85×10^{-6} moles of ozone.
    b. Determine how many grams of sodium iodide are required to process 3.00 grams of ozone.
    c. If 1,455 g NaI are available to process a sample of 250.0 g O_3 in a reaction with excess water, determine the limiting reactant and the theoretical yield of I_2.
    d. Determine the number of iodine atoms present in the theoretical yield of I_2.

GENERAL REVIEW EXERCISES

29. Melatonin is a compound secreted naturally in the brain when it is time to go to sleep. The molar mass of melatonin is 232.281 g/mol. A sample of 25.0 mg of melatonin is found to contain 16.81 mg C, 1.736 mg H, 3.015 mg N, and 3.445 mg O.
    a. Determine the percent composition of melatonin.
    b. Determine the molecular formula for melatonin.
    c. Melatonin is sold as an over-the-counter sleeping aid. Each tablet contains 3.0 mg of melatonin. Determine the number of carbon atoms in each tablet of melatonin.
30. Mercury-vapor lamps are widely used for general lighting in gymnasiums, exhibit halls, and other large spaces. One of the wavelengths in the emission spectrum of mercury is a bright green line with an energy of 2.271 eV. Determine the wavelength of this line.
31. Write the condensed electron configuration and draw the orbital diagram for yttrium, tin, chromium, and iodine.
32. Write explanations in one or two sentences describing the Aufbau principle, the Madelung rule, and Hund's Rule.
33. Use the information from Table 1.5 to calculate the atomic mass of silicon.
34. Place the following isoelectronic series of atoms and ions in order by size, from largest to smallest: Rb^+, Y^{3+}, Br^-, Kr, and Sr^{2+}.
35. Based on your knowledge of general trends in the periodic table, place the following in order of increasing electronegativity: Fe, F, Se, Fr, Ba, and Cl.
36. Draw the Lewis structures for the following molecules and polyatomic ions:

a. CCl_2S	b. PO_4^{3-}	c. NO_2^+	d. N_2H_2
e. O_3	f. SF_6	g. CO	h. H_3O^+

37. Identify the electron domain geometry and molecular geometry for each molecule listed in the previous exercise.

Chapter 6

Kinetic Theory and States of Matter

This researcher is exploring a technology known as *chemical vapor deposition* (CVD). This process is used in the manufacturing of semiconductors (computer chips) for use in computers, mobile devices, digital signal processing, and other digital technologies. In CVD processes, a thin layer of metallic (conducting) or nonmetallic (nonconducting) material is deposited directly from the gas state onto a silicon substrate. Often the gaseous material undergoes a reaction at the surface to form a coating of pure metal (such as tungsten), pure glass, or other material. Sometimes plasmas are used to assist the process, in which case the process is called *plasma-enhanced chemical vapor deposition* (PECVD).

Objectives for Chapter 6

After studying this chapter and completing the exercises, you should be able to do each of the following tasks, using supporting terms and principles as necessary.

SECTION 6.1

1. Describe the relationship between temperature and molecular energy.
2. Describe the Maxwell-Boltzmann velocity distribution and relate it to the particle velocities in gases.
3. Describe in detail the basic principles of the kinetic-molecular theory of gases.
4. Explain the cause of gas pressure.
5. State the name, symbol, and definition for the official SI system unit of pressure.
6. Write a paragraph describing how a mercury barometer works.

SECTION 6.2

7. State the four basic states of matter and give a detailed description of each at the molecular level.
8. Describe the forces that hold substances together in the solid and liquid states, distinguishing between forces acting in ionic crystals, metals, and molecular substances.
9. Explain the causes for the following properties of solids: high melting point (crystals and metals), low melting point (molecular substances), high density, incompressibility, and definite shape.
10. Define and explain the cause of *surface tension* and use surface tension to explain capillary action and wicking.
11. Define and give examples of *diffusion* in liquids and gases.
12. State common examples of plasmas.
13. State and describe the six basic phase transitions that occur between solids, liquids, and gases.
14. On a P-T phase diagram, locate the phase transition points for given pressures or temperatures.
15. Define *triple point*, and locate the triple point on a P-T phase diagram.
16. Define *critical point*, *critical temperature*, and *critical pressure*, and locate the critical point on a P-T phase diagram.
17. Describe the unique feature of the P-T phase diagram for water.
18. Given appropriate thermal constants, calculate the heat that must be added or removed to raise or lower the temperature in a given mass of solid, liquid, or gas substance.
19. Given appropriate thermal constants, calculate the heat that must be added or removed to effect a phase transition in a given mass of material.
20. Define and explain the causes of *evaporation* and *vapor pressure*.
21. Explain why water boils at lower temperatures in high-altitude locations.

6.1 Temperature, Kinetic-Molecular Theory, and Pressure

6.1.1 Temperature and Molecular Energy

So far we have encountered several different physical mechanisms by which substances can possess energy. In Chapter 1, we saw that the energy atoms possess can be modeled as kinetic energy in the motion of the atoms, or as the energy of electrons when excitation pushes electrons into excited states. In more recent chapters, we have seen that the energy in bulk substances resides in the lattice energy of crystals or bond energy of molecules or metals. In this chapter, we look more closely at the first of these—the energy of molecular motion.

According to *kinetic-molecular theory*, we understand the temperature of a substance to be directly related to the kinetic energy in the motion of atoms and molecules. In solids, atoms are not free to move around, so atomic motion is manifest as atomic vibrations. When an object vibrates back and forth its energy continuously changes from kinetic energy to potential energy, but we don't need to get into that much detail here. We can just think of vibrating atoms as each having an average kinetic energy defined as

$$E = \tfrac{1}{2}mv^2$$

where v represents some kind of average velocity. In fact, the average velocity of a mass vibrating back and forth is zero, because velocity is directional (that is, it is a vector). The mass spends equal amounts of time moving in opposite directions, and if we sum up its velocities over time (which requires calculus, since the velocity is continuously changing) the sum is zero. To get a meaningful sum instead of a zero sum, we use the so-called root-mean-square (rms) velocity, v_{rms}. Again, we don't need to get into the details here because we are not going to perform the actual calculations in this course. But in short, an rms calculation involves the square root of the mean of the squares of the individual velocity contributions. Squaring all the velocities eliminates negative signs so the sum adds up to a non-zero value for v_{rms} that gives a correct average energy. So our energy calculation for the vibrating atom becomes

$$E = \tfrac{1}{2}mv_{\text{rms}}^2$$

In fluids—liquids and gases—particles are free to move around, so they possess translational kinetic energy. In molecular fluids, molecules also possess rotational kinetic energy because they tumble as they move. In these cases, molecules are continuously colliding and ricocheting all over the place. Just as with the vibrating atoms in a solid, the average velocity of a molecule of gas is close to zero because as it bounces around, it spends roughly equal amounts of time moving in all different directions. As before, we deal with this by using an rms velocity and an energy calculation based on v_{rms}.

In a given substance, the sum of all the kinetic energies of all the particles in the substance due to all their motions is called the *internal energy*. The variable we call *temperature* is directly related to the internal energy of the substance. A temperature measurement is not an energy measurement; nevertheless, the temperature of a substance is strictly due to the internal energy of the substance. In fact, the internal energy of a substance can be expressed as a function of temperature, and vice versa.

6.1.2 Velocity Distribution of Gases

A very important application of our theory of molecular motion is in the motion of the molecules in a gas. (I use the term *molecules* because at room temperature, all gases are composed of molecules except for the noble gases. The noble gases do not form chemical bonds, and thus

Figure 6.1. Scottish physicist James Clerk Maxwell (1831–1879).

consist of individual atoms. So when I write "molecules" of gas, I mean also to include the individual atoms in noble gases.) Gas molecules are in constant motion and are constantly colliding with each other. In room temperature air for example, each molecule of nitrogen and oxygen collides with another molecule about five billion times per second. In these collisions, energy is transferred from one molecule to another. Just as with pool balls on a pool table, sometimes a molecule speeds up in a collision, a gain in kinetic energy, and sometimes a molecule slows down, a loss of kinetic energy.

At a given temperature, the molecules of a particular gas possess a predictable average (rms) velocity, but the speeds of individual molecules are spread out in a well-known distribution called the *Maxwell-Boltzmann distribution*. This speed distribution was derived from mathematical theory in 1860 by James Clerk Maxwell (Figure 6.1), the same Scottish physicist who developed Maxwell's equations explaining electricity and magnetism. Ludwig Boltzmann (Figure 6.2) was an Austrian physicist whose name is now synonymous with theory in thermodynamics and statistical mechanics. In the late 19th century, Boltzmann developed much of the explanatory theory supporting Maxwell's speed distribution curves.

Figure 6.2. Austrian physicist Ludwig Boltzmann (1844–1906).

The Maxwell-Boltzmann speed distributions for the noble gases at 25°C are shown in Figure 6.3. Notice that the peak of the xenon curve is at around 200 m/s and the peak of the helium curve is at around 1,100 m/s. For reference, recall that the speed of sound in air at this temperature is around 342 m/s.

The first thing to get from the distribution curves is the fact that the speeds lie in a *distribution*. Corresponding to a given temperature is a specific average speed, but any particular atom can have a speed lying anywhere in the distribution at any given time. For example, an atom of neon at 25°C can have a speed ranging anywhere from near zero to about 1,300 m/s, with most neon atoms having speeds close to 500 m/s.

The second important thing to notice from the distributions is that *the heavier the atom, the lower the average speed*. Connect this to what we have just been considering about molecular energy. At a given temperature, the particles of any gas have the same average energy because the temperature of the gas depends only on the en-

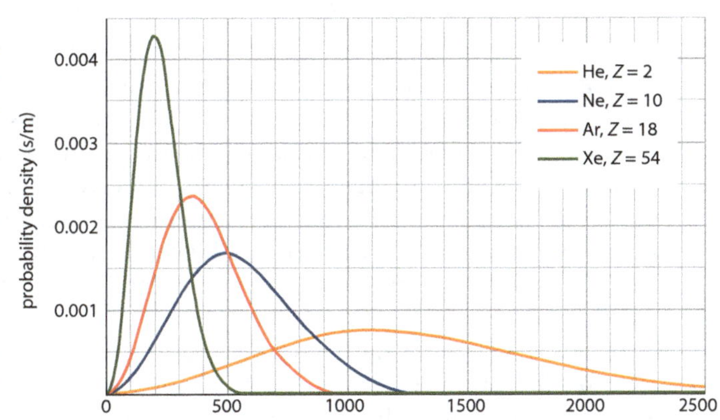

Figure 6.3. The Maxwell-Boltzmann speed distribution for noble gases at 25°C. Interpret the units of measure on the vertical axis on this graph this way: A section of area under the curve represents the proportion of atoms moving with speeds in that speed range. That area is calculated by multiplying the horizontal width of the section by its height. But a proportion is a unit-less number, thus the units of measure on the vertical axis must be the reciprocal of those on the horizontal axis so that they cancel out when multiplied.

ergy, and vice versa. The kinetic energy of an atom, $\frac{1}{2}mv^2$, is proportional to the mass and speed of the atom. So on average, heavier atoms move at slower speeds.

6.1.3 The Kinetic-Molecular Theory of Gases

The *kinetic-molecular theory of gases* is the major theory supporting our present view of gases, pressure, and the gas laws. The theory as we state it today comes directly from the work of Maxwell and Boltzmann. (Unlike many other scientists at that time, Maxwell and Boltzmann both accepted the actual existence of atoms.) It is important here to remind ourselves what theories are and what role they play in scientific inquiry. As discussed briefly in the Introduction, a scientific theory is a *model* describing and accounting for some aspect of the natural world. The kinetic-molecular theory accounts very well for the behavior of gases, with some limitations described below. The following is a basic outline of the kinetic-molecular theory of gases.

1. A gas consists of a very large number of molecules (or atoms, in the case of noble gases).
2. The total volume occupied by all the molecules of gas in a container is negligibly small compared to the volume of the container. This means we can model the molecules as point particles that occupy no space.
3. When molecules collide, they experience *elastic collisions*. An elastic collision is one in which the total kinetic energy of the objects after the collision is the same as the total kinetic energy prior to the collision. In other words, no energy is converted to other forms of energy in the collision. A molecule might slow down or speed up in an interaction with another molecule, but the total kinetic energy of the colliding molecules is unaffected by the collision and any transfer of energy that occurs.
4. Electrical forces of attraction and repulsion between molecules are negligible. This is due to a couple of factors. One is that neutral molecules have no net charge, so there are no ionic attractions. Another is that on average gas molecules are relatively far apart—something like a thousand times farther apart than the molecules in a liquid are—and thus intermolecular forces are so weak as to be negligible.
5. The average kinetic energy of the molecules in a gas is proportional to the temperature. At any given temperature, the gas molecules all have the same average energy.

One of the immediate results of the kinetic-molecular theory is an explanation for gas pressure (our next topic). Gases that behave according to the assumptions listed here are called *ideal gases*. It turns out that under certain conditions, many gases do exhibit ideal gas behavior. For example, at low pressure there are fewer gas particles in a given amount of space, so the second assumption listed above becomes an accurate description of the conditions of the gas. Also, the more energy the gas particles have (i.e., higher temperature) the less any intermolecular forces affect the particle motion, which means the third and fourth assumptions listed above are accurate descriptions.

In light of the fourth statement above, gases whose molecules are nonpolar exhibit ideal gas behavior to a larger degree than gases whose particles are polar because polar molecules are attracted to one another by Keesom forces, the strongest of the Van der Waals forces. Intermolecular forces between nonpolar molecules are limited to the much weaker London dispersion forces (see Section 4.3.4).

6.1.4 Gas Pressure

According to the kinetic-molecular theory, pressure in gases is caused by the collisions of molecules with each other and with the walls of the container. Higher internal energy means faster moving molecules, resulting in harder collisions, and consequently, higher pressure. This is why heating a gas confined in a container causes the pressure to increase. The heat increases the internal energy and thus also increases the average velocity of the gas particles.

The official SI System unit for pressure is the *pascal* (Pa), named after French scientist, mathematician, and philosopher Blaise Pascal (Figure 6.4). In the 17th century, Pascal developed many of the general principles pertaining to pressure in both liquids and gases. To help demonstrate how pressure is defined and how much pressure a pascal represents, Figure 6.5 depicts a chamber with a freely sliding side wall. The sliding wall is one meter square, so its area is 1 m^2. Imagine that the chamber contains ordinary air, and that a force of one newton (1 N, which is about 1/4 lb) is applied to the sliding wall. The wall slides in until the force of the air pushing back on the wall also equals 1 N. The forces on both sides of the wall are then equal at 1 N each. But the force on the inside of the wall is due to the gas pressure inside (all those molecules of gas colliding with the inside of the wall). The force of 1 N applied by the gas molecules inside the chamber to the area of 1 m^2 is defined as a pressure of 1 pascal (1 Pa). This is how the pascal is defined. In other words,

Figure 6.4. French scientist Blaise Pascal (1623–1662).

$$1\ \text{Pa} = 1\ \frac{\text{N}}{\text{m}^2}$$

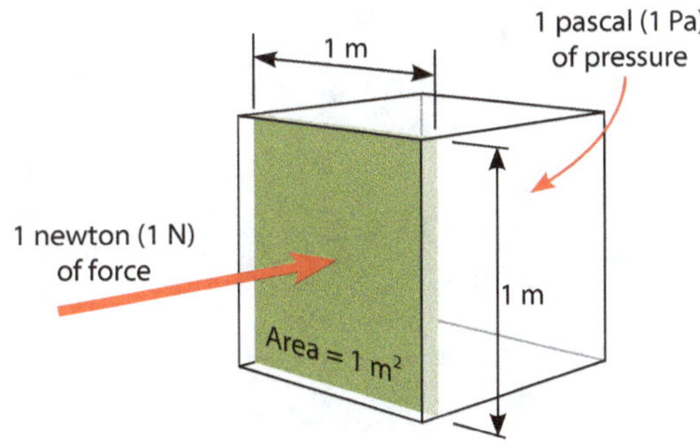

Figure 6.5. One pascal of pressure is one newton of force per square meter.

Scientists have been experimenting with pressure for a long time, and as a result there are many different units for pressure in use, more than for just about any other variable you can think of. Some of the most common units are listed in Table 6.1 along with their unit symbols, definitions, and the average atmospheric pressure in each unit. In the U.S., pounds per square inch (psi) is the most common pressure unit for general use. Weather reporters like to use a scale based on "inches of mercury." The liquid pressure 29.9 inches beneath the surface in a column of mercury with a vacuum on top is equivalent to atmospheric pressure, as explained in the box on the next two pages. Because of the historical significance of barometers and their continued use for measuring atmospheric pressure, I've included some of the history in the box.

6.2 States of Matter

6.2.1 The Four Basic States of Matter

No doubt you are already familiar with the basic states (or phases[1]) of matter, or at least three of them. Our objective here is for you to be able to distinguish the four states of matter and their characteristics at the atomic or molecular level.

1 Physics texts often refer to "phases" of matter, while chemistry texts tend to use the phrase "states" of matter. These phrases are equivalent. When discussing transitions between states, the phrase *phase transition* is commonly used by everyone.

Hmm... Interesting. How barometers work

The air of the earth's atmosphere is obviously not in a container, but nevertheless air pressure is still caused by the collisions of gas molecules. A helpful way to think about atmospheric pressure is in terms of the earth's gravity pulling the gas molecules toward the earth—the molecules are held against the earth by the force of gravity. The pressure at a given altitude is due to the molecules at that altitude colliding with the molecules just above. The pressure can be modeled in terms of the weight of the gas above. This weight in turn depends on the density of the molecules and the frequency with which they collide with each other. At higher elevations, the earth's gravitational attraction is weaker, so the molecules are more spread out. This lowers the density, reduces the frequency of molecular collisions, and thus reduces the pressure.

Atmospheric pressure varies with the weather from about 2% above normal to about 5% below normal. These variations are the cause of wind, as pressure differences between one region and another cause air to flow toward regions of lower pressure. Low pressures occur when the wind is moving rapidly, so when atmospheric pressure drops, it means stormy weather is headed your way. Atmospheric pressure is measured with a *barometer*, so weather reporters often refer to the value of atmospheric pressure as the *barometric pressure*.

Italian physicist and mathematician Evangelista Torricelli (1608–1647).

The mercury barometer for measuring air pressure was invented in 1643 by Italian physicist and mathematician Evangelista Torricelli. The diagram below shows how we might make one of these devices today, using a vacuum pump to remove the air from a glass tube set into a bowl of mercury. (Torricelli didn't have a vacuum pump, so he accomplished this a different way, but the result is the same.) As the air is removed from the glass tube, the mercury rises into the tube up to a height of approximately 760 mm. The tube is then capped off and the device registers changes in atmospheric pressure

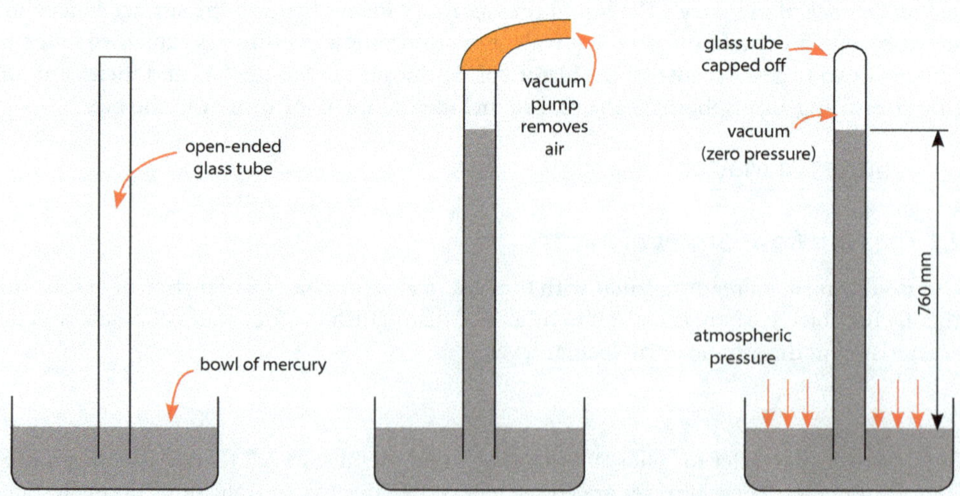

caused by the weather. The pressure of the atmosphere pushes the mercury up into the tube, and because of the vacuum, there is no gas pressure on top of the mercury column to push it back down.

As atmospheric pressure changes, the height of the mercury column in the glass tube rises and falls. At normal atmospheric pressure at sea level, the height of the mercury is 760 mm, which is about 29.9 inches, and sometimes weather reporters refer to atmospheric pressure in terms such as "thirty inches of mercury." Normal barometric pressure variation in the U.S. is from about 725 mm to 775 mm of mercury, which corresponds to a range of 28.5 to 30.5 inches. The higher the atmospheric pressure is, the farther it pushes the mercury up into the barometer tube. So if the local barometric pressure is 30.4 inches of mercury, that is a high-pressure day and the weather is likely to be clear, sunny, and calm. When the barometric pressure falls, it means a change of weather as high-pressure regions push wind and moisture toward the now low-pressure region. You might be interested to know that the world record low barometric pressure (other than the pressure inside a tornado, which is extremely low) is 25.7 inches, measured in 1979 during a typhoon in the Pacific Ocean. The pressure unit *torr* (symbol: Torr) is named after Evangelista Torricelli. Normal atmospheric pressure is equal to 760 torr.

A vacuum is a complete void in space in which there is no matter of any kind. In 1643, scientists did not believe that a vacuum could exist (because of things Aristotle had written long before), so at the time the barometer was invented, no one could formulate a correct explanation for how it worked. The big question was about the mysterious force that holds the mercury up in the tube. Blaise Pascal explained how the barometer works in 1647, theorizing that a vacuum does exist in the glass tube above the mercury. On the basis of his own experiments, Pascal explained that there is no air pressure above the mercury column in the tube, and that the pressure of the atmosphere acting on the surface of the mercury in the bowl pushes the mercury up into the tube into the vacuum region. This explanation is correct, but since at the time no one accepted the possibility of a vacuum, scientists did not want to accept Pascal's bold theory. But as with all scientific theories, a theory gains strength when it repeatedly leads to successful hypotheses, and this one did.

Pascal's explanation is easy to understand if we calculate the pressure in the tube 760 mm below the surface of the mercury. As with the pressure in the atmosphere, the pressure under a stationary liquid can be modeled in terms of the density and depth of the liquid, and the acceleration of gravity, g. If you work out the mathematics of the weight of liquid applied to a certain amount of area you get the standard equation for the pressure at a specific depth under the surface of a liquid:

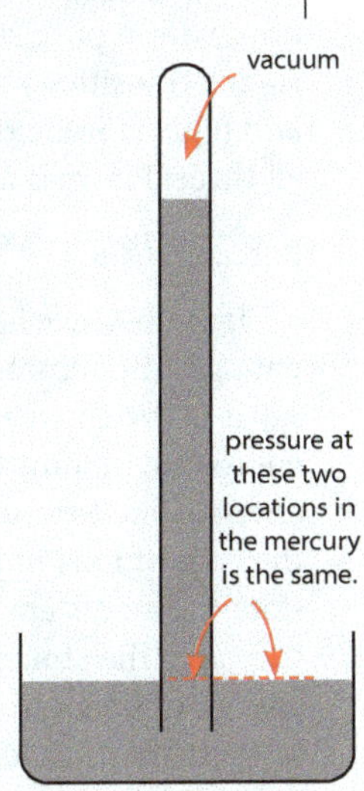

$$P = \rho g h$$

where P is the pressure in pascals, ρ is the liquid density in kg/m^3, and h is the depth below the surface in meters. The density of mercury is 13,600 kg/m^3. Using the pressure equation above, the pressure at a depth of 760 mm or 0.76 m comes out to 101,300 Pa, which agrees with the value for atmospheric pressure in Table 6.1 to four significant digits. Thus, the atmospheric pressure outside the tube is equal to the pressure due to the weight of the liquid inside the tube. This calculation demonstrates that the pressure at the same height in the mercury inside and outside the tube is the same, a principle discovered by Pascal.

Pressure Unit Name (unit symbol)	Definition	Average Atmospheric Pressure at Sea Level in these Units
pascal (Pa)	1 Pa = 1 N/m^2	101,325 Pa
kilopascal (kPa)	1 kPa = 1,000 Pa	101.325 kPa
atmosphere (atm)	1 atm = 101,325 Pa	1 atm
bar (bar)	1 bar = 100,000 Pa	1.01325 bar
pounds per square inch (psi)	1 psi = 6894.757 Pa*	14.696 psi
torr (Torr)	1 Torr = 1/760 atm	760 Torr
inches of mercury (in Hg)	height, in inches, of mercury column in a barometer	29.9 in Hg
millimeters of mercury (mm Hg)	height, in millimeters, of mercury column in a barometer (1 mm Hg = 1 Torr)	760 mm Hg

Table 6.1. Common pressure units and their definitions. In the definitions column, the one labeled with an asterisk (*) is approximate; all others are exact.

The four basic states of substances are *solid*, *liquid*, *gas* (or *vapor*), and *plasma*. The gas state may be called gas or vapor, depending on whether a substance normally exists in nature as a gas or a liquid. If a substance exists at ordinary temperature and pressure as a gas, we refer to a substance in this state as a gas. But if a substance is a liquid or a solid at ordinary temperature and pressure, we conventionally refer to the substance in the vapor state as a vapor. Steam, for example, is water *vapor* because at room temperature water is a liquid.

The internal energy of a substance is what distinguishes the states from one another. Beginning with a pure substance in the solid state, if the substance is continuously heated, the substance transitions from one state to the next as the kinetic energy of its molecules gets higher and higher. Some substances can exist as solids or liquids, some as liquids or gases, and some as all three. Plasmas are a completely unique state of matter, which we briefly address later.

6.2.2 Solids

If a substance has a low internal energy, this means its temperature is low. If the energy is low enough, the substance will be in the solid state. For a substance normally in the liquid state, we say it is frozen when it is in the solid state. Since the atoms in the solid state have a low internal energy, the atoms' kinetic energies are low and they are vibrating relatively slowly. This low energy state for the atoms means they cling strongly to one another by electrical attraction (in the case of ions in crystals) or metallic bonding (in the case of metals) forming a solid. In this state the atoms are not free to move around; they are stuck in place and can only vibrate where they are. The atoms *do* have kinetic energy, so they are not at rest. But they do not have enough energy to break free of the electrical attractions or metallic bonding holding them all together, so their kinetic energy forces them to vibrate in place. This is what is going on with the atoms in any solid. Figure 6.6 depicts the atoms in an ionic crystal solid.

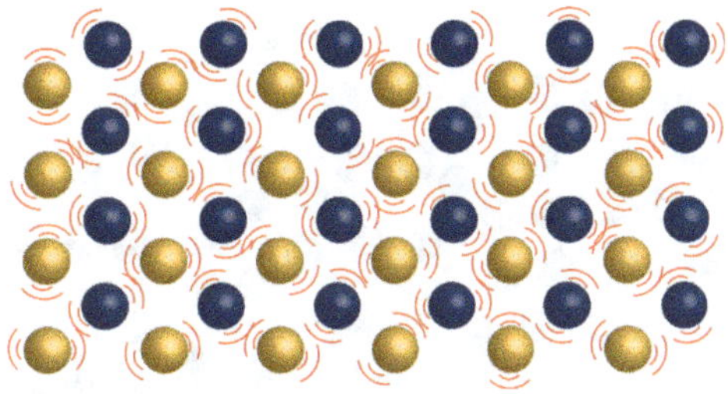

Figure 6.6. Atoms in a solid crystal vibrating in place.

The close packing of atoms in metallic and ionic solids explains their high density relative to gases. Forces of electrical attraction and repulsion in the crystal lattice, or metallic bonding in the metallic lattice, account for the rigidity and *incompressibility* (tendency to withstand being compressed into a smaller volume) of these solids.

Of course, molecules can form solids as well—ice, for example. Many molecular solids such as glass and plastics are amorphous because the molecules are not organized in an orderly fashion the way the atoms are in crystals (ice being a notable exception). Because of their relatively low energy, the molecules in a molecular solid are more closely packed than in molecular liquids. As a result, intermolecular forces (hydrogen bonding and Van der Waals forces) in molecular solids are much greater than in liquids or gases, and these intermolecular forces hold the molecules together in a solid.

The intermolecular forces in molecular solids are much weaker than the covalent bonds holding the individual molecules together. For this reason, molecular substances have relatively low melting points compared to the much higher melting points of metals and ionic crystals.

We think of solids as having a definite, permanent shape. For the most part, this is correct. However, amorphous solids such as glass can "flow" at an extremely slow rate over an extended period of time. Windows in homes built around 100 years ago are noticeably thicker at the bottom than they are at the top because of the very slow downward flow of the glass due to gravity.

Table 6.2 summarizes a few of the physical properties we have touched on in this section.

Property	Cause
high melting point in metals and ionic crystals	high bonding forces between ions or atoms
relatively low melting point of molecular solids	intermolecular forces in molecular solids are much lower than ionic forces in crystals or metallic bonding forces in metals
high density	close packing of particles due to low kinetic energies
incompressibility (definite volume)	close packing and high forces between particles (relative to gases) resist compression
definite shape	close packing and high forces between particles cause particles to hold their position, maintaining the shape of the substance

Table 6.2. Some physical properties of solids.

6.2.3 Liquids

Liquids are common here on the earth, but liquid is the least common state of matter in the universe. The range of temperatures and pressures where liquids can exist is very narrow, and is generally only found on relatively temperate planets.

At higher internal energies—higher temperatures—the kinetic energies in atoms or molecules of a substance can be high enough to overcome the attractions that bind them together in a solid. Every pure substance has a particular temperature at which the higher kinetic energy in the atoms overcomes the electrical attraction between the atoms, allowing the atoms or molecules to come apart. At this temperature—the *melting point*—a solid transitions to the liquid state. The atoms in a liquid move quickly, and there are a lot of them. The result is that the

Figure 6.7. Particles in the liquid state.

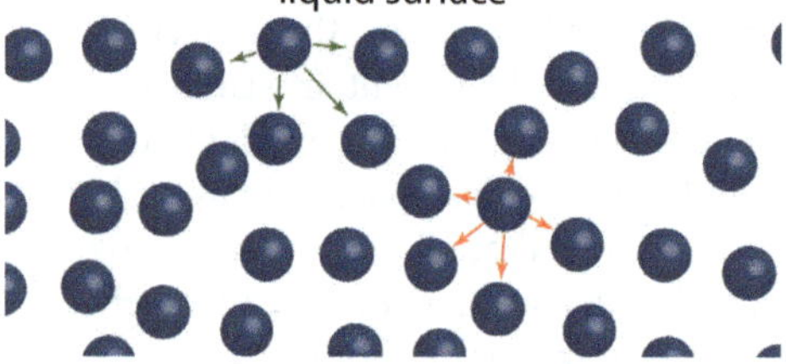

Figure 6.8. Surface tension is due to the fact that the combined effect of the attractive forces on a molecule at the surface of a liquid tends to pull the molecules toward the inside of the liquid. Molecules in the liquid interior tend to be pulled equally in all directions.

particles in a liquid, depicted in Figure 6.7, are continuously colliding with each other.

In the liquid state, the atoms or molecules of a substance are still held together loosely by ionic attractions or intermolecular forces, causing liquids to stay together in an open container, puddles, or drops. As mentioned back in the box on page 6 of the Introduction, small liquid drops are nearly spherical because this shape minimizes the surface area of the drop, and thus minimizes the potential energy between molecules or atoms at the surface. This is yet another instance of energy minimization.

Figure 6.8 depicts the unbalanced forces on a molecule at the liquid surface compared to the more balanced forces present on a molecule in the liquid interior. At the surface, the forces tend to hold the surface molecules together with each other and with the liquid interior. The unbalanced forces at the surface give rise to a liquid characteristic called *surface tension*. The surface tension holding surface molecules together is what allows insects to walk on the surface of water. The combination of surface tension and forces between surface molecules and a polar surface such as glass gives rise to several effects, including the ability of small water droplets to cling to a vertical glass surface, the meniscus that forms on the surface of water in a glass container (Figure 6.9), and *capillary action*.

Capillary action occurs in small spaces such as the inside of thin glass tubes or the open spaces in porous rocks or sponges. In a small space, the forces between the surface molecules in the liquid and the polar surface of the solid can pull the liquid up into the space. Examples are seen in the way water flows upward against gravity in a thin glass tube and in porous bricks (Figure 6.10). Back in Section 4.3.3, we discussed the fact that hydrogen bonding between water molecules and the cellulose molecules in cotton is the cause for the high absorbency of cotton. When that absorbency is combined with tiny channels in a fabric that allow liquid surface molecules to be pulled by polar molecules in the fabric, wicking occurs. Wicking is seen in the way that liquid candle wax or lamp oil travels up a wick to be burned, and in the fabrics used in athletic clothing that wick moisture away from the skin.

Figure 6.9. A liquid meniscus, the bowl-shaped curve on a liquid surface, is a result of surface tension and attraction between surface molecules and the polar surface of the glass.

Diffusion, the tendency for molecules spontaneously to mix and mingle, is a phenomenon exhibited by both liquids and gases.[2] In these states, the particles are energetic and free to move around. Their random motions result in complete mixing of a soluble liquid in another liquid, forming a homogeneous mixture (a solution). Diffusion is illustrated by the food coloring added to water in Figure 6.11. With no shak-

2 Interestingly, diffusion occurs in solids, too. For example, foreign atoms in a crystal lattice, such as the carbon atoms in the iron lattice of steel alloys, migrate through the lattice. (Alloys are discussed in Section 8.3.5).

ing or stirring at all, the molecules of the dye thoroughly mix with the water molecules after short time.

Intermolecular forces hold the molecules together in a liquid, so the particles stay together in a container. But high particle energies (due to the temperature) mean that a liquid has no fixed shape, so it takes the shape of its container.

In the liquid state, the molecules of a substance are not much farther apart from each other than they are in the solid state. Thus, liquid densities are very close to solid densities. Further, liquids are essentially incompressible. Even enormous pressure does not decrease the volume of a liquid more than a few percent.

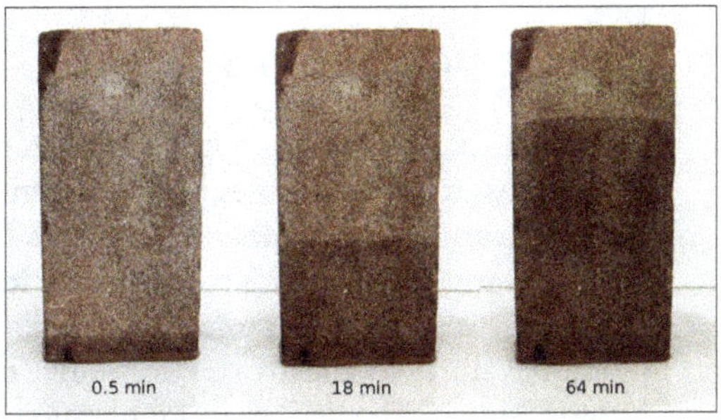

Figure 6.10. Capillary action draws water up into the pores in a brick.

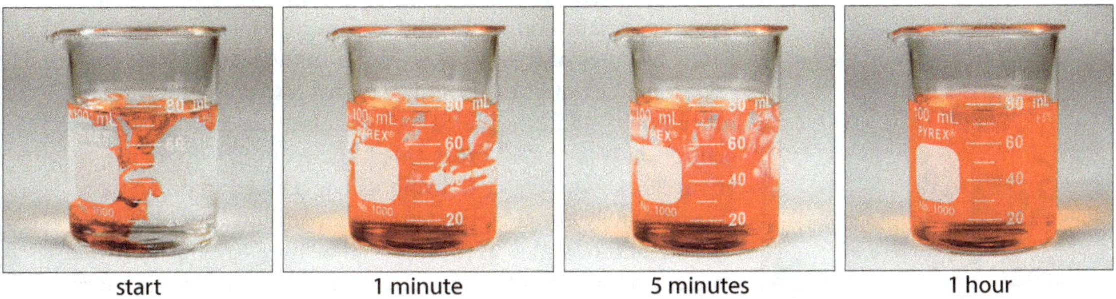

Figure 6.11. Diffusion leads to 100% mixing of a soluble liquid solute, resulting in a homogeneous mixture.

6.2.4 Gases

In gases, the internal energy of the atoms or molecules is so high that it completely overcomes the forces of attraction that hold the particles together in solids and liquids. Instead of staying together, if the particles are not contained they have so much energy they fly away at great speed. Because of their higher speeds, the atoms or molecules spread apart, still colliding with one another, but spreading out as much as they are able. If they are contained in a container of some kind, they still fly around at great speed, but instead of flying off into the atmosphere, they furiously collide with each other and with the walls of the container, creating pressure inside the container.

The average distance between particles in gases is on the order of 1,000 times that of particles in liquids. In Figure 6.12, in order for the gas particles to be visible in the figure, the gas particles are shown only about 10 times as far apart as the liquid particles in Figure 6.8. In fact, to make the picture more to scale, the tiny spheres representing the gas particles would have to be 100 times smaller than they are in the figure.

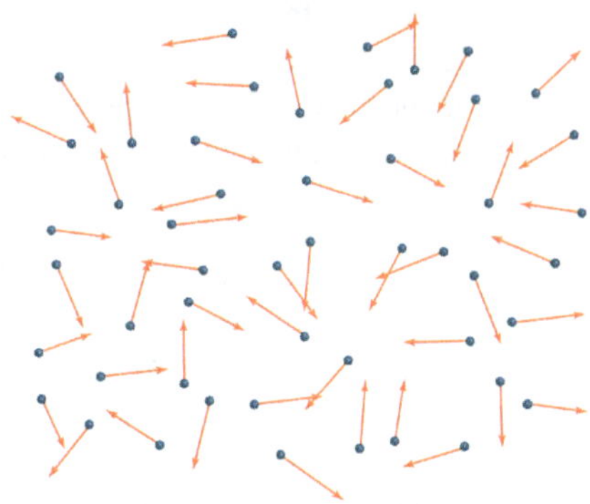

Figure 6.12. Particles in gases are far apart (on average) and move at very high speeds compared to particles in liquids and solids.

We have already seen (Figure 6.3) that even at ordinary temperatures, the average velocity of gas particles is in the same ballpark as the speed of sound in air. At room tem-

Hmm... Interesting. — *Gas diffusion*

Here is a simple experiment you can perform to observe gas diffusion and compare rates of diffusion for molecules with different masses. We are using two household substances, white (distilled) vinegar and ammonia, both available down at the grocery store.

The molecular formula for acetic acid (vinegar) is CH_3COOH, which means its molecular mass is 60.2 u. The molecular mass of ammonia, NH_3, is 17.0 u. Since vinegar molecules are 3.5 times as massive as ammonia molecules, we expect them to diffuse through air more slowly. The lighter ammonia molecules move at higher speeds, bounce more rapidly between molecules of oxygen and nitrogen in the air, and thus we expect them to diffuse through the air more rapidly.

To make the contest fair, you need to make sure the concentrations of your ammonia and vinegar are about the same. Both products are dilute aqueous solutions. When I tried this, the label on the vinegar said it was "5% acidity," which I took to mean 5% by volume acetic acid. The ammonia label did not specify the concentration. A few minutes of online research led to the Safety Data Sheet (SDS) for the ammonia brand (which is also sold under many other brand names). The SDS stated that the ammonia concentration was "<3%." From this I knew to mix some vinegar about 50-50 with water to bring its concentration down to be approximately equal to the ammonia concentration. (Experimental methods such as *titration* can be used to determine the concentrations of both the acetic acid and the ammonia more precisely. We discuss titration when we address acids and bases in Chapter 9.)

For the experiment, use two small bowls with tight-fitting plastic lids. To prevent odors from contaminating the room where your experiment is conducted, take the bowls to a separate room, fill each one with one of the solutions, and put on the lids to seal the bowls.

Perform the experiment in a room with still air (no fans, windows closed, etc.) You need two people to conduct the experiment. One person can be the odor sniffer and operate the stopwatch. The other person can be the one to open the bowls. Do each solution separately. Place one bowl on a table about 24 inches in front of the sniffer. As one person removes the lid, the sniffer starts the timer then sits quietly, breathing deeply, waiting for the fumes to arrive. As soon as the sniffer detects the odor, he or she stops the timer. Wait a few minutes for the air to clear and repeat the test with the other solution. When I tried this, my sniffer smelled the vinegar in 34 seconds and the ammonia in 11 seconds, just as we expected.

perature, molecules in the air collide with each other around five billion times per second! Because of such high energies, the molecules in a gas diffuse much faster than liquid molecules do. Recall from Figure 6.3 that the lighter the molecules are, the higher their average velocities. One can conduct a simple diffusion experiment with vinegar and ammonia to test this, as described in the box above.

Since the particles in a gas are spread so far apart, the density of gases is orders of magnitude lower than liquid and solid densities. As we will see, the volume of a gas is a function of both the temperature and pressure. As a result, gases are quite compressible: doubling the pressure on a gas reduces its volume by half. We return to study gases in more detail in the next section and in the next chapter.

6.2.5 Plasmas

A *plasma* is an ionized gas. At the very high energies where ionization occurs, molecules break apart, so it is not really meaningful to think of molecules in the context of plasmas. Instead, think in terms of single-atom gases, such as the noble gases. Neon signs, for example, are illuminated by the high-energy neon plasma inside the glass tube. As you recall from Chapter 1, each atomic species has its own emission spectrum, and the colors of light emitted by a plasma depend on the atomic species involved. The neon atoms ionized in a neon sign produce an intense orange light, shown in the lettering of the top image in Figure 6.13. Nitrogen ions emit a beautiful violet light, shown in the nitrogen gas tube in the center image of the figure. During a lightning strike, O_2 and N_2 molecules in the air are ripped apart and atoms of nitrogen and oxygen are ionized to produce a plasma. Since air is 78% nitrogen, the bright violet glow of a lightning strike at the bottom of the figure is similar in color to the light emitted by the nitrogen gas tube.

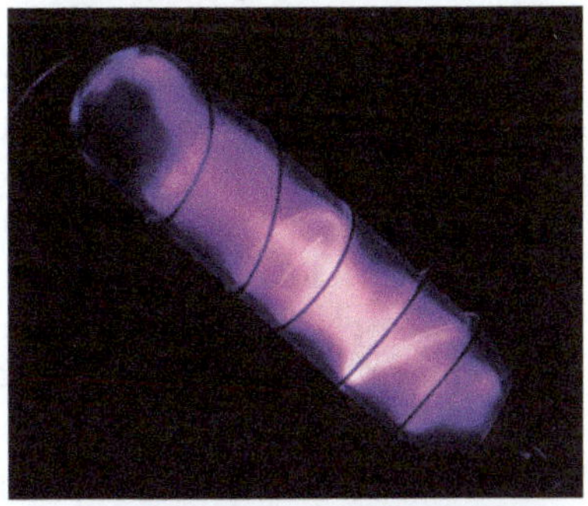

Figure 6.13. Plasmas produced by a neon sign (top), a tube of energized nitrogen gas (middle), and lightning (bottom).

The energy necessary to ionize a gas can be supplied by intense heat, as in the colored flames of a fire or welder's cutting torch—also plasmas—or by high-voltage electricity. Heat, as in a fire, excites the electrons in the atoms so much that they are broken free from the atoms; high voltage, as in neon signs and lightning, attracts the electrons so forcefully that they are torn away from the atoms. Either way, the result is a gas of ions, electrons, protons, and neutrons. Since there are so many charged particles in a plasma, plasmas conduct electricity and are affected by magnetic fields. The night-time sky colors of the *aurora borealis* or "northern lights" (Figure 6.14) are the result of a plasma. Ions from the solar wind emitted by the sun shower the earth constantly. The magnetic field of the earth, called the *magnetosphere*, shields life of the earth from the harmful effects of all this so-called *ionizing radiation*. But the ions interact with the magnetic field in the earth's upper atmosphere, resulting in the beautiful light displays visible at night in far northern and southern latitudes.

Figure 6.14. The sky light produced by the plasma in one of the polar auroras.

Plasmas are exciting to see (pardon the pun) and fascinating to study. Unfortunately, they play no significant role in the phenomena we study at this level in chemistry, so I leave them for you to study more about later and move on to more detailed analysis of the other three states of matter.

6.2.6 Phase Transitions and Phase Diagrams

As described in the previous few pages, for a given pure substance, there is a distinct temperature at which a solid substance melts, depending on the energy necessary for the atoms or molecules to break free of the forces of the bonds holding them together. The change of state or phase we call *melting* is an example of a *phase transition*. When studying phase transitions, a useful graphical aid is the *phase diagram*. The phase diagram for each substance is unique, but a typical, generic phase diagram is shown in Figure 6.15. This graph indicates the states of a substance for various combinations of pressure and temperature, so to distinguish this diagram from another type of phase diagram we will see shortly, we will call this a "P-T phase diagram."

The axes on the P-T phase diagram are pressure and temperature, and the large regions on the diagram represent the conditions of temperature and pressure that result in each of the different phases or states. Notice the dashed horizontal line at atmospheric pressure. The temperature where that line crosses the boundary from solid to liquid (point 1) represents the melting point (also the freezing point) of the substance at atmospheric pressure, and the temperature where that line crosses the boundary from liquid to vapor (point 2) represents the boiling point at atmospheric pressure. From this diagram, it is clear that at lower pressures, the melting and boiling points are lower. One consequence of this is that water boils at a much lower temperature at high altitude because the pressure at high altitude is substantially lower than at sea level. When preparing foods such as rice or vegetables that require boiling water, cooking times at high altitudes are longer than usual because the temperature of the boiling water is lower than the 100°C people are used to at low altitudes. At high altitudes, some food preparations require water to be at temperatures higher than the boiling point, which is not possible to achieve without a pressure cooker. A pressure cooker has a lid that clamps onto the cooking pot and seals it. As water boils and produces steam inside the pot, the pressure increases, which increases the boiling point so that the food can cook properly.

Looking again at Figure 6.15, notice that at the temperature and pressure of point 1, the substance can exist as a solid. If you have a solid at these conditions and hold the temperature constant while reducing the pressure, the solid first melts and then when the pressure gets low enough, the substance boils as it transitions to the vapor state (point 3).

There are six phase transitions labeled on the diagram. You are probably familiar with most of these. You have probably seen dry ice (frozen CO_2) exhibiting *sublimation*. Dry ice does this because on the phase diagram for carbon dioxide, the curves are higher up the graph and the horizontal atmospheric pressure line passes from the solid region to the vapor region. So when the temperature of dry ice is increased at atmospheric pressure, the ice undergoes the phase transition from solid to vapor—sublimation.

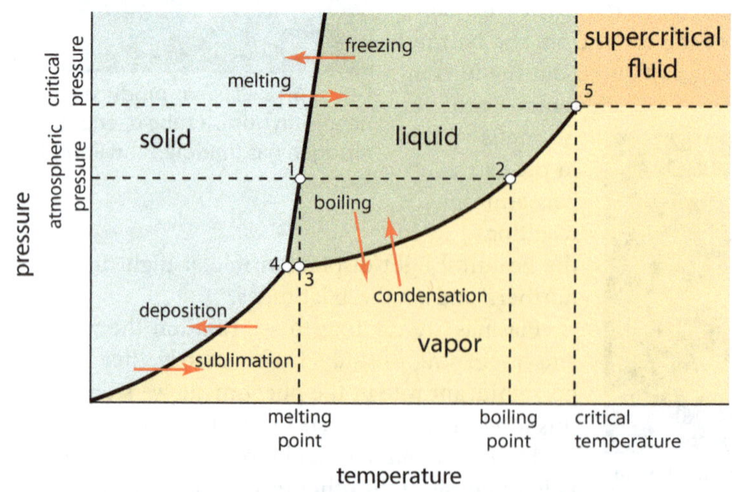

Figure 6.15. Generic P-T phase diagram.

Deposition occurs when a vapor transitions directly to the solid state, the reverse of sublimation. The formation of frost is an example of deposition, as water vapor in the air deposits onto a cold surface and freezes. This can result in some beautiful patterns, as illustrated in Figure 6.16. Semiconductor manufacturers use deposition to apply a thin layer of a metallic compound such as titanium nitride or an insulating compound such as SiO_2 to a silicon substrate. This process is called *chemical vapor deposition* (CVD). (See the photo on the opening page of this chapter.)

Figure 6.16. Deposition of water vapor can create beautiful crystal shapes.

Back again to Figure 6.15, point 4 in the diagram is called the *triple point*. This is the temperature and pressure where all three states of a substance can coexist in equilibrium. Point 5 in the diagram is called the *critical point*, located by the *critical temperature* and *critical pressure*. The critical temperature is the highest temperature where the liquid state of the substance can exist. Above this temperature, regardless of the pressure, the substance does not liquefy. The critical pressure is the lowest pressure at which the liquid state can exist at the critical temperature. At conditions above the critical temperature and critical pressure, the substance is in a state referred to as a *supercritical fluid*. In this region, distinct liquid and vapor states do not exist. At pressures close to the critical pressure, the fluid is more vapor-like. At very high pressures and close to the critical temperature, the fluid is more liquid-like. One of the most common applications of a supercritical fluid is in producing decaffeinated coffee. This process uses supercritical CO_2, and the caffeine extracted from the coffee is sold to soft drink manufacturers for use in carbonated beverages.

Critical constants for several substances are shown in Table 6.3. The critical constants are very important for engineering applications with fluids. It would be pointless, for example, to try to liquefy propane at temperatures above 97°C. Liquid propane cannot exist at such temperatures. Ammonia is easy to liquefy—room temperature is well below its critical temperature. Oxygen can be liquefied if it is first chilled to below –118.6°C.

The P-T phase diagram for water is shown in Figure 6.17. The dashed horizontal line at 1 atm represents atmospheric pressure at sea level. The two points shown on that line are the melting point and boiling point of water, at 0°C and 100°C, respectively.

The phase diagram for water is unique. Notice that the line separating the solid and liquid regions has a negative slope. In almost every other substance, this boundary has a positive slope, as shown in the generic diagram of Figure 6.15. In the case of water, the negative slope of this line is closely related to the fact that the density of liquid water decreases as the water is cooled from 4°C to 0°C. The density of other substances increases as the substance freezes. But as we have discussed before, the hydrogen bonding between water molecules causes the molecules to spread out and form hexagonal patterns just before freezing.

Substance	Critical Temperature (°C)	(K)	Critical Pressure (atm)
ammonia, NH_3	132.4	405.6	112.1
carbon dioxide, CO_2	30.9	304.1	72.8
nitrogen, N_2	–147.0	126.2	33.5
oxygen, O_2	–118.6	154.6	49.8
propane, C_3H_8	96.7	369.9	41.9
water, H_2O	373.9	647.1	217.8

Table 6.3. Critical constants for a few substances.

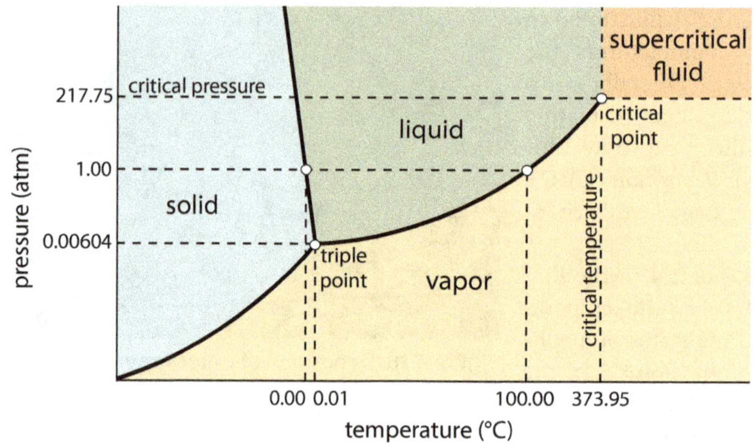

Figure 6.17. P-T phase diagram for water. (In order to make small features visible, the scales are not drawn linearly and are not in proportion.)

The phase diagrams help explain this unique behavior. In the generic diagram of Figure 6.15, if a substance is at the triple point and the pressure is increased while the temperature is held constant (a line straight up from the triple point), you see that the substance must become entirely solid. However, for water, the same change results in liquid water. Increasing the pressure packs the molecules closer together, disrupting the hexagonal hydrogen bonding in the ice and increasing the density. Notice that the same thing happens at atmospheric pressure: increasing the pressure at 0°C turns ice into water. Even if the temperature is below 0°C, if the pressure is increased enough, the ice melts. This is what makes ice skating possible. The weight of the skater is supported only by the small area of contact between the thin skate blade and the ice (and basically on only one blade at a time while the skater is moving). The small area results in a very high pressure under the skate blade. This high pressure under the blade causes a layer of water to form, which lubricates the skate so it slides easily over the ice.

6.2.7 Heat Capacity, Heat of Fusion, and Heat of Vaporization

We have seen that the difference between solids, liquids, and gases is in the internal energy of a substance. If you take a solid substance at a temperature below its melting point and heat it continuously, the added heat causes the substance to go through the following five changes:

1. The temperature of the solid increases to the melting point.
2. The substance melts, maintaining the temperature at the melting point as it does.
3. As a liquid now, the temperature of the liquid begins at the melting point and increases steadily to the boiling point.
4. The substance vaporizes, maintaining the temperature at the boiling point as it does.
5. As a vapor now, the temperature of the substance begins at the boiling point and increases.

Each of these five situations requires a specific amount of heat to accomplish, depending on the substance involved. If we begin with a vapor at a temperature above the boiling point and continuously cool the substance, it passes through the same five changes in reverse, requiring the same quantities of heat, the only difference being that heat is being removed from the substance by cooling instead of added by heating.

The situations just described are represented on a second type of phase diagram, shown in Figure 6.18. In this diagram, the axes are labeled temperature and heat, so we will call this a "T-H phase diagram." The lower graph in the figure is an enlargement of the circled region in the upper graph. The temperatures and relative heat quantities in this particular diagram are for H_2O. Similar diagrams can be drawn for other substances, but we will limit ourselves here to discussing H_2O.

The purpose of the T-H phase diagram is to show the quantities of heat that must be added to or removed from a quantity of ice, water, or steam in order to change its temperature or effect a phase transition. As you see in the upper graph, there are five distinct sections to the graph, corresponding to the five changes listed above. On the three diagonal sections of the graph, the H_2O is in the solid, liquid, or vapor state (from lower left to upper right). The horizontal sections represent the phase transitions between ice and water (the one at 0°C) and between water and steam (the one at 100°C). From our discussion in previous pages, it should be clear that a T-H phase diagram is only valid for a specific pressure. The discussion here focuses on water at atmospheric pressure.

The red horizontal arrows represent the quantities of heat required, designated as Q, to change the temperature of a quantity of water or to change its state. The variables and constants shown in the equations above the red arrows are defined in Table 6.4. It should seem intuitively obvious that the amount of heat required in any change of temperature or state depends on the quantity of water involved (the number of moles). The more water you have in a pan, the more

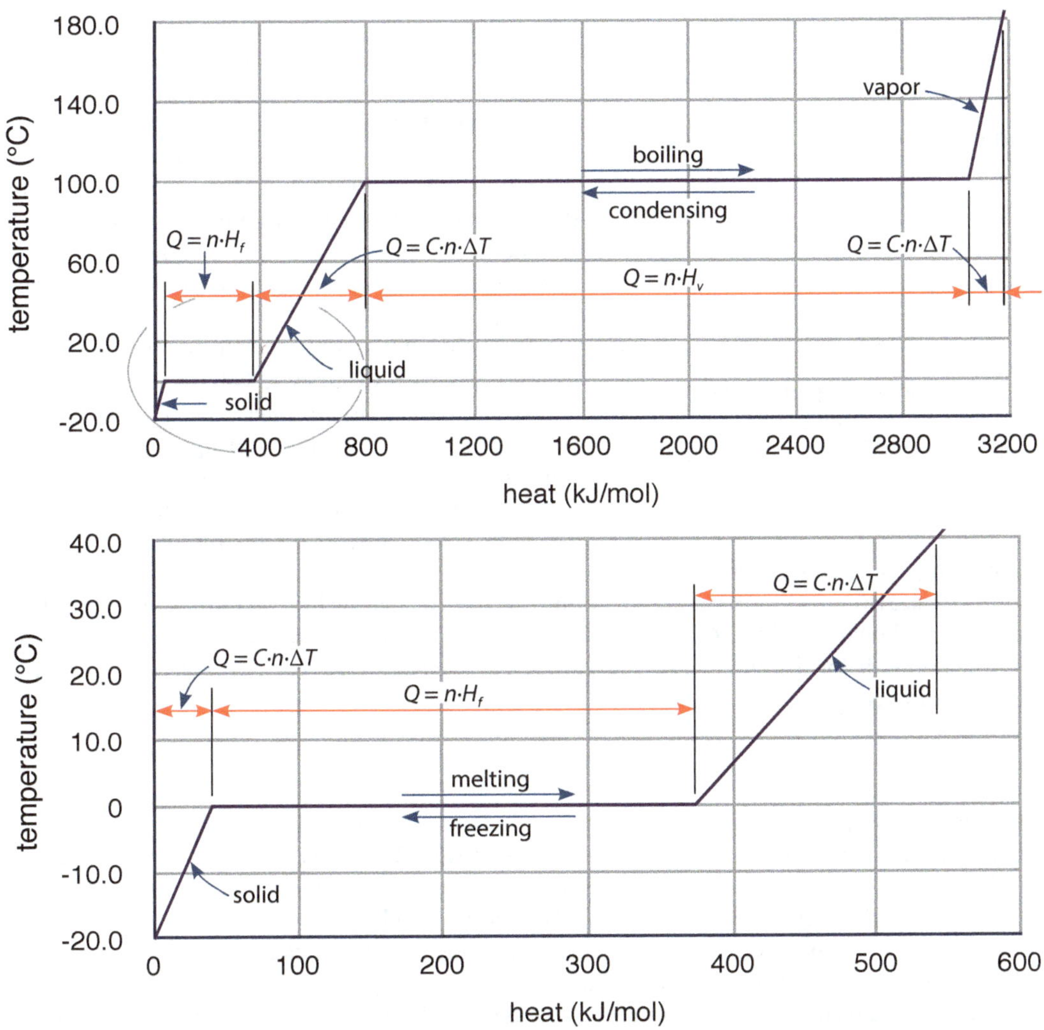

Figure 6.18. T-H phase diagram for H_2O showing quantities of heat required to change the temperature or effect a phase transition. The lower graph is an enlargement of the circled area in the upper graph. (The zero reference for the horizontal axis in both graphs is arbitrarily set to be at $T = -20°C$.)

Quantity	Units	Definition
Q	J	variable representing the quantity of heat that must be added to or removed from the substance
C	J/(mol·K)	*molar heat capacity* of the substance, defined as the quantity of heat required to raise or lower the temperature of one mole of the substance by 1 K
n	mol	variable representing the number of moles of the substance
ΔT	K	variable representing the change in temperature of the substance
H_f	J/mol	*molar heat of fusion*, defined as the quantity of heat required to melt or freeze one mole of the substance
H_v	J/mol	*molar heat of vaporization*, defined as the quantity of heat required to boil or condense one mole of the substance

Table 6.4. Thermal variables and constants for calculating quantities of heat in phase transitions and temperature changes.

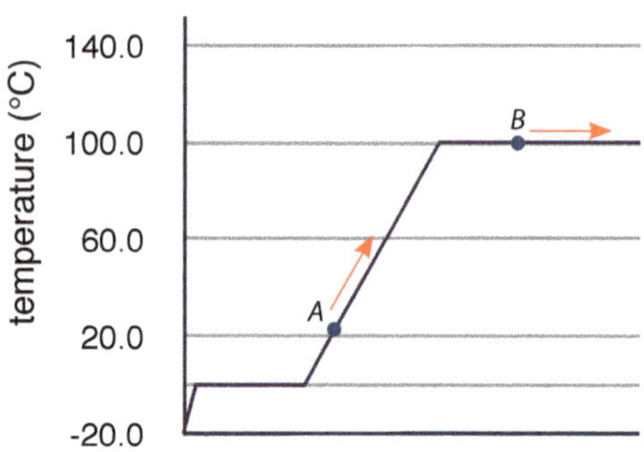

Figure 6.19. At point *A*, liquid water is at 22°C and warming up. At point *B*, a mixture of water and vapor is at 100°C and transitioning to vapor.

heat it takes to bring it to a boil. For changes of temperature, the *molar heat capacity* of a substance specifies the quantity of heat needed to change its temperature. The value of the molar heat capacity depends on what state the substance is in. For phase transitions, the constants that specify the amount of heat involved are the *molar heat of fusion* for melting or freezing, and the *molar heat of vaporization* for boiling or condensing.

To further illustrate how to read the T-H phase diagram, Figure 6.19 shows H_2O in two different situations. At point *A*, liquid water is at 22°C and warming up, as indicated by the upward pointing arrow. At point *B*, the H_2O is a mixture of water and steam at 100°C because it is in the middle of a phase transition. The arrow indicates that the mixture is on its way to becoming completely vapor.

There are two main forms of the equations used for calculating heat quantities (with variables defined as shown in Table 6.4). One is for the energy associated with a change in temperature, and the other is for the energy associated with a phase transition. For changes in temperature, the equation is

$$Q = Cn\Delta T \tag{6.1}$$

In this equation, the change in temperature, ΔT, is always defined as $\Delta T = T_f - T_i$, where T_f and T_i are the final and initial temperatures, respectively. Used this way, a positive value for Q indicates that heat is being *added* to the substance. When a substance is being cooled down, T_f is lower than T_i, so Q is a negative value. This indicates that heat is being *removed* from the substance. Note that since Celsius degrees and Kelvin "degrees" are the same size, you can calculate ΔT with either one and get the same result.

The second form of equation is used to calculate the heat associated with phase transitions:

$$Q = nH_f \tag{6.2a}$$

$$Q = nH_v \quad (6.2b)$$

You simply use the appropriate constant for the phase transition involved. To calculate the heat associated with melting or freezing, use the molar heat of fusion, H_f; to calculate the heat associated with boiling or condensing, use the molar heat of vaporization, H_v. We limit our examples here to computations involving water. The values of the thermal constants C, H_f, and H_v are physical properties of the substance under consideration. The values for H_2O are shown in Table 6.5 and in Table B.5 of Appendix B.[3]

Constant	Value
C (ice, −10°C)	$0.0364\ \frac{\text{kJ}}{\text{mol}\cdot\text{K}}$
C (water)	$0.0752\ \frac{\text{kJ}}{\text{mol}\cdot\text{K}}$
C (steam, 110°C)	$0.0369\ \frac{\text{kJ}}{\text{mol}\cdot\text{K}}$
H_f	$6.01\ \frac{\text{kJ}}{\text{mol}}$
H_v	$40.7\ \frac{\text{kJ}}{\text{mol}}$

Table 6.5. Thermal constants for water.

▼ Example 6.1

Determine the quantity of heat that must be added to 5.75 moles of ice at 0°C to melt it.

This problem is about a phase transition, melting, so we use Equation (6.2a) with the molar heat of fusion, H_f, from Table 6.5. We write the equation, insert the values, and compute the result.

$$Q = nH_f = 5.75\ \text{mol} \cdot 6.01\ \frac{\text{kJ}}{\text{mol}} = 34.6\ \text{kJ}$$

▲

▼ Example 6.2

Determine the quantity of heat released when 797 mL of water cools from 25.0°C to 10.0°C. Assume that the average density of water over this temperature range is 0.998 g/mL.

First we need to convert the quantity of water from a volume to mass, and then to number of moles. Using the density equation, we convert the volume to mass.

$$\rho = \frac{m}{V}$$

$$m = \rho V = 0.998\ \frac{\text{g}}{\text{mL}} \cdot 797\ \text{mL} = 795.4\ \text{g}$$

Now we determine the number of moles by using the molar mass as a conversion factor:

$$795.4\ \text{g} \cdot \frac{1\ \text{mol H}_2\text{O}}{18.02\ \text{g H}_2\text{O}} = 44.14\ \text{mol H}_2\text{O}$$

Finally, we calculate the heat. Our given quantities are:

3 Heat capacity values for ice and steam vary quite a bit with temperature, but within about 20°C of the melting or boiling point, the values given are reasonable approximations.

$n = 44.14$ mol

$T_i = 25.0°C$

$T_f = 10.0°C$

From Table 6.5, the molar heat capacity of water is 0.0752 kJ/(mol·K). We write down the change of temperature equation, Equation (6.1), insert the values, and compute the result.

$$Q = Cn\Delta T = Cn\left(T_f - T_i\right) = 0.0752\ \frac{\text{kJ}}{\text{mol}\cdot\text{K}} \cdot 44.14\ \text{mol} \cdot \left(10.0°\text{C} - 25.0°\text{C}\right) = -49.8\ \text{kJ}$$

Note that the temperature units cancel because a change in Celsius degrees is equal to a change in kelvins. As mentioned above, the negative sign on this quantity means that heat is being removed from the water.

▲

6.2.8 Evaporation

The kinetic-molecular theory tells us that in a liquid at a given temperature, there is a distribution of speeds among the liquid molecules, just as there is with the molecules in a gas. At ordinary temperatures, there are many molecules in a liquid that have enough energy to transition into the vapor state if they can only break free of the other molecules blocking their paths to the open air. Near the surface of the liquid, these higher energy molecules aren't blocked by other liquid molecules. Consequently, they transition to the vapor state, flying away from the liquid and joining the molecules in the air. The phase transition from liquid to vapor that occurs this way is called *evaporation*, illustrated in Figure 6.20. The warmer the liquid is, the higher the average energy of the molecules in the liquid, and the faster evaporation occurs.

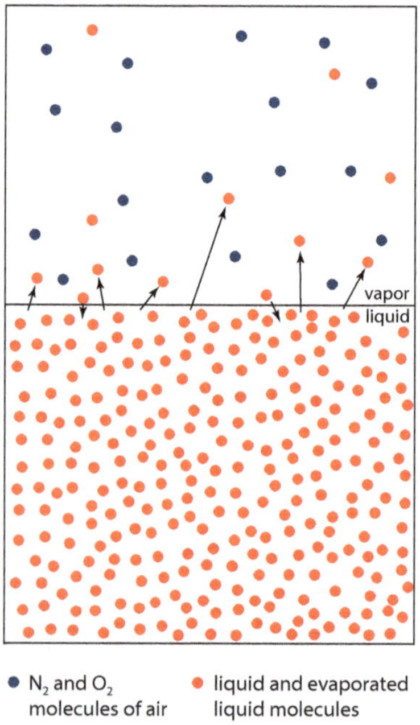

Figure 6.20. Liquid particles transitioning to the vapor state during evaporation.

In the process of evaporation, it is the highest energy molecules in the liquid that transition to the vapor state. As the molecules with higher energy depart the liquid, the average energy of the molecules left behind decreases. (This is analogous to the fact that every time a graduating class of seniors leaves a school, the average age of the students left behind drops because all the oldest students have left.) Since temperature varies with internal energy, the temperature of the liquid decreases during evaporation, an effect known as *evaporative cooling*. In warm-weather locations, evaporative cooling is used at outdoor seating areas at restaurants to keep patrons cool. Misting systems spray a fine mist of water into the air. As the molecules of water in the tiny water droplets transition to the vapor state (evaporate), the temperature of the water, and thus of the surrounding air, decreases. The amount of cooling can be calculated using H_v.

Evaporative cooling is used by the human body to keep the body cool. The human body is a thermodynamic engine, and like every other engine it constantly produces waste heat, just as boilers emit waste heat through their smoke stacks and cars through their tailpipes. In a room at 72°F, the temperature difference between the body (98.6°F) and the air is high enough that waste heat escapes and keeps the body cool. This happens by conduction between the body and air molecules near the surface of the skin, and to some extent through infrared electromagnetic radiation. But in a hot environment, where heat cannot escape from the body fast enough, the body begins secreting water onto the surface of the skin through the sweat glands, shown in Figure 6.21. The drops of moisture on the skin then evaporate, lowering the temperature of the remaining moisture and cooling the body.

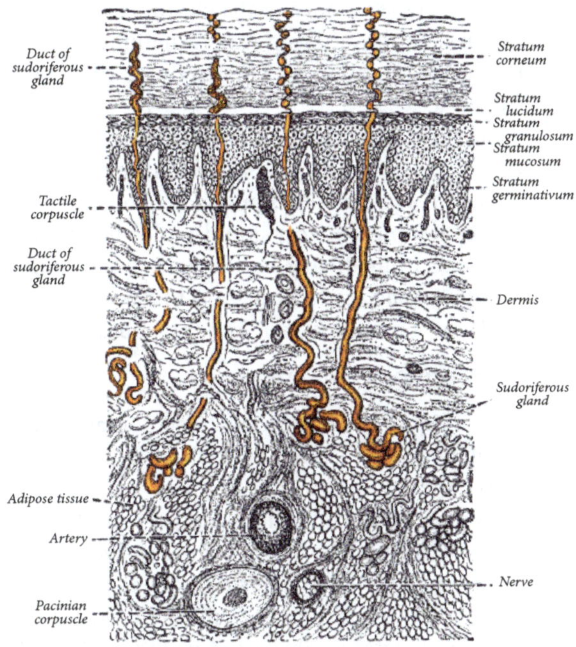

Figure 6.21. Cut-away view of the skin showing the sweat glands (orange). This drawing is from Henry Gray's classic 1858 text, *Gray's Anatomy*.

I am compelled to pause here to note *how stupendous this is.* Sweat glands (called *sudoriferous* glands in the world of anatomy) are found in our skin on almost the entire surface of our bodies. In a warm environment, an exquisite and complex control process called *thermoregulation*, controlled by the *hypothalamus* in the brain, calls upon the sweat glands to begin channeling water to the surface of the skin where evaporative cooling assists in maintaining the body's temperature. Thermoregulation, like any control process, requires a system of input sensors, a decision-making algorithm that compares current conditions to desired conditions, and an output mechanism that controls some kind of variable that has the power to change current conditions and bring them toward the desired conditions.

A diagram of the thermoregulation control system in the human body looks exactly like a sophisticated control system in a manufacturing plant. The system even includes special information channels to cause the system response to kick in more rapidly in the case of a sudden alarm—the so-called *fight or flight response*. In anticipation of a fast-approaching temperature increase due to increased exertion while contending with danger—running, rescuing people, and so on—the cooling system kicks in before the body temperature even increases. (This is why you can start sweating almost instantly when faced with a sudden danger, such as almost having a car crash.) In fact, many other systems in our bodies increase production prior to actual need in the same way, including increased heart rate in anticipation of increased oxygen demand by the cells, increased respiratory rate to bring in more oxygen, dilation of blood vessels to allow faster blood flow, and release of energy sources such as fat and glycogen in anticipation of immediate muscle use. By itself, the thermoregulation system in our bodies is stunning. Add in all these other systems, each of which is equally amazing, and the wonder of our bodies is breathtaking.

Scientists studying the body have often commented that the complexity and organization in the human body is a constant source of amazement. One famous scientist commented that the life processes in just a single bacterium were mind-boggling (and still beyond our full understanding). How much more amazing are the processes in the most complex object in the universe—the human body! You will learn more about these processes in future biology or anatomy classes. For now though, next time you break out in a sweat, you may just pause for a moment to bask in the wonder of the elegant way your body works for you and all the amazing ways it maintains itself. If you have a proper appreciation for your body, you will want to take good care of it!

Returning briefly to Figure 6.20, notice in the figure that a few of the liquid molecules that have already evaporated are recondensing into the liquid state. This might happen, for example, if a high-energy liquid molecule transitions to the vapor state, immediately collides with a molecule in the air near the liquid surface, and having lost a lot of its energy in the collision, condenses back into the liquid state.

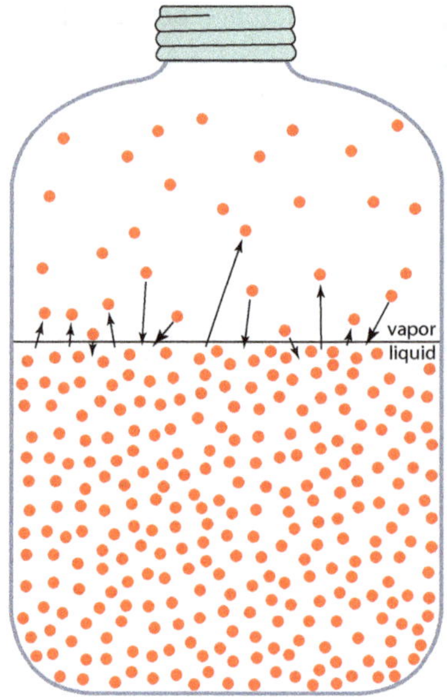

Figure 6.22. The vapor pressure of a substance in a sealed container causes an equilibrium in which vaporization and condensation are happening at equal rates.

6.2.9 Vapor Pressure

If a liquid is placed in a sealed container, molecules of the liquid begin to evaporate into the vapor space in the container above the liquid, causing the pressure in the vapor to begin increasing. This continues until an equilibrium is reached. In this equilibrium, the density of the molecules in the vapor state is high enough that the rate of molecules transitioning into the vapor state equals the rate of molecules transitioning back into the liquid state, as illustrated in Figure 6.22. The pressure at which this equilibrium is reached is called the *vapor pressure.* The vapor pressure is a measure of the rate of evaporation of a liquid and its *volatility*—its tendency to become a vapor. More volatile substances have higher vapor pressures.

You have probably noticed that if you open a container of a liquid with a high vapor pressure, such as paint thinner or gasoline, you briefly hear the sound of gas escaping from the container when you break the seal of the container's lid. This effect is due to the vapor pressure of the liquid inside the container.

The vapor pressure of liquids is strongly dependent on temperature, as indicated by the curve for the vapor pressure of water in Figure 6.23. Looking carefully at the curve of the graph, we see that at 100°C, the boiling point of water, the vapor pressure appears to be—and is—760 Torr, which is normal atmospheric pressure. This is because *the vapor pressure curve also indicates the pressure at which a liquid boils at a given temperature.* At 100°C, water boils at a pressure of 760 Torr. At 40°C (104°F) water boils at a pressure of about 55 Torr. We never see water boiling at 40°C because atmospheric pressure is never as low as 55 Torr. But when water at 40°C is placed in a vacuum chamber and the pressure is reduced to about 55 Torr, the water begins boiling, *and it is*

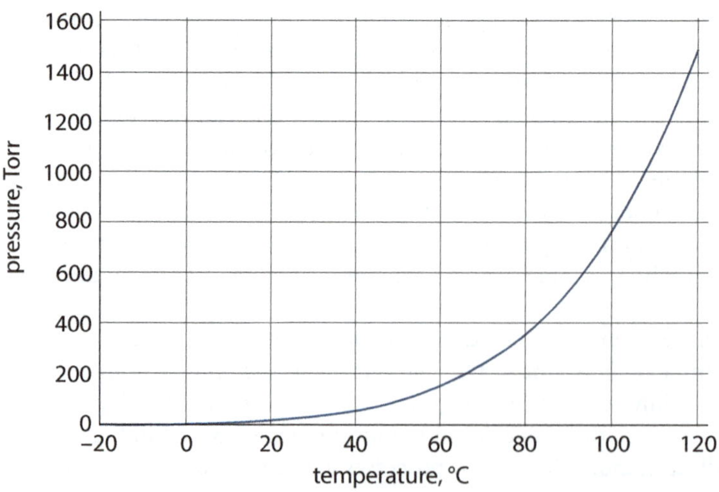

Figure 6.23. Vapor pressure curve for water.

not that hot—40°C is the temperature of hot bath water. When water is at atmospheric pressure and 100°C, it boils because atmospheric pressure is the vapor pressure of water at that temperature. At that temperature and pressure, all the water begins vaporizing at once. And by the way—the bubbles erupting from a boiling liquid are *not* air bubbles. They are bubbles of *vapor*—liquid that has transitioned to the vapor state and is on its way out.

Recall the discussion earlier about longer cooking times due to water boiling at lower temperatures at higher altitudes. In Leadville, Colorado, atmospheric pressure is about 502 Torr. From the graph in Figure 6.23, you see that water boils at that pressure at about 89°C (192°F). At the top of Mt. Elbert, the highest mountain in Colorado, atmospheric pressure is down to 418 Torr, resulting in a boiling point of 84°C.

Chapter 6 Exercises

Note: Constants and conversion factors are in Tables B.2 and B.3 in Appendix B.

SECTION 6.1

1. Give a physical description of how to interpret a temperature measurement. In other words, what does temperature *mean*?
2. What is the Maxwell-Boltzmann distribution?
3. State the approximate range of velocities and the mean velocity found in Argon atoms at 25°C (see Figure 6.3).
4. Explain what holds the mercury up in the tube of a mercury barometer.
5. As indicated in the box on page 163, the pressure under the surface of a liquid is given by $P = \rho g h$. Use this equation to calculate the height of the fluid column in a barometer that uses water instead of mercury. (Again, see Appendix B for reference data.)
6. The pressure inside an industrial cylinder of oxygen gas is 1,850 psi. State this pressure in Pa, kPa, Torr, atm, and bar.
7. Consider a typical ice hockey skate. The contact surface between the blade and the ice when the skater is gliding is 2.9 mm wide and 57 mm long. Determine the pressure on the ice under a 200.0-lb skater with one skate in contact with the ice. State your answer in atm.
8. Of the five statements given in this section outlining the kinetic-molecular theory of gases, some of them may be taken as fully accurate descriptions and some of them are based on obvious approximations. State and describe the approximations inherent in the kinetic-molecular theory of gases.
9. Explain why containers of gas, such as cans of spray paint or mosquito spray, include warning labels telling people not to leave the cans in the direct sunlight. Address this question in detail at the molecular level.
10. Describe conditions under which a real gas might exhibit ideal gas behavior.
11. Describe which of the following gases you would expect to exhibit ideal gas behavior to the largest degree, and which to the smallest degree: Ar, HF, H_2O, H_2, O_2, He, and NH_3.

SECTION 6.2

12. Write paragraphs describing each of the four states of matter.
13. Explain why solids and liquids are essentially incompressible, while gases are compressible.
14. Explain surface tension, then use surface tension to explain why capillary action and wicking occur.

15. Explain diffusion and its cause.

16. Which would you expect to diffuse faster in air: molecules of ethanol, C_2H_5OH, or butane, C_4H_{10}? Explain your response.

17. Explain evaporation and its cause.

18. Explain why evaporative cooling helps remove heat from the body in a hot environment.

19. Calculate each of the following. (Again, see Table B.5 in Appendix B for reference data.)
    a. the amount of heat required to warm 3.47 moles of water from 25°C to 55°C.
    b. the mass of steam required to release 3.65×10^5 kJ on condensation.
    c. the amount of heat required to melt 12.00 kg of ice at 273.15 K.
    d. the volume of water that can be heated from 20.0°C to 30.0°C with 7.32 kJ of heat. (Use the density at 25°C—the approximate average density of water over this temperature range.)
    e. the total amount of heat required to warm 5.00 kg of ice from –25.0°C to 0.0°C and then melt it.

20. The P-T phase diagram for carbon dioxide is shown below. Use this diagram to answer the following questions.

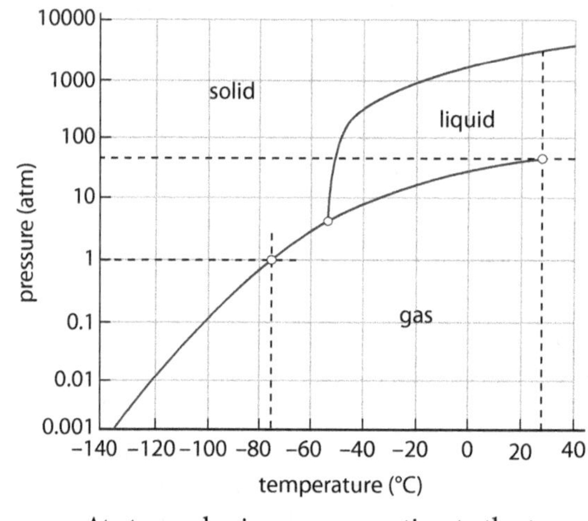

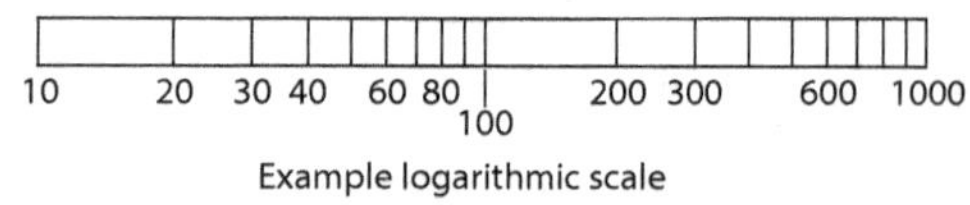

    a. At atmospheric pressure, estimate the temperature at which CO_2 sublimates.
    b. Estimate the pressure and temperature at the triple point, and state your results in both atm and Torr, and both °C and K. (Note that the pressure scale is logarithmic, not linear. See the reference scale in the figure.)
    c. What are the melting and boiling points at a pressure of 7,600 Torr?
    d. What is the lowest pressure at which liquid CO_2 can exist at 28°C?
    e. At a temperature of –20°C, what are the pressures at which CO_2 boils and freezes?
    f. At a temperature of 0°C, what is the vapor pressure?
    g. At 40°C, what happens to gaseous CO_2 if the pressure is increased to 10,000 atm?

21. When you squeeze an ice cube between your teeth—but not hard enough to crush it—the ice begins to melt away. Use the P-T phase diagram for water to account for this phenomenon.

22. If you want to store liquid water in an environment that is at 120°C, what pressure does the water have to be under? (See Figure 6.23.)

23. On a P-T phase diagram, why does the curve separating the liquid and gas regions end at the critical point?

24. At atmospheric pressure, the melting and boiling points of O_2 are –218°C and –183°C, respectively. The triple point for O_2 is at –219°C and 0.0015 atm, and the critical point is at –119°C and 49.8 atm.

    a. Sketch the phase diagram for O_2, showing the four given points and the regions of the three states.

    b. Determine whether solid O_2 floats on liquid O_2.

    c. At atmospheric pressure, when solid O_2 is heated, does it sublime or melt?

25. Cabin pressure in a commercial jet is around 594 Torr. When the flight attendants make coffee, at what temperature does the water boil? (See Figure 6.23.)

GENERAL REVIEW EXERCISES

26. Write the electron configurations and draw the orbital diagrams for bromine and yttrium.

27. The main component of gasoline is octane, C_8H_{18}. The combustion reaction when octane burns is as follows:

    $$2C_8H_{18}(l) + 25O_2(g) \rightarrow 16CO_2(g) + 18H_2O(g)$$

    a. How many moles of O_2 are required to completely burn 5.00 mol C_8H_{18}?

    b. At 20.0°C, the density of octane is 0.692 g/mL. How many grams of O_2 are required to burn 1.00 gal of octane?

28. Solid ammonium nitrate decomposes to yield nitrogen gas, oxygen gas, and water vapor. Write the formula equation for this reaction and balance it.

29. A 2.50-g strip of zinc metal is placed in an aqueous solution of silver nitrate containing 3.00 g $AgNO_3$. The following single replacement reaction occurs:

    $$Zn(s) + 2AgNO_3(aq) \rightarrow 2Ag(s) + Zn(NO_3)_2(aq)$$

    a. Which reactant is the limiting reagent?

    b. In a complete reaction, how many grams of $Zn(NO_3)_2$ form?

    c. How many grams of the excess reactant remain after the reaction?

30. Ibuprofen, the pain reliever in Advil and Motrin, contains 75.69% C, 8.80% H, and 15.51% O by mass. The molar mass is 206.3 g/mol. Determine the empirical and molecular formulas for ibuprofen.

31. Determine the oxidation state of S in H_2SO_4 and C in CO_3^{2-}.

32. Identify the following bonds as covalent, polar covalent, or ionic in character.

    a. C—N   b. I—I   c. Al—Cl

33. Explain why metals tend to be malleable.

34. For each of the following molecules, draw the Lewis structures and predict the molecular geometry:

    a. SO_4^{2-} b. H_2S c. ClO_4^-

Chapter 7

The Gas Laws

The burners in hot air balloons use liquid propane fuel to heat the air inside the balloon. According to the gas laws, if volume and pressure are held constant as the temperature of the air is raised, the number of molecules of air inside the balloon decreases. Fewer molecules inside the balloon means less dense air, so the balloon floats. The volume of a typical hot air balloon these days is in the range of 2,800 m^3. Heated to 99°C, such a balloon provides a lift force (simply called *lift*) of nearly 1,600 lb. This is about where hot air balloons operate once they have reached altitude. To get up there requires acceleration, and thus more force. To achieve this force, balloonists heat the air up to around 120°C, acquiring a lift of around 1,900 lb.

Balloons achieving lift strictly by hot air are called *Montgolfier* balloons. The Montgolfier brothers were French, and invented the hot air balloon in 1783. In October of that year, one of them went up in their balloon. Others went up that same day and over the next few weeks. The new phenomenon of manned flight caused a sensation. In recognition of their achievement, the Montgolfier brothers' *father* was elevated to the nobility by King Louis XVI of France. (Not sure if the brothers reaped any reward for themselves, but at least Dad enjoyed himself.)

Though lots of fanfare surrounded the Montgolfier's achievement, the honor of the first manned balloon flight is not theirs. They flew the first hot air balloon, but as we will see in this chapter, Jacques Charles' flight in a hydrogen balloon just two months before was the first manned balloon flight.

Objectives for Chapter 7

After studying this chapter and completing the exercises, you should be able to do each of the following tasks, using supporting terms and principles as necessary.

SECTION 7.1

1. Describe and explain Boyle's law, Charles' law, and Avogadro's law.
2. Given initial conditions of a gas, use Boyle's law and Charles' law to compute gas temperature, volume, or pressure.
3. Explain how Charles' law and Avogadro's law demonstrate some of the assumptions in the kinetic-molecular theory of gases.

SECTION 7.2

4. State the conditions associated with standard temperature and pressure (STP).
5. Compare ideal gases to real gases and explain how these concepts apply to practical chemistry.
6. Use the ideal gas law and the ideal gas constant to solve gas problems involving temperature, volume, pressure, and number of moles.
7. Use the ideal gas law, density equation, and molar mass to solve gas problems involving densities and masses of gases.

SECTION 7.3

8. Use Dalton's law of partial pressures to compute partial pressures, mole fractions, and the total pressure in a gas mixture.
9. Use the law of partial pressures and water vapor pressure data to determine the mass of a gas collected over water.

SECTION 7.4

10. Solve stoichiometric problems involving gas volumes.
11. Distinguish between diffusion and effusion.
12. Use Graham's law of effusion to determine relative velocities of gas molecules and to estimate molar mass.

7.1 Early Formulations of the Gas Laws

7.1.1 Boyle's Law

Figure 7.1. Irish-English scientist Robert Boyle (1627–1691).

The names of the gas laws are a sort of hall of fame of the great scientists whose work laid the foundations of modern science. In 1660, the Irish-English genius, inventor, and scientist Robert Boyle (Figure 7.1) first referred to the gas law that now bears his name. *Boyle's law* states that *at a given temperature, the volume of a fixed amount of gas varies in inverse proportion to the pressure.* In mathematical form, the law may be written as

$$V = \frac{k}{P} \tag{7.1}$$

where k is a constant that depends on the units of measure. Graphically, this relationship between volume and pressure is as shown in Figure 7.2. This law can be rewritten as

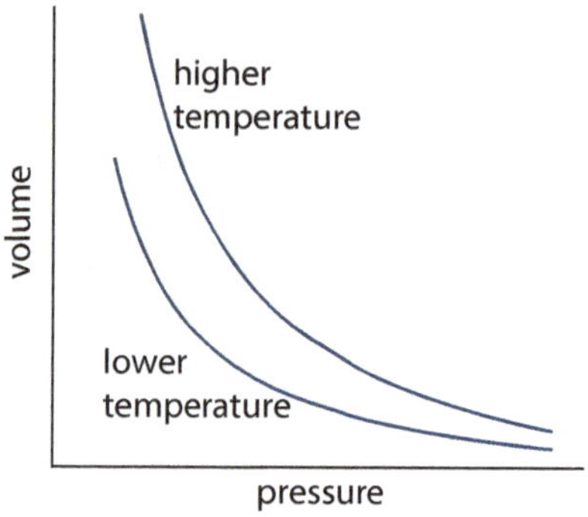

Figure 7.2. Graphical representation of the relationship between volume and pressure for a gas at constant temperature, Boyle's law.

$$VP = k \tag{7.2}$$

which means that the product of volume and the pressure is a constant. In particular, this also means that the product of volume and pressure at one set of conditions must equal the product at a different set of conditions, or

$$V_1P_1 = V_2P_2 \tag{7.3}$$

Boyles' Law

Equation (7.3) represents the most common form Boyle's law is written in today. Note that Boyle's law is simply a pair of ratios. This is clear when the law is written in the form

$$\frac{V_1}{V_2} = \frac{P_2}{P_1} \tag{7.4}$$

Since there are only ratios in this equation, we can solve problems with Boyle's law using any convenient units of measure. If we let V_1 and P_1 represent certain known initial conditions, we can use Boyle's law to determine the new volume, V_2, of the gas at a new pressure, P_2, or vice versa. (Keep in mind, though, that Boyle's law only applies when the temperature is constant.) This is illustrated by the following example.

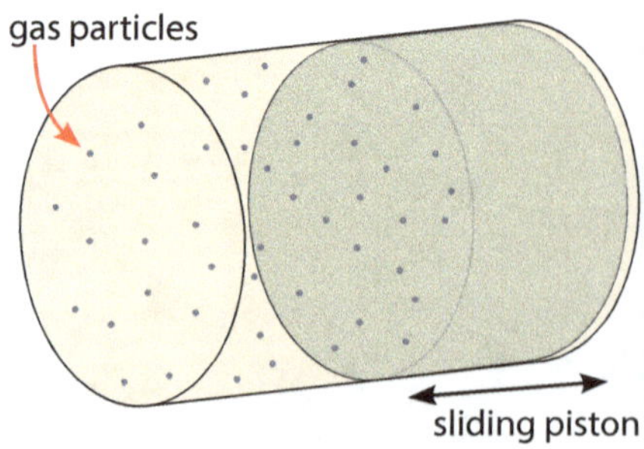

Figure 7.3. Cylinder-piston system.

▼ Example 7.1

The volume of a certain cylinder can be varied by means of an adjustable piston, as shown in Figure 7.3. The cylinder contains 0.75 L of a gas at an absolute pressure of 850 Torr. If the volume is reduced to 0.50 L at a constant temperature, determine the new gas pressure.

We let V_1 and P_1 represent the initial conditions of the gas. The new volume is V_2, and the new pressure—our unknown—is P_2. We first solve the Boyle's law equation for the unknown, P_2.

$$P_2 = P_1 \frac{V_1}{V_2}$$

Next, insert the given values and compute the new pressure. The volume units cancel, giving the new pressure value in the same units as the initial pressure.

$$P_2 = P_1 \frac{V_1}{V_2} = 850 \text{ Torr} \cdot \frac{0.75 \text{ L}}{0.50 \text{ L}} = 1275 \text{ Torr}$$

Rounding this value to the correct number of significant digits we have

$$P_2 = 1300 \text{ Torr}$$

It is important to note here that in ratio equations like Boyle's law, all values must be in absolute units. Typical pressure gauges, such as those used by mechanics to measure tire pressure, do not read absolute pressure because the "zero" of the pressure gauge is at atmospheric pressure. On an absolute pressure scale the zero of the scale is at a complete vacuum. We consider this in more detail in Section 7.2.

Figure 7.4. Jacques Charles (1746–1823) in the first manned balloon flight in 1783.

7.1.2 Charles' Law

By the late 18th century, many scientists in England and France were investigating the properties of gases. This work led to the rise of ballooning as a popular sport for scientifically minded (and fearless) individuals. In August of 1783, Jacques Charles, a French scientist, inventor, and balloonist, flew to an altitude of about 1,800 feet in the world's first manned balloon flight, an event captured in the illustration shown in Figure 7.4. The gas in the balloon was hydrogen. Hydrogen is the lightest gas there is (four times lighter than helium), but is, as you know, highly flammable. Use of hydrogen as a gas to lift manned balloons came to an end with the disastrous burning of the *Hindenburg* in 1937.

In the 1780s, Jacques Charles formulated the law now known as *Charles' law*, but his work went unpublished. In 1802, over 140 years after Boyle's law had been published, French scientist Joseph Louis Gay-Lussac (Figure 7.5) published this gas law and credited the law to Jacques Charles. Charles' law states that *at a constant pressure, the volume of a fixed amount of gas is directly proportional to its temperature*, or

Figure 7.5. French scientist Joseph Gay-Lussac (1778–1850).

$$V = kT \qquad (7.5)$$

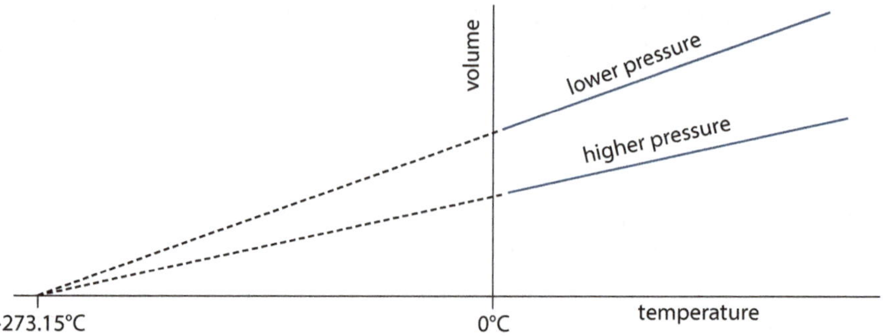

Figure 7.6. Graphical representation of the relationship between volume and temperature for a gas at constant pressure, Charles' law.

As before, k is a constant. Graphically, the relationship between volume and temperature is as shown in Figure 7.6. Charles' law states that volume is a linear function of temperature. An important and interesting feature of this law is that the lines on the graph representing different pressures all converge to the same point on the temperature axis, –273.15°C (0 K), the temperature known as *absolute zero*. This obviously suggests that the volume of a gas goes to zero at absolute zero, which may sound puzzling at first. The resolution of this puzzle lies in the kinetic-molecular theory of gases discussed in the previous chapter. There we saw that the kinetic-molecular theory of gases includes the assumption that the atoms or molecules in a gas are so far apart that the total volume of the particles is *negligible*. Obviously, the total volume of the particles in a gas is not literally zero; atoms do take up space. According to current scientific theory, at absolute zero, molecular motion ceases. So the particles in the gas do not need any room to move around, but they still take up the space of their own total volume.

The interesting thing about Charles' law is that it relates directly to the kinetic-molecular theory. Both laws include the same assumption about the total volume of the particles in a gas. This assumption is also embedded in the *ideal gas law*, which we come to soon. Under many conditions, gases behave as if the total volume of the particles in the gas really is zero, and this means that Charles' law may be used in many cases. However, under extremes of pressure or temperature, all gases depart from the simple curves of Charles' law (and Boyle's law, too).

Charles' law may also be written as

$$\frac{V}{T} = k \tag{7.6}$$

Written this way, we can see that the ratio of volume to temperature for a gas at constant pressure is always the same. This means that, as before with Boyle's law, we can equate the volume-temperature ratio at one set of conditions to the ratio at another set of conditions:

$$\frac{V_1}{T_1} = \frac{V_2}{T_2} \tag{7.7}$$

Charles' Law

With this equation, we can use V_1 and T_1 as initial conditions to compute either the volume or the temperature at a different set of conditions. The principles for using Boyle's law also apply to problems using Charles' law. First, the equation consists of two ratios, so it doesn't matter what the units of measure are. Second, the temperatures must be in absolute units, which means kelvins. (There *is* an absolute scale related to the Fahrenheit scale called the Rankine scale. If you simply must use U.S. Customary Units, you can use temperatures in °R. However, the sensible thing to do in most circumstances is to convert temperatures into kelvins, an absolute scale, and the SI system base unit for temperature.) We touch on this again in Section 7.2. Finally, keep in mind that Charles' law only applies when the pressure is constant.

Example 7.2

A special cylinder and piston arrangement is designed so that the piston is weighted and slides freely in the cylinder. This cylinder is oriented vertically, as shown in Figure 7.7. The weight of the piston provides a constant pressure on the gas, and the volume of the gas increases and decreases with changes in temperature.

A chemist observes that when 1,275 cm^3 of gas are heated to 166.0°F, the gas expands to 1,565 cm^3. Determine the original gas temperature and state the result in degrees Celsius.

We can leave the volumes in cm^3, but the temperature must be in kelvins, so we begin by executing this conversion.[1]

$$T_C = \frac{5}{9}(T_F - 32°) = \frac{5}{9}(166.0° - 32°) = 74.4°\text{C}$$

$$T_K = T_C + 273.15 = 74.4 + 273.15 = 347.6 \text{ K}$$

We are seeking to determine the original temperature of the gas, which is T_1. So we need to solve the Charles' law equation for this variable. Begin by cross multiplying and dividing by V_2:

$$\frac{V_1}{T_1} = \frac{V_2}{T_2}$$

$$V_1 T_2 = V_2 T_1$$

$$T_1 = T_2 \frac{V_1}{V_2}$$

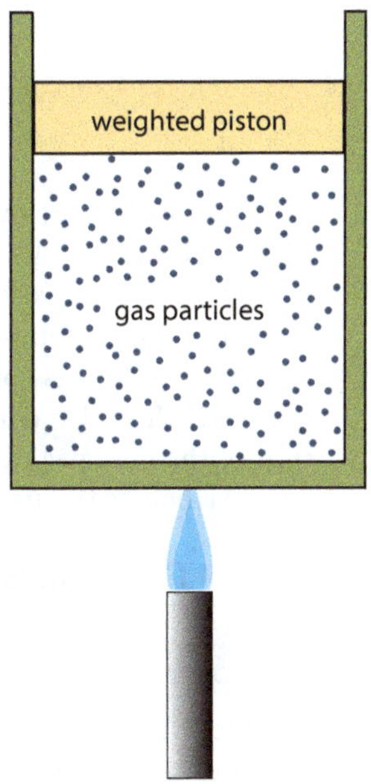

Figure 7.7. Apparatus for visualizing Charles' Law.

Next we insert the values and calculate the original temperature.

$$T_1 = T_2 \frac{V_1}{V_2} = 347.6 \text{ K} \cdot \frac{1275 \text{ cm}^3}{1565 \text{ cm}^3} = 283.2 \text{ K}$$

Finally, we convert this to °C as the problem requires, and round to three significant digits.

$$T_K = T_C + 273.15$$

$$T_C = T_K - 273.15 = 283.2 \text{ K} - 273.15 = 10.0°\text{C}$$

1 If you are rusty on unit conversions of this kind, the procedures for temperature conversions are in the unit conversions refresher in Section A.2.3 of Appendix A. The equations are listed in Appendix B, Table B.3.

7.1.3 Avogadro's Law

There are several more important historic gas laws from the 19th century for us to examine. We review one of them here, and save the others for later in the chapter. Amedeo Avogadro (Figure 7.8)[2] was an Italian scientist whose work in molecular theory was ahead of its time. In 1811, a time when relatively few scientists accepted the existence of atoms, Avogadro was already hypothesizing the existence of molecules—clusters of atoms—an idea that no one else accepted at the time. Avogadro correctly predicted that some gases, such as O_2, exist as diatomic molecules. General acceptance of Avogadro's molecular hypothesis did not occur until demonstration of it was given by Stanislao Cannizzaro in 1860, four years after Avogadro's death.

Figure 7.8. Italian physicist and chemist Amedeo Avogadro (1776–1856).

Part of Avogadro's 1811 molecular hypothesis is now known as *Avogadro's law*. This law states that *at equal temperatures and pressures, equal volumes of gas contain equal numbers of particles.* It is difficult to convey the immense importance of this discovery. The fact that equal volumes of gas contain the same number of particles means that gases can be weighed to determine relative molecular masses and atomic masses. This is how early 19th-century measurements of atomic mass and molecular mass were achieved.

One of the consequences of Avogadro's law is that the volume of a gas is directly proportional to the number of particles of gas present. Or, using the number of moles of gas, n, as a measure of particles,

$$V = kn \tag{7.8}$$

As we will see below, one mole of any gas at atmospheric pressure and 0°C has a volume of 22.4 L. This is a well-known consequence of Avogadro's law. Notice that Avogadro's law, like Charles' law, is consistent with the kinetic-molecular theory of gases. Equal quantities of all gases at the same temperature and volume exert the same pressure, regardless of the size of the gas particles. This is consistent with the assumption that the total volume of the gas particles is negligible. It is also consistent with our understanding of the internal energy of the gas as the sum of the energies of gas particles, each being $E = \frac{1}{2}mv_{\text{rms}}^2$, since larger molecules move with lower velocities (Figure 6.3).

7.2 The Ideal Gas Law

7.2.1 Standard Temperature and Pressure

Since the volume of a given amount of gas is completely dependent on the pressure (Boyle's law) and temperature (Charles' law), gases are often compared by citing volumes at *standard*

2 I have often observed that when first seeing this image, students show disrespect by laughing at Avogadro's personal appearance. I remind them that it is wrong to laugh at someone for something they have no control over. Avogadro was not only a brilliant scientist and professor, he was a faithful family man with six children. He was a human being who made the most of his natural talents—someone to admire, not to despise.

temperature and pressure, or *STP*. In chemistry, STP conditions are defined as an absolute pressure of 100 kPa (0.987 atm, 1 bar) and a temperature of 0°C (273.15 K).

In defining STP, the pressure is designated as "absolute pressure" because of the convention in industry of specifying pressures using atmospheric pressure as a zero reference rather than using a vacuum as the zero reference. For example, we would say that a flat tire on a car has no pressure (pressure = 0 psi). After the tire is repaired and pressured up, the tire might be pressurized to 35 psi. It is common to refer to pressures this way, using atmospheric pressure as a zero reference, and stating pressure values as the pressure above atmospheric pressure. A pressure value referenced this way is called a *gauge pressure*, or P_{gauge}.

A mechanic's pressure gauge for measuring tire pressure reads gauge pressure. A gauge pressure of 0 psi means atmospheric pressure, which means that gas molecules (such as air) are still present at a pressure of 14.7 psi. The term *absolute pressure*, designated as P_{abs}, is used to denote a pressure reading based on a complete vacuum (no gas particles) being the zero reference. The absolute pressure in a tire inflated to 35 psi is 35 psi + 14.7 psi = 49.7 psi. In summary, atmospheric pressure, designated P_{atm}, relates to gauge pressure and absolute pressure as:

$$P_{abs} = P_{atm} + P_{gauge} \qquad (7.9)$$

Gauge pressure values are commonly used in industry and in specifications involving consumer products such as bicycle tires. However, gas calculations in chemistry and physics require all pressure and temperature values to be in absolute units. For this reason, pressure values in chemistry are usually stated as absolute pressure values.

By contrast, temperature values are often measured or stated in Celsius degrees, which is not an absolute temperature scale. Before performing any gas calculations, temperatures in Celsius degrees must be converted to kelvins using the conversion equation

$$T_K = T_C + 273.15 \qquad (7.10)$$

7.2.2 The Ideal Gas Law

If we combine together the three gas laws we have considered so far, we have one of the most important and well-known laws in chemistry—the *ideal gas law*. We examine the ideal gas law in detail shortly, but before we do so, some words about its applicability are in order.

Chemists often refer to "ideal gases" and "real gases." The ideal gas law is a good model and useful in many circumstances when routine calculations are adequate. However, there is actually no such thing as an "ideal gas," a gas that perfectly obeys the gas laws under all conditions. As we saw previously, many gases do exhibit ideal gas behavior over a wide range of conditions, which is why the ideal gas law is so useful to chemists and physicists. But at extreme pressures—or even moderate pressures in some cases—gas particles are compressed together densely enough that their volumes are no longer negligible, and departures from ideal gas behavior are observed. Similar factors apply to temperature extremes. In this course, our calculations involve only ideal gases. But you should be aware that much more sophisticated models of gas behavior now exist. You can read a bit more about how scientists have attempted to model the behavior of "real gases" in the accompanying box.

It is important to understand how the ideal gas law emerges from the three laws we have considered so far. According to Boyle's law, expressed in a basic form in Equation (7.1), the volume of a gas is inversely proportional to the absolute pressure, or

$$V = \frac{k}{P}$$

Hmm... Interesting. *The gas laws as models*

After the ideal gas law had been worked out, the first effort at modeling the behavior of "real gases" was introduced in 1873, when Dutch physicist Johannes Diderik Van der Waals (1837–1923, shown in the photo) derived the equation that now bears his name. Van der Waals is the same scientist for whom are named the Van der Waals forces. For his work in these areas, Van der Waals was awarded the 1910 Nobel Prize in Physics. The Van der Waals equation, which applies to both liquids and gases, takes into account the particle volumes and the intermolecular forces between them. The Van der Waals equation is an improvement over the ideal gas law, but is still not adequate for highly rigorous calculations. For chemists at that level, much more complex gas models must be used.

As we have seen repeatedly, science is not about making truth claims; science is about modeling—developing theories that model nature with increasing accuracy. This may be as good a place as any to quote Gilbert N. Lewis on this subject. We met Lewis when we studied his ideas about the electron pair in covalent bonding, and his development of Lewis structures as a way of representing molecules. We encounter him again when we get to acids and bases. In 1925, Lewis was invited to give the Silliman Memorial Lectures, a series of lectures given annually by a prominent scientist. At that time, Lewis stated:

> The scientist is a practical man and his are practical aims. He does not seek the *ultimate* but the *proximate*. The theory that there is an ultimate truth, although very generally held by mankind, does not seem to be useful to science except in the sense of a horizon toward which we may proceed.

In this expression, k is a constant that depends on the units of measure involved. According to Charles' law, the volume is also directly proportional to the absolute temperature, as Equation (7.5) indicates:

$$V = kT$$

If the gas volume is inversely proportional to pressure and directly proportional to temperature, then we can combine the equations together, lumping the two constants together to be a single new constant. Doing so we have

$$V = \frac{kT}{P} \tag{7.11}$$

We have also seen from Avogadro's law that the volume of a gas is proportional to the number of particles or moles of gas present. From Equation (7.8) we have

$$V = kn$$

Combining this expression with Equation (7.11), and lumping the constants together again, we have

$$V = \frac{nkT}{P}$$

Rearranging this expression, and replacing the letter k with an R to represent the constant, we have the *ideal gas law*, one of the most famous equations in science.

$$PV = nRT \qquad (7.12)$$

Ideal Gas Law

The constant R in the ideal gas law goes by various names, including the *ideal gas constant*, the *molar gas constant*, and the *universal gas constant*. The value of the ideal gas constant depends on the units of measure associated with it, and because of all the different units in use for pressure and volume there are a lot of different values for R. A few of the most common of these are listed in Table 7.1.[3] All you have to do to solve ideal gas problems is use a value for R that incorporates the units you have for your pressure and volume values. Your values for n and T are always moles and kelvins, and all the constants listed incorporate these units.

The fourth value listed in the table uses so-called MKS units, units of measure based on the meter-kilogram-second subsystem of the SI System of units (the metric system). For this reason, this value is commonly used in physics problems, where the use of MKS units is standard practice. The numerator unit, joules (J), is derived as follows, using the fact that the unit N·m is defined as the joule, a quantity of energy:

$$m^3 \cdot Pa = m^3 \cdot \frac{N}{m^2} = N \cdot m = J$$

As long as you keep your units straight, ideal gas law problems are not difficult. Just always make sure you use absolute pressure and absolute temperature values.

R Value, with Units	Units for P, V
$8.314 \frac{L \cdot kPa}{mol \cdot K}$	kPa, L
$0.08206 \frac{L \cdot atm}{mol \cdot K}$	atm, L
$8314 \frac{L \cdot Pa}{mol \cdot K}$	Pa, L
$8.314 \frac{J}{mol \cdot K}$ *	Pa, m^3
$62.36 \frac{L \cdot Torr}{mol \cdot K}$	Torr, L
$62.36 \frac{L \cdot mm\ Hg}{mol \cdot K}$	mm Hg, L
$8.314 \times 10^{-5} \frac{m^3 \cdot bar}{mol \cdot K}$	bar, m^3
$8.206 \times 10^{-5} \frac{m^3 \cdot atm}{mol \cdot K}$	atm, m^3

**These are the MKS units.*

Table 7.1. Approximate values for the ideal gas constant.

Example 7.3

Demonstrate that at atmospheric pressure, 101.325 kPa, and 0.00°C the volume of 1.00 mole of any gas is 22.4 L.

The given information is as follows:

$$P = 101.325 \text{ kPa}$$

$$n = 1.00 \text{ mol}$$

$$T_C = 0.00°\text{C}$$

$$T_K = T_C + 273.15 = 0.00 + 273.15 = 273.15 \text{ K}$$

Now we write the ideal gas law and solve it for the unknown, the volume:

$$PV = nRT$$

$$V = \frac{nRT}{P}$$

3 Since 2019, R has an exact value. See Appendix B.

With the pressure in kPa and the required volume in L, we select a value for R from Table 7.1 of

$$R = 8.314\ \frac{\text{L} \cdot \text{kPa}}{\text{mol} \cdot \text{K}}$$

Inserting this value along with the other information into the equation for volume, we have:

$$V = \frac{nRT}{P} = \frac{1.00\ \text{mol} \cdot 8.314\ \dfrac{\text{L} \cdot \text{kPa}}{\text{mol} \cdot \text{K}} \cdot 273.15\ \text{K}}{101.325\ \text{kPa}} = 22.4\ \text{L}$$

▲

▼ Example 7.4

A certain 20.00-L cylinder of oxygen gas contains 5.00 mol O_2 at 25.00°C. Determine the pressure in atmospheres in the cylinder, the density of the gas, and the number of oxygen molecules in the cylinder.

For determining the pressure, we use the ideal gas law. The other two computations are from material covered in prior chapters.

Our given information is as follows:

$$V = 20.00\ \text{L}$$
$$n = 5.00\ \text{mol}$$
$$T_C = 25.00°\text{C}$$
$$T_K = T_C + 273.15 = 298.15\ \text{K}$$

Now we write the ideal gas law and solve it for the unknown, the pressure:

$$PV = nRT$$
$$P = \frac{nRT}{V}$$

With the volume in L and the required pressure in atm, we select a value for R from Table 7.1 of

$$R = 0.08206\ \frac{\text{L} \cdot \text{atm}}{\text{mol} \cdot \text{K}}$$

Inserting this value and the other information into the equation for P, we get

$$P = \frac{nRT}{V} = \frac{5.00\ \text{mol} \cdot 0.08206\ \dfrac{\text{L} \cdot \text{atm}}{\text{mol} \cdot \text{K}} \cdot 298.15\ \text{K}}{20.00\ \text{L}} = 6.11\ \text{atm}$$

This value is rounded to the required three significant digits. To determine the density, we need the mass and volume of the gas. We already have this volume, and the mass we obtain from the number of moles and the molar mass of O_2. Looking up the atomic mass for oxygen in the periodic table, we find that the molar mass of O_2 is 32.00 g/mol. Thus the mass is

$$m = 5.00 \text{ mol } O_2 \cdot 32.00 \ \frac{\text{g } O_2}{\text{mol } O_2} = 160.0 \text{ g } O_2$$

Since this is an intermediate calculation, I keep one extra significant digit. Using this value and the volume, we use the density equation to determine the gas density:

$$\rho = \frac{m}{V} = \frac{160.0 \text{ g}}{20.00 \text{ L}} = 8.00 \ \frac{\text{g}}{\text{L}}$$

Finally, the number of oxygen molecules is determined from the number of moles and the Avogadro constant.

$$5.00 \text{ mol } O_2 \cdot \frac{6.022 \times 10^{23} \text{ molecules } O_2}{\text{mol } O_2} = 3.01 \times 10^{24} \text{ molecules } O_2$$

The last two values are stated with three significant digits, as the problem requires.

▲

▼ Example 7.5

Figure 7.9 illustrates a flask with a device called a *manometer* attached. A manometer is a U-shaped glass tube with a liquid inside that is used to measure small pressure differences. As shown in the figure, if the pressure on both sides of the manometer is the same, the two liquid columns are the same height (indicated by the arrows on the left figure). If the pressure inside the flask increases, the higher pressure pushes the liquid toward the low-pressure side of the manometer, and the pressure inside the flask is calculated from the difference in the heights of the columns.

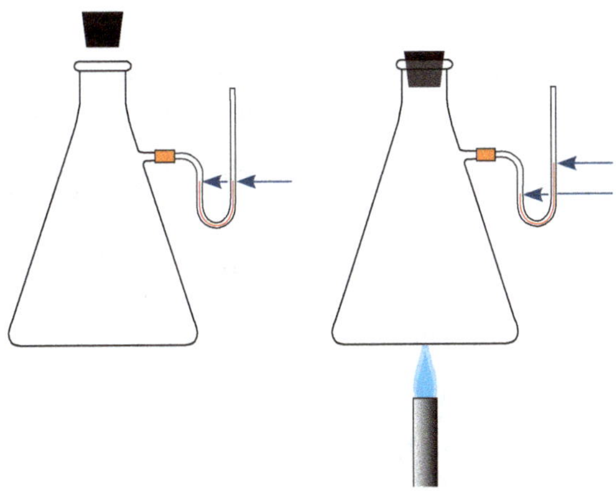

Figure 7.9. Using a manometer to measure small pressure changes at constant volume.

A sealed flask contains 1.000 L of nitrogen gas at STP. The gas is then heated until a manometer on the flask indicates the pressure has increased by 10.19 Torr. How many moles of gas are present in the flask, and what is the temperature of the gas at the higher pressure?

We are given a complete list of the initial conditions of the gas, so we use those to determine the number of moles of N_2 present in the flask. Once we know the quantity of gas present, we use the ideal gas law to find the new temperature. The given initial conditions are:

$V = 1.000 \text{ L}$

$P = 100.0 \text{ kPa}$

$T_C = 0.00°\text{C}$

$T_K = T_C + 273.15 = 0.00 + 273.15 = 273.15 \text{ K}$

From Table 7.1 we select the most convenient form of the ideal gas constant,

$$R = 8.314\ \frac{\text{L}\cdot\text{kPa}}{\text{mol}\cdot\text{K}}$$

Now we write the ideal gas law equation, solve for the number of moles, insert the values, and compute the result.

$$PV = nRT$$

$$n = \frac{PV}{RT} = \frac{100.0\ \text{kPa}\cdot 1.000\ \text{L}}{8.314\ \frac{\text{L}\cdot\text{kPa}}{\text{mol}\cdot\text{K}}\cdot 273.15\ \text{K}} = 0.044034\ \text{mol}$$

I keep an extra significant digit because we use this result in the next calculation. But we report this value as 0.04403 mol.

To determine the new temperature, we are working with pressures in torr, so we select this ideal gas constant from Table 7.1:

$$R = 62.36\ \frac{\text{L}\cdot\text{Torr}}{\text{mol}\cdot\text{K}}$$

To determine the new pressure in torr, we convert the original pressure to torr, then add the pressure increase to obtain the new pressure:

$$100.0\ \text{kPa}\cdot\frac{1\ \text{atm}}{101.325\ \text{kPa}}\cdot\frac{760\ \text{Torr}}{1\ \text{atm}} = 750.06\ \text{Torr}$$

$$P = 750.06\ \text{Torr} + 10.19\ \text{Torr} = 760.25\ \text{Torr}$$

The information we have now for the temperature calculation is as follows:

$$V = 1.000\ \text{L}$$

$$P = 760.25\ \text{Torr}$$

$$n = 0.044034\ \text{mol}$$

We write the ideal gas equation, solve for the temperature, insert the values, and compute the result.

$$PV = nRT$$

$$T = \frac{PV}{nR} = \frac{760.25\ \text{Torr}\cdot 1.000\ \text{L}}{0.044034\ \text{mol}\cdot 62.36\ \frac{\text{L}\cdot\text{Torr}}{\text{mol}\cdot\text{K}}} = 276.9\ \text{K}$$

▲

▼ Example 7.6

An engineer fills a small steel sample cylinder with a sample of process gas. The sample is taken at an absolute pressure of 766 Torr and 28°C. That night, the engineer leaves the sample in her truck outside, where the temperature drops to –4°C. Determine the pressure in the sample cylinder at the new temperature. Assume the volume of the cylinder does not change.

In this case, we do not have complete information about the gas, but we do know that one of the variables T, P, or V is a constant (V in this case). We use the ideal gas law to construct a set of proportions, like those of Boyle's law and Charles' law. (In fact, you don't really need to remember those two laws explicitly because you can work them out from the ideal gas law using the technique shown here.) Starting with the ideal gas law equation, let's get all the constants for this problem on one side and everything else on the other side:

$$PV = nRT$$

$$\frac{P}{T} = \frac{nR}{V}$$

In this problem, everything on the right side of this expression is a constant, so any ratio of temperature and pressure is equal to the same value. This means the ratios of T and P are all equal to each other:

$$\frac{P_1}{T_1} = \frac{P_2}{T_2}$$

Now, we let T_1 and P_1 represent the initial conditions, and T_2 and P_2 represent the new conditions at the colder temperature. This means that P_2 is the unknown we seek. Solving for that variable, we have

$$P_2 = P_1 \frac{T_2}{T_1}$$

Now we are ready to solve for the new pressure. As always with these problems, we must make sure our temperatures and pressures are in absolute units. Here are the data and unit conversions:

$$T_1 = 28°\text{C}$$

$$T_{1,K} = 28 + 273.15 = 301.2\ \text{K}$$

$$T_2 = -4°\text{C}$$

$$T_{2,K} = -4 + 273.15 = 269.2\ \text{K}$$

$$P_1 = 766\ \text{Torr}$$

Putting these values into the expression for P_2, we find the new pressure:

$$P_2 = P_1 \frac{T_2}{T_1} = 766\ \text{Torr} \cdot \frac{269.2\ \text{K}}{301.2\ \text{K}} = 685\ \text{Torr}$$

▲

7.2.3 Using the Ideal Gas Law to Find Molar Mass and Density

Suppose that you use the ideal gas law to determine the number of moles, n, present in a sample of unknown gas. If you also determine the mass, m, of the gas sample, then you can use the number of moles and the mass to compute the molar mass, M, because

$$n = \frac{m}{M} \tag{7.13}$$

Thus, the molar mass, M, can be calculated as

$$M = \frac{m}{n} \tag{7.14}$$

If we insert Equation (7.13) into the ideal gas law equation, we have

$$PV = nRT$$

$$PV = \frac{mRT}{M}$$

Since density is mass over volume, we can solve this equation for the density

$$PV = \frac{mRT}{M}$$

$$\frac{m}{V} = \frac{MP}{RT}$$

or

$$\rho = \frac{MP}{RT} \tag{7.15}$$

To save yourself time deriving equations, you may want to go ahead and commit Equation (7.15) to memory. This form of the ideal gas law is very useful with problems involving density or molar mass. Note that the density of a gas is directly proportional to the molar mass and to the pressure.

▼ Example 7.7

A gas sample with a volume of 1.50 L has a mass of 7.74 g at 28.0°C and 0.974 atm. Determine the molar mass of this gas.

Since the mass of the gas is given, we can calculate the molar mass from Equation (7.14) if we have the number of moles. So we use the ideal gas law to solve for the number of moles first, and then use Equation (7.14) to calculate the molar mass.

The given information, with the necessary temperature unit conversion, is as follows:

$m = 7.74$ g

$V = 1.50$ L

$P = 0.974$ atm

$T_C = 28.0°C$

$T_K = 28.0°C + 273.15 = 301.2$ K

Selecting the appropriate ideal gas constant from Table 7.1, we write the ideal gas law equation, solve for n, insert values, and compute the number of moles.

$$PV = nRT$$

$$n = \frac{PV}{RT} = \frac{0.974\ \text{atm} \cdot 1.50\ \text{L}}{0.08206\ \frac{\text{L} \cdot \text{atm}}{\text{mol} \cdot \text{K}} \cdot 301.2\ \text{K}} = 0.05911\ \text{mol}$$

Now we solve for the molar mass:

$$M = \frac{m}{n} = \frac{7.74\ \text{g}}{0.05911\ \text{mol}} = 131\ \frac{\text{g}}{\text{mol}}$$

▲

In the next example, we pile it on. There are a lot of data, and the solution involves calculating the density as an intermediate step so that we can use Equation (7.15) to obtain the molar mass. But don't get lost in the details. The procedure simply uses the given information to determine the mass of an unknown gas, then its volume, then its density. Then we use the density form of the ideal gas law, Equation (7.15), to calculate the required molar mass. There is nothing new here, but the problem is a nice example of how one actually goes about making the necessary measurements to determine the molar mass of an unknown gas.

▼ Example 7.8

A chemist uses a sequence of measurements as a pathway to determining the molar mass of an unknown gas. First, a flask is evacuated (that is, there is a vacuum inside) and its mass is found to be 133.785 g. The flask is then filled with the gas to a pressure of 735 Torr at 31.0°C and weighed again. This time the mass is 137.982 g. Finally, the flask is filled completely with water at 31.0°C and weighed again. This time the mass is 1,066.8 g. Use these data to determine the molar mass of the gas. At 31°C, the density of water is 0.997 g/mL.

The difference between the first two mass measurements is the mass of the gas itself:

$$m = 137.982\ \text{g} - 133.785\ \text{g} = 4.197\ \text{g}$$

The first and third mass measurements allow us to determine the mass of water that fills the flask, and from this we determine the volume. The mass of the water is

$$m_{\text{water}} = 1066.8\ \text{g} - 133.785\ \text{g} = 933.0\ \text{g}$$

With the mass of the water, we use the density equation to determine the volume of the flask. The ideal gas constant we use below is in liters, so we know we want the volume/density to be in terms of liters. Accordingly, we go ahead and convert the volume to liters while we are at it.

$$\rho = \frac{m}{V}$$

$$V = \frac{m}{\rho} = \frac{933.0 \text{ g}}{0.997 \ \frac{\text{g}}{\text{mL}}} = 938.5 \text{ mL} \cdot \frac{1 \text{ L}}{1000 \text{ mL}} = 0.9385 \text{ L}$$

Now we have both the mass and volume of the unknown gas, so we compute its density.

$$\rho = \frac{m}{V} = \frac{4.197 \text{ g}}{0.9385 \text{ L}} = 4.472 \ \frac{\text{g}}{\text{L}}$$

Finally, we are ready to use Equation (7.15). First, let's list our data:

$$\rho = 4.472 \ \frac{\text{g}}{\text{L}}$$

$$P = 735 \text{ Torr}$$

$$T_C = 31.0°\text{C} + 273.15 = 304.15 \text{ K}$$

Now we write the equation, solve it for the molar mass, insert values, and compute the result.

$$\rho = \frac{MP}{RT}$$

$$M = \frac{\rho RT}{P} = \frac{4.472 \ \frac{\text{g}}{\text{L}} \cdot 62.36 \ \frac{\text{L} \cdot \text{Torr}}{\text{mol} \cdot \text{K}} \cdot 304.15 \text{ K}}{735 \text{ Torr}} = 115 \ \frac{\text{g}}{\text{mol}}$$

▲

7.3 The Law of Partial Pressures

7.3.1 Dalton's Law of Partial Pressures

No doubt you remember John Dalton (Figure 7.10), the English scientist who formulated the first detailed atomic model in 1803. Also in 1803, just one year after Gay-Lussac published Charles' law, Dalton formulated the law now known as *Dalton's law of partial pressures*. This law states that *for a mixture of non-reacting gases, the total pressure of the mixture, P_T, is equal to the sum of the pressures each gas would exert if it were present alone*, or

$$P_T = P_1 + P_2 + P_3 \ldots \quad (7.16)$$

Dalton's Law

where P_1, P_2, P_3, and so on are the so-called *partial pressures* for gases 1, 2, 3, etc.

Essentially, Dalton's law of partial pressures says that at a given set of conditions (T and V), the only factor that determines the total pressure, P_T, of a mixture of gases is the total number of moles of gas that are present, and that each component gas in the mixture contributes its portion of moles to the total number of moles and an identical portion to the total pressure. The portion contributed by each gas is called the *partial pressure*, and the total pressure of the mixture is the sum of the partial pressures.

Notice that Dalton's law of partial pressures is another application of the kinetic-molecular theory of gases. If the total pressure of combined gases is simply the sum of the individual partial pressures, then we are assuming that the gases do not interact with each other (no intermolecular forces), and that the gas particles to do not take up any space.

In a gas mixture, each component gas contributes a portion of the total moles of gas present. The portion one gas contributes is called the *mole fraction*, symbolized by X. Let's let the number of moles of each component gas be represented by n_1, n_2, n_3, etc., and the total number of moles present by n_T. Then the mole fraction of gas 1 is

Figure 7.10. English scientist John Dalton (1766–1844).

$$X_1 = \frac{n_1}{n_T} \quad (7.17)$$

The partial pressure of gas 1 is the same fraction of the total pressure as the mole fraction is of the total number of moles, or

$$\frac{P_1}{P_T} = X_1 = \frac{n_1}{n_T}$$

Thus, we can write

$$P_1 = \frac{n_1}{n_T} P_T = X_1 P_T \quad (7.18)$$

For example, oxygen represents 20.95% of the earth's atmosphere, or a mole fraction of $X = 0.2095$. The pressure of the atmosphere at sea level is 101,325 Pa. Using Equation (7.18), we can compute the partial pressure of oxygen in the atmosphere:

$$P_{O_2} = X_{O_2} \cdot P_T = 0.2095 \cdot 101{,}325 \text{ Pa} = 21{,}230 \text{ Pa}$$

There are two ways to find the partial pressure of a gas in a mixture. First, as in the oxygen illustration above, if you know the total pressure and the mole fraction of the gas, you can just multiply these to get the partial pressure. Second, if you know the number of moles present for each of several gases in a mixture, you can calculate the partial pressure for each component gas using the ideal gas law, and then add the partial pressures to get the total pressure. The following example illustrates the second type of calculation.

Example 7.9

A mixture of gases is made from 6.00 g O_2 and 10.0 g CH_4. The gas mixture is placed in an 18.0-L container at a temperature of 22.0°C. Determine the total pressure in the container and the partial pressure for each gas. State the pressures in torr.

We begin by computing the number of moles present for each gas. To do this, we calculate the molar masses for O_2 and CH_4, which are 32.00 g/mol and 16.04 g/mol, respectively. The numbers of moles are thus

$$n_{O_2} = 6.00 \text{ g } O_2 \cdot \frac{1 \text{ mol } O_2}{32.00 \text{ g } O_2} = 0.1875 \text{ mol } O_2$$

$$n_{CH_4} = 10.0 \text{ g } CH_4 \cdot \frac{1 \text{ mol } CH_4}{16.04 \text{ g } CH_4} = 0.6234 \text{ mol } CH_4$$

Now we use the ideal gas law to calculate each partial pressure. Solving for the pressure from the gas law, we have

$$PV = nRT$$

$$P = \frac{nRT}{V}$$

For the oxygen, we now have these data:

$$n = 0.1875 \text{ mol}$$

$$V = 18.0 \text{ L}$$

$$T_C = 22.0°\text{C}$$

$$T_K = T_C + 273.15 = 22.0 + 273.15 = 295.2 \text{ K}$$

Selecting the appropriate gas constant for liters and torr from Table 7.1, the partial pressure for oxygen is

$$P_{O_2} = \frac{nRT}{V} = \frac{0.1875 \text{ mol} \cdot 62.36 \dfrac{\text{L} \cdot \text{Torr}}{\text{mol} \cdot \text{K}} \cdot 295.2 \text{ K}}{18.0 \text{ L}} = 191.8 \text{ Torr}$$

Rounding to three significant digits,

$$P_{O_2} = 192 \text{ Torr}$$

For the methane, all the values are the same except the number of moles. Using the number of moles for methane, the partial pressure is

$$P_{CH_4} = \frac{nRT}{V} = \frac{0.6234 \text{ mol} \cdot 62.36 \dfrac{\text{L} \cdot \text{Torr}}{\text{mol} \cdot \text{K}} \cdot 295.2 \text{ K}}{18.0 \text{ L}} = 637.6 \text{ Torr}$$

Rounded to three significant digits,

$$P_{CH_4} = 638 \text{ Torr}$$

Finally, the total pressure in the container is the sum of the partial pressures:

$$P_T = 191.8 \text{ Torr} + 637.6 \text{ Torr} = 829.4 \text{ Torr}$$

Rounding this value to three significant digits, we have

$$P_T = 829 \text{ Torr}$$

▲

Notice from the preceding example that the mole fractions and total pressure give us the same partial pressures. The total number of moles present is 0.8109 mol, so the mole fractions are

$$X_{O_2} = \frac{n_{O_2}}{n_T} = \frac{0.1875 \text{ mol}}{0.8109 \text{ mol}} = 0.2312$$

$$X_{CH_4} = \frac{n_{CH_4}}{n_T} = \frac{0.6234 \text{ mol}}{0.8109 \text{ mol}} = 0.7688$$

Using these values with the total pressure of 829.4 Torr and Equation (7.18), the partial pressures are

$$P_{O_2} = X_{O_2} P_T = 0.2312 \cdot 829.4 \text{ Torr} = 192 \text{ Torr}$$

$$P_{CH_4} = X_{CH_4} P_T = 0.7688 \cdot 829.4 \text{ Torr} = 638 \text{ Torr}$$

7.3.2 Collecting a Gas Over Water

In chemistry labs, experiments are often performed in which a gas is collected in a glass vessel sealed from the atmosphere in a pan of water. As illustrated in Figure 7.11, the collecting vessel is first filled with water and inverted into a pan of water so that there is no air in the top of the collecting vessel. Then the gas from a reaction is directed into the collection vessel and collected there, pushing the water level in the collection vessel down. After gas collection, the collection vessel is adjusted up or down until the water levels inside and outside the collecting vessel are the

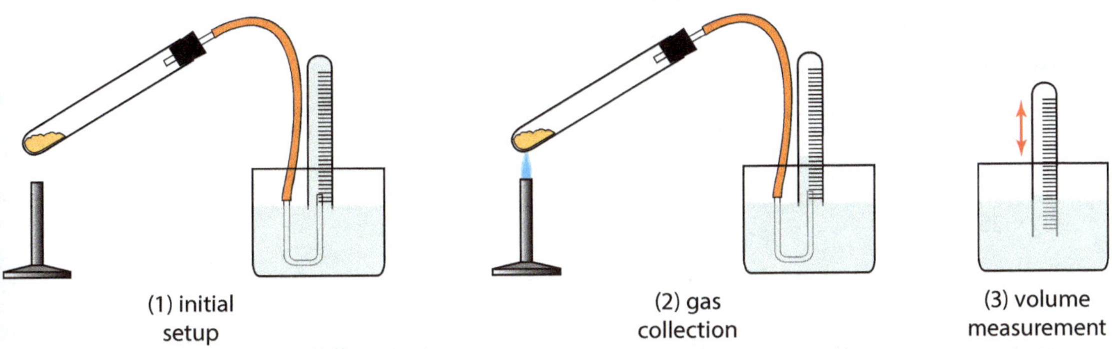

Figure 7.11. Collecting gas over water.

same. In this position, the pressure inside the collecting vessel is equal to atmospheric pressure, and the gas volume in the collecting vessel at atmospheric pressure may be read from the scale on the collecting vessel.

The gas collected in the collecting vessel is a mixture of the gas from the reaction and water vapor, which evaporates up into the collection vessel along with the desired gas from the reaction. To determine the quantity of gas collected, the experimenter first waits for the temperature and pressure in the collecting vessel to reach equilibrium with the environment in the lab. Then the experimenter measures the volume of the gas mixture in the collecting vessel, along with the barometric pressure and temperature in the lab. The partial pressure of the collected gas is then determined from a partial pressure calculation, using the vapor pressure of water at the measured temperature. (Water vapor pressures at different temperatures are listed in Appendix B, Table B.4.) Finally, the quantity of gas collected is calculated using the ideal gas law with the volume, temperature, and partial pressure for the collected gas. The following example illustrates this technique.

Example 7.10

When potassium chlorate is heated, it decomposes into potassium chloride and oxygen gas (a standard technique for producing laboratory oxygen). A sample of $KClO_3$ is decomposed to produce 0.22 L of gas, which is collected over water. Data from measurement instruments in the lab indicate a temperature of 26°C and barometric pressure of 754 Torr. Determine the number of moles of oxygen gas collected and the mass in grams of the $KClO_3$ that is decomposed.

We begin by using the law of partial pressures and the ideal gas law to determine the amount of O_2 produced. After we have that, then we use stoichiometry to work out the mass of $KClO_3$.

We must determine the partial pressure of the oxygen in the collecting vessel. From Table B.4 in Appendix B, at 26°C the vapor pressure of water—equal to its partial pressure—is 25.23 Torr. The total pressure inside the collection vessel is equal to atmospheric pressure. Dalton's law of partial pressures applied to the gas mixture in the collection vessel gives us

$$P_{\text{atm}} = P_{\text{O}_2} + P_{\text{H}_2\text{O}}$$

$$P_{\text{O}_2} = P_{\text{atm}} - P_{\text{H}_2\text{O}}$$

From this, we calculate the partial pressure of the O_2 (applying the significant digits rule for addition as we do):

$$P_{\text{O}_2} = P_{\text{atm}} - P_{\text{H}_2\text{O}} = 754\ \text{Torr} - 25.23\ \text{Torr} = 729\ \text{Torr}$$

Now we have the following data for oxygen:

$$P = 729\ \text{Torr}$$

$$V = 0.22\ \text{L}$$

$$T_C = 26°\text{C}$$

$$T_K = T_C + 273.15 = 26 + 273.15 = 299\ \text{K}$$

With this information, we solve the ideal gas law equation for n and compute the number of moles of O_2 in the collection vessel:

$PV = nRT$

$$n_{O_2} = \frac{PV}{RT} = \frac{729 \text{ Torr} \cdot 0.22 \text{ L}}{62.36 \frac{\text{L} \cdot \text{Torr}}{\text{mol} \cdot \text{K}} \cdot 299 \text{ K}} = 8.60 \times 10^{-3} \text{ mol } O_2$$

We report this result with two significant digits as

$$n_{O_2} = 8.6 \times 10^{-3} \text{ mol } O_2$$

Next, to determine the amount of $KClO_3$ decomposed we need to do a stoichiometric calculation. We need to work out the formula equation and balance it. Doing so gives us

$$2KClO_3 \xrightarrow{\Delta} 2KCl + 3O_2$$

This equation tells us that the mole ratio of O_2 to $KClO_3$ is 3 : 2. Using this ratio as a conversion factor, we have

$$8.60 \times 10^{-3} \text{ mol } O_2 \cdot \frac{2 \text{ mol } KClO_3}{3 \text{ mol } O_2} = 5.73 \times 10^{-3} \text{ mol } KClO_3$$

Finally, we convert this to grams using the molar mass for $KClO_3$:

$$5.73 \times 10^{-3} \text{ mol } KClO_3 \cdot \frac{122.5 \text{ g } KClO_3}{\text{mol } KClO_3} = 0.70 \text{ g } KClO_3$$

This final result is rounded to two significant digits as the given data require.

▲

7.4 Stoichiometry of Gases and Effusion

7.4.1 Stoichiometry of Gases

We have seen that the coefficients in a chemical equation give us the ratios of molecules and the ratios of moles of the reactants and products. If we make all our measurements at the same temperature and pressure, such as STP, the same ratios of coefficients can be used to give us *volumetric ratios* of the gases involved in the reaction. Since only ratios are involved, the volume units may be any units that are convenient. The following example illustrates.

▼ Example 7.11

Assume that 3.00×10^4 m^3 of propane gas are to be completely combusted with oxygen. Determine the volumes of O_2 required and CO_2 produced. Assume all measurements are made at STP.

Since all measurements are made at the same conditions (STP, which are 0°C and 100 kPa), the coefficients from the balanced chemical equation give us the volumetric ratios we need. The equation is

$$C_3H_8 + 5O_2 \rightarrow 4H_2O + 3CO_2$$

From this equation, we see that the volumetric ratios we need are

$C_3H_8 : O_2$ 1 : 5

$C_3H_8 : CO_2$ 1 : 3

You can probably work the math out in your head, but here it is:

$$3.00\times10^4\ \text{m}^3\ C_3H_8 \cdot \frac{5\ \text{m}^3\ O_2}{1\ \text{m}^3\ C_3H_8} = 1.50\times10^5\ \text{m}^3\ O_2$$

$$3.00\times10^4\ \text{m}^3\ C_3H_8 \cdot \frac{3\ \text{m}^3\ CO_2}{1\ \text{m}^3\ C_3H_8} = 9.00\times10^4\ \text{m}^3\ CO_2$$

▲

The next example illustrates the use of the ideal gas law in conjunction with a routine stoichiometric calculation.

▼ Example 7.12

Tungsten is the metal used for the filaments inside incandescent light bulbs. Tungsten is found in nature as tungsten oxide, WO_3, and is refined with hydrogen to produce tungsten metal and water. Determine how many liters of hydrogen gas at STP are required to react completely with 1,475 g of tungsten oxide.

We first have to work out the balanced chemical equation for this reaction, which is

$$WO_3(s) + 3H_2(g) \rightarrow W(s) + 3H_2O(l)$$

We are given a mass of WO_3. The strategy is that we convert this mass to moles and use standard stoichiometry to determine the number of moles of hydrogen required. Then we use the ideal gas equation to determine the volume of this quantity in liters.

Using the molar mass of WO_3, we find that the given quantity of WO_3 is equal to

$$1475\ \text{g}\ WO_3 \cdot \frac{1\ \text{mol}\ WO_3}{231.85\ \text{g}\ WO_3} = 6.3619\ \text{mol}\ WO_3$$

From the balanced equation, the ratio of moles WO_3 to moles H_2 is 1 : 3. Using this ratio as a conversion factor, we have

$$6.3619\ \text{mol}\ WO_3 \cdot \frac{3\ \text{mol}\ H_2}{1\ \text{mol}\ WO_3} = 19.086\ \text{mol}\ H_2$$

Now that we know the quantity of H_2 involved, we use the ideal gas equation to compute its volume at STP. First let's write down the givens for this calculation, remembering that the values for STP are exact by definition:

$n = 19.086$ mol

$P = 100.0$ kPa

$T_C = 0.00°C$

$T_K = T_C + 273.15\text{ K} = 273.15\text{ K}$

Using Table 7.1 to obtain an appropriate value for *R*, we now use the ideal gas law equation to compute the volume of hydrogen:

$PV = nRT$

$$V = \frac{nRT}{P} = \frac{19.086\text{ mol} \cdot 8.314\ \frac{\text{L} \cdot \text{kPa}}{\text{mol} \cdot \text{K}} \cdot 273.15\text{ K}}{100.0\text{ kPa}} = 433.4\text{ L}$$

▲

7.4.2 Gas Diffusion and Effusion

We have already discussed the process of diffusion in liquids and gases, in which the random motion of fluid molecules causes them to mix, and over time leads to the uniform mixing of a homogeneous mixture (solution). In the process of gas diffusion, there are always two or more gases present at the same time. For example, when a person wearing a fragrance (perfume or cologne) enters a room, molecules of the fragrance diffuse throughout the molecules of oxygen, nitrogen, and other gases in the room.

A similar gas movement phenomenon is *effusion*, in which molecules of a gas pass through a very tiny opening,[4] as illustrated in Figure 7.12. Effusion occurs when molecules of gas pass singly through the opening and begin spreading into the region on the other side. A common instance of effusion occurs when the helium atoms inside a helium balloon escape through microscopic pores in the balloon into the air. The fact that this happens explains why helium balloons eventually loose their pressure and go flat.

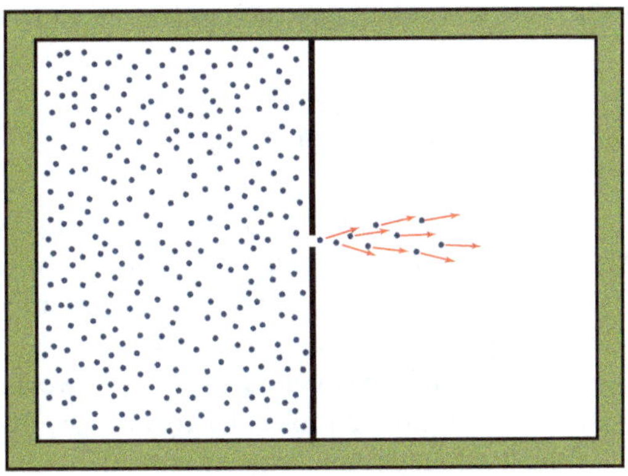

Figure 7.12. Gas effusion through a tiny opening.

The rate at which molecules of a gas effuse depends on the velocity of the molecules in the gas. You recall that the average velocity of the molecules in a gas is determined by the temperature, and thus by the kinetic energy,

$E = \frac{1}{2}mv^2$

4 For effusion, the diameter of the opening is less than the so-called *mean free path* of the particles, the average distance they travel between collisions with each other. With an opening this small, only one particle at a time passes through it. As an example of how small this is, for helium at STP the mean free path is approximately 193 nm—far smaller than the wavelengths of visible light. If the opening is large enough so that several particles can pass through simultaneously, diffusion takes place.

Let's consider the rates of effusion of two different gases at the same temperature. Suppose the molecules in one gas have mass m_a and those of the other have mass m_b. At the same temperature, the molecules of the two gases have the same average kinetic energy, which means that

$$\tfrac{1}{2}m_a v_a^2 = \tfrac{1}{2}m_b v_b^2$$

or simply

$$m_a v_a^2 = m_b v_b^2$$

We can generalize this equation by replacing the individual particle masses, m_a and m_b, with the molar masses, M_a and M_b, of the two gases:

$$M_a v_a^2 = M_b v_b^2$$

If we now collect the velocity terms on one side of the equation and the mass terms on the other, and take the square root of both sides, we have

$$\frac{v_a^2}{v_b^2} = \frac{M_b}{M_a}$$

$$\frac{v_a}{v_b} = \frac{\sqrt{M_b}}{\sqrt{M_a}} \qquad (7.19)$$

Now, the *rate* at which a gas effuses is directly proportional to the average velocity of the particles of the gas, which means we can write

$$\frac{\text{rate of effusion of gas } a}{\text{rate of effusion of gas } b} = \frac{\sqrt{M_b}}{\sqrt{M_a}} \qquad (7.20)$$

Graham's Law

This equation is known as *Graham's law of effusion*. Put into words, the law says that *the ratio of the effusion rates of two gases is equal to the inverse ratio of the square roots of their molar masses*. If the molar mass of gas *b* is four times that of gas *a*, gas *a* effuses twice as fast, as the example below helps to illustrate.

Graham's law of effusion was first worked out empirically in 1848—over a decade before the arrival of kinetic-molecular theory—by Thomas Graham, a Scottish chemist. As shown above, the law follows immediately from the kinetic-molecular theory of gases.

Graham's law is useful for comparing the effusion rates of two gases. In casual conversation, it is common for people to compare two values by looking at the difference between them: she is three years older than I am, he makes $1.25 less per hour than I do, and so on. But when scientists compare things, they do it with *ratios* not with *differences*. The difference between 10 and 20 is 10, as is the difference between 1,000 and 1,010. But 20 is 200% of 10, whereas 1,010 is only 101% of 1,000. As this example illustrates, the significance of a difference between two values usually depends on how large the values are. For this reason, scientists nearly always compare things with ratios, which do not depend on how large the values are.

Example 7.13

Consider two gases, oxygen and helium, with molecular masses of 32.00 u and 4.00 u, respectively. How much faster does helium effuse than oxygen?

As explained above, this question is asking for the *ratio* of the effusion rates of the two gases. Helium is the lighter, and thus faster, gas. Let's let He be gas *a* and O_2 be gas *b*. Using Graham's law of effusion, the ratio of effusion rates is

$$\frac{\text{rate of effusion of He}}{\text{rate of effusion of O}_2} = \frac{\sqrt{32.00 \text{ u}}}{\sqrt{4.00 \text{ u}}} = \sqrt{\frac{32.00}{4.00}} = \sqrt{8.00} = 2.83$$

Oxygen is eight times heavier than helium, which means helium effuses $\sqrt{8}$ = 2.83 times faster than oxygen.

▲

▼ Example 7.14

According to Figure 6.3, the average velocity of neon atoms at 25°C is 500 m/s. Use this value to estimate the velocity of an ammonia molecule at the same temperature. Assume that the given neon velocity is precise to three significant digits.

Hmm... Interesting. Uranium enrichment

Back in the early 1940s, scientists involved in the Manhattan Project were developing the atomic bombs eventually deployed against Japan in WWII. One of the technical challenges was the process known as *uranium enrichment*. The two main isotopes of uranium are U-238, which constitutes about 99.3% of all uranium atoms, and U-235, which makes up about 0.7% of uranium. Unfortunately for those seeking to use uranium as a nuclear fuel, only the U-235 sustains the nuclear chain reaction needed for a nuclear reactor or atomic bomb. Several different methods have been developed for increasing the percentage of U-235 by separating the two isotopes; all of them are painfully slow. Three different methods were used in succession to separate the U-235 used in the manufacture of the two atomic bombs. The first process increased the percentage of U-235 to 2%. The middle process is called *gaseous diffusion*, which uses multiple stages of effusion to increase the percentage of U-235. At the time, this process was performed at the Oak Ridge facility in Oak Ridge, Tennessee.

In the gaseous diffusion process, the uranium is first reacted with fluorine to produce uranium hexafluoride gas, UF_6, an extremely corrosive acid gas. The U-238 isotope is three neutrons heavier than the U-235 isotope (see Table 1.5). This isn't much of a difference, but it is enough that molecules of UF_6 containing U-235 atoms effuse through a tiny opening 0.4% faster than do UF_6 molecules containing U-238 atoms. By effusing the UF_6 through thousands of repeated stages one after the other, the concentration of U-235 was increased from 2% to 23%. Once separated to this level of purity, the uranium was sent through a third process that brought the concentration of U-235 up to its final level of 84%.

The name gaseous diffusion can be a bit confusing. Despite its name, the process works by effusion. The UF_6 molecules pass through openings only 10–25 nanometers in diameter, which is only about 10% of the mean free path of UF_6 molecules (see the footnote a couple of pages back). To make the confusion worse, the material with the tiny holes in it is called a *diffuser*, and the theory used to describe the rate of effusion is Graham's law of effusion.

From Equation (7.19), we solve for the particle velocity of a gas in terms of the velocity of particles for a second gas and the two molar masses. The mass of neon atoms is 20.18 u, and the molecular mass of ammonia molecules, NH_3, is 17.03 u.

Since the ammonia velocity is the unknown we seek, we will let ammonia be gas *a* and neon gas *b*. We solve Equation (7.19) for v_a, insert the values, and compute the result.

$$\frac{v_a}{v_b} = \frac{\sqrt{M_b}}{\sqrt{M_a}}$$

$$v_a = v_b \frac{\sqrt{M_b}}{\sqrt{M_a}} = 500\ \frac{\text{m}}{\text{s}} \cdot \frac{\sqrt{20.18\ \text{u}}}{\sqrt{17.03\ \text{u}}} = 500\ \frac{\text{m}}{\text{s}} \cdot \sqrt{\frac{20.18}{17.03}} = 544\ \frac{\text{m}}{\text{s}}$$

▲

Techniques for measuring the molar mass of a gas are important for identifying compounds in reaction products. We saw earlier in the chapter that the ideal gas law may be used to measure the molar mass. Graham's law can also be used to estimate a molar mass, illustrated by the following example.

▼ Example 7.15

The rate of effusion of an unknown gas is compared to the effusion rate of carbon dioxide. The CO_2 is found to effuse at a rate 2.4 times faster than the unknown gas. Estimate the molar mass of the unknown gas.

From Equation (7.20), the ratio of rates of effusion is equal to the square root of the ratio of molar masses. We let the faster gas, CO_2, be gas *a* and the unknown gas be gas *b*. This gives us

$$2.4 = \frac{\sqrt{M_b}}{\sqrt{M_{CO_2}}}$$

Solving this for the unknown molar mass, M_b, we have

$$(2.4)^2 = \frac{M_b}{M_{CO_2}}$$

$$5.76 = \frac{M_b}{M_{CO_2}}$$

$$M_b = 5.76 \cdot M_{CO_2}$$

From the periodic table, we calculate the molar mass of CO_2, which is 44.0 g/mol. Inserting this value into the equation above gives

$$M_b = 5.76 \cdot M_{CO_2} = 5.76 \cdot 44.0\ \frac{\text{g}}{\text{mol}} = 250\ \frac{\text{g}}{\text{mol}}$$

This result is rounded to two significant digits as required.

▲

Chapter 7 Exercises

SECTION 7.1

1. A fixed quantity of gas is in a 2.00-L cylinder at 22.0°C and 0.974 atm.
   a. Use Boyle's law to determine the volume the gas occupies when the pressure is reduced to 0.772 atm while the temperature is held constant.
   b. Using the original starting conditions, use Charles' law to determine the new gas temperature if the pressure is held constant and the volume increases to 3.14 L during heating.
2. A gas with a volume of 3.5 m^3 is at 0.275 bar. The volume is reduced to 1.25 m^3 while the temperature is held constant. Determine the new gas pressure.
3. A gas with a volume of 2,225 cm^3 is at a pressure of 655 Torr and a temperature of 37.00°C. In a two-stage process, the gas is first cooled until the temperature is –15°C, while the pressure is held constant. Then the pressure is decreased to 603 Torr, while the temperature is held constant. Determine the final volume of the gas.
4. Explain how the graphical depiction of Charles' law in Figure 7.6 includes assumptions that are also in the kinetic-molecular theory of gases.
5. Explain how Avogadro's law implies two of the assumptions in the kinetic-molecular theory of gases.
6. According to Boyle's law, if the volume of a gas is reduced to 1/5 of its original value at a constant temperature, what happens to the pressure?
7. According to Charles' law, if the absolute temperature of a gas is reduced by 25%, what happens to the volume?
8. According to Charles' law, the volume of a gas at 0 K is zero. Does Boyle's law predict a volume of zero at any pressure? Explain your response.

SECTION 7.2

9. What volume is required to contain 100.0 mol O_2 at STP?
10. Determine the temperature of 0.105 moles of a gas in a volume of 1,750 mL at 15.0 bar. State your answer in degrees Fahrenheit.
11. A mass of 2.10 g of methane is in a container at 145 Torr and 276 K. Determine the volume of the container holding this methane.
12. How many moles of gas are present in a cylinder with a volume of 1.75 L if the pressure is 0.922 bar and the temperature is 5°C?
13. A tank of methane holds 1,550 m^3 and is held at STP.
    a. How many moles of methane are in the tank?
    b. How many kilograms of methane are in the tank?
    c. What is the density of the methane?
    d. How many carbon atoms are in the tank?
14. You have probably seen the Goodyear blimps as they fly over sporting events. The blimps hold approximately 4,960 m^3 of helium. Assume the gas is at 17°C and 1.00 atm.
    a. What mass of helium is contained in the blimp?
    b. If the blimp were full of hydrogen instead of helium (it's not and won't be, but just suppose), what would be the mass of hydrogen in the blimp?

15. A 2.000-L tank is found containing gas at 27.0°C and 23.11 atm.
    a. Determine the number of moles of gas in the tank.
    b. If the mass of the gas is found to be 133.0 g, confirm which halogen is in the tank.
16. A weather balloon is filled with 275 L He at 29°C and a pressure of 765 mm Hg. The balloon's volume can vary, since both pressure and temperature vary with altitude.
    a. How many moles of helium are in the balloon?
    b. At an altitude of 10,000 m (a common cruising altitude for commercial aircraft), the temperature is 221 K and the pressure is 25.0 kPa. Determine the volume of the balloon at that altitude.
17. A driver fills his tires to 35 psi gauge on a day when the air is 36°C. Over the course of the next few weeks the temperature drops to –5°C. Determine the new pressure in the tires. Assume the tire volume is constant and that there are no air leaks. State your answer in psi gauge. (Note the pressure is given as gauge pressure.)
18. A SCUBA diver tank contains 12.00 L of air at a gauge pressure of 119.0 bar and a temperature of 22.00°C. (Note the pressure is given as gauge pressure.)
    a. Determine the number of moles of air in the tank.
    b. Determine the volume this much air occupies at STP.
19. Find the molar masses of the following gases, measured at the stated conditions:
    a. 0.942 g occupying 0.975 L at 26°C and 868 Torr
    b. 0.651 g occupying 888 mL at –32°C and 2.14 bar
20. Estimate the temperature at the center of a star assuming the following: a) the gas is 67.0% H and 33.0% He (this information allows you to figure the average molar mass); b) the density is 1.45 g/cm^3; c) the pressure is 1.29×10^9 atm.
21. Determine the densities of CO_2, O_2, N_2, and CH_4 at STP.
22. Distinguish between "ideal gases" and "real gases."
23. Explain why the existence of liquid and solid states of matter imply that gases, in fact, do not act like ideal gases under all conditions.
24. Is a gas more likely to act as an ideal gas at high temperature and low pressure, or vice versa? Explain why.

SECTION 7.3

25. The *Voyager 1* space probe, which is presently leaving our solar system forever, sent back data that has allowed scientists to determine the composition of the atmosphere of Titan, Saturn's largest moon. The atmosphere consists of 82.0 mol percent N_2,[5] 12.0 mol percent Ar, and 6.0 mol percent CH_4. The total atmospheric pressure at Titan's surface is 1.61 atm. Determine the partial pressure of each of these gases in Titan's atmosphere.
26. Scientists are studying the effect of a synthetic atmosphere on plant growth. The components of the synthetic atmosphere are 80.5 mol percent Ar, 18.0 mol percent O_2, and 1.5 mol percent CO_2.
    a. If the total pressure of the gas mixture is to be 748 Torr, determine the partial pressure for the argon in the mixture.

5 The specification "82.0 mol percent" means the mole fraction is 0.820.

b. If this atmosphere is to be contained in a vessel with a volume of 165 L at a temperature of 295 K, how many moles of Ar are needed?

27. A sample of 2.55 g of propane, C_3H_8, is collected in a container at a pressure of 2.55 bar. In a separate container 1.01 g of methane, CH_4, are collected. The two gases are then combined into the container the propane is in originally. Determine the mole fraction and partial pressure for each gas in the mixture, and the pressure in the container after the two gases are combined.

28. In the figure to the right, two flasks containing gases at the conditions shown are connected together with a stopcock[6] in between. The valve is then opened. (The stopcock is shown in the open position.)

   a. Determine the number of moles present for each gas.

   b. By considering the new volume available to each gas, determine the partial pressure of each gas after the stopcock is opened and the gases have mixed. (Assume the volume of the tubing is included in the volumes given, and the temperature remains unchanged.)

   c. Determine the total pressure in the system after the gases mix.

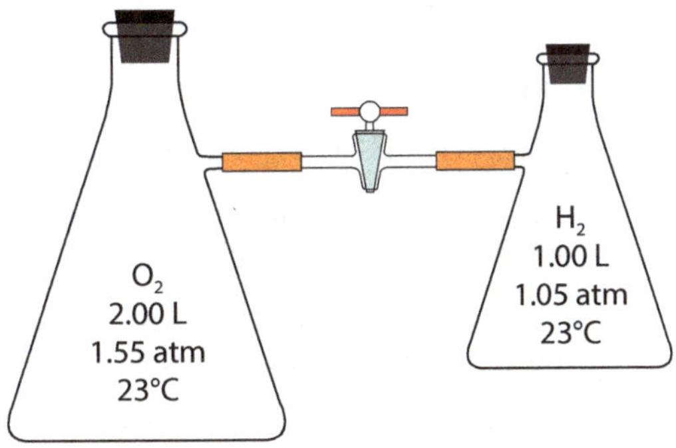

29. A sample of nitrogen is collected over water. At a total pressure of 1.00 atm and 22°C, the container volume is 285 mL. Determine the following: (Vapor pressure data for water are in Appendix B, Table B.4.)

   a. the number of moles, partial pressure, and mole fraction of the nitrogen in the sample.

   b. the number of moles, partial pressure, and mole fraction of the water vapor in the sample.

30. When zinc reacts with sulfuric acid, the reaction products are zinc sulfate and hydrogen gas according to this equation:

$$Zn(s) + H_2SO_4(aq) \rightarrow ZnSO_4(aq) + H_2(g)$$

Assume that 184 mL of H_2 is collected over water at a barometric pressure of 757 mm Hg and a temperature of 28°C. Determine how many grams of zinc are consumed in the reaction.

31. A sample of mercury oxide, HgO, is decomposed by heating to form pure mercury metal and oxygen gas. A volume of oxygen equaling 355 mL is collected over water at 18°C and 725 Torr total pressure. Determine how many grams of mercury oxide are decomposed.

SECTION 7.4

32. Carbon monoxide, CO, reacts with oxygen to produce carbon dioxide. How many liters of oxygen are required to react with 5.0 L of CO, and how many liters of CO_2 are produced by the reaction?

33. Propane, C_3H_8, combusts with O_2 to produce H_2O vapor and CO_2. If 30.00 m^3 C_3H_8 at certain conditions are consumed, determine the volumes at the same conditions for the other three compounds in the reaction.

---

6 A stopcock is a valve in a glass tube that can be opened or closed to connect or isolate the two sections of tubing.

34. Solid iron(III) hydroxide decomposes to produce iron(III) oxide and water vapor. Assume a reaction occurs in which 2.00 L of water vapor are produced at atmospheric pressure and 390 K. Determine how many grams of iron(III) hydroxide are consumed, and how many grams of iron(III) oxide are produced.

35. When Jacques Charles made his original 1783 balloon flight, the balloon contained 31,150 L H_2 at atmospheric pressure. He collected the hydrogen gas from the single replacement reaction between iron and hydrochloric acid:

$$Fe(s) + 2HCl(aq) \rightarrow FeCl_2(aq) + H_2(g)$$

Determine the mass in kilograms of the iron needed to produce this much hydrogen if the temperature is 24°C.

36. Suppose Monsieur Charles had used sulfuric acid instead of hydrochloric acid in his iron reaction to produce hydrogen. In this case, the other reaction product is iron(II) sulfate. Determine the amount of iron needed in this case, assuming the same conditions.

37. The reaction between magnesium and hydrochloric acid produces hydrogen gas. If such a reaction produces 0.015 L H_2 at STP, what mass of magnesium is used?

38. A reaction between calcium carbide and water can be used to prepare acetylene gas, C_2H_2:

$$CaC_2(s) + 2H_2O(l) \rightarrow Ca(OH)_2(s) + C_2H_2(g)$$

Calculate the volume of acetylene that is collected over water at 20.0°C by a reaction of 0.883 g CaC_2 if the total pressure of the gas is 735 Torr.

39. In a reaction between an aqueous solution of potassium iodide and chlorine gas at STP, potassium chloride in aqueous solution and 7.75 L of gaseous I_2 are produced. Determine:

   a. the number of moles of each of the reactants and each of the products.

   b. the masses in grams of each of the reactants and each of the products.

40. Solid aluminum reacts with an aqueous solution of HCl to produce an aqueous solution of $AlCl_3$ and hydrogen gas. If 110.0 g HCl are in the solution and 25.00 g Al are added, determine the following:

   a. the limiting reactant.

   b. the volume of H_2 produced at atmospheric pressure and 32.0°C.

41. Methanol, CH_3OH, is also known as wood alcohol. It is manufactured by reacting carbon monoxide and hydrogen gases at high temperature and extreme pressure. Assume 660.0 m^3 CO and 1,210.0 m^3 H_2 go into the reaction.

   a. Which reactant is present in excess?

   b. How much of the excess reactant remains after the reaction?

   c. What volume of methanol vapor is produced, assuming the same conditions?

42. Nitroglycerine, $C_3H_5N_3O_9$ or $C_3H_5(NO_3)_3$, is an oily liquid at room temperature that explodes just by being jostled too much. (It's packed with nitrates and produces lots of gas—see the box in Chapter 5 on the subject.) When it explodes, the reaction products are water vapor, carbon dioxide, nitrogen, and a bit of oxygen. Assume that in such an explosion 1.00 g H_2O is produced at 350°C and 50.0 bar. Determine:

   a. the total volume of all the reaction products.

   b. the mass of nitroglycerine consumed in the reaction.

43. Gaseous CO_2 and CCl_4 are allowed to effuse through a pinhole at identical conditions. If the rate of effusion of the CO_2 is 6.1×10^{-2} mol/s, what is the rate of effusion of CCl_4?

44. Under certain conditions, chlorine molecules have an average velocity of 0.0421 m/s. Determine the average velocity of sulfur dioxide molecules under the same conditions.

45. A sample of helium is found to effuse through a porous diffuser 7.1 times faster than an unknown gas. Determine the molar mass of the unknown mass.

46. The ratio of the molar masses of two gases is 3.4. What is the ratio of their rates of effusion?

GENERAL REVIEW EXERCISES

47. Determine the wavelength of a photon with an energy of 3.20×10^{-18} J, and state what part of the electromagnetic spectrum it is in.

48. Determine the percent composition of nitroglycerin, $C_3H_5(NO_3)_3$.

49. In Jacques Charles' hydrogen production process (exercise 35 above), if 53.5 kg Fe are used and 19,625 L H_2 are produced,

    a. what is the percent yield of the reaction?

    b. how many molecules of hydrogen are produced?

    c. what is the mass of iron(II) chloride produced (assuming 100% efficiency)?

50. Use only the periodic table to answer the following questions about sodium, bromine, strontium, plutonium, and bismuth.

    a. Which atom has the largest radius?

    b. Which element is a halogen?

    c. Which element has electrons in the 5*f* subshell?

    d. Which element is an alkaline-earth metal?

    e. Which element has the highest electronegativity?

    f. Which element has the highest electron affinity?

51. How much heat is required to melt 22.0 kg of ice at 0°C, and then warm the water to 20.0°C?

52. Carbon dioxide at –60°C and atmospheric pressure is held at a constant temperature while the pressure is increased to 10 atm. According to the phase diagram for carbon dioxide (see Chapter 6, exercise 20), what are the states the CO_2 is in at the beginning and end of this process?

Chapter 8

Solution Chemistry

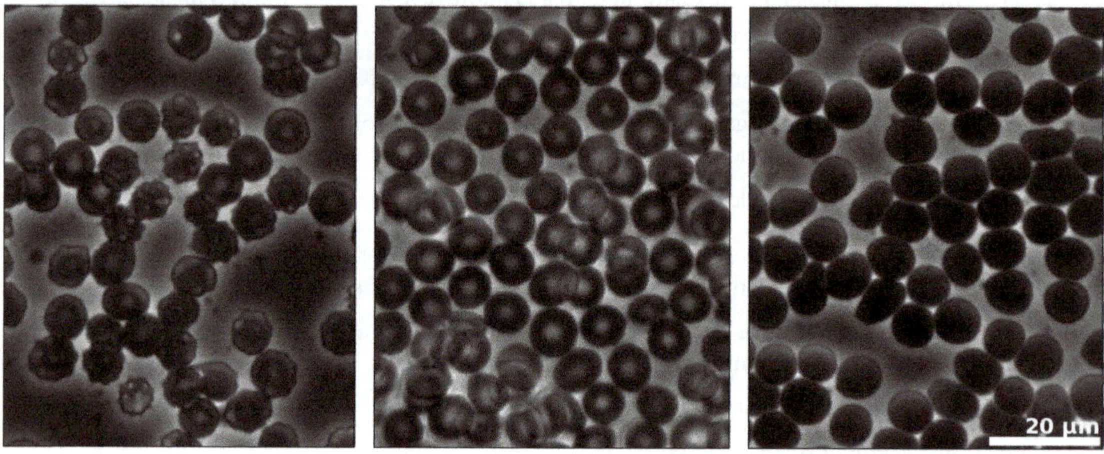

The images above are of human red blood cells, viewed by means of a technique called *phase contrast light microscopy*.

Osmosis, one of the colligative properties of solutions, is responsible for the way human blood cells maintain proper water content. On the left, blood cells are shrunken and wrinkled. Conditions leading to this situation are referred to as *hypertonic*. The shrinking is caused by water in the cells passing by osmosis through the cell wall. Osmosis happens when the concentration of dissolved solids outside the cells is greater than the concentration inside the cells. Water passes through the cell wall in a direction toward the higher concentration. In *isotonic* conditions, the cells have the proper water content. The cells in the center image look the way they are supposed to. In *hypotonic* conditions, the cells gain water by osmosis, possibly leading to rupture of the cell wall.

Chances are you have heard of osmosis before without knowing what it really is. In this chapter, you find out.

Objectives for Chapter 8

After studying this chapter and completing the exercises, you should be able to do each of the following tasks, using supporting terms and principles as necessary.

SECTION 8.1

1. Define *suspension* and *colloidal dispersion*, and describe several types and examples of colloids.
2. Describe and explain the Tyndall effect.

SECTION 8.2

3. Describe the process of dissolution of an ionic solid, acid, or base in a polar solvent, such as water.
4. Use energy concepts to explain how the enthalpy of solution might be positive, indicating an endothermic dissolution process, or negative, indicating an exothermic dissolution process.
5. Describe entropy, and use the concepts of entropy and energy to explain why an endothermic dissolution process can occur.
6. Define *electrolyte* and explain why electrolytes conduct electricity.
7. Distinguish between strong and weak electrolytes, and account for the difference.

SECTION 8.3

8. Use the solubility guidelines to classify given compounds as soluble or insoluble in water.
9. Explain what is meant by the phrase "like dissolves like" and what factors are involved.
10. Explain why solutes of ionic solids and polar molecules dissolve in polar solvents but not in nonpolar solvents, and explain why solutes of nonpolar molecules do not dissolve in polar solvents such as water.
11. Give examples and descriptions of several solid solutions.
12. State Le Châtelier's principle and Henry's law, and relate both to the equilibrium state of a gas in solution in a liquid and the effect of the gas partial pressure on the solubility.
13. Describe the effect of temperature on liquid solutions of solids and gases.

SECTION 8.4

14. Define and calculate *molarity* and *molality*.

SECTION 8.5

15. Use the solubility guidelines to determine if a precipitate forms when given aqueous solutions of ionic solids are combined.
16. Write ionic equations and net ionic equations describing precipitation reactions.
17. Define *spectator ion*.

SECTION 8.6

18. Define *colligative property* and list four colligative properties.
19. Describe, in detail, vapor pressure lowering, freezing point depression, boiling point elevation, osmosis, and osmotic pressure.
20. Give examples of freezing point depression and osmosis from everyday life.
21. Use molal freezing-point and molal boiling-point data to compute freezing points, boiling points, and molal concentrations for solutions.
22. Describe the effect of electrolytes on calculations of freezing and boiling points of solutions.

8.1 What Is and Is Not a Solution

In Section 3.1, we briefly reviewed the major types of substances. You recall that solutions are homogeneous mixtures, mixtures with a uniform composition all the way down to the molecular level. In this chapter on solutions, it is appropriate to look more closely at some classes of substances that resemble solutions, but are actually heterogeneous mixtures. These two types of substances are called *suspensions* and *colloidal dispersions*.

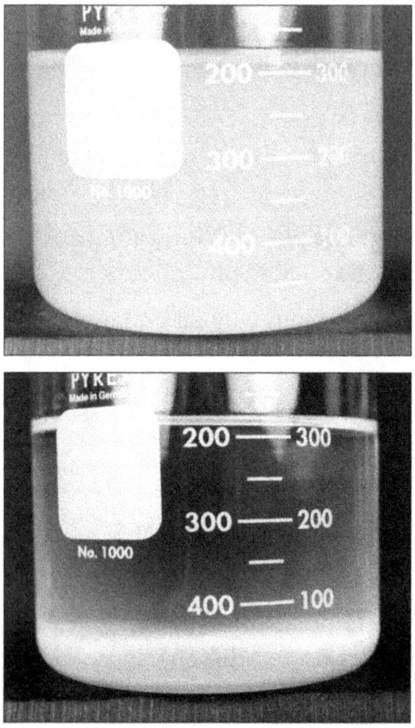

Figure 8.1. Corn starch mixed in water forms a suspension (top). The starch particles eventually settle out of the liquid and fall to the bottom of the beaker (bottom).

8.1.1 Suspensions

A *suspension* is formed when particles of size approximately 1 micrometer (1 μm) or larger are dispersed in a fluid (liquid or gas) medium. Particles this large do not remain in suspension indefinitely; they eventually settle out due to gravity.

Figure 8.1 shows an example of a suspension. At room temperature, corn starch is not soluble in water; that is, it does not dissolve in water. But if corn starch is stirred into water, a suspension is formed, as shown in the upper photo in the figure. After a few minutes, it is evident that the starch is not dissolved in the water, and after a few hours the starch particles are all at the bottom of the container, as shown in the lower photo.

Muddy water is another example of a suspension. If left to stand, the particles of sand, soil, and other organic matter eventually settle out, leaving the water quite transparent.

8.1.2 Colloidal Dispersions

Another solution look-alike is a class of substances known as *colloidal dispersions*, usually referred to simply as *colloids*. A colloid is formed when microscopic particles, ranging in size from 1 to 1,000 nm, are dispersed throughout a dispersing medium. Unlike the particles in suspensions, these particles do not settle out. Forces from *Brownian motion* (see the nearby box) keep the particles dispersed in the dispersing medium.

		Dispersing Medium State			
		Solid	**Liquid**		**Gas**
Dispersed Particle State	Solid	**solid sol** cobalt glass, cranberry glass	**sol** blood, paint	**solid emulsion** butter, cheese	**solid aerosol** smoke, fine airborne dust
	Liquid	**gel** gelatin, hair gel	**liquid emulsion** mayonnaise, milk		**liquid aerosol** fog, hair spray, insect repellent
	Gas	**solid foam** Styrofoam, foam rubber, marshmallow	**foam** shaving cream, whipped cream		**(none)**

Table 2.1. Types of colloids with examples.

Hmm... Interesting. Brownian motion

In 1827, Scottish botanist Robert Brown (1773–1858) was using a microscope to examine tiny particles from pollen grains. The particles were suspended in water, and Brown noticed that they jittered back and forth and moved through the water in an erratic and apparently random path. The motion of those particles is now known as *Brownian motion*, although no one knew what caused it until 1905. That was the year Albert Einstein published a paper demonstrating that the motion was caused by a particle being bombarded on all sides by moving water molecules. On this topic, still pictures just don't cut it. There are lots of animations online, but nothing beats the real thing. Go to commons.wikimedia.org, search under Brownian motion, and watch the clip with the file name *Brownianmotion beads in water spim video.gif* (sic).

Table 8.1 lists the different types of colloids, distinguished by the states of the dispersed particles and the dispersing medium.[1] Since most of these substances are probably familiar to you, although perhaps not by name, I will just point out a few things about the information in the table.

You may or may not have seen items made of cobalt glass or cranberry glass, both examples of *solid sols* and shown in Figure 8.2. Cobalt glass is ordinary silicon dioxide, SiO_2, with cobalt salts such as cobalt(II) aluminate, $CoAl_2O_4$, dispersed within the molten glass. Cranberry glass contains particles of gold(III) oxide in the molten glass. Once the glass hardens, the colloid that remains consists of solid particles dispersed within a solid medium.

Two different terms are used for colloids consisting of solid particles dispersed within a liquid medium, *sol* and *solid emulsion*. As the examples show, these two classes of colloids are rather different. In a *gel*, the solid particles link together in long molecules known as polymers, and these linked solids form a network or matrix throughout the substance. Finally, there are no colloids consisting of gas particles dispersed in other gas particles. This is because when no reaction occurs, all gases are *miscible* in each other, that is, they mix together in all proportions. Thus, mixed gases form solutions, not colloids.

Figure 8.2. Cobalt glass (left) and cranberry glass (right), also known as "gold ruby."

We conclude this section with two more interesting facts about colloids. First, although the dispersed particles in a colloid are smaller than the particles in a suspension, they are still not small enough to pass through a filtration medium. In a true solution, the individual particles are ions or molecules, which means their sizes are down in the sub-one nanometer range, and they will pass right through a filtration medium such as filter paper. (So don't bother trying to separate salt from water by passing saltwater through a filter.) Colloidal particles are much larger than this and will not pass through a filter.

1 Sources differ to a frustrating degree on the definitions of some of the terms and on the identifications of some of the substances. If you go digging around in different texts or online sources, prepare to get confused.

This is why people can wear dust masks to filter out pollens and dust (but not water vapor) from breathing air.

For colloids in fluid media, there is a test that distinguishes between solutions and colloids. Figure 8.3 illustrates the so-called *Tyndall effect* in a colloid. In the upper photograph, a flashlight beam is passed from right to left through a solution of saltwater. The light from the beam cannot be seen passing through the solution because the particles in the solution are so small they do not scatter light. In the lower photo, the beam is passed through a mixture of gelatin and water. The particles of gelatin form a colloidal dispersion, and are large enough that they do scatter the light. This is demonstrated by the fact that the beam is visible, by the illumination of the liquid, and by the brightness of the white markings on the beaker. The two photos were taken under the same lighting conditions. The light in the room was very dim except for the flashlight.

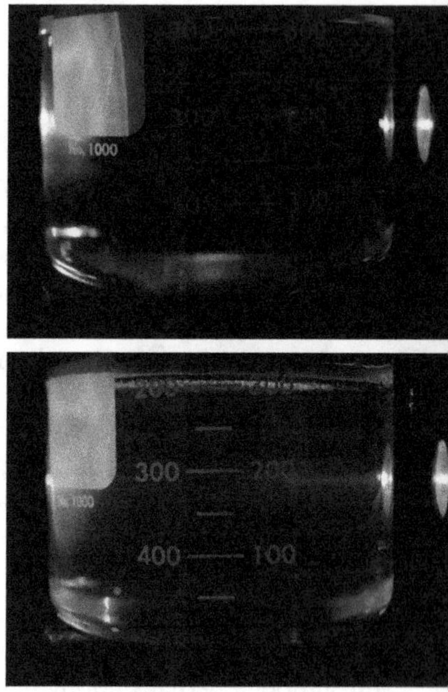

Figure 8.3. A salt water solution (top) allows the flashlight beam to pass through without scattering the light. A colloidal dispersion (bottom) scatters the light, making the beam visible, illuminating the surrounding liquid and the markings on the beaker from behind.

The Tyndall effect is what makes the beams from a car's headlights visible in foggy conditions. Fog is a colloid of liquid particles of water in air (a liquid aerosol), and the liquid particles scatter the light. And by the way, true water vapor—which is what steam is—is an invisible gas, just as the humidity in the air is invisible. If you can see the moisture, you are seeing light scattered by liquid water particles in air, and you are looking at a colloid, a mist of tiny condensed water droplets, not at steam.

8.2 Dissolution

8.2.1 The Process of Dissolving

Today it is difficult for us to imagine the intellectual climate among chemists in the 1880s. Many prominent scientists at that time were still opposed to the theory of atoms (including Dmitri Mendeleev), despite the success of John Dalton's 1803 atomic theory. At this time, a key problem was to explain why some substances dissolve in water while others do not, and why after dissolving in water, some solutions conduct electricity while others do not.

Figure 8.4. Swedish chemist Svante Arrhenius (1859–1927).

In 1887, Swedish chemist Svante Arrhenius (Figure 8.4) published his theory of ionic dissociation. Despite the combative atmosphere of the time, with scientists ridiculing each other for their belief—or their refusal to believe—in atoms, Arrhenius' theory encountered little opposition and was widely accepted by chemists within just a few years. Arrhenius and three other prominent chemists of the day became known as the "wild army of the Ionists," and they fought tooth and nail for acceptance of dissociation theory. They won the debate, and today dissociation theory is the most fundamental component of our theoretical understanding of solutions.

The process of dissolving—or *dissolution*—involves a *solvent*, often water, into which is dissolved another substance, called the *solute*. If a particular solute dissolves in a given solvent, it is said to be *soluble* in that solvent. So many things are soluble in water that water is often called the "universal solvent," although this term is quite hyperbolic. A lot of stuff *doesn't* dissolve in water, as you know if you have ever tried to rinse oil or grease off your hands with water but no soap.

The solubility of different substances in water varies over an extremely wide range. Some substances are essentially insoluble, some are soluble to a degree, and some liquids are infinitely soluble. Just as we saw with the mixing gases, substances that can dissolve in a solvent in any proportions are referred to as miscible. Ethanol and acetic acid, two compounds we discuss further below, are both miscible in water.

The term *dissociation* refers to the separation or coming apart of an ionic substance in a solvent composed of polar molecules. Common substances that dissociate are many ionic compounds, acids, and bases.

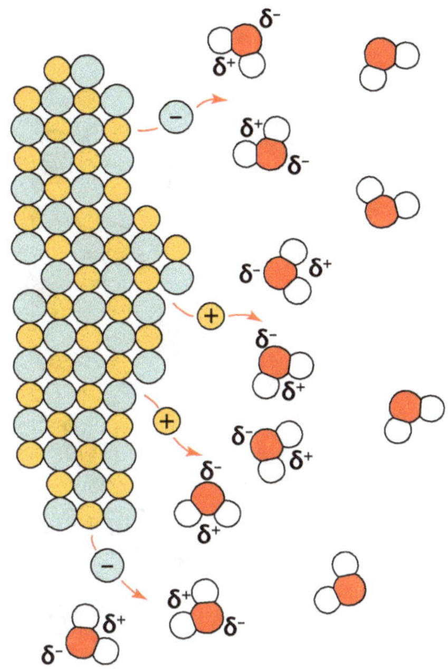

Figure 8.5. Dissociation of an ionic crystal in water. (Note that for clarity, the H_2O molecules are drawn much farther apart than they actually are.)

Water, as you know, is composed of polar molecules. Figure 8.5 illustrates what happens when an ionic compound such as sodium chloride dissolves in water. The hydrogen or δ^+ ends of the water molecules exert an electrical attraction on the negative ions in the crystal, the chloride ions in the case of sodium chloride. At the same time, the oxygen or δ^- ends of the water molecules exert an electrical attraction on the positive ions in the crystal, sodium ions in the case of sodium chloride. These attractions are strong enough to pull the ions out of the crystal lattice. This is amazing, when you think about it, because ionic bonds are quite strong. Note that because of hydrogen bonding, the H_2O molecules in water are actually much closer together than indicated by the figure. The molecules are shown spread out in the drawing for clarity.

Once again, this chemistry is all about electrical attractions. But understanding the process of dissociation also requires us to consider the energies involved in these electrical attractions. The lattice energy of a typical ionic crystal is so high that if the energy to pull the crystal apart were to come entirely from the internal energy of the water molecules (that is, the kinetic energy associated with their motion at a certain temperature), the temperature of the water molecules would have to plummet. Not only is energy required to dissociate the crystal lattice, additional energy is required to overcome the hydrogen bonding and separate the water molecules from each other to make room among the H_2O molecules for the ions.

For example, just considering the energy needed to overcome the lattice energy of 10 g NaCl, if the energy were to come from the thermal energy in the water, the temperature of 400 mL H_2O would drop by 141°F! (This is a simple calculation based on the equation from Chapter 6, $Q = Cn\Delta T$.) Obviously, when you dissolve a bit of salt in a glass of water, this doesn't happen. So where is the energy coming from that keeps the water temperature more or less constant while the salt dissolves?

The answer is given by the process of *hydration* that occurs in parallel with the dissociation. Recall that when entities of opposite charge are allowed to fly toward each other, energy is released. Figure 8.6 illustrates what happens as ions from a crystal dissociate in water. The polar

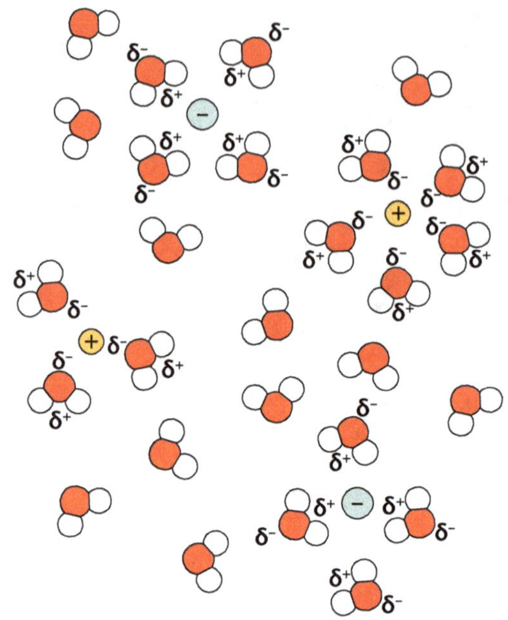

Figure 8.6. Hydration of ions in aqueous solution. (Again, H_2O molecules are drawn farther apart than they actually are.)

water molecules rush in to form cages of molecules around the ions, with either the δ^+ ends or the δ^- ends of the molecules pointing toward the ions, depending on each ion's charge. The decrease in electrical potential energy between the molecules and ions as they come together means energy is released. In the case of salt in water, this release of energy balances (more or less) the energy absorbed by the dissociation of the crystal lattice.

Before we move on, note that the term hydration refers specifically to the ion-surrounding process that occurs in water. Other polar solvents act the same way, and the more general term for solvent molecules surrounding ions in solution is *solvation*. Besides water, other polar solvents include ammonia, ethanol, and methanol.

Figure 8.7 shows the forces and energy flow involved during the process of an ionic compound dissolving in water. To dissociate a crystal and push apart the solvent molecules, energy is required, so heat flows into areas where these processes are occurring. Where hydration is occurring, heat is released. The combined effect of these energy absorptions and releases determines whether the dissolving of a solute is an endothermic process (net flow of energy is in) or an exothermic process (net flow of energy is out). If the energy required for dissociation and solvent molecule separation is greater than the energy released during hydration, then more energy is absorbed than released and the temperature of the solution drops—the complete process is endothermic. The drop in solution temperature occurs because the energy difference has to come from the thermal energy in the solution, and when some of that energy is removed, the solution temperature decreases. However, if the energy released by hydration exceeds the energy requirements of dissociation and solvent separation, then additional energy is released into the solution, and its temperature increases—so the complete process is exothermic.

When sodium chloride dissolves in water, the temperature does not change noticeably. But temperature changes can be substantial when other ionic compounds dissolve. The instant cold packs popular among sports trainers these days can be made from ammonium nitrate. This compound is separated from water inside the cold pack by a thin barrier. Twisting the cold pack ruptures the barrier, allowing the ammonium nitrate to dissolve into the water. This turns out to be a very endothermic process, so the pack gets cold and can be used to treat injuries such as sprains.

Our discussion so far has concentrated on ionic compounds. However, when acids and bases dissolve in water, the same logic applies. One significant difference, though, is that when acids or bases dissolve in water, the process is generally highly exothermic. Adding even a small amount of many different acids and bases to water generates a lot of heat. For this reason, it is standard procedure in every lab always to add a concentrated acid or base to water, and never the other way around. By adding the acid to water, the active ingredients in the process get diluted. The dissolving processes and the heat generated by this exothermic process are distributed through the water in the container. Conversely, when water is added to a concentrated acid or base, the

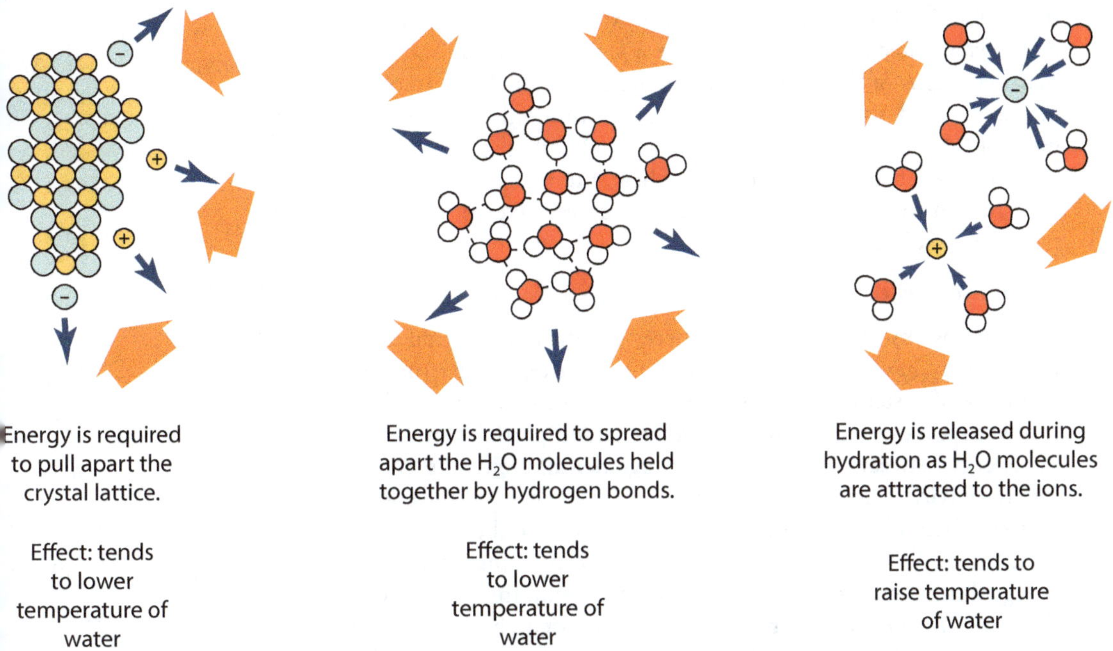

Figure 8.7. Forces and energy during dissociation and hydration. Blue arrows show directions of forces involved, orange arrows show direction of energy flow. Thermal energy from the water flows into areas where forces are pulling particles apart, and out of areas where particles are rushing together.

acid or base remains concentrated, as does the heat released, and in some cases creates explosive fumes, such as hydrogen. Concentrated heat and explosive fumes can cause dangerous boiling, hot containers, and even explosions.

8.2.2 Enthalpy of Solution

The net amount of energy we have been talking about has a technical name, which we introduce here. The net amount of heat that flows into or out of a chemical system during a reaction is called the *change in enthalpy*, ΔH. *Enthalpy*, H, is a conceptual term from the field of thermodynamics. By change in enthalpy we mean the difference between the enthalpy of the reaction products and the enthalpy of the reactants, as shown in the following equation:

$$\Delta H = H_{\text{products}} - H_{\text{reactants}}$$

More advanced studies in chemistry require the student to get deep into thermodynamics, a subject with a reputation for inducing weeping and gnashing of teeth. Happily however, at this stage what you need to know about enthalpy can be boiled down to these two statements. First, the enthalpy of the reactants or products in a reaction cannot be measured directly. Only changes in enthalpy can be measured. So you don't really need to worry about what enthalpy *is*, only what the change in enthalpy represents. Second, *for a process that occurs at constant pressure, the change in enthalpy represents the heat that flows into (is absorbed by) or out of (is released from) a system*. If ΔH is positive, heat is absorbed during the process and the process is endothermic. If ΔH is negative, heat is released during the process and the process is exothermic.

Now you see how the concept of the change in enthalpy relates directly to our previous discussion of the energy involved during the processes of dissociation and solvation. The net amount of heat energy absorbed or released during dissolution is called the *enthalpy of solu-*

tion. Table 8.2 lists enthalpy of solution values for a few common *electrolytes*, substances that ionize in aqueous solution. The large negative values of the acids and bases are due to the large amounts of heat released, which we discussed above. Ammonium nitrate, the compound in the instant cold packs, has a fairly large positive value indicating a substantial absorption of heat. But make sure you have your mittens on if you ever have to mix up a beaker of rubidium perchlorate solution!

Electrolyte	Enthalpy of Solution, ΔH (kJ/mol)
$HClO_4$	−88.76
HCl	−74.84
KOH	−57.61
NaOH	−44.51
LiCl	−37.03
HNO_3	−33.28
CH_3COOH	−1.51
NaCl	3.88
$AgNO_3$	22.59
NH_4NO_3	25.69
KNO_3	34.89
$KClO_3$	41.38
$RbClO_4$	56.74

Table 8.2. Enthalpy of solution (at 25°C and 1 atm) for some common electrolytes.

8.2.3 Entropy and Free Energy

Before we move on, let's pause here to address a question some readers may have at this point. We have seen repeatedly that the tendency of systems toward energy minimization is a rule of nature with a lot of explanatory power. Charged particles minimize the energy between them by coming together and releasing energy into the environment. But when charges are pulled apart, the energy between them *increases*, and this is what happens to the ions during dissolution. As we have seen, there is a corresponding decrease in energy during hydration. But in the case of an endothermic dissolution, there is a net gain in the energy of the particles in the solution.

How can this occur? In the case of an endothermic dissolution process, the net flow of energy is into the solution, and the energy required for dissociation exceeds the energy released during hydration. The key to understanding this puzzle is in the concept of *entropy*. Entropy plays a significant role in topics we encounter in coming chapters, so let's take time here to dive into it a bit.

We first encountered entropy back in the Introduction. There I defined entropy as a measure of the disorder present in a system. Just as minimizing energy is one of the universal principles according to which inanimate objects behave, maximizing entropy is as well. In fact, the second law of thermodynamics is basically a statement saying that in any process, the entropy of the universe increases. When entropy decreases in one location or system, it always increases even more somewhere else.

To bring home the concept of entropy, let's illustrate disorder with some examples. Figure 8.8 depicts four scenarios in which ordered and disordered states are represented side by side. In each case, the disordered state is a condition with higher entropy, entailing mixing, spreading out, randomness, and the erasing of concentrations, distinctions, or boundaries. For example, notice in the hot-cold object scenario (B), when the objects are first brought together, there is a sort of boundary or transition region between the hot object and the cold one. If heat is allowed to flow across this boundary, it moves from the hot object to the cold one. The boundary gradually disappears as thermal equilibrium is reached; at that point entropy is maximized.

The four scenarios represent three different cases with respect to the energy involved. In scenario (A), the house of cards represents a higher energy state than cards laying flat on the floor. In this scenario, nature's tendency to minimize energy and maximize entropy pull in the same direction—to bring down the house. Scenarios (B) and (C) are energy neutral. There is no gain or loss of energy in the overall system, assuming these objects are contained in ways that keep the total thermal energy or total number of gas particles isolated from the outside world. Scenario (D) represents an increase of energy. The circles represent ions, which are electrically attracted to each other and thus must be pulled apart by the input of energy. In this case, the tendencies toward energy minimization and entropy maximization are in opposite directions.

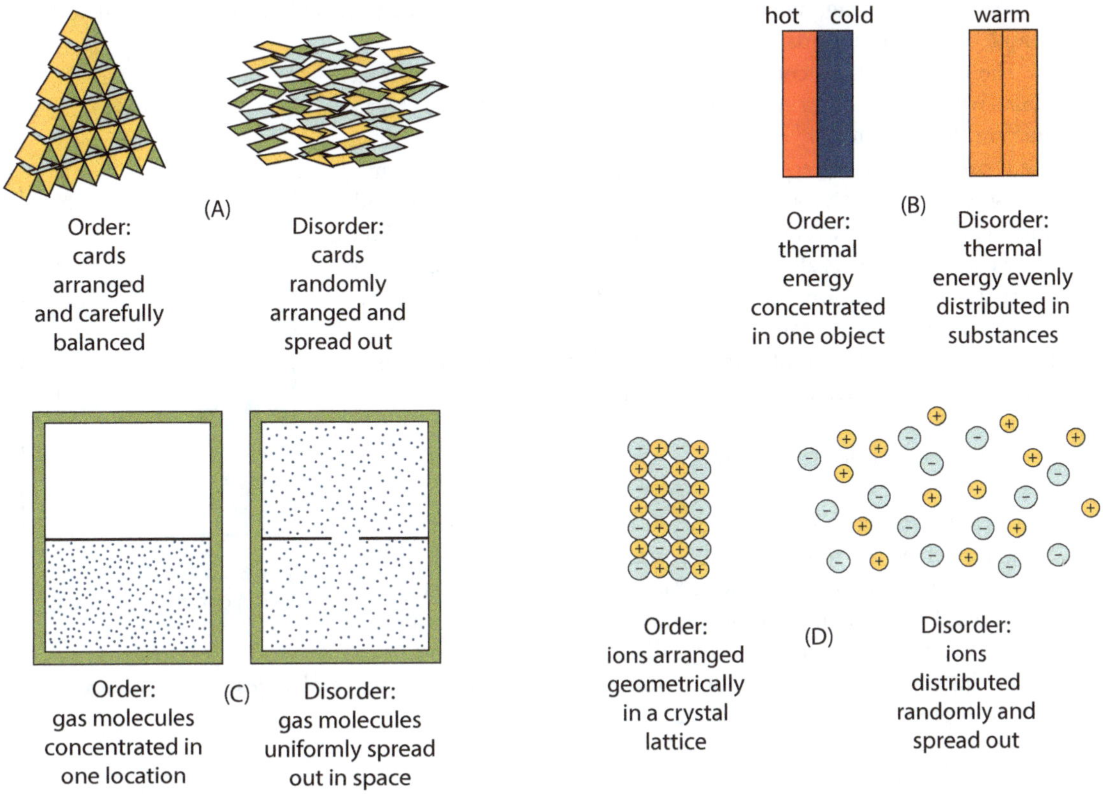

Figure 8.8. Four examples of entropy increases due to an increase in disorder. The following observations hold regarding the energy in these cases: In (A) energy is minimized; (B) and (C) are energy neutral; (D) represents an increase in system energy.

In many physical and chemical processes, the relationship between energy and entropy is like the ions in scenario (D): the tendencies to minimize energy and maximize entropy pull the system in opposite directions. In a case like this, the system can move toward a condition of minimum energy or toward a condition of maximum entropy but not both. Endothermic dissolution of an ionic crystal is one example of this situation. In such cases, the system goes the direction in which it gains the most (lower energy or more entropy), and it does so by borrowing, so to speak, from the other side. In the endothermic dissolution of a crystal, energy is borrowed from the thermal energy in the solution. But the borrowing is worth it because of the gain in entropy, so the crystal dissolves.

In any process, there is (a) an increase or decrease in the internal energy in the system, and (b) energy borrowed or expended because of the increase or decrease of entropy. Roughly speaking, and without getting too mathematical at this point, the difference between these two is called the *free energy*.[2] We come back to the free energy in a later chapter. For now simply note that *natural processes proceed in the direction that reduces the free energy in the system.*

8.2.4 Electrolytes

As I mentioned in Section 8.2.2, electrolytes are compounds that ionize in aqueous solutions. Ions in solution conduct electricity. The compounds that do this are the same as those we have

2 This is a rough definition. A more accurate definition for the free energy of a system must wait until we define these terms in more detail in Chapter 10.

just been considering, namely, compounds that ionize in water. As we have seen, these include soluble ionic compounds, acids, and bases.

The fact that these compounds ionize in aqueous solution explains why they conduct electricity. Ions have a net electrical charge, and are thus affected by electric fields. When two electrodes from the terminals of a battery are placed in the solution, as illustrated in Figure 8.9, positive ions travel toward the negative electrode and negative ions travel toward the positive electrode. "Conducting electricity" means allowing electric current to flow, and electric current is, by definition, the movement or flow of electric charge.

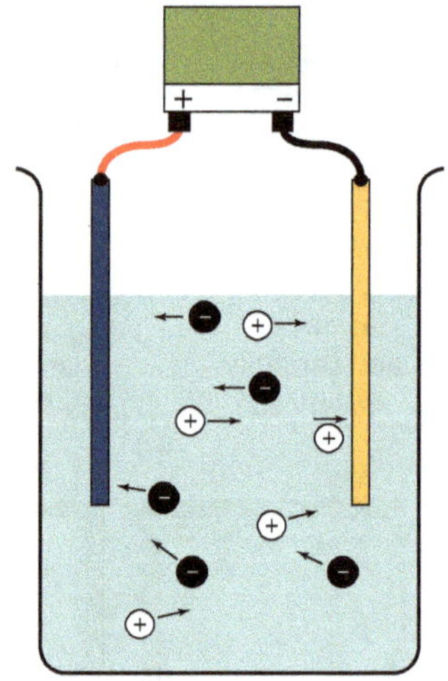

Figure 8.9. Electrolytes in solution conduct electricity.

However, as we examine in more detail next chapter, acids and bases ionize in water to differing degrees. Hydrochloric acid dissociates almost completely in water, as represented by this equation:

$$HCl(aq) \rightarrow H^+(aq) + Cl^-(aq)$$

Since HCl dissociates so thoroughly, it is called a *strong electrolyte*. Strong electrolytes ionize readily, and thus in aqueous solution conduct electricity at any concentration. By contrast, acetic acid, CH_3COOH, (the main component of vinegar other than water), is a *weak electrolyte*. When acetic acid dissolves in water, most of the molecules stay intact and do not dissociate to form ions. The few molecules that do dissociate may be represented by this equation:

$$CH_3COOH(aq) \rightleftharpoons H^+(aq) + CH_3COO^-(aq)$$

In this equation, the normal reaction "yields" arrow is replaced with a double arrow, indicating that this reaction proceeds in both directions simultaneously. When acetic acid dissolves in water, the solution quickly comes to an equilibrium state in which the rate of dissociation of CH_3COOH molecules equals the rate at which H^+ and CH_3COO^- ions are joining back together to become CH_3COOH molecules. The double arrows indicate this two-directional process.

Just because a substance is soluble in water does not mean it is an electrolyte. Sucrose, or ordinary table sugar, is quite soluble in water. A molecule of sucrose has eight different —OH groups that form hydrogen bonds with water molecules, so it readily dissolves. (Recall from Chapter 4 that an —OH group joined to a molecule is called a *hydroxyl group*.) However, molecules of sucrose are not ions; they have no net charge. Accordingly, sucrose is not an electrolyte.

States of equilibrium like acetic acid in water are very important in chemical processes. We encountered one example of equilibrium in Chapter 6—when a substance is at its triple point, all three states exist in equilibrium. Molecules are continuously freezing, melting, boiling, and condensing at the same rate. This is also happening in an insulated container of ice water. The ice water is always at 0°C, and molecules of H_2O in the water are freeing and melting at the same rate. (Of course, in a warm room, the ice all eventually melts because there is no perfect insulation, and heat from outside slowly conducts into the container, causing slightly more molecules to melt than freeze.) We encounter more examples of equilibrium in this and later chapters.

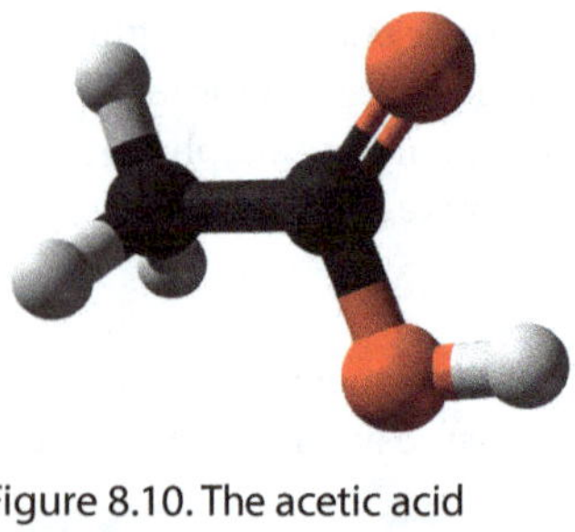
Figure 8.10. The acetic acid molecule, CH_3COOH. As usual, black = C, red = O, and white = H.

By the way, in case you are wondering about the formula for acetic acid, sometimes the formula is written as HCH_3COO so that its formula resembles those of other acids, with the ionizing H atom written first. However, these days the formula is most often written CH_3COOH because this way the formula resembles the sequence of atoms in the molecule, shown in Figure 8.10. The hydrogen atom that dissociates from the molecule is in the hydroxyl group shown at the lower right of the molecule in the figure.

8.3 Solubility

8.3.1 Ionic Solids in Water

Solids that are soluble in water are only so up to a point. There is a maximum amount of a solid solute that dissolves in a given quantity of water. A solution that contains this maximum amount of dissolved solute is said to be *saturated*. Saturation is another example of chemical equilibrium. At saturation, ions are dissociating from the crystal lattice and returning to recrystallize on the lattice at equal rates, as illustrated in Figure 8.11.

For many solutions, the saturation point depends on the temperature, and for many solutes, the solubility increases with temperature. An important exception to this common trend is NaCl. The solubility of NaCl increases only very slightly with increasing temperature.

Just about any ionic solid dissolves in water to at least a minute degree. However, as a working definition for solubility we can say that if at least 0.05 mole of a substance dissolves in one liter of water, then the substance is soluble in water; otherwise the substance is insoluble.

We don't really have a simple method for predicting whether a given ionic solid is soluble in water, but we do have lots of empirical data regarding the solubility of various compounds. These data have been generalized into a set of rules of thumb—guidelines we use to describe the solubility of certain broad classes of compounds. These guidelines are listed in Table 8.3 on the next page. It is important to notice that all ionic solids with an alkali metal or ammonium ion as the cation are soluble. Also notice that all the nitrates and acetates are soluble.

▼ Example 8.1

Classify each of these compounds as soluble or insoluble in water: (a) barium sulfate, (b) sodium carbonate.

According to Table 8.3, sulfates are generally soluble. However, sulfates with barium as the cation are listed as an exception. Thus, barium sulfate is insoluble.

Carbonates are generally insoluble. However, all ionic compounds with an alkali metal as the cation are soluble, so sodium carbonate is soluble.

▲

8.3.2 Ionic Solids in Nonpolar Solvents

Seeing is believing. If you have never done so, go into your kitchen, pour a bit of cooking oil into a glass, and sprinkle in some salt. You can stir the salt in the oil all you like, but it won't dissolve. Oils like cooking oil, lubricating oil, and motor oil are composed of nonpolar molecules. There are no forces in a solvent of nonpolar molecules strong enough

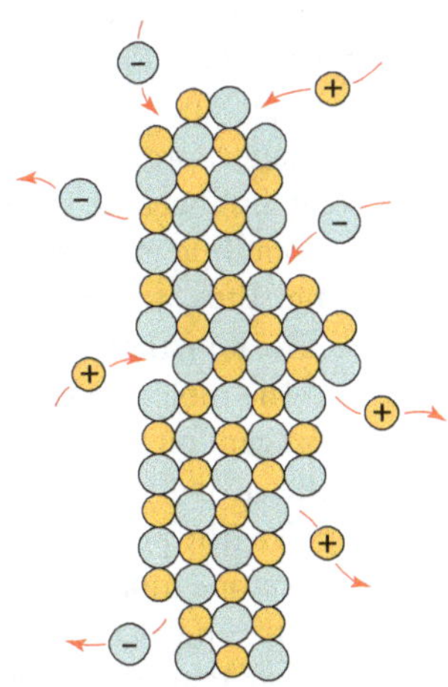

Figure 8.11. At saturation, ions dissociate and recrystallize at the same rate.

Soluble Compounds	
Compounds Containing These Anions	Cations That Are Significant Exceptions
NO_3^-	None
CH_3COO^-	None
Cl^-	Compounds of Ag^+, Hg_2^{2+}, Pb^{2+}
Br^-	Compounds of Ag^+, Hg_2^{2+}, Pb^{2+}
I^-	Compounds of Ag^+, Hg_2^{2+}, Pb^{2+}
SO_4^{2-}	Compounds of Sr^{2+}, Ba^{2+}, Hg_2^{2+}, Pb^{2+}
Insoluble Compounds	
Compounds Containing These Anions	Cations That Are Significant Exceptions
S^{2-}	Compounds of NH_4^+, alkali metal ions, Ca^{2+}, Sr^{2+}, Ba^{2+}
O^{2-}	Compounds of NH_4^+, alkali metal ions, Sr^{2+}, Ba^{2+}
CO_3^{2-}	Compounds of NH_4^+, alkali metal ions
PO_4^{3-}	Compounds of NH_4^+, alkali metal ions
OH^-	Compounds of NH_4^+, alkali metal ions, Ca^{2+}, Sr^{2+}, Ba^{2+}

Table 8.3. Solubility guidelines for ionic compounds in water.

to pull ions out of a crystal lattice, so ionic solids do not dissociate in nonpolar solvents. We return to nonpolar solvents shortly.

8.3.3 Polar Liquids

You may have heard the rule that when it comes to solvents, "like dissolves like." The like-dissolves-like rule refers to the types of molecular forces present in a liquid solvent and potential solute. Polar solutes dissolve in polar solvents. Nonpolar solutes dissolve in nonpolar solvents. But nonpolar substances do not dissolve in polar substances, and vice versa.

You are already familiar with the polarity of water and ammonia molecules. Each of the substances shown in Figure 8.12 also consists of polar molecules. These substances are all soluble in water, and each can be used as a solvent to dissolve other polar substances. The first three substances are all members of a class of substances called *alcohols*. These molecules are not only polar, but because they each have a hydroxyl group attached, they are able to engage in hydrogen bonding in water. Looking back at the acetic acid molecule in Figure 8.10, you see the same thing. Thus, as a molecule, acetic acid is polar and participates in hydrogen bonding in water.

The fourth molecule shown in Figure 8.12, acetone, does not participate in hydrogen bonding. The molecule is covered with hydrogen atoms, but those hydrogen atoms are not attached to oxygen, nitrogen, or fluorine atoms as required for hydrogen bonding. Nevertheless, because of the strongly electronegative oxygen atom on one side of the molecule, the acetone molecule is polar. The polarity allows the molecules of a polar solvent such as water to form cages around the acetone molecules, just as they do around the ions of acetic acid, and thus acetone is soluble in water and alcohols.

Note that since acetone does not participate in hydrogen bonding, it is quite volatile. Recall from our discussion of vapor pressure in Chapter 6 that volatility is a measure of how readily a liquid evaporates to the vapor state. The more volatile a liquid is, the higher its vapor pressure. Acetone molecules are attracted to one another by Keesom forces, forces between permanent dipoles. But Keesom forces, as with all Van der Waals forces, are much weaker than the forces in hydrogen bonding, and with only weak forces attracting molecules toward one another, the molecules easily break lose and transition to the vapor state.

Alcohols are about twice as volatile as water; acetone is about 10 times as volatile as water. Liquids more volatile than water are also famous for often being flammable—the more volatile, the more flammable. (This explains the use of the term volatile as a metaphor to describe a person's personality.)

As we saw before with the sucrose molecule, all these polar molecules are soluble but none of them is an electrolyte. Again, these molecules have no net charge—they are not ions—and they do not ionize in water. Alcohols and other polar liquids that do not ionize are not electrolytes.

8.3.4 Nonpolar Liquids

Nonpolar liquids such as cooking oil do not dissolve in water. If you try to mix a nonpolar liquid into water, the hydrogen bonding between the polar water molecules attracts them toward each other and squeezes the nonpolar molecules out. This is why oil and vinegar salad dressing separates after shaking. Vinegar is a solution of about 5% acetic acid and 95% water—all polar molecules that do not mix with nonpolar oil molecules.

Molecules of greases and oils are all nonpolar, including oils from plants like olive oil, peanut oil, and cottonseed oil, as well as fats from animals. Waxes are another example of nonpolar molecules. Nonpolar solvents include gasoline, mineral spirits, benzene, toluene, and carbon tetrachloride. These are insoluble in water, just as oils are.

In nonpolar liquids, the main forces holding the molecules together are London dispersion forces—the random forces between temporary dipoles. Even though nonpolar substances are not soluble in water, they are soluble in each other—another application of the like-dissolves-like rule. Nonpolar liquids dissolve in each other because mixing of the molecules is favored by the increase in entropy that occurs when the molecules mix. An impressive demonstration of this is that cooking oil is very effective at removing oil-based stain or grease from your hands. Try it—you'll be amazed.

8.3.5 Solutions of Solids

Most of our attention has been focused on liquid solutions. That is as it should be, and we return to them shortly. We also hit on solutions of gases briefly at the beginning of the chapter, where we saw that non-reacting gases are miscible. Gas molecules mix together in any proportion to make a homogeneous mixture of gases—a solution in the gas state. Solid solutions, that is, metal *alloys*, are another important class of solutions. Here we look briefly at several common metal alloys.

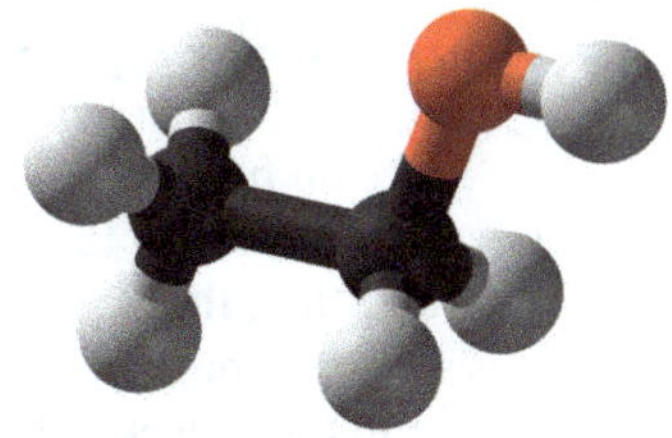

Ethanol, CH_3CH_2OH or C_2H_6O. Also called grain alcohol. This is the type of alcohol found in alcoholic beverages.

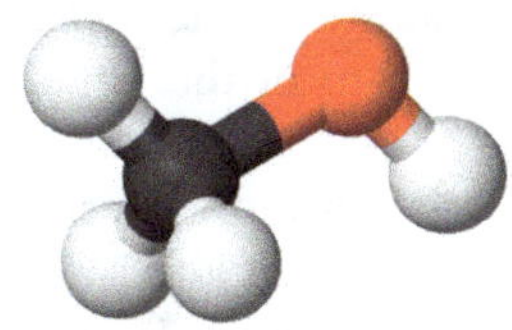

Methanol, CH_3OH, the simplest alcohol. Also called wood alcohol. This substance is toxic.

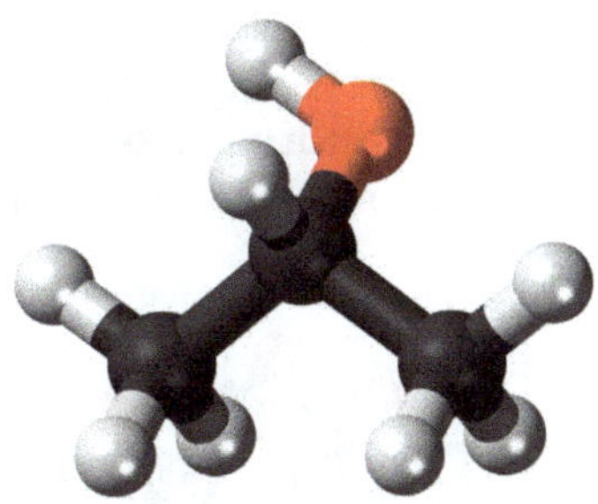

Isopropyl alcohol, written as C_3H_8O, C_3H_7OH, or $(CH_3)_2CHOH$. This is the alcohol used in rubbing alcohol.

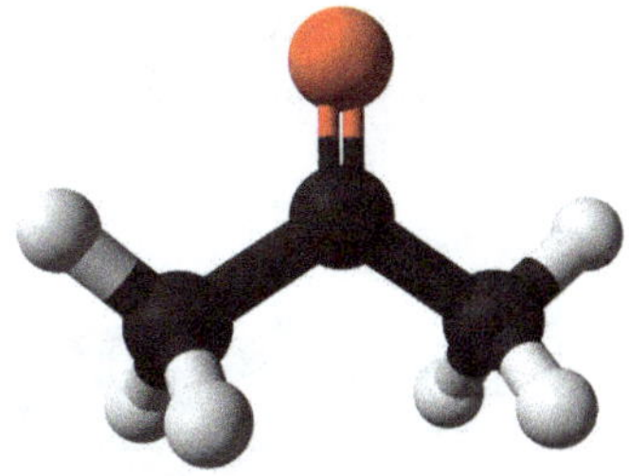

Acetone, $(CH_3)_2CO$, the solvent used in fingernail polish remover.

Figure 8.12. Examples of polar solvent molecules. These substances all dissolve in water.

Hmm... Interesting. How soap works

Be thankful you don't live back in the days before soap was discovered. The ancient Greeks used to rub themselves with sand mixed with olive oil. The olive oil removed the dirt, and then the whole mixture, along with the dead skin cells broken lose by the sand, was scraped from the body with a special tool called a *strigil*. Back then, getting clean was a pain, and the squeeky-clean feeling we are accustomed to today after a shower did not exist.

Soap is obviously soluble in water. But how can it simultaneously bond to a nonpolar substance such as oil? How does soap work? You may have been involved in a soap-making project when you were a kid; what we seek here is to understand how soap does its job.

First, let's look at how it is made. Legend has it that soap was discovered when rain runoff from animal sacrifices made at a temple on Mount Sapo flowed into the Tiber River near Rome. Women washing clothes downstream of the temple found that while it was raining the water cleaned the clothes more effectively. The reason was that the runoff from the temple contained a crude soap formed in a reaction between the fats from the burnt offerings, the wood ash from the fire, and the heat of the fire. The reaction used in making soap is called *saponification*, and the words soap and saponification both derive from the name of Mount Sapo.

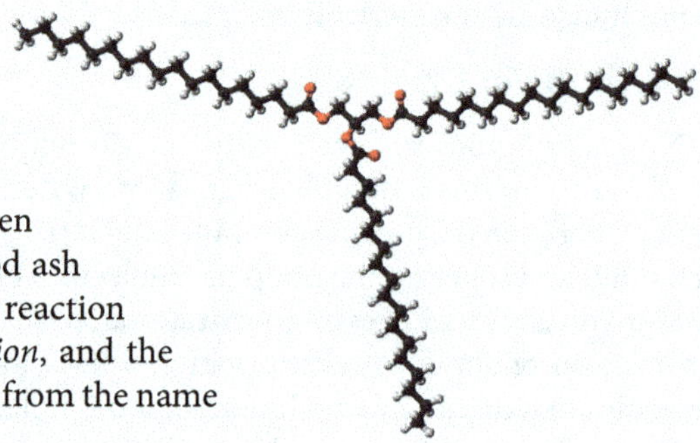

The two key ingredients that led to those clean clothes are potash, which is potassium hydroxide, from the ashes of the wood fire, and triglycerides, which are long, three-stranded fat molecules (shown in the first figure) from the animal sacrifices.

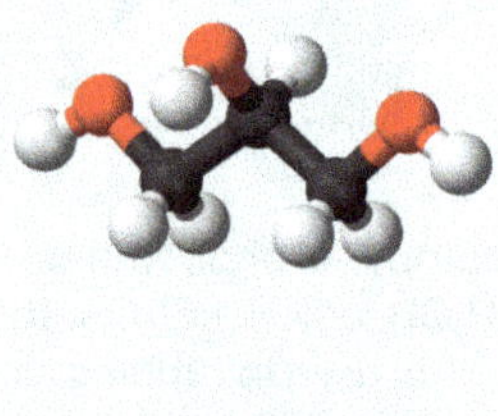

The backbone of each triglyceride strand is a carbon chain with two hydrogen atoms attached at each carbon bond. The red atoms are oxygens. With heat and the potassium hydroxide, the three long strands separate from the triglyceride molecule, leaving behind the center section of the former triglyceride, with hydroxyl groups attached in their places. This center molecule, which is *not* the soap, is called *glycerol* (second figure). The soap is the long strands, which go by various names, including glycerides, stearate ions, or, if a hydrogen atom is attached to the oxygen on the end (third figure), fatty acids. In the saponification reaction, instead of a hydrogen atom attaching to the glyceride molecule, the potassium atom from the KOH attaches there. (The arrow in the figure points to the potassium atom). *That* is a soap molecule. Or, at least it is in the case of soft soap.

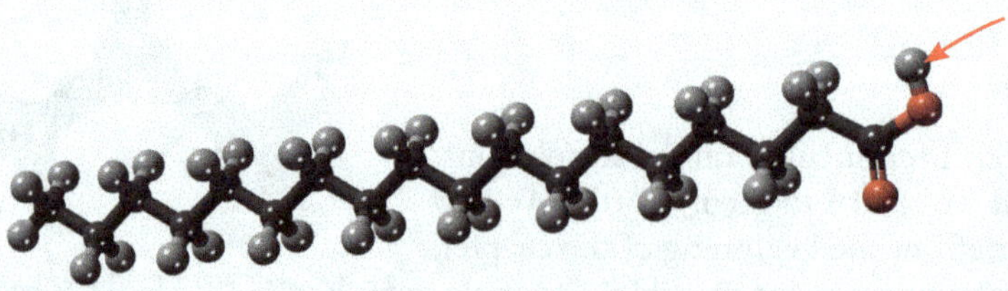

The same saponification reaction occurs with sodium hydroxide, NaOH or *lye*, instead of the KOH. (In the old days, sodium hydroxide was called soda ash. It was formed by burning seaweed instead of wood.) In a reaction with NaOH, the atom at the red arrow is a sodium atom. Making soap from NaOH is the most common industrial process for making soap, and results in bars of soap that are hard. Soft soaps are made with KOH; hard soaps are made with NaOH.

Now we come to how soap does its job. In water, the alkali metal atom ionizes (K^+ or Na^+), leaving the long stearate anion. (Actually, the stearate ion is just one possibility. There are others depending on the source of the triglycerides.) With the glyceride ionized to form a stearate ion, the oxygen where the K or Na was attached is now an O^- ion, and is attracted to the δ^+ regions of the water molecules. But the long carbon chains of these anions, being nonpolar, are not part of this action and are constantly being squeezed out of the way by the hydrogen bonding between water molecules. So the stearate carbon chains are looking for a safe place to get out of the way, and in the presence of particles of nonpolar oil or grease, the carbon chains collect together by inserting themselves into the grease particles, forming what are called *micelles* (last figure). The white spheres in the figure represent the O^- ions at the ends of the stearate ions, and the yellow strands represent the carbon chains. Once the grease particle gets completely covered with the inserted stearate strands, it is no longer being squeezed out of the way. The ball is now just a big negative ion, which becomes surrounded by a cage of water molecules (the δ^+ ends) and flushes away with water, carried along by water's own polar molecules.

There are many steel alloys; basic diagrams of the two most common are shown in Figure 8.13. *Carbon steel* contains 2.1% or less carbon by weight. The carbon atoms are much smaller than the iron atoms, and are located in the spaces or *interstices* between iron atoms. If the carbon content is greater than 2.1%, the alloy is known as *cast iron*. Cast iron is extremely strong, but also very inflexible. A lower carbon content gives the steel more flexibility, enabling it to be drawn into wire (ductility) or hammered into shapes (malleability). Cast iron is so inflexible (brittle) that it must be cast into shapes. It is not malleable at all.

On the right in Figure 8.13 is shown a typical atomic arrangement in *stainless steel*. Stainless steel is important because it resists rust. In stainless steel, atoms of chromium and nickel, which are about the same size as atoms of iron, have taken the places of iron atoms in the crystal lattice of the metal. In general, the chromium content of stainless steel is 10.5% by weight or higher. We encountered the popular 304 and 316 stainless steel alloys in a previous chapter. In these alloys, the chromium and nickel contents are about 18% and 8%, respectively.

Another alloy we all encounter is gold. The indication of purity for gold jewelry uses the *carat* system, which specifies the amount of gold out of 24 parts. With 18-carat gold, the alloy is 18 parts gold (75%) and 6 parts (25%) another metal, often silver. At a purity of 24 carat, the gold must be at least 99.9% pure. The atoms of silver in the alloy take the place of gold atoms in the metal crystal lattice, just as the chromium and nickel atoms do in stainless steel. Pure gold is a fairly soft metal. Adding some silver makes jewelry much stronger, while preserving its gold color.

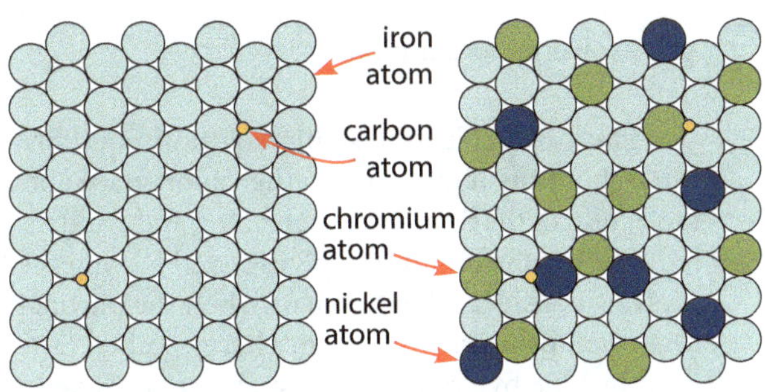

Figure 8.13. Carbon steel has 2.1% or less by weight of carbon atoms in the interstices between iron atoms (left.) Stainless steel (right) has this carbon content, plus typically about 18% chromium atoms and 8% nickel atoms in place of iron atoms in the crystal lattice.

As with gold, pure silver is very soft. A popular silver alloy is called *sterling silver*, which is 92.5% silver and 7.5% other metals, usually copper. Adding the copper increases the strength, while preserving the color and luster. Sterling silver has been popular for centuries as an alloy for making dishes, candlesticks, vases, jewelry, eating utensils, and many other items.

A few copper alloys need mentioning, both because of their historical significance and their continued use today. You may already be familiar with these. *Brass* is an alloy of copper and zinc; the proportions of each vary widely for different purposes. Brass resists corrosion, which makes it an excellent material for use outdoors. Copper has the excellent property of being lethal to bacteria. This makes brass an excellent material for doorknobs, which get touched by many dirty hands.

Figure 8.14. A 15th-century bronze door knocker.

Copper's antibacterial properties also make it highly appropriate for the alloys used for coins. (Pure copper is a good choice for water supply plumbing for the same reason.) Nowadays, pennies have the least copper of all U.S. coins. Since 1982, pennies have been 98.5% zinc, with just a thin copper plating on the outside. All other currently minted U.S. coins are alloys with at least 75% copper content.

Finally, *bronze* is a copper alloy usually containing tin. Bronze has been popular for centuries as a material for tools, machines, weapons, and statuary. As with brass, the copper content varies widely, but bronze is roughly 90% copper and 10% tin. Other metals, including zinc, are sometimes included in the alloy as well. Artisans have long preferred the rich color of bronze, as shown in the 15th-century bronze door knocker from Germany shown in Figure 8.14.

8.3.6 Gases in Liquid Solutions: Le Châtelier's Principle and Henry's Law

One more important class of solutions are those in which a gas is dissolved in a liquid. Oxygen molecules dissolve in water, and those molecules are extracted by the gills of fish. Another familiar example is the CO_2 dissolved in soft drinks. Let's use this example to make a few important points about how gases dissolve.

Unlike other solutions we have discussed, the solubility of gases in liquids is strongly affected by pressure. Recall from our discussion of vapor pressure in Chapter 6 that at a given temperature, the gas and liquid states reach an equilibrium in which molecules are evaporating and condensing at the same rate. The same holds for a solution of gas in a liquid such as the sugar solution of a soft drink and a dissolved gas like CO_2. We represent this situation with an equilibrium equation this way:

$$\text{gas} + \text{solvent} \rightleftharpoons \text{solution}$$

Figure 8.15 illustrates this solution equilibrium. Molecules of gas are shown dissolved throughout the liquid, and in the vapor space above the liquid. An equilibrium exists in which molecules of the liquid are vaporizing and condensing at the same rate, and molecules of the gas are entering the liquid (dissolving) and leaving the liquid at the same rate.

Now consider what happens when the partial pressure of the dissolved gas is increased. An important principle that applies here is called *Le Châtelier's principle*. This principle was formulated in 1888 by French chemist Henri Louis Le Châtelier (1850–1936). Le Châtelier's principle states that *when a system at equilibrium is subjected to stress, the system seeks a new equilibrium state that minimizes the stress*. This principle applies to all types of chemical equilibrium scenarios, and we have already seen several. (We briefly reviewed these a few pages back in Section 8.2.4.)

Applying Le Châtelier's principle to the gas/liquid solution in Figure 8.15, increasing the CO_2 pressure imposes a stress on the equilibrium system. When the CO_2 pressure is increased, there are more CO_2 molecules in the vapor space. This means the CO_2 dissolves into the liquid at a faster rate. With more CO_2 molecules in solution, they are also exiting the solution at a higher rate. A new equilibrium position is reached in which there is a higher concentration of dissolved CO_2 molecules.

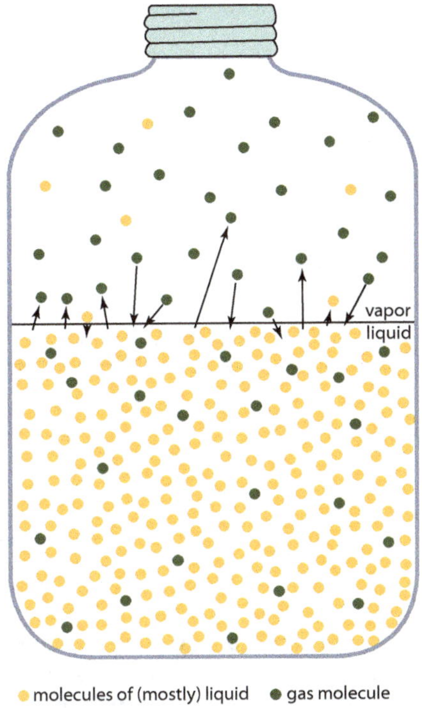

Figure 8.15. When a gas dissolves in a liquid an equilibrium is reached, and the solubility of the gas is directly proportional to the partial pressure of the gas.

When soft drinks are manufactured, the CO_2 is added to water at a pressure of 5–10 atm to form carbonated water.[3] This high pressure forces a lot of CO_2 molecules into solution. In addition to understanding this in terms of Le Châtelier's principle, we can also understand it in terms of another chemical principle called *Henry's law*, formulated in 1803 by English chemist William Henry (1774–1836). Henry's law states that *the solubility of a gas in a liquid is directly proportional to the partial pressure of the gas at the surface of the liquid*. In other words, when the CO_2 pressure is increased, the amount of dissolved CO_2 increases in proportion.

When you open a container of your favorite soft drink, the pressure in the container drops to atmospheric pressure. According to Henry's law, the solubility of CO_2 in a soft drink drops at that point by a factor of five to ten, depending on what the pressure was in the container. And according to Le Châtelier's principle, the liquid/gas system seeks a new equilibrium point to relieve this new stress: CO_2 comes out of solution until its concentration reaches its solubility at atmospheric pressure. This is why carbonated beverages begin *effervescing* (fizzing) when the container is opened. Over time, if the container is left open, the amount of dissolved CO_2 drops off dramatically. We say that the beverage becomes "flat."

Although we have focused our attention on the CO_2 dissolved in beverages, these same principles apply to all solutions of gases in liquids.

8.3.7 The Effect of Temperature on Solubility

The temperature of a solution can have a pronounced effect on the solubility of both ionic solids and gases in aqueous solution. The graphs shown in Figures 8.16 and 8.17 show solubility curves for assorted ionic solids and gases in water. These curves show the amount of dissolved solute at saturation at different temperatures. In Figure 8.16, the vertical axis indicates the percentage of the total mass of the solution that is due to the mass of the solute. A few particulars in the graph are interesting to note. First, the solubility of NaCl is almost flat; it does increase with temperature, but only increases by about 2% over the temperature range of liquid water. The solubility of Li_2SO_4 actually decreases substantially with increasing temperature, an uncommon trend. By contrast, note the strong upward trend for the solubility of KNO_3.

3 The CO_2 is the source of the carbonic acid in soft drinks. When CO_2 is dissolved in water, this chemical equilibrium is established: $CO_2(g) + H_2O(l) \rightleftharpoons H_2CO_3(aq)$

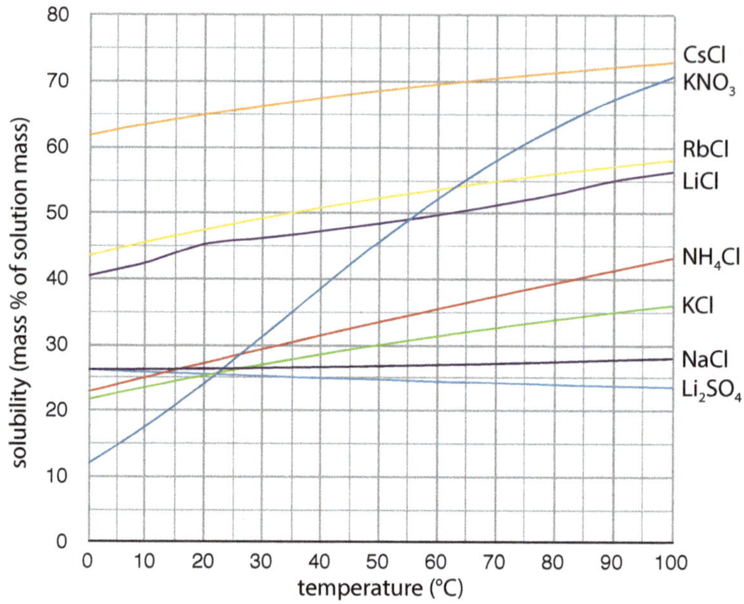

Figure 8.16. Temperature dependence of solubility for a few ionic solids in water.

In Figure 8.17 are four curves representing the solubilities of some common gases. The vertical axes for these two curves represent the mole fraction of the solution attributable to the gas, and all the data have been multiplied by 1,000.

These curves indicate solubility at atmospheric pressure, and in every case but one, the solubility decreases with temperature. The exception is oxygen, whose solubility is virtually constant over the temperature range shown. The solubility of oxygen is very low compared to the other gases, but apparently it is high enough for the fish to be happy. What's really interesting is that its solubility is relatively *constant*. It does vary, and at 0°C (not shown) it is about twice as high as it is at 20°C. Still, the concentration of oxygen in ponds does not vary with much temperature, which is good for the fish who live there. (This matters most in shallow ponds and in rivers that are alternatively fed by rain water or snow melt. In lakes and oceans, a few meters below the surface the temperature does not vary much.)

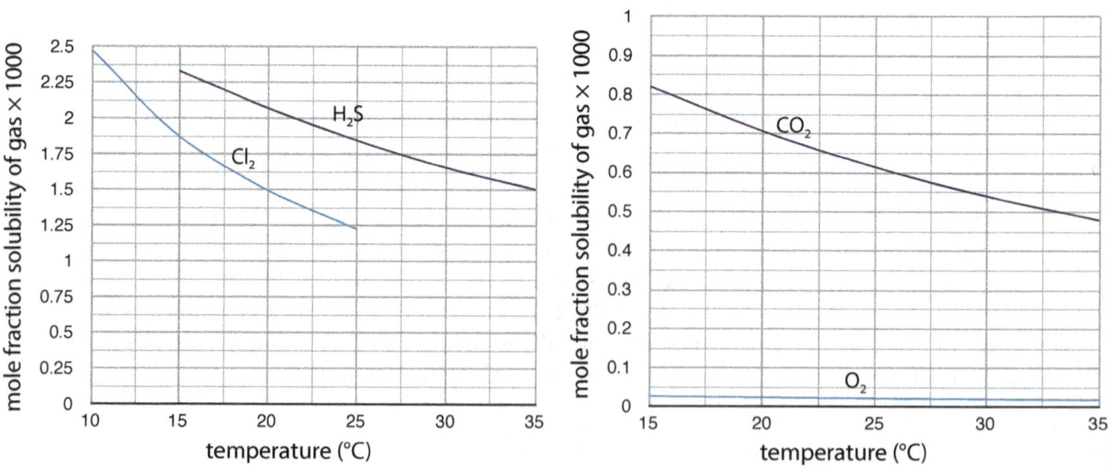

Figure 8.17. Temperature dependence of a few gases in water. The left graph shows H_2S and Cl_2. The right graph shows CO_2 and O_2. Mole fraction percentages in these figures have been multiplied by 1,000.

8.4 Quantifying Solution Concentration

8.4.1 Molarity

As you might expect, there are standard ways of referring to the concentrations of solutions. The most common is by a solution's *molarity*. Molarity, signified by *M*, is defined as the number of moles of solute per liter of solution:

$$\text{molarity }(M) = \frac{\text{moles solute}}{\text{solution volume (L)}} \tag{8.1}$$

A solution with 1.00 mole of solute in 1.00 liter of total solution is referred to as a "1.00 molar solution." This concentration is written as 1.00 *M*. Note that the denominator in Equation (8.1) is the volume of the *solution*, not the volume of the solvent by itself.

▼ Example 8.2

A solution of sodium nitrate is prepared by adding 45.55 g $NaNO_3$ to enough water to bring the solution volume to 0.100 L. Determine the molarity of the solution.

The molar mass of sodium nitrate is 84.99 g/mol. Converting the given quantity to moles we have

$$45.55\text{ g NaNO}_3 \cdot \frac{1\text{ mol NaNO}_3}{84.99\text{ g NaNO}_3} = 0.5359\text{ mol NaNO}_3$$

With this amount in moles and a total solution volume of 0.100 L, we now calculate the concentration.

$$\text{molarity }(M) = \frac{\text{moles solute}}{\text{solution volume (L)}} = \frac{0.5359\text{ mol}}{0.100\text{ L}} = 5.36\ M$$

This result is rounded to three significant figures.

▲

Since molarity is defined in terms of total solution volume, to prepare a solution with a certain molarity one must first measure out the required amount of solute, dissolve it in water, and then add water to bring the total volume up to what is needed for the specified molarity. This is usually done using a *volumetric flask*, as shown in Figure 8.18 on the following page. A volumetric flask has a line etched into the neck of the flask indicating a volume of 0.1 L, 0.25 L, 0.5 L, or 1 L. To make an aqueous solution of a particular molarity, the correct mass of solute is first calculated, based on its molar mass and the volume of the flask. The solute is then weighed out and added to the flask. The flask is filled partially with water to enable the solute to dissolve and then filled to the line with water. The following example illustrates this procedure using the same solute and quantities used for the photos in Figure 8.18.

▼ Example 8.3

A 0.25-L volumetric flask is to be used to prepare a solution of copper(II) chloride with a concentration of 0.100 *M*. Determine the mass of copper(II) chloride needed for this preparation.

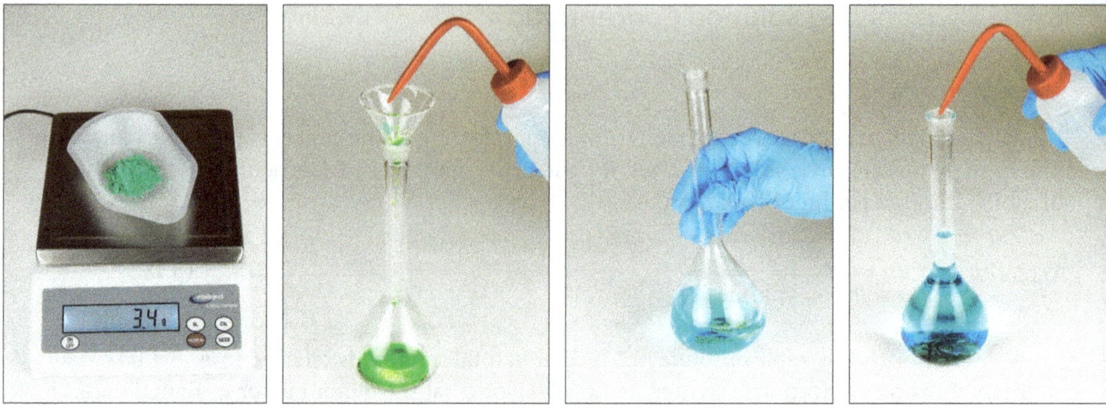

(a) (b) (c) (d)

Figure 8.18. Using a volumetric flask to prepare a standard quantity of solution with a specific concentration. First (a), the calculated mass of solute is measured out. Next (b), the solute is added to the volumetric flask and the solute particles rinsed out of the funnel and into the flask with water. Third (c), the flask is partially filled with water and the flask is swirled to dissolve the solute. Finally (d), the flask is filled to the line with water. As a final step, the flask is stoppered and inverted back and forth several times to ensure complete mixing.

The calculation involves solving the molarity equation for the moles of solute required, and then converting the number of moles to grams using the molar mass of $CuCl_2$. The molar mass calculation indicates the molar mass of $CuCl_2$ to be 134.45 g/mol.

$$\text{molarity } (M) = \frac{\text{moles solute}}{\text{solution volume (L)}}$$

We solve this for the moles of solute. However, instead of writing M for the molarity units, it is helpful to write the units for the solution concentration as mol/L (with a horizontal fraction bar) to make the unit cancellation clear.

$$\text{moles solute} = \text{ molarity} \cdot \text{ solution volume (L)} = 0.100\ \frac{\text{mol}}{\text{L}} \cdot 0.250\ \text{L} = 0.0250\ \text{mol}$$

Finally, we convert this mole quantity into grams.

$$0.0250\ \text{mol CuCl}_2 \cdot \frac{134.45\ \text{g CuCl}_2}{1\ \text{mol CuCl}_2} = 3.36\ \text{g CuCl}_2$$

▲

▼ Example 8.4

A chemist needs 50.0 g potassium chromate for a reaction. She finds that the only source of potassium chromate she has on hand is 4.5 L of 8.0 M K_2CrO_4. What volume of this solution should the chemist take in order to have the required 50.0 g K_2CrO_4?

As usual, we begin by calculating the molar mass of K_2CrO_4, which we find is 194.19 g/mol. Now we convert the required quantity from grams to moles:

$$50.0\ \text{g K}_2\text{CrO}_4 \cdot \frac{1\ \text{mol K}_2\text{CrO}_4}{194.19\ \text{g K}_2\text{CrO}_4} = 0.2575\ \text{mol K}_2\text{CrO}_4$$

We use the molarity equation to determine the quantity of 8.0 M solution needed to provide 0.2575 mol K_2CrO_4. Start with the equation and solve it for the solution volume.

$$\text{molarity }(M) = \frac{\text{moles solute}}{\text{solution volume (L)}}$$

$$\text{solution volume (L)} = \frac{\text{moles solute}}{\text{molarity}} = \frac{0.2575 \text{ mol } K_2CrO_4}{8.0 \ \frac{\text{mol } K_2CrO_4}{\text{L of solution}}} = 0.032 \text{ L solution}$$

This result is rounded to two significant digits.

▲

8.4.2 Molality

Another measure of solution concentration is *molality*, defined as follows:

$$\text{molality }(m) = \frac{\text{moles solute}}{\text{solvent mass (kg)}} \tag{8.2}$$

A solution with 1.00 moles solute in 1.00 kg of solvent is a "1.00 molal" solution, written as 1.00 m.

Take careful note of the difference between the ways molarity and molality are calculated. Both ways of specifying solution concentration entail moles of solute. But molarity uses the total volume of the solution, whereas molality uses the mass of only the solvent. While molarity is a more convenient concentration to use for stoichiometric calculations, molality is more convenient in applications of varying temperature, such as vapor pressure. The overall volume of a solution may change at different temperatures, causing a change in the molar concentration. But the molal concentration does not change because it is based on the mole or mass quantities of the substances used to prepare the solution, not on the volume of the solution.

▼ Example 8.5

A 3.0 m solution of table sugar (sucrose, $C_{12}H_{22}O_{11}$) and water is prepared from 0.75 L water. Determine the mass of sucrose present in that solution.

For this calculation we need the molar mass of sucrose, which we find to be 342.30 g/mol.

Next we use the density equation to get the mass of water, using the density of water, 0.998 g/mL at room temperature. We also convert the mass to kilograms while we are at it because the molality equation requires the solvent mass to be in kilograms.

$$\rho = \frac{m}{V}$$

$$m = \rho \cdot V = 0.998 \ \frac{\text{g}}{\text{mL}} \cdot 750 \text{ mL} = 748.5 \text{ g} \cdot \frac{1 \text{ kg}}{1000 \text{ g}} = 0.7485 \text{ kg}$$

Now we use the molality equation to determine how many moles of sucrose we have. We begin with the molality equation:

$$\text{molality } (m) = \frac{\text{moles solute}}{\text{solvent mass (kg)}}$$

We solve this equation for the moles of solute and calculate the number of moles, showing the units for molality explicitly to make it clear how the units cancel:

$$\text{moles solute} = \text{molality} \cdot \text{solvent mass (kg)}$$

$$= 3.0\ \frac{\text{mol } C_{12}H_{22}O_{11}}{\text{kg } H_2O} \cdot 0.7485\ \text{kg } H_2O$$

$$= 2.25\ \text{mol } C_{12}H_{22}O_{11}$$

Finally, we use the molar mass to convert this mole quantity into grams.

$$2.25\ \text{mol } C_{12}H_{22}O_{11} \cdot \frac{342.30\ \text{g } C_{12}H_{22}O_{11}}{1\ \text{mol } C_{12}H_{22}O_{11}} = 770\ \text{g } C_{12}H_{22}O_{11}$$

This final result is rounded to two significant digits, as required by the given information.

▲

8.5 Compounds in Aqueous Solution

8.5.1 Ionic Equations and Precipitates

In Chapter 5, we saw that double replacement reactions, also called *exchange* or *metathesis* reactions, can lead to precipitation. (Metathesis, which is the Greek word meaning "to transpose," is pronounced meh-TATH-esis.) Earlier in this chapter, in Table 8.3, we reviewed the guidelines for determining which compounds are soluble in water and which are insoluble. Now we put these together and look more closely at how we can describe what happens when compounds in solution are mixed. As we go, keep this basic principle in mind: *in a mixture of soluble ionic compounds—electrolytes—if nothing precipitates, no reaction occurs.*

Consider a possible reaction between lead(II) nitrate and potassium iodide. From Table 8.3 we see that all nitrate compounds and all compounds with alkali metal cations are soluble. This means that lead(II) nitrate and potassium iodide are both strong electrolytes. When separate solutions of these electrolytes are prepared, their dissociation reactions are as follows:

$$Pb(NO_3)_2(s) \rightarrow Pb^{2+}(aq) + 2NO_3^-(aq)$$

$$KI(s) \rightarrow K^+(aq) + I^-(aq)$$

There are three important things to notice in these dissociation reactions. First, coefficients are placed on ions in the equations where required. The law of conservation of mass in chemical reactions still applies, even when we are just dissolving things in water. The formula for lead(II) nitrate indicates two nitrate ions for each lead ion, so the NO_3^- ion has a coefficient of 2. Second, not only is mass conserved, but charge is conserved as well. The reactant compounds are neutral, so the total numbers of positive and negative charges in the ions must add up to zero. Third, the equations still show the relative mole ratios of substances involved, just as other balanced chemical equations do. In other words, one mole of lead nitrate, $Pb(NO_3)_2$, dissociates to form one mole of Pb^{2+} ions, and two moles of NO_3^- ions. Likewise, one mole of potassium iodide, KI, dissociates to form one mole of K^+ ions and one mole of I^- ions.

Now assume that we combine solutions of these two soluble compounds. The left side of the chemical equation is

$$Pb(NO_3)_2(aq) + KI(aq) \rightarrow$$

Figure 8.19. Yellow lead iodide precipitates out of solution when lead nitrate and potassium iodide react. The yellow crystals are hexagonal in shape, and the precipitation is sometimes called "golden rain."

To determine if a reaction occurs when these are combined, we consider the possible product compounds that can form. In addition to the reactants we started with, the possibilities are PbI_2 and KNO_3. From Table 8.3 we see that the KNO_3 is soluble. Compounds formed with the I^- anion are generally soluble too, but PbI_2 is an exception, shown in Figure 8.19. *When solutions of soluble compounds are mixed, if an insoluble compound can form and precipitate, it does.* The soluble products remain as ions in solution. So now we see that the formula equation is as follows:

$$Pb(NO_3)_2(aq) + KI(aq) \rightarrow PbI_2(s) + KNO_3(aq)$$

In this equation, PbI_2 is identified as a solid (*s*), which means it precipitates. The balanced equation is

$$Pb(NO_3)_2(aq) + 2KI(aq) \rightarrow PbI_2(s) + 2KNO_3(aq)$$

This familiar form of the reaction equation is called the *molecular equation*. But when studying compounds in aqueous solution, it is helpful to write an *ionic equation* showing the ions present before and after the precipitation occurs. The ionic equation is written by writing all the reactant and product ions, and showing insoluble precipitates as solids. The reactant ions for this equation are shown in the dissociation equations above. The product ions are the K^+ and NO_3^- that remain in solution. Thus the ionic equation is

$$Pb^{2+}(aq) + 2NO_3^-(aq) + 2K^+(aq) + 2I^-(aq) \rightarrow PbI_2(s) + 2K^+(aq) + 2NO_3^-(aq)$$

Notice that all the coefficients from the molecular equation have been preserved on the ions in the ionic equation. This is necessary to show that the law of conservation of mass in chemical reactions is still in force, that charge neutrality is preserved, and to indicate the relative molar quantities involved.

8.5.2 Net Ionic Equations and Spectator Ions

Look again at the ionic equation just above. Some of the ions appear on both sides of the equation, namely, NO_3^- and K^+. Since these ions are in solution before the reaction and remain in solution, they are not part of the precipitation reaction that occurs. Ions present in the same form in both the reactants and products of an ionic equation are called *spectator ions*.

Spectator ions play no direct role in the reaction. To see only the ions and substances that participate in the reaction, we eliminate the spectator ions from both sides of the ionic equation to form the so-called *net ionic equation*. For the reaction above, the net ionic equation is

$$Pb^{2+}(aq) + 2I^{-}(aq) \rightarrow PbI_2(s)$$

Notice that in the net ionic equation charge and mass are both conserved, just as in the dissociation equations.

There are many different soluble ionic compounds—electrolytes—and thus there are many different dissociation reactions. Net ionic equations are important because they make it easier to compare them. After the illustration we just worked through, we know that any time solutions of ionic compounds containing lead and iodine are mixed, we always get a lead(II) iodide precipitate. For example, magnesium iodide, MgI_2, is soluble. This means that combining aqueous solutions of lead nitrate and magnesium iodide leads to results similar to those of the case we just examined.

In summary, to develop the net ionic equation for a precipitation reaction, follow these steps:

1. Identify which potential product compounds are soluble and insoluble.
2. Write the balanced molecular equation.
3. Write the ionic equation.
4. Eliminate the spectator ions and write the net ionic equation.

Example 8.6

Consider the potential outcomes of combining aqueous solutions of ammonium carbonate and copper(II) chloride. Write the ionic equation, determine if a reaction occurs, identify any spectator ions, and write the net ionic equation if a reaction does occur.

According to Table 8.3, carbonates are generally insoluble, but an exception occurs when carbonate is coupled with the ammonium cation. Also, chlorides are soluble with a few exceptions that aren't relevant here, so both our reactant compounds form aqueous solutions. This means the left side of the formula equation is

$$(NH_4)_2CO_3(aq) + CuCl_2(aq) \rightarrow$$

We see that when these electrolytes dissociate, the ions formed are NH_4^{+}, CO_3^{2-}, Cu^{2+}, and Cl^{-}. Thus potential product compounds are NH_4Cl and $CuCO_3$. From Table 8.3, it is clear that ammonium chloride is soluble and copper carbonate is not. So a reaction does occur with the formation of a copper carbonate precipitate. The full formula equation is

$$(NH_4)_2CO_3(aq) + CuCl_2(aq) \rightarrow NH_4Cl(aq) + CuCO_3(s)$$

Balancing this equation gives us the complete molecular equation, which is

$$(NH_4)_2CO_3(aq) + CuCl_2(aq) \rightarrow 2NH_4Cl(aq) + CuCO_3(s)$$

From here we write the ionic equation, taking care to preserve quantities by introducing coefficients.

$$2NH_4^+(aq) + CO_3^{2-}(aq) + Cu^{2+}(aq) + 2Cl^-(aq) \rightarrow 2NH_4^+(aq) + 2Cl^-(aq) + CuCO_3(s)$$

The spectator ions are those that appear in the same form on both sides of the ionic equation, namely, NH_4^+ and Cl^-. Deleting these, we have the net ionic equation.

$$CO_3^{2-}(aq) + Cu^{2+}(aq) \rightarrow CuCO_3(s)$$

If you've never seen the copper carbonate precipitate, it has a lovely blue color, as you can see from Figure 8.20.

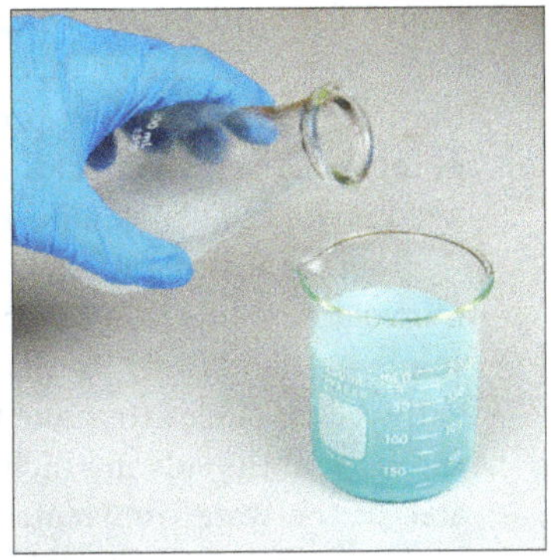

Figure 8.20. Formation of the pretty blue copper carbonate precipitate.

8.6 Colligative Properties of Solutions

Several important properties of solutions depend on the concentration of solute in a solution but not on the particular chemical species present in the solution. These are called *colligative properties*. The first three relate to the vapor pressure, so we begin there.

8.6.1 Vapor Pressure Lowering

As you recall from Chapter 6, the vapor pressure of a liquid at a given temperature is the pressure at which that temperature is the boiling point. The curve on a P-T phase diagram separating the vapor and liquid states is the curve representing the vapor pressure of the liquid. In Figure 8.21, the generic phase diagram we examined in Chapter 6 is shown again to illustrate the effect of a nonvolatile solute such as sugar or salt in solution in the liquid. As before, the black curves separate the states of the pure solvent. The red curves in the figure represent the new positions of the curves separating the states when the solute is present in solution.

The upper horizontal dashed line represents atmospheric pressure. This line crosses the phase boundaries at the freezing point (f.p.) and boiling point (b.p.) of the pure solvent. The lower horizontal dashed line indicates the vapor pressure of the solution at the boiling point of the pure solvent. The change in pressure, ΔP, is the amount the vapor pressure (v.p.) has been lowered at the boiling point of the pure solvent. Not surprisingly, this effect is called *vapor pressure lowering*.

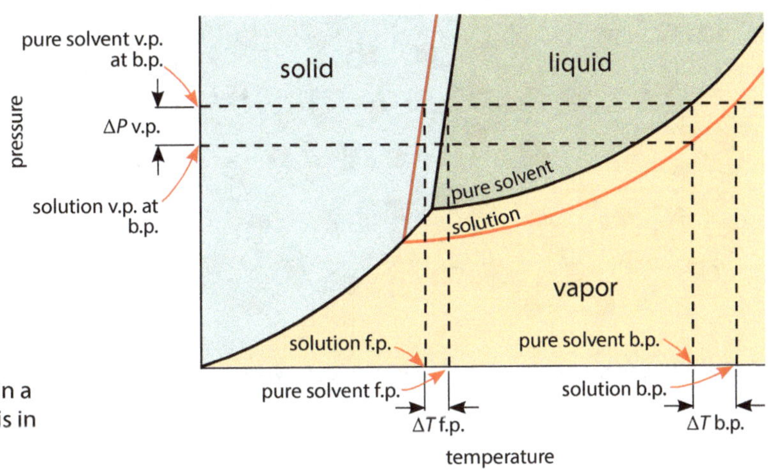

Figure 8.21. Changes to vapor pressure (v.p.), freezing point (f.p.), and boiling point (b.p.) that occur in a solvent when a nonvolatile solute is in solution.

The P-T phase diagram clearly shows that the lowering of the vapor pressure curve has an effect on the freezing point and boiling point of the solution. Two of the vertical dashed lines in Figure 8.21 indicate the freezing point and boiling point of the pure solvent. The other two are drawn from the points where the atmospheric pressure line crosses the new phase boundaries (shown in red). The changes in temperature, ΔT, between the two sets of lines are the amounts the freezing and boiling points changed with the solute in solution. As you see, the freezing point decreases, an effect known as *freezing point depression*. The boiling point increases, an effect known as *boiling point elevation*.

There is a straightforward physical explanation for the lowering of vapor pressure when a nonvolatile solute is in solution in a liquid, illustrated in the aqueous solution of Figure 8.22. The solute particles (ions in this instance) with their little cages of water molecules are highlighted with yellow circles to make them more easily visible. These clusters of particles are distributed throughout the solvent, including at the liquid surface, where they reduce the surface area available to the water molecules for evaporation. Simply put, they are *in the way* of some of the higher-energy water molecules that might otherwise make it to the surface and evaporate.

The illustration in Figure 8.22 is based on ions in water, but the same effect occurs with nonelectrolyte solutes such as sugar, just so long as the solute is nonvolatile so that it is not itself trying to evaporate. As mentioned at the very beginning of Section 8.6, the factor governing colligative properties like vapor pressure lowering is the concentration of the solute, not the identity of the solute. The species of solute doesn't matter; what matters is how many particles there are—the molar concentration. The more particles there are, the more the surface area available to the solvent molecules for evaporation is effectively reduced.

Notice one more important detail in Figure 8.22. When we refer to the concentration of solute particles as the key factor producing the colligative properties of solutions, we are referring to the total number of moles of *particles* in solution, regardless of what those particles are. Sugar molecules do not ionize in water, so one mole of sugar in solution results in one mole of

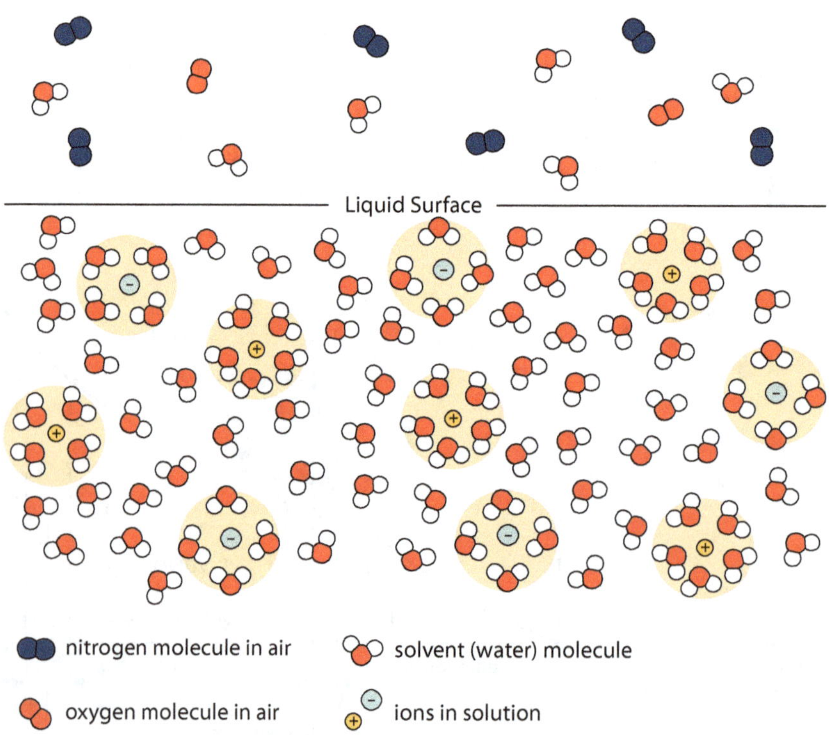

Figure 8.22. Ions in solution reduce the amount of liquid surface available to the solvent molecules for evaporation. This slows the evaporation rate and lowers the vapor pressure.

particles. However, NaCl ionizes to form Na^+ ions and Cl^- ions—a pair of them for every unit of NaCl. This means that one mole of NaCl produces two moles of solvent particles in solution. Lithium sulfate, Li_2SO_4, ionizes to form two Li^+ ions and one SO_4^{2-} ion. So every mole of Li_2SO_4 produces three moles of particles in solution. We might expect that a solute that produces two or three moles of particles in solution would have two or three times the effect on the vapor pressure compared to a solute that produces one mole of particles.

This is, in fact, *approximately* what happens. However, there are factors that prevent this analysis from working out so neatly in the real world. For example, the more clustered the solute particles are, the more their motion in the solvent is hindered. Such clustering could be caused by hydration, as in the illustration of Figure 8.22, or by attractions to other ions of the opposite polarity. These clusters encumber the ions, making them less mobile, and if the mobility of particles is impaired due to such clustering, then the solvent particles get around them more easily and evaporate. Thus, we should expect to see something less than the doubling or tripling of the effect that our simple analysis leads us to predict.

What happens in practice depends on whether hydrogen bonding is present, the precise nature of the other intermolecular forces involved, and in the case of ions, their charge. As far as intermolecular forces go, the net result in a solution could be close to, higher than, or lower than what we would predict. As far as ions are concerned, the charge on the ions has a noticeable effect on what happens in the solution. Since intermolecular forces cause the clustering of particles that reduces the effect, we expect that ions with higher charges, and thus greater attractions for oppositely charged species, exhibit less of an effect than ions with lower amounts of charge. As shown below, this is indeed the case.

8.6.2 Freezing Point Depression and Boiling Point Elevation

In Figure 8.21, we see that two side effects of vapor pressure lowering are freezing point depression and boiling point elevation. The graph in Figure 8.21 indicates generally why these effects follow from the vapor pressure lowering, and the discussion of Figure 8.22 provides a physical explanation for why vapor pressure lowering occurs. In this section, we examine a simple quantitative model chemists developed to help predict the size of the effect on the freezing and boiling points for different solvents and different molal concentrations of solute.

Before we proceed to the technical material, I will describe two very common examples of freezing point depression—examples that many students are already familiar with. The first is the fact that in icy winter conditions, many cities use trucks to spread salt on the roads to prevent ice formation or to melt ice that has already formed. Likewise, consumers can purchase salt to sprinkle on their porches and sidewalks. Figure 8.23 shows the effect on some steps in cold weather in the northern U.S.

To be effective, the salt must dissolve in liquid water; salt laying atop solid ice does nothing to melt the ice. But on the surface of the ice there is usually a very thin layer of water. Also, recall from Chapter 6 that if the temperature of the ice is not far below the freezing point, an increase of pressure causes the ice to melt. If the temperature is not too low, the pressure of cars and

Figure 8.23. Salt for consumer use on icy steps and walkways. The bare places on the steps were salted just a few minutes before.

trucks on roadways causes a thin film of water on the surface of the ice. Once the salt dissolves in this liquid film, the freezing point of the water drops. At the boundary between the water layer and the ice beneath, the water and ice are close to phase transition equilibrium, except that ice molecules are entering the liquid state at a greater rate than the other way around because once liquid, the dissolved ions interfere with movement back toward the ice layer. So the salt water on top of the ice gradually leads to the melting of all the ice, allowing the water to evaporate and leading to a dry road surface.

Figure 8.24. Simple ice cream maker for home use.

Salting roads is only effective down to the new freezing point of the ice water solution. Sodium chloride works down to about 15°F (–9°C). Magnesium chloride produces more ions than sodium chloride, and can be used down to about 5°F (–15°C), and calcium chloride works down to about –20°F (–29°C). The salt product used for the photo in Figure 8.23 is shown at the top of the figure—calcium chloride, the one that provides the most freezing point depression.

Another example of freezing point depression with salt water is fun to be a part of: making ice cream outside at home. A typical home ice cream maker is shown in Figure 8.24. The mixture of cream, sugar, and flavorings is in a metal canister inside the wooden bucket. A mixture of ice and rock salt (sodium chloride) is placed in the space between the canister and the walls of the wooden bucket. The salt mixed in with the ice lowers the freezing point, forcing the ice to melt. For solid H_2O to become liquid, energy must be supplied (the molar heat of fusion). This energy comes from the surrounding ice and water, lowering the temperature of the salt-ice-water mixture. Thus, the ice cream mixture inside the metal canister is sitting in a bath that is colder than 0°C, so the ice cream becomes very cold and freezes. The fun comes in taking turns turning the crank to mix the ice cream as it freezes, and eating the ice cream afterwards! If your family has never made ice cream—try it!

Now we do the math. Using ΔT_f to represent the change in freezing point, and ΔT_b to represent the change in boiling point, the two mathematical models are as follows:

$$\Delta T_f = K_f m \tag{8.3a}$$

$$\Delta T_b = K_b m \tag{8.3b}$$

In these two equations, m is the molality of the solution, the molal concentration of solute in moles of solute per kilogram of solvent. K_f and K_b are called the *molal freezing-point constant* and *molal boiling-point constant*, respectively. The units for these constants are °C/m, so that when multiplied by the molality (with units m) the result is the temperature change in °C. The presence of a solute always lowers the freezing point and raises the boiling point, and the results of calculations based on Equations (8.3) are interpreted accordingly.

Table 8.4 lists freezing and boiling points and molal constants for a few common solvents. As we go through the following examples, keep in mind that ionizing solutes produce more than one mole of particles in solution for each mole of solute. So to use Equations (8.3) with ionizing solutes, one must multiply the molality, which gives moles of solute per kilogram of solvent, by a factor relating moles of solute to moles of dissolved particles.

Solvent	Freezing Point (°C)	Molal f.p. Constant, K_f (°C/m)	Boiling Point (°C)	Molal b.p. Constant, K_b (°C/m)
acetic acid (p)	16.6	3.63	117.9	3.22
acetone (p)	–94.7	2.67	56.1	1.80
benzene (np)	5.5	5.07	80.1	2.64
ethanol (p)	–114.4	1.959	78.3	1.23
methanol (p)	–97.5	2.56	64.6	0.86
phenol (p)	40.9	6.84	181.9	3.54
toluene (np)	–95.0	3.55	110.6	3.40
water (p)	0.00	1.86	99.974	0.513

Table 8.4. Molal freezing point and boiling point constants for a few solvents (at 1 atm). (p) = polar solvent; (np) = nonpolar solvent

▼ Example 8.7

Determine the freezing point of a solution composed of 50.0 g sugar, $C_{12}H_{22}O_{11}$, in 2.00 kg of water.

To calculate the molality of the solution, we first determine the molar mass of sugar, which is 342.3 g/mol. Converting the given mass of sugar to moles, we have

$$50.0 \text{ g } C_{12}H_{22}O_{11} \cdot \frac{1 \text{ mol } C_{12}H_{22}O_{11}}{342.3 \text{ g } C_{12}H_{22}O_{11}} = 0.1461 \text{ mol } C_{12}H_{22}O_{11}$$

With this we calculate the molality:

$$m = \frac{\text{mol solute}}{\text{kg solvent}} = \frac{0.1461 \text{ mol } C_{12}H_{22}O_{11}}{2.00 \text{ kg } H_2O} = 0.07305\ m$$

Now we are ready to calculate ΔT_f, using Equation (8.3a) and the K_f value for water from Table 8.4.

$$\Delta T_f = K_f m = 1.86\ \frac{°\text{C}}{m} \cdot 0.07305\ m = 0.136°\text{C}$$

Finally, this ΔT_f value tells us how much the freezing point of water was lowered. To get the freezing point of the solution, we need to subtract this value from the freezing point for the pure solvent.

$$T_{f,\text{solution}} = T_{f,\text{normal}} - \Delta T_f = 0.00°\text{C} - 0.136°\text{C} = -0.14°\text{C}$$

▲

In the next example, we treat a case of an ionizing solvent that produces more than one mole of particles in solution for each mole of solute.

▼ Example 8.8

A solution of 180.0 g LiCl in 1.25 L water is prepared. Determine the boiling point of this solution.

We first determine the number of moles of LiCl, based on the molar mass of 42.39 g/mol.

$$180.0 \text{ g LiCl} \cdot \frac{1 \text{ mol LiCl}}{42.39 \text{ g LiCl}} = 4.246 \text{ mol LiCl}$$

Now we need to determine the mass of the water in kilograms, for which we use the density equation and the density of water, 0.998 g/mL.

$$\rho = \frac{m}{V}$$

$$m = \rho V = 0.998 \ \frac{\text{g}}{\text{mL}} \cdot 1.25 \text{ L} \cdot \frac{1 \text{ kg}}{1000 \text{ g}} \cdot \frac{1000 \text{ mL}}{\text{L}} = 1.248 \text{ kg}$$

Now we can compute the molality of the solution:

$$m = \frac{\text{mol solute}}{\text{kg solvent}} = \frac{4.246 \text{ mol}}{1.248 \text{ kg}} = 3.402 \ m$$

This molality is for the solute LiCl. But when LiCl ionizes, two ions are produced (Li^+ and Cl^-) for each unit of LiCl. We form a mole ratio from this information to compute the molality of the ions in solution. (We are simply doubling the molal concentration.)

$$3.402 \ m \text{ LiCl} \cdot \frac{2 \ m \text{ ions}}{1 \ m \text{ LiCl}} = 6.804 \ m \text{ ions}$$

This is the molality we need to use for calculating ΔT_b. Next, we compute the ΔT_b value for this solution, using the value of K_b from Table 8.4.

$$\Delta T_b = K_b m = 0.513 \ \frac{°\text{C}}{m} \cdot 6.804 \ m = 3.49°\text{C}$$

Finally, we calculate the new boiling point by adding the ΔT_b value to the boiling point of water.

$$T_{b,\text{solution}} = T_{b,\text{normal}} + \Delta T_b = 99.97°\text{C} + 3.49°\text{C} = 103.46°\text{C}$$

Notice here that the correct number of significant digits for the ΔT_b value is three. But once this value is added to the boiling point of water using the addition rule, we end up with five significant digits in our final result.

▲

▼ Example 8.9

A solution of benzene and an unknown nonpolar, non-ionizing solute is found to have a freezing point of 1.1°C. Determine the molal concentration of the solution.

To solve this problem, we again use Equation (8.3a). Instead of solving for ΔT_f, we use the given freezing point and the information from Table 8.4 to determine the ΔT_f value. From there we calculate m.

ΔT_f is the difference between the given freezing point and the one in Table 8.4:

$$\Delta T_f = T_{f,\text{normal}} - T_{f,\text{solution}} = 5.5°\text{C} - 1.1°\text{C} = 4.4°\text{C}$$

We use this value with the molal freezing point constant from Table 8.4 to solve for m. Starting with Equation (8.3a), we solve for m, insert the values, and compute the result.

$$\Delta T_f = K_f m$$

$$m = \frac{\Delta T_f}{K_f} = \frac{4.4°\text{C}}{5.07\ \frac{°\text{C}}{m}} = 0.87\ m$$

▲

We have seen that for electrolytes, the degree of freezing point depression or boiling point elevation depends on the molal concentration of ions, and not on the molality of the solute itself. If the solute produces two moles of ions, we expect the ΔT to be twice as large; if three moles, we expect the ΔT to be three times as large. Table 8.5 lists data that demonstrate how closely these predictions agree with empirical results. Three solutes are shown, each producing either two or three moles of ions in solution, and four different concentrations are shown for each. The right column of the table shows the ratio of the experimental ΔT_f to the predicted ΔT_f value for

Solute	Concentration (m)	Experimental ΔT_f (°C)	ΔT_f Prediction for Nonionizing Solute (°C)	Ratio of Experimental value to non-ionizing prediction
HCl	0.05	0.18	0.093	1.9
	0.25	0.90	0.47	1.9
	0.50	1.86	0.93	2.0
	0.75	2.90	1.40	2.1
$MgSO_4$	0.05	0.13	0.093	1.4
	0.25	0.55	0.47	1.2
	0.50	1.01	0.93	1.1
	0.75	1.50	1.40	1.1
$CaCl_2$	0.05	0.25	0.093	2.7
	0.25	1.27	0.47	2.7
	0.50	2.66	0.93	2.9
	0.75	4.28	1.40	3.1

Table 8.5. Empirical f.p. depression results compared to predictions using molal f.p. constant.

a non-ionizing solute such as sugar (where one mole of solute produces one mole of particles in solution).

HCl produces two moles of ions for each mole of solute, so we expect to see a doubling of the ΔT_f value compared to that predicted for a non-ionizing solute. As you see, the experimental data agree quite well with this prediction, and the data are consistent at various concentrations. Similarly, we expect to see a tripling of the ΔT_f value for $CaCl_2$, and this also agrees fairly well with the empirical data.

However, the $MgSO_4$ values are significantly less than the doubling we expect for $MgSO_4$. One explanatory factor is in the charges on the ions: the charge on the cation (Mg^{2+}) is higher than the H^+ cation from the HCl, and the charge on the anion (SO_4^{2-}) is higher than the Cl^- anion from both the other solutes listed. In our discussion a few pages back, we noted that higher ionic charges tend to reduce the freezing-point depression and boiling-point elevation effects because of greater intermolecular forces and resulting particle clustering. Another causal factor may be the size of the SO_4^{2-} ion. This ion is a molecule with five atoms in it, compared to the single-atom Cl^- ions. We expect larger ions to form larger clusters during hydration, becoming less mobile and offering less interference to higher-energy H_2O molecules working their way towards the liquid surface.

8.6.3 Osmotic Pressure

A fourth colligative property of solutions is an effect known as *osmotic pressure*, an extremely important factor for cell biology. Osmotic pressure arises from a process called *osmosis* that occurs when a *semipermeable membrane* separates a solution with a relatively high solute concentration from a solution with a lower solute concentration.

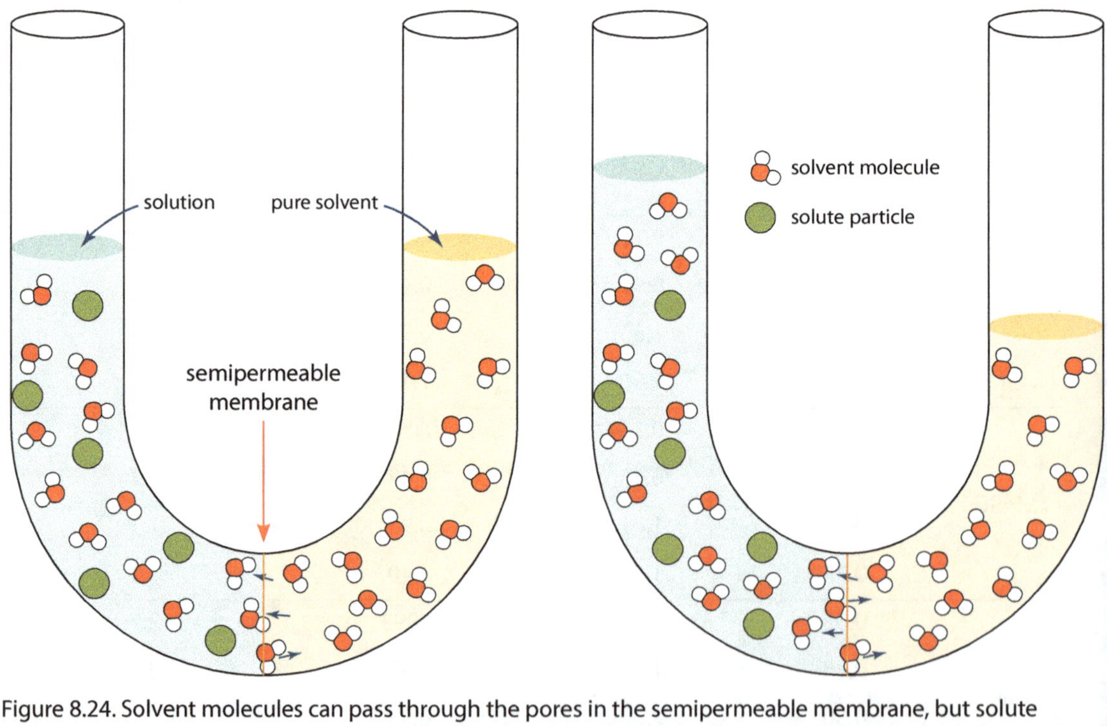

Figure 8.24. Solvent molecules can pass through the pores in the semipermeable membrane, but solute particles cannot. At first (left), the flow rate of solvent molecules through the membrane is higher from solvent into solution, increasing the amount of liquid on the solution side of the membrane. When the pressure on the solution side of the membrane is high enough, flow rates through the membrane become equal in both directions.

Figure 8.24 illustrates how osmosis works. A U-shaped tube contains a semipermeable membrane separating the two sides. The semipermeable membrane is designed to allow solvent molecules to pass through, but not solute particles. One side of the tube contains a solution and the other contains the pure solvent. Since the solvent molecules can pass through the membrane, they pass through the membrane in both directions. But because of the solute particles in the liquid on the solution side of the membrane, the route for the solvent particles to reach the membrane is partially blocked, so the flow rate of solvent molecules is greater from the pure solvent into the solution than it is in the other direction. This causes the liquid to build up on the solution side, causing a higher liquid column and a correspondingly greater pressure on the membrane on that side. The increased pressure increases the flow rate of solvent particles from the solution back into the pure solvent. When the pressure on the solution side of the membrane is high enough, an equilibrium is achieved in which the flow rates in both directions are the same.

The net flow of solute molecules from the pure solvent side to the solution side of a semipermeable membrane is called *osmosis*. *Osmotic pressure* may be defined as the external pressure required to stop osmosis and establish equilibrium.

The role of osmosis in biological systems is huge. Biological membranes, such as cell walls, human skin, and the skins of vegetables, are semipermeable, so the process of osmosis is happening all the time in biological systems. Plant roots absorb water by osmosis, and plant cells maintain correct water content and thus proper stiffness (called *turgor*) by osmotic pressure. The cells in our bodies maintain correct concentrations of dissolved solids by osmotic pressure. If the concentration of solutes in the water inside a cell is higher than the concentration outside, water flows into the cell. If the cell has a lower concentration of dissolved solids than the fluid outside the cell, then the excess water inside the cell flows out of the cell through the cell wall. Images of human blood cells that illustrate these processes are shown on the opening page of this chapter.

When you are swimming or in the bath, the water in the cells of your skin has a higher concentration of dissolved solids than the water outside your skin. The result is water flowing into your skin cells by osmosis. The excess water causes the cells to swell and your skin to wrinkle.

The process of pickling involves osmosis. Cucumbers are placed in a briny solution, and the high salt concentration in the brine causes some of the water in the cucumbers to pass by osmosis through the skin of the cucumber and into the brine. At the same time, salts and vinegar enter the cucumbers by diffusion, preserving the cucumbers and giving them flavor. This diffusion occurs at a slower rate than osmosis.

Chapter 8 Exercises

SECTION 8.1

1. Distinguish between suspensions, colloidal dispersions, and solutions.
2. Why are there no colloids composed of two different gases?
3. Explain the Tyndall effect and describe an example of it.
4. Describe how the Tyndall effect can be used to determine if an unknown mixture is (a) a true solution or (b) a suspension or colloid.
5. What is Brownian motion?

SECTION 8.2

6. Describe in detail the process of dissolution of an ionic crystal in a polar solvent, such as water.

7. What determines whether the dissolution of a given ionic solute is exothermic or endothermic? Explain the different *sources* or *sinks* of energy in such processes. (A sink is the opposite of a source. Energy is supplied by a source and taken in by a sink.)
8. Why should one never add water to a concentrated acid? Explain the details behind why this action is hazardous.
9. Define enthalpy of solution.
10. Explain the concept of entropy.
11. Consider the four processes described in the accompanying table and address these questions:
    a. Which process always happens spontaneously?
    b. Which process never happens spontaneously?
    c. Two of the processes may or may not happen in a given instance. Which two are these, and what determines whether the processes occur?

1. Energy increases; entropy increases	2. Energy increases; entropy decreases
3. Energy decreases; entropy increases	4. Energy decreases; entropy decreases

12. Explain what an electrolyte is and distinguish between strong and weak electrolytes.
13. How does the concept of equilibrium apply to solutions of weak electrolytes such as acetic acid?

SECTION 8.3

14. Classify the following compounds as soluble or insoluble in water:

a. cobalt(II) hydroxide	b. calcium nitrate	c. ammonium phosphate
d. potassium sulfate	e. iron(III) oxide	f. magnesium hydroxide
g. lead(II) acetate	h. strontium oxide	i. copper(II) chloride

15. Imagine that you have some sort of sticky residue on the side of your arm and want to clean it off. You try rinsing it off in water, but the water doesn't remove the substance. Explain why it would be pointless to try cleaning off the sticky substance with rubbing alcohol, and suggest a common product that will remove it (other than soap and water).
16. Ammonia is commonly mixed with water as a cleaning agent for mopping floors. Explain why ammonia dissolves in water.
17. Explain why oil and vinegar salad dressing separates into layers of oil and vinegar.
18. Consider the equilibrium that exists in a solution of pressurized gas in a liquid.
    a. Describe the equilibrium.
    b. Apply Le Châtelier's principle to this equilibrium and describe what happens to the equilibrium if some of the gas leaks out of the container.
    c. Apply Henry's law to this equilibrium and describe what happens to the amount of dissolved gas if the partial pressure of the gas is increased by 50%.
19. Describe the equilibrium that exists in a saturated solution of an ionic solid with excess solid at the bottom of the container. Apply Le Châtelier's principle to this equilibrium and describe what happens to the equilibrium if additional water is added to the solution.
20. Referring to Figures 8.16 and 8.17, respond to these questions:

a. At 50°C, what is the maximum percentage of an aqueous solution's mass that can be attributed to dissolved cesium chloride?

b. For a solution of chlorine dissolved in water at 20°C, what are the mole fractions in the solution of chlorine and of water?

c. Compare the solubilities of KCl and NH_4Cl as functions of temperature.

d. By what percentage does the solubility of CO_2 decrease if the temperature is changed from 20°C to 30°C?

21. The accompanying table contains solubility data for $Al_2(SO_4)_3$ in aqueous solution. As an example of how to read the solubility values, at a temperature of 40°C, the mass percent of $Al_2(SO_4)_3$ in a saturated solution is 29.2%. This means that if the entire solution has a mass of 100.0 g, there are 29.2 g $Al_2(SO_4)_3$ and 70.8 g H_2O in the solution. Plot the data from the table on graph paper with the solubility on the vertical axis (resolution of vertical scale = 0.1), and connect the data points with a smooth curve. Then address the following questions:

Temperature (°C)	$Al_2(SO_4)_3$ Solubility (mass percent of solution mass)
30	28.2
40	29.2
50	30.7
60	32.6
70	34.9
80	37.6

a. Describe the relationship between $Al_2(SO_4)_3$ solubility and temperature.

b. Estimate the solubility of $Al_2(SO_4)_3$ at 55°C.

c. In a saturated solution with a mass of 250.0 g at 70°C, how many grams of $Al_2(SO_4)_3$ are present?

d. In a saturated solution with a mass of 45 g at 30°C, how many moles of both $Al_2(SO_4)_3$ and water are present? Compare the mass percentages to the mole percentages for $Al_2(SO_4)_3$ and water at this temperature.

SECTION 8.4

22. Determine the mass of solute required to make each of the following solutions:

a. an 8.00 *M* solution of NaOH, with water added to bring the solution volume to 2.25 L.

b. a 15.0 *M* solution of acetic acid, CH_3COOH, with water added to bring the solution volume to 1.00 L.

23. A solution is prepared by dissolving 65.11 g $(NH_4)_2SO_4$ in enough water to bring the solution volume to 125 mL. Determine the molarity of this solution.

24. Determine the mass in grams required to make a 1.25 *M* solution of $Ca(NO_3)_2$, with a solution volume of 3.10 L.

25. To extract 22.5 g silver nitrate from a 1.2 *M* aqueous solution of silver nitrate, what volume of solution is required?

26. If 54.0 mL of a $CuSO_4$ solution reacts with enough iron to produce 7.7 g Cu in a single replacement reaction, what is the molarity of the $CuSO_4$ solution?

27. Solutions of 0.60 *M* Na_2SO_4 and 1.00 *M* $Ba(NO_3)_2$ are mixed.

a. Write the balanced equation for the reaction that occurs.

b. If 475 mL Na_2SO_4 solution are combined with excess $Ba(NO_3)_2$ solution, what is the mass of the precipitate that forms?

28. Determine the volume in mL of 12.0 *M* H_2SO_4 required for a complete reaction with 655 mL of 6.00 *M* NaOH. The products of this reaction are Na_2SO_4 and water.

29. Excess iron is reacted with 2.00×10^3 L of 2.0 *M* HCl to produce hydrogen gas and iron(II) chloride. Determine the volume of H_2 produced at STP.

30. Determine the mass of solute required in order to make each of the following solutions:

    a. a 10.0 *m* solution of HCl in 1.00 kg H_2O

    b. a 5.25 *m* solution of NaOH in 1.75 L H_2O.

31. If you prepare a solution of 225 g sucrose, $C_{12}H_{22}O_{11}$, in 5.00×10^2 mL water, what is the molality of the solution?

32. Determine the quantity in milliliters of water that must be added to 235 g of $Ca(NO_3)_2$ to produce a 2.0 *m* solution.

SECTION 8.5

33. Solutions of Na_2CO_3 and $Ca(NO_3)_2$ are combined.

    a. Use Table 8.3 to identify the precipitate that forms.

    b. Write the balanced molecular equation.

    c. Write the ionic equation and the net ionic equation for the process.

    d. Identify the spectator ions.

34. In a reaction between aqueous solutions of $AgNO_3$ and LiCl, identify the spectator ions and the precipitate compound in the reaction.

35. Aqueous solutions of magnesium chloride and lead(II) nitrate are combined. The magnesium chloride solution is 0.50 molar and 40.0 mL of it are used.

    a. Write the balanced molecular equation, the ionic equation, and the net ionic equation.

    b. Identify the spectator ions and the precipitate.

    c. If all the magnesium chloride is used in the reaction, determine the mass of the precipitate that forms.

36. Aqueous solutions of sodium nitrate and copper(II) chloride are combined. Describe in detail what occurs.

SECTION 8.6

37. Why is the vapor pressure of an aqueous solution of calcium chloride lower than the vapor pressure of water?

38. How does osmosis work?

39. Determine the boiling point of a solution formed by dissolving 0.4000 kg of table sugar, $C_{12}H_{22}O_{11}$, in 1.100 L of water.

40. Glycerol, $C_3H_8O_3$, is a soluble nonelectrolyte used widely in the food industry. (Also, recall that glycerol is the molecule that remains after the three strands of a triglyceride have been separated during saponification.) Determine the molal concentration of a glycerol and water solution for boiling points of 101.75°C, 100.99°C, and 103.17°C.

41. Ethylene glycol, $C_2H_4(OH)_2$, is the compound in automotive antifreeze. It is a nonelectrolyte, is nonvolatile, and is miscible in water. Determine the molality of ethylene glycol in water required to prevent the solution from freezing at –18°C.

42. The freezing point of a solution of urea, $CO(NH_2)_2$, a nonelectrolyte, in water is found to be −2.93°C. If the mass of the water in the solution is 775 g, determine the mass of the urea in the solution.

43. The *specific gravity* of a substance is defined as the ratio of the substance's density to the density of water. For example, a substance with a specific gravity of 0.750 is 75.0% as dense as water, and thus has a density of $0.750 \cdot 0.998$ g/mL = 0.749 g/mL. A solution is prepared by adding carbon tetrachloride, CCl_4, to 2.50 L of benzene. The specific gravity of benzene is 0.878. If the boiling point of the solution is found to be 85.34°C, determine the mass in grams of the CCl_4 in the solution.

44. Predict the freezing point of a solution of 155 g LiCl in 575 g water.

45. Predict the freezing point of a solution of 456 g ethanol and 75.0 g KCl.

46. Assume you are given 0.040 *m* solutions in acetic acid of KI, $MgCl_2$, and $CaSO_4$. Determine the expected change to the boiling point of each.

47. Aqueous solutions with a molality of 0.020 *m* are prepared from Na_3PO_4, KCl, $C_6H_{12}O_6$ (glucose), and $CaCl_2$. Determine the expected freezing point of each solution.

48. The *CRC Handbook of Chemistry and Physics* indicates that the freezing point of a 2.00 *m* solution of nitric acid is lowered by 8.34°C compared to pure water. Compute the expected value of the freezing point lowering for this solution. Comment on whether the value listed in the *CRC* is consistent with the models we have studied in this chapter.

GENERAL REVIEW EXERCISES

49. Determine the molar mass of a gas, given that 1.30 g of the gas occupies 2.24 L at 6.85°C and 1.14 atm.

50. If 17.5 L of methane, CH_4, combust completely with excess oxygen at STP, determine the volumes of each of the products formed.

51. What is surface tension and how does it relate to capillary action?

52. As illustrated to the right, a steel support column in a building is supporting a load of 12,500 lb and has a steel plate at the bottom that is 10.0 inches square. What is the pressure in atm under this plate? Note that necessary conversion factors are in Appendix B.

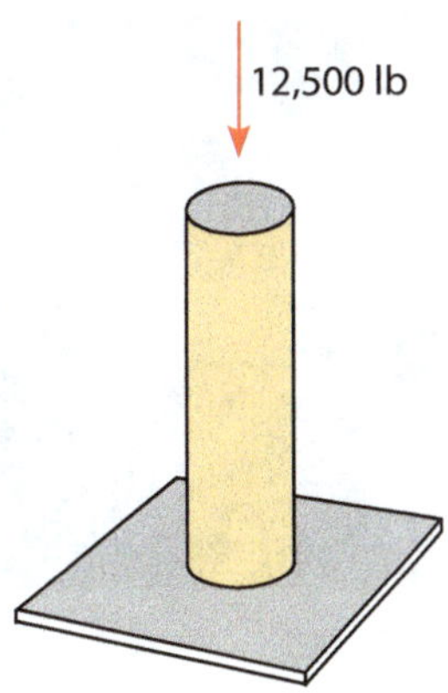

53. The process of coal gasification is used to convert coal, which is pure carbon, into methane gas according to the following reaction:

$$2C(s) + 2H_2O(l) \rightarrow CH_4(g) + CO_2(g)$$

If the reaction has 82% yield, determine the mass of methane produced from 100.0 kg coal.

54. Salicylic acid, $C_7H_6O_3$, is reacted with acetic anhydride, $C_4H_6O_3$, to produce aspirin, $C_9H_8O_4$, and acetic acid in this reaction:

$$C_7H_6O_3(s) + C_4H_6O_3(l) \rightarrow C_9H_8O_4(s) + CH_3COOH(l)$$

If 125.0 mol $C_7H_6O_3$ are used in this reaction, determine the mass in kg of aspirin produced.

55. Determine the percentage composition of salicylic acid.

56. Determine the oxidation state of chlorine in the chlorate ion.

Chapter 9

Acids and Bases

The beautiful flowers of the hydrangea are a natural indicator of soil pH. The flowers are blue in acidic soil and pink in alkaline (basic) soil. Unfortunately for chemistry students, this color palette is just the opposite of *litmus*, one of the oldest laboratory pH indicators. Litmus is a mixture of various dyes extracted from lichens, and turns red in acids and blue in bases. Although other pH indicators are used more commonly now, the use of litmus was so common in chemistry labs for so long that we now associate its colors red and blue with acids and bases, respectively.

Objectives for Chapter 9

After studying this chapter and completing the exercises, you should be able to do each of the following tasks, using supporting terms and principles as necessary.

SECTION 9.1

1. List six general properties of acids and five general properties of bases.
2. Name at least four acids that are important industrial chemicals, stating at least two industrial uses for each.
3. Define *binary acid*, *oxyacid*, and *carboxylic acid*, and give at least three examples of each.
4. Given chemical formulas, state correct names for binary acids and oxyacids.

SECTION 9.2

5. State definitions for *acid*, *base*, and *acid-base neutralization reaction* according to the Arrhenius, Brønsted-Lowry, and Lewis acid-base theories.
6. Explain why water is amphiprotic under the Brønsted-Lowry acid-base theory.
7. Explain why NH_3 acts as a base according to all three acid-base theories.
8. Given a chemical equation, identify compounds or ions that act as acids or bases under each of the three acid-base theories.
9. Define *conjugate acid* and *conjugate base*, give examples of each, and identify each in a given chemical equation.
10. State the names of at least three of each of the following: strong acid, strong base, weak acid, weak base.
11. Explain the factor that determines whether an acid or base is weak or strong.
12. Write ionic and net ionic equations for neutralization reactions.
13. Identify which direction the equilibrium lies in a Brønsted-Lowry acid-base reaction.
14. Relate acid strength to the number of oxygen atoms bound to the central atom in an acid molecule, and explain why acid strength goes up with the number of oxygen atoms.

SECTION 9.3

15. Explain the concept of the self-ionization of water.
16. Use the ionization constant of water to compute $[OH^-]$ from $[H_3O^+]$, and vice versa.
17. Given $[H_3O^+]$ or $[OH^-]$, compute both pH and pOH.
18. Define *pH* and *pOH*, and compute one given the other.
19. Given pH or pOH, compute both $[H_3O^+]$ and $[OH^-]$.
20. Describe the features of the pH scale, and give five examples of common substances and their pH values.
21. Explain what pH indicators are and what their two main purposes are.
22. Explain how pH indicators function in aqueous solution to produce different colors.
23. State the names and approximate transition intervals for three different pH indicators, one each with its transition interval below 7, including 7, and above 7.
24. Explain the purpose of titration, and give a general explanation of titration procedure.
25. Determine $[H_3O^+]$ and $[OH^-]$ from titration data.

9.1 Properties and Nomenclature of Acids and Bases

9.1.1 Introduction

Acids and bases are of supreme importance in chemistry—a great deal of chemistry concerns reactions between acids and bases or their reactions with other substances. We are surrounded by compounds that act as acids or bases in daily life, in the foods we eat and the products we use. There are thousands of compounds that exhibit acidic or basic properties, and in this chapter we only scratch the surface of this vast subject.

Figure 9.1. Acids in common foods.

Acids often give flavors to foods, including the sour or tart flavors we are familiar with in fruits. Several examples of these are displayed in Figure 9.1. Many bases are used in products commonly found in the home. Several of these products are shown in Figure 9.2. Tums, the antacid shown in the figure, contains only $CaCO_3$. But $Al(OH)_3$ and $Mg(OH)_2$ are also bases commonly found in antacids.

Acids are also of crucial importance in industry, where their uses are limitless. Some of the major industrial acids and their uses are summarized in Table 9.1.

9.1.2 Properties of Acids and Bases

It is important to be familiar with the well-known properties of acids and bases. For acids, these include the following:

1. In concentrated form, acids are often corrosive, which means they react vigorously with other substances. Because of this property, acids in the laboratory must always be treated with great care. Always wear eye protection, gloves, and—if you value your clothing—a laboratory apron.
2. Acids have a sour taste. This is why several of them appear so often in tart foods. However,

Figure 9.2. Common household bases.

Acid	Industrial Uses and Other Notes
hydrochloric acid, HCl historical name: *muriatic acid*	• manufacture of PVC pipe and polyurethane • swimming pool pH maintenance • leather processing • metal cleaning and descaling (pickling) • the acid in the human stomach, known as gastric acid
nitric acid, HNO_3 historical name: *aqua fortis*	• manufacture of ammonium nitrate fertilizer by reaction with ammonia • manufacture of explosives • precursor to manufacture of other compounds, including nylon
sulfuric acid, H_2SO_4 historical name: *oil of vitriol*	• most produced industrial chemical in the world • manufacture of fertilizer • metal cleaning and descaling (pickling) • precursor to manufacture of other compounds • the acid used in car batteries
phosphoric acid, H_3PO_4	• rust removal • food additive for delivering tangy taste • used in dentistry and toothpastes to clean teeth
acetic acid, CH_3COOH	• main ingredient in vinegar (other than water) • source for acetate in photographic film • manufacture of synthetic fabrics • food additive

Table 9.1. Important industrial acids.

note that you should never taste acids in a laboratory. They can be poisonous and extremely damaging to human tissue.

3. Acids react with bases to produce a *salt* and water. This type of chemical reaction is called *neutralization* because the corrosive properties of acids and bases are neutralized when they exchange ions to form the salt. A chemical salt is defined as a compound formed in an acid-base reaction, composed of the cation from the base and the anion from the acid.
4. Acids that form ions in aqueous solution are electrolytes. As we have seen, this means they conduct electricity.
5. Acids cause a color change in *pH indicators*. A pH indicator is a substance used to detect whether a solution is acidic or basic by its color. We study indicators in detail later in this chapter, including the use of *pH* as a measure of the strength of acids and bases. The juice from red cabbage is a natural pH indicator. Its amazing range of color is shown in Figure 9.3.

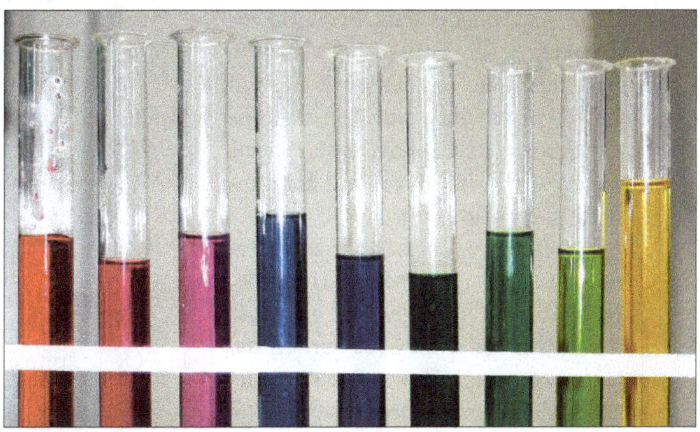

Figure 9.3. The colors exhibited by red cabbage in solutions with the following pH values (L to R): 1–3–5–7–8–9–10–11–13.

6. Some acids react vigorously with metals, releasing hydro-

gen gas in the process. For this reason, acids are not stored in metal containers or metal cabinets. The photos in Figure 5.17 of magnesium reacting with hydrochloric acid are a typical example of this property.

Bases also have some common properties, many of which correlate to the properties of acids. These properties include:

1. In concentrated form, bases are often corrosive, which means they react vigorously with other substances. Bases in the laboratory must always be treated with great care. Always wear eye protection and gloves.
2. Bases have a bitter taste and slippery feel. Soap exhibits both these characteristics, as you know. However, note that you should never taste bases in a laboratory. They can be poisonous and extremely damaging to human tissue.
3. Bases react with acids to produce a salt and water.
4. Bases that form ions in aqueous solution are electrolytes. As with acids, this means they conduct electricity.
5. Bases cause a color change in pH indicators.

Formula	Name of Pure Compound	Name in Aqueous Solution
HF	hydrogen fluoride	hydrofluoric acid
HCl	hydrogen chloride	hydrochloric acid
HBr	hydrogen bromide	hydrobromic acid
HI	hydrogen iodide	hydroiodic acid
H_2S	hydrogen sulfide	hydrosulfuric acid.

Table 9.2. Binary acids.

9.1.3 Acid Names and Formulas

We briefly touched on naming acids in the discussion of polyatomic ions in Chapter 3. Now it is time to become more familiar with these terms.

The first major class of acids are the *binary acids,* formed between hydrogen and one of the more electronegative nonmetals. Several of these are *hydrogen halides*—hydrogen with a halogen—HF, HCl, HBr, and HI. Another binary acid is H_2S.

In their pure forms, all these compounds are colorless gases with names that begin with *hydrogen* and end with *–ide.* In this state, they are sometimes referred to as *acid gases.* But in aqueous solution they are proper acids, and their names begin with *hydro–* and end

Formula	Oxyacid Name	Oxyanion
HIO_3	iodic acid	IO_3^-, iodate
HClO	hypochlorous acid	ClO^-, hypochlorite
$HClO_2$	chlorous acid	ClO_2^-, chlorite
$HClO_3$	chloric acid	ClO_3^-, chlorate
$HClO_4$	perchloric acid	ClO_4^-, perchlorate
$HMnO_4$	permanganic acid	MnO_4^-, permanganate
HNO_2	nitrous acid	NO_2^-, nitrite
HNO_3	nitric acid	NO_3^-, nitrate
H_2CO_3	carbonic acid	CO_3^{2-}, carbonate
H_2CrO_4	chromic acid	CrO_4^{2-}, chromate
H_2SO_3	sulfurous acid	SO_3^{2-}, sulfite
H_2SO_4	sulfuric acid	SO_4^{2-}, sulfate
H_3PO_3	phosphorous acid	PO_3^{3-}, phosphite
H_3PO_4	phosphoric acid	PO_4^{3-}, phosphate

Table 9.3. Common oxyacids and oxyanions.

with *acid.* This information is summarized in Table 9.2.

The second major class of acids is the *oxyacids*. An oxyacid contains one or more oxygen atoms, one or more hydrogen atoms, and at least one other element, usually a nonmetal. We have already encountered many of these in this text. Table 9.3 lists only a *few* of the most common ones; there are many more.

Acid Formula	Acid Name	Anion
$HClO_4$	perchloric acid	ClO_4^-, perchlorate
$HClO_3$	chloric acid	ClO_3^-, chlorate
$HClO_2$	chlorous acid	ClO_2^-, chlorite
$HClO$	hypochlorous acid	ClO^-, hypochlorite

Table 9.4. Illustration of oxyacid naming convention.

As a brief review of the naming convention for the oxyacids, Table 9.4 lists the four Cl—O oxyacids. The convention for naming the acids is to retain anion prefixes and change *–ate* and *–ite* suffixes to *–ic* and *–ous* in the acid names. This convention is reflected in the names of the oxyacids listed in Table 9.3.

A third major category of acids is the *carboxylic acids.* These contain the *carboxyl group*, —COOH. Acetic acid, CH_3COOH, which we have encountered before, is a simple carboxylic acid. The number of carboxylic acids is enormous, and they are of great importance in organic chemistry and biochemistry. We do not consider the carboxylic acids further here, except to mention two of the more well-known ones. Formic acid, the simplest of all carboxylic acids, is famous for being in the venom of many ant species. These species bite the victim and spray the formic acid on the wound.[1] Benzoic acid is widely used as a food preservative. The Lewis structures for these three carboxylic acids are shown in Figure 9.4.

Finally, note that some acids do not fit any of the broad acid classes described here. An example of this that we have already seen is hydrocyanic acid, HCN, formed from the cyanide ion, CN^-.

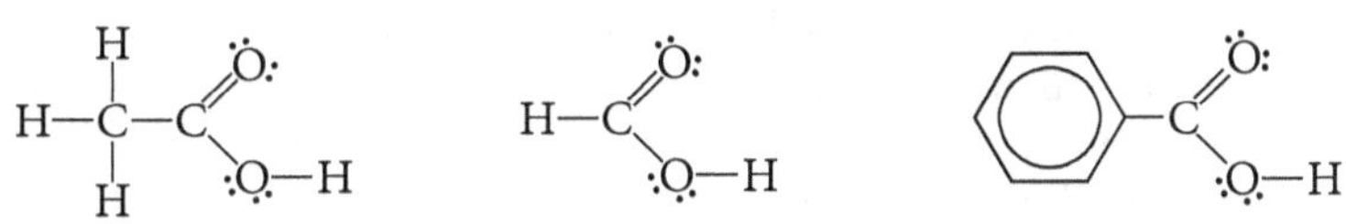

(a) (b) (c)

Figure 9.4. Lewis structures for acetic acid (a), formic acid (b), and benzoic acid (c).

9.2 Acid-Base Theories

Recall once again our discussion in the Introduction that chemistry is all about modeling—developing theories that describe the natural world as accurately as possible. This frame of mind is particularly valuable in the study of acids and bases. The reason for this is that when we try to define what acids and bases are, and then understand their behavior in light of those definitions, nature does not much want to cooperate. The result is three major acid-base theories, each one defining acids and bases differently from the others. The theories do overlap, and substances like the acid H_2SO_4 and the base NaOH are identified as acid or base in all three theoretical systems. But there are differences as well, and in order to engage in the conversation about what happens when acids and bases react, students must be familiar with all three theoretical systems.

One of the first acid theories was developed by French chemist Antoine Lavoisier. Recall from Chapter 5 that in 1774 Lavoisier first identified oxygen and gave the element its name. The

1 However, fire ants, one of the downsides of living in Texas, are different. They only bite to get a grip, and then sting the victim with poison from the abdomen, like a wasp.

name *oxygen* means "acid former," and Lavoisier chose the name because at the time scientists thought that all acids contained oxygen. As it became clear that many acids did not contain oxygen, the task of defining exactly what acids and bases are became pressing.

9.2.1 Arrhenius Acids and Bases

As you recall from the previous chapter, Swedish chemist Svante Arrhenius was a groundbreaking contributor to our understanding of the behavior of substances in solution. The acid-base theory put forward by Arrhenius in 1884 is the first theory to learn. We begin with definitions of acids and bases according to Arrhenius' model.

An Arrhenius acid is a substance that increases the concentration of hydrogen ions, H^+, in aqueous solution. Acids do this because they are electrolytes containing hydrogen, and when they ionize, they contribute H^+ cations to the solution. The solution becomes electrically conductive as a result.

An Arrhenius base is a substance that increases the concentration of hydroxide ions, OH^-, in aqueous solution. Thus, like acids, bases are electrolytes, compounds that ionize in aqueous solution. Bases contain the OH^- anion (or produce them in solution, as NH_3 does). When they ionize, the presence of the OH^- anions in solution makes the solution electrically conductive.

Arrhenius' definitions are very powerful. Thinking about his definition for acids, computer models of three common acids are shown in Figure 9.5. Hydrogen atoms are shown in white in these models, and the hydrogens that ionize in solution are the ones connected to oxygen atoms (red). As you see, nitric acid, HNO_3, contributes one H^+ cation when it ionizes. Such acids are referred to as *monoprotic* (one proton). Sulfuric acid, H_2SO_4, first ionizes as H^+ and HSO_4^-. Then, some of the HSO_4^- anions dissociate further, producing more H^+ cations and SO_4^{2-} anions in solution. The resulting solution contains all these species simultaneously. Acids that can contribute more than one H^+ cation in solution are referred to as *polyprotic*. Those like H_2SO_4 that can contribute two hydrogen ions are termed *diprotic*; an acid such as phosphoric acid, H_3PO_4, that can contribute three hydrogen ions is *triprotic*.

The third acid in Figure 9.5, acetic acid, contains four hydrogen atoms, but those attached to the carbon atom do not dissociate in solution; they stay with the molecule. Only the hydrogen atom connected to the oxygen atom dissociates.

As mentioned in Chapter 8, some acids ionize completely in water and some ionize to a very limited degree. Hydrochloric acid ionizes completely:

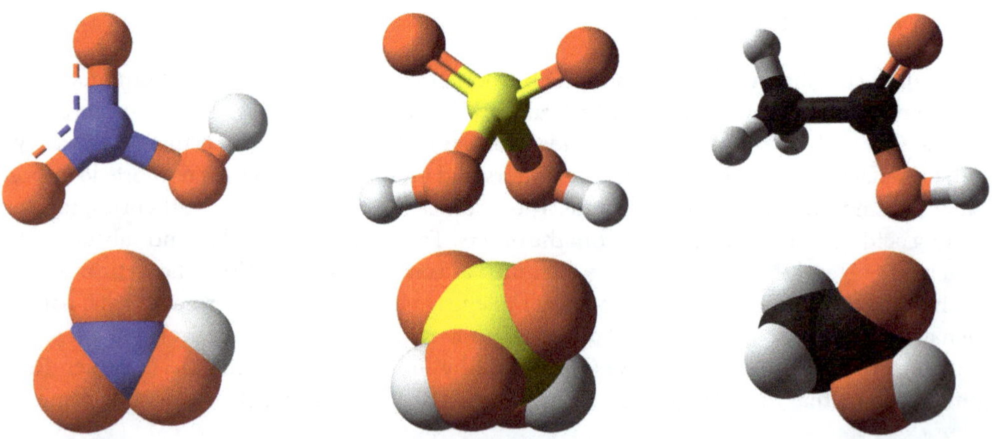

Figure 9.5. Ball-and-stick models (top) and space-filling models (bottom) of three acids: nitric acid (left), sulfuric acid (middle), and acetic acid (right).

$$HCl(aq) \rightarrow H^+(aq) + Cl^-(aq)$$

Acetic acid, on the other hand, ionizes very little in water; most of the molecules remain intact in the solution. Accordingly, the equation shows the ionization of acetic acid as a bidirectional process:

$$CH_3COOH(aq) \rightleftharpoons H^+(aq) + CH_3COO^-(aq)$$

In solution, CH_3COOH reaches an equilibrium in which molecules of CH_3COOH are dissociating at the same rate that H^+ and CH_3COO^- ions are recombining into molecules.

The more completely an acid ionizes, the stronger an electrolyte it is, and the stronger an acid it is. The same holds for bases: when a base ionizes completely, it is a strong electrolyte and a strong base. Those that ionize only to a limited extent are weak electrolytes and weak bases. Table 9.5 identifies a number of common acids and bases, classifying them as strong or weak.

Strong Acids	Weak Acids	Strong Bases	Weak Bases
HCl	CH_3COOH	NaOH	NH_3
HNO_3	HCN	KOH	$NaCH_3COO$ (sodium acetate)
H_2SO_4	HF	RbOH	CH_3NH_2 (methylamine)
$HClO_4$	H_2CO_3	CsOH	
HBr	H_3PO_4	$Ca(OH)_2$	
HI	H_2S	$Sr(OH)_2$	
	HSO_4^-	$Ba(OH)_2$	
	HCO_3^-		

Table 9.5. Classification of some common acids and bases.

While Arrhenius' definition of an acid is very helpful for understanding acids, we now know that the H^+ ions—protons—contributed to a solution by an acid do not remain in solution as individual protons. Instead, each H^+ ion attaches to one of the nonbonding pairs on a water molecule, forming the H_3O^+ *hydronium ion*. For this reason, it is now common to say that *an Arrhenius acid is a substance that increases the concentration of hydronium ions in solution*. The concepts of increasing H^+ ions (Arrhenius' original concept) and increasing H_3O^+ ions (which is where the H^+ ions end up) are used interchangeably.

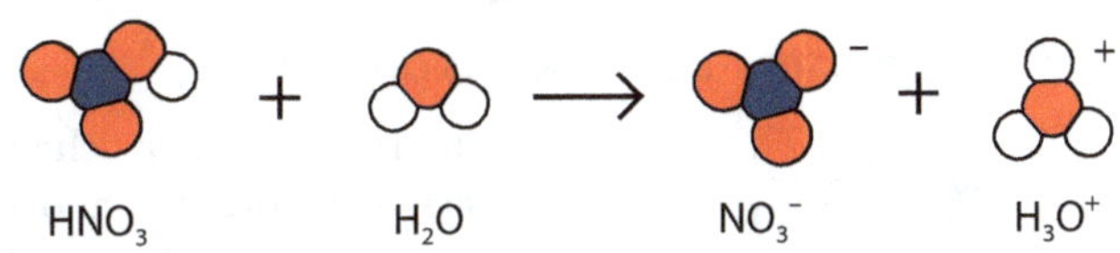

Figure 9.6. Nitric acid dissociating in water to produce nitrate and hydronium ions.

As an illustration, the formation of hydronium ions when nitric acid dissolves in water is represented by the following equation:

$$HNO_3(l) + H_2O(l) \rightarrow H_3O^+(aq) + NO_3^-(aq)$$

This process is further illustrated by Figure 9.6.

We show the ionization of bases in equations in the same way. For example, the ionization of the strong base sodium hydroxide in water is represented this way:

$$NaOH(s) \xrightarrow{H_2O} Na^+(aq) + OH^-(aq)$$

The weak base ammonia, NH_3, also dissolves in water, producing OH^- ions in the process. However, ammonia doesn't dissociate. To show what happens in this reaction, we have to show the H_2O molecules in the equation. Ammonia does not contain the OH^- ion, but OH^- ions are produced when hydrogen atoms leave water molecules to bond to ammonia molecules. Recall that the electron domain geometry of the ammonia molecule is tetrahedral, with one nonbonding electron pair on the nitrogen atom. This is where a hydrogen atom can attach, producing an ammonium ion, NH_4^+, as shown in this equation:

$$NH_3(aq) + H_2O(l) \rightarrow NH_4^+(aq) + OH^-(aq)$$

Not very many hydrogen atoms are going to make the jump from water molecules to ammonia molecules. For this reason, not many OH^- ions are produced, and thus ammonia is a weak base.

Arrhenius gave us a good starting place for defining acids and bases. But it gradually became apparent that there are many cases his definitions don't address, primarily because they were articulated in terms of ions in aqueous solution. Many acid-base reactions can occur apart from aqueous solution. For example, HCl and NH_3 can react as gases to produce solid particles of NH_4Cl, as this equation shows:

$$HCl(g) + NH_3(g) \rightarrow NH_4Cl(s)$$

Another example is the acid-base reaction between acetic acid and ammonia:

$$CH_3COOH + NH_3 \rightarrow CH_3COO^- + NH_4^+$$

In this reaction, the acetic acid molecule donates a proton, but the reaction cannot be described in terms of Arrhenius model because the proton does not form hydronium. Instead, it is transferred to the ammonia molecule to form ammonium.

As scientists learned about occasions when substances act as acids or bases apart from aqueous solutions, the door was open for a new, more all-encompassing definition.

9.2.2 Brønsted-Lowry Acids and Bases

Figure 9.7. Danish chemist Johannes Brønsted (1879–1947).

In 1923, Danish chemist Johannes Brønsted (Figure 9.7) and English chemist Thomas Martin Lowry (Figure 9.8) independently recognized that an acid-base reaction can be modeled as a *proton transfer*, and they developed definitions for acids and bases based on this fundamental concept. The Brønsted-Lowry theory is now the one most commonly used for general analysis of acid-base interactions.

The definitions for acids and bases according to the Brønsted-Lowry model are as follows: *A Brønsted-Lowry acid is a proton donor, and a Brønsted-Lowry base is a proton acceptor.* These definitions sound

Figure 9.8. English chemist Thomas Martin Lowry (1874–1936).

Hmm... Interesting. What is an alkali?

The beautiful photo below is an image from the salt flats at Badwater Basin in Death Valley, California—the lowest place in North America (282 feet below sea level). Salt flats are sometimes called *alkali flats* because of the carbonates of alkali metals or alkaline-earth metals (chiefly Na_2CO_3) deposited there. Outside the field of chemistry, the term *alkali salt* is sometimes applied to hydroxide compounds, such as NaOH (called *lye* or *caustic soda*), KOH (called *caustic potash*), $Mg(OH)_2$, and $Ca(OH)_2$. These compounds can form when the carbonate compounds react with water. However, chemists refer to the hydroxide compounds as *bases*, reserving the term *salt* for the compounds that form when a base reacts with an acid.

There are also many *alkali lakes* around the world, where living things are rare because of the high concentrations of salts, often seen encrusted along their shores. But the salts in alkali lakes are typically carbonates such as Na_2CO_3, not hydroxides.

The meanings of the terms *alkali* and *alkaline* are tough to chase down because of varying usages in different contexts. When describing natural features like flats and lakes, soils and waters are typically referred to as *alkaline* any time the soil or water pH is high, regardless of the particular base that is causing it.

In the context of chemistry, we also find the terms used in several different ways. The most general use is in referring to any base or basic solution. But some sources consider this usage obsolete, and other sources specifically distinguish between an *alkali* and a *base*, and between *alkalinity* and *basicity*. Other usages apply the term *alkali* to a subset of the bases, but there are several different ways to identify the subset the term applies to, including:

- to describe only bases that include the OH^- anion
- to describe any basic salt of the alkali metals or alkaline earth metals
- to describe any water-soluble base that forms OH^- ions (which would include NH_3 but not $Mg(OH)_2$, because magnesium hydroxide is not very water soluble)

The term *alkali* derives from the Arabic *al-qaly*, and originally referred to wood ashes containing K_2CO_3 and other potassium compounds, known generally as *potash*. *Caustic potash*, KOH, is a much stronger base, and is the substance English scientist Sir Humphry Davy used when he first isolated potassium in 1807. The chemical symbol K for potassium is from the Medieval Latin word *kalium*—plant ashes—which derives from the same Arabic term that gave us *alkali*. All this is interesting, but doesn't really help us much in establishing a specific contemporary meaning for the term *alkali*.

With so many different definitions in current use, my hunch is that there is no hope of the term regaining any specific meaning, if it ever had one. So use it as you see fit.

simple, but they have far-reaching implications for what we call an acid and what we call a base, as we will now see. (As we proceed, keep in mind that a proton is identical to a hydrogen ion.)

To begin, let's first look again at the reaction between acetic acid and ammonia:

$$CH_3COOH + NH_3 \rightarrow CH_3COO^- + NH_4^+$$

In this reaction, acetic acid is a proton donor, so it is a Brønsted-Lowry acid; ammonia is a proton acceptor, so it is a Brønsted-Lowry base.

However, let's now look again at the dissolving of nitric acid in water:

$$HNO_3(l) + H_2O(l) \rightarrow H_3O^+(aq) + NO_3^-(aq)$$

Here the nitric acid donates a proton, so it is a Brønsted-Lowry acid. But the proton acceptor in this case is water, so according to the Brønsted-Lowry model, water is a Brønsted-Lowry base.

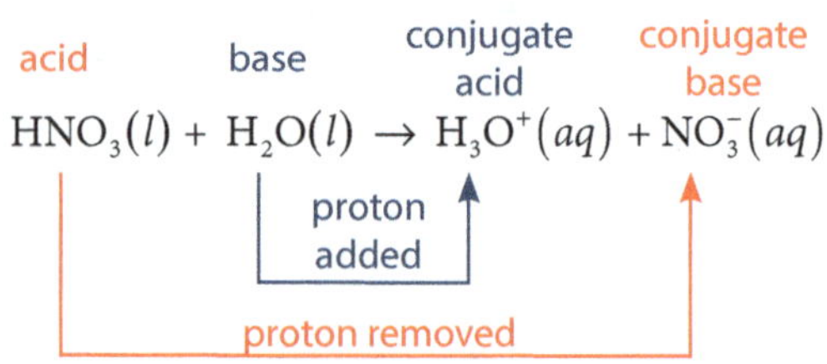

Figure 9.9. Conjugate acids and bases in the Brønsted-Lowry model.

After the reaction, notice that the HNO_3 has become NO_3^-, and as such it can now *accept* a proton. This makes it a base according to the Brønsted-Lowry model. Likewise, H_2O has become H_3O^+, so it is now a potential proton *donor*. This situation is always the case with the Brønsted-Lowry model: once a proton has been donated, the species that remains is able to accept a proton, and once a proton has been accepted, the new species is able to donate it. Special terms are used to describe these chemical species, illustrated in Figure 9.9.

When a Brønsted-Lowry acid donates a proton, the new species created is called the *conjugate base* of that acid. Likewise, when a Brønsted-Lowry base accepts a proton, the new species created is called the *conjugate acid* of that base.

As another example, let's look again at the reaction between ammonia and water:

$$NH_3(aq) + H_2O(l) \rightarrow NH_4^+(aq) + OH^-(aq)$$

Figure 9.10 indicates the proton transfer and identifies the acid, base, conjugate base, and conjugate acid for this reaction. As before, the Brønsted-Lowry base, ammonia, accepts a proton, and the new species created, NH_4^+, is the conjugate acid of that base. However, now we see that in this reaction, the H_2O molecule is the proton donor, which means that in this reaction water is a Brønsted-Lowry acid. The hydroxide ion formed is the conjugate base of that acid.

In the two examples we have just reviewed, water is a Brønsted-Lowry base in one of them and a Brønsted-Lowry acid in the other. We use the term *amphoteric* to describe any species that can act as either an acid or a base in a general sense. A subset of amphoteric species are molecular compounds that can act as either a proton donor or a proton acceptor. The term *amphiprotic* is used to describe these. Water is an amphiprotic (and amphoteric) compound, and can act as either a Brønsted-Lowry acid or a Brønsted-Lowry base.

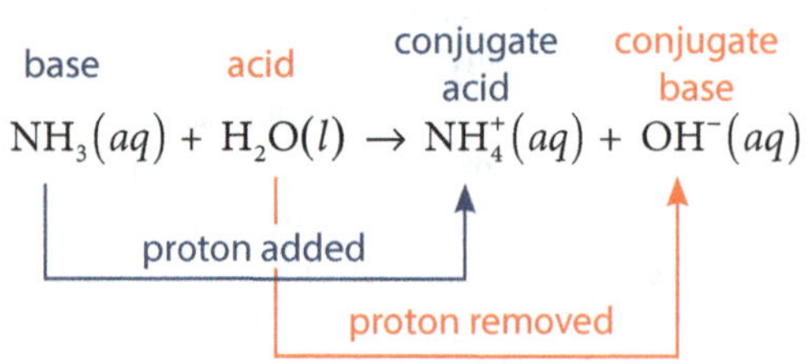

Figure 9.10. Another example of conjugate acids and bases in the Brønsted-Lowry model.

The Brønsted-Lowry model is certainly helpful in understanding what goes on when an acid or base dissolves in water. Let's now see what happens when we attempt to analyze a standard Arrhenius acid-base reaction in terms of the Brønsted-Lowry model. Consider the acid-base neutralization reaction between hydrochloric acid and sodium hydroxide:

$$HCl(aq) + NaOH(aq) \rightarrow NaCl(aq) + H_2O(l)$$

The reaction is often written this way, and demonstrates a widely understood fact about acid-base reactions: they produce a salt and water.

But HCl is a strong acid, and dissociates completely in water. Likewise, NaOH is a strong base and dissociates completely in water. So to analyze this reaction further in terms of both the Arrhenius and Brønsted-Lowry models, it is more helpful to write this equation in terms of the ions that appear in solution:

$$HCl(aq) + NaOH(aq) \xrightarrow{H_2O} Na^+(aq) + Cl^-(aq) + H_3O^+(aq) + OH^-(aq)$$

Now it is clear that in terms of the Arrhenius model, HCl is an Arrhenius acid because the hydrogen ion from HCl goes to create a new hydronium ion. Likewise, NaOH is an Arrhenius base because it contributes an OH^- ion to the solution.

HCl is also identified as an acid in terms of the Brønsted-Lowry model because it donates the proton that forms the hydronium ion. However, NaOH does not fit the definition of a base according to the Brønsted-Lowry model because it does not accept the proton. The water molecule accepts the proton, and so H_2O is the Brønsted-Lowry base in this reaction. The OH^- ion created in the process is now able to receive a proton, and thus is ready to act as a Brønsted-Lowry base in the future. Even though the Brønsted-Lowry model does not identify NaOH as a base, hydroxides such as NaOH are commonly referred to as bases, so one has to get used to mentally switching back and forth from the Arrhenius model to the Brønsted-Lowry model, depending on the reaction in question. Note also that NaOH *produces* a Brønsted-Lowry base (OH^-) just as HCl *produces* a Brønsted-Lowry acid (H_3O^+). Thus, we can call NaOH a base and HCl an acid in this sense as well.

As mentioned previously, there are also reactions that do not take place in water where the Brønsted-Lowry model is very helpful. The reaction between ammonia and hydrochloric acid is one of these:

$$HCl(g) + NH_3(g) \rightarrow NH_4Cl(s)$$

To see what is happening in this case, the Lewis structures are helpful:

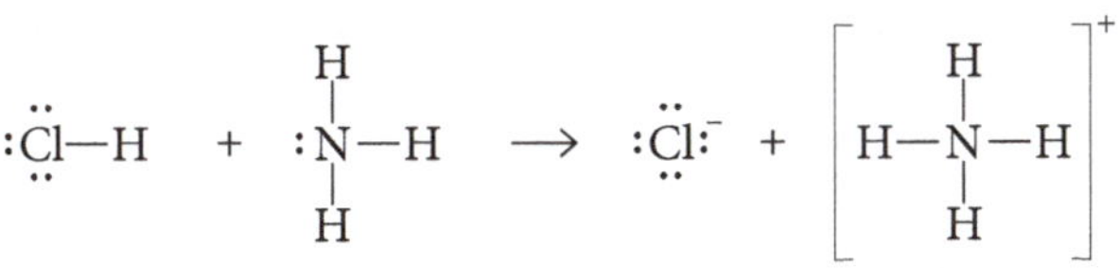

Here, HCl is a Brønsted-Lowry acid, donating a proton and creating a Cl^- ion, which is the conjugate base. NH_3 is a Brønsted-Lowry base, accepting the proton to form the NH_4^+ ion, which is the conjugate acid. As mentioned above, this reaction can take place in the vapor state, in

Figure 9.11. Vapors from HCl (left) and NH_3 (right) react in the air to form solid particles of NH_4Cl.

the air, as you see from the photo in Figure 9.11. When it does, the Cl^- and NH_4^+ ions combine to form solid, airborne particles of the salt NH_4Cl.

9.2.3 Lewis Acids and Bases

In 1923, the great American chemist G.N. Lewis proposed a third acid-base model. As you know, Lewis originally conceptualized the covalent bond, describing it in terms of the formation of a shared electron pair. Lewis' acid-base theory is expressed in the same terms—electron pairs. Since Lewis expressed his model in terms of electron pairs rather than protons, a substance without any hydrogen atoms can act as a Lewis acid.

A Lewis acid is any species that accepts an electron pair to form a covalent bond. A Lewis base is any species that donates an electron pair to form a covalent bond. A Lewis acid-base reaction occurs when one or more covalent bonds are formed between an electron pair donor and an electron pair acceptor.

The vapor-state reaction between boron trifluoride and ammonia is commonly used to illustrate the Lewis acid-base reaction:

$$BF_3(g) + NH_3(g) \rightarrow F_3BNH_3(g)$$

Represented with Lewis structures, this reaction is as follows:

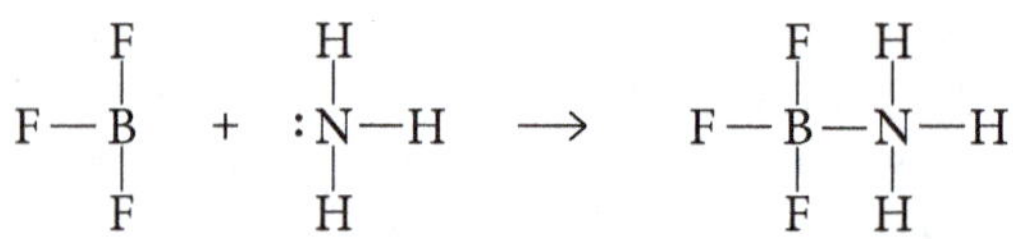

Recall that boron acts in exception to the octet rule and forms compounds such as BF_3 that furnish boron with six electrons in its valence shell rather than eight. As the Lewis structures show, the lone pair on the ammonia molecule attaches to the boron atom, giving boron a full octet. In this reaction, BF_3 is a Lewis acid because it accepts the electron pair from NH_3 to form a covalent bond. NH_3 is a Lewis base because it donates the electron pair to form the covalent bond.

Another Lewis acid-base reaction is between BF_3 and the fluorine anion, F^-:

$$BF_3(aq) + F^-(aq) \rightarrow BF_4^-(aq)$$

Each of the three acid-base theories is useful. The Lewis theory is especially useful in organic chemistry, where it is one of the dominant tools for modeling chemical reactions. As we have seen, sometimes the three theories classify substances in different ways. In other cases, all three models classify a compound the same way. We have seen that NH_3 acts as a base under all three models. It is an Arrhenius base because it causes OH^- ions to form in water, it is a Brønsted-Lowry base because it accepts a proton to form NH_4^+, and it is a Lewis base because it donates an electron pair to form a covalent bond.

Example 9.1

For each of the reactions shown, identify the Arrhenius, Brønsted-Lowry, and Lewis acids and bases.

$$HBr(aq) + H_2O(l) \rightarrow Br^-(aq) + H_3O^+(aq)$$

$$CH_3COOH + NaHCO_3 \rightarrow CH_3COONa + H_2O + CO_2$$

In the first reaction, HBr ionizes and donates a proton, so it is an Arrhenius acid and a Brønsted-Lowry acid. The H_2O accepts the proton, making it a Brønsted-Lowry base. There is no Arrhenius base. The H^+ from the HBr acts as a Lewis acid, receiving the electron pair from H_2O, which acts as a Lewis base, to form a covalent bond in H_3O^+.

The second reaction does not have to occur in solution. Pure acetic acid (called glacial acetic acid) is a liquid, and sodium bicarbonate is a solid. The acetic acid donates a proton, so it is a Brønsted-Lowry acid. An OH^- ion from the dissociated $NaHCO_3$ acts as the Brønsted-Lowry base, receiving the proton and forming a molecule of water. In the process, we could say that the H^+ from the CH_3COOH acts as a Lewis acid, receiving the electron pair from the OH^- ion, which is acting as a Lewis base, to form a covalent bond. This reaction can also occur in solution. In that case, the CH_3COOH contributes H^+ ions (and thus H_3O^+ ions) to the solution, and thus acts as an Arrhenius acid. The $NaHCO_3$ contributes OH^- ions to the solution and thus acts as an Arrhenius base.

9.2.4 Strength of Acids and Bases

Here we use the Brønsted-Lowry model to elaborate further on the concept of acid and base strength. We find that the Brønsted-Lowry model allows us to make general predictions about the equilibrium point when an acid or base ionizes in water.

Table 9.6 lists several Brønsted-Lowry acids and bases in conjugate pairs. In each row, an acid is listed side-by-side with its conjugate base. Water appears twice, and the two placements of water divide the list into three major sections. The strong acids are compounds that ionize 100% in water. The conjugate base for each of these exhibits negligible properties as a base. For example, HCl ionizes completely, and is a strong acid.

Acid	Base
HCl	Cl^-
$HClO_4$	ClO_4^-
H_2SO_4	HSO_4^-
HNO_3	NO_3^-
$HClO_3$	ClO_3^-
H_3O^+	H_2O
HSO_4^-	SO_4^{2-}
$HClO_2$	ClO_2^-
H_3PO_4	$H_2PO_4^-$
HF	F^-
CH_3COOH	CH_3COO^-
H_2CO_3	HCO_3^-
H_2S	HS^-
$H_2PO_4^-$	HPO_4^{2-}
HClO	ClO^-
NH_4^+	NH_3
HCO_3^-	CO_3^{2-}
HPO_4^{2-}	PO_4^{3-}
H_2O	OH^-
OH^-	O^{2-}
H_2	H^-
CH_4	CH_3^-

Acids: ionize 100% in water (HCl through $HClO_3$) — strong; weak; negligible; increasing acid strength (upward).
Bases: negligible; weak; strong — protonate 100% in water (O^{2-}, H^-, CH_3^-); increasing base strength (downward).

Table 9.6. Relative strengths for some conjugate acid and base pairs.

But Cl^- ions in solution show little tendency to accept a proton, so Cl^- is classified as having negligible base strength.

In the center of the chart are the weak acids and bases. These are weak because they dissociate only partially in water. In acetic acid, for example, only about 1.4% of the molecules in 0.10 *M* CH_3COOH ionize. But the more an acid dissociates, the stronger it is and the higher up it is in the scale of acid strength.

Notice the pattern among the diprotic and triprotic acids. H_2SO_4 is a strong acid, and ionizes in steps. The first step occurs immediately as almost 100% of the H_2SO_4 ionizes to become HSO_4^-. But HSO_4^- is a weak acid, and only partially ionizes. This second step of ionization only occurs to about 29% of the molecules in 0.10 *M* H_2SO_4. All polyprotic acids ionize in steps like this, and after each step of ionization, the acid that remains is weaker and ionizes less readily.

In the lower section of the chart are substances that contain hydrogen but display negligible tendency to donate a hydrogen ion, so these compounds have negligible acid strength. Their conjugate bases are strong bases, which means 100% of these molecules accept a proton in aqueous solution, a process called *protonation*.

The first thing we notice when the conjugate acid-base pairs are arranged this way is that *the stronger an acid is, the weaker its conjugate base is, and the stronger a base is, the weaker its conjugate acid is.* In the Brønsted-Lowry model, acid and base strength are determined by a compound's readiness to donate or accept a proton. The strongest acids are those most ready to donate a proton, and the strongest bases are those most ready to accept a proton.

In any Brønsted-Lowry acid-base reaction, there are two conjugate pairs: an acid and its conjugate base, a base and its conjugate acid. In the reaction, a weak base does not compete with a strong base for protonation. Likewise, a weak acid does not compete with a strong acid for donating protons. These considerations lead to this important generalization: *In a Brønsted-Lowry acid-base reaction, the reaction favors the transfer of protons from the strong acid to the strong base, producing larger quantities of the weak acid and weak base.* The way we describe this verbally is to reference the written reaction equation and say that the reaction "favors the right (or left) side," or "the equilibrium point lies to the right (or left)."

To illustrate, consider once again the dissolution of acetic acid in water:

$$CH_3COOH(aq) + H_2O(l) \rightleftharpoons CH_3COO^-(aq) + H_3O^+(aq)$$

In this reaction, acetic acid is a Brønsted-Lowry acid, and water is a Brønsted-Lowry base. The conjugate base is the acetate ion, CH_3COO^-, and the conjugate acid is the hydronium ion, H_3O^+. Referring to Table 9.6, we find that H_3O^+ is a stronger acid than CH_3COOH. We also find that CH_3COO^- is a stronger base than H_2O. So in the equation, the stronger acid and the stronger base are on the right and the weaker acid and weaker base are on the left. Since the reaction favors the production of the weaker acid and the weaker base, we say that the equilibrium lies to the left in this reaction. This is simply a way of saying that the equilibrium point in the solution is such that the production of CH_3COOH and H_2O molecules is favored more than the production of H_3O^+ and CH_3COO^- molecules. In the solution, concentrations of CH_3COOH and H_2O molecules are much higher than the concentrations of H_3O^+ and CH_3COO^- molecules.

Now consider again the dissociation of nitric acid:

$$HNO_3(aq) + H_2O(l) \rightleftharpoons NO_3^-(aq) + H_3O^+(aq)$$

In this reaction, nitric acid is a Brønsted-Lowry acid, and water is a Brønsted-Lowry base. The conjugate base is the nitrate ion, NO_3^-, and the conjugate acid is the hydronium ion, H_3O^+. Referring to Table 9.6, we find that HNO_3 is a stronger acid than H_3O^+. We also find that H_2O is a stronger base than NO_3^-. So in the equation, the stronger acid and the stronger base are on the left and the weaker acid and weaker base are on the right. Since the reaction favors the production of the weaker acid and the weaker base, we say that the equilibrium lies to the right in this reaction. That is, the equilibrium favors the production of NO_3^- and H_3O^+ in the solution.

Example 9.2

For the proton-transfer reaction shown, identify the Brønsted-Lowry conjugate pairs, and predict whether the equilibrium point lies to the right (favoring the production of SO_4^{2-} and HCO_3^-) or to the left (favoring the production of HSO_4^- and CO_3^{2-}).

$$HSO_4^-(aq) + CO_3^{2-}(aq) \rightleftharpoons SO_4^{2-}(aq) + HCO_3^-(aq)$$

We begin by identifying the conjugate acid-base pairs, as shown in the next equation.

$$HSO_4^-(aq) + CO_3^{2-}(aq) \rightleftharpoons SO_4^{2-}(aq) + HCO_3^-(aq)$$

acid	base	conjugate base	conjugate acid

The two acids are HSO_4^- and HCO_3^-. Referring to the acid column in Table 9.6, HSO_4^- is a stronger acid than HCO_3^-. We also see in the table that CO_3^{2-} is a stronger base than SO_4^{2-}. Thus, the stronger acid and the stronger base are on the left side of the equation. The reaction favors the transfer of the proton from the stronger acid to the stronger base, resulting in the production of the species that are the weaker acid and the weaker base. Thus, the equilibrium in this reaction lies to the right.

We have seen that according to both Arrhenius' model and the Brønsted-Lowry model, the strength of an acid depends directly on how readily the acid molecule donates a proton, H^+. When we look at the actual structure of oxyacid molecules, there is a straightforward physical basis for understanding what makes one molecule more ready to donate a proton than another. Figure 9.12 shows space-filling molecular models for the four chlorine oxyacids, arranged in order of increasing acidity. As you see, the only difference between these molecules is the number of oxygen atoms attached to the central chlorine atom.

As you know, oxygen is strongly electronegative. As such, it pulls the electron density of the bonding electron pair of the hydrogen atom toward itself, away from the hydrogen atom. This makes the O—H bond polar and enables the hydrogen atom to separate more easily from the rest of the molecule. This polarity becomes magnified with each additional oxygen atom in the

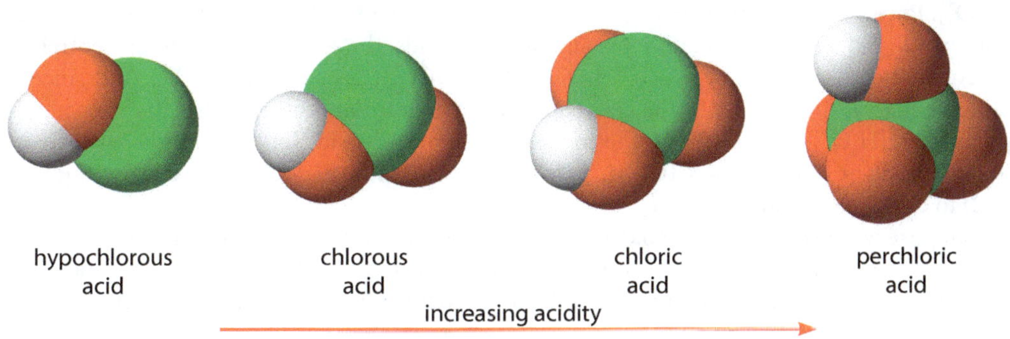

Figure 9.12. The more oxygen atoms there are on the central atom in an acid, the stronger the acid is.

molecule. The more there are, the more strongly the electron density of the electron pair bonding the hydrogen atom is pulled toward the oxygen atoms and away from the hydrogen atom. This makes the molecule more polar, so that less and less energy is required to separate the hydrogen atom from the molecule. As indicated in Table 9.6, the chlorine oxyacids with three and four oxygen atoms are strong acids because they ionize 100% in water.

9.3 Aqueous Solutions and pH

9.3.1 The Self-ionization of Water

We have seen that water is amphiprotic; it can donate or accept a proton in a Brønsted-Lowry acid-base reaction. So it may come as no surprise that pure water can and does ionize all by itself, transferring protons from one water molecule to another to create hydronium and hydroxide ions in the water. This process is called the *self-ionization of water*. Any time one water molecule donates a proton to become an OH^- ion, another water molecule accepts the proton to become an H_3O^+ ion. This is represented in equation form as follows:

$$H_2O(l) + H_2O(l) \rightleftharpoons H_3O^+(aq) + OH^-(aq)$$

The molar concentrations of OH^- and H_3O^+ ions in pure water, represented as $[OH^-]$ and $[H_3O^+]$, are the same. At 25°C, these values are

$$\left[H_3O^+\right]=\left[OH^-\right]=1.0\times10^{-7}\ M \tag{9.1}$$

The product of these two concentrations is called the *ionization constant of water, K_w*. The ionization constant K_w is equal to

$$K_w=\left[H_3O^+\right]\cdot\left[OH^-\right]=1.0\times10^{-7}\ M\cdot1.0\times10^{-7}\ M=1.0\times10^{-14}\ M^2 \tag{9.2}$$

At 25°C, the product of $[OH^-]$ and $[H_3O^+]$ is always equal to this value, even when other ions are present in solution. The ionization constant of water does vary with temperature: the higher the temperature, the more water ionizes, causing K_w to increase. But at any given temperature, K_w is a constant. For our purposes, you may assume that at any ordinary laboratory temperature, K_w is equal to $1.0 \times 10^{-14}\ M^2$.

9.3.2 Calculating $[H_3O^+]$ and $[OH^-]$

Strong acids are those that readily donate protons to form H_3O^+ ions. Thus, the measure of the acidity of an aqueous solution is the molar concentration of these ions in the solution, $[H_3O^+]$. Pure water, in which $[H_3O^+] = 1.0 \times 10^{-7}\ M$, is considered to be neither acidic nor basic but *neutral*. In fact, in any solution if Equation (9.1) holds, the solution is said to be neutral.

Any aqueous solution in which $[H_3O^+] > 1.0 \times 10^{-7}\ M$ is said to be acidic. And since K_w always equals $1.0 \times 10^{-14}\ M^2$, if the concentration of hydronium ions is higher than its value in pure water, $[H_3O^+] > 1.0 \times 10^{-7}\ M$, the concentration of hydroxide ions must be lower than it is in pure water, or $[OH^-] < 1.0 \times 10^{-7}\ M$. If we know one of these two values, we can use Equation (9.2) to calculate the other.

Suppose we have a solution of 0.010 M NaOH. NaOH is a strong base and thus ionizes 100% in water. As a result, when NaOH dissolves in water, each mole of NaOH in the solution supplies one mole of sodium ions and one mole of hydroxide ions in the solution, as shown by the coefficients in the equation:

$$NaOH(s) \xrightarrow{H_2O} Na^+(aq) + OH^-(aq)$$

This implies that the molar hydroxide ion concentration, $[OH^-]$, is the same as the molar concentration of NaOH. Thus, the concentration of hydroxide ions in a 0.010 *M* NaOH solution is $[OH^-] = 1.0 \times 10^{-2}$ *M*. This value for $[OH^-]$ is five orders of magnitude greater than it is for pure water, so this solution is very basic.

From the $[OH^-]$ value we can compute $[H_3O^+]$ using Equation (9.2). The product of $[H_3O^+]$ and $[OH^-]$ is 1.0×10^{-14} M^2. So

$$[H_3O^+]\cdot[OH^-]=1.0\times10^{-14}\ M^2$$

$$[H_3O^+]=\frac{1.0\times10^{-14}\ M^2}{[OH^-]}=\frac{1.0\times10^{-14}\ M^2}{1.0\times10^{-2}\ M}=1.0\times10^{-12}\ M$$

The following example illustrates a slightly more complex situation.

Example 9.3

Calcium hydroxide, $Ca(OH)_2$, is a strong base that dissociates completely in water. Determine the values of $[H_3O^+]$ and $[OH^-]$ for a 0.0012 *M* solution of $Ca(OH)_2$.

When $Ca(OH)_2$ dissociates, two moles of OH^- ions are produced for each mole of $Ca(OH)_2$. Thus, $[OH^-]$ is

$$[OH^-]=2\cdot0.0012\ M=0.0024\ M$$

From this, we calculate $[H_3O^+]$ using the ionization constant of water.

$$[H_3O^+]\cdot[OH^-]=1.0\times10^{-14}\ M^2$$

$$[H_3O^+]=\frac{1.0\times10^{-14}\ M^2}{[OH^-]}=\frac{1.0\times10^{-14}\ M^2}{0.0024\ M}=4.2\times10^{-12}\ M$$

▲

The calculation technique used in the preceding example depends on the knowledge that $Ca(OH)_2$ dissociates completely in water. The same procedure may be applied to solutions of $Ba(OH)_2$ and $Sr(OH)_2$ because these hydroxides are also strong bases that dissociate completely. (The solubility of these compounds in water is quite limited, but as much as dissolves also dissociates.) You may be tempted to assume that the same technique could be applied to H_2SO_4 to determine $[H_3O^+]$. However, as Table 9.6 indicates, only the first step in the ionization of H_2SO_4 is complete. The second step involves HSO_4^-, which is a weak acid and does not ionize completely. For this reason, we cannot simply double the molar concentration for H_2SO_4 and obtain $[H_3O^+]$. There are methods of determining $[H_3O^+]$ for polyprotic acids, but they lie beyond the scope of this text.

9.3.3 pH as a Measure of Ion Concentration and Acidity

As we have seen in these examples, molar concentrations of H_3O^+ and OH^- are very small values. To save ourselves the trouble of always having to write concentrations using scientific notation, the convenient *pH scale* was developed. The pH scale is used widely for specifying the acidity or basicity of a solution. The pH of a solution is defined as

$$pH = -\log[H_3O^+] \tag{9.3}$$

Now, I have a hunch that at least a few of the students reading this text may be a bit fuzzy about how logarithms work and what they mean. So let's pause here and work on this equation for a moment. *A logarithm is an exponent.* To demonstrate what this means, consider the expression

$$\log X = Y$$

In this expression, log$X$ is an exponent, and Y is an exponent. (They are equal—if one is an exponent, the other is as well.) So exactly what exponent is log$X$? Log$X$ *is the exponent you have to put on a 10 to get X.* (See footnote.[2]) In other words, the expression log$X$ = Y means the same thing as

$$10^Y = X$$

So, the expression "log[H_3O^+]" is an exponent. Let's say the numerical value of [H_3O^+] is 1×10^{-4}, which is simply 10^{-4}. The expression log[H_3O^+] means log(10^{-4}). This expression is a logarithm, and a logarithm is an exponent. So what exponent is this logarithm? It is the exponent you have to put on a 10 to get 10^{-4}. Obviously, that value is –4. That is how logarithms work. (Note: There is an issue here with significant digits, which we will get to in a moment.)

Now let's look again at the definition of pH. By definition, pH is equal to the negative of the log of the hydronium ion concentration. So consider again the illustration in the preceding paragraph. In that illustration, the numerical value of [H_3O^+] is 1×10^{-4}. The value of log(10^{-4}) is –4. The pH is the negative of this value, or pH = 4.

When dealing with the mathematics of logarithms, we have to be careful with the significant digits. When we calculate expressions like log(1×10^{-4}), the number to the left of the decimal in the result does not count in the significant digits; it is an exponent and only *locates the decimal.* The significant digits in a logarithm are the digits to the right of the decimal. This means we must have the same number of significant digits *to the right of the decimal* in the result as we have in the original value that we found the logarithm of. The value 1×10^{-4} has one significant digit, which means we must have one digit to the right of the decimal in the logarithm of it, so –log(1×10^{-4}) = 4.0.

▼ Example 9.4

Determine the pH of pure water.

From Equation (9.1), for pure water, [H_3O^+] = 1.0×10^{-7} *M*. Thus,

$$pH = -\log[H_3O^+] = -\log(1.0 \times 10^{-7}) = 7.00$$

Note that since the given hydronium ion concentration has two significant digits, we must have two digits to the right of the decimal in the logarithm, that is, in our result.

▲

2 This is the way *common* logarithms are defined. The common logarithm, symbolized log$X$, is a base-10 logarithm. The *natural* logarithm, symbolized ln$X$, is a base-*e* logarithm. So the expression ln$X$ means "the exponent you have to put on *e* to get X." The number *e* is, of course, a transcendental number, like π. The numerical value of *e* begins 2.71828....

Figure 9.13 is a representation of the pH scale, with a number of common substances located on it. Solutions with pH values below 7 are acidic, and the lower the pH, the more acidic the solution. Notice that since pH values are powers of 10, a solution with a pH of 4.0 has 10 times the concentration of hydronium ions compared to a solution with a pH of 5.0. Thus, the solution with pH = 4.0 is 10 times as strong an acid as the solution with pH = 5.0. A pH of 7.0 represents the same hydronium ion concentration (and hydroxide ion concentration) as pure water, and is a neutral solution, neither acidic nor basic. Solutions with pH values above 7.0 are basic. Each higher value means $[H_3O^+]$ is 10 times less, and that means $[OH^-]$ is 10 times more because the product of the two is always 1.0×10^{-14} for aqueous solutions at room temperature.

Sources differ on the original meaning of the term "pH." The term was coined in 1909 by a chemist in Copenhagen, Denmark. This was in Arrhenius' day, when scientists assumed that an ionizing acid added H^+ ions to a solution. That was before we knew that hydrogen ions didn't exist in water, and that they formed hydronium ions instead. So because the term arose in the time when scientists thought that acidity was all about having hydrogen ions in the water, some say the term derives from one of the French phrases *pouvoire hydrogène* or *piussance hydrogène*, both of which mean "power of the hydrogen." Others speculate that the letters derive from the Latin *pondus hydrogenii* (weight of hydrogen) or *potentia hydrogenii* (hydrogen potential). There are other possibilities as well. The fact is, we

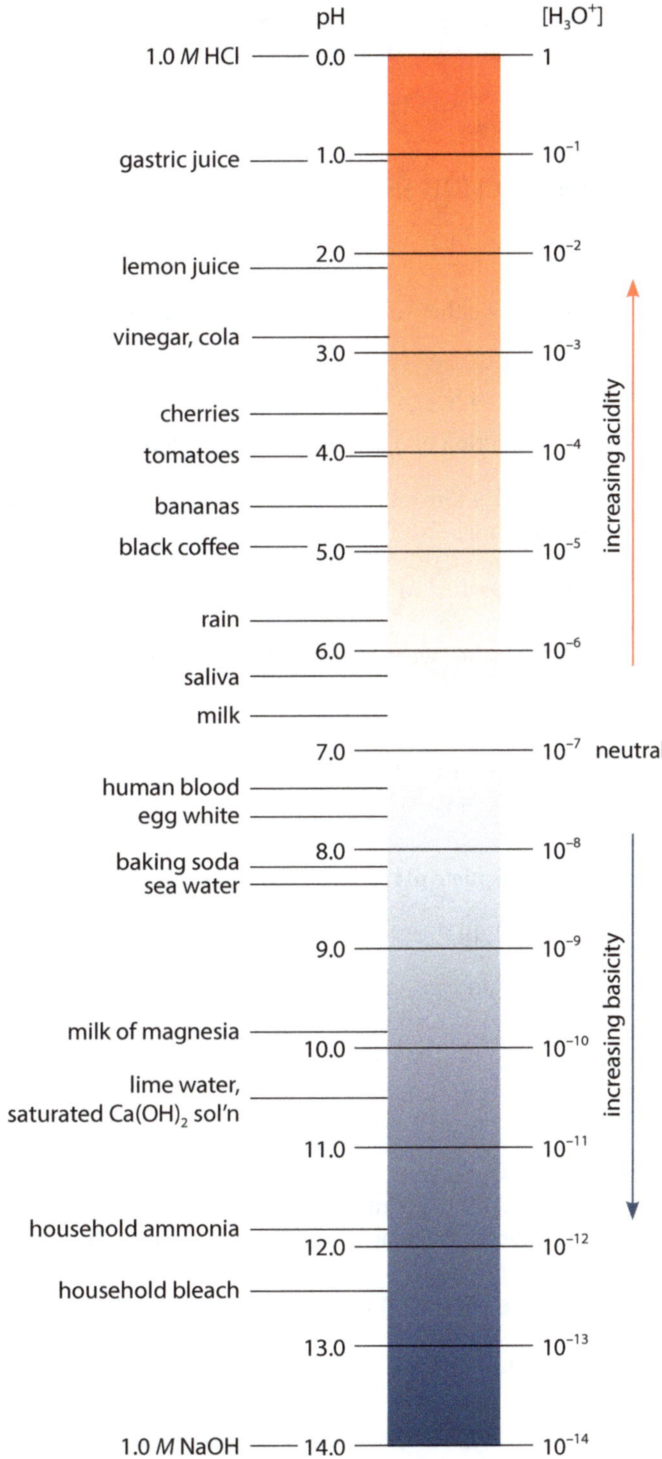

Figure 9.13. The pH scale with approximate values for common substances.

are not sure what the designation pH meant when it was first used. We do know that in its first use in Copenhagen it was written p_H.

In the same way that pH is defined, we can define the quantity pOH as

$$\text{pOH} = -\log\left[\text{OH}^-\right] \tag{9.4}$$

Since $[H_3O^+]\cdot[OH^-] = 1.0 \times 10^{14}$, the rules of logarithms give us

$$\text{pH} + \text{pOH} = 14.0 \tag{9.5}$$

If you know either the pH or the pOH of a solution, you can calculate the other from Equation (9.5).

Example 9.5

Given a solution of 3.0×10^{-3} *M* NaOH, determine $[H_3O^+]$, $[OH^-]$, the pH, and the pOH.

When NaOH ionizes in water, one mole of hydroxide ions are formed for each mole of NaOH in the solution. This means

$$[\text{OH}^-] = 3.0 \times 10^{-3}\ M$$

From this value, we use the ionization constant of water to determine $[H_3O^+]$:

$$[\text{H}_3\text{O}^+]\cdot[\text{OH}^-] = 1.0 \times 10^{-14}\ M^2$$

$$[\text{H}_3\text{O}^+] = \frac{1.0 \times 10^{-14}\ M^2}{[\text{OH}^-]} = \frac{1.0 \times 10^{-14}\ M^2}{3.0 \times 10^{-3}\ M} = 3.3 \times 10^{-12}\ M$$

Now we calculate pH and pOH using the definitions for each shown in Equations (9.3) and (9.4).

$$\text{pH} = -\log[\text{H}_3\text{O}^+] = -\log(3.3 \times 10^{-12}) = 11.48$$

$$\text{pOH} = -\log[\text{OH}^-] = -\log(3.0 \times 10^{-3}) = 2.52$$

Note that pH + pOH = 14.00.

When working in a research laboratory, pH is typically measured with a *pH meter*, illustrated in Figure 9.14. Once the pH is known, the chemist can then calculate the values of $[H_3O^+]$ and $[OH^-]$. Such a calculation requires using the definition of a logarithm discussed previously to rearrange the logarithmic Equations (9.3) and (9.4) into base-10 power functions.

To illustrate, if you know $[H_3O^+]$ and wish to compute the pH, you use Equation (9.3):

$$\text{pH} = -\log\left[\text{H}_3\text{O}^+\right]$$

This is as simple as entering the molar hydronium ion concentration into your calculator and hitting the log*x* button. But now we are talking about the inverse calculation: given the pH, we desire to calculate $[H_3O^+]$.

To do this, we must re-express the logarithmic equation as a base-10 power function. Remember, the expression

$$\log X = Y$$

means the same thing as

$$10^Y = X$$

This is a base-10 power function.[3] In the same way, the equation

$$\text{pH} = -\log\left[\text{H}_3\text{O}^+\right]$$

can be re-expressed as a base-10 power function, as follows:

$$\text{pH} = -\log[\text{H}_3\text{O}^+]$$

$$\log[\text{H}_3\text{O}^+] = -\text{pH}$$

$$10^{-\text{pH}} = [\text{H}_3\text{O}^+]$$

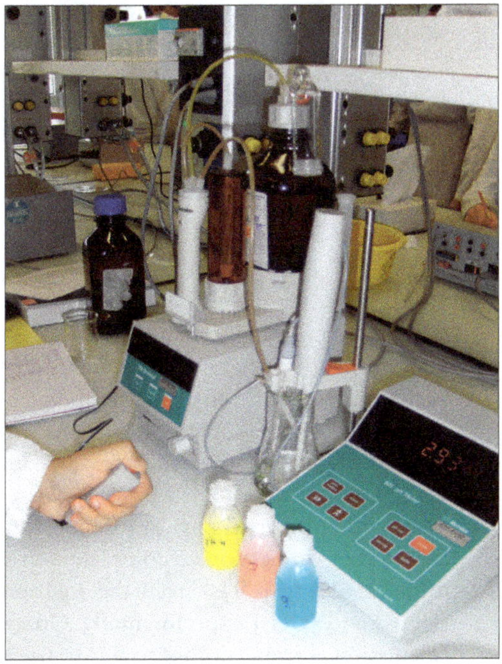

Figure 9.14. Using an electronic laboratory pH meter to measure the pH of various solutions.

This gives us

$$[\text{H}_3\text{O}^+] = 10^{-\text{pH}} \quad (9.6)$$

A similar expression applies for calculating $[\text{OH}^-]$ from the pOH value:

$$[\text{OH}^-] = 10^{-\text{pOH}} \quad (9.7)$$

We now illustrate the use of this technique in the following example.

Example 9.6

A technician measures the pH of a solution with a pH meter and finds that pH = 4.72. Determine the molar hydronium ion concentration and molar hydroxide ion concentration for this solution.

We use the pH to find $[\text{H}_3\text{O}^+]$ using Equation (9.6):

$$[\text{H}_3\text{O}^+] = 10^{-\text{pH}} = 10^{-4.72} = 1.9 \times 10^{-5}\ M$$

Note here that the two digits to the right of the decimal in the pH value allow us two significant digits in the molar hydronium concentration. Now that we have $[\text{H}_3\text{O}^+]$, we use the ionization constant of water (Equation (9.2)) to calculate $[\text{OH}^-]$.

3 The base-10 power function is sometimes referred to as an *antilog*.

$$[H_3O^+]\cdot[OH^-]=1.0\times10^{-14}\ M^2$$

$$[OH^-]=\frac{1.0\times10^{-14}\ M^2}{[H_3O^+]}=\frac{1.0\times10^{-14}\ M^2}{1.9\times10^{-5}\ M}=5.3\times10^{-10}\ M$$

We note that this solution is acidic, since the pH is less than 7.0.

▲

There is a significant reason why this type of calculation is important. If you prepare a solution of a strong acid such as HCl or HNO_3, the molar concentration of hydronium ions can be determined directly from the molarity of the solution. The fact that the acid completely dissociates in water means that the molar hydronium ion concentration is equal to the molarity of the acid (for monoprotic acids). As Example 9.3 shows, alkaline-earth hydroxides dissociate completely, and the molar hydroxide concentration of an alkaline-earth hydroxide is twice the molar concentration of the base.

However, referring again to Table 9.6, most of the acids and bases listed are weak, which means they do not ionize completely in water. Without complete ionization, we cannot determine $[H_3O^+]$ or $[OH^-]$ from the solution molarity. Instead, we must determine the values of $[H_3O^+]$ and $[OH^-]$ experimentally. One way to do this is to measure the pH and perform the calculation above. Another way is to use the technique known as *titration* to determine $[H_3O^+]$ and $[OH^-]$. We consider titration procedure in detail soon. First though, we need to become familiar with the world of pH measurement and pH indicators, our next topic.

9.3.4 pH Measurement, pH Indicators, and Titration

A *pH indicator* is a substance that changes color in solutions of different pH. Some pH indicators, such as *litmus*, exhibit two basic colors. Litmus is an extract from lichens, and is one of the oldest known pH indicators. Illustrated in Figure 9.15, litmus paper turns red when it comes in contact with an acidic solution, and blue when in contact with a basic solution. The reason we associate red with acidity and blue with basicity (see again Table 9.6 and Figure 9.13) is because of litmus paper. Litmus paper has been around a long time.

We have already encountered two other natural pH indicators. The opening page of the chapter illustrates the two colors the hydrangea takes on in acidic and basic soils (and as the caption says, the colors are the opposite of litmus paper). Another natural pH indicator is the juice from red cabbage, which exhibits an entire spectrum of colors across the range of the pH scale, shown in Figure 9.3. A fourth type of indicator, popular now in school science laboratories, is indicator paper that uses various colors to indicate the pH more precisely. One brand of this type of indicator paper is shown in Figure 9.16. Indicator strips like this are convenient to use and inexpensive.

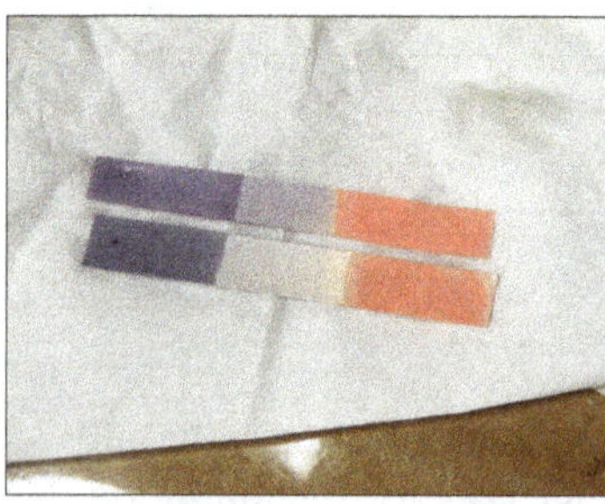

Figure 9.15. Strips of litmus paper. The paper turns red when in contact with an acidic solution, and blue when in contact with a basic solution. In a neutral solution, the paper remains a dull color.

Electronic pH meters are now rapidly rendering paper-strip type indicators obsolete. Figure 9.14 shows an expensive bench pH meter, for use in a research laboratory. Figure 9.17 shows one of the inexpensive portable models available today. pH meters work by inserting two electrodes into the solution and reading the voltage between them. Acids and bases are electrolytes, and the meter uses the voltage measurement to determine the conductivity of the solution. The conductivity depends on ion concentration, and so does the pH.

There are two basic reasons for using a pH indicator. The first is to estimate the pH of a solution. The pH indicator strips shown in Figure 9.16 are suitable for this purpose. But again, inexpensive pH meters are rapidly taking over this task from indicator papers.

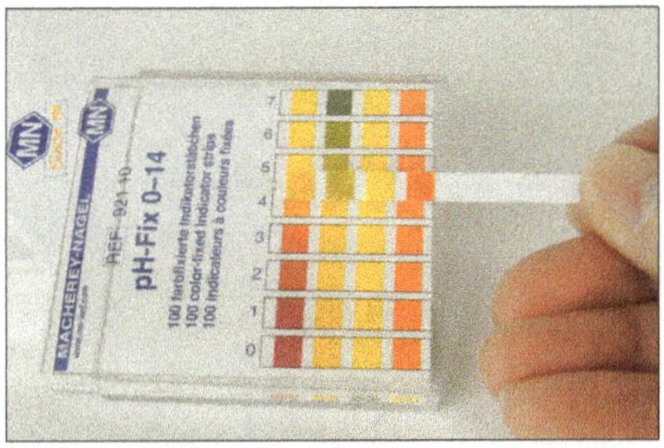

Figure 9.16. With four colored bars on each strip, this indicator paper can give a fairly accurate indication of pH. The strip shown has been dipped in a solution with a pH of 5, or possibly a bit lower as evidence by the color on the left end. The second color spot is much closer to a pH of 5 than a pH of 4.

A second use of indicators arises in a titration, a procedure used to ascertain the molar concentration of an unknown acid or base solution based on the molar quantity of solution of a known molarity required to achieve neutralization. I am reserving the details of titration procedure for Section 9.3.5, but in general, here's how a titration works: during titration, an acid of precisely known concentration, called a *standard solution*, is gradually added to a base of unknown concentration (or vise versa) in a measured fashion until the point is reached when neutralization occurs, indicating that chemically equivalent amounts of the acid and base exist in the solution. From the molar amount of the standard solution and the volumetric amount of the solution of unknown concentration, the molar concentration of the unknown may be calculated.

A liquid indicator in the solution changes color when chemically equivalent amounts of acid and base are present. There are a number of different liquid indicators that may be used for this purpose, and a few of these are shown in Figure 9.18 on the next page. At the top is *bromothymol blue* (also called *bromthymol blue*). This indicator changes color in the pH range of 6.0–7.6. This range is called the indicator's *transition interval.*

The next indicator shown is *bromophenol blue* (also called *bromphenol blue*), with a transition interval in the pH range of 3.0–4.6. The third indicator shown is a solution called *phenolphthalein*, which has its transition interval in the pH range of 8.2–10.0. There are many other indicators as well, with transition intervals in different regions of the pH scale. A few of the most common indicators are listed in Table 9.7.

It may appear to you that the transition intervals for the pH indicators in Figure 9.18 are rather wide, and that an accurate pH measurement from the color would be next to impossible, but that's not really the point. First, in a titration, we are not actually trying to measure the pH; we only want to know when chemically equivalent amounts of acid and base are present in the solution. This point in the titration is called the *equivalence point*. The indicator changes color at a point called its *endpoint*, which we use as an approximation for the equivalence point. Second, when the equivalence point is reached in a titration, the pH and the indicator color both change very fast with only a minute amount of additional acid or base in the solution. So even though the color change takes place over a relatively wide range of pH, this does not really affect the accuracy with which the equivalence point is known. We will return to these matters shortly when we discuss titration procedure.

Figure 9.17. An inexpensive, portable electronic pH meter.

There are three different basic scenarios in a titration, summarized in Table 9.8. These three scenarios depend on whether the acid and base in the titration are strong or weak. As

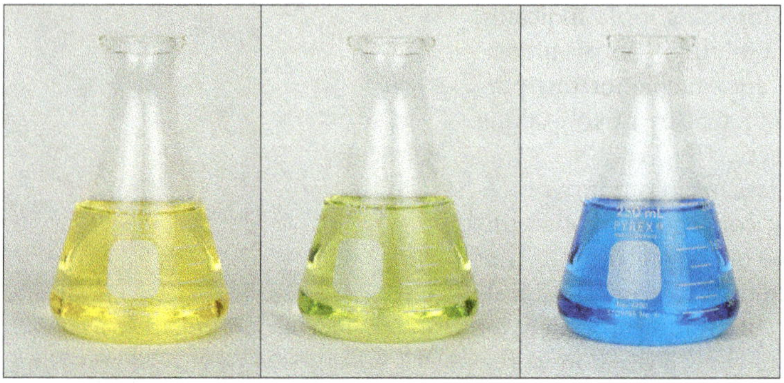

4.5 5.9 8.2

bromothymol blue (transition interval 6.0–7.6)

2.5 3.0 4.0

bromophenol blue (transition interval 3.0–4.6)

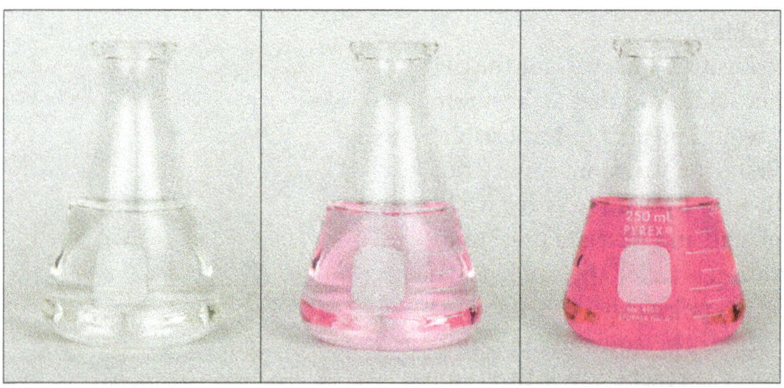

7.9 8.4 9.0

phenolphthalein (transition interval 8.2–10.0)

Figure 9.18. Colors for three pH indicators at different pH values. In each case, colors on the left and right are the colors just below and above the transition. The color in the center is right at the transition. The difference between each photo in each series is a single drop of the standard solution added to the solution being titrated.

before, the terms strong and weak refer to how fully the acid or base ionizes in water, per Table 9.6.

When a strong acid is neutralized with a strong base, the resulting salt is neutral. This means the entire solution is neutral at the equivalence point, which occurs close to pH = 7.0. The indicator used in the titration needs to have its transition interval in the pH region where the equivalence point is expected to occur. Of the three indicators pictured in Figure 9.18, the transition interval for bromothymol blue is in the interval where the equivalence point is expected for a strong acid/strong base titration, so this indicator is an appropriate choice for this titration.

When a strong acid is neutralized by a weak base, the resulting salt solution is slightly acidic.[4] This means that the equivalence point occurs at a pH below 7.0. For this titration, an indicator such as bromophenol blue, methyl red, or malachite green should be used, since their transition intervals are below 7.0. Malachite green has two color transitions, one on the very acidic end of the pH scale and one on the very basic end of the scale.

When a weak acid is neutralized with a strong base, the resulting salt is slightly basic, so the equivalence point occurs above pH 7.0. For a titration like this, an indicator such as phenolphthalein should be used, since its

4 We address the reason for this in Chapter 11.

transition interval is above 7.0. As Table 9.7 shows, malachite green might also be used. Finally, the weak acid/weak base combination is not used for titration because the equivalence point can occur at just about any pH, depending on the specific compounds in solution. The relationships between acid and base strength and their impact on titration are summarized in Table 9.8.

Indicator	Transition Interval
malachite green (first transition)	0.0–2.0
bromophenol blue	3.0–4.6
methyl orange	3.1–4.4
methyl red	4.4–6.2
bromothymol blue	6.0–7.6
phenol red	6.8–8.2
phenolphthalein	8.2–10.0
malachite green (second transition)	11.6–14.0

Table 9.7. pH indicators and their transition intervals.

pH indicators are weak acids or weak bases. In a solution, an indicator partially ionizes and is present in equilibrium. For example, an acidic indicator in solution is represented by this equation:

$$HX \rightleftharpoons H^+ + X^-$$

In this equation, X represents the anion of the indicator (usually a large, complicated molecule). The anion of an indicator like this acts as a Brønsted-Lowry base in solution. If an acid is present, the X^- ions accept protons from the acid to form HX molecules, and thus the indicator is mostly in its non-ionized form in the solution. In this form, the indicator takes on its acid-indicating color, illustrated in Figure 9.19. In a basic solution, there are more OH^- ions present than H^+ ions, prompting the indicator to ionize and give up its proton. In this case, the indicator is present in its ionized form, and the solution takes on a different color.

Titration	Resulting Salt	Example Indicator
strong acid—strong base	neutral	bromothymol blue
strong acid—weak base	acidic	bromophenol blue
weak acid—strong base	basic	phenolphthalein

Table 9.8. The three types of titrations lead to different types of salts, and require different pH indicators.

As mentioned in the caption to Figure 9.18, the color transition at the titration end point occurs with the addition of a single drop or so of the standard solution. The addition of another drop drives the color change to completion. Again, the endpoint in a titration is when the color change occurs, and this is used as an estimate for the equivalence point—the moment when chemically equivalent amounts of acid and base are present in the solution.

As the equivalence point is approached, the pH of the combined solution during a titration becomes very sensitive to additional amounts of the standard solution. Figure 9.20 illustrates how the pH changes when a strong acid of unknown concentration is titrated with a standard solution of a strong base. At first the pH increases very gradually. Then addition of smaller and smaller quantities of

Figure 9.19. The color of a pH indicator in solution depends on whether the indicator is in its non-ionized form or its ionized form.

the base causes the pH to increase by larger and larger amounts. Then when the equivalence point is reached, just a drop of the base increases the pH dramatically and initiates the color change of the indicator. One more drop of the standard solution again increases the pH dramatically, and the color transition of the indicator goes to completion. (See the pH values listed under the colored solutions in Figure 9.18.)

The fact that the pH of the combined solution is so sensitive at the equivalence point allows the equivalence point to be determined with high precision. The pH indicator in the solution is a critical factor and provides the experimenter with a visual notification of when the equivalence point is being approached and when it has been achieved.

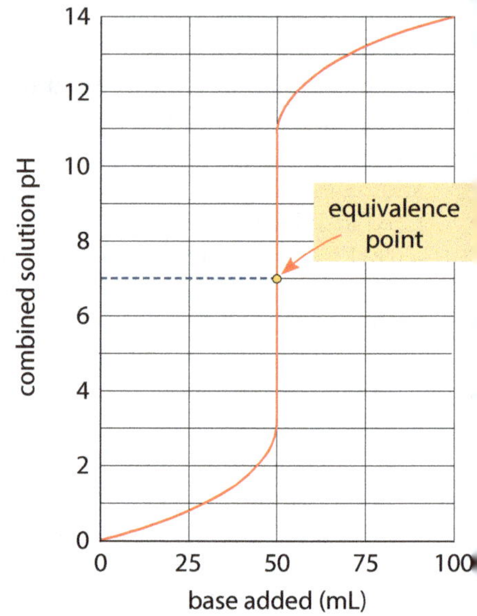

Figure 9.20. Change in the pH of the combined solution as a strong acid of unknown concentration is titrated with a standard solution of strong base.

9.3.5 Titration Procedure

Titration makes use of an item of laboratory apparatus called a *buret*, pictured in Figure 9.21. Two burets are required, along with a stand, a dual buret clamp, and an *Erlenmeyer flask*.

Before the titration, a standard solution of precisely known molarity must be prepared. This is done by first preparing a volume of the standard solution with the desired concentration. The standard solution is then titrated with a *primary standard*, a solid, highly purified compound that is used to determine the precise concentration of the standard solution. Then the standard solution is ready to be used for titrating the solution of unknown concentration.

The procedure for titration is described below. This procedure describes the titration of an acid solution of unknown concentration with a basic standard solution.

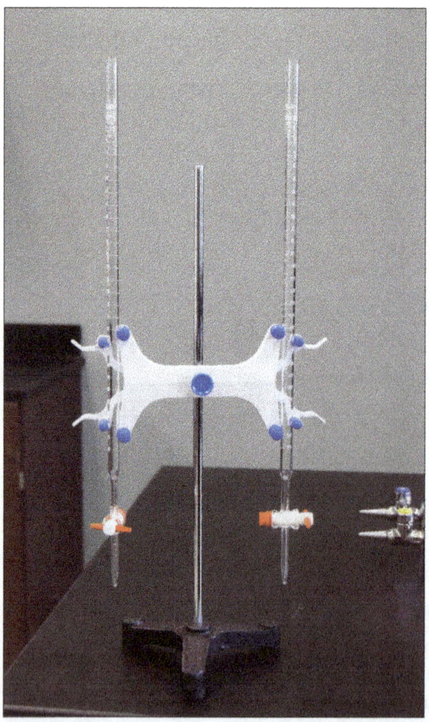

Figure 9.21. Dual buret setup for titration.

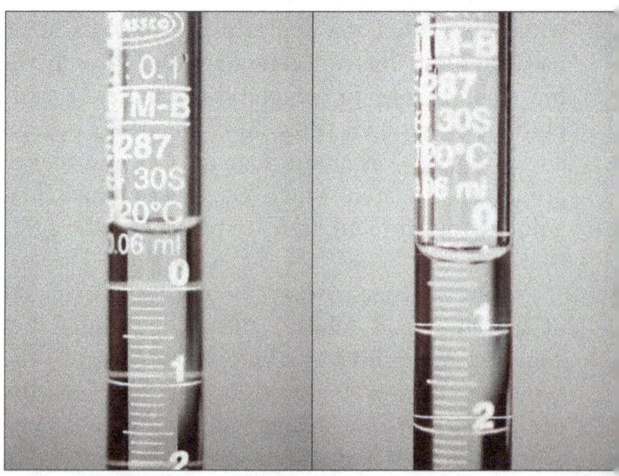

Figure 9.22. Filling the acid above the calibrated scale (left), then releasing acid until the liquid is within the calibrated scale (right).

1. Set up the stand with two clean burets and assign one of the burets to hold the acid and the other to hold the base. Rinse the acid buret three times with the acid solution to be titrated. Rinse the base buret three times with the base standard solution. Note that these rinses cannot be performed with water because even a few drops of water remaining in the buret compromises the accuracy of the titration.
2. Fill the acid buret to above the calibrated scale with the solution to be titrated. To eliminate any air bubbles in the stopcock, release some acid into a waste beaker until the top of the liquid falls down into the region of the calibrated scale on the buret. Record the reading in the buret to the nearest 0.01 mL. (Remember to read the buret at the bottom of the meniscus.) These steps are illustrated in Figure 9.22.
3. Release acid into a clean, dry Erlenmeyer flask. Your instructor or lab procedure will indicate the approximate volume of acid to use. Record the new buret reading to the nearest 0.01 mL. The difference between the two buret readings is the precise amount of acid solution in the flask.
4. Add three drops of the pH indicator to the flask. The rest of this description assumes phenolphthalein is being used.
5. Fill the base buret with the standard solution to above the calibrated scale. To remove air bubbles from the stopcock, release some base into a waste beaker until the buret level falls to within the calibrated scale. Record the buret reading to the nearest 0.01 mL.
6. Place the Erlenmeyer flask containing the acid solution under the base buret. Adjust the height of the buret so the tip of the buret extends slightly into the flask, as shown in Figure 9.23.
7. Slowly release standard base solution into the flask, swirling the flask as you do, as illustrated in Figure 9.24. As the concentration of base in the flask increases, flares of pink color

Figure 9.23. The tip of the buret should extend slightly into the flask.

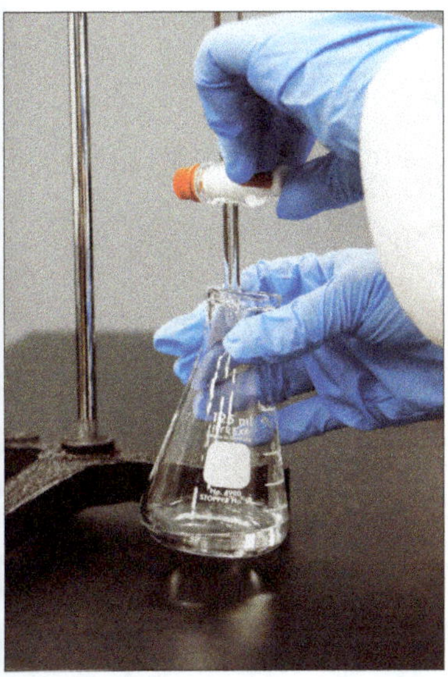

Figure 9.24. Titration.

Figure 9.25. Flares of pink color begin lasting for a second or two when the indicator endpoint is approaching.

Figure 9.26. A single drop of the standard solution eventually turns the indicator pink, indicating that the equivalence point has been reached.

appear where the base solution enters the titration solution. The colored liquid dissipates and becomes colorless as the flask is swirled.

8. When the pink color begins to stay in the solution for a second or two before disappearing, the endpoint is being approached, shown in Figure 9.25. After this, begin adding the standard base slowly, drop by drop.
9. When a single drop of base causes the solution to become light pink and stay that way even after 30 seconds of swirling, the endpoint has been reached, as shown in Figure 9.26. Record the reading on the standard base buret to the nearest 0.01 mL. The difference between the initial and final base buret readings is the amount of standard solution required to neutralize the volume of acid solution in the flask.

9.3.6 Determining $[H_3O^+]$ or $[OH^-]$ from Titration Data

Determining the concentration of the solution of unknown concentration with titration data is essentially a stoichiometric calculation. The starting point is the chemical equation showing the mole ratios in the reaction. For example, if titrating HCl with NaOH, the equation is

$$\text{HCl}(aq) + \text{NaOH}(aq) \rightarrow \text{NaCl}(aq) + \text{H}_2\text{O}(l)$$

The coefficients on the HCl and NaOH in this equation tell us that 1 mol HCl neutralizes 1 mol NaOH. Thus, chemically equivalent amounts of acid and base are equal mole quantities.

Next, determine the number of moles of standard solution present at the titration endpoint. For example, if 24.50 mL are used of a standard solution of 1.100×10^{-3} *M* NaOH, then the mole quantity of NaOH is

$$24.50\ \text{mL} \cdot \frac{1\ \text{L}}{1000\ \text{mL}} \cdot \frac{1.100 \times 10^{-3}\ \text{mol}}{\text{L}} = 0.02695\ \text{mol}$$

Since the mole ratio of HCl to NaOH is 1 : 1, there are 0.02695 mol HCl in the solution as well. Now supposing that 21.00 mL HCl are initially released into the flask, the concentration of the acid solution is computed as

$$\frac{0.02695\ \text{mol}}{21.00\ \text{mL}} \cdot \frac{1000\ \text{mL}}{\text{L}} = 1.283\ M$$

The following example illustrates a case when the mole ratios of $[OH^-]$ and $[H_3O^+]$ in the titration solution are not the same.

Example 9.7

Suppose 22.73 mL of a standard solution of 0.0995 *M* $Sr(OH)_2$ are required to neutralize 23.21 mL of HNO_3 solution. Determine the concentration of the HNO_3 solution.

Recall from Example 9.3 that the three soluble alkaline-earth hydroxides each dissociate completely in water. For $Sr(OH)_2$, there are two OH^- ions in each formula unit of $Sr(OH)_2$, and thus the molarity of the OH^- is twice the molarity of the $Sr(OH)_2$. From this information we know that

$$[\text{OH}^-] = 2 \cdot 0.0995\ M = 0.199\ M$$

Using this with the volume given for the standard solution, we find the number of moles of OH^- to be

$$22.73 \text{ mL } Sr(OH)_2 \cdot \frac{1 \text{ L}}{1000 \text{ mL}} \cdot \frac{0.199 \text{ mol } OH^-}{\text{L } Sr(OH)_2} = 4.523 \times 10^{-3} \text{ mol } OH^-$$

The ionic equation for this neutralization reaction is

$$2HNO_3(aq) + Sr(OH)_2(aq) + 2H_2O(l)$$
$$\rightarrow 2NO_3^-(aq) + Sr^{2+}(aq) + 2OH^-(aq) + 2H_3O^+(aq)$$

This equation indicates that the mole ratio of hydroxide ions to hydronium ions is 2 : 2, which is the same as 1 : 1. This is also the mole ratio of OH^- to HNO_3. Thus, the number of moles of HNO_3 is 4.523×10^{-3} mol.

Now we calculate the HNO_3 concentration with the number of moles and the volume:

$$\frac{4.523 \times 10^{-3} \text{ mol } HNO_3}{23.21 \text{ mL } HNO_3} \cdot \frac{1000 \text{ mL}}{\text{L}} = 0.195\ M$$

▲

Chapter 9 Exercises

SECTION 9.1

1. List the general properties of acids and bases.
2. What properties do acids and bases have in common?
3. Distinguish between oxyacids and carboxylic acids.
4. Name the following acids, and identify each as monoprotic, diprotic, or triprotic.

   a. HF
   b. HClO
   c. HNO_2
   d. H_2SO_3
   e. HBr
   f. H_3PO_4
   g. H_2S
   h. H_2CrO_4
   i. CH_3COOH

5. Is formic acid a binary acid? Explain your response.
6. Write the formulas for the following acids:

   a. chlorous acid
   b. perbromic acid
   c. hydrofluoric acid
   d. hydroiodic acid

7. Describe three industrial uses for three different major industrial acids.
8. Draw the Lewis structures for formic acid and benzoic acid.
9. Explain why acetic acid is not a "tetraprotic" acid.

SECTION 9.2

10. Describe the three acid-base theories in detail.

11. Give examples of two acids and two bases according to each acid-base theory. For each one, write a short explanation of why the species acts as you have described under the given theory.
12. Explain why water is amphiprotic under the Brønsted-Lowry acid-base model.
13. For the reactions shown, identify the Brønsted-Lowry acid-base conjugate pairs.

a. $HNO_2(aq) + CN^-(aq) \rightleftharpoons NO_2^-(aq) + HCN(aq)$

b. $HCO_3^-(aq) + H_2O(l) \rightleftharpoons H_2CO_3(aq) + OH^-(aq)$

c. $HSO_4^-(aq) + C_2O_4^{2-}(aq) \rightleftharpoons SO_4^{2-}(aq) + HC_2O_4^-(aq)$

d. $NH_4^+(aq) + HPO_4^{2-}(aq) \rightleftharpoons NH_3(aq) + H_2PO_4^-(aq)$

14. Explain why ammonia is classified as a base under all three acid-base theories.
15. Explain the factor that determines whether an acid or base is classified as strong or weak.
16. Explain why the chlorine oxyacids get more acidic as the number of oxygen atoms in the molecule increases.
17. Based on your knowledge of molecular bonding, formulate a reason for the stepwise ionization and subsequent weakening of the acid for polyprotic acids.
18. Sodium carbonate is an important industrial base, even though it contains no hydroxide anion. Write an equation that demonstrates the character of sodium carbonate as a base.
19. For each of the following acid-base neutralization reactions, complete the formula equation and balance it. Then write the ionic equation and the net ionic equation.

a. $HCl(aq) + KOH(aq) \rightarrow$

b. $HNO_3(aq) + NaOH(aq) \rightarrow$

c. $H_2SO_4(aq) + Ca(OH)_2(aq) \rightarrow$

d. $H_2SO_3(aq) + KOH(aq) \rightarrow$

20. In each of the following reactions, an acid reacts with a metal to form a salt, releasing hydrogen gas. Complete the formula equation and balance it. Then write the ionic equation and the net ionic equation.

a. $Al(s) + H_2SO_4(aq) \rightarrow$

b. $Zn(s) + HCl(aq) \rightarrow$

21. In each of the following reactions, an acid reacts with a carbonate base. Complete the formula equation and balance it.

a. $HNO_3(aq) + CaCO_3(aq) \rightarrow$

b. $HCl(aq) + BaCO_3(aq) \rightarrow$

c. $H_2SO_4(aq) + Na_2CO_3(aq) \rightarrow$

d. $HNO_3(aq) + MgCO_3(aq) \rightarrow$

22. For each acid-base reaction given, complete the proton-transfer equation for the transfer of a single proton. Then, using Table 9.6, predict the direction in which the equilibrium is favored.

a. $H_3PO_4(aq) + F^-(aq) \rightarrow$

b. $HSO_4^-(aq) + CO_3^{2-}(aq) \rightarrow$

c. $ClO_2^-(aq) + H_3O^+(aq) \rightarrow$

d. $HClO_3(aq) + HCO_3^-(aq) \rightarrow$

e. $H_3O^+(aq) + HS^-(aq) \rightarrow$

f. $HPO_4^{2-}(aq) + NH_3(aq) \rightarrow$

23. A 196-g sample of sodium carbonate is dissolved in nitric acid. Assuming excess acid, determine the following:

a. the mass and volume at STP of dry CO_2 gas produced by the reaction.

b. the volume of 6.00 *M* nitric acid required for this reaction.

24. According to the activity series of metals (Table 5.2), copper does not react with sulfuric acid. However, if the acid is hot enough and concentrated enough, copper reacts with H_2SO_4 in a single-replacement reaction. If excess copper reacts with 125 mL of 8.0 *M* H_2SO_4, determine the following:

a. the mass in grams of copper sulfate produced.

b. the volume in liters of hydrogen gas released at STP.

25. Some forms of marble are made of calcium carbonate. In a reaction with hydrochloric acid, $CaCl_2$, CO_2, and H_2O are produced. Determine the following:

a. the mass of marble required to produce 2.5×10^4 mL CO_2 at STP.

b. the corresponding volume of 2.00 *M* HCl required.

SECTION 9.3

26. Explain the significance of the small value of the ionization constant of water, K_w, and relate this to the electrical conductivity of water.

27. Explain what the pH of a solution indicates. Select three different pH values to use as examples in your explanation.

28. For each of the following, indicate whether the solution is acidic, basic, or neutral.

a. $[OH^-] = 1.0 \times 10^{-4}\ M$

b. $[H_3O^+] = 1.0 \times 10^{-6}\ M$

c. $[H_3O^+] = 1.0 \times 10^{-9}\ M$

d. $[H_3O^+] = [OH^-]$

e. $pH = 4.91$

f. $[H_3O^+] < [OH^-]$

g. $pOH = 4.91$

h. $[OH^-] = 1.0 \times 10^{-7}\ M$

29. Develop a mathematical explanation of why a very small amount of standard solution can cause the pH in a titrated solution to change dramatically near the equivalence point. (Refer to Figure 9.20.)

30. Distinguish between equivalence point and endpoint.

31. There are four possible combinations of strong/weak acid/base that could be used in a titration. For each combination, state in general what the expected pH is at the equivalence point, and suggest an appropriate pH indicator.

32. Calculate $[H_3O^+]$ and $[OH^-]$ for each of the following:

a. 1.56×10^{-3} *M* $HClO_4$

b. 0.0110 *M* $Sr(OH)_2$

c. 2.50 *M* NaOH

d. 8.911×10^{-5} *M* $HClO_3$

33. Calculate the pH and pOH for each of the following:

a. 0.0250 *M* NaOH

b. 6.0 *M* HNO_3

c. 12 *M* HCl

d. 2.03×10^{-2} *M* $Ba(OH)_2$

34. For each of the following values, determine $[H_3O^+]$ and $[OH^-]$.

a. pH = 1.00

b. pH = 3.5

c. pOH = 10.0

d. pH = 11.88

e. pOH = 2.1

f. pOH = 7.0

g. pOH = 4.00

h. pH = 14.0

35. In each of the following items, one of these values is given: pH, pOH, $[H_3O^+]$, or $[OH^-]$. In each case, calculate the other three.

a. pH = 5.6

b. $[OH^-] = 5.5 \times 10^{-10}$ *M*

c. $[H_3O^+] = 1.0 \times 10^{-9}$ *M*

d. pOH = 11.5

e. $[H_3O^+] = 4.5 \times 10^{-3}$ *M*

f. pOH = 1.00

g. pOH = 7.22

h. pH = 13.50

i. pH = 2.0

j. $[OH^-] = 3.2 \times 10^{-2}$ *M*

36. Listed below are several acid-base combinations that could be used in a titration. Determine the molarity of the first substance listed that is chemically equivalent with the amount given for the second substance.

a. 18.55 mL HNO_3 with 24.05 mL of 1.65×10^{-3} *M* NaOH

b. 23.78 mL KOH with 21.00 mL of 1.500 *M* $HClO_3$

c. 22.94 mL $Ca(OH)_2$ with 20.87 mL of 4.077×10^{-2} *M* $HClO_4$

d. 22.00 mL HCl with 20.05 mL of 3.050 *M* LiOH

37. A chemist uses 23.09 mL of a 5.06×10^{-3} *M* standard solution of HCl to neutralize 20.07 mL of KOH. Determine the molar concentration of the KOH solution.

38. In a titration, a student adds 22.01 mL of 0.0231 *M* $Ba(OH)_2$ to 27.0 mL of an HCl solution of unknown concentration. Determine the molarity of the acid solution.

39. A 2.5×10^{-3} *M* standard solution of NaOH is used to titrate a 0.04098 *M* standard solution of a monoprotic weak acid. If 23.90 mL of the base are required to neutralize 20.00 mL of the acid, what is the percent ionization of the acid?

GENERAL REVIEW EXERCISES

40. The solubility of strontium hydroxide is 1.77 g $Sr(OH)_2$/100 mL water. With this information, verify that it is possible to prepare the $Sr(OH)_2$ solution referred to in Example 9.7.

41. A solution is prepared by dissolving 38.95 g sucrose, $C_{12}H_{22}O_{11}$, in enough water to bring the solution volume to 250.0 mL. Determine the molar concentration of the solution.

42. An ice cube with a mass of 37.05 g is in equilibrium at −18.0°C. Determine the amount of heat required to raise the ice cube's temperature to 0°C, melt it, and warm the resulting water to 20.0°C. (The properties of water are in Appendix B.)

43. An aqueous solution is prepared with 35.55 g NaOH. To this is added 157.0 mL of 4.777 *M* HNO_3.

    a. Determine the limiting reactant.

    b. Determine the amount of sodium nitrate produced.

    c. Determine the amount of water produced.

    d. Determine how many moles of the excess reactant remain unconsumed by the reaction.

44. Draw the Lewis structures for the following molecules and polyatomic ions:

a. CCl_4	b. CO_3^{2-}	c. NO_2^-	d. C_2H_2
e. O_3	f. H_2S	g. CO_2	h. H_3O^+

45. Aqueous solutions of sodium carbonate and barium nitrate are combined. The sodium carbonate solution is 0.050 molar and 25.50 mL of it are used.

    a. Write the balanced molecular equation, the ionic equation, and the net ionic equation.

    b. Identify the spectator ions and the precipitate.

    c. If all the sodium carbonate is used in the reaction, determine the mass of the precipitate that forms.

46. Gaseous CO_2 and HCl are allowed to effuse through a pinhole at identical conditions. If the rate of effusion of the CO_2 is 2.1×10^{-3} mol/s, what is the rate of effusion of HCl?

47. When copper reacts with sulfuric acid, the reaction products are copper sulfate and hydrogen gas. Assume that 109 mL of H_2 is collected over water at a barometric pressure of 765 Torr and a temperature of 25°C. Determine how many grams of copper are consumed in the reaction.

48. A 2.500-L tank contains oxygen gas at 22.0°C and 13.11 atm.

    a. Determine the number of moles of gas in the tank.

    b. Determine the mass of the gas.

49. A 25.00-g sample of iron metal is placed in an aqueous solution containing 35.8 g $CuSO_4$.

    a. Determine the limiting reactant.

    b. Determine the mass of iron(II) sulfate produced.

    c. Determine the mass of the excess reactant that remains.

50. At 30.0°C, a 2.00-L container contains a gas sample with a mass of 8.755 g at a pressure of 1.05 bar. Determine the molar mass of the gas.

51. Determine the energy in electron volts of an X-ray photon with a wavelength of 0.22 nm.

52. A certain unknown gas is found to effuse at a rate 1.91 times slower than the effusion rate of methane, CH_4. Determine the molar mass of the unknown gas.

Chapter 10

Thermochemistry and Kinetics

In the 19th century, physicists learned how to quantify the amount of heat released or absorbed in chemical reactions, such as in the beautiful exothermic reaction pictured above. For reactions that occur at constant pressure like the one pictured above, this quantity of heat is called the *enthalpy of reaction*. In the 1870s, American scientist Josiah Willard Gibbs combined enthalpy, entropy, and temperature together in an equation that defines a quantity called the *Gibbs free energy*. The Gibbs free energy is a powerful theoretical tool used to predict if reactions will occur at stated conditions.

After studying this chapter, you will be able to look at the reaction pictured above and state whether the Gibbs free energy for the reaction is positive or negative.

Objectives for Chapter 10

After studying this chapter and completing the exercises, you should be able to do each of the following tasks, using supporting terms and principles as necessary.

SECTION 10.1

1. Define *enthalpy of reaction* and relate this term to the energy pathways for exothermic and endothermic reactions.
2. Interpret enthalpy diagrams and thermochemical equations.
3. Use the *First Law of Thermodynamics* to distinguish between *enthalpy* and *energy*.
4. Define *enthalpy of combustion* and *enthalpy of formation*.

SECTION 10.2

5. State Hess's law.
6. Use enthalpy of combustion and enthalpy of formation data with the principle of Hess's law to calculate the enthalpy of reaction and enthalpy of formation for given reactions.

SECTION 10.3

7. Describe the relationships between entropy and temperature, entropy and phase, and entropy and the number of moles or particles of a substance.
8. Define the *Gibbs free energy*.
9. Outline the four possibilities for ΔH and ΔS in a chemical reaction, describe the resulting possibilities for ΔG, and describe the conditions under which a reaction occurs spontaneously.
10. Calculate the Gibbs free energy two different ways and use the result to determine if a reaction occurs spontaneously.

SECTION 10.4

11. Describe collision theory and identify the two requirements for a collision between molecules of reactants to be effective in leading to the formation of product species.
12. Describe four factors scientists can control that influence the rate of reactions.
13. Explain the related concepts of *reaction mechanism* and *intermediates*.
14. Explain the related concepts of *activation energy* and the *activated complex*.
15. Using the Haber-Bosch process as an example, explain how metal catalysts work.
16. Distinguish between heterogeneous and homogeneous reactions.
17. Given a reaction mechanism with reaction speeds labeled, identify the rate-determining step and determine the rate law for the reaction.
18. Given a rate law, determine the effect on the reaction rate of altering the concentrations of reactants by specific proportions.
19. Explain how an experimentally determined rate law can provide clues to help chemists determine a reaction's mechanism.
20. Identify the order of a rate law and the order of a reaction with respect to the concentrations in the rate law.

10.1 Energy in Chemical Reactions

10.1.1 Introduction

Energy plays a huge role in chemical reactions. The purpose for many chemical reactions is specifically to deliver energy. Examples of reactions like this are metabolism in animals, electrochemical reactions in batteries to power electronic devices, and burning of fuels to power machines. Figure 10.1 shows the test firing of one of the Space Shuttle main engines. The fuel for this engine is liquid hydrogen and liquid oxygen, which means there is nothing coming out of this engine except water vapor and heat. And yet there is enough energy released by this reaction to power a vehicle weighing over 2,000 tons.

Figure 10.1. The chemical reaction between liquid hydrogen and liquid oxygen is the energy source used by the Space Shuttle main engine. The white formation under the engine at the bottom is water vapor that has condensed to liquid droplets, allowing the water vapor to become visible.

Throughout this text, we have seen that one of the major principles of nature is that processes tend to go in directions that minimize energy. But in Section 8.2.1 we saw that some processes, such as the dissolution of an ionic crystal, entail an energy increase. The melting of an ice cube is another example of a process that spontaneously goes in a direction leading to higher particle energy.

In Section 8.2.3, we saw that the reason such processes can occur is that there is another factor involved in natural processes besides energy—entropy. The entropy of a system is a measure of the disorder present in the system. Just as spontaneous processes tend to go in a direction that minimizes energy, they also tend to go in a direction that maximizes entropy. Several of these examples are discussed in Section 8.2.3. Together, the energy and the entropy pathways open to a substance or combination of substances determine what the substances do. Applying these topics to the study of chemical reactions is the subject of a branch of chemistry known as *thermochemistry*. A tour through the basic principles of thermochemistry will occupy us for the first three sections of this chapter.

We begin with a study of energy in reactions. In Section 10.3.3, we come back to the topic of entropy, and we combine energy and entropy together into a single reaction-governing principle known as the *Gibbs free energy*. Before proceeding further in this chapter, you may wish to reread the paragraphs in Section 8.2.3 on entropy.

10.1.2 Enthalpy

Recall from Section 8.2.2 that the *enthalpy* in a chemical system is related to the heat flow that occurs during chemical processes. The net amount of heat that flows into or out of a chemical system during a reaction at constant pressure is called the *change in enthalpy*, ΔH. By change in enthalpy we mean the difference between the enthalpy of the reaction products and the enthalpy of the reactants, as shown in the following equation:

$$\Delta H = H_{\text{products}} - H_{\text{reactants}} \tag{10.1}$$

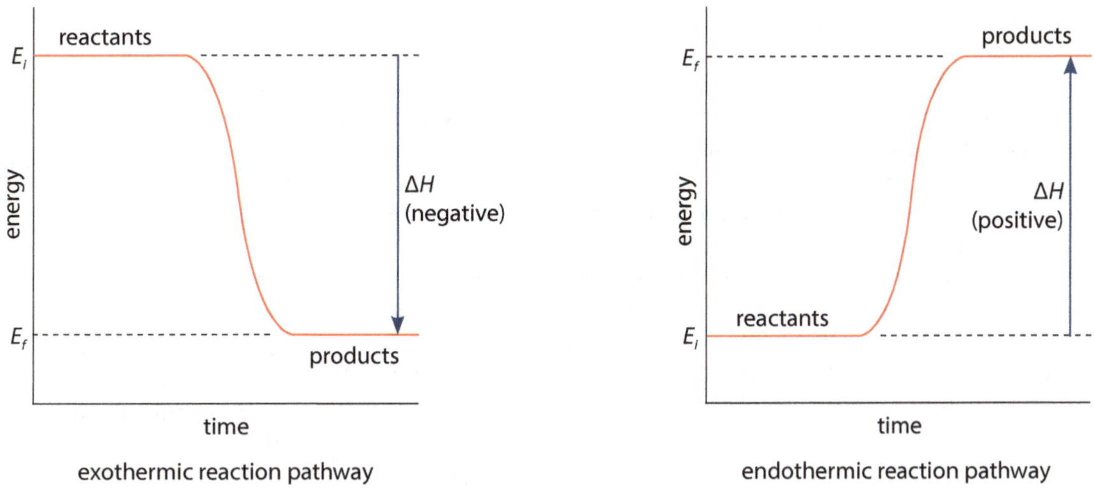

Figure 10.2. Energy pathways for exothermic and endothermic reactions.

The change in enthalpy represents the heat that flows into (is absorbed by) or out of (is released from) a system during a process occurring at a constant pressure. During any process, the change in enthalpy is either positive or negative, as illustrated in Figure 10.2. In this figure, the red curves illustrate the *energy pathway* during the course of a reaction. As the reaction proceeds, the substances involved either release heat or absorb heat.

As shown on the left side of the figure, if ΔH is negative, heat is released during the process, and the process is exothermic. In this case, the energy in the reaction products, E_f, is less than the energy in the reactants, E_i. For processes at constant pressure, this change in energy is equal to the change in enthalpy, ΔH. If ΔH is positive, as shown on the right, heat is absorbed during the process and the process is endothermic. In this case, the chemical potential energy in the reaction products is higher than the energy in the reactants.

The enthalpy, H, of the reactants or products in a reaction cannot be measured directly. Only changes in enthalpy can be measured. As mentioned in Section 8.2.2, we are not going to concern ourselves too much with what enthalpy itself is, although in Section 10.1.3 we briefly explore how enthalpy relates to the internal energy in a system and the ways energy can flow into or out of a system. Our main concern here is with what the change in enthalpy represents.

The general term for the change in enthalpy during a chemical reaction is *enthalpy of reaction*, also called the *heat of reaction*. The enthalpy of reaction is often listed with a chemical equation to form a *thermochemical equation*. For example, we have regularly encountered the reaction between hydrogen gas and oxygen gas to form water vapor:

$$2H_2(g) + O_2(g) \rightarrow 2H_2O(g)$$

In this reaction, two moles of hydrogen react with one mole of oxygen to produce two moles of water vapor, and in the process 483.6 kJ of heat are released. We write the thermochemical equation this way:

$$2H_2(g) + O_2(g) \rightarrow 2H_2O(g) \qquad \Delta H = -483.6 \text{ kJ} \qquad (10.2)$$

Again, ΔH is negative because the energy in the reaction products is less than the energy in the reactants and heat is released during the reaction. The information in Equation (10.2) can also be represented on an *enthalpy diagram*, as shown in Figure 10.3. The horizontal lines in this diagram indicate the enthalpies of the reactants and the products. The reaction goes in the direction of the red arrow, releasing 483.6 kJ of heat at constant pressure.

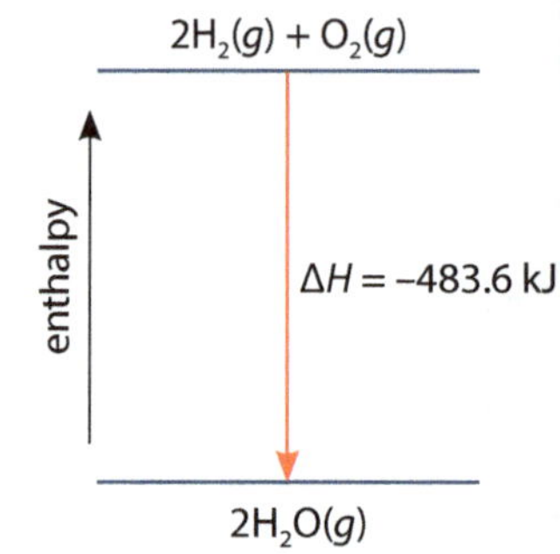

Figure 10.3. Enthalpy diagram showing the enthalpy of reaction for the combustion of hydrogen.

Several comments are in order about thermochemical equations such as Equation (10.2). First, enthalpy is an extensive property. This means that ΔH depends on the amount of the substances present in the reaction. For this reason, when interpreting thermochemical equations we always take the coefficients in the equation to represent numbers of moles rather than numbers of atoms. In fact, sometimes fractional coefficients are used in thermochemical equations. (We see why in Section 10.2.1.) Thus, for the reaction of hydrogen and oxygen involving half as many moles as shown in Equation (10.2), we write the thermochemical equation this way:

$$H_2(g) + \frac{1}{2}O_2(g) \rightarrow H_2O(g) \qquad \Delta H = -241.8 \text{ kJ}$$

When half as many moles are involved in the reaction, the enthalpy of reaction is half as much.

Second, it is conventional to state enthalpy values in thermochemical equations at 25°C (and, again, constant pressure). This means that the reactants begin at 25°C and after the reaction are returned to 25°C. Equation (10.2) is written according to this convention, even though the H_2O is shown as a vapor in the equation. As written, the equation states that when the reaction occurs and the H_2O vapor is returned to 25°C, the total energy released is 483.6 kJ.

Third, as you can see from the previous comment, it is important always to indicate the states for each reactant and product in thermochemical equations. In the hydrogen reaction shown in Equation (10.2), the enthalpy of reaction is different if H_2O is in the liquid state instead of the vapor state shown in the equation. This is because energy must be removed from H_2O vapor in order to condense it to liquid. At 25°C, the molar heat of vaporization for water is 44.0 kJ/mol, so to condense two moles of H_2O, 88.0 kJ of heat must be removed. The thermochemistry of this condensation can be represented as

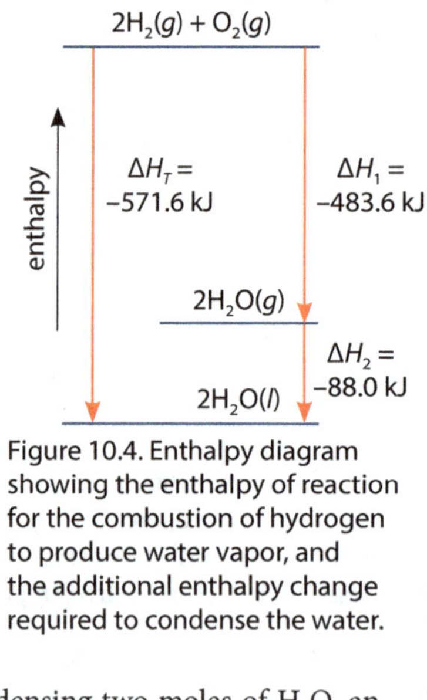

Figure 10.4. Enthalpy diagram showing the enthalpy of reaction for the combustion of hydrogen to produce water vapor, and the additional enthalpy change required to condense the water.

$$2H_2O(g) \rightarrow 2H_2O(l) \qquad \Delta H = -88.0 \text{ kJ}$$

So then, to write Equation (10.2) showing water in the liquid state, we must include the enthalpy change associated with condensing two moles of H_2O, an additional decrease of 88.0 kJ. Including this quantity of heat in ΔH gives:

$$2H_2(g) + O_2(g) \rightarrow 2H_2O(l) \qquad \Delta H = -571.6 \text{ kJ} \qquad (10.3)$$

The combined process of combustion and condensation is represented in the enthalpy diagram of Figure 10.4. In this diagram, the enthalpy change in the combustion step is labeled ΔH_1 and the enthalpy change for the condensation step is labeled ΔH_2. The total enthalpy change is denoted by ΔH_T.

Finally, thermochemical equations are often written in reverse, as we see in Section 10.2.1. When we do this, the enthalpy change for the reverse reaction has the opposite sign. For example, the thermochemical equation for the combustion of methane is:

$$CH_4(g) + 2O_2(g) \rightarrow CO_2(g) + 2H_2O(l) \qquad \Delta H = -890.8 \text{ kJ} \qquad (10.4)$$

This is an exothermic reaction that releases 890.8 kJ of heat for each mole of methane burned. If we imagine the reverse reaction of combining water and carbon dioxide to form methane, we have

$$CO_2(g) + 2H_2O(l) \rightarrow CH_4(g) + 2O_2(g) \qquad \Delta H = 890.8 \text{ kJ}$$

This reverse reaction is endothermic, and the same amount of energy has to be supplied as is released in the forward reaction. Accordingly, the sign on ΔH is reversed to show that energy is absorbed by the reactants. ΔH is now a positive quantity, as it always is for an endothermic reaction.

Example 10.1

Hydrogen peroxide, H_2O_2, is a compound that decomposes to water and oxygen according to the following reaction:

$$2H_2O_2(l) \rightarrow 2H_2O(l) + O_2(g) \qquad \Delta H = -196 \text{ kJ}$$

Answer the following questions about this process:

1. Is this process exothermic or endothermic?
2. What is the change in enthalpy for each mole of hydrogen peroxide that decomposes?
3. If 25.0 g of H_2O_2 decomposes at 25°C, determine the net amount of heat released or absorbed.

First, since ΔH is a negative quantity, the products of the reaction have less energy than the reactants, so heat is released during the reaction. The reaction is exothermic.

Second, the heat of reaction shown is for two moles of H_2O_2, as the coefficient in the thermochemical equation indicates. The change in enthalpy for one mole—half this amount—is half as much, or –98.0 kJ.

Third, we begin by determining how many moles of H_2O_2 we have. The molecular mass is 34.01 g/mol. Calculating the number of moles from this, we have

$$25.0 \text{ g} \cdot \frac{1 \text{ mol}}{34.01 \text{ g}} = 0.7351 \text{ mol}$$

Finally, the given equation tells us that 196 kJ are released for every two moles of H_2O_2 involved. We use this ratio as a conversion factor and apply it to the number of moles we have.

$$0.7351 \text{ mol} \cdot \frac{196 \text{ kJ}}{2 \text{ mol}} = 72.0 \text{ kJ}$$

We find that 72.0 kJ of heat is released by the decomposition of 25.0 g H_2O_2.

10.1.3 Understanding Enthalpy and Energy

Before we move on, you may be wondering why we need to use the term *enthalpy change* to describe the heat being absorbed or released by a chemical reaction. The term *enthalpy* is probably unfamiliar to you and may sound strange or conceptually difficult. So why can't we just use a term such as *energy change* or *internal energy change*, since we are talking about heat energy entering or leaving the thermochemical system? The answer has to do with the fact that there is more than one way for energy to enter or leave a thermochemical system. Thus, in general, the change in enthalpy may or may not be equal to the change in the internal energy of the substances in the reaction or the heat absorbed or released. Let's consider these energy possibilities.

First, energy can enter or leave the system through the absorption or release of heat. This can occur by means of an exothermic chemical reaction releasing heat into the system or an endothermic reaction absorbing heat from the system. Heat exchange can also occur by means of heating or cooling the system.

Next, energy can enter or leave by means of work being done by or on the system. For example, imagine a chemical reaction—such as an explosive combustion of fuel—that occurs inside a cylinder, as depicted in Figure 10.5. This is, in fact, exactly what happens when a mixture of gasoline and air is ignited in one of the cylinders of a car engine. The explosion products are gas molecules at high pressure. This high pressure pushes the piston, which is mechanically connected to the machinery that makes the wheels of the car turn. Pushing the piston is an example of *mechanical work*: force is applied to an object and moves it a certain distance. Work is energy, and when the expanding gases push the piston, the energy involved—the work—leaves the thermochemical system (the gases in the cylinder) and is converted into the kinetic energy of the moving car. This is called *expansion work*, and is energy that leaves the thermochemical system by a mechanism other than heat.

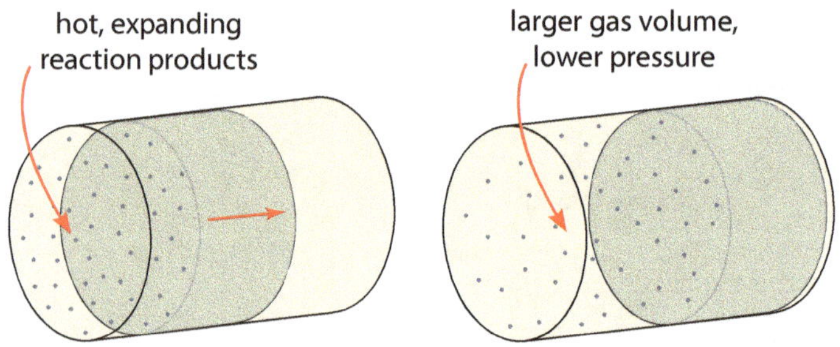

Figure 10.5. Energy leaves a system when a gas does mechanical work by expanding inside a cylinder and pushing a piston.

Third, there are other forms of work by which energy can enter or leave the system, such as electromagnetic interactions, or mechanical parts driven from the outside that stir (accelerate) the gases in the system and thus increase their internal energy. For the rest of this discussion, we assume these play no role in our system and that only expansion work is involved.

In general, the change in enthalpy in a system is a function of pressure and entropy. But *if the pressure is held constant*, then the mathematics allows us to describe the change in enthalpy more simply as follows:

$$\Delta H = \Delta U + P\Delta V \tag{10.5}$$

In this equation, ΔU is the change in internal energy in the system (see Section 6.1.1), P is the system pressure, and ΔV is the change in the system volume. Again, Equation (10.5) can only be written because we are holding the pressure constant.[1]

Now to see how this change in enthalpy can be equal to the heat absorbed or released, we start with the so-called *first law of thermodynamics*, which is:

1 An example of a reaction at constant pressure is a chemical reaction that occurs in the open atmosphere. The pressure is atmospheric pressure throughout the reaction.

$$Q = \Delta U + W \tag{10.6}$$

In this equation, Q is the heat entering or leaving the system, ΔU is the change in the internal energy of the molecules as before, and W is the work done by (or on) the system. Now, since we are considering only the expansion work, you may recall from an earlier physics class that the work associated with pushing an object is defined as

$$W = F \cdot d \tag{10.7}$$

In this equation, F represents a force (in newtons) and d represents a distance (in meters). We can easily relate this expression for work to the pressure and volume of a gas. Recall from Section 6.1.4 that pressure is defined as force per unit area, or

$$P = \frac{F}{A} \tag{10.8}$$

Solving this equation for the force, we have

$$F = P \cdot A \tag{10.9}$$

Substituting this expression into Equation (10.7) gives us

$$W = P \cdot A \cdot d \tag{10.10}$$

Figure 10.6 relates these quantities to a piston in a cylinder. The force applied to the end of the piston is equal to the pressure in the cylinder times the surface area of the end of the piston. The force causes the piston to move a distance d. (Again, we assume this happens in a way that allows the pressure to remain constant.) When this happens, the volume change, ΔV, in the cylinder, shown in blue, is equal to the surface area of the end of the piston, A, times the distance it moves, d, or

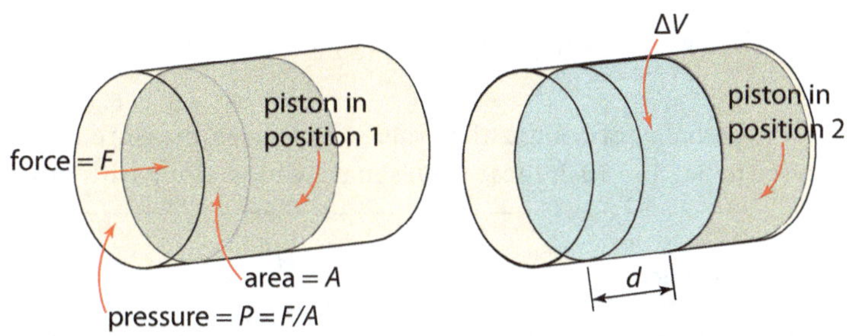

Figure 10.6. Given $W = F \cdot d$, $F = P \cdot A$, and $A \cdot d = \Delta V$, we get $W = P\,\Delta V$.

$$\Delta V = A \cdot d \tag{10.11}$$

Substituting this expression into Equation (10.10) gives us

$$W = P\Delta V \tag{10.12}$$

Finally, substituting this expression for the expansion work into Equation (10.6) gives:

$$Q = \Delta U + P\Delta V \tag{10.13}$$

Comparing this equation to Equation (10.5), you see that $\Delta H = Q$, and thus the change in enthalpy equals the heat entering or leaving the system. This result depends on the pressure being constant and on the expansion work being the only form of work affecting the system. As a quick application, the chemical reaction of an outdoor explosion releases heat and produces expansion work as the hot gases formed by the explosion push back the atmosphere. In this case, the change in enthalpy of the chemical system is equal to the heat released by the reaction.

10.1.4 Enthalpy of Combustion

The heat released by the complete combustion of one mole of a substance at a constant pressure is called the *molar enthalpy of combustion* (or *heat of combustion*). The enthalpy of combustion is a special case of the enthalpy of reaction we have been discussing. Conventionally, enthalpy of combustion values are listed for controlled conditions in which reactants and products are in their so-called *standard states* at standard conditions ($T = 25°C$ and $P = 1$ bar). The standard state is the state compounds are normally in at standard conditions. Table 10.1 lists enthalpy of combustion values for a number of representative substances. Notice in the table that the value for hydrogen is listed as –285.8 kJ/mol. For two moles, this is –517.6 kJ, which agrees with Equation (10.3).

We use a special notation to designate enthalpy of combustion. Using hydrogen as an example, the enthalpy of combustion is

$$\Delta H_c^\circ = -285.8 \ \frac{\text{kJ}}{\text{mol}}$$

The subscript *c* designates *combustion*, and the superscript ° indicates that the value pertains to the reactants and products in their standard states. As another example, the thermochemical equation for the combustion of butane is

$$C_4H_{10}(g) + \frac{13}{2}O_2(g) \rightarrow 4CO_2(g) + 5H_2O(l) \qquad \Delta H_c^\circ = -2877.6 \ \frac{\text{kJ}}{\text{mol}}$$

Enthalpy of combustion values have been measured for many common substances. As we see in Section 10.2, these values may be used along with other data to perform thermochemical

Substance	Formula	ΔH_c° (kJ/mol)	Substance	Formula	ΔH_c° (kJ/mol)
acetic acid (*l*)	CH_3COOH	-874.2	methane (*g*)	CH_4	-890.8
acetone (*l*)	C_3H_5OH	–1789.9	ethyne (acetylene) (*g*)	C_2H_2	–1301.1
ammonia (*g*)	NH_3	–382.8	ethene (ethylene) (*g*)	C_2H_4	–1411.2
carbon (graphite) (*s*)	C	–393.5	ethane (*g*)	C_2H_6	–1560.7
carbon monoxide (*g*)	CO	–283.0	propane (*g*)	C_3H_8	–2219.2
ethanol (*l*)	C_2H_5OH	–1366.8	butane (*g*)	C_4H_{10}	–2877.6
ethylene glycol (*l*)	$C_2H_6O_2$	–1189.2	pentane (*l*)	C_5H_{12}	–3509.0
formic acid (*l*)	CH_2O_2	–254.6	benzene (*l*)	C_6H_6	–3267.6
glucose (*s*)	$C_6H_{12}O_6$	–2805	hexane (*l*)	C_6H_{14}	–4163.2
hydrogen (*g*)	H_2	–285.8	toluene (*l*)	C_7H_8	–3910.3
methanol (*l*)	CH_3OH	–726.1	heptane (*l*)	C_7H_{16}	–4817.0
sucrose (*s*)	$C_{12}H_{22}O_{11}$	–5640.9	naphthalene (*s*)	$C_{10}H_8$	–5156.3

Table 10.1. Standard molar enthalpy of combustion values for representative substances at $T = 25°C$ and $P = 1$ bar. Common hydrocarbons are shown in the right column.

calculations that can be used to predict how chemical reactions would proceed without the need for additional experimentation. Being able to predict the results of a reaction without actual experimental data is a powerful analytical tool.

Scientists measure the enthalpy of combustion using a device called—believe it or not—a *bomb calorimeter*. In general, a *calorimeter* is an insulated container with accommodations for measuring the temperature of the contents. Substances at different temperatures are placed inside the calorimeter, and allowed to reach thermal equilibrium without any heat flowing into or out of the calorimeter from the outside. Combined with the known masses of the substances inside the calorimeter, the temperature at thermal equilibrium may be used to determine thermal properties (such as the molar heat capacity) of one of the substances inside. Calculations like this are based on the heat calculations we encountered in Chapter 6 and are not difficult. For simple experiments, a calorimeter can be as rudimentary as an insulated coffee cup with a lid and a thermometer.

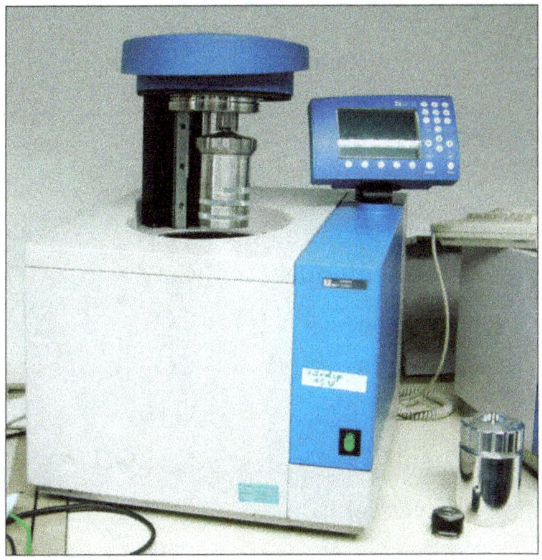

Figure 10.7. A bomb calorimeter.

The bomb calorimeter used for measuring enthalpy of combustion values is significantly more sophisticated, as the photo in Figure 10.7 shows. The sample to be burned is placed in the steel canister, called a *bomb*. Inside the canister is an electronic ignition source. After being sealed, the air in the canister is replaced with oxygen and the canister is lowered into the larger chamber shown in the figure. The larger chamber contains a precisely measured amount of water at a precisely measured temperature and is insulated to prevent heat flowing in from or out to the outside. The ignition source is used to ignite the sample under test, which burns completely in the presence of pure oxygen. The energy released by the reaction is calculated from the temperature rise of the water in the outer chamber.

10.1.5 Enthalpy of Formation

The reaction shown in Equation (10.2) describes the formation of water from its elements, hydrogen and oxygen. In a manner similar to the enthalpy of combustion, the *molar enthalpy of formation* is defined as *the amount of heat released or absorbed when one mole of a substance is formed from its elements*. As with enthalpy of combustion, enthalpy of formation values are stated assuming that the reactants and products are at standard conditions (25°C, 1 bar) and in their standard states. The notation used for designating enthalpy of formation is

$$\Delta H_f^\circ$$

In this notation, the subscript *f* denotes *formation*.

For most substances, the formation of the substance from its elements is an exothermic reaction, resulting in a negative heat of formation. The enthalpy of formation for water is the same as the enthalpy of combustion of hydrogen (since hydrogen and oxygen are the elements composing water), and so for water

$$\Delta H_f^\circ = -285.8 \frac{\text{kJ}}{\text{mol}}$$

Substance	Formula	ΔH_f° (kJ/mol)	Substance	Formula	ΔH_f° (kJ/mol)
acetylene (*g*)	C_2H_2	227.4	hydrogen chloride (*g*)	HCl	-92.3
ammonia (*g*)	NH_3	-45.9	hydrogen fluoride (*g*)	HF	-273.3
benzene (*l*)	C_6H_6	49.1	hydrogen iodide (*g*)	HI	26.5
calcium chloride (*s*)	$CaCl_2$	-795.4	methane (*g*)	CH_4	-74.6
carbon monoxide (*g*)	CO	-110.5	methanol (*l*)	CH_3OH	-239.2
carbon dioxide (*g*)	CO_2	-393.5	propane (*g*)	C_3H_8	-103.8
ethane (*g*)	C_2H_6	-84.0	sodium chloride (*s*)	NaCl	-411.2
ethanol (*l*)	C_2H_5OH	-277.6	sucrose (*s*)	$C_{12}H_{22}O_{11}$	-2226.1
ethylene (*g*)	C_2H_4	52.4	sulfur dioxide (*g*)	SO_2	-296.8
glucose (*s*)	$C_6H_{12}O_6$	-1273.3	water (*l*)	H_2O	-285.8
hydrogen bromide (*g*)	HBr	-36.3	water (*g*)	H_2O	-241.8

Table 10.2. Standard molar enthalpy of formation values for representative substances at $T = 25°C$ and $P = 1$ bar. See Appendix B, Table B.7 for a more extensive list.

Considering all we have discussed so far in this text about energy, it should make sense to you that substances with a large negative value for the enthalpy of formation are very stable because a lot of energy is required to undo the formation of the substance. Substances with a small negative value for the enthalpy of formation are relatively unstable, since only a small amount of energy can disrupt the structure of the atoms in such a substance.

Table 10.2 contains enthalpy of formation values for a number of representative substances. As you see, some substances have a positive value for ΔH_f°. The formation of such compounds is endothermic, and the compounds are quite unstable. These compounds can spontaneously decompose or react violently. As examples, hydrogen iodide spontaneously decomposes when stored at room temperature. Acetylene is the gas used with oxygen in the cutting torches welders use to cut through steel. A lot of energy has to be supplied to form acetylene, so the molecules are unstable and combust very violently, releasing a lot of energy as heat in the process. Although it is not listed in the table, mercury(II) fulminate, $Hg(CNO)_2$, has a high positive enthalpy of formation, $\Delta H_f^\circ = 270$ kJ/mol. This substance is very sensitive to shock, which makes it a great detonator to set off larger charges of more environmentally safe explosives (safer, that is, because they don't contain mercury). Interestingly, although mercury(II) fulminate was first prepared in 1800, its crystal structure was not known until 2007. A space-filling model of a portion of the crystal lattice for this compound is shown in Figure 10.8. This compound has made an appearance in a number of popular movies and TV series.

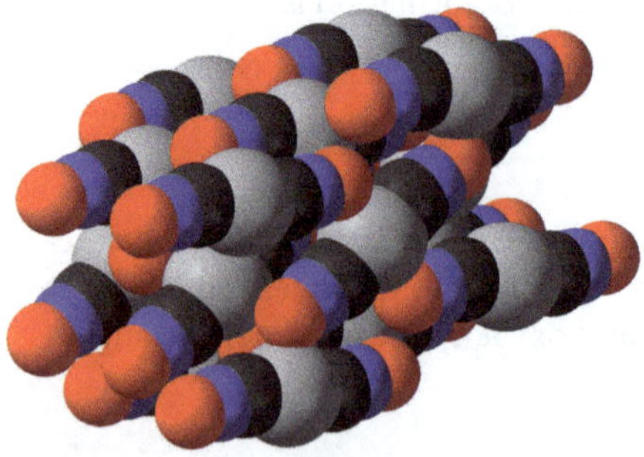

Figure 10.8. Crystal structure of mercury(II) fulminate. Mercury atoms are shown in gray.

The enthalpy of formation for pure elements in their standard states is defined as $\Delta H_f^\circ = 0$. This value applies to molecules for elements that are naturally molecular at standard conditions. Thus, for H_2, O_2, and the other diatomic elements, $\Delta H_f^\circ = 0$. However, the enthalpy of formation for atomic hydrogen, H, is $\Delta H_f^\circ = 218.0$ kJ/mol. To separate an H_2 molecule, energy must be supplied to break the covalent bond in

H_2. If you refer back to Table 3.9, you see that the bond energy for one mole of H_2 is –436.0 kJ. Since two atoms of H are in each H_2 molecule, the heat of formation for a mole of individual H atoms is half this amount (and positive).

The implication of defining the enthalpy of formation for pure elements to be zero is that substances with a negative value for ΔH_f° are more stable than the elements from which they are formed. Compounds with a positive value for the molar enthalpy of formation are less stable than the elements from which they are formed.

10.2 Calculating Enthalpy of Reaction and Enthalpy of Formation

10.2.1 Hess's Law

Hess's law[2] states that *the enthalpy change in a chemical reaction is independent of the pathway the reaction takes*. Another way of stating this principle is that the enthalpy change in a reaction is the same whether the reaction takes place in one step or in many steps. So if we can imagine a sequence of reaction steps that leads, in principle, from the same reactants to the same products as for a single-step reaction, the sum of the enthalpy changes for the individual steps must equal the enthalpy change for the single-step reaction. This important principle can be used to determine the enthalpy of reaction in cases where measuring ΔH is difficult or even impossible. We can use Hess's law to solve for any unknown enthalpy of reaction so long as we know either the enthalpy of combustion values for all the reactants and products, or the enthalpy of formation values for all the reactants and products.

To illustrate the use of Hess's law for such calculations, we calculate the enthalpy of formation for methane gas, CH_4. The elements composing methane are solid carbon (graphite) and hydrogen gas. This means the equation showing the formation of methane from its elements is

$$C(s) + 2H_2(g) \rightarrow CH_4(g) \qquad \Delta H_f^\circ = ? \qquad (10.12)$$

For this illustration, *we derive the enthalpy of formation for methane using only enthalpy of combustion values* for the three reactants and products (C, H_2, and CH_4) in Equation (10.12). These three thermochemical equations, with enthalpy values from Table 10.1, are as follows:

$$C(s) + O_2(g) \rightarrow CO_2(g) \qquad \Delta H_c^\circ = -393.5 \text{ kJ/mol}$$

$$H_2(g) + \frac{1}{2}O_2(g) \rightarrow H_2O(l) \qquad \Delta H_c^\circ = -285.8 \text{ kJ/mol}$$

$$CH_4(g) + 2O_2(g) \rightarrow CO_2(g) + 2H_2O(l) \qquad \Delta H_c^\circ = -890.8 \text{ kJ/mol}$$

Notice that the equation for the combustion of hydrogen is written to represent the combustion of one mole of H_2 because the enthalpy of combustion value from Table 10.1 is for a single mole. Writing the equation this way requires the use of a fractional coefficient on O_2.

Our plan now is essentially to *add these equations together* to generate the formation equation for methane, Equation (10.12). To do this, we must arrange the equations so that the desired compounds are on the correct side of the equation in each case. Unwanted terms in the addition cancel out if we have the same term (with the same coefficient) appearing on both sides.

2 American orthography would normally have us write Hess' law. However, the British orthography with the final *s* added is used universally when referring to Hess's law.

To assure proper cancellation, it may be necessary to multiply one or more of the equations by a constant. So we are about to make these two preparatory maneuvers before adding the equations: (1) reverse equations and signs on the enthalpy as necessary to get compounds on the correct side according to Equation (10.12), and (2) multiply equations by constants as necessary so that when the equations are added together Equation (10.12) results without any extra terms.

In Equation (10.12), carbon and hydrogen are on the left side. In the first two combustion equations, carbon and hydrogen are on the left, so these equations are fine in this respect. However, methane is on the right in Equation (10.12), so we need to reverse the methane combustion equation to show methane on the right. Recall from earlier in the chapter that when we reverse a thermochemical equation, the sign on the enthalpy of reaction changes.

$$CO_2(g) + 2H_2O(l) \rightarrow CH_4(g) + 2O_2(g) \qquad \Delta H^\circ = 890.8 \text{ kJ/mol}$$

(Note that the enthalpy is now written ΔH°, without the subscript, because written in reverse this is no longer a combustion reaction.)

Now let's list our equations in preparation for adding them together.

$$C(s) + O_2(g) \rightarrow CO_2(g) \qquad \Delta H_c^\circ = -393.5 \text{ kJ/mol}$$

$$H_2(g) + \frac{1}{2}O_2(g) \rightarrow H_2O(l) \qquad \Delta H_c^\circ = -285.8 \text{ kJ/mol}$$

$$CO_2(g) + 2H_2O(l) \rightarrow CH_4(g) + 2O_2(g) \qquad \Delta H^\circ = 890.8 \text{ kJ/mol}$$

When we add these together, we need all the CO_2, H_2O, and O_2 terms to cancel out because none of these compounds appears in Equation (10.12). Let's look at what it takes for this to happen in the case of O_2. We see that if we multiply the entire second equation by 2, the $\frac{1}{2}O_2$ term becomes just O_2, and the O_2 terms on the left add to be equal the $2O_2$ shown on the right, enabling them to cancel out. It turns out that doubling the coefficients on the second equation enables everything to cancel out that needs to cancel out. Accordingly, we multiply this entire equation by 2, *including the enthalpy*. Doing so and canceling out terms that appear on both sides gives us the following set of equations. These we now add together to give the equation at the bottom, which is identical to Equation (10.12).

$$C(s) + \cancel{O_2(g)} \rightarrow \cancel{CO_2(g)} \qquad \Delta H_c^\circ = -393.5 \text{ kJ/mol}$$

$$2H_2(g) + \cancel{O_2(g)} \rightarrow \cancel{2H_2O(l)} \qquad 2\Delta H_c^\circ = 2(-285.8 \text{ kJ/mol})$$

$$\cancel{CO_2(g)} + \cancel{2H_2O(l)} \rightarrow CH_4(g) + \cancel{2O_2(g)} \qquad \Delta H^\circ = 890.8 \text{ kJ/mol}$$

$$C(s) + 2H_2(g) \rightarrow CH_4(g) \qquad \Delta H_f^\circ = -74.3 \text{ kJ/mol}$$

Adding together the enthalpy values gives us the desired result, the enthalpy of formation of methane, –74.3 kJ/mol. Comparing this value to the one given in Table 10.2, we see a slight difference. This is probably due to rounding in the values stated in Tables 10.1 and 10.2.

Figure 10.9 is an enthalpy diagram that compares the artificially constructed multistep reaction pathway we used in our calculation (Path 1) to the single-step reaction pathway we sought to compute (Path 2). The two pathways are indicated in the figure by yellow circles. Hess's law says that the net enthalpy change is the same for all such paths. This is why we can use the ΔH values for the sequence of three reactions in Path 1 to determine the ΔH for the desired single-step reaction in Path 2. The diagram looks complicated—but ignore the dashed arrows for the moment as we walk through it.

Both pathways begin with solid carbon, C(*s*), at a reference enthalpy of 0 kJ/mol. If you follow the red arrows and yellow circles for each path, you see they both end up at $CH_4(g)$ at $H = -74.3$ kJ/mol.

The simplest path is Path 2. The enthalpy change for the path is what we sought to compute (the unknown). We computed it using Hess's law applied to the sequence of known reactions with known enthalpy of combustion values in Path 1. To follow any path, keep this in mind: the reactants in the reaction we sought to compute are C(*s*) and $2H_2(g)$. We can make use of these reactants in the path whenever we need to. But whenever a path requires other reactants that have not been produced yet, we "borrow" them from the environment for the purpose of our calculation, knowing that we can pay them back later in the reaction sequence. (Since this computation is a mathematical model, we should not be bothered if we need to "borrow" some oxygen for a moment.)

Path 1 begins with a reaction between C(*s*) and $O_2(g)$ because this is the first reaction listed in our computation. (We could take these reactions in any sequence, but I am just taking them in the same order as they are listed in the computation.) So, we borrow one mole of O_2 to begin. This reaction produces $CO_2(g)$, and the change in enthalpy is −393.5 kJ/mol. The second reaction is between $2H_2$(g) and O_2(g). Thus, we now borrow another mole of O_2 in order to proceed. We don't yet need our CO_2 from the previous reaction, so we will simply bring it with us for later use. This is indicated by the dashed blue arrow. The product of the second reaction is $2H_2O(l)$, with an additional enthalpy change of −571.6 kJ/mol, bringing the total enthalpy change so

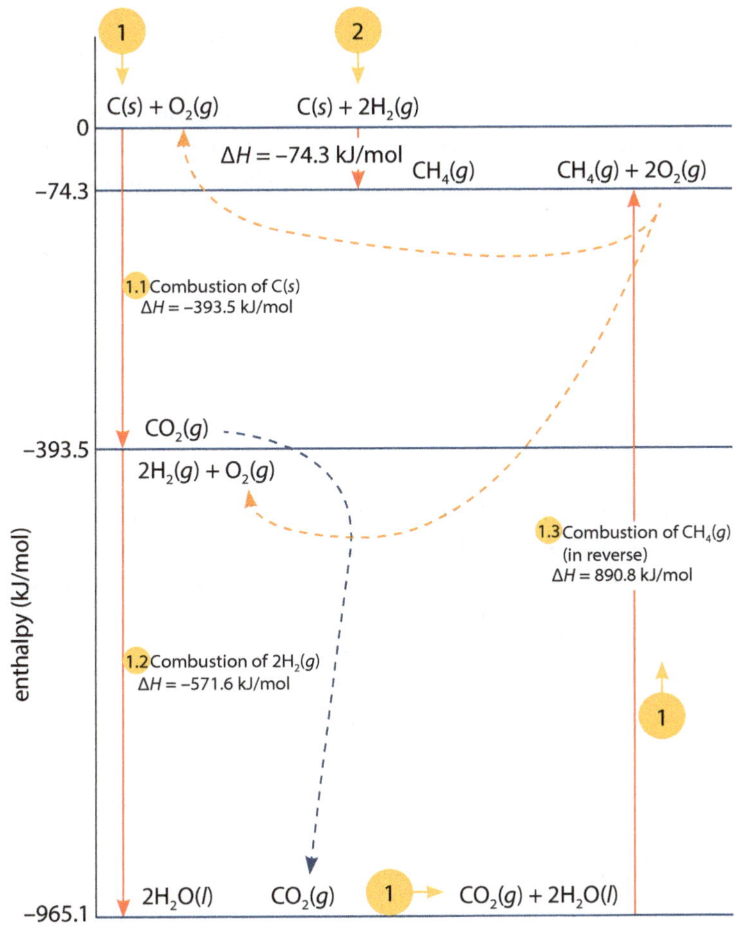

Figure 10.9. Enthalpy diagram comparing the net enthalpy of reaction for two separate pathways leading from C(*s*) + $2H_2(g)$ to $CH_4(g)$. According to Hess's law, the net enthalpy change is the same for all such paths.

far to –965.1 kJ/mol. For the third reaction in Path 1, we take the $CO_2(g)$ and $2H_2O(l)$ produced by the first two reactions, and together they undergo an endothermic reaction to produce $CH_4(g)$ and $2O_2(g)$. This reaction entails an increase in enthalpy of 890.8 kJ/mol, bringing us up to an enthalpy level of –74.3 kJ/mol. Here we are at the end of Path 1. We arrived at $CH_4(g)$, and created two moles of $O_2(g)$ along the way. As shown by the orange dashed arrows, we use these to pay back what we borrowed, leaving us with $CH_4(g)$ and a net enthalpy change of

$$\Delta H_f^\circ = -74.3 \text{ kJ/mol}$$

Now notice what we just did. We calculated the *enthalpy of formation* for CH_4 using only *enthalpy of combustion* data. In the next example, we calculate an *enthalpy of combustion* using only *enthalpy of formation data.*

Example 10.2

Use Hess's law to determine the enthalpy of combustion for propane, given the enthalpy of formation for each of the reactants and products in the thermochemical equation for the combustion of propane:

$$C_3H_8(g) + 5O_2(g) \rightarrow 3CO_2(g) + 4H_2O(l) \qquad \Delta H_c^\circ = ? \qquad (10.13)$$

The thermochemical equations for the formation of the other three reactants and products are as follows. There is no formation equation for O_2 because the enthalpy of formation for pure elements is zero. Enthalpy of formation values are from Table B.7 in Appendix B.

$$3C(s) + 4H_2(g) \rightarrow C_3H_8(g) \qquad \Delta H_f^\circ = -103.8 \text{ kJ/mol}$$

$$C(s) + O_2(g) \rightarrow CO_2(g) \qquad \Delta H_f^\circ = -393.5 \text{ kJ/mol}$$

$$H_2(g) + \frac{1}{2}O_2(g) \rightarrow H_2O(l) \qquad \Delta H_f^\circ = -285.8 \text{ kJ/mol}$$

To prepare for adding these together, we must get each compound on the correct side and adjust coefficients so the addition cancels out all unwanted terms. In Equation (10.13), propane is on the left, so we reverse the first equation and change the sign on the enthalpy. Carbon dioxide and water are both on the right in Equation (10.13) and they are also on the right in the formation equations. Reversing the first equation gives us

$$C_3H_8(g) \rightarrow 3C(s) + 4H_2(g) \qquad \Delta H^\circ = 103.8 \text{ kJ/mol}$$

$$C(s) + O_2(g) \rightarrow CO_2(g) \qquad \Delta H_f^\circ = -393.5 \text{ kJ/mol}$$

$$H_2(g) + \frac{1}{2}O_2(g) \rightarrow H_2O(l) \qquad \Delta H_f^\circ = -285.8 \text{ kJ/mol}$$

Looking now at the coefficients, Equation (10.13) has no $C(s)$ or $H_2(g)$ in it. Multiplying the second equation by three enables the carbons to cancel, and multiplying the third equation by four enables the hydrogens to cancel. Remember, when multiplying an equation by a constant, the enthalpy must be multiplied as well. The following equations show the new coefficients and enthalpies after the multiplications, the cancellations of terms, and the summed equation at the bottom.

$$C_3H_8(g) \rightarrow \cancel{3C(s)} + \cancel{4H_2(g)} \qquad \Delta H^\circ = 103.8 \text{ kJ/mol}$$

$$\cancel{3C(s)} + 3O_2(g) \rightarrow 3CO_2(g) \qquad 3\Delta H_f^\circ = 3(-393.5) \text{ kJ/mol}$$

$$\cancel{4H_2(g)} + 2O_2(g) \rightarrow 4H_2O(l) \qquad 4\Delta H_f^\circ = 4(-285.8) \text{ kJ/mol}$$

$$C_3H_8(g) + 5O_2(g) \rightarrow 3CO_2(g) + 4H_2O(l) \qquad \Delta H_c^\circ = -2219.9 \text{ kJ/mol}$$

The bottom equation is identical to Equation (10.13), as required. We calculate the enthalpy of combustion for propane to be –2219.9 kJ/mol. Comparing this to the value listed in Table 10.1, we again see a small difference in the last digit due to rounding.

▲

Notice this important tip from this example: CO_2 and H_2O are compounds that appear in the final equation and each appears only once in the formation equations. When this happens, simply multiply their equations by the coefficients you need in the final equation. Everything else works out. (This tip will come in handy in the exercises.)

An enthalpy diagram for the reaction sequence in Example 10.2 is shown in Figure 10.10. Path 2 represents the combustion of propane, the enthalpy of combustion we sought to compute. Path 1 is the sequence of three reactions we put together using Hess's law. Reactants we can use whenever we need them are the $C_3H_8(g)$ and $5O_2(g)$. As before, the blue dotted arrows indicate reac-

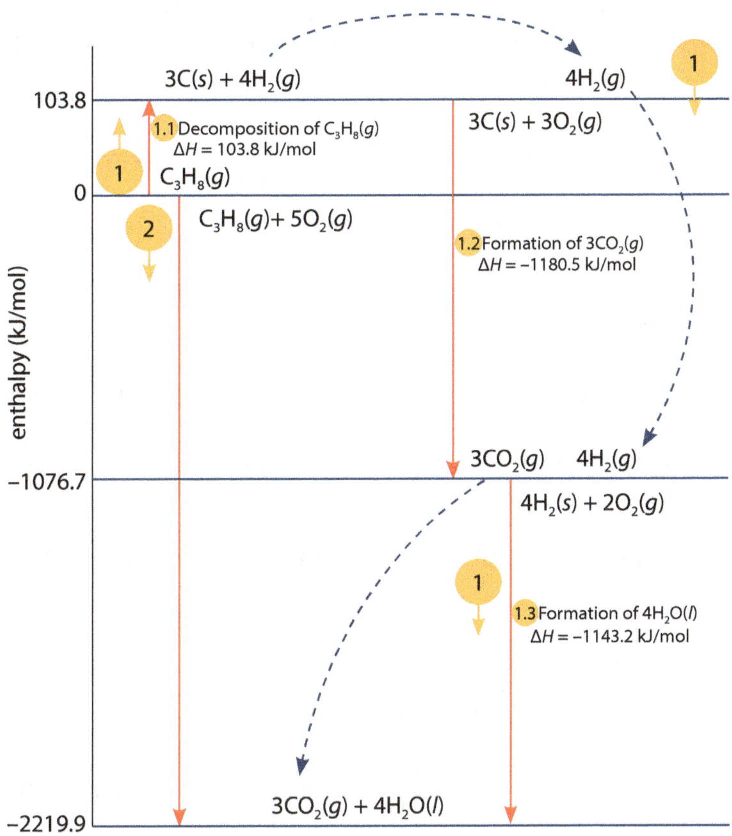

Figure 10.10. Enthalpy diagram comparing the net enthalpy of reaction for two separate pathways leading from $C_3H_8(g) + 5O_2(g)$ to $3CO_2(g) + 4H_2O(l)$.

tion products that we carry along from one step to another until they are needed. No "borrowing" is needed this time.

10.2.2 Hess's Law and the General Enthalpy Change Equation

Recall that Equation (10.1) defines the change of enthalpy for a reaction as the difference between the enthalpy of the products and the enthalpy of the reactants. This relationship may be expressed using the enthalpy of formation values, ΔH_f°, for products and reactants as follows:

$$\Delta H^\circ = \Sigma \Delta H^\circ_{f,\text{products}} - \Sigma \Delta H^\circ_{f,\text{reactants}} \tag{10.14}$$

In this expression, the upper case Greek letter sigma, Σ, means "sum." So this expression tells us that *we can calculate any enthalpy of reaction by summing the enthalpy of formation values for the products, summing the enthalpy of formation values for the reactants, and then taking the difference between these two sums*. This is a powerful computational tool. The entire process of manipulating reactions we just described boils down to the difference between two sums of enthalpy of formation values.

To illustrate, we will revisit Example 10.2, noting that Table B.7 in Appendix B contains an extended list of enthalpy of formation values. The goal is to calculate the enthalpy of reaction for the following reaction:

$$C_3H_8(g) + 5O_2(g) \rightarrow 3CO_2(g) + 4H_2O(l)$$

We tabulate the ΔH_f° values for the reactants and for the products, recalling that the enthalpy of formation for a pure element in its standard state (such as O_2) is zero. We need to take into account the number of moles for each reactant and product in the equation. Pulling the ΔH_f° values from Appendix B, Table B.7, we have:

	Reactants		Products	
		ΔH_f° (kJ/mol)		ΔH_f° (kJ/mol)
	$C_3H_8(g)$	−103.8	$3CO_2(g)$	3(−393.5) = −1180.5
	$5O_2(g)$	0	$4H_2O\ (l)$	4(−285.8) = −1143.2
Sum (Σ)		−103.8		−2323.7

Now we use Equation (10.14) to compute this enthalpy of reaction:

$$\Delta H^\circ = \Sigma \Delta H^\circ_{f,\text{products}} - \Sigma \Delta H^\circ_{f,\text{reactants}} = \left(-2323.7\ \frac{\text{kJ}}{\text{mol}}\right) - \left(-103.8\ \frac{\text{kJ}}{\text{mol}}\right) = -2219.9\ \frac{\text{kJ}}{\text{mol}}$$

This is the same value we calculated before in Example 10.2.

Example 10.3

Use Equation (10.14) to calculate the enthalpy of reaction for the following reaction:

$$SiCl_4(l) + 2H_2O(l) \rightarrow SiO_2(s) + 4HCl(g)$$

Tabulating the enthalpy of formation values from Appendix B, we have:

	Reactants		Products	
		ΔH_f° (kJ/mol)		ΔH_f° (kJ/mol)
	$SiCl_4(l)$	-687.0	$SiO_2(s)$	-910.7
	$2H_2O(l)$	2(−285.8) = −571.6	$4HCl(g)$	4(−92.3) = −369.2
Sum (Σ)		−1258.6		−1279.9

Inserting these two sums into Equation (10.14), we have the following:

$$\Delta H^\circ = \Sigma\Delta H^\circ_{f,\text{products}} - \Sigma\Delta H^\circ_{f,\text{reactants}} = \left(-1279.9\ \frac{\text{kJ}}{\text{mol}}\right)-\left(-1258.6\ \frac{\text{kJ}}{\text{mol}}\right) = -21.3\ \frac{\text{kJ}}{\text{mol}}$$

▲

10.3 Free Energy

10.3.1 Brief Review of Enthalpy and Entropy

We are now ready to put enthalpy and entropy together into a single law that can predict whether a process spontaneously occurs. As we saw in Chapter 8, the tendency in nature is for processes to go in a direction that decreases potential energy and increases entropy. But these tendencies may or may not point in the same direction, so we need a way to weigh them against each other so we can determine if a reaction occurs or not. The goal of this section is to present the equation that does this. Let's first summarize what we have seen so far about enthalpy and entropy. For enthalpy, we know the following:

Enthalpy Summary

1. For processes at constant pressure, the change in enthalpy, ΔH, is equal to the heat absorbed or released during the process.
2. For exothermic reactions, ΔH is negative; for endothermic reactions, ΔH is positive.
3. Most reactions that occur in nature are exothermic.
4. When an exothermic reaction occurs, heat is released. This means that the potential energy in the molecules of the products is lower than in those of the reactants, so the product molecules are more stable.
5. Nature favors processes in which the enthalpy decreases, but processes in which the enthalpy of a system increases also occur.
6. The First Law of Thermodynamics for constant-pressure processes may be written as $\Delta H = \Delta U + P\Delta V$.

And now, a summary of what we have seen about entropy, along with two new pieces of information added for completeness:

Entropy Summary

1. The entropy of a system is a measure of the disorder or randomness in the system.
2. For every process, the entropy of the universe increases. If the entropy decreases in a certain system, this can only happen by a larger increase in entropy somewhere else. (This is one way of stating the Second Law of Thermodynamics.)

3. Nature favors processes in which the entropy increases, but processes in which the entropy of a system decreases also occur.
4. (New) Entropy is denoted by S, and a change in entropy by ΔS.
5. (New) The entropy of a pure crystalline substance at absolute zero (0 K) is equal to zero. As temperature increases, entropy does as well.

10.3.2 More on Entropy

The last point above about entropy deserves comment. I have described entropy as a measure of the disorder in a system. The mathematical way physicists characterize entropy is by assessing the number of different possible states that are available for a system to be in. In this statement, the term "states" refers to every possible combination of position and velocity available to all the atoms in the system. Since we are talking about the states of atoms, the term scientists use for all these states is *microstates*.[3]

At absolute zero, molecular motion ceases. A perfectly ordered crystal in which there is no movement at all can only be in one state, so there is no randomness at all—no uncertainty at all about where an atom might be or how fast it might be moving. Thus, the entropy of a pure crystal at 0 K is zero.

Now, if the temperature of the crystal is a bit above 0 K, the atoms in the crystal are vibrating. There is not much randomness in this crystal because the atoms are not free to translate, but since there is now a bit of uncertainty associated with the position and velocity of each of the vibrating atoms, there is entropy in the system. In general, the higher the temperature of a system gets, the more entropy there is because there is more randomness—more microstates available to the particles in the system. A large increase in entropy occurs when the substance melts to become a liquid because then the particles are free to move around a bit. And when vaporization occurs, an even larger increase in entropy takes place. In the gaseous state, the velocities of particles are high and the particles are spread out, so the number of microstates available to all the particles in the system is huge. These generalizations about the entropy of a system with respect to temperature are depicted graphically in Figure 10.11.

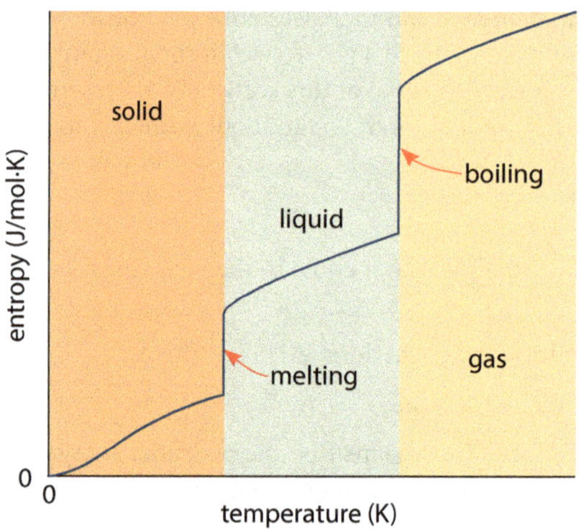

Figure 10.11. Variation of entropy with temperature and phase transitions in a substance.

One more observation we can make about entropy pertains to the numbers of moles indicated by chemical equations. The Haber-Bosch process for the manufacture of ammonia goes according to this equation:

$$N_2(g) + 3H_2(g) \rightarrow 2NH_3(g) \qquad (10.15)$$

3 It was Ludwig Boltzmann's work in statistical mechanics in the 1870s that led to the equation that quantifies the entropy of a system: $S = k \log W$, where W is the number of microstates available to the particles in a system, and k is the Boltzmann constant, 1.3806×10^{-23} J/K. At Ludwig Boltzmann's grave in Vienna, Austria, this equation is engraved in the marble above a bust of Boltzmann. Nowadays, the equation is typically written $S = k_B \log \Omega$.

We have seen this reaction a few times before. Note that for every four moles of reactants, only two moles of products are produced. This reaction entails a reduction in entropy since there are fewer moles of gas in the products than in the reactants. The next equation represents a process called *steam reforming*, which is used to produce a gas mixture called *syngas*.

$$CH_4(g) + H_2O(g) \rightarrow CO(g) + 3H_2(g) \qquad (10.16)$$

Syngas is a mixture of carbon monoxide and hydrogen used as a fuel and as a precursor for the manufacture of methanol and a number of other industrial substances. The steam reforming reaction is strongly endothermic, but there is also an increase in entropy, since every two moles of reactants produce four moles of gas products. An increase in the number of moles of gas means more microstates are available to the particles in the gas, and thus indicates an increase in entropy.

10.3.3 Gibbs Free Energy

There are three names associated with the foundations of statistical mechanics that you should know. We have already encountered James Clerk Maxwell and Ludwig Boltzmann (Section 6.1.2), who gave us the Maxwell-Boltzmann speed distributions for gas particles. The third is the great American scientist Josiah Willard Gibbs (Figure 10.12). In the 1870s, Gibbs introduced a concept he called "available energy," a quantity based on both the enthalpy and entropy in a system. Today, this quantity is called the *Gibbs free energy* or simply *free energy*. The Gibbs free energy, G (J/mol), in a system is defined as

$$G = H - TS \qquad (10.17)$$

Figure 10.12. American scientist Josiah Willard Gibbs (1839–1903).

In this equation, H, T, and S are the molar enthalpy (J/mol), temperature (K), and molar entropy [J/(mol·K)] of the system. As with enthalpy, the Gibbs free energy cannot be measured directly, but changes in the free energy can be measured. Thus, we work with the change to the Gibbs free energy, ΔG. For processes at constant temperature involving the standard states of reactants and products, Equation (10.17) may be re-expressed as follows:

$$\Delta G^\circ = \Delta H^\circ - T\Delta S^\circ \qquad (10.18)$$

Spontaneous processes in nature favor energy decreases and entropy increases. The Gibbs free energy equation relates them together, and *spontaneous processes in nature always go in the direction that reduces the Gibbs free energy of the system*. In other words, if ΔG° for a process is negative, the process happens spontaneously. Also, any process in which ΔG° increases at the stated conditions does not occur spontaneously.

Since the absolute temperature is always positive, the terms in Equation 10.17 that determine if ΔG° is negative are ΔH° and ΔS°. Either of these can be positive or negative, and this gives us four possibilities to consider when seeking to determine if a reaction will occur. These are summarized in Table 10.3. First, any exothermic process in which the entropy increases occurs spontaneously. In this case, nature's preferences for minimizing energy and increasing entropy are both being satisfied, and such a process always occurs.

The use of the term "spontaneous" here is the traditional language used to describe a process with a negative value for ΔG°, but the term can be misleading and requires some explanation.

ΔH°	ΔS°	ΔG°	Example
negative (exothermic)	positive (increases disorder)	always negative (spontaneous)	$2O_3(g) \rightarrow 3O_2(g)$
negative (exothermic)	negative (decreases disorder)	negative at lower temperatures (spontaneous at lower temperatures)	$H_2O(l) \rightarrow H_2O(s)$
positive (endothermic)	positive (increases disorder)	negative at higher temperatures (spontaneous at higher temperatures)	$H_2O(s) \rightarrow H_2O(l)$
positive (endothermic)	negative (decreases disorder)	never negative (never spontaneous)	$3O_2(g) \rightarrow 2O_3(g)$

Table 10.3. Four possible combinations of ΔH and ΔS lead to the different possibilities for the sign of ΔG.

The fact that the Gibbs free energy is negative means that a chemical reaction will occur *so long as the activation energy is present to initiate the reaction.* (We address activation energy in just a few pages.) Reactions such as fuel combustion always have a negative value for ΔG°, but fuels generally don't ignite "spontaneously" (for which we are thankful). A spark or flame is required to initiate the reaction; after that it goes by itself.

The last row in the table is for the opposite case—enthalpy increases and entropy decreases. Such a process never occurs spontaneously because it goes against the natural tendencies of both the enthalpy and entropy. (Such process can be made to occur only by supplying enough energy to drive the process, which increases entropy even more somewhere else.) The two middle rows deal with the cases in which the enthalpy and entropy pull in opposite directions. The deciding factor in each of these scenarios is the temperature because it multiplies the change in entropy and thus determines how significant the entropy contribution is to the Gibbs free energy. Note that in cases such as the phase transition examples shown in the table, the process does occur literally spontaneously. No activation energy is needed to make an ice cube melt; it just happens.

Example 10.4

For the steam reforming reaction to make syngas, shown in Equation (10.16), ΔH° = 206.1 kJ/mol and ΔS° = 0.215 kJ/(mol·K). Determine if this reaction will occur at standard temperature, 298 K.

We use Equation (10.18) to compute the change in the Gibbs free energy for this process, ΔG°. If it is negative, then the reaction occurs.

$$\Delta G^\circ = \Delta H^\circ - T\Delta S^\circ$$

$$\Delta G^\circ = 206.1 \frac{\text{kJ}}{\text{mol}} - 298 \text{ K} \cdot 0.215 \frac{\text{kJ}}{\text{mol} \cdot \text{K}} = 142.0 \frac{\text{kJ}}{\text{mol}}$$

At standard temperature, the Gibbs free energy increases for this process. Thus, the reaction does not occur spontaneously.

Example 10.5

The equation for the decomposition of ammonium chloride to ammonia and hydrogen chloride is as follows:

$$NH_4Cl(s) \rightarrow NH_3(g) + HCl(g)$$

Given ΔH° = 176 kJ/mol and ΔS° = 285 J/(mol·K), address the following questions about this process:

1. Will the reaction occur spontaneously at standard conditions?
2. What is the effect of increasing the temperature?
3. Is the given value for ΔS° consistent with the reaction equation shown above?

To answer the first question, we need to calculate the change in the Gibbs free energy. The temperature for standard conditions is 25°C, or 298 K. Note that the value for ΔS° is given in J/(mol·K). This must be changed to 0.285 kJ/(mol·K) to make the energy units consistent.

Using Equation (10.18) and the given information, we calculate the change in the Gibbs free energy for this process as follows.

$$\Delta G^\circ = \Delta H^\circ - T\Delta S^\circ$$

$$\Delta G^\circ = 176\ \frac{\text{kJ}}{\text{mol}} - 298\ \text{K} \cdot 0.285\ \frac{\text{kJ}}{\text{mol} \cdot \text{K}} = 91\ \frac{\text{kJ}}{\text{mol}}$$

The change in the Gibbs free energy is positive, indicating that this reaction does not occur spontaneously at 25°C.

A note about the significant digits in the calculation may be in order here. The product of 298 K and 0.285 kJ/(mol·K) is 84.93 kJ/mol. When this value is subtracted from 176 kJ/mol, the decimal places are all lost because of the addition rule for significant digits. Thus, the result is 91 kJ/mol.

Regarding the second question, we note that although this reaction is endothermic, the entropy change is positive. Thus, at a high enough temperature, the product $T\Delta S$ will be larger than ΔH, making ΔG° negative and enabling the reaction to occur.

Regarding the third question, the reaction equation indicates that one mole of a solid reactant forms two moles of gaseous products, indicating an increase in entropy. Since ΔS° is positive, it is consistent with the equation.

▲

Three more important points remain to be made in this section. First, the mere fact that the Gibbs free energy is negative says nothing about the rate at which a reaction occurs. For the reaction below, $\Delta G^\circ = -3.3\ \text{J}/(\text{mol} \cdot \text{K})$, and thus the reaction is favorable:

$$\text{C}(s, \text{diamond}) \rightarrow \text{C}(s, \text{graphite})$$

However, the reaction takes place so slowly it is never observed.

Second, make note of this: *the general principle embedded in Equation (10.14) may be used to calculate* ΔS° *for any reaction*. Data for S° are tabulated in standard reference sources for thousands of standard compounds, including Table B.8 in Appendix B. Thus, ΔS° for any reaction may be calculated as

$$\Delta S^\circ = \Sigma S^\circ_{\text{products}} - \Sigma S^\circ_{\text{reactants}}$$

The method for calculating ΔS° is the same as that illustrated in Example 10.3. You simply tabulate standard molar entropy values for products and reactants, multiply each by the number of moles in the equation, determine the two sums, and take the difference of the two sums. Thus, one can easily determine whether the reaction entails an increase or decrease in entropy. Make note, though, that *standard entropy values are usually given in J/(mol·K) rather than kJ/ (mol·K)*, so you have to watch out for the need to convert the units.

Third, the same idea applies to calculating the change in the Gibbs free energy. Data for the standard *free energy of formation*, ΔG°_f, are tabulated in reference sources for common compounds, including Table B.9 in Appendix B. Just as with the change in enthalpy in Equation (10.14), the change in the Gibbs free energy, ΔG°, for any reaction may be calculated as

$$\Delta G^\circ = \Sigma \Delta G^\circ_{f,\text{products}} - \Sigma \Delta G^\circ_{f,\text{reactants}}$$

Of course, ΔG° may also be calculated according to Equation (10.18), where ΔH° and ΔS° are calculated from standard enthalpy of formation and entropy data as described above.

10.4 Reaction Kinetics

One of the goals of research in chemistry is to understand more precisely what actually happens during a chemical reaction. Such research seeks answers to questions like these: How exactly does chemical species *A* get transformed into chemical species *B*? Does the transformation happen all at once or in a sequence of steps? What factors govern how fast a reaction occurs? What are the physical mechanisms of such factors? The subject of the physical mechanisms of reactions and factors that govern the rates of chemical reactions is known as *reaction kinetics*. In this section, we take a brief tour of these topics.

10.4.1 Collision Theory

Collision theory is a general term that refers to our present understanding of chemical reactions at the molecular level. The essential idea behind collision theory is that chemical reactions occur when molecules of different chemical species collide. However, just because molecules jostle around together does not mean a chemical reaction occurs. You can pour gasoline on the ground but it does not begin reacting with the oxygen in the air (combustion) without a source of ignition.

When molecules collide, the clouds of electrons in the orbitals surrounding the atoms repel each other with a great deal of force. In order for a chemical reaction to occur, a molecular collision must involve enough energy to force the molecules close enough together that existing bonds between atoms can be broken and new bonds formed. Not only does this require enough energy; it also requires that the molecules have the right orientation in space at the moment they collide. For example, the steam reforming reaction shown in Equation (10.16) is depicted in Figure 10.13. On the left side of the figure, the CH_4 and H_2O molecules collide and bounce apart because the orientation of the H_2O molecule is not favorable relative to that of the CH_4 molecule. On the right side of the figure, the H_2O molecule is oriented so that the oxygen atom can disrupt the bonds in both molecules and bond with the carbon atom in CH_4 to produce one molecule of CO.

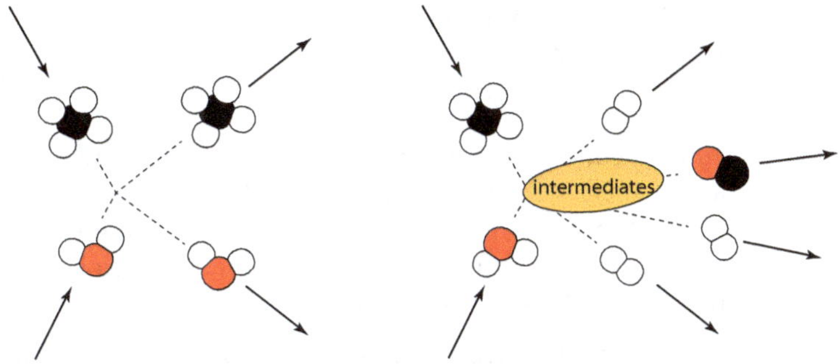

Figure 10.13. If molecular orientation is not favorable, a molecular collision may not result in a reaction (left). A different orientation of molecules may enable a reaction to occur (right). During the reaction, intermediate species form as part of the steps in the reaction mechanism.

The right side of Figure 10.13 also indicates that during the reaction, intermediate species form that are not part of the product species. We return to considering *intermediates* shortly.

10.4.2 Factors Influencing Reaction Rate

The *reaction rate* of a chemical reaction is the rate at which the concentrations of reactants and products are changing. Based on collision theory, there are four important factors we can control that can influence the rate of chemical reactions. These are as follows:

1. *Temperature*

   As you know, the higher the temperature of the reactants, the higher the kinetic energy of the moving particles. At higher speeds, molecules collide more frequently, providing more opportunities for the favorable orientations that lead to the breaking of bonds and formation of new bonds. Note that increasing the temperature only increases the reaction rate if the temperature is high enough for the reaction to commence. We discuss this issue more in Section 10.4.4.

   On rare occasions, higher temperatures can decrease a reaction rate. This can occur, for example, if the higher temperature increases the reverse reaction rate at the expense of the forward reaction rate. We see an example of this in the next chapter, Section 11.1.4.

2. *Surface Area*

   As a factor in chemical reactions, the surface area of solid reactant applies to *heterogeneous reactions*—reactions in which there is more than one state of matter represented among the reactants. A solid reactant that is to react with a liquid or gas has a certain surface area exposed to the liquid or gas reactant. According to collision theory, reactions happen through collisions, so the larger the surface area of the solid reactant, the more the atoms in the solid are exposed to collisions from the fluid reactant molecules.

   A 1-cm cube of solid material has a surface area of 6 cm^2. If the same amount of material is ground to small particles with grains that are also cubes but with side lengths of 0.1 mm, the total surface area of the solid is 600 cm^2. The finer the grains in the powder, the larger the total surface area of the solid.

   Sawdust, coal dust, and powderized aluminum are all examples of substances that may not sound all that dangerous but which can explode violently in a so-called *dust explosion*. In all these cases, the explosion happens because of the enormous surface area provided by a reactant consisting of small particles. If the dust forms a suspension in air, explosive conditions result. Sawdust is essentially powdered carbon, as is coal dust. A spark in a dusty woodworking shop or coal mine has often led to a tragic explosion. Dust from dry grain, flour, powdered milk, and other sources of carbon can produce the same explosive suspension in air. Both aluminum and titanium metals will also form explosive suspensions in air when in powdered form.

3. *Concentration*

   Since chemical reactions occur by means of molecular collisions, the more particles of reactant species there are available, the faster a reaction occurs. When one or more of the reactants are in a solution of liquid or gas, the higher the concentrations of those reactants are, the more molecules there are in the solution to react.

   As an example, oxygen accounts for about 21% of the gas molecules in the air. At the same pressure, the concentration of oxygen molecules in pure oxygen gas is nearly five times the concentration in air. As a result, a combustion reaction fueled by pure oxygen occurs much faster than the same reaction in air, releasing heat energy at the same increased rate. We have seen that acetylene is an unstable molecule with a large positive enthalpy of formation (Table 10.2). When this gas is combined with pure, high-pressure oxygen and ignited, we have a combustion reaction that gives off so much heat so rapidly that the flame can be used for cutting steel, as illustrated in Figure 10.14. The oxyacetylene torch shown has two separate gas tubes arriving at the nozzle, one for the C_2H_2 and one for the O_2. The two gases are stored in separate steel storage tanks.

Figure 10.14. The flame from an oxyacetylene torch cuts right through steel plate.

   The reaction between oxygen and acetylene is an example of a *homogeneous reaction*—one in which all the reactants are in the same state (gas in this case).

4. *Catalysis*

   A *catalyst* is a substance that expedites or facilitates a chemical reaction without itself being consumed by the reaction. The process by which a catalyst influences the rate of a chemical reaction is known as *catalysis*. Metals often act as catalysts because of a phenomenon known as *adsorption*. Adsorption is the collection of molecules on a surface, attracted there by electrostatic attraction or Van der Waals forces. When reactant molecules adsorb onto the surface of a catalyst, the surface forces cause reactant molecules to separate into individual atoms, at which time they are free to bond to other atoms to form new molecules.

   The presence of iron as a catalyst is the key that allows the formation of NH_3 in the Haber-Bosch process. After the N_2 and H_2 molecules adsorb onto the surface of the iron, the diatomic molecules of nitrogen and hydrogen come apart and individual atoms of nitrogen and hydrogen form on the metal surface. After that, hydrogen atoms begin bonding with nitrogen atoms, building up NH_3 molecules by adding one hydrogen atom at a time. The catalyst works by providing a completely different energy pathway, significantly reducing the energy required to make the reaction occur. We will look at this energy pathway more closely shortly.

   A catalyst that is in the same state as the reactants in a chemical system is referred to as a *homogeneous catalyst*. A catalyst in a different state, such as the metal catalyst used with the gases in the Haber-Bosch process reaction, is called a *heterogeneous catalyst*. Note that in reversible reactions such as equilibrium systems, a catalyst increases the reaction rate equally in both directions.

10.4.3 Reaction Mechanisms

Experimental evidence now strongly supports the theory that chemical reactions almost never take place in one simple step. Instead, reactions typically require several separate steps in which the bonds of reactant molecules are rearranged into the bonds of the product molecules. The sequence of steps in which a reaction takes place is called the *reaction mechanism*. A reaction mechanism is a *model* depicting a sequence of steps and *intermediates* involved in a chemical reaction. As mentioned above, intermediates are chemical species that form temporarily during the reaction, but which are not part of the net reaction equation.

As an example, when hydrogen gas, H_2, and iodine gas, I_2, are combined in a sealed container, the gases form the following equilibrium:

$$H_2(g) + I_2(g) \rightleftharpoons 2HI(g)$$

Hydrogen and hydrogen iodide are both colorless gases; iodine gas is purple. At equilibrium, all three gases are present in the container. By means of some ingenious experimental techniques—which include hitting the molecules with certain wavelengths of light—scientists discovered that the H_2 and I_2 molecules do not simply come apart and recombine to form HI gas. Instead, various intermediates are involved.

Figure 10.15 shows three different reaction mechanisms that have been proposed as models for this reaction. All three begin with the formation of individual atoms of iodine, an intermediate. In the first mechanism, the dissociated iodine atoms interact with H_2 to form HI. In the second mechanism, the intermediate I atoms form H_2I, another intermediate, before finally forming HI. In the third mechanism, the dissociated I atoms attach to either side of the H_2 molecule to form yet another intermediate, IH_2I. The attached iodine atoms then cause the breaking of the H_2 bond, leaving separate molecules of HI.

As another example, consider again the Haber-Bosch process. The reaction mechanism for this process, as presently understood, is shown in Figure 10.16 on the next page. This diagram is a combination enthalpy diagram, such as we saw in Figures 10.9 and 10.10, and energy pathway diagram, such as those shown in Figure 10.2. The figure ties together many of the concepts we have already studied in this chapter, as well as a couple of ideas we haven't even gotten to yet, so let's take the time to look it over carefully.

The diagram begins on the left with N_2 and H_2 gas and ends on the right with NH_3. Two different energy pathways are shown. Notice that as Hess's law predicts, the total enthalpy change for the reaction (ΔH = –46 kJ/mol) is the same regardless of the energy pathway. It just happens that one pathway is feasible for an industrial process and the other is not.

The upper path models the way the reaction might proceed without the metal catalyst. The first step is a huge en-

$$I_2 \rightleftharpoons 2I$$
$$2I + H_2 \rightleftharpoons 2HI$$

$$I_2 \rightleftharpoons 2I$$
$$I + H_2 \rightleftharpoons H_2I$$
$$H_2I + I \rightleftharpoons 2HI$$

$$I_2 \rightleftharpoons 2I$$
$$2I + H_2 \rightleftharpoons IH_2I$$
$$IH_2I \rightleftharpoons 2HI$$

Figure 10.15. Three proposed reaction mechanisms for the formation of HI gas from H_2 and I_2.

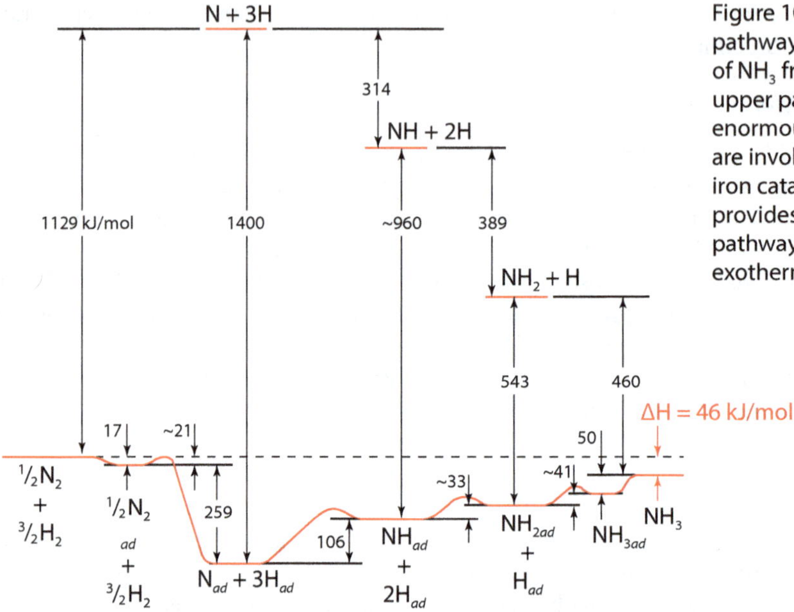

Figure 10.16. Energy pathways for the formation of NH_3 from N_2 and H_2. The upper pathway indicates the enormous enthalpies that are involved without the iron catalyst. The catalyst provides a different energy pathway which is initially exothermic.

dothermic reaction, in which 1,129 kJ/mol of heat must be supplied to break the N_2 triple bond and the H_2 bonds. This produces separate atoms of N and H, which are intermediates. Afterwards, a sequence of exothermic steps leads to the net enthalpy change of $\Delta H = -46$ kJ/mol. In these exothermic steps, hydrogen atoms are added one at a time to the nitrogen to form the intermediates NH and NH_2, and then finally NH_3.

Look now at the fascinating energy pathway at the bottom of the figure, which shows the mechanism for the catalyzed process. The first step is the adsorption (*ad*) of the N_2 and H_2 molecules on the surface of the metal catalyst. Next, while adsorbed on the metal surface, the diatomic molecules separate and this separation is actually *exothermic* under these conditions (!). So instead of having to supply the reactants with 1,129 kJ/mol of heat to break apart the N_2 and H_2 molecules, only about 21 kJ/mol have to be supplied to get over the little energy hump there (more on that soon) and then the molecules release 259 kJ/mol as they dissociate while adsorbed on the catalyst surface. This forms the adsorbed N and H intermediates. From here, several small endothermic steps lead to the intermediates NH and NH_2, and then finally to NH_3. Notice that all these steps occur while the molecules are still adsorbed on the catalyst. Only after the NH_3 molecules are formed is a final boost of heat supplied (50 kJ/mol) to remove the NH_3 molecules from the catalyst and form them into NH_3 gas. The net enthalpy change for this pathway is the same as for the other energy pathway, and the same intermediate species are formed. But with the intermediates adsorbed onto the catalyst, the energies involved are quite different.

10.4.4 Activation Energy and the Activated Complex

Look again at the lower energy pathway in Figure 10.16. Before each energy transition, the reactants need a small energy boost in order for the next step in the reaction to commence. This small energy boost is seen for both exothermic transitions and endothermic transitions.

Figure 10.17 shows an enlarged view of generic energy pathways for both exothermic and endothermic reactions. To commence the reaction, an amount of energy labeled E_a or E_a' must be supplied. This energy is the *activation energy*, E_a (or E_a'), roughly defined as *the amount of energy required to initiate the chemical reaction.* Depending on the specifics of the reaction, the activation energy may be supplied by heating, a spark, a flame, an electrical current, or even by mechanical means such as physical shock or compression (which supplies energy to a gas by

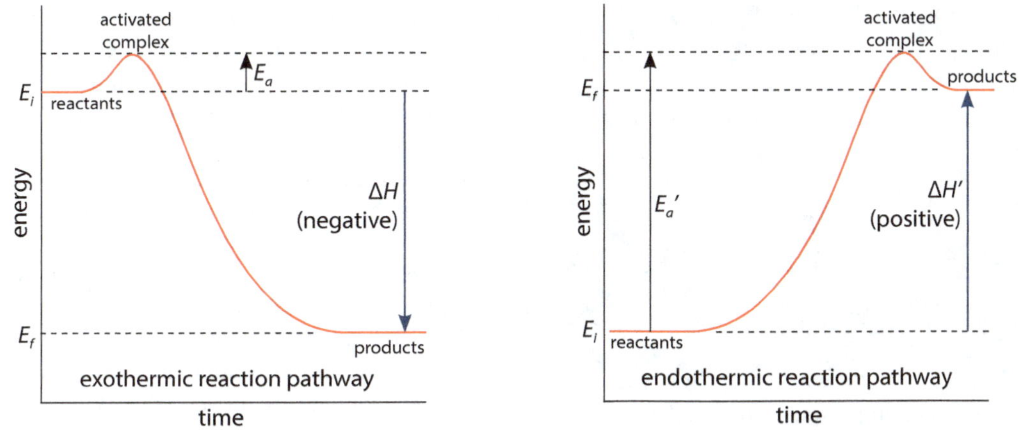

Figure 10.17. To initiate the exothermic or endothermic reaction, the activation energy E_a or E_a' must first be supplied. E_a is the energy required to form the activated complex. Once the activated complex is formed, the reaction commences.

doing work on the gas). However the activation energy is supplied, its effect is to increase the temperature, giving the molecules the kinetic energy they need so that effective molecular collisions occur with enough force to break molecular bonds. Current theory holds that when these effective, energetic collisions occur, a so-called *activated complex* results. An activated complex is a range of transitional molecular structures of partially formed bonds that results from an effective molecular collision with adequate energy. The activated complex lasts only while bonds are in the process of breaking and reforming. The formation of the activated complex is shown in Figure 10.17 at the peak of the energy curve; once this peak is reached and the activated complex forms, the reaction commences. The formation of the activated complex gives us a more nuanced definition for the activation energy: *the activation energy is the minimum energy necessary to transform the reactant molecules into the activated complex for a given reaction.*

Figure 10.17 displays separate diagrams for exothermic and endothermic reactions. However, these diagrams can also be interpreted as representing the energy pathways for the forward and reverse directions in a single reversible reaction. In the exothermic reaction, the total energy change from the energy of the activated complex to the energy of the reaction products is

$$E_a - \Delta H = E_a + |\Delta H|$$

This amount of energy becomes the activation energy, E_a', for the reverse reaction.

$$E_a' = E_a + |\Delta H|$$

The energy required to form the activated complex is the same in either direction. But the energy that must be supplied to the reactants, E_a or E_a', depends on the energy the reactants are starting with.

In some cases, the reactants already have enough energy at room temperature to form an activated complex. Just pour a bit of vinegar into some baking soda and watch what happens—no additional heating is necessary. But in a great many cases, an extra amount of energy must be supplied. As an example, when there is a natural gas leak, why doesn't the gas spontaneously combust when coming in contact with the air? The reason is that although the reactants (methane and oxygen) are together, the activation energy has not been supplied. An open flame might be enough to do the job. Another example is illustrated in Figure 10.18. I hope you have seen

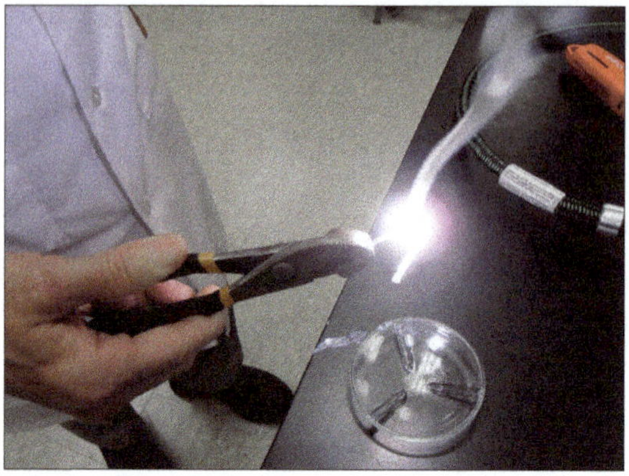

Figure 10.18. Burning magnesium, which lights with a butane lighter, but not with a match.

the bright light of burning magnesium before because the picture doesn't do it justice. One thing I always point out to my students when performing this demonstration is that the magnesium does not ignite with a match, but it does light with a butane gas-grill lighter. The flame of the burning match is not as hot as the butane flame, so it does not produce enough heat to supply the activation energy. I also like to note that the flame always goes out when it is about 1/4 inch from the pliers I use to hold the strip of magnesium metal. The pliers are metal and conduct heat pretty well, cooling the magnesium metal strip. When the flame gets close enough to the pliers, the pliers conduct heat away from the flame so fast that the energy supplied by the flame falls below the activation energy needed to keep the reaction going. Without the requisite amount of energy, the activated complex cannot form and the reaction does not happen.

Now that we have introduced the concepts of reaction mechanism and activation energy, let's return for a moment to the general topic of how catalysts work. Figure 10.19 compares the energy pathway for a single-step exothermic reaction to the alternative, three-step energy pathway provided by the presence of a catalyst. Instead of the large activation energy, E_a, required to initiate the single-step process, the catalyst lowers the activation energy by providing an alternative reaction mechanism. In the generic case shown in the figure, the alternative mechanism consists of three separate reaction steps, each with its own activation energy. But since none of these activation energies is as great as the activation energy in the pathway without the catalyst, the catalyst allows the desired reaction to occur more expeditiously. We saw in the case of the iron catalyst used in the Haber-Bosch process that surface attractions between metal atoms and adsorbed particles reduce the energy required to break apart molecular bonds in the reactants, so much so that the reactant dissociation actually becomes exothermic. Endothermic steps occur after that as the product molecules begin to form, but again energy requirements are substantially lower than the energy required to break apart reactant molecules without the catalyst.

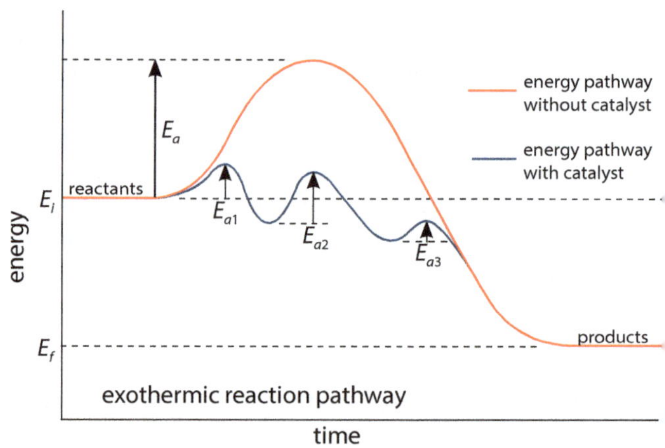

Figure 10.19. The catalyst provides an alternative energy pathway for the reaction with lower activation energy. The alternative path shown entails a three-step reaction.

10.4.5 Reaction Rate Laws

A few pages back, we saw that the concentration of reactants in solution affects the rate of the reaction. In referring to reaction rate, we are referring to the rate at which the molarity of the reactants and products are changing per unit time. This means that the reaction rate, *R*, of a chemical reaction has units of *M*/s, molarity per second.

Consider the reaction between ammonium and nitrite ions to form nitrogen and water:

$$NH_4^+(aq) + NO_2^-(aq) \rightarrow N_2(g) + 2H_2O(l) \tag{10.19}$$

This reaction has two reactants. One of the ways to investigate the effect of concentration on the reaction rate is to vary the concentration of one reactant while holding the concentration of the other reactant constant. For the reaction above, experiments have shown that the reaction rate, *R*, is directly proportional to the NH_4^+ concentration, $[NH_4^+]$. This means that if $[NH_4^+]$ is doubled or tripled, the rate of the reaction doubles or triples, respectively. We model this relationship algebraically as:

$$R = k\left[NH_4^+\right]$$

As we have seen before, *k* is a constant of proportionality. Thus, this expression says that the reaction rate, *R*, is directly proportional to the ammonium ion concentration.

It has also been found experimentally that for the reaction in Equation (10.19), the following holds:

$$R = k\left[NO_2^-\right]$$

Or, in words, the reaction rate is directly proportional to the nitrate ion concentration. Combining these two expressions gives

$$R = k\left[NH_4^+\right]\left[NO_2^-\right] \tag{10.20}$$

Equation (10.20) is an example of a *rate law*—a mathematical expression that characterizes reaction rates in terms of the concentrations of the reactants. The constant *k* is known as the *rate constant*. The units for the rate constant are such as to allow the reaction rate to have units of *M*/s. (The units for *k* are illustrated in the next example.)

In general, we can consider a generic chemical equation such as the following:

$$aA + bB \rightarrow cC + dD \tag{10.21}$$

The reaction rate law for such a reaction is of the following form:

$$R = k[A]^m[B]^n \tag{10.22}$$

In Equation (10.22), the terms in square brackets are the concentrations of the reactants. The exponents *m* and *n*, as well as the rate constant *k*, must be determined experimentally. The values of the exponents *m* and *n* are typically 0, 1, or 2. In Equation (10.20), the exponent on both concentrations is 1, so we say that "the reaction rate is first-order in $[NH_4^+]$" and "the reaction rate is first-order in $[NO_2^-]$." When two first-order terms are multiplied in a rate law, the overall rate law is second order—the orders on the individual terms add together to make the overall rate law order.

Note that the rate constant, *k*, is a *constant* at a given temperature, but the concentrations and reaction rate are clearly not. From the initial concentrations, these vary exponentially (at least, for first-order reactions), decreasing from their initial concentrations and finally stabilizing at their equilibrium concentrations. But once we know the rate law, we can calculate the rate for any given set of concentrations.

Example 10.6

Use the rate data in Table 10.4 to derive the rate law and rate constant for the reaction of ammonium and nitrate ions in water.

Trial	$[NH_4^+]$, initial (M)	$[NO_2^-]$, initial (M)	R, Initial (M/s)
1	0.0050	0.100	1.35×10^{-7}
2	0.0100	0.100	2.70×10^{-7}
3	0.0200	0.100	5.40×10^{-7}
4	0.100	0.0100	2.70×10^{-7}
5	0.100	0.0200	5.40×10^{-7}
6	0.100	0.0300	8.10×10^{-7}

Table 10.4. Rate data for the reaction of ammonium and nitrite ions in aqueous solution (25°C).

Looking at trials 1 and 2, we see that when the initial value of $[NH_4^+]$ is doubled while holding $[NO_2^-]$ constant, the initial reaction rate doubles. Trials 1 and 3 show us that when $[NH_4^+]$ is increased by a factor of four ($[NO_2^-]$ constant), the initial rate increases by the same factor. From these data, we see that the reaction rate is first order in $[NH_4^+]$, or

$$R = k\left[NH_4^+\right]$$

Looking at trials 4 and 5, we see that when $[NO_2^-]$ is doubled ($[NH_4^+]$ constant), the initial rate doubles. Trials 4 and 6 show that when $[NO_2^-]$ is tripled ($[NH_4^+]$ constant), the initial rate increases by the same factor. From these data, we see that the reaction rate is first order in $[NO_2^-]$, or

$$R = k\left[NO_2^-\right]$$

We now know that the rate law is of the form

$$R = k\left[NH_4^+\right]\left[NO_2^-\right]$$

To determine the rate constant, we solve the rate law for k and insert experimental data from one of the trials. Data from any trial gives the same value for k. We will use data from trial 2 so that all values have three significant digits.

$$k = \frac{R}{\left[NH_4^+\right]\left[NO_2^-\right]} = \frac{2.70\times10^{-7}\ \dfrac{M}{\text{s}}}{(0.0100\ M)(0.100\ M)} = 2.70\times10^{-4}\ M^{-1}\text{s}^{-1}$$

The complete rate law is found to be

$$R = 2.70\times10^{-4}\ M^{-1}\text{s}^{-1}\left[NH_4^+\right]\left[NO_2^-\right]$$

10.4.6 Rate Laws and Reaction Mechanisms

The form and order of a reaction rate law is determined by the reaction mechanism. For a single-step reaction, the powers m and n on the reactant concentrations are equal to the coefficients on the reactants. Thus, if Equation (10.21) represents a single-step reaction, the rate law is

$$R = k[\text{A}]^a[\text{B}]^b$$

However, if a reaction takes place by means of a multistep mechanism, the steps occur at different rates and one of the steps occurs more slowly than the others. This step is referred to as the *rate-determining step*. The rate-determining step determines the rate law, and the coefficients on the reactants in the rate-determining step form the powers on the terms in the rate law. This is a powerful clue that helps chemists figure out the actual reaction mechanism.

For example, consider the following reaction:

$$CO(g) + NO_2(g) \rightarrow CO_2(g) + NO(g) \qquad (10.23)$$

It has been found experimentally that the rate law for this reaction is

$$R = k[NO_2]^2$$

This rate law is second order in $[NO_2]$ and doesn't depend on $[CO]$ at all. The fact that the rate law in this case is second order in $[NO_2]$ implies that the reaction in Equation (10.23) is a multistep reaction, and that its rate-determining step begins as follows:

$$2NO_2 \rightarrow$$

We conclude this because, as mentioned above, the rate law for a single-step reaction is formed by multiplying the reactant concentrations together and raising each one to a power that is the same as the coefficient on that reactant. In fact, the mechanism proposed for the reaction in Equation (10.23) is the following:

step 1: $2NO_2 \rightarrow NO_3 + NO$ (slow)

step 2: $NO_3 + CO \rightarrow NO_2 + CO_2$ (fast)

▼ Example 10.7

Nitrogen dioxide reacts with fluorine as follows:

$$2NO_2(g) + F_2(g) \rightarrow 2NO_2F(g)$$

A proposed mechanism for the reaction is the following two-step sequence:

step 1: $NO_2 + F_2 \rightarrow NO_2F + F$ (slow)

step 2: $F + NO_2 \rightarrow NO_2F$ (fast)

Identify the rate-determining step and propose a rate law for the reaction of nitrogen dioxide and fluorine.

The rate-determining step in a multistep reaction is the slowest step. Thus, the first step in the proposed mechanism is the rate-determining step. From this reaction step, the rate law is

$$R = k[NO_2][F_2]$$

▲

Example 10.8

Consider the reaction between nitric oxide and hydrogen:

$$2NO(g) + 2H_2(g) \rightarrow N_2(g) + 2H_2O(g)$$

From experiments, the rate law for this reaction has been found to be

$$R = k[NO]^2[H_2]$$

Address these questions:

1. What is the reaction order for this rate law?
2. If the concentration of H_2 is doubled, what is the effect on the reaction rate?
3. If [NO] is tripled, what is the effect on the reaction rate?

The rate law is second order in [NO] and first order in $[H_2]$. We add these together to get the overall order of the rate law. This rate law is third order.

The rate law is first order in $[H_2]$. Thus, doubling $[H_2]$ doubles the reaction rate.

The rate law is second order in [NO]. Tripling [NO] increases the reaction rate by a factor of $(3)^2 = 9$.

Chapter 10 Exercises

SECTION 10.1

1. Describe how a thermochemical system might do mechanical work. Consider (a) a gas in a cylinder, and (b) reaction products in the open air.
2. Why is ΔH negative for an exothermic reaction?
3. Use the first law of thermodynamics to distinguish between the change in enthalpy and the change in internal energy of a thermochemical system during a reaction.
4. Use the enthalpy diagram shown on the next page to address the following questions.
   a. Is the reaction $CO_2(g) + 2H_2O(l) \rightarrow CO(g) + 2H_2O(l) + \frac{1}{2}O_2(g)$ exothermic or endothermic?
   b. What is the enthalpy change for $CH_4(g) + 2O_2(g) \rightarrow CO_2(g) + 2H_2O(l)$?
   c. Write the thermochemical equations, including the enthalpy change, for the three endothermic processes represented in the diagram.
5. Distinguish between enthalpy of combustion and enthalpy of formation.
6. Why is it important to specify the states of the compounds in a thermochemical equation?

SECTION 10.2

7. Use the enthalpy diagram shown below as an example to explain Hess's law.
8. What practical advantages are there to being able to use Hess's law to compute the enthalpy of reaction for a given process?

9. For each of the following compounds, write the reaction illustrating the formation of the compound from its elements. Show the enthalpy of formation with each equation (see Appendix B, Table B.7 for enthalpy of formation data).

   a. $C_2H_2(g)$

   b. $Li_2SO_4(s)$

   c. $SO_3(g)$

   d. $NH_3(g)$

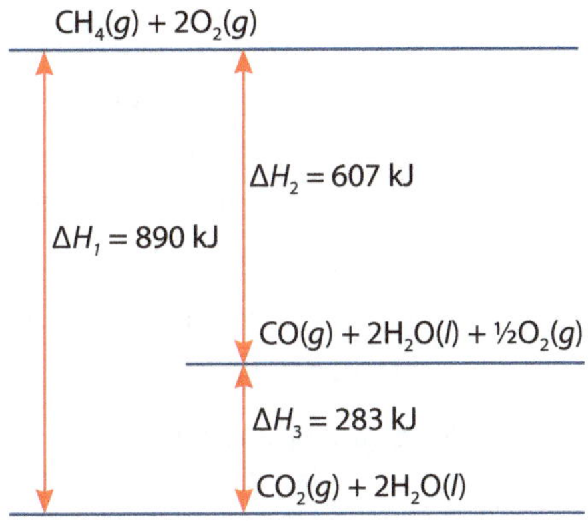

10. Use the method illustrated in Example 10.2 to determine the enthalpy of reaction for each of the following reactions. In each case, solve by using enthalpy of formation data from Appendix B.

    a. $NO(g)+\frac{1}{2}O_2(g) \rightarrow NO_2(g)$

    b. $Fe_2O_3(s) + 3CO(g) \rightarrow 2Fe(s) + 3CO_2(g)$

    c. $Ca(OH)_2(s) \rightarrow CaO(s) + H_2O(g)$

    d. $CaCO_3(s) \rightarrow CaO(s) + CO_2(g)$

    e. $Mg(OH)_2(s) \rightarrow MgO(s) + H_2O(l)$

11. Use Equation (10.14) and the method illustrated in Example 10.3 to verify your results for the five reactions in the previous item.

12. Determine the enthalpy of formation for the following compounds. Solve by using enthalpy of combustion data from Table 10.1. (No combustion equation is required for O_2.)

    a. ethylene, $C_2H_4(g)$

    b. formic acid, $CH_2O_2(l)$

    c. benzene, $C_6H_6(l)$

13. In a combustion reaction at standard conditions, each of the following compounds produces $CO_2(g)$ and $H_2O(l)$. Determine the enthalpy of combustion for one mole of each by using enthalpy of formation data.

    a. methanol, CH_3OH

    b. ethanol, C_2H_5OH

SECTION 10.3

14. For the reactions listed below, inspect the equation and predict whether ΔS° is positive or negative.

    a. $2SO_2(g) + O_2(g) \rightarrow 2SO_3(g)$

    b. $CO(g) + 2H_2(g) \rightarrow CH_3OH(l)$

    c. $Ba(OH)_2(s) \rightarrow BaO(s) + H_2O(g)$

15. Describe how the entropy of a substance relates to the state (solid, liquid, or gas) and temperature of the substance.

16. Why is the entropy of a pure crystalline substance at absolute zero equal to zero?

17. Consider a reaction for which $\Delta H^\circ > 0$ and $\Delta S^\circ < 0$. Explain why this reaction does not happen spontaneously in nature.

18. The boiling point for benzene, C_6H_6, is 80°C. At 60°C, which do you expect to be greater for the vaporization of benzene, ΔH or $T\Delta S$? Explain your response.

19. Determine the change in the standard Gibbs free energy for each of the following reactions at 25°C. For each, state whether the reaction is spontaneous at the given conditions.

    a. $N_2(g) + O_2(g) \rightarrow 2NO(g)$, $\Delta H^\circ = 182.6$ kJ and $\Delta S^\circ = 24.8$ J/K

    b. $C(s) + O_2(g) \rightarrow CO_2(g)$, $\Delta H^\circ = -393.5$ kJ and $\Delta S^\circ = 2.70$ J/K

20. For the following reactions, ΔH and ΔS are given for the temperature indicated. Calculate ΔG for each and indicate whether the reaction is spontaneous at the given conditions.

    a. $\Delta H = 108$ kJ/mol, $\Delta S = 0.0375$ kJ/(mol·K), $T = 292$ K

    b. $\Delta H = -93.1$ kJ/mol, $\Delta S = 0.105$ kJ/(mol·K), $T = 135$°C

    c. $\Delta H = -278$ kJ/mol, $\Delta S = 0.375$ kJ/(mol·K), $T = 775$ K

21. Using enthalpy and entropy data in Appendix B (Tables B.7 and B.8), determine the change in the standard Gibbs free energy for the Haber-Bosch process, in which ammonia is formed from nitrogen and hydrogen gases. First calculate ΔG° at 25°C, then calculate ΔG at 500°C, assuming that the values for ΔH_f and S do not change. (In fact, they do change a little, but this calculation is a good approximation.) Comment on what these two values tell you about the prospect of running the Haber-Bosch process at a pressure of 1 bar (standard pressure).

22. Using data from Appendix B, calculate ΔG°, ΔH°, and ΔS° for each of the following reactions. Then compare your ΔG° values to the values obtained from $\Delta G^\circ = \Delta H^\circ - T\Delta S^\circ$ for each reaction. (You will find the values are virtually the same.)

    a. $6Cl_2(g) + 2Fe_2O_3(s) \rightarrow 4FeCl_3(s) + 3O_2(g)$

    b. $H_2(g) + F_2(g) \rightarrow 2HF(g)$

    c. $NO_2(g) + N_2O(g) \rightarrow 3NO(g)$

    d. $SO_2(g) + 2H_2(g) \rightarrow S(s) + 2H_2O(g)$

    e. $MgCl_2(s) + H_2O(l) \rightarrow MgO(s) + 2HCl(g)$

SECTION 10.4

23. Explain the two factors that determine whether a molecular collision will result in the breaking and reforming of chemical bonds.

24. What is a reaction mechanism?

25. In a certain demonstration, a teacher fills a metal can with non-dairy coffee creamer and makes it explode. The demonstration works by placing a small amount of explosive black powder in the bottom of the metal can, covering the black powder with a paper towel, and filling up the can with coffee creamer. The black powder is ignited with a fuse and this blows the coffee creamer out of the can. The instant the coffee creamer is out of the can it explodes. Explain why coffee creamer, which is basically carbon, explodes as described. (Do not try this demonstration without expert supervision and full attention to safety procedures.)

26. The change in Gibbs free energy for the combustion of gasoline is a large negative number. Explain why gasoline doesn't immediately burst into flames when it is spilled on the ground.

27. A certain reaction is found to occur in a single step, with the following mechanism:

    $A + 2B \rightarrow AB_2$

Determine the rate law for this reaction and use it to predict the effect on the reaction rate of tripling [B].

28. Given the following reaction mechanism:

step 1: $A_2 + A_2 \rightarrow C + D_3$ slow
step 2: $B + D_3 \rightarrow A_2 + E_2$ fast

Determine the overall balanced equation for the reaction, identify the rate determining step, identify the intermediates, and propose an appropriate rate law for the reaction.

29. Draw energy diagrams similar to those shown in Figure 10.18 to fit the following reactions:

a. $\Delta H = -20$ kJ/mol, $E_a = 15$ kJ/mol

b. $\Delta H = 30$ kJ/mol, $E_a = 50$ kJ/mol

c. $\Delta H = -5$ kJ/mol, $E_a = 45$ kJ/mol

30. Consider a reversible reaction. Compare the activation energy and enthalpy change in the forward direction to those of the reverse direction.

31. A certain chemical reaction is described by the single-step mechanism $2A + B \rightarrow C$. Rate data are collected for this reaction, indicating the reaction rates for different concentrations of reactants. These data are shown in the chart.

a. Determine the rate law for the reaction.

b. Determine the value of the rate constant.

c. Determine the initial value of the reaction rate if the initial concentrations of A and B are both 0.25 *M*.

Trial	Initial [A]	Initial [B]	Initial Rate of Forming C
1	0.10 *M*	0.20 *M*	1.15×10^{-4} *M*/s
2	0.10 *M*	0.40 *M*	2.30×10^{-4} *M*/s
3	0.20 *M*	0.40 *M*	9.20×10^{-4} *M*/s

GENERAL REVIEW EXERCISES

32. Determine the volume in liters of 3.55 g of SO_3 gas at STP.

33. Aluminum oxidizes to form Al_2O_3. Assume 0.046 mol Al is combined with 0.028 mol O_2. Determine the limiting reactant and the mass of Al_2O_3 formed.

34. Fluorine gas reacts violently with water to produce HF and O_3. What volumes of HF and O_3 are produced if 5.500 L of F_2 reacts with water? Gas volumes are all at STP.

35. Determine the molarity of a solution of citric acid, $C_6H_8O_7$, if 132.7 g of citric acid are diluted in water and the resulting solution is diluted to a volume of 2500.0 mL.

36. Determine the pOH, $[H_3O^+]$, and $[OH^-]$ for a solution with a pH of 9.81.

37. Determine the expected boiling and freezing points of brine solution formed by dissolving 175.00 g KBr in 450.0 g H_2O.

38. A volume of 17.00 mL of 0.650 *M* NaOH is required to reach the equivalence point in a titration of vinegar. Determine the molarity of acetic acid in the vinegar if the vinegar volume is 18.50 mL.

Chapter 11

Chemical Equilibrium

The interesting formation in this test tube is not a bean sprout. It is lead iodide precipitating out of solution. Although lead iodide is considered an insoluble compound, at low enough concentrations it is soluble. The same can be said for just about any insoluble ionic compound. A saturated solution of an insoluble compound is an example of an equilibrium system. Chemists have developed techniques that enable quantitative predictions to be made about potential precipitations in equilibrium systems. This is one of the topics we address in this chapter.

Objectives for Chapter 11

After studying this chapter and completing the exercises, you should be able to do each of the following tasks, using supporting terms and principles as necessary.

SECTION 11.1

1. Explain why both physical and chemical equilibria are dynamic rather than static.
2. Characterize the direction an equilibrium lies for different values of the equilibrium constant.
3. Write the expression for the equilibrium constant for a given reaction.
4. Given equilibrium concentration data, calculate the value of the equilibrium constant.
5. Use Le Châtelier's principle qualitatively to describe how an equilibrium system responds (that is, in which direction the equilibrium shifts) to changes in temperature, pressure, or concentration.
6. State three examples of circumstances that cause reactions to "go to completion."
7. Explain the common-ion effect.

SECTION 11.2

8. Calculate the value of an acid dissociation constant from concentration data and from the pH of a solution.
9. Use K_a and K_b values to compare the strengths of acids and bases.
10. Explain in detail how a buffered solution is able to maintain the pH unchanged when small amounts of an acid or base are added.
11. Describe the processes of anion and cation hydrolysis.
12. Use hydrolysis to explain why acid-base neutralization reactions and titrations can be acidic or basic at the endpoint.

SECTION 11.3

13. Given concentration data for a solubility equilibrium, calculate the solubility product constant, K_{sp}.
14. Compare solubilities of compounds by comparing the values of their solubility products.
15. Determine the solubility of a compound from its solubility product.
16. Use the solubility product to predict whether precipitation will occur when low-concentration ionic solutions are mixed.

11.1 Physical and Chemical Equilibrium

11.1.1 Equilibria We Have Seen So Far

The concept of equilibrium plays an important role in chemistry. In this chapter, we look at chemical equilibrium in some detail. But before diving into new material, let's review the numerous instances of equilibrium we have seen so far and tie them all together. As we do, it is important to distinguish between *physical equilibrium* and *chemical equilibrium*. Physical equilibrium does not involve chemical reactions; chemical equilibrium does.

In Section 6.2.6, we encountered phase transitions and phase diagrams. There we noted that the triple point of a substance is the set of temperature and pressure conditions that allows all three states of a substance to exist together in equilibrium. This is an instance of physical equilibrium. At the triple point, solid, liquid, and vapor forms of a substance can coexist indefinitely in equilibrium. Under these conditions, there is a continuous process of solid particles transitioning to liquid and vice versa, and liquid particles transitioning to vapor and vice versa. These phase transition rates are all equal. Note that equilibrium is a *dynamic* rather than a *static* condition. It is not as if certain quantities of solid, liquid, and vapor are together with nothing happening. Instead, at the triple point, molecules are continuously engaged in phase transitions—a dynamic situation. But since the phase transition rates are equal, equilibrium is established between the states.

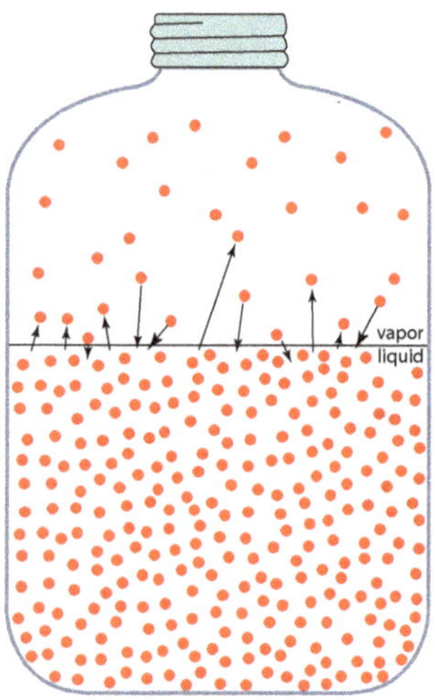

Figure 11.1. The vapor pressure of a substance in a sealed container causes an equilibrium in which vaporization and condensation are happening at equal rates.

In Section 6.2.9, we encountered another instance of physical equilibrium: the vapor pressure of a substance in a sealed container. Figure 6.22 depicts a sealed container containing a liquid, shown again here as Figure 11.1. Evaporation of the liquid occurs until the pressure inside equals the vapor pressure of the liquid at that temperature. At this point, an equilibrium is established in which molecules of the liquid are transitioning from liquid to vapor and vice versa at equal rates. Once again, the equilibrium is dynamic.

A third instance of equilibrium we have seen is that of weak acids and bases in solution. In Sections 8.2.4 and 9.2.4, we examined acetic acid as an example of a weak acid that partially ionizes in water. For example, at a concentration of 0.1 *M*, acetic acid ionizes 1.4%, leaving most of the CH_3COOH molecules non-ionized and establishing the following chemical equilibrium:

$$CH_3COOH(aq) \rightleftharpoons H^+(aq) + CH_3COO^-(aq) \qquad (11.1)$$

Once again, this equilibrium is dynamic: acetic acid molecules are ionizing in solution at the same rate that ions are recombining into molecules.

We encountered a fourth example of equilibrium—saturated solutions—in Section 8.3.1. Figure 8.11, shown again here as Figure 11.2, illustrates an ionic crystal in a saturated solution. This, too, is a dynamic situation: particles are dissolving from the crystal into solution and recrystallizing at the same rate.

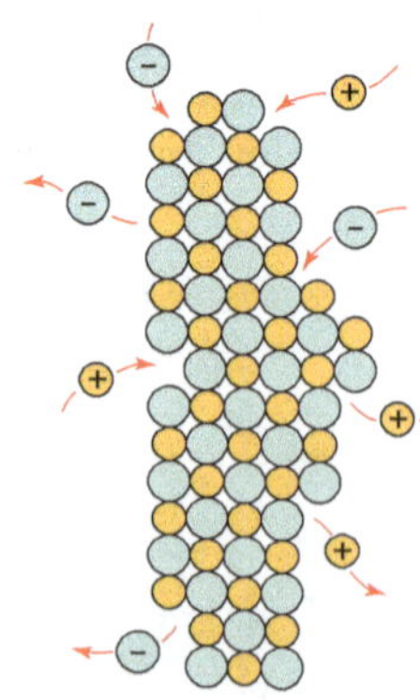
Figure 11.2. At saturation, ions dissociate and recrystallize at the same rate.

I hope you are now wondering whether the last two examples are instances of physical equilibrium or chemical equilibrium. This is not an easy issue to resolve, but you can read about it in the accompanying box.

We saw a fifth example of equilibrium in Section 8.3.6 when we considered Henry's law and gases in solution. Figure 8.15, shown again here as Figure 11.3, represents a gas in liquid solution. Henry's law states that in the equilibrium achieved inside the container the amount of gas dissolved in the liquid is directly proportional to the partial pressure of the gas in the container. This is a physical equilibrium.

In Section 9.2.4, we looked at the equilibrium between conjugate acid-base pairs. There we saw that in a Brønsted-Lowry acid-base reaction, the reaction favors the transfer of protons from the strong acid to the strong base, and thus favors the

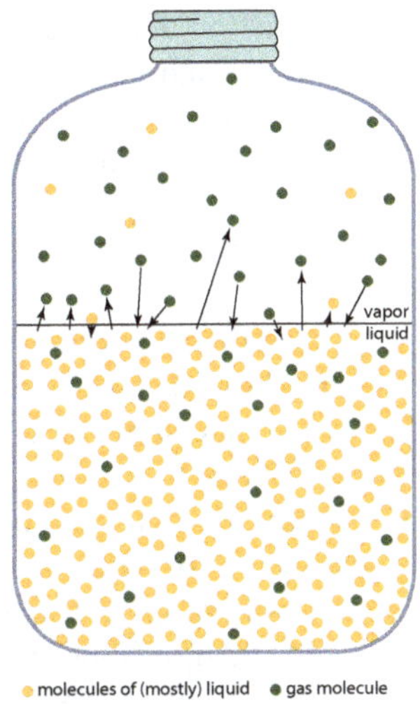

Figure 11.3. When a gas dissolves in a liquid, an equilibrium is reached and the solubility of the gas is directly proportional to the partial pressure of the gas.

Hmm... Interesting. Is dissolution a chemical change?

Sources differ widely on whether the dissolution of an ionic solid in water is a physical change or a chemical change. The nature of this change determines whether the equilibrium established is a physical equilibrium or a chemical equilibrium. The same issue applies to the acetic acid equilibrium. Let's pause here and see where the conflict lies.

Mixtures—such as solutions—are often distinguished from compounds this way: the components in a mixture can be separated by physical means, such as filtration or evaporation. In contrast, the elements in a compound are chemically bonded together and can only be separated by chemical means—a chemical reaction that breaks the chemical bonds.

If salt or sugar is dissolved in water, the salt or sugar can be recovered from the water by allowing the water to evaporate away—a physical change, and a physical means of separating the solute from the water. A physical process can also be used to separate acetic acid from water. The boiling point of acetic acid is about 118°C, so a solution of acetic acid in water can be separated by heating the solution to 100°C and boiling off the water. According to these considerations, dissolution is a physical process, not a chemical one, and thus the equilibrium established in a solution is a physical equilibrium.

However, there is another way of looking at this. One way of distinguishing chemical processes from physical processes is this: in chemical processes new chemical species are formed. According to this definition, the dissolution of sugar must still be regarded as physical. Sugar is a non-electrolyte and does not ionize in water. The sugar molecules are intact and no new chemical species are formed when sugar dissolves. But with salt and acetic acid, the situation is different. When NaCl dissolves, the species Na^+ and Cl^- are formed as the NaCl crystal dissociates. Likewise, when CH_3COOH dissolves, the species H^+ and CH_3COO^- are formed (and the H^+ ions immediately form H_3O^+ ions). Considered this way, the dissolution of salt and acetic acid are chemical processes, and thus the resulting equilibria are chemical equilibria. (However, compare this to the first paragraph of Section 8.5.1.)

Regretfully, there is no easy resolution to these conflicting definitions. Half the sources out there classify the dissolution of salt as a physical process; the other half classify it as chemical. Remember, chemistry is all about *modeling*. We try to make our definitions as robust and distinctive as possible, but our models aren't perfect, and sometimes the lines established by scientific definitions are a bit fuzzy. Such is life.

production of the weak acid and the weak base. As a result, the equilibrium is shifted to the left or to the right.

In Section 10.4.3, we saw another chemical equilibrium, the reversible reaction between hydrogen and iodine to form hydrogen iodide:

$$H_2(g) + I_2(g) \rightleftharpoons 2HI(g) \tag{11.2}$$

Finally, in Section 8.3.6 we encountered an important governing principle that applies to equilibria, Le Châtelier's principle. Le Châtelier's principle states that if a stress is placed on a system in equilibrium, the system shifts the equilibrium so as to remove or reduce the stress.

Whether physical or chemical, equilibrium systems are dynamic, and the equilibrium of a system may "shift to the right" or "shift to the left," depending on the conditions. For example, at higher concentrations, acetic acid ionizes more. So if the concentration of CH_3COOH molecules in a solution is increased, the equilibrium represented by Equation (11.1) shifts to the right and more ions form. On the other hand, if CH_3COOH molecules are removed from the solution, the equilibrium shifts to the left, and ions combine to form more CH_3COOH molecules.

11.1.2 Dynamic Chemical Equilibrium

Although any reaction may theoretically be considered to be reversible, in practice there are some reactions that are in fact reversible and others that we normally do not think of as reversible. An explosion is generally not considered reversible. On the other hand, the Haber-Bosch process is reversible, and molecules of H_2 and N_2 are just as likely to form from the NH_3 as vice versa. When discussing chemical equilibrium, we are generally discussing chemical reactions that are reversible this way in practice.

In some chemical equilibria, production of reactants and products is equally favored. An example is the reaction Joseph Priestley used (Section 5.1.2) to decompose mercury(II) oxide:

$$2HgO(s) \rightleftharpoons 2Hg(l) + O_2(g)$$

If the mercury oxide is heated in a sealed container, the forward reaction separating HgO into Hg and O_2 reaches an equilibrium with the reverse reaction in which Hg and O_2 recombine to form HgO, with the two reactions being equally favored.

In other cases, the equilibrium formed favors one side of the reaction equation or the other. We saw an example of this in the Brønsted-Lowry acid-base reaction of Example 9.2. In that example, the reactants are the stronger acid and base, so production of the products is favored and the equilibrium lies to the right. In this case, the reaction arrows may be drawn unevenly to indicate that the forward reaction is favored, as follows:

$$HSO_4^-(aq) + CO_3^{2-}(aq) \rightleftharpoons SO_4^{2-}(aq) + HCO_3^-(aq)$$

Similarly, the equilibrium can also lie to the left, indicated in the equation by the left-pointing arrow being longer than the right-pointing arrow.

As with the other equilibria we reviewed in the previous section, a chemical equilibrium is a dynamic system in which forward and reverse reactions are happening simultaneously and at the same rate. Consider the following generic chemical equation:

$$A + B \rightleftharpoons C + D$$

Figure 11.4 illustrates how the forward and reverse reaction rates change over time. If this reaction is initiated with reactants A and B, the forward reaction rate starts out high and decreases as reactants are converted to products. The reverse reaction rate is zero at the beginning because the product species do not yet exist. As the concentrations of the product species increase, the rate of the reverse reaction increases. The rate of the forward reaction continues to decrease and the rate of the reverse reaction continues to increase until the two rates are equal. At this point, chemical equilibrium is achieved.

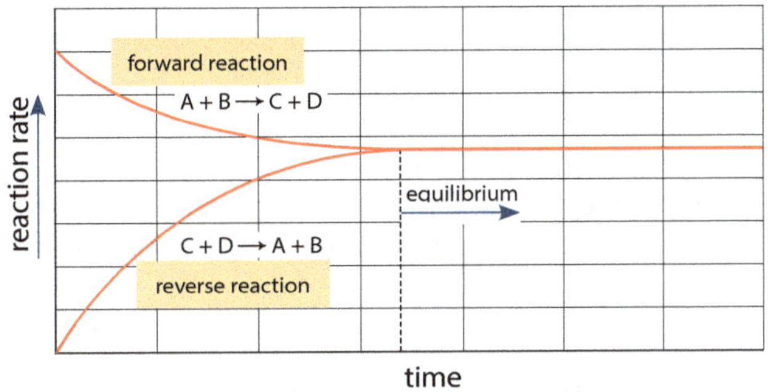

Figure 11.4. Initially, the forward reaction rate is high and the reverse reaction rate is zero. At equilibrium, the rates are equal.

11.1.3 The Law of Chemical Equilibrium and the Equilibrium Constant

As we consider chemical equilibrium now in more depth, we examine some actual data for the reaction between hydrogen and iodine to form hydrogen iodide, Equation (11.2). The data we consider are based on a reaction occurring in a 1.000-L reaction vessel at 440°C. Concentrations of H_2, I_2, and HI for several different reaction trials are shown in Figure 11.5.

The red curves in the upper graph show the concentrations over time when 1.000 mole of H_2 and 1.000 mole of I_2 are introduced into the vessel. The concentrations of H_2 and I_2 both fall from 1.000 to 0.218 *M*, while the concentration of HI rises from 0 to 1.564 *M*. When equilibrium is reached, the concentrations remain constant.

The blue curves in the upper graph show the concentrations if we begin with the reverse reaction by introducing 2.000 moles of HI into the reaction vessel. The concentration of HI falls from 2.000 *M* to the same level as before, and the concentrations of H_2 and I_2 rise from 0 to the same level as before. For both the reactions shown in the upper graph, if the reaction is initiated with initial amounts that are half as much, the final concentrations are half as great.

The lower graph shows how the concentrations proceed if we begin the reaction with unequal amounts of H_2 and I_2: 1.000 mole H_2 and 0.500 mol I_2.

Regardless of how this reaction is conducted or the initial concentrations, we find that the equilibrium concentrations always satisfy this equation:

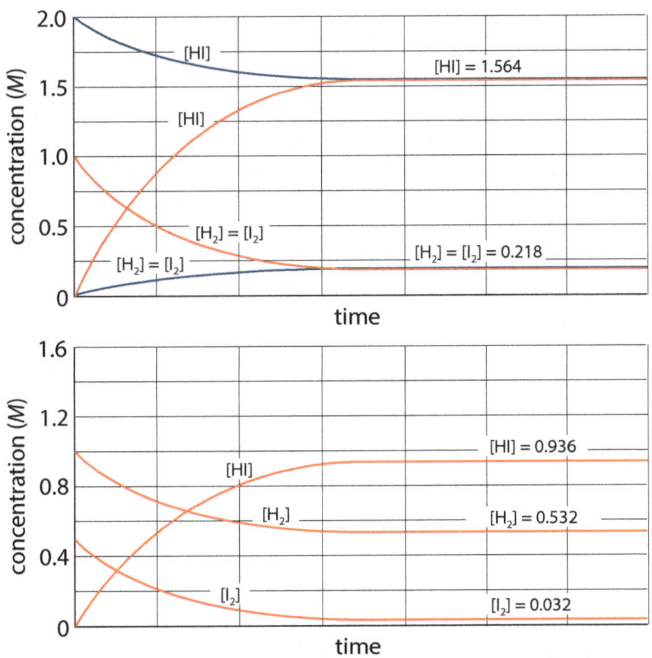

Figure 11.5. Changes in reactant and product concentrations during the development of chemical equilibrium for different starting conditions.

$$\frac{[HI]^2}{[H_2][I_2]} = 51.5$$

The value for this expression is called the *equilibrium constant* for this reaction, symbolized by *K*. It turns out that all chemical equilibria display the same kind of regularity. So, consider now a generic equation representing a reversible chemical reaction in equilibrium:

$$a\text{A} + b\text{B} \rightleftharpoons c\text{C} + d\text{D}$$

The equilibrium constant, *K*, for the reaction is given by this expression:

$$K = \frac{[C]^c[D]^d}{[A]^a[B]^b} \quad (11.3)$$

Technically, the units of measure for *K* vary based on the terms in the equilibrium expression. In practice, the units for *K* are ignored and *K* is treated as a dimensionless number.

The value of the ratio *K* is a constant at a given temperature for any chemical equilibrium. This fact is known as the *law of chemical equilibrium*: *At a given temperature, the product of the concentrations of reaction products raised to the powers of their coefficients divided by product of the concentrations of reactants raised to the powers of their coefficients is a constant.*

When writing expressions for the equilibrium constant for a given reaction, only species whose concentrations are variable are included. Pure substances in condensed phases (solid or liquid) are not included. For example, water is not included for reactions in aqueous solution. Even though some water molecules do form H_3O^+ and OH^- ions, the concentration of H_2O molecules is vastly greater than any other species and remains virtually unchanged. In the case of all solids, and other pure substances in the liquid phase, the "concentration" is not really even a defined quantity. Thus, these substances are not included in the expression for the equilibrium constant.

Notice in Equation (11.3) that the value of *K* implies a lot about the nature of the equilibrium. A large value for *K* indicates that the equilibrium lies to the right—the reaction products are favored. A small value for *K* indicates that the concentrations of the reactants are greater, and the equilibrium lies to the left. A moderate value for *K* indicates that significant quantities of all species are present in the equilibrium. There are no fixed boundaries for what constitutes a "moderate value," but a range of 0.01 to 100 (10^{-2} to 10^2) may be used as a rough guide.

Example 11.1

Write an expression for the equilibrium constant for each of the reactions shown:

1. $2\text{Cl}_2(g) + 2\text{H}_2\text{O}(g) \rightleftharpoons 4\text{HCl}(g) + \text{O}_2(g)$
2. $\text{CaCO}_3(s) \rightleftharpoons \text{CaO}(s) + \text{CO}_2(g)$
3. $\text{HF}(aq) + \text{H}_2\text{O}(l) \rightleftharpoons \text{H}_3\text{O}^+(aq) + \text{F}^-(aq)$
4. $4\text{H}_2\text{O}(g) + 3\text{Fe}(s) \rightleftharpoons 4\text{H}_2(g) + \text{Fe}_3\text{O}_4(s)$

In the first reaction, H_2O is included in the equilibrium constant because it is a gas in the reaction. Writing products in the numerator and reactants in the denominator and raising each to the power of its coefficient, we have:

$$K = \frac{[HCl]^4[O_2]}{[Cl_2]^2[H_2O]^2}$$

In the second reaction, both $CaCO_3$ and CaO are solids and are not included. Thus, the expression for the equilibrium constant is:

$$K = [CO_2]$$

In the third reaction, water is a liquid, so it is not included.

$$K = \frac{[H_3O^+][F^-]}{[HF]}$$

In the fourth reaction, both Fe and Fe_3O_4 are solids and are omitted from the equilibrium constant expression.

$$K = \frac{[H_2]^4}{[H_2O]^4}$$

▲

▼ Example 11.2

In an equilibrium mixture of N_2, O_2, and NO gases at 1500 K, the concentrations are found to be the following:

$[N_2] = 3.2 \times 10^{-3}\ M$

$[O_2] = 8.5 \times 10^{-4}\ M$

$[NO] = 5.5 \times 10^{-6}\ M$

Determine the equilibrium constant for this reaction and judge from its value where the equilibrium lies.

We begin by writing the balanced equation:

$$N_2(g) + O_2(g) \rightleftharpoons 2NO(g)$$

From this we write the expression for the equilibrium constant:

$$K = \frac{[NO]^2}{[N_2][O_2]}$$

Next, we insert the values for the equilibrium concentrations and compute the value of K:

$$K = \frac{[NO]^2}{[N_2][O_2]} = \frac{(5.5\times10^{-6}\ M)^2}{(3.2\times10^{-3}\ M)(8.5\times10^{-4}\ M)} = 1.1\times10^{-5}$$

From the low value for K, we see that the equilibrium lies to the left, favoring the production of N_2 and O_2. This should be evident from the values of the equilibrium concentrations given in the problem. The equilibrium concentrations of N_2 and O_2 are approximately 100 to 1,000 times higher than the concentration of NO.

▲

One more important instance of an equilibrium constant needs to be mentioned here. Recall the equilibrium equation for the self-ionization of water from Section 9.3.1:

$$H_2O(l) + H_2O(l) \rightleftharpoons H_3O^+(aq) + OH^-(aq)$$

In this equation, H_2O is a liquid, and thus does not appear in the expression for the equilibrium constant. So, the expression for the equilibrium constant for this reaction is:

$$K = \left[H_3O^+\right]\left[OH^-\right]$$

This expression for the equilibrium constant for the self-ionization of water is identical to the expression given in Chapter 9 for the ionization constant for water, K_w:

$$K_w = \left[H_3O^+\right]\left[OH^-\right] = 1.0\times10^{-7}\ M\cdot1.0\times10^{-7}\ M = 1.0\times10^{-14}\ M^2$$

The point here is simply to draw your attention to the fact that the ionization constant for water is an equilibrium constant. (However, unlike other equilibrium constants, it is customary to show the units of measure for K_w.) We soon encounter other examples of equilibrium constants.

11.1.4 Le Châtelier's Principle and Equilibrium Shifts

Le Châtelier's principle may be used to predict the consequences of a change introduced into an equilibrium system. We will use the Haber-Bosch reaction to illustrate several possibilities. In this reaction, nitrogen and hydrogen gases are combined in the presence of a metal catalyst to produce ammonia:

$$N_2(g) + 3H_2(g) \rightleftharpoons 2NH_3(g)$$

Changes in Temperature Consider an exothermic process in which heat is released. The colder the environment is, the faster heat flows out of an exothermic system and into the environment. For example, if a person runs a few miles outdoors, the person cools off faster after the run on a cool day than on a warm day.

Every reversible reaction is exothermic in one direction and endothermic in the opposite direction. As in the case of the runner, exothermic chemical reactions release heat. (In fact, your body's metabolic processes *are* exothermic reactions.) Lowering the temperature of the reaction environment increases the rate at which heat flows out of the reaction, and thus shifts the equilibrium in the direction of the exothermic reaction.

In the same way, the reactants in an endothermic reaction absorb heat. Raising the temperature increases the rate at which heat flows into the compounds in the reaction, and thus favors the endothermic reaction, shifting the equilibrium in that direction.

Often the heat released in an exothermic reaction is treated as a reaction product and written in the reaction equation. As we saw in Section 10.4.3, the enthalpy of reaction for the Haber-Bosch process is –46 kJ/mol (see also Table 10.2), so the forward reaction in which ammonia is produced is exothermic. The balanced equation for the reaction shows two moles of ammonia being produced, and can be written this way:

$$N_2(g) + 3H_2(g) \rightleftharpoons 2NH_3(g) + 92\text{ kJ}$$

Now let's consider this reaction and apply Le Châtelier's principle to determine how the equilibrium is affected by a change in temperature. Imagine what happens if the reaction takes place in a thermally insulated vessel so that no heat flows in or out. The forward reaction produces heat—46 kJ per mole of ammonia to be precise. In a sealed vessel, this heat cannot escape, so the temperature inside the reaction vessel increases. Le Châtelier's principle says that the equilibrium shifts in a direction to relieve this stress. If the exothermic production of ammonia produces heat and warms things up, then the reverse reaction absorbs heat and cools things down. Another way to put this is to say that higher temperatures favor the decomposition of ammonia into nitrogen and hydrogen, and thus drive the reaction to the left. Lowering the temperature favors the release of heat, and thus shifts the reaction to the right.

In Section 10.4.2, we saw that reaction rates are higher at higher temperatures. In reactions involving diatomic nitrogen, high temperatures are even more important because the nitrogen molecule is held together by a triple bond that requires a lot of energy to break. But in this case, the reaction favored by higher temperatures is not the desired one, at least not if the purpose is to manufacture ammonia. This presents a problem because at lower temperatures, the forward reaction happens too slowly to be commercially useful. Thus, a compromise temperature of about 500°C is used—low enough to allow the forward reaction to proceed, but high enough to make the reaction commercially useful. The real key to the reaction is the pressure, which we consider shortly.

An example of an endothermic reaction is the roasting of calcium carbonate to produce quick lime (CaO), the main ingredient in cement. In endothermic reactions, the reactants absorb heat, and the amount of heat absorbed by the reactants can be shown on the left side of the equation, as if the heat itself were a reactant in the reaction. In the roasting of one mole of $CaCO_3$, 556 kJ of heat are absorbed. Thus we have:

$$556\text{ kJ} + CaCO_3(s) \rightleftharpoons CaO(s) + CO_2(g)$$

The expression is equivalent to writing

$$CaCO_3(s) \rightleftharpoons CaO(s) + CO_2(g) \qquad \Delta H = 556\text{ kJ}$$

This reaction occurs above 825°C. Increasing the temperature further favors the endothermic reaction, which is the decomposition of calcium carbonate into calcium oxide and carbon dioxide.

As a final reminder, the equilibrium constant is temperature dependent. The equilibrium constant depends on the reaction temperature.

Changes in Pressure Three important facts you have learned must be borne in mind as we consider how pressure variations affect systems in chemical equilibrium. First, practically speaking, solids and liquids are incompressible. Thus, changes in pressure only affect gas-phase equilibrium systems. Second, recall that the pressure of a gas mixture is equal to the sum of the partial pressures of the individual gases. And third, the partial pressure of a gas is dependent on the number of moles of that gas that are present. An implication of these facts is that attempting to influence a reaction by increasing the number of moles of a gas that is not a reactant does nothing. If, for example, inert helium gas is pumped into the reaction vessel where nitrogen and hydrogen are reacting to produce ammonia, the pressure in the vessel increases, but the reaction is not affected because the partial pressures of nitrogen, hydrogen, and ammonia are not changed. Thus, to affect the equilibrium of a gas system by changes in pressure, the number of moles of reactant gases must be manipulated.

Look again at the equation for the Haber-Bosch process:

$$N_2(g) + 3H_2(g) \rightleftharpoons 2NH_3(g)$$

Suppose that the pressure is increased by increasing the amounts of nitrogen and hydrogen gas in the reaction vessel. Increasing the N_2 and H_2 in the vessel increases the partial pressures of these species. Le Châtelier's principle says that the equilibrium shifts to relieve the stress of this change. In the reaction, four moles of gas are converted into two moles of gas, which means that the pressure in the reaction vessel decreases if molecules of nitrogen and hydrogen are converted into ammonia. Thus, an increase in pressure caused by the addition of more nitrogen and hydrogen drives the equilibrium to the right, favoring the production of ammonia.

We saw before that the temperature of the Haber-Bosch reaction is a compromise between the requirements of the forward reaction and the desire to prevent the reverse reaction; so manipulating the temperature to produce more ammonia is not an option. But manipulating the pressure has a substantial effect on the reaction. It should now be apparent why the Haber-Bosch process operates at an extremely high pressure, in the range of 300 to 1,000 atm.

Our analysis here of the consequences of increasing N_2 and H_2 quantities is conducted in terms of the partial pressures of the reactants and the effect of this change in pressure on the equilibrium system. But we can easily analyze this in terms of concentration instead. As you recall, increasing the concentration of reactants increases the reaction rate. In the next segment, we use the concentration of ammonia as our analytical tool.

Changes in Concentration Consider the effect of removing molecules of NH_3 from the reaction vessel. This reduces the concentration of ammonia in the reaction vessel since the concentration of ammonia is equal to the number of moles of ammonia per liter of volume in the reaction vessel. If ammonia is removed, Le Châtelier's principle implies that the system responds by producing more molecules of ammonia, again shifting the equilibrium to the right.

Removing the ammonia from the reaction vessel is relatively simple since ammonia condenses at a much higher temperature (240 K, –33°C) than N_2 (77 K, –196°C) and H_2 (20 K, –253°C). The entire process is interesting enough to warrant showing the overall process flow diagram, Figure 11.6. The left half of the flow diagram shows the process of producing the necessary N_2 and H_2. The Haber-Bosch reaction occurs in the vessel labeled *Reactor*. The pipe exiting the top of the reactor contains a mixture of N_2, H_2, and NH_3. This hot gas line is first passed through a boiler to produce steam, which recovers heat that otherwise would be wasted. Then the pipe goes through a cooler where the ammonia condenses out and is removed. The cooled nitrogen and hydrogen are recompressed (which also heats them up) and returned to the reactor.

Example 11.3

Apply Le Châtelier's principle to an equilibrium system consisting of acetic acid in water to determine the effect of increasing the concentration of acetate ions in the solution.

The equation describing the equilibrium is as follows:

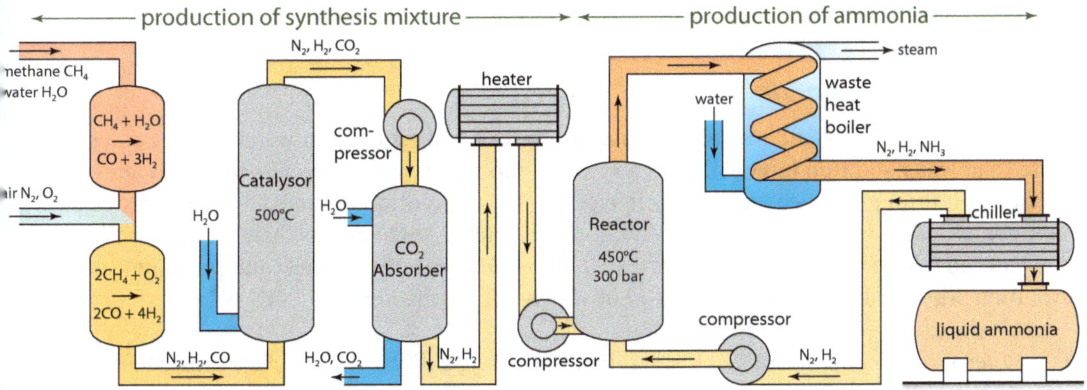

Figure 11.6. Process flow diagram for the Haber-Bosch process. At the right, the condensation unit for removing liquid ammonia is shown.

$$CH_3COOH(aq) + H_2O(l) \rightleftharpoons H_3O^+(aq) + CH_3COO^-(aq)$$

Hydronium and acetate ions are produced when acetic acid ionizes in water. An increase in the concentration of acetate ions in the solution introduces a stress on the equilibrium system. Accordingly, the equilibrium system shifts to the left, converting some of the added CH_3COO^- ions into molecules of CH_3COOH.

We see this very process in action below when we discuss the *common-ion effect*.

11.1.5 Reactions That Go To Completion

Obviously, not every chemical reaction develops an equilibrium in which a reversible reaction is occurring in both directions simultaneously at equal rates. In particular, there are three cases in which a reaction "goes to completion," producing reaction products in a form that removes them from the action.

The first is the case of a reaction in aqueous solution open to the atmosphere in which a gas evolves. An example of this that is probably familiar to you is the decomposition of carbonic acid to form water and carbon dioxide:

$$H_2CO_3(aq) \rightarrow H_2O(l) + CO_2(g) \tag{11.4}$$

A sealed soft drink container is an equilibrium system containing carbonic acid, H_2CO_3, in solution (along with sugar and other flavorings). When the container is opened, the carbonic acid spontaneously decomposes into water and carbon dioxide, and the carbon dioxide leaves the solution—that is, the gas fizzes out of the beverage. Once the gas has evolved, it is no longer present in the system. Thus, the reaction shown in Equation (11.4) goes to completion.

A second event that removes species from a solution is the formation of a precipitate, which we addressed in Section 8.5.1. Clearly, the formation of a precipitate removes the insoluble species from the solution and is another instance of a reaction going to completion.

A third example is acid-base neutralization reactions. In a neutralization reaction, the cation from the base and the anion from the acid remain in solution. But the OH^- ion from the base and the H^+ ion from the acid mostly form water molecules. A few of these ions remain in solution as H_3O^+ and OH^-. But as you know, the concentration of these ions in water is very low

(1×10^{-7} *M* each in a neutral solution). Thus, the formation of water molecules is essentially a reaction that goes to completion.

11.1.6 The Common-Ion Effect

Hydrogen chloride and sodium chloride are both highly soluble in water. Consider a saturated salt water solution (an equilibrium system) into which hydrogen chloride gas is added by bubbling it through the water. What will happen? The dissolution of the HCl produces an overabundance of Cl^- ions, a stress on the equilibrium system. Applying Le Châtelier's principle, the system can relieve this stress if some of the Cl^- ions combine with Na^+ ions and precipitate out of solution. The result is a new equilibrium in which $[Cl^-]$ is greater and $[Na^+]$ is lower. However, the product $[Cl^-]\cdot[Na^+]$ remains constant.

When a soluble solute is added to a solution containing a different solute, if the two solutes have an ion in common, the result is either precipitation or reduced ionization. This phenomenon is called the *common-ion effect.*

The HCl–NaCl illustration is an instance of precipitation occurring to relieve the stress on the equilibrium caused by the addition of ions that are already abundantly present in a solution. In other cases, the system response is to reduce ionization rather than cause precipitation. This happens when one of the ions from a weak electrolyte is added to the solution to excess. For example, if sodium acetate, $NaCH_3COO$, is added to a solution of acetic acid, CH_3COOH, the acetate ion appears in excess when the $NaCH_3COO$ is added.

This circumstance is illustrated in Figure 11.7. Beginning on the left, the first two panels in the figure illustrate the formation of the acetic acid solution. As you know, the CH_3COOH only partially ionizes, so CH_3COOH molecules are in the solution along with H_3O^+, CH_3COO^-, and H_2O molecules. This is the situation in the second panel of the diagram. In the third panel, $NaCH_3COO$ is added to this solution, and the $NaCH_3COO$ dissociates to form Na^+ and CH_3COO^- ions. At this point, there is an excess of CH_3COO^- ions. As we saw in Example 11.3, the system relieves the stress of this overabundance by some of the CH_3COO^- ions recombining with protons from H_3O^+ ions to form acetic acid molecules. This is illustrated in the right-most panel of the graphic.

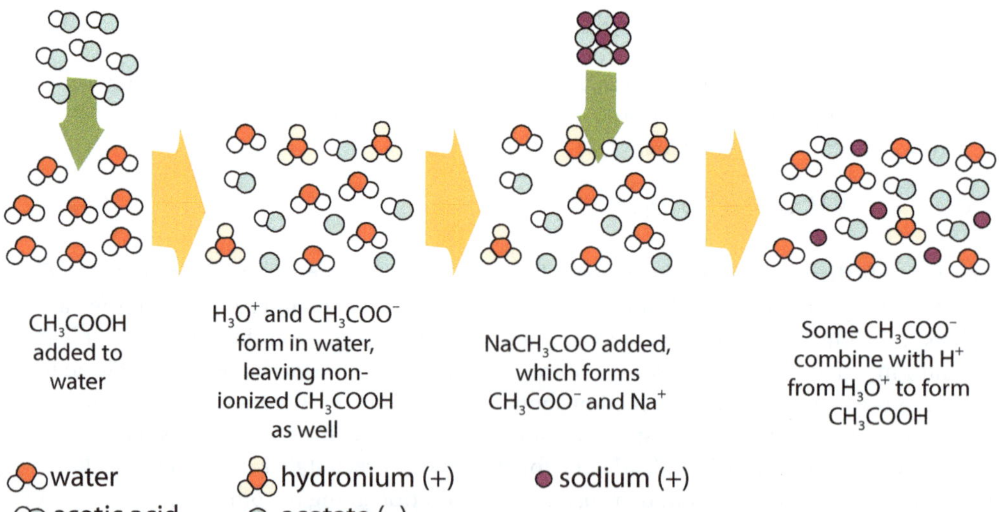

Figure 11.7. Illustration of the common ion effect. Adding $NaCH_3COO$ to an acetic acid solution increases $[CH_3COO^-]$, which in turn increases non-ionized $[CH_3COOH]$. (Hydrogen atoms in H_3O^+ molecules have been lightly colored to help distinguish H_3O^+ from H_2O.)

The result of the common-ion effect in the case of a weak acid and a salt containing the same anion is to reduce the ionization of the acid. Since the formation of acetic acid molecules requires protons from hydronium molecules, the result of the common ion effect in the case of a weak acid is to decrease $[H_3O^+]$ and thus increase the pH. In summary, the addition of the $NaCH_3COO$ to the solution initially increases $[CH_3COO^-]$, but the increase is relieved by a decrease in $[CH_3COO^-]$ and $[H_3O^+]$, an increase in $[CH_3COOH]$, and a corresponding increase in pH.

11.2 Acid-Base Equilibrium

11.2.1 The Acid Dissociation Constant

We have repeatedly used acetic acid to illustrate the dissociation of a weak acid in water:

$$CH_3COOH(aq) + H_2O(l) \rightleftharpoons H_3O^+(aq) + CH_3COO^-(aq)$$

The equilibrium constant for this dissociation is:

$$K = \frac{[H_3O^+][CH_3COO^-]}{[CH_3COOH]}$$

Let's use this expression as a model for a general equilibrium constant for the dissociation of an acid. We will use HA to represent a generic acid with ions H^+ and A^-. The equilibrium equation for the dissociation of this generic acid is:

$$HA(aq) + H_2O(l) \rightleftharpoons H_3O^+(aq) + A^-(aq) \qquad (11.5)$$

From this equation, we obtain the equilibrium constant:

$$K_a = \frac{[H_3O^+][A^-]}{[HA]} \qquad (11.6)$$

Equation (11.6) describes the equilibrium constant for an important class of equilibria—the equilibrium of an acid in water. Because of its importance, this equilibrium constant is denoted with a subscript—*a* for acid. The term K_a is called the *acid dissociation constant* (or *acid ionization constant*). As with other equilibrium constants, K_a is constant at a given temperature.

As mentioned in Section 11.1.3, when the value of an equilibrium constant is below about 10^{-2}, the equilibrium lies to the left. You already know that weak acids do not ionize very much in solution. So in the case of a weak acid, there are plenty of HA molecules in the solution and relatively few H_3O^+ and A^- molecules. Thus, the equilibrium in Equation (11.5) lies to the left, and weak acids tend to have K_a values between 10^{-3} and 10^{-10}. Table 11.1 lists a few weak acids and their dissociation constants. Several of these are also listed in Table 9.6, and you may wish to compare these two tables.

There are several ways to calculate K_a from experimental data. An obvious possibility is to measure concentrations precisely and use Equation (11.6). For weak acids, an interesting alternative approach that is easy to perform in a lab is to use a couple of approximations that enable a simple computation of K_a from the pH of a solution with a known molarity. The first approximation is to ignore the contribution to $[H_3O^+]$ due to the self-ionization of water, which is quite small (10^{-7} *M*). If the pH measurement gives a value for $[H_3O^+]$ that is on the order of 100 times

Table 11.1. Representative weak acids and dissociation constants. The hydrogen atom that dissociates is shown in red in the Lewis structure.

Acid	Lewis Structure	Conjugate Base	K_a
hydrofluoric (HF)	H—F	F^-	6.3×10^{-4}
nitrous (HNO_2)	H—O—N=O	NO_2^-	5.6×10^{-4}
hypochlorous (HClO)	H—O—Cl	ClO^-	4.0×10^{-8}
hydrocyanic (HCN)	H—C≡N	CN^-	6.2×10^{-10}
acetic (CH_3COOH)	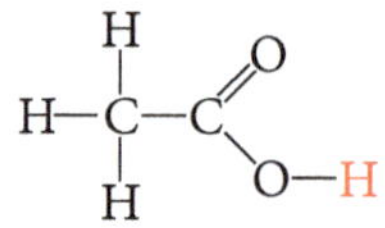	CH_3COO^-	1.7×10^{-5}
formic (HCOOH)	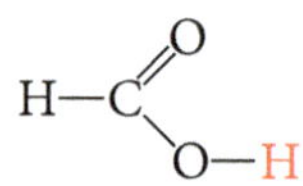	$HCOO^-$	1.8×10^{-4}
phenol (HOC_6H_5)	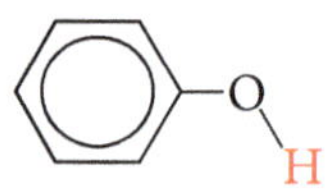	$C_6H_5O^-$	1.0×10^{-10}
benzoic (C_6H_5COOH)	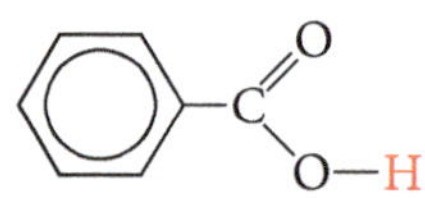	$C_6H_5COO^-$	6.3×10^{-5}

greater (10^{-5} or more), then the error introduced by ignoring the hydronium production due to the self-ionization of water is less than 1%.

The second approximation is to use the molarity of the acid solution as an approximation for [HA] at equilibrium. At equilibrium, the concentration of non-ionized acid molecules is the original molarity of the acid solution less the number of acid molecules that have ionized. Each HA that ionizes produces one H_3O^+ ion, so

$$[HA] = (\text{original molarity}) - [H_3O^+]$$

For weak acids, $[H_3O^+]$ is typically 10^{-3} or less, so we can use the molarity of the acid solution as an approximation for [HA] without much loss of accuracy.

With these two approximations, the basic procedure is to determine the value of $[H_3O^+]$ from the pH, just as we did in Section 9.3.2. Then we notice that each acid molecule that dissociates produces one H_3O^+ molecule and one A^- molecule, so the values for $[H_3O^+]$ and $[A^-]$ are the same.[1] The following example illustrates the details.

Example 11.4

A student prepares a 0.10 *M* solution of formic acid, HCOOH. She measures the pH at 25°C and finds it to be 2.38. Determine the value of K_a for 0.10 *M* HCOOH at 25°C.

Start by writing the equation for the equilibrium reaction, which is as follows:

$$HCOOH(aq) + H_2O(l) \rightleftharpoons H_3O^+(aq) + HCOO^-(aq)$$

1 This is the case for monoprotic acids, which is all we consider here. Clearly, such is not the case for polyprotic acids.

From the equilibrium equation and Equation (11.6), we see that the acid dissociation constant is calculated as

$$K_a = \frac{[H_3O^+][HCOO^-]}{[HCOOH]}$$

We need to determine the values of $[H_3O^+]$, $[HCOO^-]$, and $[HCOOH]$.

We determine $[H_3O^+]$ from the pH as follows:

$$pH = -\log[H_3O^+]$$

$$\log[H_3O^+] = -pH$$

$$10^{-pH} = [H_3O^+]$$

$$[H_3O^+] = 10^{-pH} = 10^{-2.38} = 4.2 \times 10^{-3}\ M$$

Before we proceed, we should verify that this value is large compared to the concentration of H_3O^+ due to the self-ionization of water, 10^{-7} *M*. Calculating the ratio we find that our $[H_3O^+]$ value is 42,000 times the $[H_3O^+]$ from the self-ionization of water, so we can safely ignore the contribution to $[H_3O^+]$ from the self-ionization of water and still obtain a reasonably accurate result. (Another way to verify that 10^{-7} is small compared to $[H_3O^+]$ is to compute the percentage:

$$\frac{1 \times 10^{-7}\ M}{4.2 \times 10^{-3}\ M} = 2.4 \times 10^{-5} = 0.0024\%$$

Since the percentage of H_3O^+ due to the self-ionization of water is minuscule, we can ignore it.)

As described above, each dissociated molecule of HCOOH produces one molecule of H_3O^+ and one molecule of $HCOO^-$, so $[H_3O^+]$ and $[HCOO^-]$ are equal. Next, to obtain [HCOOH] we use our second approximation, giving us [HCOOH] = 0.10 *M*.

Again, this is because

$$[HCOOH] = (\text{original molarity}) - [H_3O^+] = 0.10\ M - 4.2 \times 10^{-3}\ M \cong 0.10\ M$$

Now we insert values into Equation (11.6) and compute K_a:

$$K_a = \frac{[H_3O^+][HCOO^-]}{[HCOOH]} = \frac{4.2 \times 10^{-3} \cdot 4.2 \times 10^{-3}}{0.10} = 1.8 \times 10^{-4}$$

▲

11.2.2 The Base Dissociation Constant

The principles from the preceding section can be applied to the equilibrium of a basic solution. As a typical case considered from a Brønsted-Lowry perspective, consider a solution of carbonate ion, CO_3^{2-}, and water. Carbonate acts as a Brønsted-Lowry base, accepting a proton from H_2O, which acts as a Brønsted-Lowry acid. The proton transfer creates HCO_3^- and OH^- ions.

$$CO_3^{2-}(aq) + H_2O(l) \rightleftharpoons HCO_3^-(aq) + OH^-(aq)$$

The formation of OH^- ions continues until equilibrium is achieved. As $[OH^-]$ increases, $[H_3O^+]$ decreases, but their product stays at the ionization constant of water, 1.0×10^{-14}. Since $[H_3O^+]$ decreases, the pH is greater than 7 and this solution is basic.

We can write a generic equation describing this equilibrium:

$$B(aq) + H_2O(l) \rightleftharpoons HB^+(aq) + OH^-(aq)$$

From this generic equation we can write the following expression for the equilibrium constant:

$$K_b = \frac{[HB^+][OH^-]}{[B]} \tag{11.7}$$

The expression in Equation (11.7) is called the *base dissociation constant*, and always refers to an equilibrium that occurs when a Brønsted-Lowry base reacts with water to form hydroxide ions and the Brønsted-Lowry conjugate acid. Table 11.2 lists a few representative weak Brønsted-Lowry bases and their K_b values.

Finally, although deriving this goes beyond our treatment here, it is worth noting that for Brønsted-Lowry conjugate pairs, the following relationship holds:

$$K_a \cdot K_b = K_w \tag{11.8}$$

11.2.3 Buffered Solutions

Buffering is an important phenomenon observed in acid-base equilibrium systems. *When a solution is buffered, it resists changes to its pH that otherwise would be caused by the addition of another acid or base to the solution.*

Base	Lewis Structure	Conjugate Acid	K_b
ammonia (NH_3)	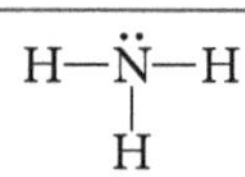	NH_4^+	1.8×10^{-5}
hydrosulfide ion (HS^-)	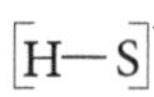	H_2S	1.8×10^{-7}
carbonate ion (CO_3^{2-})	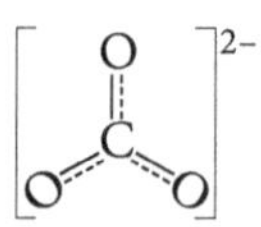	HCO_3^-	1.8×10^{-4}
hypochlorite ion (ClO^-)	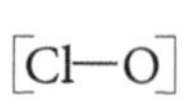	HClO	3.3×10^{-7}
methylamine (NH_2CH_3)	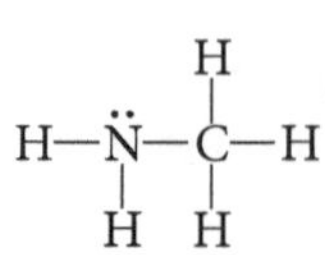	$NH_3CH_3^+$	4.6×10^{-4}

Table 11.2. Representative weak bases and dissociation constants.

The effect is similar to the common-ion effect addressed in Section 11.1.6. There we saw that when an acid and a salt of that acid are in solution, reduced ionization of the acid occurs. Buffering is a bit more complex, but it works in a similar fashion.

A buffered solution can be acidic or basic. A buffered acidic solution contains a weak acid and a salt of that acid, which means they have their anions in common. A buffered basic solution contains a base and a salt of that base, which means they have their cations in common. Figures 11.8 and 11.9 illustrate the way buffering works in the case of a buffered acid solution. To make these diagrams easier to follow, I have arranged the species in the starting solution in compartments around a central area where the action will take place. Of course, in an actual solution all these species are evenly mixed together; they aren't compartmentalized.

Figure 11.8 illustrates the response of a buffered acid solution when a small amount of another acid is added. In the left circle of the figure, we have a solution of acetic acid and sodium acetate. These both contain the acetate ion, CH_3COO^-. There are five separate species in the solution (including the H_2O molecules), but I have placed the acetate ions from the sodium acetate in a separate compartment from the acetate ions contributed by the acetic acid. (This is just to help keep track of what happens; all the acetate ions are identical and mixed together in an actual solution. Also, there is a small amount of OH^- in the solution due to the self-ionization of water, but these molecules are not shown.) On the left side of the first circle are the species associated with the acid, and on the right of that circle are the species from the salt, and the water. The second circle illustrates the addition of a small amount of a second acid to the solution (shown in the middle of the circle). This acid dissociates, contributing new anions (in yellow) and forming new hydronium ions. (This doesn't noticeably affect the abundance of H_2O molecules because there are so many.) The third circle illustrates how the equilibrium shifts in light of this new situation. The increase in H_3O^+ concentration causes acetate ions in the solution to react with these new hydronium ions, forming more non-ionized acetate ions and bringing the H_3O^+ concentration back down nearly to where it was before.

In the buffered acid solution, the excess anions from the first acid and salt react with any new hydronium ions that form. The result is to hold $[H_3O^+]$ steady. And since the pH of the solution is a measure of $[H_3O^+]$, the pH is unchanged. In short, $[CH_3COOH]$ increases, $[CH_3COO^-]$ decreases, and $[H_3O^+]$ remains nearly the same.

Figure 11.9 illustrates the response of the same buffered acid solution to the addition of a small quantity of base. The first circle shows our buffered starting solution, as before. In the sec-

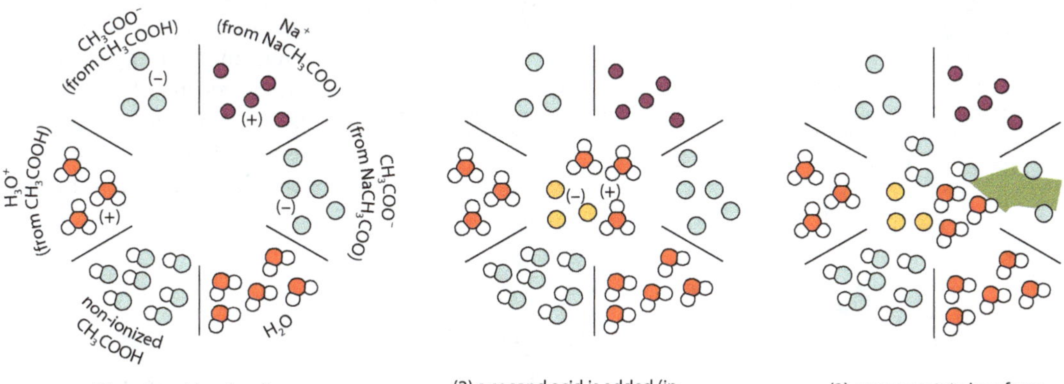

Figure 11.8. Buffering in a solution of a weak acid and a salt of that acid. When small amount of a second acid is added, the pH of the solution remains nearly unchanged.

ond circle (upper right) a base is added, which dissociates into cations (in blue) and hydroxide ions. The third circle shows hydronium ions from the solution reacting with the new hydroxide ions to produce water molecules. (Remember, $[H_3O^+]\cdot[OH^-]$ is a constant, K_w, so when $[OH^-]$ increases from the dissociation of the added base, the equilibrium shifts in response.) Finally, the concentration of H_3O^+ is now lower than what the K_a for the acetic acid requires, so some acetic acid molecules ionize to bring it back up. This restores $[H_3O^+]$ to its previous level and increases $[CH_3COO^-]$ as well. (Again, the formation of new H_3O^+ molecules has no noticeable effect on the H_2O.) In summary, when a small amount of base is added to the buffered acid solution, $[CH_3COOH]$ decreases, $[CH_3COO^-]$ increases, and $[H_3O^+]$ remains the same.

The buffering effect, which steadies the pH of a solution regardless of whether an acid or base is added to the solution, is extremely important in biological systems, as described below.

A buffered base solution contains a weak base and a salt of the weak base, which means the base and the salt share the same cation. Such a buffered solution responds to additions of small amounts of acid or base in a fashion similar to that of the buffered acid solution, with the effect of holding $[H_3O^+]$ steady. As a quick illustration, consider a buffered base solution composed of our favorite weak base, ammonia, and ammonium chloride. If a base is added, additional hydroxide ions form. Ammonium ions in the solution then donate protons to these new hydroxide ions to form water molecules. This action holds $[OH^-]$ nearly constant, and since $[H_3O^+]\cdot[OH^-]$ is a constant, the pH is held steady as well.

If an acid is added to this solution, hydroxide ions in the solution accept protons from the added acid. This lowers $[OH^-]$, and ammonia molecules in the solution ionize (accept protons from water molecules) to form new hydroxide ions. The result is to hold $[OH^-]$ to a steady value, and thus $[H_3O^+]$ and the pH as well.

The biochemical systems in living organisms make use of many different buffering agents to maintain pH levels. In the world of biochemistry, the catalysts that expedite the chemical reactions are called *enzymes*. Many enzymes function only within a narrow pH range. Thus, maintaining pH levels by means of buffering is a critically important aspect of living organisms. The buffering in human

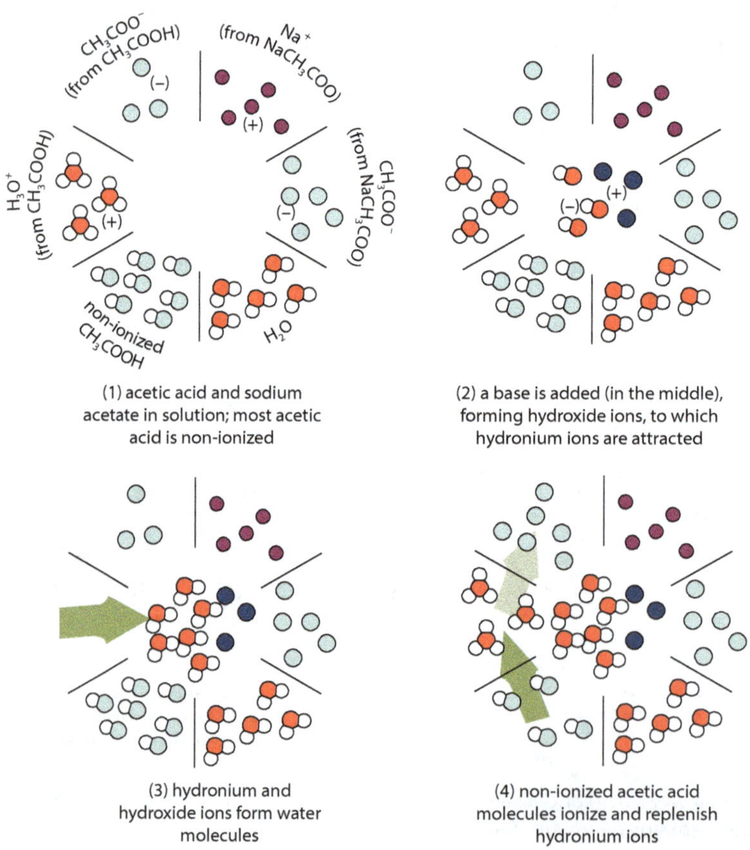

Figure 11.9. Buffering in a solution of a weak acid and a salt of that acid. When a small amount of base is added, the pH of the solution remains nearly unchanged.

Hmm... Interesting. **Buffering in blood**

The most celebrated example of buffering in human physiology is the buffering in human blood. As you probably know, blood transports oxygen in the arteries from the lungs to the cells in all parts of the body, and carbon dioxide in the veins from the cells back to the lungs. The photo shows two red blood cells. The one on the left is oxygenated. The one on the right is deoxygenated, and would be heading back to the lungs to unload CO_2 and reload O_2.

The buffering in blood is provided by carbonic acid, H_2CO_3, and bicarbonate, HCO_3^-, in the blood plasma. Since carbonic acid also decomposes to H_2O and CO_2, a three-tiered equilibrium results:

$$H^+(aq) + HCO_3^-(aq) \rightleftharpoons H_2CO_3(aq) \rightleftharpoons H_2O(l) + CO_2(g)$$

The buffering system controls the pH of the blood to stay within the very narrow range of 7.35 to 7.45—a difficult task given all the different molecules (such as bodily wastes) that are transported in the blood. Severe problems such as disruption to proteins, cell membranes, and enzyme function occur if blood pH strays outside this range.

The ability of the human body to control blood pH this tightly is another example of just how wonderful our bodies are.

blood, described in the accompanying box, is a well-known example of buffering in the human body.

Another common example of buffering is "buffered" aspirin. Aspirin is acetylsalicylic acid. Although widely used as a pain reliever, aspirin can cause the fluids in the stomach to become too acidic, and this can lead to bleeding of the stomach lining. The gastric juice in the stomach contains hydrochloric acid, and when the additional acid from the aspirin arrives, the trouble can start. Buffered aspirin contains a salt to prevent the aspirin from causing the pH in the stomach to change too much.

11.2.4 Hydrolysis of Salts

In Section 9.3.4, I mention that the salt solution formed when a strong acid is neutralized by a weak base is slightly acidic. Likewise, when a strong base is neutralized with a weak acid, the resulting salt solution is slightly basic. To explain these facts, we cannot naïvely say that this happens because the "strong" acid is stronger than the base, and thus makes the solution acidic. In an acid-base neutralization, the acid and base are present in chemically equivalent amounts; there is no acid "left over" to make the solution acidic. To explain a neutralization reaction that produces an acidic solution, we must account for the extra hydronium ions that make it acidic. In the case of a solution that remains basic, we must account for the production of hydroxide ions because an increase in $[OH^-]$ causes a decrease in $[H_3O^+]$ (their product being a constant, K_w) and this is what makes a solution basic.

The reason these "neutralized" solutions are not pH neutral (pH = 7) even though the acid and base are present in chemically equivalent amounts is due to *hydrolysis*, a chemical reaction between water molecules and the ions of a dissolved salt. The production of H_3O^+ or OH^- ions after a neutralization reaction is a consequence of the ions from the salt.

Consider an aqueous solution of sodium acetate salt:

$$NaCH_3COO(s) \rightleftharpoons Na^+(aq) + CH_3COO^-(aq) \qquad (11.9)$$

The sodium ions show no tendency to react further and simply remain completely in solution. But consider the potential Brønsted-Lowry acid-base reaction between the acetate ion and water:

$$\underset{\text{base}}{CH_3COO^-(aq)} + \underset{\text{acid}}{H_2O(l)} \rightleftharpoons \underset{\text{conjugate acid}}{CH_3COOH(aq)} + \underset{\text{conjugate base}}{OH^-(aq)} \qquad (11.10)$$

Recall from Chapter 9 that in a Brønsted-Lowry acid-base reaction, the equilibrium favors transfer of a proton from the strong acid to the strong base and thus lies in the direction of the weaker acid and weaker base. Referring again to Table 9.6, we see that the equilibrium in this case lies to the left, because H_2O and CH_3COO^- are the weaker acid and base. But will the equilibrium lie *completely* to the left? That is, is this a case of the reaction going to completion toward the left? The answer to this question is no.

To see why, consider the acid dissociation constant for acetic acid:

$$K_a = \frac{[H_3O^+][CH_3COO^-]}{[CH_3COOH]}$$

The lower the value for K_a, the higher $[CH_3COOH]$ is, and this corresponds to a stronger attraction between CH_3COO^- molecules and protons relative to the attraction OH^- molecules have for protons. (This is closely related to the fact that the acid doesn't completely ionize.) Table 11.1 lists the dissociation constant for acetic acid as $K_a = 1.7 \times 10^{-5}$, which is quite low. The result is that CH_3COO^- molecules are strong enough bases to accept some protons from H_2O molecules and make CH_3COOH molecules, producing OH^- molecules in the process. So the equilibrium in Equation (11.10) does not lie completely to the left; some of the CH_3COO^- molecules react with H_2O to form OH^-. And when $[OH^-]$ goes up, $[H_3O^+]$ goes down and pH goes up.

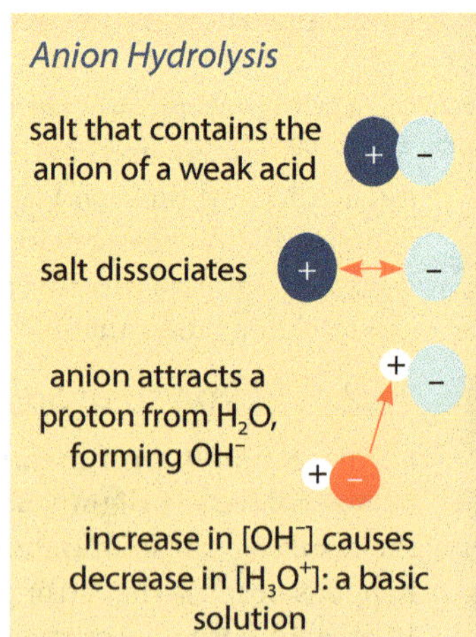

Figure 11.10. Summary of the process of anion hydrolysis.

This whole process begins with Equation (11.9), the simple dissociation of sodium acetate in water. We begin with water, pH = 7, and end up with a pH greater than 7—a basic solution. The cause is the fact that the anion of the salt reacts with water to produce hydroxide ions and a basic solution. This process is called *anion hydrolysis*. The process described here is summarized in Figure 11.10.

Now you can see why the neutralization reaction between a weak acid like CH_3COOH and a strong base like NaOH ends up with a solution that is basic. The CH_3COO^- ions produced by the neutralization produce some OH^- ions of their own, reducing $[H_3O^+]$ and increasing the pH.

Now let's consider a solution of ammonium chloride:

$$NH_4Cl(s) \rightleftharpoons NH_4^+(aq) + Cl^-(aq)$$

In this case, the chlorine ions remain happily in solution. Not so with the ammonium ions, which can engage in this Brønsted-Lowry acid-base reaction:

$$NH_4^+(aq) + H_2O(l) \rightleftharpoons NH_3(aq) + H_3O^+(aq)$$

Using the same analysis as before, we conclude that this reaction does occur to some extent in the solution, even though H_3O^+ is a stronger acid than NH_4^+ and NH_3 is a stronger base than H_2O. So the dissolution of ammonium chloride ends up producing hydronium ions, and the resulting solution is acidic. This is *cation hydrolysis*—the reaction between water and the cation of a salt to produce hydronium ions and an acidic solution. This process is summarized in Figure 11.11.

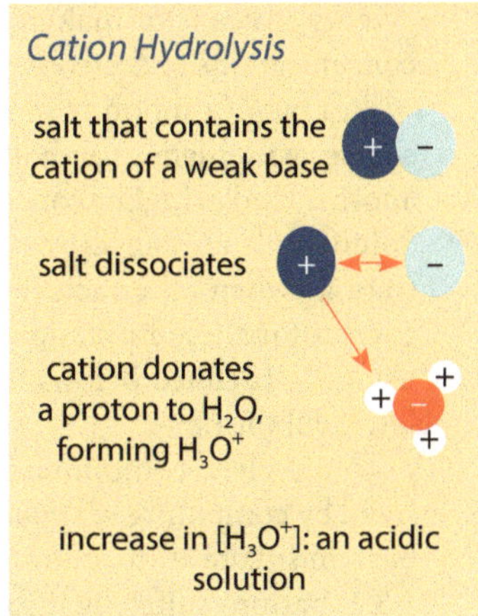

Figure 11.11. Summary of the process of cation hydrolysis.

We now apply this logic to the neutralization reaction between a strong acid and a weak base. If HCl is neutralized by NH_3, the salt produced is NH_4Cl, which remains as ions in solution. But the ammonium ions produce some hydronium ions, making the resulting solution acidic.

Now let's briefly consider what happens when a strong acid is neutralized with a strong base. The most common example is the neutralization between HCl and NaOH:

$$HCl(aq) + NaOH(aq) \rightarrow Na^+(aq) + Cl^-(aq) + H_2O(l)$$

The Cl^- ion is the conjugate base of the strong acid HCl. In Table 9.6, the Cl^- ion is the weakest base listed, so in solution Cl^- exhibits no tendency to accept protons. The result is that both the Na^+ and Cl^- ions remain in solution without reacting with H_2O at all, and the resulting solution remains at $pH = 7$.

To round out this discussion, we should consider what happens when a weak acid is neutralized with a weak base. The fact is, we cannot make general predictions about what happens. It all depends on what ions are produced, and the result could be a solution that is acidic, basic, or neutral.

11.3 Solubility Equilibria

11.3.1 The Solubility Product

In Chapter 8, we considered general rules that allow us to make qualitative predictions about the solubility of compounds formed in metathesis reactions, and about whether those reaction products are likely to precipitate. Speaking generally, we typically refer to a compound as *soluble* if its solubility is approximately 1 g per 100 mL or higher. Compounds such as calcium hydroxide with solubilities in the range of 0.2 g/100 mL are also regarded as soluble (if barely), but compounds with solubilities below 0.1 g/100 mL are regarded as *insoluble*. However, ionic compounds that are insoluble by these measures do still dissolve in water to some extent, even if it is at the level of milligrams or micrograms per 100 mL of water. Such compounds are said to be *sparingly soluble*. Many chemical systems involve only very dilute solutions, and it is possible to form "insoluble" reaction products at very low concentrations that do not precipitate. The techniques we develop in this section apply to compounds with extremely low solubilities and give us a means of making quantitative predictions about potential precipitations at very low concentrations in aqueous solution.

A note of caution is in order before we proceed. The use of the solubility product for substances with higher solubilities can easily lead to errors. This is because ions are attracted to one another, and at higher concentrations, these attractions have a significant effect on the solubility. Solubility is also affected by ion hydrolysis, which occurs at greater rates with higher concentrations when other acid-base equilibria are going on in the same solution. There are ways to correct for these limitations, which we do not explore here. You should simply be aware that using K_{sp} (defined below) in the simple way that we do here is limited to substances with very low solubilities.

A dissolved compound in a saturated solution forms an equilibrium system and can be analyzed by computing *yet another* equilibrium constant (our last one). Consider a quantity of solid barium sulfate at the bottom of a container of water. According to the rules outlined in Section 8.3.1, barium sulfate is insoluble. But over time, a saturated solution forms. This solution is an equilibrium system with very small amounts of Ba^{2+} and SO_4^{2-} ions in solution. At equilibrium, ions from the solid crystal dissociate into solution and recrystallize at the same rate. Since the $BaSO_4$ at the bottom of the container is in solid form, this is a heterogeneous equilibrium system. This situation is called a *solubility equilibrium.*

The equilibrium established is as follows:

$$BaSO_4(s) \rightleftharpoons Ba^{2+}(aq) + SO_4^{2-}(aq)$$

As we have done with previous equilibrium systems, we can calculate an equilibrium constant for this equilibrium, using the same definitions and rules. Since the $BaSO_4$ is a solid, its concentration does not appear in the expression for the equilibrium constant. Thus, the equilibrium constant is

$$K_{sp} = [Ba^{2+}][SO_4^{2-}]$$

The equilibrium constant for solubility equilibria like this one is denoted with the subscript *sp*, and is called the *solubility product constant*, or simply the *solubility product*. As before, the concentrations in K_{sp} are raised to the power of their coefficients in the equilibrium equation. As another illustration, the equilibrium equation for calcium fluoride is

$$CaF_2(s) \rightleftharpoons Ca^{2+}(aq) + 2F^-(aq)$$

The resulting solubility product is

$$K_{sp} = [Ca^{2+}][F^-]^2$$

Solubility product values for many substances have been catalogued and are available for scientists to use. Table B.10 in Appendix B lists values for a number of representative substances.

11.3.2 Calculating K_{sp} from Concentration or Solubility Data

The value of K_{sp} can be calculated from concentration measurements made on saturated solutions in equilibrium. The following example illustrates this type of computation.

▼ Example 11.5

A quantity of silver chromate is placed in a container of pure water. The mixture is stirred for several days to insure that a saturated solution at equilibrium is achieved. The concentration of silver ions in the solution is then measured and found to be $[Ag^+] = 1.3 \times 10^{-4}$ *M*. Determine the solubility product for silver chromate. Assume that silver chromate dissociates completely and that no other equilibria involving Ag^+ or CrO_4^{2-} ions that would affect their concentrations are occurring in the solution.

Begin by writing the equilibrium equation and the expression for K_{sp}. The equation is

$$Ag_2CrO_4(s) \rightleftharpoons 2Ag^+(aq) + CrO_4^{2-}(aq)$$

From this equation, we see that the expression for K_{sp} is

$$K_{sp} = \left[Ag^+\right]^2\left[CrO_4^{2-}\right]$$

We are given a value for $[Ag^+]$, and we need to determine $[CrO_4^{2-}]$. Notice from the stoichiometry of the equilibrium equation that for every two moles of silver ions in solution, there is one mole of chromate ions in solution. Thus, the chromate ion concentration is half the silver ion concentration. Or, to compute this formally using mole ratios,

$$\left[CrO_4^{2-}\right] = 1.3\times10^{-4}\ \frac{\text{mol Ag}^+}{\text{L}}\cdot\frac{1\ \text{mol CrO}_4^{2-}}{2\ \text{mol Ag}^+} = 6.5\times10^{-5}\ \frac{\text{mol}}{\text{L}}$$

Now we compute K_{sp}:

$$K_{sp} = \left[Ag^+\right]^2\left[CrO_4^{2-}\right] = \left(1.3\times10^{-4}\right)^2\left(6.5\times10^{-5}\right) = 1.1\times10^{-12}$$

▲

A procedure similar to the one used in Example 11.5 can be used to compute a solubility product from solubility data. We simply convert the solubility into a molarity, and then work out the value of K_{sp} as shown in the previous example. There is a small approximation in this computation that does not affect the accuracy of the result, but which you should be aware of to avoid confusion. The calculation involves first determining the molarity of a saturated solution and then converting this molarity to a solubility stated in grams per liter. The definition of molarity is moles of solute per liter of *solution*, whereas the solubility is expressed in grams of solute per liter of *water*. But we are dealing with extremely low concentrations, so we expect the volume of water to be essentially identical with the volume of the solution. Further, calculations of this sort are always limited to two or three significant digits because measuring very low concentrations is difficult and tabulated K_{sp} values are limited to a precision of only two or three digits. So, since our precision is limited and the concentrations are low, the approximation of using the volume of water to be equal to the volume of solution does not affect the accuracy of our results.

The following example illustrates the use of a solubility value to compute K_{sp}.

▼ Example 11.6

Iron(II) hydroxide is sparingly soluble, with a solubility of 2.07×10^{-5} grams per 100 mL. Determine the solubility product constant for this compound.

We begin by converting the given solubility into a molar concentration, making use of the approximation mentioned above.

$$\frac{2.07\times10^{-5}\text{ g}}{100\text{ mL}}\cdot\frac{1000\text{ mL}}{1\text{ L}}\cdot\frac{1\text{ mol}}{89.96\text{ g}}=2.30\times10^{-6}\ \frac{\text{mol}}{\text{L}}$$

Using the approximation discussed above, we have

$$\left[\text{Fe(OH)}_2\right]=2.30\times10^{-6}\ M$$

Now we write the equilibrium equation and the expression for the solubility product.

$$\text{Fe(OH)}_2(s) \rightleftharpoons \text{Fe}^{2+}(aq) + 2\text{OH}^-(aq)$$

$$K_{sp}=\left[\text{Fe}^{2+}\right]\left[\text{OH}^-\right]^2$$

From the stoichiometry of the equilibrium equation, we see [Fe^{+}] is the same as [$Fe(OH)_2$]. Also, since two moles of OH^- are in solution for each mole of dissolved $Fe(OH)_2$, [OH^-] is double [$Fe(OH)_2$]. We insert these concentrations into the expression for K_{sp} and compute:

$$K_{sp}=\left[\text{Fe}^{2+}\right]\left[\text{OH}^-\right]^2=\left(2.30\times10^{-6}\right)\left(4.60\times10^{-6}\right)^2=4.87\times10^{-17}$$

▲

11.3.3 Calculating Solubility from K_{sp}

The solubility of a substance can be determined from its solubility product by using a clever bit of algebra. The following example illustrates computations of this type.

▼ Example 11.7

Determine the solubility of calcium fluoride at 25°C (Figure 11.12) and express your result in grams per liter.

We begin by looking up the solubility product for calcium fluoride in Table B.10 in Appendix B and we find $K_{sp} = 3.45 \times 10^{-11}$. Next we write the equilibrium equation and the expression for K_{sp}:

$$\text{CaF}_2(s) \rightleftharpoons \text{Ca}^{2+}(aq) + 2\text{F}^-(aq)$$

$$K_{sp}=\left[\text{Ca}^{2+}\right]\left[\text{F}^-\right]^2$$

Let's now use x to represent [Ca^{2+}]:

$$\left[\text{Ca}^{2+}\right]=x$$

From the equilibrium equation, we see that the concentration of fluorine ions in the solution is twice the concentration of calcium ions (as in the previous example), so

$$[F^-]=2x$$

Now we substitute these into the expression for K_{sp}:

$$K_{sp}=[Ca^{2+}][F^-]^2=(x)(2x)^2=3.45\times10^{-11}$$

Next, we solve for x:

$$(x)(2x)^2=3.45\times10^{-11}$$

$$(x)(4x^2)=3.45\times10^{-11}$$

$$4x^3=3.45\times10^{-11}$$

$$x^3=8.625\times10^{-12}$$

$$x=(8.625\times10^{-12})^{1/3}=2.05\times10^{-4}\ M$$

This is our value for [Ca^{2+}]. Looking again at the equilibrium equation, we see that for each mole of calcium ions in solution, one mole of CaF_2 dissolved. Thus, the value of [CaF_2] is the same.

The last step is to convert this molarity into a solubility in grams per liter, using the same approximation we used in Example 11.6. First, we calculate the molar mass of CaF_2. Then we use this as a conversion factor to convert the molarity into a solubility.

$$2.05\times10^{-4}\ \frac{\text{mol}}{\text{L}}\cdot\frac{78.08\ \text{g}}{\text{mol}}=0.0160\ \frac{\text{g}}{\text{L}}$$

▲

Figure 11.12. Calcium fluoride (upper photo) occurs naturally as the mineral fluorite (lower photo), which is the source of most of the world's fluorine.

11.3.4 Using K_{sp} to Predict Precipitation

Now that you are familiar with calculations involving the solubility product, we can use K_{sp} to make quantitative predictions about precipitations that may occur in metathesis reactions. *In a solution, the product of ion concentrations (raised to powers, if appropriate) cannot exceed the solubility product for the compound formed by those ions.* Consider a general compound AB with K_{sp} = [A][B]. If a quantity of AB is dissolved, then the concentrations of A^+ and B^- ions in the solution are equal. But if the A^+ and B^- ions appear in the solution from separate sources—such as from mixing two solutions together—[A] and [B] may not be equal.

When using K_{sp} to make predictions about precipitation, it doesn't matter whether the concentrations of A^+ and B^- ions are equal. What matters is whether the product [A][B] is lower or higher than K_{sp}. If it is higher, then precipitation occurs until the product of concentrations has been reduced to the solubility product. At that point, a saturated equilibrium solution exists.

The following example illustrates how to predict precipitation when two solutions are mixed.

Example 11.8

A chemist plans to mix 20.0 mL of 0.0010 *M* $BaCl_2$ with 20.0 mL of 0.0050 *M* Na_2SO_4. Will a precipitate form when these solutions are combined?

The two compounds that can form when these solutions are mixed are NaCl and $BaSO_4$. Since the molarity of both the initial solutions is low and the solubility of NaCl is quite high, the only compound we need to consider is $BaSO_4$. According to the rules in Table 8.3, $BaSO_4$ is insoluble. But this simply means that $BaSO_4$ is sparingly soluble and its solubility is very low. To determine if precipitation will occur, we must calculate the product of the concentrations of Ba^{2+} and SO_4^{2-} and compare the result to the solubility product of $BaSO_4$.

From Table B.10, the solubility product for barium sulfate is $K_{sp} = 1.08 \times 10^{-10}$. This is the figure we compare to.

We begin, as always, by writing the equilibrium equation and the formula for K_{sp}.

$$BaSO_4(s) \rightleftharpoons Ba^{2+}(aq) + SO_4^{2-}(aq)$$

$$K_{sp} = \left[Ba^{2+}\right]\left[SO_4^{2-}\right]$$

The next step is to determine the mole quantities present in the mixture for the ions we are interested in. For the barium, each mole of $BaCl_2$ contributes one mole of Ba^{2+} to the solution. Thus, the number of moles of Ba^{2+} is the same as the number of moles of $BaCl_2$ in the $BaCl_2$ solution. Calculating this mole quantity, we have

$$0.0010\ \frac{\text{mol}}{\text{L}}\ Ba^{2+} \cdot 0.0200\ \text{L} = 2.00\times10^{-5}\ \text{mol}\ Ba^{2+}$$

Similarly, the mole quantities of Na_2SO_4 and SO_4^{2-} are equal. Thus we have

$$0.0050\ \frac{\text{mol}}{\text{L}}\ SO_4^{2-} \cdot 0.0200\ \text{L} = 1.00\times10^{-4}\ \text{mol}\ SO_4^{2-}$$

Now we calculate the two ion concentrations. Note that when the solutions are mixed, the total volume is 40.0 mL, which is 0.0400 L.

$$\left[Ba^{2+}\right] = \frac{2.00\times10^{-5}\ \text{mol}}{0.0400\ \text{L}} = 5.00\times10^{-4}\ M$$

$$\left[SO_4^{2-}\right] = \frac{1.00\times10^{-4}\ \text{mol}}{0.0400\ \text{L}} = 2.50\times10^{-3}\ M$$

With these values, we calculate the product of the concentrations:

$$\left[Ba^{2+}\right]\left[SO_4^{2-}\right] = \left(5.00\times10^{-4}\ M\right)\left(2.50\times10^{-3}\ M\right) = 1.3\times10^{-6}$$

As low as it is, this value is still much larger than $K_{sp} = 1.08 \times 10^{-10}$. Therefore, we expect a barium sulfate precipitate to form. After that, the solution will still be saturated in Ba^{2+} and SO_4^{2-}.

▲

Chapter 11 Exercises

SECTION 11.1

1. State which reaction in an equilibrium (forward or reverse) is favored under each of the following conditions:
   a. K is very small (much less than 1)
   b. K is very large (much greater than 1)
   c. $K = 1$
2. Write the expression for the equilibrium constant for each of the following equilibrium equations:
   a. $4H_2(g) + CS_2(g) \rightleftharpoons CH_4(g) + 2H_2S(g)$
   b. $Cd(OH)_2(s) \rightleftharpoons Cd^{2+}(aq) + 2OH^-(aq)$
   c. $CO(g) + H_2O(g) \rightleftharpoons CO_2(g) + H_2(g)$
   d. $HNO_2(aq) + H_2O(l) \rightleftharpoons H_3O^+(aq) + NO_2^-(aq)$
   e. $2SO_2(g) + O_2(g) \rightleftharpoons 2SO_3(g)$
   f. $Ag(CN)_2^-(aq) \rightleftharpoons Ag^+(aq) + 2CN^-(aq)$
   g. $HNO_2(aq) \rightleftharpoons H^+(aq) + NO_2^-(aq)$
   h. $C(s) + H_2O(g) \rightleftharpoons CO(g) + H_2(g)$
3. Consider again the Haber-Bosch reaction for manufacturing ammonia, and answer the following questions:
   a. When is the forward reaction rate at a maximum—at the beginning, after equilibrium has been reached, or at some time in between?
   b. When is the reverse reaction rate at a maximum—at the beginning, after equilibrium has been reached, or at some time in between?
   c. Describe what happens to each of the three species from the time the reaction begins until equilibrium is achieved.
4. For the reaction shown below, predict the direction the equilibrium shifts under the given condition. For each response, justify your prediction in terms of Le Châtelier's principle.

   $N_2O_3(g) \rightleftharpoons N_2O(g) + O_2(g)$

   a. The concentration of N_2O is increased.
   b. The pressure is decreased.
   c. Argon is pumped into the reaction vessel.

5. For the reaction shown below, predict the direction the equilibrium shifts under the given condition. For each response, justify your prediction in terms of Le Châtelier's principle.

   $$CO(g) + H_2O(g) \rightleftharpoons CO_2(g) + H_2(g)$$

   a. The temperature is decreased.

   b. The concentration of CO is decreased.

   c. The pressure is increased.

6. For each of the following reactions, specify the relative conditions (high or low) of temperature, pressure, and reactant concentration that maximizes production of the highlighted species.

   a. $4HCl(g) + O_2(g) \rightleftharpoons 2H_2O(g) + 2Cl_2(g)$

   b. $2HI(g) + 12.6\text{ kJ} \rightleftharpoons H_2(g) + I_2(g)$

   c. $H_2O(l) + 42\text{ kJ} \rightleftharpoons H_2O(g)$

7. For the reaction shown in exercise 6(a), at a certain temperature the equilibrium concentrations are found to be: 2.4×10^{-3} M HCl, 7.6×10^{-8} M O_2, 1.2×10^{-1} M H_2O, and 1.2×10^{-1} M Cl_2. Determine the equilibrium constant for this system.

8. At 450°C, the value of the equilibrium constant for the Haber-Bosch ammonia process is $K = 6.59 \times 10^{-3}$. If $[H_2] = 2.61 \times 10^{-3}$ M and $[NH_3] = 1.17 \times 10^{-4}$ M at equilibrium, determine $[N_2]$ at equilibrium.

9. At 100°C, the equilibrium constant for the reaction below is $K = 2.19 \times 10^{-10}$.

   $$COCl_2(g) \rightleftharpoons CO(g) + Cl_2(g)$$

   For each of the following mixtures, determine if the system is at equilibrium. If it is not, indicate the direction in which the reaction must proceed to achieve equilibrium.

   a. $[COCl_2] = 2.00 \times 10^{-3}$ M, $[CO] = 3.3 \times 10^{-6}$ M, $[Cl_2] = 6.62 \times 10^{-6}$ M

   b. $[COCl_2] = 4.50 \times 10^{-2}$ M, $[CO] = 1.1 \times 10^{-7}$ M, $[Cl_2] = 2.25 \times 10^{-6}$ M

   c. $[COCl_2] = 0.0100$ M, $[CO] = [Cl_2] = 1.48 \times 10^{-6}$ M

10. At 25°C, the equilibrium constant for the reaction shown below is $K = 1.7 \times 10^{13}$. At equilibrium, concentration data are $[N_2O] = 0.0029$ M and $[O_2] = 0.0023$ M. Determine the value of [NO] at equilibrium.

    $$N_2O(g) + \frac{1}{2}O_2(g) \rightleftharpoons 2NO(g)$$

11. For the following reactions, identify the species that are most abundant and least abundant at equilibrium.

    a. $N_2(g) + O_2(g) \rightleftharpoons 2NO(g);\ K = 1.5 \times 10^{-10}$

    b. $2HBr(g) \rightleftharpoons H_2(g) + Br_2(g);\ K = 5.8 \times 10^{-18}$

12. At a certain temperature, the equilibrium constant for the hydrogen iodide reaction in Equation (11.2) is $K = 40.0$.

a. What does the value of the equilibrium constant tell you about the relative concentrations of reactants and products at equilibrium?

b. What is the value of the equilibrium constant for the reverse reaction?

13. Drawing from example reactions used in this text or reactions you are familiar with in ordinary life, describe three examples of reactions going to completion. Select examples that go to completion for different reasons, and explain why each example you choose goes to completion.

14. Explain the common-ion effect.

SECTION 11.2

15. List the acids in Table 11.1 in order from weakest to strongest.

16. Given a monoprotic acid, what is implied by an acid dissociation constant of $K_a = 1.0$?

17. Lactic acid, $CH_3CH(OH)COOH$, ionizes by donating one proton (the last hydrogen in the formula). If a pH of 2.44 is measured for a 0.10 *M* solution of lactic acid, determine K_a.

18. A 0.0500 *M* solution of pyruvic acid, $CH_3COCOOH$, has a pH of 1.85. Determine K_a for this acid.

19. At certain conditions, a 0.100 *M* solution of bromoacetic acid, $BrCH_2COOH$, is 13.2% ionized. This means that $[H_3O^+]$ is 0.132(0.100 *M*) = 0.0132 *M* and $[BrCH_2COOH]$ is 0.868(0.100 *M*) = 0.0868 *M*. Determine the pH of the solution and the value of K_a for this acid.

20. At certain conditions, a 0.100 *M* solution of chloroacetic acid, $ClCH_2COOH$, is 11.0% ionized. From this information, determine $[H_3O^+]$, $[ClCH_2COO^-]$, $[ClCH_2COOH]$, and K_a for chloroacetic acid.

21. The pH of a certain sample of vinegar is 2.92. Given the dissociation constant for acetic acid (see Table 11.1), determine the concentration of acetic acid in the solution. (Assume acetic acid is the only acid in the vinegar.)

22. Calculate the percent ionization for hydrocyanic acid at concentrations of 0.150 *M*, 0.0800 *M*, and 0.0200 *M*. See Table 11.1 for K_a for hydrocyanic acid.

23. Determine K_b for the base OH^- at 25°C.

24. Explain why a solution containing both acetic acid and sodium acetate has a higher pH than a solution containing the same concentration of acetic acid but no sodium acetate.

25. Some nasal sprays include a decongestant called ephedrine, a nervous system stimulant. In solution, this weak organic base accepts a proton and creates a hydroxide ion as follows:

$$C_{10}H_{15}ON(aq) + H_2O(l) \rightleftharpoons C_{10}H_{15}ONH^+(aq) + OH^-(aq)$$

A pH of 11.33 is measured for a 0.035 *M* solution of ephedrine. Determine the following:

a. The equilibrium concentrations of $C_{10}H_{15}ON$, $C_{10}H_{15}ONH^+$, and OH^-.

b. The value of K_b for ephedrine.

26. The K_b values for ammonia and another weak base, hydroxylamine, NH_2OH, are $K_b = 1.8 \times 10^{-5}$ and $K_b = 8.7 \times 10^{-9}$, respectively. Address the following questions.

a. Which is the stronger base?

b. Which of their conjugate acids is the stronger acid, ammonium ion (NH_4^+) or hydroxylammonium ion (NH_3OH^+)?

c. What are the K_a values for NH_4^+ and NH_3OH^+?

27. The main buffering agents in human blood are H_2CO_3 and HCO_3^-. Explain how this buffering system is able to maintain blood pH at a steady level when a small amount of a generic strong acid HA (with ions H^+ and A^-) is added.

28. Explain the processes of cation and anion hydrolysis.

29. When a strong acid is titrated with a weak base, the resulting solution is acidic at the titration endpoint. Explain why this is the case.

SECTION 11.3

30. If the concentration of a saturated solution of CaF_2 is 1.24×10^{-3} *M* at 35°C, determine the value of K_{sp} at this temperature.

31. At 25°C, a 1.00-L solution saturated with calcium sulfate, $CaSO_4$, contains 0.956 g of $CaSO_4$. Determine the solubility product for calcium sulfate at 25°C.

32. The solubility of copper(I) chloride is 4.11×10^{-3} g/100 mL. From this information, determine the solubility product constant.

33. Lead(II) chloride has a solubility at 25°C of 0.45 g/100 mL. Determine K_{sp} at 25°C for this salt.

34. A chemist measures the solubility of strontium fluoride, SrF_2, at 25°C and finds it to be 1.29×10^{-2} g per 100 mL of aqueous solution. Determine K_{sp} at this temperature for SrF_2.

35. Use the K_{sp} of yttrium fluoride at 25°C to determine the molar solubility of YF_3 at this temperature. Solubility product constant data are listed in Table B.10 in Appendix B.

36. Determine the solubility of each of the following compounds in g/100 mL at 25°C. Use the K_{sp} values found in Appendix B.

    a. copper(I) iodide

    b. lead(II) sulfate

    c. mercury(I) bromide

    d. mercury(II) bromide

    e. cadmium hydroxide

    f. calcium phosphate

    g. thallium(I) chromate

37. List the following carbonates in order of solubility: $MgCO_3$, $PbCO_3$, $SrCO_3$, $BaCO_3$, $CdCO_3$, Ag_2CO_3.

38. Determine if a precipitate forms when the following substances are combined in solution at 25°C:

    a. 0.37 L of 0.0034 *M* $Ca(NO_3)_2$ and 0.25 L of 0.0065 *M* NaOH.

    b. 100.0 mL of 0.050 *M* silver nitrate and 10.0 mL of 0.040 *M* sodium sulfate.

    c. 1.70 g of solid silver nitrate and 19.5 g solid sodium chloride in 175 mL of water.

    d. 50.0 mL of 0.0103 *M* $Fe(NO_3)_2$ and 45 mL of 1.1×10^{-9} *M* NaOH.

    e. 0.97 g of solid sodium carbonate and 0.19 g of barium bromide in a 12.0-L solution.

GENERAL REVIEW EXERCISES

39. The molar mass of a certain compound is found to be 168.19 g/mol. Its composition is found to be as follows: 64.27% carbon, 7.19% hydrogen, and 28.54% oxygen. Determine the empirical and molecular formulas for this compound.

40. Potassium permanganate and glycerin react in an explosion as follows:

$$14KMnO_4 + 4C_3H_5(OH)_3 \rightarrow 7K_2CO_3 + 7Mn_2O_3 + 5CO_2 + 16H_2O$$

a. If 5.00 mol of $KMnO_4$ are consumed, how many moles of CO_2 are produced?

b. If 2.66 kg of glycerine are consumed, what mass of H_2O is produced?

c. How many grams of $KMnO_4$ are required to react with 500.0 g of glycerin?

41. When acetic acid and ethanol are heated in the presence of a sulfuric acid catalyst, ethyl acetate is produced. Ethyl acetate is a sweet-smelling solvent used in the manufacturing of glue and nail polish remover. (It is also an additive in cigarettes. Think about that.) The reaction is as follows:

$$CH_3COOH + CH_3CH_2OH \rightarrow CH_3COOCH_2CH_3 + H_2O$$

Determine the percent yield for this reaction in the following cases:

a. 4.36 moles of ethanol are reacted with excess acetic acid and 2.97 moles of ethyl acetate are produced.

b. 2.90 moles of ethanol are reacted with 4.10 moles of acetic acid and 2.30 moles of ethyl acetate are produced.

42. A sample of argon gas occupies a volume of 4.99 L at 2.15 atm. What is the volume of this gas at a pressure of 1.00 bar?

43. A chemist prepares a sample of H_2S gas by collecting it over water at a temperature of 28°C. The collected gas has a volume of 15.3 mL and a total pressure of 205.15 kPa. Determine the following:

a. the partial pressure of the H_2S collected.

b. the number of moles of H_2S collected.

c. the mass of H_2S collected.

44. Dinitrogen pentoxide, N_2O_5, is a solid of colorless crystals that sublimes at room temperature, decomposing into nitrogen dioxide, NO_2, and oxygen. Determine the volume of oxygen produced at STP from the decomposition of 100.0 g of N_2O_5.

45. Determine the freezing point of a solution prepared by dissolving 5.62 g of calcium chloride in 100.0 mL of water.

46. Determine the oxidation state of nitrogen in each of the following molecules and ions: NO, NO_2, N_2O, N_2, NO_2^-, HCN, HNO_3, and NH_3.

47. Determine the change in the Gibbs free energy for the reaction between sodium hydroxide and sodium metal to produce sodium oxide and hydrogen gas.

Chapter 12

Redox Chemistry

This is a 1940s-vintage photo of a phosphate smelting furnace in Alabama. The smelter produces elemental phosphorus. The specific redox reaction to produce phosphorus depends on the particular mineral from which the phosphorus is being extracted. Calcium phosphate mined in Florida is one possibility, for which the redox reaction is $2Ca_3(PO_4)_2 + 6SiO_2 + 10C \rightarrow 6CaSiO_3 + 10CO + P_4$. The calcium phosphate is mixed with sand (SiO_2) and coke (carbon) and heated to around 1,500°C. The carbon is oxidized to form carbon monoxide; the phosphorus is reduced to form elemental phosphorus vapor (P_4).

In the process shown above, the heat is being supplied by electricity, and the large cables are electrical conductors carrying the current into the furnace. You can be confident that where the worker is standing, the heat is so intense it is difficult to remain there.

Objectives for Chapter 12

After studying this chapter and completing the exercises, you should be able to do each of the following tasks, using supporting terms and principles as necessary.

SECTION 12.1

1. Define *oxidation* and *reduction* and write example reaction equations to represent each.
2. Identify the oxidation states of elements in elemental molecules and in compounds.
3. Compare relative strengths of oxidizing and reducing agents, and use a table of such agents to predict if one substance is able to oxidize another.

SECTION 12.2

4. Given a substance and its oxidized or reduced form, complete and balance the half-reaction and indicate whether the half-reaction is an oxidation or reduction.
5. Given substances in a redox reaction, write the oxidation and reduction half-reactions.
6. Balance redox reaction equations using the half-reaction method.
7. After balancing a redox reaction equation, combine and/or add ions to show the original compounds used to produce the reaction.

SECTION 12.3

8. Explain the essential principles of electrochemistry.
9. Describe the basic components of an electrochemical cell.
10. Describe and differentiate between voltaic and electrolytic cells.
11. On a diagram of a voltaic or electrolytic cell, identify the anode, cathode, direction of electron flow, salt bridge or porous barrier, and direction of ion movement.
12. Draw a diagram of a voltaic or electrolytic cell, indicating the anode, cathode, direction of electron flow, salt bridge or porous barrier, and direction of ion movement.
13. Explain the purpose of a salt bridge and describe how a salt bridge is physically constructed.
14. Describe the processes of oxidation and reduction in electrochemical cells, including where they occur and what kinds of species (atoms or ions) are involved in each.
15. Define *electrode potential* and explain how standard reduction potential values are established.
16. Calculate cell potentials from a table of standard electrode potentials.
17. Identify several technologies that use electrochemical redox reactions.

12.1 Oxidation and Reduction

12.1.1 Introduction to Redox Reactions

Chemical reactions in which substances are oxidized and reduced are *oxidation-reduction reactions*, or *redox reactions*. These reactions form a very large class, including many of the reactions we have studied. All combustion reactions are redox reactions, illustrated by the beautiful launch of Apollo 15 in Figure 12.1. All reactions in which coatings such as rust or verdigris form on metals are also redox reactions, illustrated by the verdigris on a copper bay window in Figure 12.1.

Reactions between metals and acids or bases that result in corrosion or destruction of the metal are also redox reactions. Alkali batteries, such as those shown in Figure 12.2, contain potassium hydroxide, a strong base. When the battery seals fail and the KOH leaks out, a redox reaction corrodes the metal. The batteries themselves also supply electrical power by means of redox reactions, a topic we address later in this chapter.

The single replacement reactions we studied in Section 5.2.4 are also redox reactions. The sequence of photos in Figure 12.3 illustrates several redox reactions. The top three images show an aluminum wire placed into a solution of copper(II) chloride. The aluminum replaces the copper in the solution, and the copper comes out of solution, forming a fluffy coating on the aluminum. In the lower image, copper hydroxide (blue) and copper carbonate (blue green) have formed on the copper—both of these being redox reactions between the copper and the air. The copper compounds then leached back into the water containing the aluminum ions.

Figure 12.1. The combustion of rocket fuel at the Apollo 15 launch (above) and the formation of verdigris on a copper-clad bay window in Brooklyn (below) are examples of redox reactions.

Oxidation reactions and reduction reactions always occur together in pairs—when one substance is oxidized, another substance is reduced. Recall from Section 5.1.5 that redox reactions may be generally thought of as electron transfer reactions, particularly in the case of the oxidation or reduction of elements in ionic compounds. (And remember the mnemonic: *LEO the lion says GER*—Lose Electrons Oxidize; Gain Electrons Reduce.)

In the case of molecular compounds, it is not very helpful to think in terms of electron transfer. Instead it is better to think in terms of the oxidation states of the elements involved in the reaction. We review this information below, but it would be a good idea at this point for you to reread Section 5.1.5 if you have not done so recently.

Figure 12.2. Metal corrosion from the strong base potassium hydroxide contained inside alkali batteries.

12.1.2 Oxidation States

When a metal atom loses one or more electrons, it becomes an ion with a positive charge. Its oxidation state as a pure metal is 0. As an ion, its oxidation state is the same as its ionic charge: +1 for alkali metals, +2 for alkaline-earth metals, and so on.

Figure 12.3. More redox: Aluminum wire (top); aluminum wire in $CuCl_2$ solution (second from top); copper metal out of solution on the aluminum (third from top); $Cu(OH)_2$ and $CuCO_3$ form on the copper, then leach into the water (bottom).

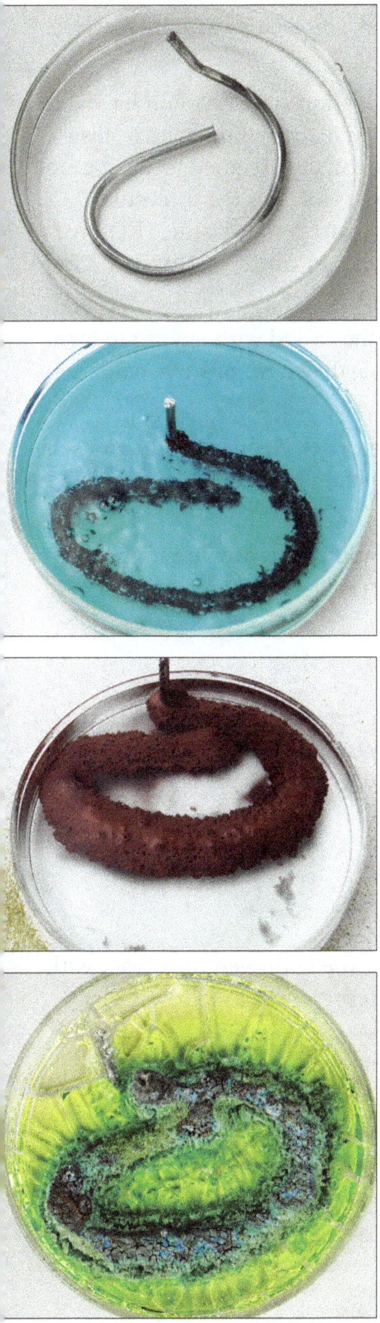

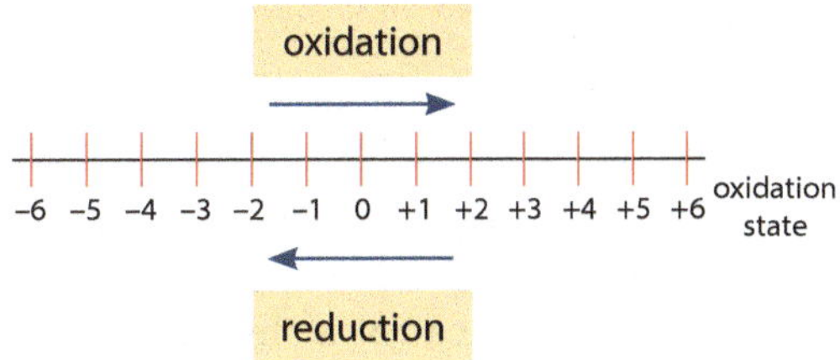

Figure 12.4. Oxidation occurs when the oxidation state increases; reduction occurs when the oxidation state decreases.

When a nonmetal atom gains one or more electrons, it becomes an ion with a negative charge. Its oxidation state as a pure element is 0 (including elements that form diatomic molecules). As an ion, its oxidation state is equal to its ionic charge: −1 for the halogens, −2 for the chalcogens, and so on.

In general, *any time the oxidation state of an element increases, that element is said to have been oxidized.* Conversely, *any time the oxidation state of an element decreases, that element is said to have been reduced.* These definitions are illustrated in Figure 12.4. Thinking in terms of electron transfer, an element that loses electrons is oxidized. The electrons must have somewhere to go, and the element they go to is the element that is reduced. The element that is oxidized causes the reduction of the element that is reduced. For this reason, the oxidized element is called the *reducer* or *reducing agent* or *reductant*. Conversely, the element that is reduced causes the oxidation of the element that is oxidized. For this reason, the reduced element is called the *oxidizer* or *oxidizing agent* or *oxidant*. These relationships are illustrated in Figure 12.5.

For monatomic ions and elements in ionic compounds, the oxidation state is equal to the value of the ion's charge. However, in molecules, the oxidation state of an element in the molecule is a hypothetical or artificial number assigned to each element in the molecule as if bonding electrons were held completely by one atom or another. In molecules, the oxidation state is associated with the strength of the element's attraction to the bonding electrons, and thus is based on the element's electronegativity relative to the other elements in the molecule. The oxidation state in a molecule is not based on any actual charge.

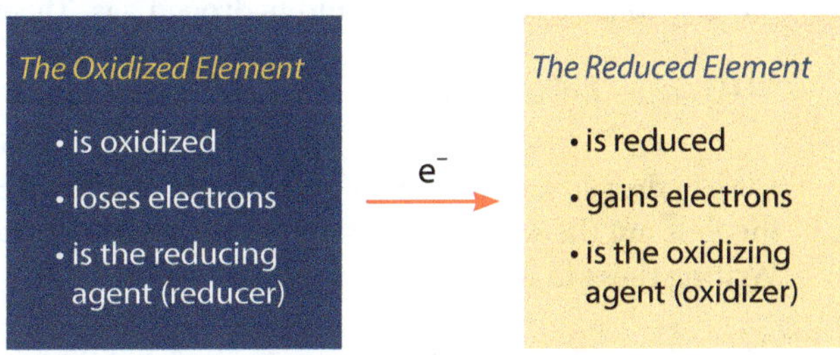

Figure 12.5. Relationships and nomenclature for the oxidizing agent and the reducing agent.

Analyzing redox reactions requires knowledge of how to establish the oxidation state of any element in any compound. As indicated above, this is easy for ions. For molecules, we have to go by the set of rules you learned in Section 5.1.5. For your convenience, here again are the rules we follow to determine the oxidation state of elements in covalent compounds:

1. The operating rule is this: *in a molecular compound, the oxidation states of the elements involved add up to the charge on the molecule.* In a neutral molecule, the oxidation numbers add up to zero. In a polyatomic ion, they add up to the charge on the ion.
2. In pure elements in their natural state, the oxidation state is 0. This also applies to diatomic molecules such as H_2 and the other elements that naturally make molecules with themselves, S_8 and P_4.
3. The most electronegative element in the molecule has the oxidation state it would have if it were an anion. Since fluorine has the highest electronegativity, its oxidation state is always −1. Oxygen has the second-greatest electronegativity, so it is usually assigned an oxidation state of −2. One exception is when oxygen forms peroxides in its form as the molecule O_2^{2-}, in which case its oxidation state is −1. A common peroxide is hydrogen peroxide, H_2O_2. Another exception is when oxygen bonds with fluorine to form OF_2, in which case its oxidation state is +2 (since F is always −1).
4. The oxidation state for halogens is −1. The exception is when they bond with oxygen to form polyatomic ions with negative charge (oxyanions). In that case, they have positive oxidation states (except for fluorine).
5. When hydrogen forms a compound with an element that is more electronegative than itself (a nonmetal), its oxidation state is +1. In compounds with metals, the oxidation state of H is −1.

Let's look at some examples of reactions and identify the oxidation states we find in the reactants and products. The equation for the combustion of hydrogen is

$$2H_2(g) + O_2(g) \rightarrow 2H_2O(g)$$

The elements in pure gases, hydrogen and oxygen, have oxidation states of 0. In water, the oxidation state of oxygen and hydrogen are −2 and +1, respectively. Writing in the oxidation states, we have:

$$\overset{0}{2H_2}(g) + \overset{0}{O_2}(g) \rightarrow 2\overset{+1}{H_2}\overset{-2}{O}(g)$$

Hydrogen is oxidized and oxygen is reduced in this reaction.

When zinc metal is placed in a solution of hydrochloric acid, the hydrogen ions react with the zinc to produce zinc ions and hydrogen gas. This reaction is represented as follows:

$$2H^+(aq) + Zn(s) \rightarrow H_2(g) + Zn^{2+}(aq)$$

Here hydrogen ions are forming hydrogen gas and zinc is being ionized. The oxidation states on the ions are the same as their ionic charges; the oxidation states of the pure elements are zero. Writing these in, we have:

$$\overset{+1}{2H^+}(aq) + \overset{0}{Zn}(s) \rightarrow \overset{0}{H_2}(g) + \overset{+2}{Zn^{2+}}(aq)$$

Now we see that hydrogen is reduced and zinc is oxidized. Since hydrogen is reduced, it is the oxidizing agent—hydrogen oxidizes the zinc. The zinc is oxidized, so it is the reducing agent—zinc reduces the hydrogen.

As one more example, from Table 5.2 we see that zinc is a more active metal than copper. This means zinc replaces copper in solution in a single replacement reaction, such as the following:

$$\text{Zn}(s) + \text{Cu}(\text{NO}_3)_2(aq) \rightarrow \text{Cu}(s) + \text{Zn}(\text{NO}_3)_2(aq)$$

When we write in the oxidation states on this more complex equation, we have:

$$\overset{0}{\text{Zn}}(s) + \overset{+2}{\text{Cu}}(\underset{+5}{\text{N}}\overset{-2}{\text{O}}_3)_2(aq) \rightarrow \overset{0}{\text{Cu}}(s) + \overset{+2}{\text{Zn}}(\underset{+5}{\text{N}}\overset{-2}{\text{O}}_3)_2(aq)$$

Note here that the charge on the nitrate ions is −1, and the sum of the oxidation states of the elements in that polyatomic ion add up to the charge on the ion. As usual, the oxidation state of oxygen is −2. Since there are three oxygen atoms, the oxidation state of nitrogen is +5 so that $5 + 3(-2) = -1$. In this reaction, the oxidation states of the elements in the nitrate ions do not change. The redox reaction is the oxidation of zinc and the reduction of copper.

Example 12.1

"Nicad" batteries are common now as AA and AAA size rechargeable dry cell batteries. These batteries contain nickel and cadmium, hence their name. The redox reaction that allows the batteries to produce electricity is shown below. Write the oxidation states above each element in the reaction. Identify the substances that are oxidized and reduced, and the oxidizing and reducing agents.

$$\text{Cd}(s) + \text{NiO}_2(s) + \text{H}_2\text{O}(l) \rightarrow \text{Cd(OH)}_2(s) + \text{Ni(OH)}_2(s)$$

Using the guidelines for assigning oxidation states, we write in these numbers as follows:

$$\overset{0}{\text{Cd}}(s) + \overset{+4}{\text{Ni}}\overset{-2}{\text{O}}_2(s) + \overset{+1}{\text{H}}_2\overset{-2}{\text{O}}(l) \rightarrow \overset{+2}{\text{Cd}}(\overset{-2}{\text{O}}\overset{+1}{\text{H}})_2(s) + \overset{+2}{\text{Ni}}(\overset{-2}{\text{O}}\overset{+1}{\text{H}})_2(s)$$

Cadmium is oxidized, since its oxidation state increases from 0 to +2. Nickel is reduced, since its oxidation state decreases from +4 to +2. Since cadmium is oxidized, cadmium is the reducing agent. Since nickel is reduced, nickel is the oxidizing agent.

Example 12.2

When sulfur dioxide gas is dissolved in water, small amounts of sulfurous acid form according to the reaction shown below. Write the oxidation states above each element in the reaction. Identify the substances that are oxidized and reduced, and the oxidizing and reducing agents.

$$\text{SO}_2(g) + \text{H}_2\text{O}(l) \rightarrow \text{H}_2\text{SO}_3(aq)$$

Using the guidelines for assigning oxidation states, we write in these numbers as follows:

$$\underset{+4}{\mathrm{S}}\overset{-2}{\mathrm{O}_2}(g) + \overset{+1}{\mathrm{H}_2}\overset{-2}{\mathrm{O}}(l) \rightarrow \overset{+1}{\mathrm{H}_2}\underset{+4}{\mathrm{S}}\overset{-2}{\mathrm{O}_3}(aq)$$

None of the oxidation states change. Thus, this is not a redox reaction.

12.1.3 Strengths of Oxidizing and Reducing Agents

In Section 9.2.4, we studied Table 9.6, which charts the relative strengths of acids and bases. By comparing their relative strengths, we are able to predict the direction of the equilibrium in an acid-base reaction. In a similar way, Table 12.1 lists a few representative substances according to their relative strengths as oxidizing and reducing agents. Some of the information in this table anticipates topics we have not yet addressed. So for now, attend only to the first two columns. The significance of the "half-reactions" in the third column will become clear in the next section, and we discuss the numerical "electrode potentials" in the fourth column later in the chapter.

From our studies so far in this text, you already know that lithium is an extremely active metal. It reacts vigorously to lose its single 2*s* electron, which is very loosely held in the second shell. When lithium loses its electron in reactions, it is oxidized, increasing its oxidation state from 0 to +1. The substance lithium is reacting with is the oxidizing agent. In such a reaction, lithium is the reducing agent, and the other substance in the redox reaction is reduced, accepting the electron from lithium.

Each substance listed in Table 12.1 appears in two columns side-by-side. The red column shows the relative strength as an oxidizing agent; the green column shows the relative strength as a reducing agent. As you see, lithium is the strongest reducing agent. And once lithium loses its electron, it shows no tendency to want to take it back, which makes Li^+ the weakest oxidizing agent.

Another strong reducer is sodium metal. If you have never seen the reaction that occurs when a lump of sodium metal is tossed into a basin of water, you need to—it's fantastic. (Ask your chemistry teacher about demonstrating it.)[1]

The strongest oxidizing agent is fluorine gas. Fluorine oxidizes just about anything, which makes it a very dangerous substance. It is interesting to compare the relative strengths of some of the oxidizing agents you are familiar with. The strength of ozone, O_3, is just below that of fluorine. In the discussion of "Why nitrates and nitros blow up" (page 135), we saw that hydrogen peroxide is a very aggressive oxidizer that is used for rocket fuel. Here you can see that it is the third strongest oxidizer listed. And note that oxygen, the most familiar oxidizer of all, is a fairly strong oxidizer but not as aggressive as some others. This has some important practical consequences. Oxygen is readily available in the atmosphere. As a strong oxidizer, it reacts readily with fuels to power the machines in the industrial world we live in. This is obviously of immense value for contemporary society. But O_2 is not such a strong oxidizer that it reacts with everything in sight the way F_2 does. If it did, the world would be a radically different place—everything would be destroyed by being oxidized. So you might say that oxygen's strength as an oxidizer is in the "Goldilocks zone": not too hot, not too cold, but just right. The downside of oxygen's strength as an oxidizer is that it oxidizes many metals, causing rust on iron and steel and corrosion of other unprotected metals.

1 Note to teachers: Instructions on how to perform the sodium demonstration safely are in my book, *Favorite Experiments for Physics and Physical Science*. See centripetalpress.com.

Oxidizing Agent	Reducing Agent	Reduction Half-Reaction	$E°$ (V)
Li^+ (weakest)	Li (strongest)	$Li^+ + e^- \rightarrow Li(s)$	−3.0401
K^+	K	$K^+ + e^- \rightarrow K(s)$	−2.931
Ca^{2+}	Ca	$Ca^{2+} + 2e^- \rightarrow Ca(s)$	−2.868
Na^+	Na	$Na^+ + e^- \rightarrow Na(s)$	−2.71
Mg^{2+}	Mg	$Mg^{2+} + 2e^- \rightarrow Mg(s)$	−2.372
Al^{3+}	Al	$Al^{3+} + 3e^- \rightarrow Al(s)$	−1.662
Zn^{2+}	Zn	$Zn^{2+} + 2e^- \rightarrow Zn(s)$	−0.7618
Fe^{2+}	Fe	$Fe^{2+} + 2e^- \rightarrow Fe(s)$	−0.447
Cd^{2+}	Cd	$Cd^{2+} + 2e^- \rightarrow Cd(s)$	−0.4030
Co^{2+}	Co	$Co^{2+} + 2e^- \rightarrow Co(s)$	−0.28
Ni^{2+}	Ni	$Ni^{2+} + 2e^- \rightarrow Ni(s)$	−0.257
Sn^{2+}	Sn	$Sn^{2+} + 2e^- \rightarrow Sn(s)$	−0.1375
Pb^{2+}	Pb	$Pb^{2+} + 2e^- \rightarrow Pb(s)$	−0.1262
H^+ (H_3O^+)	H_2	$2H^+ + 2e^- \rightarrow H_2(g)$	0
Cu^{2+}	Cu	$Cu^{2+} + 2e^- \rightarrow Cu(s)$	0.3419
I_2	I^-	$I_2(s) + 2e^- \rightarrow 2I^-$	0.5355
Fe^{3+}	Fe^{2+}	$Fe^{3+} + e^- \rightarrow Fe^{2+}$	0.771
Ag^+	Ag	$Ag^+ + e^- \rightarrow Ag(s)$	0.7996
NO_3^-	NO	$NO_3^- + 4H^+ + 3e^- \rightarrow NO(g) + 2H_2O$	0.957
Br_2	Br^-	$Br(l) + 2e^- \rightarrow 2Br^-$	1.066
O_2	H_2O	$O_2(g) + 4H^+ + 4e^- \rightarrow 2H_2O$	1.229
$Cr_2O_7^{2-}$	Cr^{3+}	$Cr_2O_7^- + 14H^+ + 6e^- \rightarrow 2Cr_3^+ + 7H_2O$	1.36
Cl_2	Cl^-	$Cl_2 + 2e^- \rightarrow 2Cl^-$	1.3583
MnO_4^-	MnO_2	$MnO_4^- + 4H^+ + 3e^- \rightarrow MnO_2(s) + 2H_2O$	1.679
H_2O_2	H_2O	$H_2O_2(aq) + 2H^+ + 2e^- \rightarrow 2H_2O$	1.776
O_3	O_2	$O_3(g) + 2H^+ + 2e^- \rightarrow O_2(g) + H_2O$	2.076
F_2 (strongest)	F^- (weakest)	$F_2(g) + 2e^- \rightarrow 2F^-$	2.866

Table 12.1. Relative strengths of oxidizing and reducing agents, with reduction half-reactions and standard reduction potentials.

Which metals will be oxidized by oxygen? This table answers that question. *Any reducing agent will be oxidized by any of the oxidizing agents below it.* And the farther apart the oxidizer and reducer are in the table, the likelier it is that a redox reaction will take place between them.

You may be wondering why aluminum, which is far above oxygen, doesn't exhibit the same tendency to be corroded (oxidized) by the oxygen in the atmosphere as other common metals do. After all, we are surrounded by aluminum products nowadays—just look around. It is probable that the window frames in the building you are in (and maybe the outer case of your laptop or tablet) are made of aluminum. The reason is that aluminum *is* oxidized by oxygen, and very quickly. However, in aluminum's case the oxide formed remains as a thin surface layer, protecting the interior metal from reacting further with oxygen. This process is called *passivation*, because the rest of the metal becomes "passive" (i.e., it doesn't react) because of its protective coating. This handy property makes aluminum an attractive structural material.

Notice that chlorine gas is one of the strongest oxidizing agents. Chlorine gas was the first chemical weapon used in World War I, deployed by Germany against the French in 1915. Chlorine gas is devastating to human tissue. In large water treatment facilities, chlorine gas is dissolved in water to produce hypochlorous acid (HOCl), a relatively weak acid. Both Cl_2 and HOCl kill microorganisms and algae (by oxidizing the carbon in them, and thus destroying them), which is why chlorine is so important for water treatment. Because Cl_2 is such a strong oxidizer, it is too dangerous for consumer use. Thus, treatment for home swimming pools uses HOCl only. The HOCl forms when solid cakes of $Ca(ClO)_2$ or NaClO are added to the water—an example of anion hydrolysis (see Section 11.2.4).

Finally, if the list of metals in the column of reducing agents looks familiar, it should. This is where the ordering of activity series of metals comes from (Table 5.2). The explanation for the activity series lies in the strengths of the metals as reducers and oxidizers. Why are copper ions in solution replaced by aluminum ions in a single replacement reaction (Figure 12.3)? Because aluminum metal is a strong reducing agent that is above Cu^{2+} in the table. So a redox reaction occurs between these two, the Al metal being oxidized to become Al^{3+} ions, and the Cu^{2+} ions being reduced to metallic copper. Figure 5.16 shows that iron metal and copper ions do the same thing. Table 12.1 helps us to see why.

We return to Table 12.1 later in the chapter after we learn more about redox chemistry. But as we go through the examples of redox reactions in the coming pages, you should refer back to the table to see how the strengths of the oxidizing agent and reducing agent in the reactions compare.

12.2 Redox Reaction Equations

12.2.1 Redox Half-Reactions

Redox reactions may be thought of as electron-transfer reactions. Just as acid-base reactions may be thought of as a process of proton transfer, redox reactions may be thought of as a process of electron transfer. The analytical methods used for redox reactions amount to a bookkeeping system that keeps track of the electrons. The first tool you need in order to do this is to be able to establish elements' oxidation states. The second tool is to be able to write separate equations for the oxidation and reduction processes. These separate equations are called *half-reactions*.

Consider the reaction between solid sodium metal and chlorine gas to form sodium chloride. If we write the equation with the oxidation states above the elements in the compounds, we have:

$$2\overset{0}{Na}(s) + \overset{0}{Cl_2}(g) \rightarrow 2\overset{+1}{Na}\overset{-1}{Cl}(s) \tag{12.1}$$

In this reaction, sodium is oxidized because sodium loses one electron and its oxidation state increases from 0 to +1. Chlorine is the oxidizing agent, so it is reduced; its oxidation state goes from 0 to −1. The half-reactions for this redox reaction are two separate ionic equations showing the oxidation and reduction processes. The oxidation half-reaction is written as follows:

$$\overset{0}{2\mathrm{Na}}(s) \rightarrow \overset{+1}{2\mathrm{Na}^+}(s) + 2e^- \quad (12.2)$$

This equation shows the two atoms of sodium ionizing to become Na^+ ions. Since their oxidations states increase from 0 to +1, the sodium atoms are oxidized. In the half-reaction, the electrons the sodium atoms give up are shown as a reaction product.

Like all reaction equations, half-reactions like Equation (12.2) must include coefficients to indicate the conservation of mass. But since half-reactions show electron transfer, they have a second requirement: they must also indicate the conservation of charge. The two sodium ions with charges of +1 are accompanied by two electrons with charges of −1. Thus, the total charge on each side of the equation is zero.

The reduction half-reaction is as follows:

$$\overset{0}{\mathrm{Cl}_2}(g) + 2e^- \rightarrow \overset{-1}{2\mathrm{Cl}^-}(s)$$

Sodium enters the reaction as individual atoms of sodium leaving the crystal lattice of the metal. But chlorine enters the reaction as molecules of gas which decompose to become separate Cl^- ions. As before, the charges on both sides of the equation must balance. Since this is a reduction half-reaction, the electrons are shown on the reactant side because they are combining with the chlorine atoms. The total charge of −2 on the left (from the two electrons) balances the total charge of −2 on the right (from the two Cl^- ions).

The two half-reactions may be added together to produce the complete net equation for the reaction:

$$\begin{array}{lr} \textit{oxidation} & 2\mathrm{Na}(s) \rightarrow 2\mathrm{Na}^+(s) + \cancel{2e^-} \\ \textit{reduction} & \mathrm{Cl}_2(g) + \cancel{2e^-} \rightarrow 2\mathrm{Cl}^-(s) \\ \hline & 2\mathrm{Na}(s) + \mathrm{Cl}_2(g) \rightarrow 2\mathrm{NaCl}(s) \end{array}$$

Notice that in the sum, the electrons cancel out because there are two electrons on each side.

Many redox reactions occur between ions in aqueous solution. Two important features of reactions in solution need mentioning here because of their bearing on how the half-reactions are written. First, to make it easier to keep track of the ions, hydrogen ions from acids are written simply as H^+ rather than as the hydronium ions they actually are in solution. Writing the ions as H_3O^+ is more convenient for acid-base reactions, but in redox reactions, it is more convenient to treat ionized hydrogen atoms as H^+ ions.

A second feature of redox reactions in solution is that once compounds ionize in solution, spectator ions appear that are not actually part of the redox reaction process. Since they are not part of the redox reaction, the spectator ions are not included in either half-reaction. In such cases, adding the half-reactions together shows the redox part of the reaction that occurs, but it does not show all the ions that came from the original compounds in solution unless they are added back in. We see in the next section how we handle this in the more general scheme of things. For now, just note that when we write the half-reactions for a redox reaction that occurs in solution, the sum of the two half-reactions does not include the spectator ions.

To illustrate these two points, consider the reaction shown in Figure 5.17 between magnesium and a solution of hydrochloric acid. When we are not concerned with the ions and the redox chemistry, we write this reaction this way:

$$Mg(s) + 2HCl(aq) \rightarrow MgCl_2(aq) + H_2(g) \tag{12.3}$$

Adding the oxidation states, we have the following:

$$\overset{0}{Mg}(s) + 2\overset{+1}{H}\overset{-1}{Cl}(aq) \rightarrow \overset{+2}{Mg}\overset{-1}{Cl_2}(aq) + \overset{0}{H_2}(g)$$

In this reaction, magnesium metal is oxidized to become Mg^{2+} ions, and hydrogen ions are the oxidizing agent. Hydrogen ions are reduced to become H_2 gas and magnesium is the reducing agent.

Now, from our studies in prior chapters, you know that the HCl and $MgCl_2$ shown in this equation are actually ions in solution. Let's rewrite the equation to show the ions explicitly. As we do so, we leave the hydrogen molecule intact because it represents an evolving gas in the reaction, not an ionized substance in solution.

$$\overset{0}{Mg}(s) + 2\overset{+1}{H^+}(aq) + 2\overset{-1}{Cl^-}(aq) \rightarrow \overset{+2}{Mg^{2+}}(aq) + 2\overset{-1}{Cl^-}(aq) + \overset{0}{H_2}(g)$$

With the equation written as ions this way, it is easier to pick out the oxidation and reduction half-reactions. When magnesium is oxidized to become the Mg^{2+} ion, two electrons are set loose. Thus, the half reaction for the oxidation of magnesium is as follows:

$$\overset{0}{Mg}(s) \rightarrow \overset{+2}{Mg^{2+}}(aq) + 2e^-$$

The two electrons on the right are necessary to make the total charge on the right zero, as it is on the left. The reduction of hydrogen involves two H^+ ions receiving the two electrons to become hydrogen gas:

$$2\overset{+1}{H^+}(aq) + 2e^- \rightarrow \overset{0}{H_2}(g)$$

The two electrons on the left are necessary to make the total charge on the left zero, as it is on the right.

If we now add these two equations together, we have the complete redox reaction:

oxidation $$Mg(s) \rightarrow Mg^{2+}(aq) + \cancel{2e^-}$$

reduction $$2H^+(aq) + \cancel{2e^-} \rightarrow H_2(g)$$

$$Mg(s) + 2H^+(aq) \rightarrow Mg^{2+}(aq) + H_2(g)$$

As you can see, the reaction we obtain by adding the half-reactions together is the actual redox part of the original reaction equation. The chlorine ions go into solution and remain there as spectator ions; they do not participate in the redox reaction itself.

Example 12.3

The single replacement reaction shown in the first three images of Figure 12.3, and again in Figure 12.6, is as follows:

$$2\text{Al}(s) + 3\text{CuCl}_2(aq) \rightarrow 3\text{Cu}(s) + 2\text{AlCl}_3(aq)$$

For this reaction,

a. Write the oxidations states above the elements in the given equation.

b. Write the ionic equation with oxidation states shown.

c. Write the two half-reactions, showing the oxidation states.

d. Add the two half-reactions together to show the redox reaction that occurs.

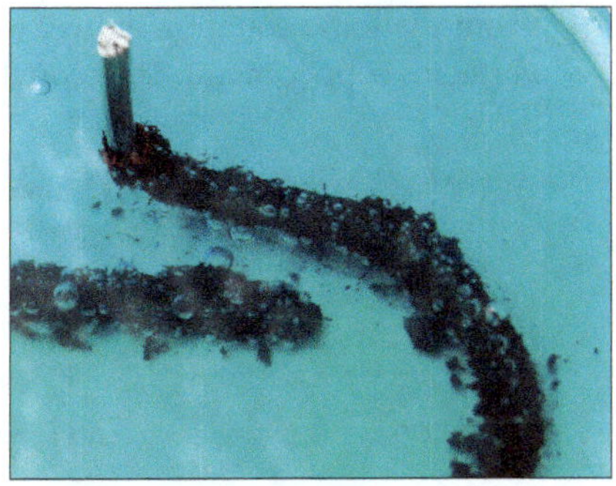

Figure 12.6. Redox reaction between aluminum metal and aqueous copper(II) chloride.

From the chlorine subscripts, we infer the oxidation states on the copper and aluminum in the two salts. In an ionic compound, chlorine's oxidation state is −1. The aluminum and copper metals (*s*) have oxidation states of 0, as all pure elements do. This gives us the following for the answer to part (a):

$$2\overset{0}{\text{Al}}(s) + 3\overset{+2}{\text{Cu}}\overset{-1}{\text{Cl}_2}(aq) \rightarrow 3\overset{0}{\text{Cu}}(s) + 2\overset{+3}{\text{Al}}\overset{-1}{\text{Cl}_3}(aq)$$

The copper chloride and aluminum chloride are both in aqueous solution. Dividing these compounds into aqueous ions gives us the answer to part (b). Notice that there are six Cl^- ions on each side of the ionic equation.

$$2\overset{0}{\text{Al}}(s) + 3\overset{+2}{\text{Cu}^{2+}}(aq) + 6\overset{-1}{\text{Cl}^-}(aq) \rightarrow 3\overset{0}{\text{Cu}}(s) + 2\overset{+3}{\text{Al}^{3+}}(aq) + 6\overset{-1}{\text{Cl}^-}(aq)$$

The aluminum metal is oxidized to Al^{3+} ions. Two aluminum atoms are shown in the equation. Thus we write the half reaction as

$$2\overset{0}{\text{Al}}(s) \rightarrow 2\overset{+3}{\text{Al}^{3+}}(aq) + 6e^-$$

Six electrons are added to the right side so that the total charge on the right is zero, as it is on the left.

(A comment is in order here about half-reactions such as this one. It is equally legitimate to write the half-reaction without the coefficients, as follows:

$$\overset{0}{\text{Al}}(s) \rightarrow \overset{+3}{\text{Al}^{3+}}(aq) + 3e^-$$

I am writing the half-reactions with the coefficients simply because they are given in the original equation. If we were working with an unbalanced equation, we would not know what coefficients to use. In such a case, the half-reaction is written without coefficients. Don't worry about this just yet. We sort out these coefficients in the next section.)

From the ionic equation above, we see that the copper is being reduced as Cu^{2+} ions are converted to copper metal. The reduction half-reaction is written as

$$\overset{+2}{3Cu^{2+}}(aq) + 6e^- \rightarrow \overset{0}{3Cu}(s)$$

For the answer to item (d), we add the half-reactions together, cancelling out the electrons as follows:

$$\begin{array}{lrcl} \textit{oxidation} & 2Al(s) & \rightarrow & 2Al^{3+}(aq) + \cancel{6e^-} \\ \textit{reduction} & 3Cu^{2+}(aq) + \cancel{6e^-} & \rightarrow & 3Cu(s) \\ \hline & 2Al(s) + 3Cu^{2+}(aq) & \rightarrow & 2Al^{3+}(aq) + 3Cu(s) \end{array}$$

From here we can see the actual redox reaction that takes place: aluminum is oxidized and copper is reduced. As a final note about the coefficients, note that the coefficients in the two half-reactions are different. If we had written the half-reactions without coefficients, we would have to apply coefficients here for the sum. We would simply select the coefficients so that the electrons cancel out when the half-reactions are added together. The coefficients of 2 and 3 give us the least common multiple of 6 for the number of electrons involved.

▲

12.2.2 Balancing Redox Equations

Balancing simpler equations may be done by inspection, as we learned in Chapter 5 and have been doing throughout this text. However, balancing the more complex equations involved in redox reactions is much trickier. For this reason, a methodical step-by-step procedure is necessary. The method we present here for balancing redox reaction equations is called the *half-reaction method*. The general idea builds on what you learned in the previous section: we write an unbalanced ionic equation for the reaction, extract the half-reactions from it, and add them together to produce the balanced equation.

To use this method, there are two important items to add to what you learned in the previous section. First, in the examples we encountered there, we didn't have to worry about balancing the masses in the half-reactions. All the elements we needed were present and the only issue we encountered was the coefficients. However, in the more general cases we consider now, the half-reactions are often missing oxygen or hydrogen atoms necessary to balance the equation. This occurs when reactions take place in acidic or basic solutions. As shown in the examples coming up, we handle this as follows:

- *In acidic solutions, we add water molecules as necessary to one side to balance the oxygen, then we add H^+ ions as necessary to one side to balance the hydrogen.*
- *In basic solutions, we add water molecules as necessary to one side to balance the oxygen, then we add H^+ ions as necessary to one side to balance the hydrogen, then we add OH^- ions as necessary to both sides to "neutralize" the H^+ ions.*

The second new item is that we must make sure that the number of electrons set loose in the oxidation half-reaction matches the number of electrons taken in by the reduction half-reaction. Doing this may require multiplying one or both of the half-reactions by coefficients.

We now go through the method. For this example, we use the redox reaction that occurs when hydrogen sulfide gas is combined with nitric acid in solution. The products of the reaction are sulfuric acid, nitrogen dioxide, and water.

Step 1 *Write the formula equation and the ionic equation.*

formula equation:

$$H_2S(g) + HNO_3(aq) \rightarrow H_2SO_4(aq) + NO_2(g) + H_2O(l)$$

ionic equation:

$$H_2S(g) + H^+(aq) + NO_3^-(aq) \rightarrow 2H^+(aq) + SO_4^{2-}(aq) + NO_2(g) + H_2O(l)$$

Step 2 *Assign oxidation states to each element. Identify all species containing only elements that do not change oxidation state in the reaction, and delete them from the equation.*

$$\overset{+1\,-2}{H_2S}(g) + \overset{+1}{H^+}(aq) + \overset{+5\,-2}{NO_3^-}(aq) \rightarrow 2\overset{+1}{H^+}(aq) + \overset{+6\,-2}{SO_4^{2-}}(aq) + \overset{+4\,-2}{NO_2}(g) + \overset{+1\,-2}{H_2O}(l)$$

Sulfur and nitrogen both change oxidation states, so species containing S and N are retained. All others are deleted.

$$\overset{+1\,-2}{H_2S}(g) + \overset{+5\,-2}{NO_3^-}(aq) \rightarrow \overset{+6\,-2}{SO_4^{2-}}(aq) + \overset{+4\,-2}{NO_2}(g)$$

Step 3 *Write the oxidation reaction terms, then balance the atoms, then balance the charges.*

This is the reaction involving sulfur, which is oxidized in the reaction.

$$H_2\overset{-2}{S}(g) \rightarrow \overset{+6}{S}O_4^{2-}(aq)$$

Sulfur is already balanced. This is an acidic solution, so to balance the oxygen and hydrogen we use H_2O and H^+. We first add four H_2O molecules to the left side to balance the oxygen.

$$H_2S(g) + 4H_2O(l) \rightarrow SO_4^{2-}(aq)$$

Now we see that we have 10 hydrogen atoms on the left, so we must add 10 H^+ ions to the right to balance the hydrogen.

$$H_2S(g) + 4H_2O(l) \rightarrow SO_4^{2-}(aq) + 10H^+(aq)$$

Now we have to add electrons to make the charges balance. On the right side, the sulfate ion has a charge of –2. This is added to 10 positive charges on the hydrogen ions, giving a total charge on the right side of $-2 + 10(+1) = +8$. The total charge on the left side is zero. To make the charges on the two sides match, we must add eight electrons to the right side to cancel out the +8 charge on that side.

$$H_2S(g) + 4H_2O(l) \rightarrow SO_4^{2-}(aq) + 10H^+(aq) + 8e^-$$

This is the complete oxidation half-reaction.

Step 4 *Write the reduction reaction terms, then balance the atoms, then balance the charges.*

Nitrogen is reduced. These terms are:

$$\overset{+5\,-2}{NO_3^-}(aq) \rightarrow \overset{+4\,-2}{NO_2}(g)$$

Nitrogen is already balanced. Again, this is an acidic solution, so we balance oxygen and hydrogen with H_2O and H^+. We first add one water molecule to the right side to balance the oxygen.

$$NO_3^-(aq) \rightarrow NO_2(g) + H_2O(l)$$

Now we must add two H^+ ions to the left side to balance the hydrogen.

$$NO_3^-(aq) + 2H^+(aq) \rightarrow NO_2(g) + H_2O(l)$$

Next, we have to add electrons to make the charges balance. The charges shown on the left are −1 from the nitrate ion and two +1 charges from the H^+ ions for a total charge of $-1 + (2)(+1) = +1$. The total charge on the right side is zero, so to make the left side balance with the right side, we must add one electron to the left side to cancel out the +1 charge there.

$$NO_3^-(aq) + 2H^+(aq) + e^- \rightarrow NO_2(g) + H_2O(l)$$

This is the complete reduction half-reaction.

Step 5 *Add the half-reactions together. Before adding, multiply one or both equations by a constant as necessary so that the numbers of electrons in the two half-reactions are equal. (The appropriate number of electrons is the least common multiple from the coefficients on the electrons in the two half-reactions.)*

First, we write the two half-reactions.

$$H_2S(g) + 4H_2O(l) \rightarrow SO_4^{2-}(aq) + 10H^+(aq) + 8e^-$$

$$\underline{NO_3^-(aq) + 2H^+(aq) + e^- \rightarrow NO_2(g) + H_2O(l)}$$

With eight electrons in the first reaction and one electron in the second reaction, we multiply the second reaction by eight to make the numbers of electrons in the two equations match. The goal here is that the electrons cancel out when the two half-reactions are added together.

$$H_2S(g) + 4H_2O(l) \rightarrow SO_4^{2-}(aq) + 10H^+(aq) + 8e^-$$

$$\underline{8[NO_3^-(aq) + 2H^+(aq) + e^- \rightarrow NO_2(g) + H_2O(l)]}$$

Now we multiply this value into the second equation and add the two equations together. The electrons cancel out. H_2O and H^+ appear on both sides, so some of these cancel out. H_2O and H^+ each appear on only one side of the sum.

$$\textit{oxidation} \qquad H_2S(g) + 4H_2O(l) \rightarrow SO_4^{2-}(aq) + 10H^+(aq) + 8e^-$$

$$\textit{reduction} \qquad 8NO_3^-(aq) + 16H^+(aq) + 8e^- \rightarrow 8NO_2(g) + 8H_2O(l)$$

$$H_2S(g) + 8NO_3^-(aq) + 6H^+(aq) \rightarrow SO_4^{2-}(aq) + 8NO_2(g) + 4H_2O(l)$$

Some texts regard the problem as complete at this point. The equation as it stands shows the ions and molecules involved in the redox chemistry in their proper proportions. However, I like to add the next step because it allows us to see again the compounds the ions all came from.

Step 6 *Combine the ions to form the compounds in the original formula equation. If you need extra ions of any sort, add the same number to both sides of the equation.*

Nitric acid, HNO_3, is in the formula equation on the left side, so we need to combine the H^+ ions with the NO_3^- ions to form HNO_3. However, we only have six H^+ ions. Thus, we add two more H^+ ions to each side of the equation. The two added to the right side combine with the SO_4^{2-} ion to make the H_2SO_4 that appears on the right side of the formula equation. First, we'll add the two H^+ ions:

$$H_2S(g) + 8NO_3^-(aq) + 8H^+(aq) \rightarrow SO_4^{2-}(aq) + 8NO_2(g) + 4H_2O(l) + 2H^+(aq)$$

Now we'll combine the ions:

$$H_2S(g) + 8HNO_3(aq) \rightarrow H_2SO_4(aq) + 8NO_2(g) + 4H_2O(l)$$

This is the complete, balanced equation showing the original compounds that provided the species in the redox reaction.

Step 7 *Check the final equation to make sure that all elements are correctly balanced and that there is no common multiple that can be divided out of all the coefficients.*

A final check of our equation indicates all is correct.

▼ Example 12.4

Potassium permanganate, $KMnO_4$, reacts with hydrocyanic acid, HCN, in aqueous solution to form fulminic acid, HCNO, manganese oxide, MnO_2, and potassium hydroxide, KOH. Note that manganese oxide is insoluble and precipitates out of solution. Because of the production of excess hydroxide ions, this reaction occurs in basic solution.

As a side note, you may recall the compound mercury fulminate we encountered back in Section 10.1.5 (and Figure 10.8). Mercury fulminate is a highly unstable compound used in detonators. Mercury fulminate is a salt of fulminic acid. A space-filling model of the fulminic acid molecule is shown in Figure 12.7.

Figure 12.7. Fulminic acid, HCNO. (H and O are white and red, as usual. C is black and N is blue.)

Determine the complete, balanced equation for this redox reaction, showing the original compounds involved.

Step 1 From the description above, the formula equation is

$$KMnO_4(aq) + HCN(aq) \rightarrow HCNO(aq) + MnO_2(s) + KOH(aq)$$

Each of the compounds is soluble and ionizes in solution, except MnO_2. Even without the tip-off in the problem statement, we know that manganese oxide is insoluble from the solubility rules of Table 8.3. So leaving MnO_2 as a non-ionized solid, the ionic equation is

$$K^+(aq) + MnO_4^-(aq) + H^+(aq) + CN^-(aq) \rightarrow$$
$$H^+(aq) + CNO^-(aq) + MnO_2(s) + K^+(aq) + OH^-(aq)$$

Step 2 Our next task is to write in the oxidation states.

$$\overset{+1}{K}{}^+(aq) + \overset{+7}{Mn}\overset{-2}{O}{}_4^-(aq) + \overset{+1}{H}{}^+(aq) + \overset{+2}{C}\overset{-3}{N}{}^-(aq) \rightarrow$$
$$\overset{+1}{H}{}^+(aq) + \overset{+4}{C}\underset{-3}{N}\overset{-2}{O}{}^-(aq) + \overset{+4}{Mn}\overset{-2}{O}{}_2(s) + \overset{+1}{K}{}^+(aq) + \overset{-2}{O}\overset{+1}{H}{}^-(aq)$$

Carbon is oxidized and manganese is reduced. The manganese and carbon compounds contain these elements. Eliminating everything else, we have

$$\overset{+7}{Mn}\overset{-2}{O}{}_4^-(aq) + \overset{+2}{C}\overset{-3}{N}{}^-(aq) \rightarrow \overset{+4}{C}\underset{-3}{N}\overset{-2}{O}{}^-(aq) + \overset{+4}{Mn}\overset{-2}{O}{}_2(s)$$

Step 3 Next we write the terms involved in the oxidation, then balance the atoms, and then balance the charges.

The carbon is oxidized, so writing the carbon terms gives

$$CN^-(aq) \rightarrow CNO^-(aq)$$

Carbon and nitrogen are already balanced. We add one H_2O to the left to balance the oxygen:

$$CN^-(aq) + H_2O(l) \rightarrow CNO^-(aq)$$

Now we add two H^+ ions to the right to balance the hydrogen:

$$CN^-(aq) + H_2O(l) \rightarrow CNO^-(aq) + 2H^+(aq)$$

We have one extra step to perform as part of the mass balancing. Recall that for redox reactions in basic solution, we must "neutralize" the H^+ ions by adding an equal number of OH^- ions to each side of the equation. Adding two OH^- ions to each side gives us

$$CN^-(aq) + H_2O(l) + 2OH^-(aq) \rightarrow CNO^-(aq) + 2H_2O(l)$$

Before we continue, notice that we have water molecules on both sides. We can simplify the equation here a bit by subtracting one H_2O molecule from each side. This is not strictly necessary; the cancellations occur when we add the half-reactions. So if you forget this step, it's no problem. But I like to keep things simplified, so now we have:

$$CN^-(aq) + 2OH^-(aq) \rightarrow CNO^-(aq) + H_2O(l)$$

Next we balance the charges. The total charge on the left side is −3; on the right side it is −1. To make these balance, we add two electrons to the right side:

$$CN^-(aq) + 2OH^-(aq) \rightarrow CNO^-(aq) + H_2O(l) + 2e^-$$

This is the completed oxidation half-reaction.

Step 4 Now we write the terms involved in the reduction, then balance the atoms, and then balance the charges.

The manganese is reduced. Writing the manganese terms gives

$$MnO_4^-(aq) \rightarrow MnO_2(s)$$

Manganese is already balanced. We add two H_2O molecules to the right side to balance the oxygen:

$$MnO_4^-(aq) \rightarrow MnO_2(s) + 2H_2O(l)$$

Now we add four H^+ ions to the left to balance the hydrogen:

$$MnO_4^-(aq) + 4H^+(aq) \rightarrow MnO_2(s) + 2H_2O(l)$$

Again, this reaction occurs in basic solution. So we have the extra step here of adding enough OH^- ions to both sides to neutralize those H^+ ions. Adding four OH^- ions to each side we have

$$MnO_4^-(aq) + 4H_2O(l) \rightarrow MnO_2(s) + 2H_2O(l) + 4OH^-(aq)$$

As before, there are now water molecules on both sides. We can subtract two from each side to simplify a bit:

$$MnO_4^-(aq) + 2H_2O(l) \rightarrow MnO_2(s) + 4OH^-(aq)$$

The last piece of this step is to balance the charges. The total charge on the left is −1; on the right it is −4. We add three electrons to the left side to make these match.

$$MnO_4^-(aq) + 2H_2O(l) + 3e^- \rightarrow MnO_2(s) + 4OH^-(aq)$$

This is the completed reduction half-reaction.

Step 5 We can now write the two half-reactions in preparation for adding them. But before we add them together, we need to make sure the electrons will cancel.

oxidation $$CN^-(aq) + 2OH^-(aq) \rightarrow CNO^-(aq) + H_2O(l) + 2e^-$$

reduction $$MnO_4^-(aq) + 2H_2O(l) + 3e^- \rightarrow MnO_2(s) + 4OH^-(aq)$$

There are two electrons in the oxidation half-reaction and three in the reduction half-reaction. To make the electrons cancel, we need six electrons in each half-reaction (six being the least common multiple of two and three). We need to multiply the oxidation half-reaction by three and the reduction half-reaction by two. Multiplying in these constants gives us the equations below. I have also marked out the items that cancel and added the equations together.

oxidation $$3CN^-(aq) + \cancel{6OH^-}(aq) \rightarrow 3CNO^-(aq) + \cancel{3H_2O(l)} + \cancel{6e^-}$$

reduction $$2MnO_4^-(aq) + \overset{1}{\cancel{4}}H_2O(l) + \cancel{6e^-} \rightarrow 2MnO_2(s) + \overset{2}{\cancel{8}}OH^-(aq)$$

$$3CN^-(aq) + 2MnO_4^-(aq) + H_2O(l) \rightarrow 3CNO^-(aq) + 2MnO_2(s) + 2OH^-(aq)$$

This is the completed redox equation. However, to show the original compounds involved as the problem requires, we need to add in some ions in the next step.

Step 6 We need three H^+ ions on each side to join with the carbon-containing ions. We also need to add two K^+ ions to make potassium permanganate and potassium hydroxide. Adding these ions, we have:

$$3H^+(aq) + 3CN^-(aq) + 2K^+(aq) + 2MnO_4^-(aq) + H_2O(l) \rightarrow$$
$$3H^+(aq) + 3CNO^-(aq) + 2MnO_2(s) + 2K^+(aq) + 2OH^-(aq)$$

And combining the ions gives us finally

$$3HCN(aq) + 2KMnO_4(aq) + H_2O(l) \rightarrow 3HCNO(aq) + 2MnO_2(s) + 2KOH(aq)$$

Step 7 Our final step is to double check the mass balance by counting elements on both sides. This checks.

▲

12.3 Electrochemistry

12.3.1 Copper and Zinc Redox

We have seen several single replacement reactions with metals in copper sulfate. You now know that these are all redox reactions. Here is another one, this time involving zinc. The plating known as *galvanizing* that is often used to protect steel from rusting is made of zinc. Figure 12.8 shows a beaker of copper sulfate solution and a galvanized steel plate purchased from a hardware store. (I used a galvanized plate for this demo because I didn't have any solid zinc metal handy. The galvanized part from the hardware store is completely covered in zinc, so it works

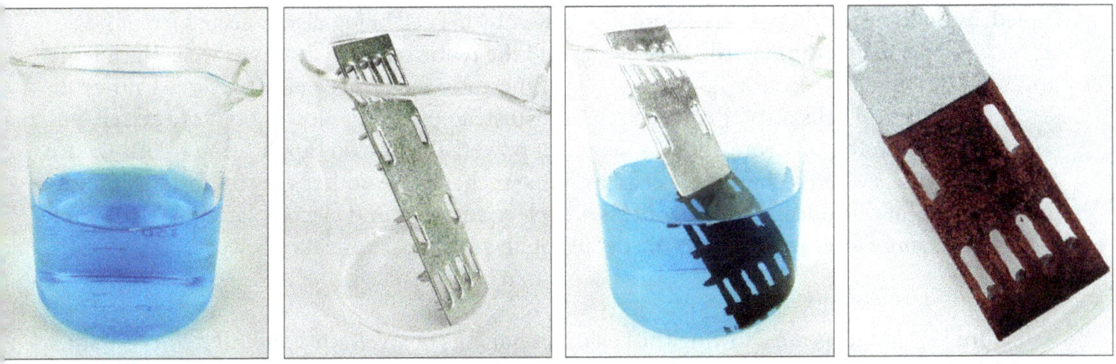

Figure 12.8. The spontaneous redox reaction between zinc and copper. Copper ions in the blue $CuSO_4$ solution are reduced by the zinc, so the copper ions come out of solution. The copper metal adheres to the zinc plate. (The plate was removed from the solution before the reaction was complete, so the solution was still blue.)

just fine.) The zinc-coated plate is placed into the copper sulfate solution, and five minutes later the zinc plate has a nice copper coating.

This redox reaction occurs spontaneously. In the reaction, the electron transfer from the atoms of zinc metal (which are being oxidized) to the copper ions (which are being reduced) occurs directly between the zinc atoms at the surface of the metal and the copper ions in solution right next to the zinc metal.

The introduction to electrochemistry in this section involves an apparatus using the same redox reaction, but we are going to arrange things so the copper and zinc are not in direct contact. For the redox reaction to occur, the electrons take a different route to get from the zinc to the copper. We provide this electron pathway in the form of an electrical conductor (a copper wire) connected from the zinc to the copper. So before we go any further with the redox reaction, let's discuss the benefits of doing this.

12.3.2 Electricity Instead of Heat

Every chemical reaction involves a transfer of energy. As you now know, heat flows into a chemical system during an endothermic reaction and out of a chemical system in an exothermic reaction. But since redox reactions involve electron transfer, there is another way energy can flow into or out of a chemical system. Electric current is the flow of electrons. Since we have electrons in motion anyway in redox reactions, we use the electric current of the electrons being transferred as a means for getting useful electrical energy out of a reaction, or as a means of forcing reactions to occur by putting electrical energy in. This is what electrochemistry is all about.

Instead of being released as heat, the energy released during an exothermic redox reaction is released in the form of an electric current. All we have to do is rig up a system so that the electrons being transferred in the redox reaction flow through an external electric circuit. When we do this, we have made what everyone calls a *battery*.

The importance of redox chemistry should now be obvious to you. We are surrounded now by battery-operated devices. Not only do we have millions of different products that run on standard batteries such as the common AAA, AA, C, and D cells (and others), but we now also have rechargeable batteries everywhere—every mobile phone, mobile device, and laptop computer uses a rechargeable battery.

A battery—actually, the proper term is *cell*; a *battery* is a set of two or more cells connected together—is an electrochemical system that uses a redox reaction to create an electric current composed of the electrons being transferred in the reaction. When the cell is connected to an external circuit, the electrons flow and the reaction occurs. When the external circuit is interrupted, by opening a switch in the external circuit or by the cell being disconnected from the device it is connected to, the electron flow ceases and the redox reaction ceases as well. The energy produced by the redox reaction flows as electrical energy to the device being powered by the cell.

Electrochemical cells also operate in reverse—instead of energy flowing out of the cell, energy can flow into the cell from another source of power. The obvious application of this we are all familiar with is when rechargeable cells are being recharged. Another very important application is the use of cells for industrial processes such as smelting and electroplating. We will look at these a bit more after we dive into the details of how electrochemical cells work.

12.3.3 Electrochemical Cells

Figure 12.9 is a diagram showing the basic components of an *electrochemical cell*, a container in which the two half-reactions of an electrochemical redox reaction take place. The cell shown makes use of zinc and copper metal as the two *electrodes*, the sites of the oxidation and reduction half-reactions.[2] The zinc electrode is immersed in a zinc sulfate solution. This solution conducts electricity, and is thus an electrolyte.[3] The copper electrode is immersed in a solution of copper sulfate, also an electrolyte.

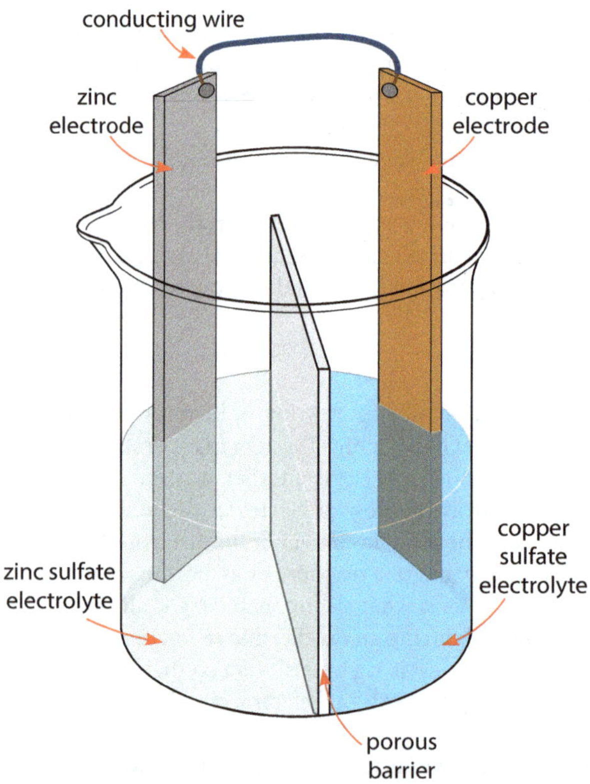

Figure 12.9. The essential parts of an electrolytic cell. The cell shown uses zinc and copper for the electrodes.

The zinc and copper solutions are divided into separate *half-cells* where the redox half-reactions take place. The zinc and copper solutions are separated so that the half-reactions occur in separate compartments, but there must be a means for ions to migrate from one cell to another. (We are setting up an electric circuit in which charges—electrons or ions—flow in a closed loop. If charges can't get from one half-cell to the other, the cell won't work.) In the simplified cell pictured, ion migration occurs through a *porous barrier* separating the half cells. This barrier keeps the solutions from randomly mixing, but it is made of a material with microscopic holes large enough to allow ions to migrate through from one side of the cell to the other. A better cell design allows for ion transfer while keeping the zinc and copper ions completely separated from each other. Such a cell allows ion transfer to occur through a *salt bridge*. We look more closely at these details a bit later.

The final component of the electrochemical cell is a conductor connected between the two electrodes. With this

2 Many other metals may be used in electrochemical cells. The copper/zinc cell is the usual one used for demonstration and discussion of the basic principles of the cell's operation.

3 In our studies in previous chapters, we use the term *electrolyte* to refer to a compound that ionizes in water to produce an electrically conductive solution. In the context of electrochemistry, it is customary to refer to the solution itself as the electrolyte.

conductor between the electrodes, we now have a complete closed electrical circuit. Electrons flow up one electrode, through the wire, and down the other electrode. To complete the circuit, ions flow through the porous barrier. The movement of charge represented by the electrons in the wire is balanced by ion transfer through the barrier. The direction of the electron flow depends on whether the cell is producing electrical energy or consuming it, as I explain below. Note that a simple wire connecting the electrodes wouldn't actually allow the cell to be used for anything. Instead, that connecting wire needs to connect to an outside device if the cell is producing electricity, or have a power source inserted in it if the cell is consuming electricity. Details on these connections are also part of the discussion below.

The cell shown in Figure 12.9 contains liquid, and is accordingly called a *wet cell*. The cells we use to power portable electronics contain moist solids (pastes or gels), but no liquids, and are called *dry cells*. Cells powering mobile phones, laptops, and flashlights are dry cells; conventional car batteries are wet cells.

There are two basic configurations of the electrochemical cell, based on whether the cell is producing electricity or consuming it, and equally based on the direction of electron flow in the redox process. The *voltaic cell* or *galvanic cell* (named after Alessandro Volta or Luigi Galvani, depending on what you call it) is used to provide electrical power to an external device. Cells in mobile devices, power tools, and flashlights are voltaic cells. The other configuration is the *electrolytic cell*. Electrolytic cells are those that consume electricity instead of producing it. We will now examine the details of what is going on in each type of cell.

Voltaic Cells The basic arrangement of the voltaic cell is shown on the left side of Figure 12.10. If you have not yet studied DC circuits in one of your science classes, just focus on the diagram on the left. But for those students who are familiar with DC circuits, the electrochemical cell is represented by the "battery" (cell) in the electrical schematic diagram shown on the right side of the figure. If you are familiar with DC circuits, the schematic will help you understand how the cell functions electrically. Recall from your study of DC circuits that the way the flow of positive current (I) is defined, the current, I, and the electrons flow in opposite directions in the circuit.

The figure shows a generic electrical device (the light bulb) being powered by the voltaic cell. The connections between the voltaic cell and the electrical device are shown so you can get an idea of how a voltaic cell is put to use. Naturally, we would not actually use a wet cell made of two beakers to power anything; we would use the dry cells and batteries we are all familiar with. But the voltaic cell shown might be used in a laboratory for experimental purposes and is convenient for demonstrating how electrochemical cells work.

We are going to examine this diagram in detail, but first, here's the big picture in brief: The voltaic cell is divided into two half-cells containing the two electrodes, each immersed in a salt solution of the same metal. For our demonstration example, the two electrolytes are zinc sulfate and copper sulfate. The cell and the device it is connected to form a complete electrical circuit—a closed path in which electrons or other forms of charge flow in a closed loop. Beginning at the bottom of the zinc electrode, the electrons flow up the zinc electrode, out of the cell in the blue wire, through the electrical device (the light bulb or whatever it is), through another blue wire, into the copper electrode, and down the copper electrode to where it is submerged in the electrolyte. At this point the movement of charge around the circuit is taken over by ions in the two half-cells and in the salt bridge connecting them together.

We begin our tour around the loop in the zinc half-cell. At the lower end of the zinc electrode, down in the electrolyte, oxidation of the zinc metal is occurring: zinc atoms are ionizing and the Zn^{2+} ions are separating from the zinc metal and entering the zinc sulfate electrolyte. (Thus, the zinc metal is gradually dissolving.) Note this: *In any electrochemical cell, the electrode*

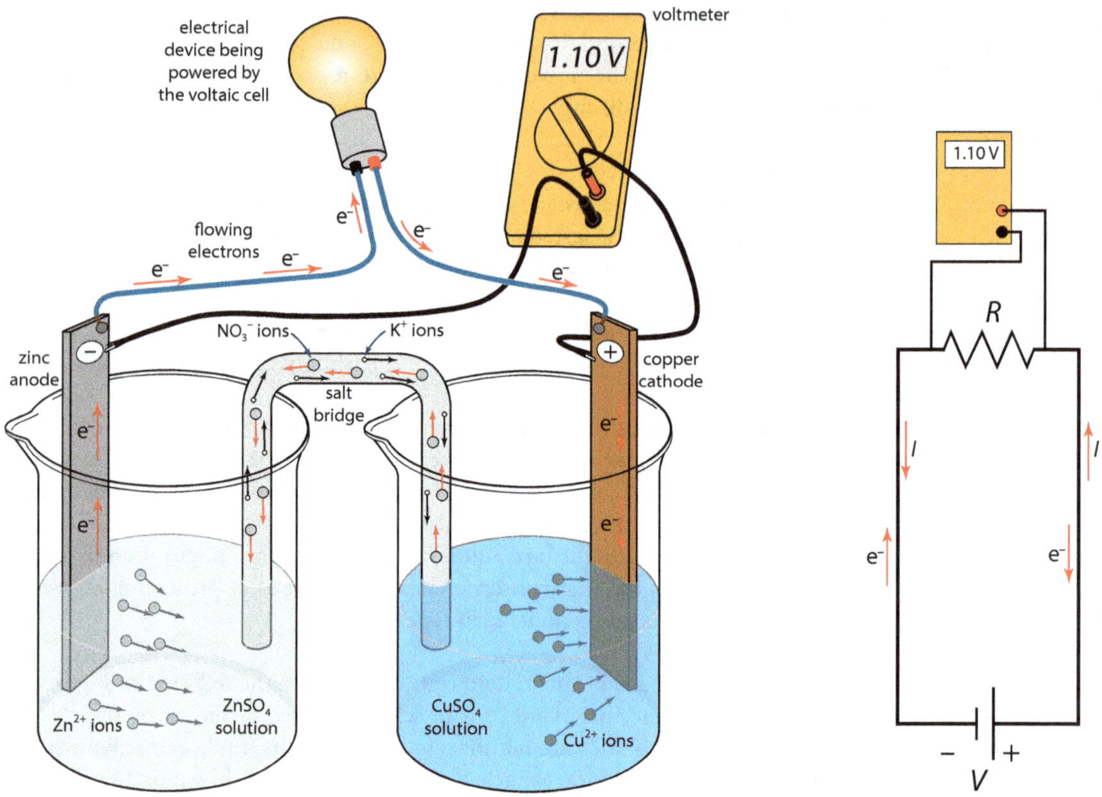

Voltaic Cell

The cell is powering an electrical device, while a voltmeter measures the voltage the cell is producing.

Schematic

An electrical representation of the voltaic cell. The cell is driving current into an external device, represented in the schematic by a resistor.

Figure 12.10. Diagram of the voltaic cell showing flow of charge (left). The cell is providing electrical energy to a generic electrical device shown. A voltmeter is connected to the cell to show the voltage the cell is producing. On the right is the equivalent electrical schematic representation of the system. (Note that the cell could only be used this way if the 1.10 V it produces is a high enough voltage to run the device.)

where the oxidation is taking place is called the anode. Thus, in our demonstration voltaic cell, the zinc electrode is the anode. The half-reaction for the oxidation of the zinc is

$$\mathrm{Zn}(s) \rightarrow \mathrm{Zn}^{2+}(aq) + 2\mathrm{e}^-$$

The electrons turned loose by all the oxidizing zinc atoms (two per atom, according to the half-reaction) flow up the anode and out of the cell in the blue wire, as indicated by the red arrows. This flow of electrons goes through an electrical device, providing energy to operate the device. The redox reaction is releasing energy, and this is where the energy is going. The electrons returning to the cell from the electrical device enter the copper electrode. Down under the surface of the $CuSO_4$ electrolyte, Cu^{2+} ions in the electrolyte are being reduced and are attaching to the crystal lattice of the copper electrode. Thus the copper electrode is where the reduction is taking place. *In any electrochemical cell, the electrode where the reduction is taking place is called the cathode*. Thus, the copper electrode in this voltaic cell is the cathode. The electrons flowing into the upper end of the cathode travel down the copper metal and reduce the copper ions as they attach to the cathode and become copper metal. (As a result, the copper ions are gradually being depleted from the electrolyte, and the cathode is getting larger and heavier as additional copper atoms continue to attach to it.) The reduction half-reaction is

$$Cu^{2+}(aq) + 2e^- \rightarrow Cu(s)$$

(To aid you in remembering which electrode goes with which half-reaction, note that *anode* and *oxidation* both begin with vowels; *cathode* and *reduction* both begin with consonants.)

We return to Figure 12.10 to discuss the salt bridge in a moment. But first, Figure 12.11 shows hypothetical extreme close-up depictions of the two electrodes. (Remember that atoms and ions are much smaller than the wavelength of light, and no amount of magnification would allow us actually to see these things. These images are visual models of entities that cannot be perceived with our eyes.) At the anode, oxidation of zinc atoms sends them into the $ZnSO_4$ electrolyte as Zn^{2+} ions, while the electrons from the oxidized atoms flow up the anode and out of the cell. The electrons return to the cell at the cathode, where they travel down to where the reduction of Cu^{2+} ions is occurring, which adds copper metal to the cathode.

Now we return to our discussion of Figure 12.10. For an electric circuit to work, charge must be able to flow in a path that forms a complete, closed loop. Without a way for charge to flow between the two half-cells, nothing happens and no current flows. In a demonstration cell like the one we are considering here, the connection allowing charge to flow between the cells is

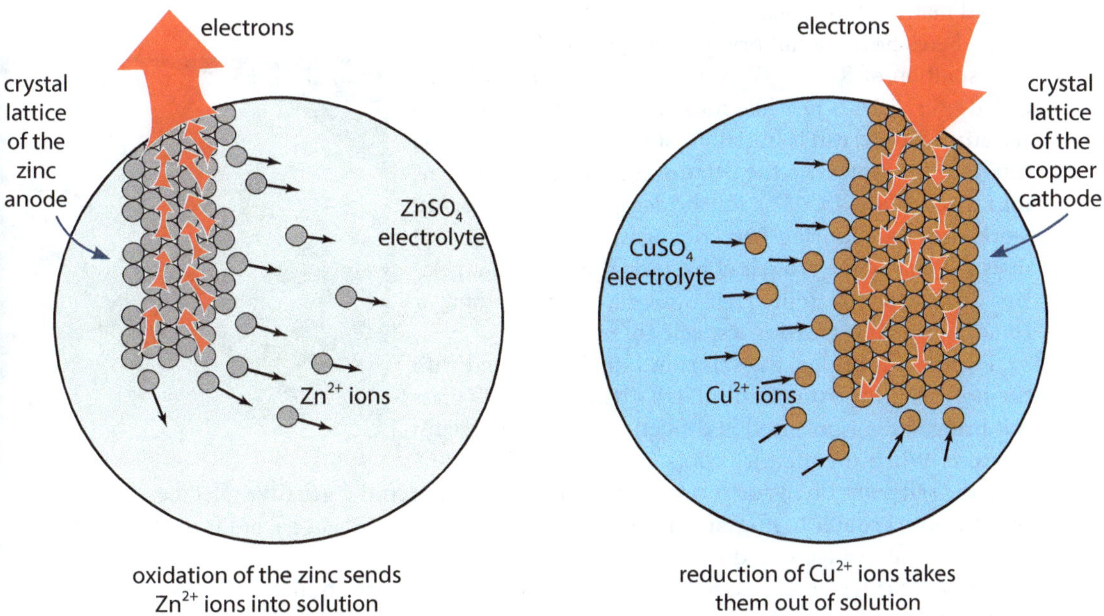

Figure 12.11. A close-up view of the oxidation occurring at the anode (left) and the reduction occurring at the cathode (right) in the voltaic cell. The red arrows represent the electrons streaming away from the oxidized zinc ions on the left, and flowing down to reduce the copper ions on the right. (Note that the ions are shown on a colored background for clarity. If we could really "see" ions, we would not see the rest of the solution as a colored background. We would instead see the particles of which the solution is composed—a dense mixture of water molecules, metal ions, sulfate ions, and of course, a few hydronium and hydroxide ions as well.)

often provided by a pathway called a *salt bridge*. The salt bridge is essentially a tube filled with concentrated salt solution. The salt is chosen so that the particular cations and anions, which are dissolved in solution, do not react with the substances of which the electrodes and electrolytes are composed. Potassium nitrate is a salt commonly used for this purpose. The anions in the salt solution, NO_3^- in our example, are negatively charged just like the electrons are. Thus, they move around the electrical circuit in the same direction the electrons are moving. This means that the salt anions migrate through the salt bridge toward the half-cell containing the anode. The cations in the salt solution (K^+) migrate in the opposite direction. To summarize, in the electrodes and in the external electric circuit, the flowing charge is electrons. In the salt bridge connecting

Hmm... Interesting. How are salt bridges made?

If you are the curious type, you want to know what a salt bridge *is*—that is, what they are *made of*. We know the function of the salt bridge in the electrochemical cell—to allow charge migration between the half-cells. This can be done with a simple salt solution. But the trick is, we need some way to keep the salt solution inside the glass tube of the salt bridge. This is done in a clever way that makes use of the porous properties of gels. At the beginning of Chapter 8, we looked briefly at some types of heterogeneous mixtures, one of which was gels. I mention there that in a gel, the solid particles link together in long molecules known as polymers, and these linked solids form a network or matrix throughout the substance. What this means is that a gel is like a sponge, but with pores too small to see. The solid material is the network of polymer strands and between these strands is a connected matrix of micro-sized passageways.

So here's how the salt bridge is prepared. You mix up a 1-*M* solution of KNO_3. Then you mix up a batch of a gel called *agar*. This is just like making Jell-O, which is a gel (gelatin). The gelatin is initially a solid powder, which you mix into hot water. As the gelatin mixture cools, it turns into a gel, like Jell-O. For salt bridges, and for biological work such as culturing bacteria, agar is used because it does not enable the growth of bacteria the way other gelatins do. The photo to the right shows a petri dish of agar (blue) in which is growing a species of moss.

For the salt bridge, the salt solution is combined with the hot gelatin mixture and mixed. Then the empty glass salt bridge tube is inverted and filled with the salt/gelatin mixture. When the mixture cools, it solidifies like Jell-O, with the salt solution inside it occupying the spaces between the gelatin molecules. In this way, the ions are able to migrate through the salt bridge, and the agar gel holds the salt solution in the salt bridge so it doesn't run out.

The salt bridge described here is typically used for demonstration cells like the one described in this chapter. But a simple strip of paper towel soaked in salt water and draped between the beakers also works. And instead of a salt bridge, some demonstration cells have their compartments connected together at the bottom by a tube containing some kind of porous barrier. The porous material is chosen to allow the sulfate and zinc ions to migrate directly from one half-cell to another without the use of an extra salt. In such a case, the negatively charged SO_4^{2-} ions migrate toward the zinc anode and the Zn^{2+} ions migrate toward the cathode.

the two half-cells, the flowing charge is the ions from the salt solution. The interesting details of how a salt bridge is actually made are in the accompanying box.

With the salt bridge connecting the two half-cells of the cell, charge can flow all the way around the circuit. As the cell operates, nitrate ions from the salt bridge become concentrated in the anode half-cell with the zinc sulfate solution. This solution also becomes increasingly concentrated with Zn^{2+} ions from the decomposing anode. The potassium cations in the salt bridge migrate into the cathode half-cell and become concentrated there in the copper sulfate solution. The concentration of copper ions in this solution is decreasing, as copper ions come out of solution and join the other copper metal atoms in the cathode. For each electron that leaves the anode, another electron from the external circuit enters the cathode. Together, these charge movements make the anode one unit more positive and the cathode one unit more negative. For each of these electron exchanges, one nitrate ion enters the anode compartment from the salt bridge, replacing the negative charge lost as the electron leaves. Also, one potassium ion enters the cathode compartment from the salt bridge, its positive charge canceling out the negative charge of the electron that enters. The result is that charge remains balanced between the two half-cells as the two half-reactions proceed.

Finally, note that the reaction in a voltaic cell occurs spontaneously (*really* spontaneously[4]) and delivers energy to the outside world. This means that the change in the Gibbs free energy for the reaction is negative—the reaction happens all by itself once the outside circuit is connected to the cell.

Electrolytic Cells

Electrolytic cells are essentially voltaic cells running in reverse, with all reactions proceeding in the opposite direction. But the change in the Gibbs free energy for a system operating this way is positive, so the process is not spontaneous. An energy source must be supplied to force the electrical current to flow in the opposite direction.

The components and arrangement in the electrolytic cell are the same as for a voltaic cell, with one difference. Figure 12.12 is a diagram showing the electrode arrangement for the electrolytic cell. An electrical power source is now connected into the external circuit and electron flow is in the opposite direction. Instead of the cell producing power to operate an external device, an external source of power drives the redox reaction in the cell. This is essentially what is going on when "batteries" are recharged. The forward redox reaction concentrates ions in one region until they are so thick the cell can no longer operate. The recharging puts all these ions back in the other direction so the forward reaction can begin afresh.

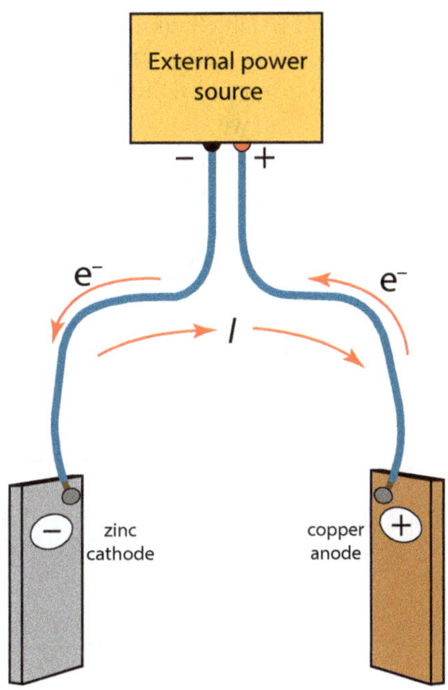

Figure 12.12. Connection of a power source to the electrodes of an electrolytic cell. The power supply drives the redox reaction occurring in the cell. (The rest of the cell is not shown.)

It is important for you to note in Figure 12.12 that *the zinc electrode is now the cathode*. If you refer again to Figure 12.10, you see that in the voltaic cell the cathode is the copper electrode. The electrons are flowing *into* this electrode because this electrode is the site of the reduction half-reaction. For reduction to occur, a continuous electron supply is needed. When reduction

4 Recall my comments about the use of the term "spontaneous" in Section 10.3.3.

occurs, ions are combining with electrons, reducing their oxidation states, coming out of solution, and becoming metal atoms in solid form. The incoming electrons are joining with the ions, enabling the reduction to happen.

Compare that situation to the sketch here in Figure 12.12. With the power source connected between the electrodes, the electrons are being forced by the power source to flow in the opposite direction. The electrode where the electrons are entering is the one where the reduction is taking place, and thus it is the cathode.

To identify the anode and cathode in any electrochemical cell, simply remember this: when reduction occurs, electrons are being supplied to metal ions to convert them from ions to atoms of pure metal. So to identify the cathode, you can simply look at the direction of the electron flow. The electrode where the electrons are entering the cell is the electrode where the reduction is taking place, and is thus the cathode. And again, as a final reminder to those who have studied DC circuits, in any electrical circuit, the direction of the flow of positive current, *I*, is the opposite of the direction of electron flow. The same applies to the electrolytic cell.

Figure 12.13 summarizes the nomenclature and arrangement of the two types of electrochemical cells. In this sketch, the two types of cells are shown side-by-side to aid you in studying and remembering the details. The voltaic cell has a resistor in the external circuit. This resistor represents whatever the cell is connected to (even if it is only connected to a voltmeter). The electrolytic cell has an external source of DC current in the wire between the electrodes.

12.3.4 Electrode Potentials

Look once again at Figure 12.10. In the diagram, the voltmeter connected to the electrodes reads 1.10 volts (V). The voltage produced by a voltaic cell depends on the materials the two electrodes are made of. (It also depends on the concentration of the electrolytes, a detail we will not concern ourselves with here). Each different electrode material exhibits a certain tendency to engage in a reduction half-reaction. For a given reducing agent in a particular reduction half-reaction, we quantify this tendency with a constant called the *standard reduction potential*, symbolized as E° and measured in volts. The standard reduction potential for a material used as an electrode is usually just called the *electrode potential*.

The overall potential for a complete cell made of two electrodes is referred to as the *cell potential*. The cell potential is calculated as

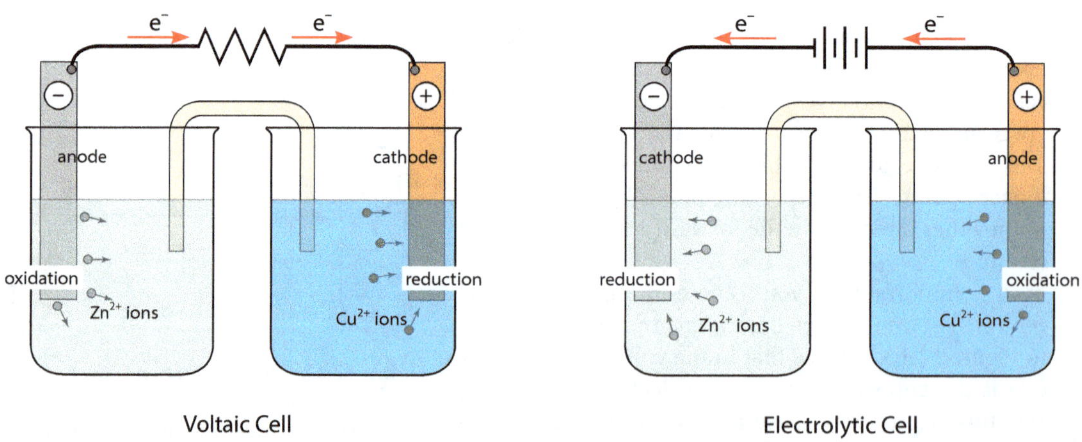

Figure 12.13. Summary of configurations and ion flow for voltaic and electrolytic cells. The key to remembering everything is that the electrode where the electrons are entering is where the reduction is taking place, and is thus the cathode.

$$E^{\circ}_{cell} = E^{\circ}_{cathode} - E^{\circ}_{anode} \qquad (12.4)$$

For the redox reaction in any cell to occur spontaneously, the cell potential must be greater than zero. Ideally, the voltage produced by a voltaic cell is equal to this cell potential. However, nothing is ever ideal, so in practice the actual voltage the cell produces while electrical current is flowing between the electrodes is a bit less and drops slowly over time as the concentrations of the electrolytes change.

It is now time to refer back to Table 12.1, which lists the standard reduction potentials for a number of different reduction half-reactions.[5] The standard reduction potential is a measure of the readiness of the reducing agent to engage in a particular reduction half-reaction. The more negative the standard electrode potential for the half-reaction is, the stronger the reducing agent is. The table begins with the most negative half-reaction potential, which is associated with the strongest reducing agent, and ends with the most positive potential, associated with the strongest oxidizing agent. For any pair of substances selected to be the electrodes in a voltaic cell, *the substance with the most positive electrode potential is the cathode.*

With the data from Table 12.1 and Equation (12.4), we can calculate the cell potential for the copper-zinc cell (denoted as Cu | Zn) we have been considering. The electrode potential for copper is listed as 0.3419 V and the electrode potential for zinc is listed as −0.7618 V. Inserting these values into Equation (12.4), the cell potential for our Cu | Zn cell is

$$E^{\circ}_{cell} = E^{\circ}_{cathode} - E^{\circ}_{anode} = 0.3419\text{ V} - (-0.7618\text{ V}) = 1.1037\text{ V}$$

This is why the voltmeter in Figure 12.10 reads 1.10 V.

If you look once again at Figure 12.10, you see that the copper cathode is labeled positive (⊕) and the zinc anode is labeled negative (⊖). In a voltaic cell, the electrode with the positive (or more positive) electrode potential, which is the cathode, is labeled positive. Now you know what the + and – symbols on a dry cell or battery actually mean! In an electrolytic cell, represented by Figure 12.12, the reaction processes operate in reverse, and the electrode potentials are reversed as well. This makes the anode the more positive electrode in an electrolytic cell.

Example 12.5

A voltaic cell is to be constructed using cadmium and tin electrodes. The standard reduction half-reactions for these metals are as follows:

$$Cd^{2+} + 2e^- \rightarrow Cd(s)$$

$$Sn^{2+} + 2e^- \rightarrow Sn(s)$$

Identify the anode and cathode, write the two redox half-reactions for this cell, and determine the cell potential for the resulting voltaic cell.

The standard reduction potentials listed in Table 12.1 for Cd^{2+}/Cd and Sn^{2+}/Sn are as follows:

5 Any oxidation potential for an oxidation half-reaction (the reverse of the reduction half-reaction) is the negative of the reduction potential. Thus, there is no need to list both in tables such as Table 12.1. Chemists deal in standard reduction potentials by common consent. There is no reason to prefer reduction potentials over oxidation potentials, except for the convenience of having a standard way to refer to data.

Cd^{2+}/Cd: −0.4030 V
Sn^{2+}/Sn: −0.1375 V

Tin is the material with the most positive electrode potential, and thus tin is the cathode in this cell. Since reduction occurs at the cathode, the standard reduction half-reaction given above for tin applies. The cadmium electrode is the anode, where the oxidation occurs. Accordingly, the reduction half-reaction for cadmium must be reversed to convert it to an oxidation half-reaction. This gives us the following two half-reactions for the cell:

oxidation $Cd(s) \rightarrow Cd^{2+} + 2e^-$

reduction $Sn^{2+} + 2e^- \rightarrow Sn(s)$

The cell potential is obtained by inserting the electrode potentials into Equation (12.4):

$$E^\circ_{\text{cell}} = E^\circ_{\text{cathode}} - E^\circ_{\text{anode}} = -0.1375\text{ V} - (-0.4030\text{ V}) = 0.2655\text{ V}$$

The cell potential is positive, so when this cell is assembled and connected to an outside circuit, the redox reaction commences and current flows in the external circuit. The voltage of the cell is ideally 0.2655 V, but in an actual cell is lightly less.

▲

At this point, the meaning of the standard reduction potentials listed in Table 12.1 may still be a bit of a mystery to you. To understand what the standard reduction potential means, it helps to know how it is measured. Standard reduction potentials are relative voltage values, as are all electrical voltages. This is analogous to the fact that the height of any object is relative to our reference for measuring height, which could be the ground, or sea level, or the floor in a room. The reference standard for measuring reduction potentials is the reduction of hydrogen ions to form hydrogen gas. In this half-reaction, hydrogen ions are the oxidizing agent:

$$2H^+ + 2e^- \rightarrow H_2(g) \qquad (12.6)$$

A standard reduction potential is the voltage a voltaic cell reads when a given reducing agent is being oxidized in the anode half-cell while a special reference electrode reduces H^+ to H_2 in the cathode half-cell.

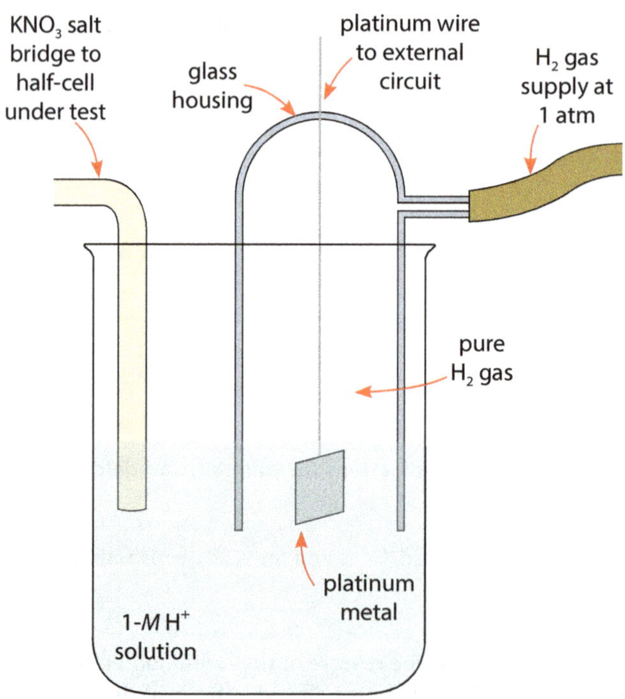

Figure 12.14. A half-cell containing a standard hydrogen electrode for testing standard reduction potentials of other electrode materials.

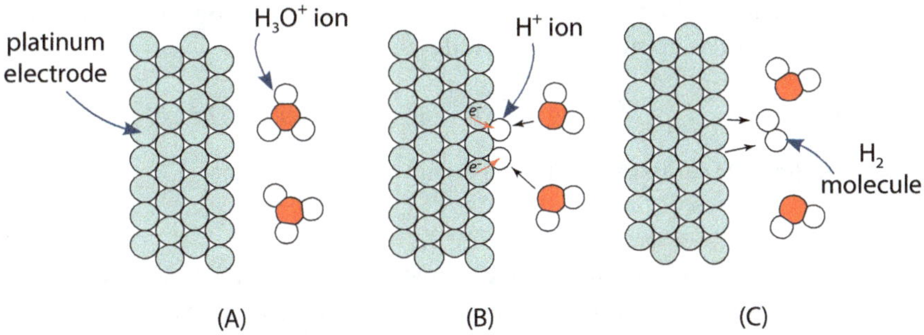

Figure 12.15. The sequence A–B–C shows how the platinum electrode acts as a catalyst for the reduction of H^+ ions to H_2. H^+ ions separate from H_3O^+ ions, adsorb onto the platinum where they are reduced, and then combine to form H_2.

The reference electrode is called a *standard hydrogen electrode* (SHE), or *normal hydrogen electrode* (NHE). The standard reduction potential for the half-reaction shown in Equation (12.6) is defined as 0 V and has been established as the reference standard for measuring all other standard reduction potentials. All other standard reduction potentials are measured relative to an SHE.

Figure 12.14 shows generally how an SHE is constructed. A platinum electrode is lowered into a pure solution with an H^+ concentration of $[H^+] = 1\ M$ (pH = 0). Platinum is used because it is a fairly unreactive metal. The platinum electrode is suspended by a platinum wire that serves as the electrical connection to the external circuit. The entire platinum assembly is contained inside a glass housing filled with H_2 gas at a pressure of 1 atm. A conventional KNO_3 salt bridge connects the hydrogen half-cell to the half-cell containing the electrode under test.

The mechanics of the reduction half-reaction that forms the H_2 gas in the SHE are illustrated in Figure 12.15. As you know, the H^+ ions come from H_3O^+ ions in the water. The platinum electrode acts as a catalyst for the reduction half-reaction. The H^+ ions separate from H_3O^+ ions and adsorb onto the platinum surface. There they are reduced and combine to form molecules of H_2 gas.

To summarize, the standard reduction potentials are relative values, as are all electrical voltages. The SHE has been established as the reference-standard half-cell. The electrode potentials of all other reducing agents are measured by installing them in an anode half-cell connected to an SHE cathode half-cell. The overall potential of the voltaic test cell—and thus the electrode potential of the reducing agent under test—is the difference between the E° of the SHE cathode (0 V) and the E° of the anode, according to Equation (12.4).

12.3.5 Electrochemical Applications

We conclude this chapter with a brief look at a few of the hundreds of real-world applications of electrochemistry in technology.

Dry Cells Beginning with the most familiar application of all, Figure 12.16 is a diagram of a typical alkaline dry cell. The materials shown as "zinc/KOH anode" and "manganese oxide cathode" are typically made of a gel or paste that allows movement of ions throughout the material. The anode is actually a zinc powder mixed with KOH, which is where the cell gets the name *alkaline*. The cathode material is in direct contact with the positive terminal of the cell at the top. Electrical connection to the negative terminal at the bottom is made with a metal rod

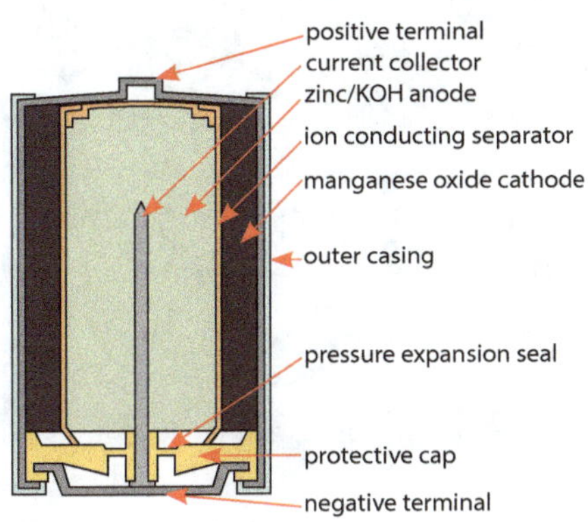

Figure 12.16. Typical alkaline dry cell construction.

that extends up into the anode material. The half-reactions for alkaline dry cells are quite complex, but can be approximated as follows:

oxidation $$Zn(s) + 2OH^-(aq) \rightarrow Zn(OH)_2(s) + 2e^-$$

reduction $$2MnO_2(s) + 2H_2O(l) + 2e^- \rightarrow 2MnO(OH)(s) + 2OH^-(aq)$$

Wet Cell Batteries Figure 12.17 shows the inside of a small six-cell automotive battery. The battery shown came out of a four-wheeler, but larger batteries for cars are similar. Typical automotive batteries use plates made of lead and lead oxide immersed in an H_2SO_4 electrolyte. For this reason they are often called *lead-acid* batteries. There are six cells connected in series inside the battery, so this device is truly a *battery*, unlike dry cells. Automotive batteries are designed to deliver a very large current (hundreds of amps) for a very brief time (typically under one second, just long enough to start the car). This requirement is quite different from the way dry cells typically operate, which is to provide a very small current to an electronic device for an extended period of time.

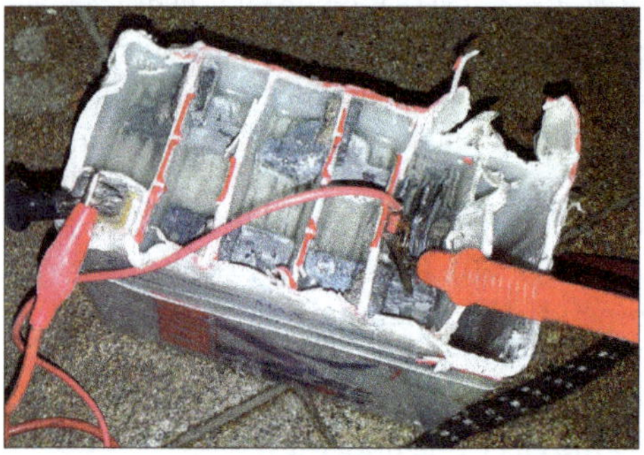

Figure 12.17. A small, six-cell automotive battery cut open to show the cells inside.

Each of the six cells has a cell potential of just over 2.0 V, and six of them connected in series produces an overall voltage for the battery of 12 V. The half-reactions for this common battery are as follows:

oxidation $$Pb(s) + HSO_4^-(aq) \rightarrow PbSO_4(s) + H^+(aq) + 2e^-$$

reduction $$PbO_2(s) + HSO_4^-(aq) + 3H^+(aq) + 2e^- \rightarrow PbSO_4(s) + 2H_2O(l)$$

Electroplating *Electroplating* is a very widely used instance of electrochemistry. The basic idea behind plating is to form a tightly adhering metallic coating on a metal or plastic object. In electroplating, the part being plated is the cathode. The anode is made of the metal that is plating the part. The reaction takes place with anode and cathode together in the same electrolyte, which contains a high concentration of ions of the same metal as the anode. Once the current is switched on, the ions in the electrolyte begin plating onto the cathode piece.

Figure 12.18. Chrome plating on a Velocette Viper, a British motorcycle manufactured in the 1950s and 60s.

Electroplating technology is used for a wide variety of purposes. One use is to coat an object made of a metal that rusts or corrodes (such as steel) with a metal that is more resistant to corrosion, such as chrome, nickel, or gold. A second purpose is for aesthetic appeal. The chrome plating on the motorcycle parts in Figure 12.18 protects the steel parts from rust, and at the same time makes the machine shiny and attractive. Some coins are made this way as well. Nearly all coins are made of one metal plated onto another. Since 1982, the U.S. penny has been made of a zinc disk with an electroplated copper coating. The penny is 97.5% zinc and 2.5% copper.

A third example of electroplating technology is to adhere a copper layer to a fiberglass board in the manufacturing of printed circuit boards (PCBs). Electronic devices such as computers, televisions, and microwave ovens all contain printed circuit boards where the electronic components are mounted and interconnected. Instead of being connected together with individual wires, the electronic components on the PCB are connected together with copper "traces" plated onto the board. Figure 12.19 shows some PCBs hanging in an electroplating machine during the manufacturing process. The plated copper is visible on the PCBs.

Figure 12.19. Printed circuit boards (PCBs) hanging in an electroplating machine.

Anodizing *Anodizing* is an electrochemical process commonly used on aluminum parts, although other metals may be anodized as well. Recall from Section 12.1.3 that aluminum parts form a passivation layer of aluminum oxide that protects the aluminum from further corrosion. The purpose of anodizing is to increase the thickness of the autopassivated layer (the layer the aluminum part forms on itself) by an electrochemical redox reaction. The anodizing also allows the metal part to be dyed. Anodized aluminum objects are quite common these days. Figure 12.20 displays a couple of familiar examples. The purpose of the anodizing on the cooking pan at the top is to protect the aluminum. With the inexpensive "not-for-climbing" carabiners below, the purpose is clearly for the colorful dye that makes the objects attractive.

In anodizing, the part to be anodized is the anode—hence the name of the process. Recall from Figure 12.12 that in electrolytic cells, the anode is the positive electrode. The electrons flow through the electrolyte from the cathode to the anode. The electric current releases hydrogen at the cathode and oxygen at the anode. The oxygen-rich environment at the anode is what causes the aluminum oxide layer to form.

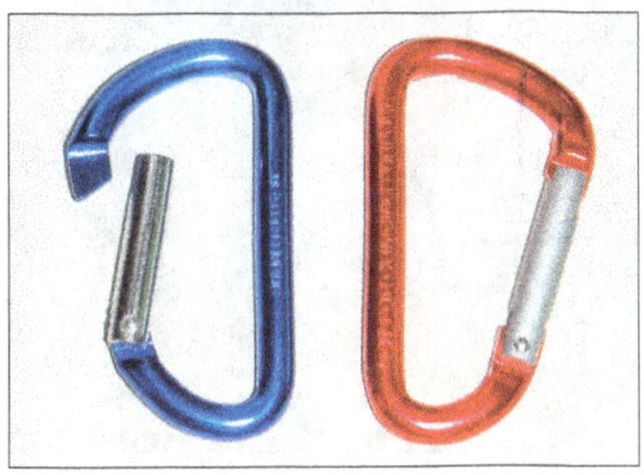

Figure 12.20. Examples of anodized aluminum objects.

Cathodic Protection Underground steel pipes and ships with steel hulls or steel propellers have this in common: they are subject to the merciless processes of corrosion—oxidation—that destroy the metal. If not controlled, corrosion can damage steel pipes so badly that they rupture. The steel hulls of ships can simply corrode away to the point where only costly repair work can save the ship from the scrap yard.

Figure 12.21. At the top, the lozenge-shaped object on the bottom of this boat is a sacrificial anode. The middle image shows what the sacrificial anode looks like after it has corroded during use. The metallic rectangles on the large ship in the bottom image are zinc sacrificial anodes. Metal propellers are also protected from corrosion as long as there is a solid electrical contact between the propeller and the cathode, which is usually the steel hull of the ship.

This type of passive corrosion of steel parts is a redox reaction. Recall from our discussion of Figure 12.10 that the anode in a voltaic cell dissolves away during the redox reaction. This means that it is the anode that is corroded in these corrosion reactions—when active metals like steel (which is mostly iron) are buried in the ground or pushed around in salt water, the salts in the ground and in the water allow the steel parts to become the anodes—as if they were in voltaic cells—in redox reactions that gradually destroy those parts.

Cathodic protection, illustrated in Figure 12.21, makes use of a *sacrificial anode* to protect other parts, such as pipes, ship hulls, and ship propellers. The idea is to convert the parts to be protected from anodes to cathodes by attaching a chunk of metal that is a more active reducing agent than the steel parts are. This chunk of active metal is called the sacrificial anode, because it is specifically chosen to corrode (i.e., sacrifice itself) in lieu of the corrosion of the protected parts.

For cathodic protection to work, the anode must have a more negative electrode potential than the metal to be protected. Typical metals for this purpose are zinc, aluminum, or magnesium, or alloys of these. The electrode potentials of these metals are all more negative than the electrode potential of steel.

Referring again to Figure 12.10, keep in mind that electrochemical processes that run by themselves are basically voltaic cells. In a voltaic cell, the electrons flow out of the anode into the external circuit, and charge in the form of ions flows from the cathode half-cell toward the anode half-cell. This is essentially what is happening when a pipe corrodes underground or a ship hull corrodes in the water. In the case of a pipe without cathodic protection, oxidation and reduction both occur all along the pipe. Iron atoms in the pipe oxidize to form iron oxide—rust. Atmospheric oxygen is reduced at the same time.

But a sacrificial anode changes the scenario, as illustrated in Figure 12.22. With the addition of the sacrificial anode, a metal that is a stronger reducing agent than the iron in the steel is, the oxidation all occurs at the sacrificial anode, so it gets corroded instead of the pipe. The electrons

Figure 12.22. Instead of oxidation occurring all along the pipe, the sacrificial anode alone is destroyed by oxidation. Reduction still occurs all along the pipe, as atmospheric oxygen is reduced to form H_2O.

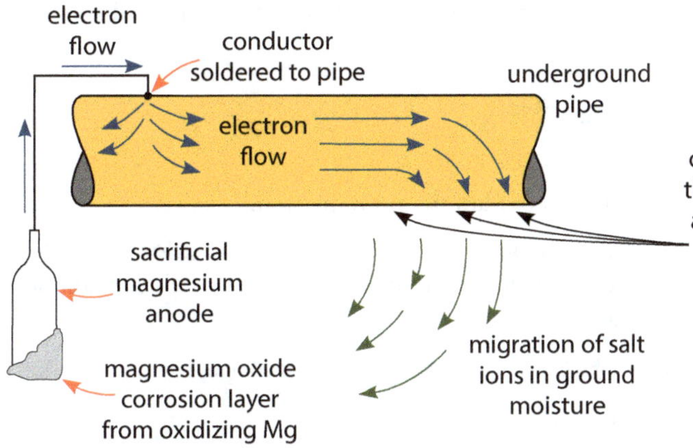

from the oxidation flow into the pipe, and reduction still occurs all along the pipe walls, wherever there is oxygen to be reduced to form water. But the reduction doesn't hurt the pipe. The corrosion is not the result of the reduction half-reaction; it is from the oxidation half-reaction at the anode. With a sacrificial anode, the anode can simply be replaced from time to time after it corrodes away, and the pipe remains in good condition. The same idea works in the case of ship hulls underwater.

Smelting I have mentioned a few times in this text that *smelting* is another important example of redox technology at work. Of all the technologies we have reviewed here, smelting is *by far* the oldest. Evidence now persuades historians and anthropologists that lead smelting was in use in Asia Minor by 6,500 BCE.[6] Apparently, copper smelting was known by 5,500 BCE in the region of the Balkans.

The point of smelting is to reduce the metal ions in ores to pure metal. You can't simply heat the ore up and melt the metal out. (At least, not unless you heat the ore to the extremely high temperature needed to melt the rock.) The metal ions are bonded in the crystal lattice of the mineral, so to convert them into metal, they must be reduced.

Pure metals have always been valuable as tools, weapons, sculptural materials, and structural materials, so people have smelted metals for a long time. In the old days, the redox reactions required for smelting were generated strictly from heat. Thus, smelters prior to the 20th century were essentially large furnaces. The classic reducing agent for smelting is carbon monoxide, CO, which is easily produced by an oxygen-starved fire. In the thermal process, metal ores are heated to the necessary temperature in the presence of CO. The carbon in the CO is oxidized and forms CO_2 (which increases the oxidation state of carbon from +2 to +4). In the process, the metal from the ore is reduced and forms a pure liquid metal, which is cast into ingots.

Once electrochemical theory was developed and industrial-grade electrical supplies were available, smelting by electrochemical processes became as common as smelting by strictly thermal processes. Today, electrochemical smelting of some metals (zinc, for example) is more common than thermal processes. The facilities for smelting metals like aluminum and zinc can have cells the size of large automobiles connected in series circuits of over 100 cells. The electrical currents involved can be well over 100,000 amps. Alas, the manufacturers all guard their technologies closely, so photos of such facilities are not to be had. However, I can tell you that the magnetic fields surrounding conductors carrying these kinds of currents are huge. If you wear a watch, carry credit cards, or take a mobile phone into such an area, any of them will

6 BCE stands for *before common era*.

cease functioning—either permanently erased or magnetized so strongly that they won't work. (*Hmm... Interesting.*)

Chapter 12 Exercises

SECTION 12.1

1. Distinguish between oxidation and reduction.
2. For the reaction equations listed below, write the oxidation states for each element in each compound, identify elements that are oxidized or reduced, and identify all oxidizing agents and reducing agents.
   a. $PBr_3(l) + 3H_2O(l) \rightarrow H_3PO_3(aq) + 3HBr(aq)$
   b. $2KClO_3(s) \rightarrow 2KCl(s) + 3O_2(g)$
   c. $2H_2SO_4(aq) + 2NaBr(s) \rightarrow Br_2(l) + SO_2(g) + Na_2SO_4(aq) + 2H_2O(l)$
   d. $NaCl(aq) + AgNO_3(aq) \rightarrow AgCl(aq) + NaNO_3(aq)$
   e. $NaI(aq) + 3HClO(aq) \rightarrow 3HCl(aq) + NaIO_3(aq)$
   f. $3SO_2(g) + 2HNO_3(aq) + 2H_2O(l) \rightarrow 3H_2SO_4(aq) + 2NO(g)$
3. Use Table 12.1 to identify the strongest and weakest reducing agent in each of the following groups of substances:
   a. I^-, Ca, NO, Zn
   b. Fe^{2+}, Li, Br^-, Cu,
   c. Al, Mg, Cr^{3+}, Ag
   d. Sn, O_2, F^-, MnO_2
4. Use Table 12.1 to address the following questions:
   a. Can I_2 be used to oxidize nickel metal?
   b. Will cadmium metal reduce nitrate ions?
   c. Can silver metal act as a reducing agent to reduce Ca^{2+} ions?
   d. Will Fe^{2+} ions oxidize copper metal?
5. What is the significance of one reducer being stronger than another?

SECTION 12.2

6. For each of the following half-reactions, complete and balance the half-reaction and indicate whether it is an oxidation or reduction.
   a. $SO_3^{2-}(aq) \rightarrow SO_4^{2-}(aq)$ (basic solution)
   b. $N_2(g) \rightarrow NH_3(g)$ (basic solution)
   c. $N_2(g) \rightarrow NH_4^+(aq)$ (acidic solution)
   d. $Sn^{2+}(aq) \rightarrow Sn^{4+}(aq)$ (acidic or basic solution)
   e. $ClO_3^-(aq) \rightarrow Cl^-(aq)$ (acidic solution)

f. $OH^{-}(aq) \rightarrow O_2(g)$ (basic solution)

g. $H_2SO_3(aq) \rightarrow SO_4^{2-}(aq)$ (acidic solution)

h. $O_2(g) \rightarrow H_2O(l)$ (basic solution)

7. For each of the following redox reactions, complete and balance the equation using the half-reaction method. Identify the oxidizing and reducing agents. Combine and/or add ions to produce a balanced equation showing the reactants and products given.

   a. In aqueous solution, ferric chloride, $FeCl_3$, combines with hydrogen sulfide, H_2S, to form ferrous chloride, $FeCl_2$, hydrochloric acid, and solid sulfur. Everything dissociates except the solid sulfur, and the reaction takes place in acidic solution.

   b. Solid iodine, I_2, combines with aqueous hypochlorous acid, HClO, to produce aqueous iodic acid, HIO_3, and aqueous hydrochloric acid, HCl. In this reaction, everything ionizes except the I_2. The reaction takes place in acidic solution.

   c. Aqueous potassium permanganate, $KMnO_4$, combines with aqueous methanol, CH_3OH, to produce aqueous manganese(II) hydroxide, aqueous formic acid, HCOOH, and aqueous potassium hydroxide, KOH. Everything ionizes except the methanol, and the reaction takes place in acidic solution. (Tip 1: $KMnO_4$ ionizes to K^+ and permanganate, MnO_4^-; formic acid ionizes to H^+ and $COOH^-$. Tip 2: You need to add in potassium and hydroxide ions at the very end.)

   d. Aqueous hydrogen peroxide, H_2O_2, and aqueous chlorine dioxide, ClO_2, combine to produce chlorous acid, $HClO_2$, and oxygen gas. Only the acid dissociates. The reaction takes place in basic solution.

   e. Solid aluminum combines with aqueous lithium chloride and aqueous lithium nitrite to produce aqueous lithium aluminate, $LiAlO_2$, and aqueous ammonium chloride. Dissociations should be obvious. The reaction takes place in basic solution.

   f. In aqueous solution, hydrochloric acid combines with permanganic acid, $HMnO_4$, to produce chlorine gas and manganese(II) chloride.

   g. Solid sulfur combines with aqueous nitric acid to produce aqueous sulfurous acid and nitrous oxide gas, N_2O. (N_2O is sometimes called *laughing gas* because inhaling it induces euphoria. Formerly, it was used as an anesthesia. Today, it is added to racing fuel because it is a powerful oxidizer.)

   h. In aqueous solution, hydrobromic acid, HBr, combines with permanganic acid, $HMnO_4$, to produce bromic acid, $HBrO_3$, and solid manganese(IV) oxide.

8. Why do free electrons always appear on the left side of a reduction half-reaction?

9. A *thermite reaction* occurs when a metal fuel powder is combined with a powdered metal oxide. When heated, the powdery mixture briefly emits intense light and heat. The reaction is not explosive, but generates enough heat that it is used for a process called *thermite welding*. Thermite reactions

are redox reactions. The following thermite reaction is used for welding steel rails together in train tracks:

$$Fe_2O_3(s) + 2Al(s) \rightarrow 2Fe(s) + Al_2O_3$$

In the image above, a small quantity (about 110 g) of the iron(III) oxide/aluminum powder is heated. The sparks are molten particles of pure iron. The reaction is taking place in a cast iron skillet (destroyed during the reaction). Address the following questions about this thermite reaction:

a. Explain why this reaction occurs. (Use data from Table 12.1 to support your answer.)

b. Identify the oxidizing agent and reducing agent.

c. Explain the significance of the fact that the reactants are both in powdered form and thoroughly mixed together.

d. What mass of powdered aluminum should be mixed with 65 g of powdered Fe_2O_3 to make the thermite powder described by the reaction above?

e. If 65 g Fe_2O_3 is consumed in the reaction, what mass of iron is produced?

SECTION 12.3

10. Explain the basic principles of electrochemistry. In formulating your response, consider addressing the question of why electrons spontaneously flow in an external electric circuit connected to a voltaic cell.

11. Explain the difference between the anode and the cathode in an electrochemical cell. Does your answer hold for both voltaic and electrolytic cells?

12. Draw a diagram of a voltaic cell made with cobalt and iron metals, $CoCl_2$ and $FeCl_2$ electrolytes, and sodium chloride salt bridge. On your diagram, identify the following: anode, cathode, positive electrode, negative electrode, direction of electron flow, direction of chloride ion migration in the salt bridge, direction of sodium ion migration in the salt bridge. Finally, determine the cell potential.

13. Explain why electrolytic cells require an outside source of electrical power to drive the reaction.

14. Draw a diagram of a voltaic cell made with Ag in $AgNO_3$ and Ni in $NiSO_4$, with a potassium nitrate salt bridge. On your diagram, identify the following: anode, cathode, positive electrode, negative electrode, direction of electron flow, direction of nitrate ion migration in the salt bridge, direction of potassium ion migration in the salt bridge. Finally, determine the cell potential.

15. Describe what you expect to happen if an aluminum wire is dipped into a solution of tin(II) sulfate. Base your response on the values of standard reduction potentials.

16. When a voltaic cell operates with a potassium nitrate salt bridge, potassium ions are drawn from the salt bridge into the cathode half-cell. Explain why this happens.

17. Explain why the copper electrode in a Cu | Zn electrolytic cell is the positive electrode.

18. Explain why E°_{cell} must be greater than zero for a voltaic cell to work.

19. For the following pairs of half-cells, determine which is the anode and which is the cathode if the redox reaction is to occur spontaneously. Then determine the overall cell potential, E°_{cell}.

a. Sn^{2+}/Sn, Ag^+/Ag

b. Co^{2+}/Co, Fe^{3+}/Fe^{2+}

c. Na^+/Na, Br_2/Br^-

d. I_2/I^-, Br_2/Br^-

20. Explain why some standard reduction potentials are positive and some are negative.

21. Explain the general idea behind cathodic protection. In your response, address the question of why cathodic protection works—that is, why the protected object is protected from corrosion.

22. Explain what is meant by the phrase, "reduction occurs at the cathode." Include the details about what is actually happening at the atomic level at the cathode.

23. Explain how a rechargeable dry cell incorporates the electrochemistry of both the voltaic cell and the electrolytic cell.

GENERAL REVIEW EXERCISES

24. Nitrogen dioxide reacts with ozone to produce dinitrogen pentoxide and oxygen according to the following equation:

$$2NO_2(g) + O_3(g) \rightarrow N_2O_5(s) + O_2(g)$$

a. Determine the percent yield for a reaction in which 1.27 g NO_2 reacts and 1.20 g N_2O_5 is recovered.

b. Determine the mass of oxygen produced if 4.0 mol ozone react and the percent yield is 76.0%.

c. Determine the limiting reagent if 4.304 g NO_2 are available to react with 2.268 g O_3.

25. A steel tank of O_2 gas is filled to a gauge pressure of 62.4 atm at 22.00°C. The temperature then increases to 35.00°C. Determine the new pressure in the tank and state your result as a gauge pressure in atm. Assume the volume of the tank does not change.

26. A chemist collects a sample of ethane by water displacement at a temperature of 16.0°C and a total pressure of 102,360 Pa. The collection bottle used to draw the sample has a volume of 275 mL. Determine the number of moles of ethane collected in the sample.

27. The molecular formula for aspirin, acetylsalicylic acid, is $C_9H_8O_4$. How many carbon atoms are contained in a bottle of aspirin, if the bottle contains exactly 200 aspirin tablets with a mass of 325 mg per tablet?

28. Ethanol burns to produce water and CO_2. Determine the volume of ethanol consumed in producing 1.00 L of water. The density of ethanol is 0.789 g/cm^3.

29. Determine the change in the Gibbs free energy for the combustion of ethane, C_2H_6, producing $CO_2(g)$ and $H_2O(l)$ at 25°C.

Chapter 13

Organic Chemistry—An Introduction

Just about everything in this photograph is composed of organic molecules—molecules containing carbon. Not only are the plants and grasses composed of organic molecules, but the greenhouse is as well. This plastic tunnel—called a *hoop house* or *polytunnel*—is composed entirely of an organic molecule called *polyethylene*. Polyethylene is an example of a *polymer*, a long molecule composed of smaller units, called *monomers*, chained together. Polyethylene molecules can take on three different forms, and these forms determine how dense and how flexible the plastic is. Low density polyethylene (LDPE) is very flexible, and is used to make the plastic sheeting that covers the hoop house (as well as many of the plastic bags we use). High density polyethylene (HDPE) is much stronger and stiffer and is used to make the structural members of the hoop house (as well as many of the plastic bottles we use).

Hoop houses like the one shown here are becoming increasingly popular with farmers as a means of controlling temperature and humidity, and sheltering delicate crops from harm due to wind and hail. Because of their low cost, huge hoop houses can be constructed at minimal expense, and enable farmers to grow crops at times normally considered off season. In 2009, the U.S. Department of Agriculture announced a new program to help farmers install hoop houses.

Objectives for Chapter 13

After studying this chapter and completing the exercises, you should be able to do each of the following tasks, using supporting terms and principles as necessary.

SECTION 13.1

1. Define *organic chemistry*.
2. Describe the carbon bonds that do and do not allow for part of a molecule to rotate relative to the other part.
3. Describe several of the allotropes of carbon and relate their atomic structures to their properties.

SECTION 13.2

4. Describe four examples of uses for aliphatic hydrocarbons of different lengths.
5. State the names and formulas of the first 10 alkanes.
6. Define *hydrocarbon* and *isomer*, and give three example pairs of hydrocarbon compounds that are isomers with the same formulas but different structures.
7. Distinguish between saturated and unsaturated hydrocarbons and relate the concept of saturation to the types of carbon bonds present in hydrocarbon molecules.
8. Explain substitution reactions using halogenation as an example.
9. Explain addition reactions using hydrogenation as an example.
10. Define *alkyl group* and state the formulas for the methyl, ethyl, and propyl groups.
11. Given structural formulas or condensed structural formulas, use IUPAC naming conventions to determine names of compounds; and given the names, draw the formulas.
12. Draw the structural formulas for each of the functional groups listed in Table 13.5.
13. Distinguish between *cis*- and *trans*- isomers in alkenes, and incorporate these prefixes into the naming of alkenes and the drawing of structural formulas.
14. Describe the *cyclo*– compounds and distinguish them from aromatic hydrocarbons.

SECTION 13.3

15. Identify the following from their structural formulas: alcohols, ethers, aldehydes, keytones, carboxylic acids, esters, amines, and amides.
16. Describe how the structures of the functional groups relate to the properties of substances.
17. Describe condensation and elimination reactions and give examples of each.

SECTION 13.4

18. Define *monomer* and *polymer*.
19. State four examples of common polymers, describe the monomers of which they are composed, and give examples of their use.

SECTION 13.5

20. Describe the functional group structure that defines amino acids.
21. Describe the condensation reaction between amino acids that forms a peptide.
22. Draw the structural formula of a peptide bond.
23. Distinguish between peptides and proteins.
24. Describe step-growth and chain-growth polymerization and give examples of each.

13.1 The Chemistry of Carbon

13.1.1 Introduction

You are growing up in a time when the term *organic* is used frequently to describe products made from natural ingredients or farm products grown without synthetic chemicals. But the term has a different meaning in the study of chemistry. The branch of chemistry known as *organic chemistry* is devoted to the study of compounds containing carbon. There are a few carbon-containing compounds that are considered *inorganic*, such as the carbonates (compounds containing the carbonate ion, CO_3^{2-}), the oxides (CO, CO_2, etc.), and a few others. With these exceptions, compounds containing carbon are organic compounds. The term *organic* arose because in the early days of this discipline, chemists believed that the compounds being studied were the compounds involved in living organisms. However, later it became clear that defining organic chemistry in terms of the compounds in or from living organisms was too narrow. For example, the atmosphere of the planet Saturn contains trace amounts of methane—definitely an organic compound—but there are no organisms on Saturn to our knowledge.

Our objectives in this introductory chapter are somewhat different than in previous chapters. Instead of working with equations to analyze chemical reactions, we focus on describing some of the most common classes of organic molecules and the rules governing how they are named. This is going to make this chapter seem like a big parade of technical vocabulary that you must memorize. We can't really get around this—when you first step into organic chemistry it *is* a big parade of technical vocabulary. However, there is one reward you can look forward to. Along the way, we encounter a few terms that you have heard before, but probably did not know what they actually referred to. For example, you have probably heard of polymers, saturated fats and *trans*-fats. Soon you will know what these terms denote.

As we go through different types of organic compounds, we also encounter four important classes of chemical reactions. These are underlined in the Chapter Objectives. To highlight them in the text and make them easy for you to find as you study, they are denoted by green markers like the one shown to the left.

Reaction

13.1.2 Carbon Molecular Geometry

The molecular bonds and chemical reactions we study in organic chemistry are the same as those you have learned about in our study already, which has focused primarily on inorganic compounds. But since all organic compounds contain carbon, the peculiar bond structure exhibited by carbon has a major impact on the shapes of organic molecules. As you know, carbon atoms make four bonds in molecular compounds. When all these bonds are single bonds, the resulting molecular geometry is tetrahedral, as illustrated by the methane molecule, CH_4, modeled in Figure 13.1.

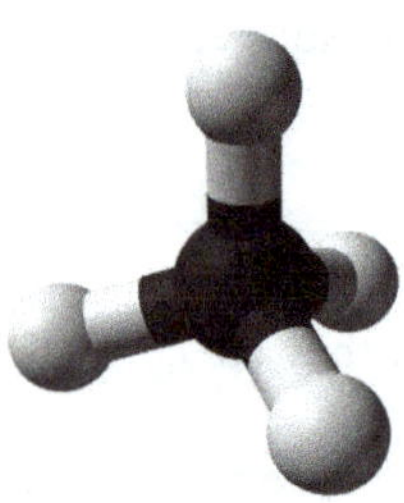

Figure 13.1. Methane illustrates the tetrahedral molecular geometry exhibited by a carbon atom when bonded with four single bonds.

One of the most significant features of organic chemistry is the frequent occurrence of carbon atoms connected together in chains, as illustrated by the hexane molecule, C_6H_{14}, modeled in Figure 13.2. The tetrahedral geometry of carbon single bonds results in

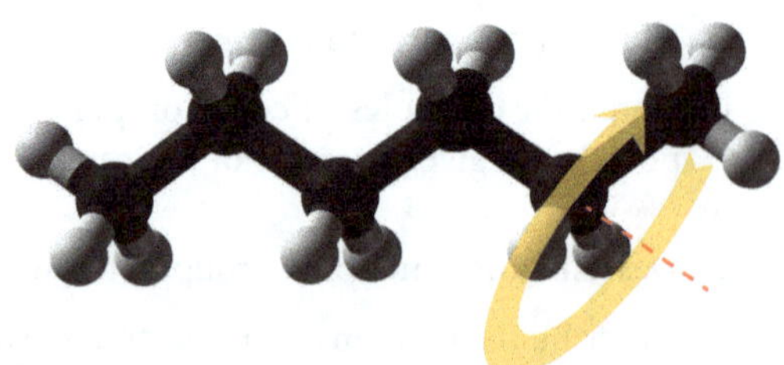

Figure 13.2. The tetrahedral geometry of carbon results in a zigzag molecular shape when carbon atoms are chained together. Molecular groups can rotate around single bonds.

zigzag-patterned chains, often with two hydrogen atoms connected to carbon atoms in the interior of the chain. An important aspect of the carbon single bond is that the two sections of a molecule on either side of a single bond can rotate relative to each other. This is illustrated by the yellow arrow in the figure, which shows how the last five hydrogen atoms and the last carbon atom can rotate around the axis indicated by the dashed red line.

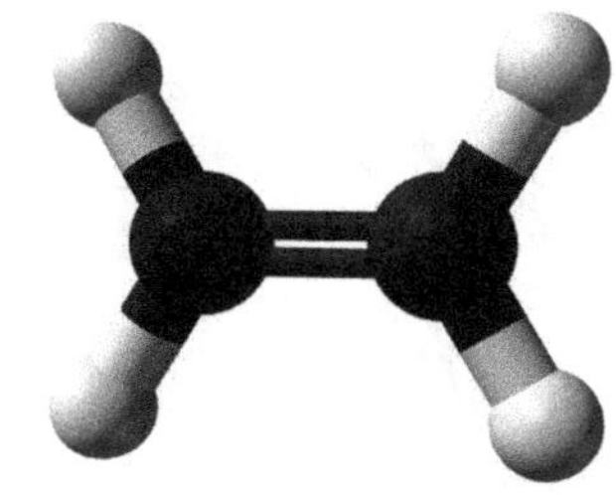

Figure 13.3. Sections of molecules connected by a double bond are not free to rotate relative to one another.

When a carbon atom forms a double bond with one other carbon atom, things are different. Figure 13.3 shows a model of an ethene molecule, C_2H_4, (usually called ethylene) in which the carbon atoms are connected by a double bond. There are four hydrogen atoms in the molecule, instead of the six found in the molecule when the carbons are joined by a single bond. Additionally, the sections of the molecule on either side of the double bond are not free to rotate relative to each other (at least not easily). Thus, the atoms in the molecule shown all lie in a plane.

All the other factors addressed in Chapter 4 that govern molecular geometries apply to organic compounds as well. The most important of these are summarized in Table 13.1.

13.1.3 Carbon Allotropes

For some elements, there is more than one way the atoms can be arranged in the pure substance. These different forms are called *allotropes*. Since we are entering into a study of carbon compounds, let's take a moment here to examine a few of the allotropes of carbon, a subject we haven't hit on yet. The structural geometries of the different allotropes results in unique properties for each.

Figure 13.4 depicts eight different carbon allotropes. Image (a) depicts the crystal structure of *diamond*, the hardest material known. Its hardness makes diamond ideal for use on the tips of the bits and blades of industrial drilling and cutting tools. Image (b) shows several layers

Element	Maximum Number of Single Bonds	Bond Geometry			
		Single Bond	Double Bond	Double Bond	Triple Bond
carbon	4	C (tetrahedral)	>C= (trigonal)	=C= (linear)	—C≡ (linear)
hydrogen	1	H—			
halogens	1	:X:—			
oxygen	2	:O: (bent)	:O=		
nitrogen	3	:N (pyramidal)	:N= (bent)		:N≡

Table 13.1. Summary of bond geometries for the common elements and bonds in organic compounds.

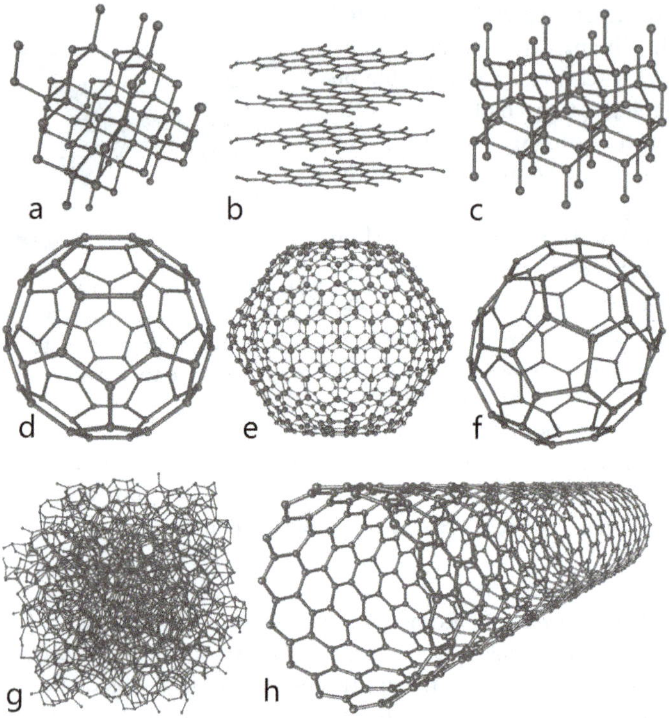

Figure 13.4. Eight allotropes of carbon.

of *graphene*, each one atom thick. A sheet of graphene is an extraordinarily strong substance. When many small sections of these layers are compacted together, we have a third allotrope, *graphite*, the mineral used in pencils for making marks on paper. (Everyone calls that material lead, but it is not. I wish we didn't call it that.) The individual sections of graphene can slide past each other. This allows pencils to write because the sections of graphene are easily scraped off and deposited onto a rough surface such as paper. And the slipperiness of graphene sheets makes graphite useful as a solid lubricant.

Image (c) depicts a carbon allotrope called *Lonsdaleite*, named in honor of Kathleen Lonsdale, a British crystallographer. Lonsdaleite is also called "hexagonal diamond." In nature, Lonsdaleite has been found only in the carbon in meteorites, where the heat and pressure of a meteor's impact with earth results in this structure.

Image (d) depicts one molecule of *buckminsterfullerene*, a 60-atom sphere of pure carbon with the atoms arranged in the same pattern as the corners on the surface of a soccer ball. C_{60}, as it is also called, was named after American architect Buckminster Fuller, who invented the geodesic dome. In case you haven't seen a geodesic dome, Figure 13.5 is a photo of the Montreal Biosphère, designed by Fuller in 1967. C_{60} molecules are also called *buckyballs*. The study of buckyballs is part of an entire research discipline devoted to the study of *fullerenes*, the allotropes of carbon shaped as balls, ellipsoids, tubes, or other shapes. Scientists have discovered a host of interesting properties for C_{60}, but so far no practical applications have made it to the market. Images (e) and (f) are two more fullerenes, C_{540} and C_{70}. Image (g) is a depiction of *amorphous carbon*, an allotrope with very little structural strength. The term *amorphous* derives from Greek and means "no shape." The term is used to describe carbon that is not structured in a regular way as the other allotropes are. The carbon in coal and soot is amorphous, which is why coal is crumbly. Finally, the tubular allotrope in image (h) is called a *carbon nanotube*. A nanotube is basically a rolled section of graphene; its walls are one atom thick. Relative to their diameter, the tensile strength of nanotubes is higher than any other material yet discovered, and many applications for nanotubes are currently being researched.

Figure 13.5. Biosphère Montréal, designed by Buckminster Fuller in 1967.

13.2 Hydrocarbons

We begin our study of organic chemistry with a class of organic compounds known as *hydrocarbons* because they are composed entirely of carbon and hydrogen. The most significant sources of hydrocarbons are *petroleum* and *natural gas*. Natural gas is composed primarily of methane. Petroleum, also called *crude oil*, is a mixture of hydrocarbon molecules with carbon chains ranging from one to scores of carbon atoms in length. Common uses for hydrocarbon molecules of different lengths are listed in Table 13.2.

The entire global industrial economy came to be because of fuels from petroleum and natural gas and the other compounds hydrocarbons can be used to produce. Hydrocarbons continue to be the major fuel source for industrial nations today. Needless to say, hydrocarbons are important. They also happen to be the simplest organic molecules since they only contain carbon and hydrogen. This makes the study of hydrocarbons a good place to begin the study of organic chemistry.

Molecule Size	Use
C_1 to C_4	Fuel gases such as natural gas, propane, and butane
C_5 to C_{12}	Gasoline, lighter fluid
C_5 to C_{15}	Jet fuel
C_{12} to C_{18}	Kerosene, heating oil
C_{16} and up	Lubricating oils
C_{20} and up	Paraffin wax, candles
C_{36} and up	Asphalt

Table 13.2. Common uses for hydrocarbons from petroleum.

Two broad classes of hydrocarbons we review here are the *aliphatic* compounds and the *aromatic* compounds. The aliphatic compounds consist of three basic types: alkanes, alkenes, and alkynes.

13.2.1 Alkanes

The alkanes are hydrocarbon chains in which all the carbon bonds are single bonds. The alkane names all end in *–ane*. The alkanes are the simplest hydrocarbons and the simplest organic molecules of all, and many of the other organic compounds we address are combinations or derivatives of alkanes.

Alkanes are our fuels. As I wrote above, the primary component in natural gas is methane, and gasoline is basically a mixture of alkanes with five to twelve carbon atoms, with some other compounds added in. These organic compounds have been the energy backbone of the world's industries since the beginning of the 20th century. The alkanes are also the basic building blocks for a huge array of industrial chemicals.

Prefix	Number of Carbon Atoms	Alkane Molecular Formula	Alkane Name	Boiling Point (°C)
meth–	1	CH_4	methane	–161
eth–	2	C_2H_6	ethane	–89
prop–	3	C_3H_8	propane	–42
but–	4	C_4H_{10}	butane	0
pent–	5	C_5H_{12}	pentane	36
hex–	6	C_6H_{14}	hexane	69
hept–	7	C_7H_{16}	heptane	98
oct–	8	C_8H_{18}	octane	126
non–	9	C_9H_{20}	nonane	151
dec–	10	$C_{10}H_{22}$	decane	174

Table 13.3. Hydrocarbon prefixes, and the names and formulas for the first 10 alkanes.

The naming conventions for the first ten alkanes are shown in Table

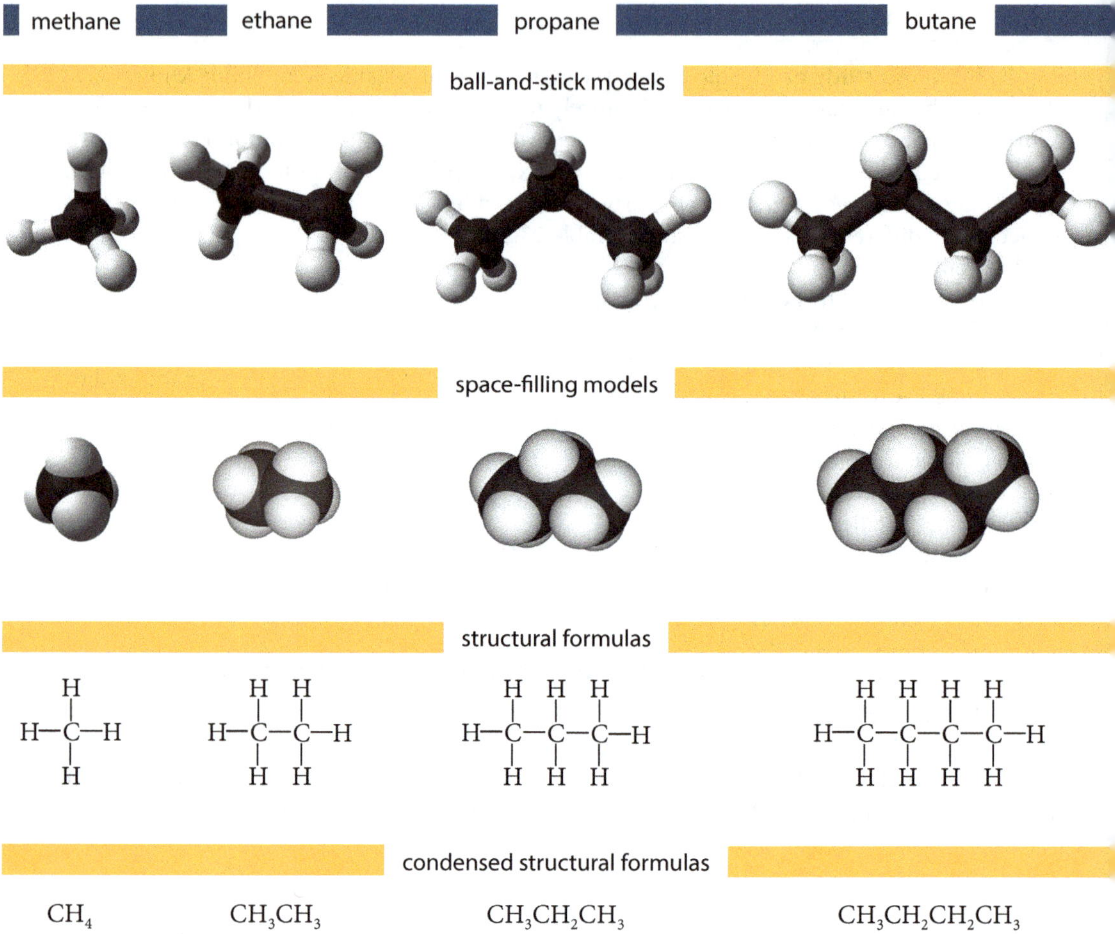

Figure 13.6. Ball-and-stick models, space-filling models, structural formulas, and condensed structural formulas for the first four alkanes.

13.3. Beginning with pentane, the prefixes are similar to those used for naming binary compounds (Table 3.8). Also shown in the table are the boiling points for these compounds. Note the steady increase in boiling point as the size of the molecule increases. This is due to the fact that larger molecules exhibit greater attractions to each other because there are more opportunities for Van der Waals forces to develop.

The smallest alkane is methane, with only one carbon atom. Models of the first four alkanes—molecules containing one, two, three, or four carbon atoms—are shown in Figure 13.6. The ball-and-stick models are useful for making the molecular geometry clear. However, a better idea of the actual shape and compactness of organic molecules is given by space-filling models, also shown in the figure.

Beneath the models are diagrams called *structural formulas*, which are similar to Lewis structures. Structural formulas are a common way of representing organic compounds when working with molecular shapes on paper. When working with hydrocarbons, it is common to draw structural formulas in a rectangular fashion rather than trying to represent the zigzag shape the molecules actually take on. Later, we will see that when we add atoms other than carbon and hydrogen to the basic hydrocarbon units, bond angles are commonly shown as other than 90°.

At the bottom of the figure, the *condensed structural formulas* are shown. In a condensed structural formula, each carbon atom is listed and after each one are listed the elements attached to it. Subscripts are used to indicate multiples in the elements attached to each carbon atom. Condensed structural formulas are often used as a shorthand notation.

Each of the alkanes differs from the one before it by the addition of one $—CH_2$ unit in the middle of the chain. A series of compounds like this in which each compound differs from the one before it by the addition of the same structural unit is called a *homologous series*. Recall that hydrocarbon molecules can rotate around single carbon bonds. Alkanes do this constantly and fairly rapidly, even at room temperature, so even though we draw them as nice, straight chains, it is more accurate to think of them as wiggling around like a piece of metal chain that is being shaken.

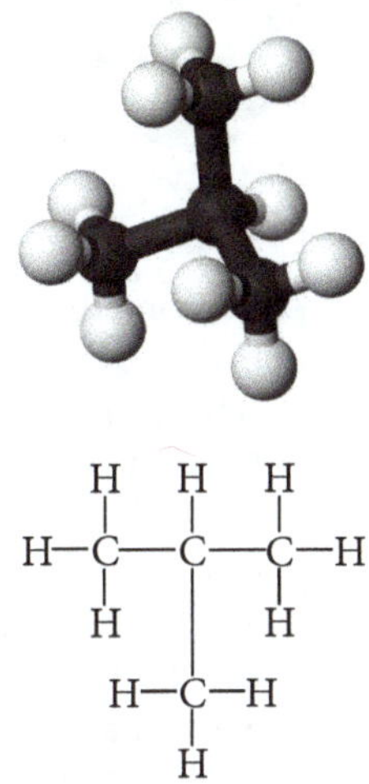

Figure 13.7. Isobutane.

There are no double or triple carbon bonds in alkanes, and for this reason they are known as *saturated hydrocarbons*. When carbon chains contain only single bonds and hydrogen atoms, they have the maximum number of hydrogen atoms possible, which is what the term *saturated* denotes in this context.

The name of an alkane denotes the number of carbon atoms in the molecule. But beginning with butane, there is more than one way for an alkane molecule to form. Figure 13.7 shows this alternate structure in the case of butane. As you see, instead of the four carbons being connected one after the other in a chain, the molecule has a side branch extending off the middle carbon atom. Thus, there are two different molecules with the formula, C_4H_{10}, both entitled to the name butane. This phenomenon—where several different forms exist of the same set of atoms with the same formula—is called *isomerism*, and the different forms of a given set of atoms are called *isomers*. Isomerism is ubiquitous in organic chemistry. Isomers may look similar to one another, but they are distinctly different compounds and their physical and chemical properties can be quite different.

The number of possible isomers increases dramatically as the number of carbon atoms in the molecule increases. For butane there are two. The one shown in Figure 13.6 is called *n*-butane. (The *n*- is for "normal.") The isomer shown in Figure 13.7 goes by the common name *isobutane*. There are five isomers of hexane, nine heptanes, and 75 decanes.

Substitution Reactions

An important feature of alkanes and other hydrocarbons is that hydrogen atoms can be replaced with halogens in a *substitution reaction* called *halogenation*. In substitution reactions, one or more atoms in a molecule are replaced with others. The compounds resulting from halogenation are called *haloalkanes* or *alkyl halides*. Thousands of *haloalkanes* are possible and industrial researchers explored them extensively during the 20th century. Many of these became hugely important in industry and are manufactured for use as propellant in spray cans, refrigerant in air conditioning systems, gases in fire extinguishers, and many other uses. The compounds known as the *chlorofluorocarbons* or *CFCs* are a subgroup of the haloalkanes. The famous story of the CFCs is told in the box on the next page.

One halogenation process is called *free radical halogenation*. The chlorination of methane is a simple example of how the free radical halogenation reaction proceeds. The reaction requires several steps. First, ultraviolet light is used to split a chlorine molecule, creating two chlorine free radicals:

$$Cl_2 \rightarrow :\underset{..}{\ddot{Cl}}\cdot + :\underset{..}{\ddot{Cl}}\cdot$$

Hmm... Interesting. The end of CFCs (almost)

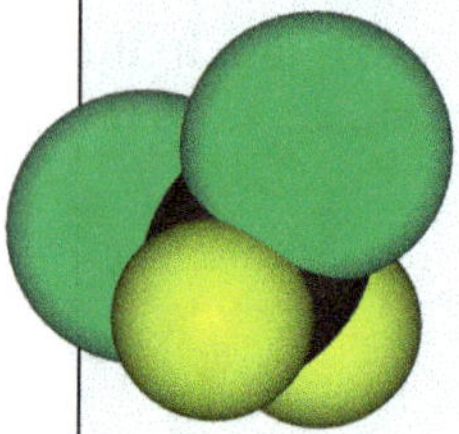

The *chlorofluorocarbons*, or *CFCs* for short, are molecules derived from the first three alkanes—methane, ethane, and propane. CFCs are produced by chemical reactions in which all the hydrogen atoms are replaced with chlorine or fluorine atoms. CFCs are highly stable, unreactive, and relatively nonflammable, so they are safe to work with.

CFCs have been commercially produced since the late 19th century. Their properties make them ideal for use as the gas in fire extinguishers, as the propellent in spray cans such as hair spray and mosquito repellent, and as the refrigerant in air conditioning systems. For decades, the main refrigerant used in automotive air conditioning systems was dichlorodifluoromethane, CCl_2F_2, shown in the first image. This molecule is a methane molecule with the four hydrogen atoms replaced with two chlorine atoms and two fluorine atoms. This refrigerant was popularly known as R-12 and was marketed under the trade name Freon by the DuPont Corporation.

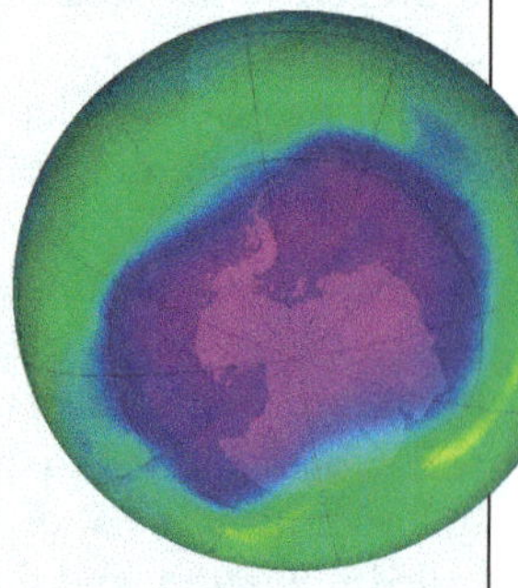

In the early 1970s, a new gas detection technology led to the discovery that CFCs persist in the atmosphere for a very long time and were found to be widespread in the atmosphere. In 1974, scientists discovered that the CFCs in the atmosphere were connected with an alarming phenomenon that had been observed: the overall depletion of the earth's ozone layer, and the appearance of seasonal holes in the ozone layer. The ozone layer is an accumulation of ozone in the earth's stratosphere that absorbs ultraviolet light, shielding life on Earth from this harmful solar radiation.

It turned out that because CFCs are so unreactive, once released into the atmosphere they remain there without breaking down. When these molecules diffuse up into the stratosphere, the sun's ultraviolet light begins breaking them down. Chlorine atoms cut loose from CFC molecules react with ozone to convert ozone into O_2. The second image (above), produced by NASA, shows the hole in the ozone layer that appeared over Antarctica in 2006, the largest yet.

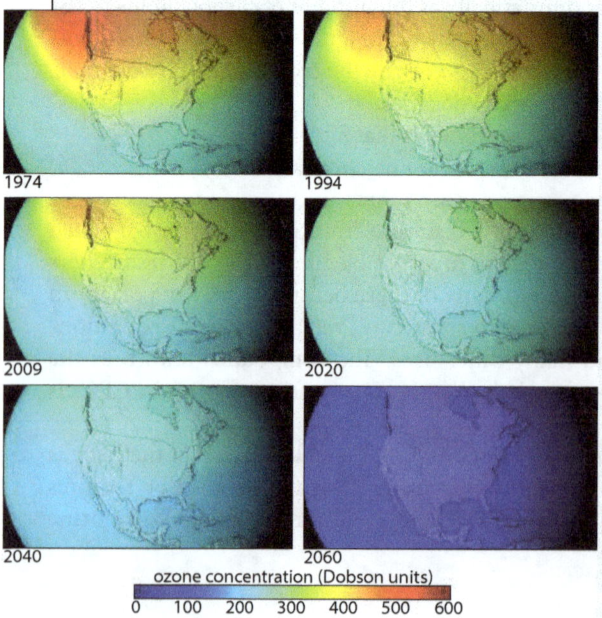

In a 1989 treaty called the Montreal Protocol, industrial nations agreed to reduce and eventually eliminate production of CFCs. Production has been reduced and consumer products no longer use CFCs, but the efforts to get rid of them are still underway. The images to the left, also prepared by NASA, are a computer model showing the predicted reduction in atmospheric ozone concentration that would have occurred had the CFC ban not been put into effect.

Next, a so-called *chain propagation* occurs in two steps. First, a methyl free radical is produced when one of the chlorine free radicals pulls a hydrogen atom from a methane molecule to form a methyl free radical and a molecule of HCl:

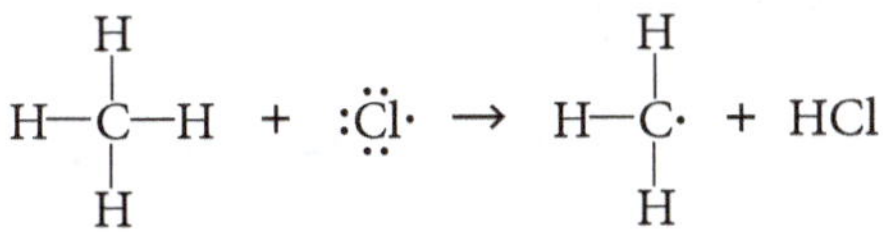

Then the methyl free radical pulls a chlorine ion from a chlorine molecule, forming chloromethane—a haloalkane—and leaving a chlorine free radical:

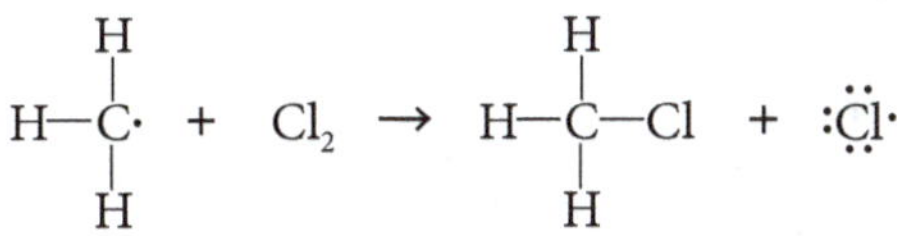

The chain propagation can continue, and further steps can be used to produce ethane, C_2H_6, by combining two methyl free radicals, or other haloalkanes containing more than one chlorine atom.

The alkanes are the basis for a lot of organic chemistry. Thus, it may be helpful to summarize the take-away points on the topic. These are in Table 13.4.

Alkanes: Take-Away Points
• Single carbon bonds in zigzag chains
• Saturated hydrocarbons
• The main components in our most common fuels and lubricating oils
• Hydrogen atoms often replaced with halogens in substitution reactions to form haloalkanes (free radical halogenation)
• The structural backbone for thousands of important industrial chemicals, including CFCs

Table 13.4. Take-away points on alkanes.

13.2.2 Alkyl groups

There is a systematic naming system for all the different hydrocarbons and their isomers. But before getting into naming, you need to know a bit more about the side branches that occur in hydrocarbon isomers. Many alkanes have a hydrocarbon side branch attached in place of one of the hydrogen atoms that would normally be in the molecule. Side branches like this that are themselves derivable from alkanes are called *alkyl groups*.

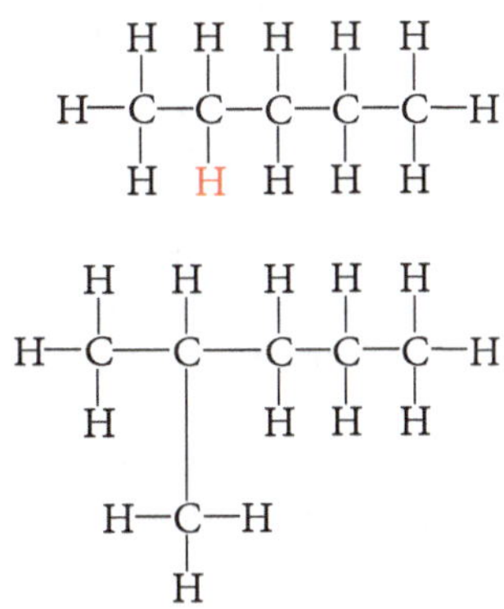

Figure 13.8. Pentane (top), and 2-methylpentane (bottom).

In Figure 13.8, the upper structural formula is that of pentane. One of the hydrogen atoms is shown in red. In the lower formula, the red hydrogen atom is replaced by a branch that is equivalent to a methane molecule with the hydrogen atom removed from one end. This side branch is an alkyl group derived from methane, so it is called the *methyl group*. The methyl group is attached to the pentane molecule at the site where the red hydrogen atom was. As I explain in the next section, the name of this molecule is 2-methylpentane.

Figure 13.9 shows several common alkyl groups. Alkyl groups are named by using the prefix of the parent molecule and adding the suffix *–yl*.

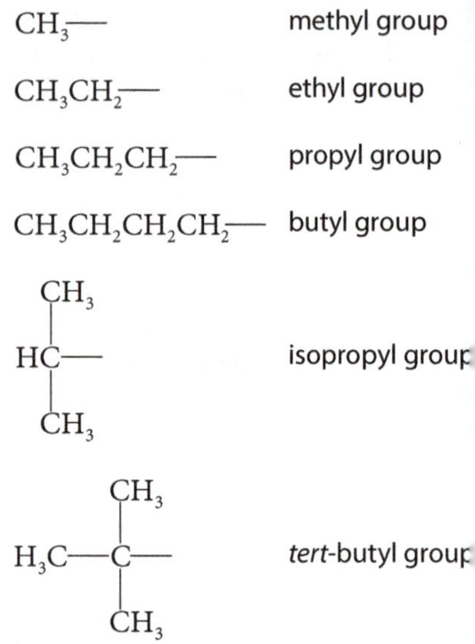

Figure 13.9. Common alkyl groups and condensed structural formulas.

13.2.3 IUPAC Naming Standards

In the early days of chemistry, common names without much systemization became associated with many compounds. Some compounds, such as isobutane, acetylene, and toluene, still go by these common names. But as the number of known organic molecules began to explode—there are now millions of known compounds—it became clear that a new systematic procedure for naming organic compounds was essential.

In 1860, August Kekulé, a very prominent European chemist, headed a committee to address this need. (You may recall that Kekulé was the chemist who discovered the structure of the benzene ring through his dream of a snake eating its tail.) The organizational effort started by Kekulé eventually led to the founding of the International Union of Pure and Applied Chemistry (IUPAC) in 1919. The IUPAC is best known for its role as the organization that superintends the naming of compounds.

Our task in this section is to address the basic conventions for naming the isomers of hydrocarbons and the haloalkanes. It is simplest to describe the naming convention as a sequence of steps with examples.

Step 1 *Identify the number of carbon atoms in the parent chain—the chain with the longest carbon sequence. The alkane associated with this number of carbon atoms forms the root of the compound name.*

This step is illustrated as follows:

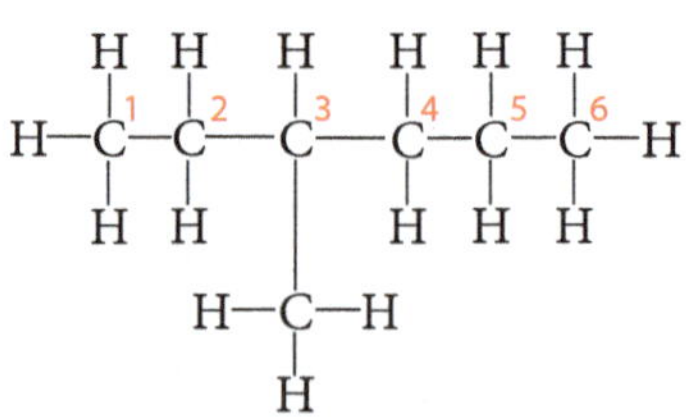

The longest carbon sequence in this molecule is numbered in the structural formula above and contains six carbon atoms (There are two shorter chains, one containing four carbon atoms and one with five.) Note, however, that the longest chain in a structural formula drawn on paper may not consist of carbons atoms in a straight line. Recall again that hydrocarbon molecules can rotate around single bonds. For this reason, the following structural formula is equivalent to the one above:

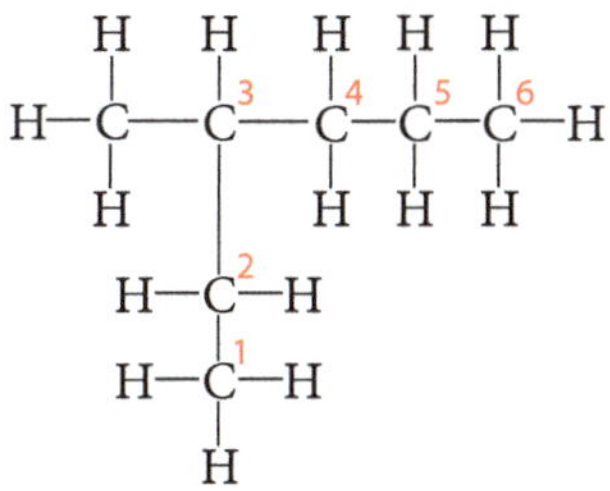

These look different on paper, but they are the same. The carbon atom labeled 3 can rotate around the single bond to its right, so this molecule can be represented by these two equivalent structural formulas (as well as others). The longest carbon chain in this molecule contains six carbon atoms, so this molecule is a hexane.

The parent chain is the one with the most *straight-chain branches*. A straight-chain branch is a branch without side branches. Consider the following two possibilities, each of which leads to a count of eight carbons in the parent chain (the hydrogens are left off for clarity):

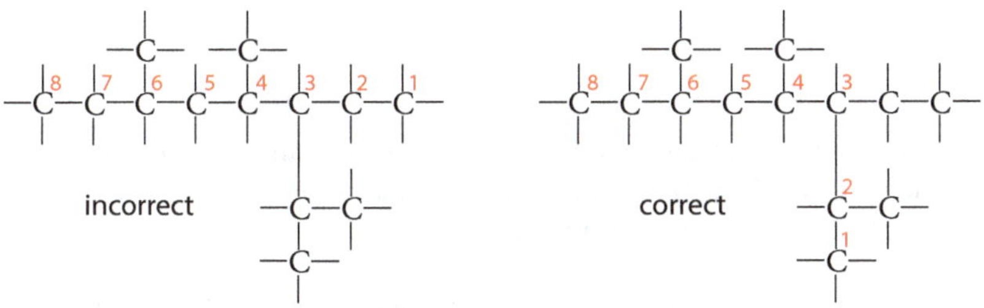

In the counting scheme on the left, the side branch attached at the third carbon atom is not a straight-chain branch, so the number of straight-chain branches on the parent chain is two. In the numbering scheme on the right, all side branches are straight-chain branches, which gives four straight-chain branches. The counting scheme on the right is correct.

Step 2 *Identify the numbers of the carbon atoms where alkyl groups or non-hydrogen species are attached. Begin counting from the end of the parent chain that results in the lowest values for the attachment points.*

In the "correct" structural formula just above, if we count the carbon atoms as shown by the red numerals, the attachment points for the side branches are carbon atoms 2, 3, 4, and 6. If we were to begin counting at the left end, the attachment points would be carbon atoms 3, 5, 6, and 7. The set of numbers we get from the counting sequence shown in the diagram has lower values than the other numbering possibility, and is thus the correct one.

Step 3 *Identify the attached alkyl groups or other species. If the same species occurs more than once, indicate so with the prefixes di–, tri–, tetra–, and so on. The names of the attached groups with their prefixes become prefixes on the root name.*

In the following molecule, there are six carbon atoms in the parent chain, and two methyl groups attached to the parent chain:

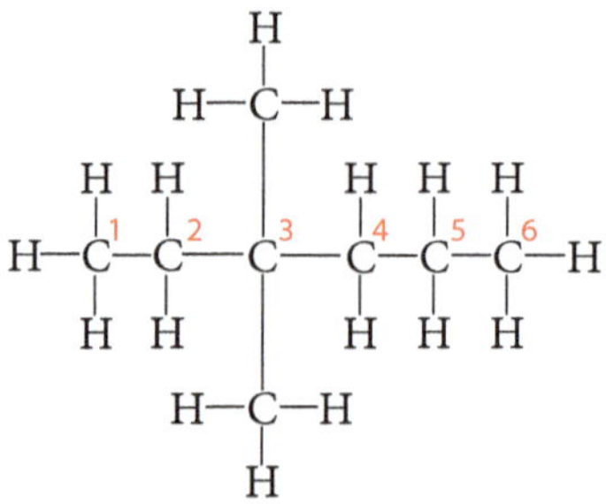

With a parent chain of six carbons, the root of the compound name is hexane. Since two methyl groups are attached, the prefix dimethyl is added to the root name to give dimethylhexane.

As another example of the use of prefixes in naming, the haloalkane molecule below has a parent chain consisting of eight carbons, with three chlorine atoms attached in place of hydrogen atoms:

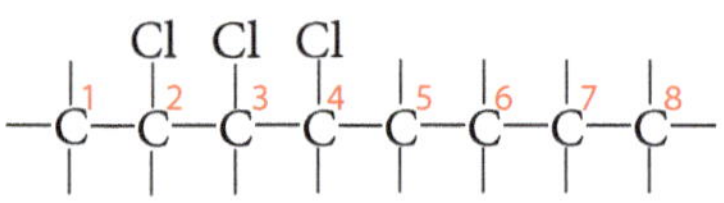

The prefixes used for halogens in haloalkanes end in *-o-*, as in *chloro-* and *bromo-*. Thus, this compound is called trichlorooctane. Note that the parent chain has been numbered so that the attachment points for the chlorine atoms are 2, 3, and 4.

Step 4 *Write the compound name using digits separated by commas for the attachment points of each branch. Digits are separated from names of attached species with hyphens. Attached species are ordered by increasing group size for alkyl groups or alphabetical order for halogens.*

Applying this to the trichlorooctane compound just above, the full IUPAC name for the compound is 2,3,4-trichlorooctane. Applying the rules to the dimethylhexane compound in step 3, the full name is 3,3-dimethylhexane.

The following molecule has chlorine atoms attached at carbon atoms 2 and 3. Combining the rules outlined in the previous steps, the full IUPAC name of the compound is 2,3-dichlorohexane.

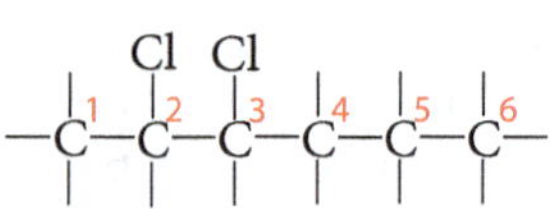

When more than one halogen is present, they are listed in the compound name in alphabetical order, without considering any prefixes on the halogens. The following compound is 2,2-dibromo-4-chlorohexane:

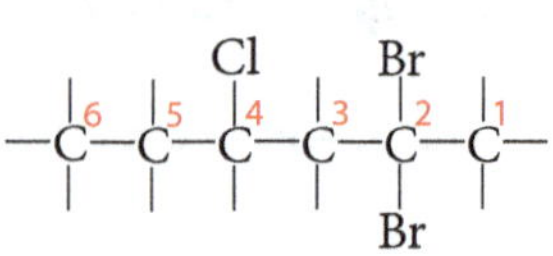

Alkyl groups are listed in order of increasing size: methyl first, ethyl second, and propyl third. In the following compound, there are nine carbons in the parent chain, which is a nonane. There are methyl groups at positions 3 and 5, and an ethyl group at position 3. So the name of the compound is 3,5-dimethyl-3-ethylnonane.

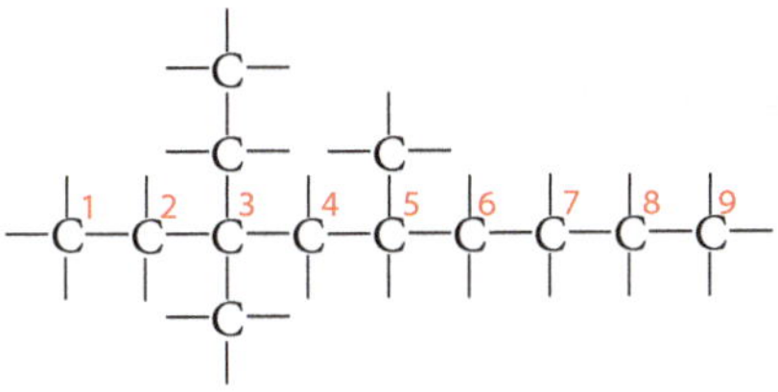

Example 13.1

Determine the IUPAC name for isobutane.

From Figure 13.7, the structural formula for this compound is as follows:

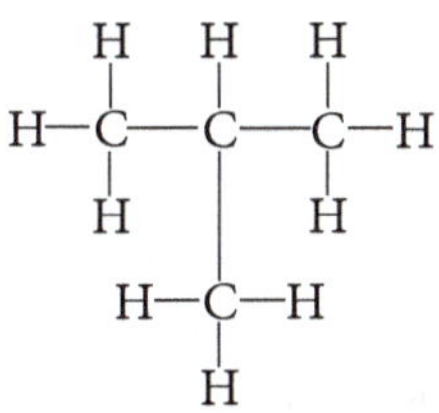

According to the IUPAC standards, this molecule consists of a parent chain of three carbon atoms with a methyl group attached. Thus, this molecule is methylpropane. Numerals are optional in this name because the only place a branch can attach to a propane molecule is the second carbon atom. (If the methyl group were attached to the end, the molecule would no longer be propane; it would be *n*-butane.)

Example 13.2

Draw the structural formula for 2,2,3-triethylheptane. For clarity, show where the hydrogen atoms attach, but leave off the hydrogen atoms themselves.

The parent molecule is a heptane, so begin by drawing a seven-carbon chain:

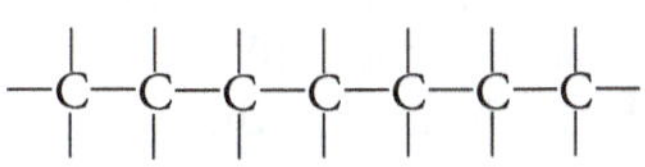

There are three ethyl groups attached to this chain, each one having this structure:

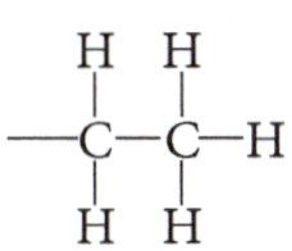

Without the hydrogen atoms, we can represent the ethyl group as

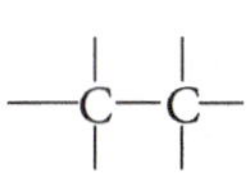

Two of the ethyl groups attach to the second carbon in the parent chain, and the third attaches to the third carbon in the chain. To draw these ethyl groups at two adjacent carbon atoms, we

need to stretch out the connection between the second and third carbon atoms so there is room to draw the branches without them overlapping. Thus, the structural diagram is as follows:

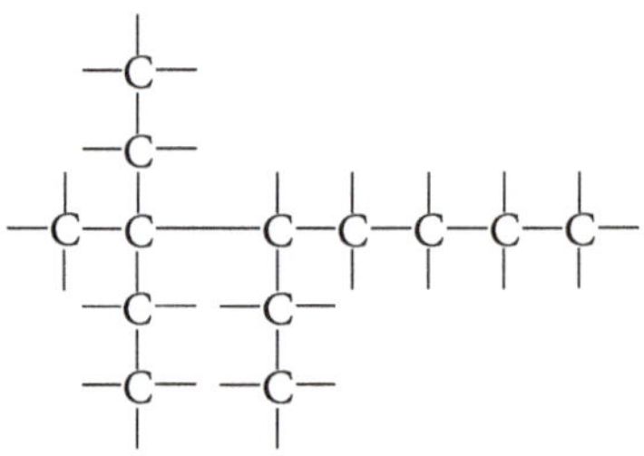

Note that the ethyl group attached to the third carbon atom may be shown above or below the parent carbon chain. These represent the same structure.

▲

13.2.4 Functional Groups

A *functional group* is an atom or group of atoms that establishes a class of compounds with specific chemical properties. The carbon double bond and the carbon triple bond act as functional groups within the hydrocarbons. Hydrocarbons possessing a carbon double bond are called alkenes, and those possessing the triple bond are called alkynes. These two classes of compounds are our next topic.

When we start getting into compounds containing elements other than carbon and hydrogen, there are many other functional groups that identify important classes of compounds. There are scores of functional groups, but in this short introductory chapter, we only introduce a few of them, namely, those shown in Table 13.5.

Some functional groups give rise to more than one class of compounds, depending on what is attached to the functional group. In the fourth column of the table, note the use of the letters R, R′ (pronounced "R-prime"), and R″ (pronounced "R-double-prime") holding positions in the molecule. At such points, any molecular group can attach and the R is a conventional placeholder representing any such group. The designations R′ and R″ are used generically to indicate that the groups connected there may be different molecular groups than the one designated by R.

Table 13.5 probably seems a bit overwhelming—an ocean of names and structures that are similar but different. I present them all together here in the table so you have them all together in one place. This should make it easier to compare them as you study the details that differentiate them from one another.

I don't have any magic bullets that can make memorizing the names any easier. You just have to do it. But as I briefly comment on each of these classes of compounds in the coming sections, I point out some details in their structures that may make remembering them a bit easier. With that, we now continue.

13.2.5 Alkenes and Alkynes

If we start with an alkane and allow one or more of the C—C bonds to be a double bond, we have the class of hydrocarbons known as the *alkenes*. In the petrochemical industry, these compounds are also known as *olefins*. If we start with an alkane but allow the compound to have one or more triple bonds, we have the class known as the *alkynes*. The names for the alkenes and alkynes use the same prefixes as the alkanes, and the compound names end in *–ene* or *–yne*, respectively. Because of the multiple bonds in alkenes and alkynes, these hydrocarbons are unsaturated.

Functional Group Name	Functional Group Structure	Compound Class	Representative Class Structure
alkyl		alkane	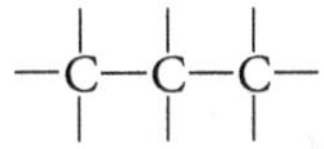
alkenyl	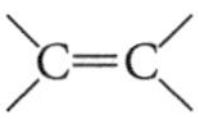	alkene	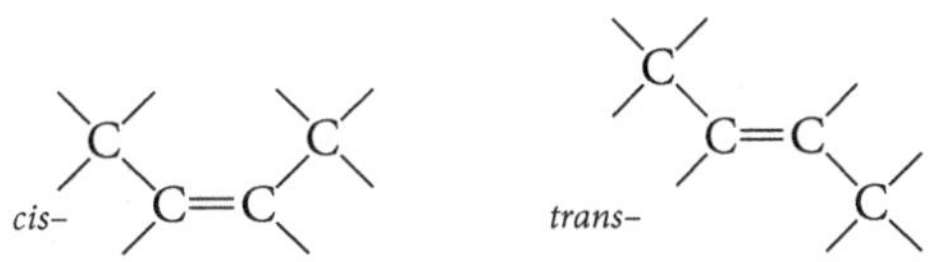
alkynyl	—C≡C—	alkyne	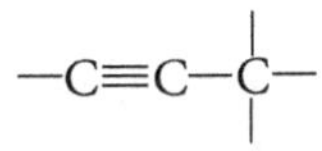
hydroxyl	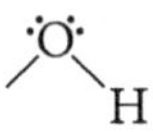	alcohol	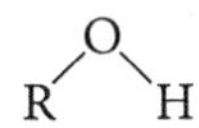
ether		ether	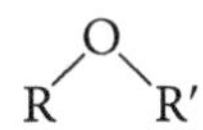
carbonyl		aldehyde, keytone	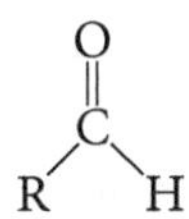 aldehyde 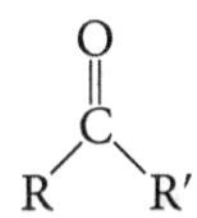 keytone
phenyl	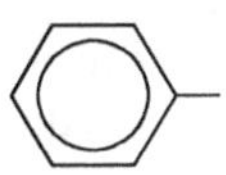	benzene derivative	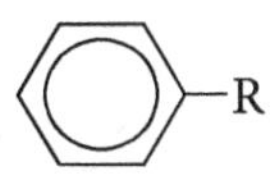
carboxyl	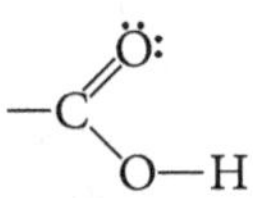	carboxylic acid	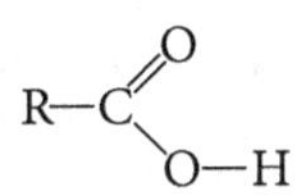
ester	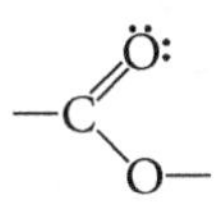	ester	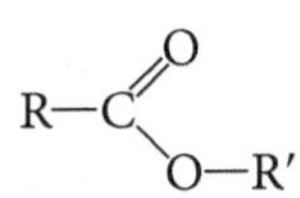
amine	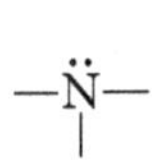	amine	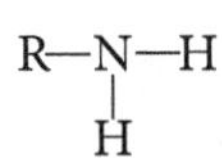 primary amine 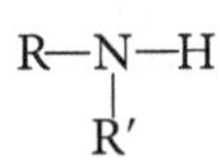 secondary amine 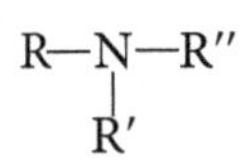 tertiary amine
carboxamide	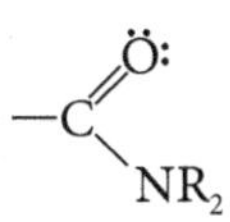	amide	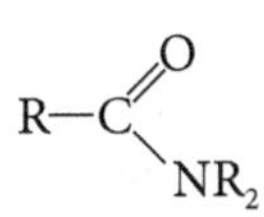

Table 13.5. A few of the most common functional groups and the classes of compounds associated with them.

The IUPAC name for the simplest alkene is ethene, but this compound always goes by its common name, ethylene. This compound is shown back in Figure 13.3. The three-carbon alkene, propene, also goes by a common name—propylene. The IUPAC name for the simplest alkyne is ethyne, but this compound always goes by its common name, acetylene. This compound is shown in Figure 13.10.

Figure 13.10. Ethyne, aka acetylene.

Ethylene is the most heavily produced organic compound in the world. It is the *precursor* or starting chemical for the production of an enormous number of compounds, and for compounds which are precursors for other compounds. Ethylene is used to make polyethylene, the material many of our plastic bottles and packages are made of. We will discuss that application later when we get to polymerization.

In the petrochemical industry, ethylene is produced by a process called *steam cracking*. In this process, larger hydrocarbons are heated to around 800–900°C, and this heat generates an abundance of hydrocarbon free radicals. After they form, these free radicals are hit with a steam "quench" that cools them down so they join together into smaller, unsaturated molecules. The ethylene is then separated from the other compounds formed in the steam cracking reaction by a process called *fractional distillation*. This is an important process in the petrochemical industry because it is also used to separate the different components in crude oil from one another. A brief look at the details is included in the accompanying box.

Another interesting thing to know about ethylene is that it is a plant hormone that stimulates or regulates the ripening of fruit. This happens naturally in the plant while it grows, but ethylene can also be used to ripen fruit at home after the fruit are purchased. Fruit are often sold before they are ripe. Unripened fruit such as pears or peaches need to ripen before they are good to eat. You may know that if you place the unripened fruit in a paper bag with a piece of ripe fruit, the unripened fruit ripen faster. This occurs because ripe fruit release ethylene as a gas, and the ethylene gas in the bag stimulates the unripened fruit to ripen. Damaged fruit release even more ethylene, hence the (correct) saying that "one bad apple spoils the whole bushel."

Concerning uses for the alkynes, the main one to know about is the smallest alkyne, acetylene (Figure 13.10). The triple carbon bond is unstable, and when high pressure acetylene burns in oxygen, the flame temperature is over 3,300°C, hot enough to cut through steel with ease. The oxyacetylene torch used for this purpose is depicted back in Figure 10.14.

Naming alkenes and alkynes requires that we identify the location of the multiple bond in the molecule. (Molecules can have more than one multiple bond, but I do not treat the naming for those cases in this text.) We always begin counting the carbon atoms in the parent chain at the end of the molecule closest to the multiple bond, and the parent chain always includes the multiple bond. Additionally, in naming the isomers of alkenes we must recall that the sections of a molecule on either side of a double (or triple) carbon bond are not free to rotate relative to one another. For the alkenes, this means that there are two ways for the carbon chain to continue on either side of the double bond: on the same side of the molecule or on the opposite side of the molecule. This type of isomerism is called *cis/trans* isomerism because the Latin designations *cis-* (meaning "on the same side") and *trans-* ("on the opposite side," or "across") are used to distinguish between the two isomers. The *cis-* and *trans-* labels are usually written in italics. The *cis-* and *trans-* isomers are illustrated in Figure 13.11 using condensed structural formulas.

Consider the four isomers of C_4H_8 shown in Figure 13.12. In the first case, the parent chain has three carbon atoms in it, no matter how you count it. This means the root of the compound name is propene and the extra CH_3 attached at the lower left is a methyl group. There is only one way to place the double bond in a propene molecule. It cannot be placed between carbon atoms 2 and 3 because carbon atom 3 would then have five bonds. Put another way, when the double

Hmm... Interesting. Fractional distillation

Fractional distillation is a process widely used in the petrochemical industry to separate mixtures of hydrocarbon molecules that have different molecular weights. The process works because the higher the molecular weight of the compound, the higher its boiling point/condensation temperature, as Table 13.3 shows.

The hydrocarbon mixture is first vaporized and then fed into a *fractionating column*, such as one of those shown to the left. Inside the column, the temperature is highest at the bottom and gradually decreases from bottom to top.

Inside the column, vapors flow up through the middle and condensed vapors (liquids) flow down along the sides. Trays placed in the column, as shown in the diagram below, facilitate the separation. When vapors arrive at the region in the column that is at their condensation temperature, they condense.

When separating two molecular components, the lighter component is collected at the top of the column as a vapor and the heavier component is collected as a liquid at the bottom. When separating mixtures such as crude oil that contain a wide range of molecular weights, the different weights separate into many different trays—heavier molecules condensing in trays near the bottom, lighter molecules condensing in trays near the top, and so on—where they are removed from the column. In the diagram, the upper tray is at a lower temperature than the lower tray, so the molecules condensing in the upper tray have a lower molecular weight (smaller molecules). Typically, there are anywhere from five to 50 trays in a fractionation column like this.

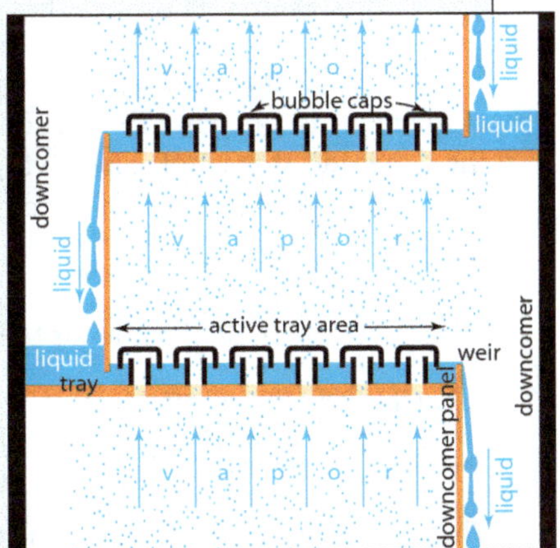

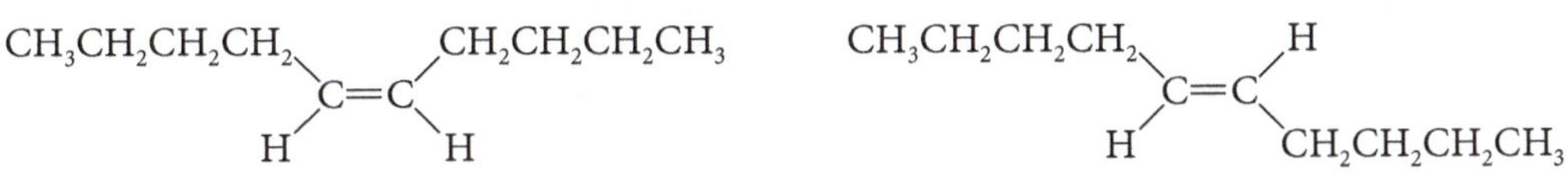

cis- *trans-*

Figure 13.11. The *cis-* and *trans-* forms of an alkene molecule, *cis*-5-decene and *trans*-5-decene.

bond is at the end of the molecule, the first carbon atom has only two hydrogen atoms attached to it (just like an intermediate CH_2 unit in a long chain). There are thus no other options, and the molecule is named methylpropene.

In the second case, the parent chain has four carbon atoms, so we are dealing with butene. The double bond is at the end of the molecule, right after the first carbon atom, and this is des-

methylpropene 1-butene *cis*-2-butene *trans*-2-butene

Figure 13.12. The four isomers of C_4H_8.

ignated with a numeral indicating the number of the carbon atom the bond comes after, giving us 1-butene. There are no more considerations for this molecule.

In the third case, the double bond comes after the second carbon atom, no matter how you look at it, so this is 2-butene. The CH_3 groups on either side of the double bond are on the same side of the molecule, so this molecule is *cis*-2-butene. The final case is identical to the third case except that the CH_3 groups on either side of the double bond are on opposite sides of the molecule, so this molecule is *trans*-2-butene.

The alkenes and alkynes can have alkyl groups attached just as the alkanes can. Let's consider the naming of this molecule:

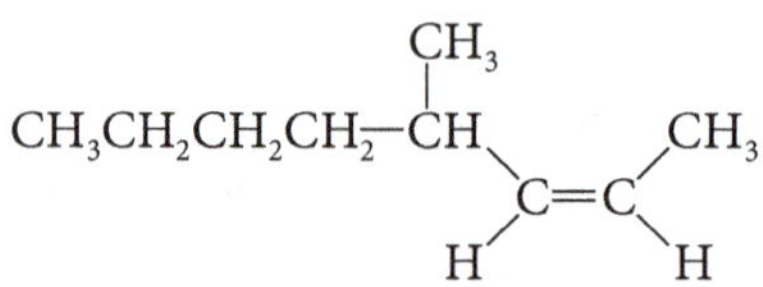

The parent chain has eight carbons, and the double bond is after the second one, so the root of the name is 2-octene. The carbon chains continuing on either side of the double bond are on the same side of the molecule, giving us *cis*-2-octene. Finally, there is a methyl group attached at carbon 4, so the full name of the compound is 4-methyl-*cis*-2-octene.

Example 13.3

Determine the IUPAC name for the following molecule:

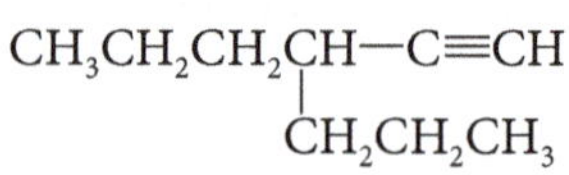

The longest carbon chain including the triple bond has six carbon atoms, so the root of the name is hexyne. The triple bond occurs after the first carbon atom, giving 1-hexyne. There is a propyl group attached at carbon 3, so the full name is 3-propyl-1-hexyne.

Example 13.4

Draw the condensed structural formula for 4-ethyl-2-pentyne.

Pentyne contains five carbons and a triple bond in the parent chain. The 2- tells us that the triple bond comes after the second carbon atom in the parent chain. With this information as a starting point, we have

$CH_3{-}C{\equiv}C{-}CHCH_3$

When drawing formulas like this, it is very tricky to get the carbons right in the vicinity of the triple bond. You should always check and recheck the carbons to make sure four bonds are associated with each of them (counting the triple bond as three). In the structure drawn so far, the fourth carbon from the left has only one hydrogen attached, and three bonds total. This is because the ethyl group attaches here. Attaching it, we have

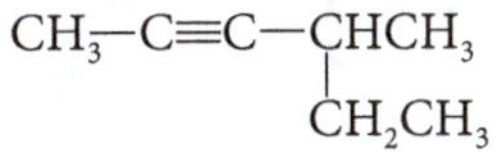

Addition Reactions

Alkenes and alkynes are processed in *addition reactions* to increase their saturation. As the name implies, atoms are added to a molecule in an addition reaction. A common addition reaction is *hydrogenation*, a reaction in which hydrogen atoms are added to alkene or alkyne molecules, converting multiple bonds into single bonds and increasing the saturation of the molecule.

The hydrogenation of an alkene may be modeled as follows:

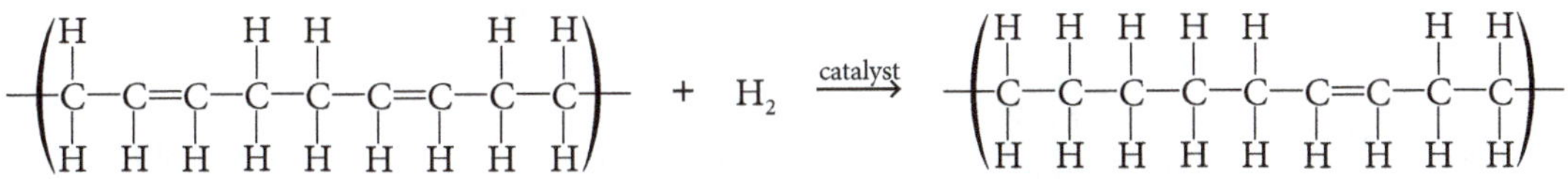

The brackets around the molecules indicate that the structural formulas only show part of the molecules and that the carbon chains continue past the brackets. To enable the reaction to occur at lower temperatures, this hydrogenation reaction usually takes place in the presence of a heterogeneous metal catalyst such as platinum or nickel. The reaction shown in the equation above represents a *partial* hydrogenation because after the reaction, double bonds still exist in the molecule. The saturation of the molecule has been increased, but the molecule is not fully saturated.

If you have spent any time preparing baked goods in a kitchen, you may already know that vegetable shortening is composed of partially hydrogenated vegetable oil. If you are health conscious, you may also know that partial hydrogenation is associated with the production of "trans fats" which are associated with health risks. People talk about trans fats all the time without understanding what they are or how they arise. The issue is fairly easy to understand once you know about the basic molecular structures involved. We get into more detail on this issue later after you have become familiar with some other molecular structures.

13.2.6 Aliphatic Cyclo– Hydrocarbons

Together, the alkanes, alkenes, and alkynes discussed so far are called *open-chain hydrocarbons*. But there is one more variation on the aliphatic compound theme that we are not going to go into much but will at least mention: the *cyclo–* compounds. Since hydrocarbon molecules can rotate around single bonds, it is possible for compounds with at least three carbon atoms to wind around into a ring shape and make a closed loop. These are called *cycloalkanes*, *cycloalkenes*, and *cycloalkynes*.

I will illustrate the structure of *cyclo–* compounds with cyclohexane, C_6H_{12}. I have chosen this compound for a specific reason: it has six carbon atoms in a ring, somewhat like benzene, C_6H_6. However, there are two key differences between the cycloalkane and benzene. Because of these differences, cycloalkanes—which are aliphatic—do not have the aromatic smell that aromatic compounds have.

Figure 13.13 depicts the ball-and-stick model of cyclohexane. Note that this compound—like other cycloalkanes—has two hydrogen atoms attached at each carbon site, and each carbon has four single bonds. This makes this compound completely different from benzene, which is a resonance structure possessing a single hydrogen atom at each carbon site. Also notice that *cyc-*

lo– molecules still possess the zigzag structure common to all aliphatic compounds, so the rings of carbon do not lie in the same plane—the molecules aren't flat the way benzene is. In Figure 13.13, the upper image is the ball-and-stick model, and the lower image depicts the shape of the carbon ring in cyclohexane. This shape is called a "chair" for obvious reasons.

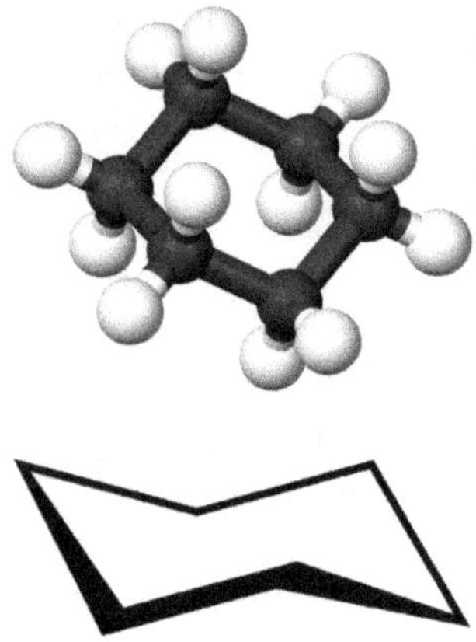

Figure 13.13. Cyclohexane, C_6H_{12}.

13.2.7 Aromatic Hydrocarbons

Have you wondered about the source of the sweet smell in natural substances such as vanilla, cloves, and cinnamon? The smell of these aromatic substances is due to the benzene ring, depicted in Figure 13.14. Hydrocarbons containing the benzene ring are called *aromatic hydrocarbons*. Benzene itself, a highly flammable liquid, has a sweet smell and used to be common in hundreds of products, from after-shave lotion to solvents and fuels. However, during the 20th century it became clear that benzene was a carcinogen—a cancer-causing agent, so all the consumer uses of benzene came to a stop. But benzene is still an important industrial compound, playing a major role in the production of hundreds of chemicals.

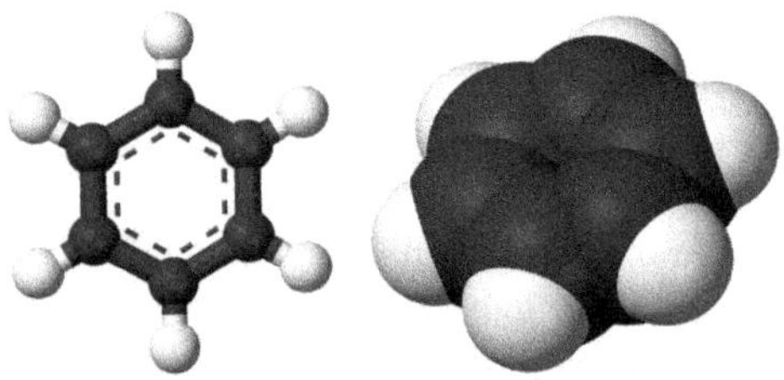

Figure 13.14. Ball-and-stick (left) and space-filling (right) models of the benzene molecule.

As mentioned above, the benzene ring is flat. (This was finally demonstrated in 1929 by Kathleen Lonsdale, after whom the carbon allotrope Lonsdaleite is named. See Figure 13.4.)

The full structural formula for benzene is as follows:

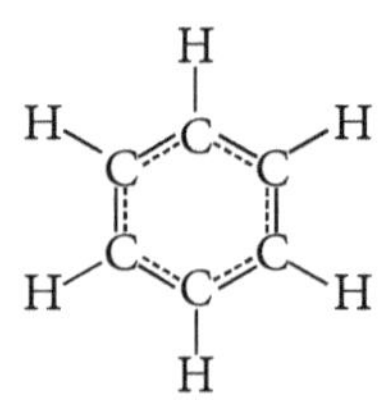

However, it is conventional to show benzene this way:

The circle inside the hexagon is there to remind us of benzene's resonant structure. The bond number of each carbon atom is the same ($\frac{3}{2}$), and the three electrons involved in the resonance structure are said to be *delocalized*—spread evenly among the six carbon atoms in the ring.

Benzene rings can join together to produce *polycyclic aromatic hydrocarbons* such as those shown in Figure 13.15. Naphthalene is a great example of the sweet odors produced by aromatic hydrocarbons because you have probably smelled it—it is the compound moth balls are composed of. Anthracene and pyrene are each used in the production of multitudes of compounds,

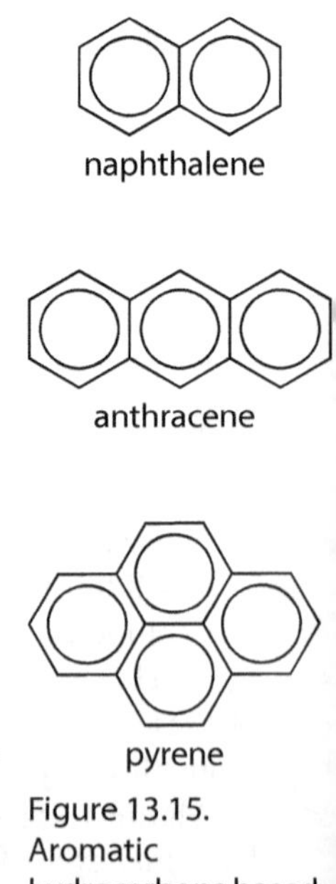

Figure 13.15. Aromatic hydrocarbons based on benzene.

many of them dyes. Anthracene is used to make alizarin, shown in Figure 13.16. Alizarin is the compound in the dyes known as Mordant Red 11 and Turkey Red. It is also the compound used to make pigments known to painters as Rose madder and Alizarin crimson. Pyrene is used to make pyranine, which is used to make the colored inks in highlighters.

Figure 13.16. A sample of alizarin.

Benzene is nonpolar—soluble in other nonpolar solvents but not in water. Benzene is regarded as highly unsaturated since only six hydrogen atoms are attached to its six carbon atoms instead of the possible 12. Benzene readily undergoes substitution reactions, replacing one of the hydrogen atoms at the corners with a functional group, halogen, or other carbon chain. The number of compounds that can be formed is essentially limitless. We will take a quick look at a few of them and the reactions used to produce them.

Halogenation of benzene occurs by substitution, as in the following chlorination reaction that uses iron(III) chloride as a catalyst:

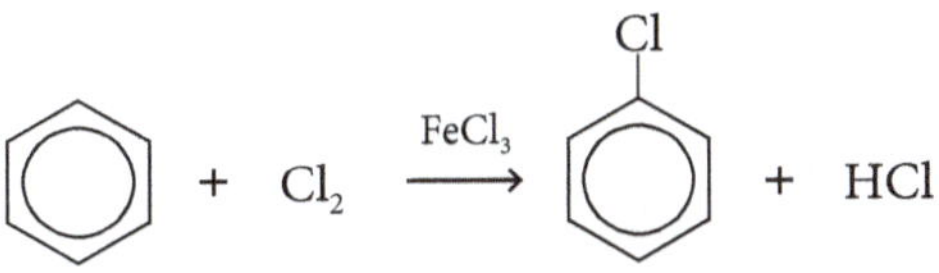

The compound formed in this reaction is known as chlorobenzene. The same reaction can be performed using iron(III) bromide as the catalyst to add bromine atoms to benzene rings. If more than one halogen atom is added, IUPAC naming of the isomers must be considered. If a single halogen is attached, the compound is simply called chlorobenzene, bromobenzene, and so on. If two identical halogens are attached, there are three isomers, illustrated with bromine in Figure 13.17. Each of these compounds has two bromine atoms attached, and is accordingly called dibromobenzene. One of the attached halogens is used as a starting point for the counting of carbon atoms around the ring, and the locations of the two attached groups are denoted with the lowest possible numerals, just as we have done before. Using the bromine at the top as the starting point, the isomer on the left is 1,2-dibromobenzene. The names of the isomers in the center and on the right are 1,3-dibromobenzene and 1,4-dibromobenzene, respectively. These compounds also have common names that are still in wide use. The common names use the Greek prefixes *ortho*–, *meta*–, and *para*–, which are always written in italics. These names and their common abbreviations are shown in the figure.

If benzene is heated in a mixture of nitric acid and sulfuric acid, nitro groups can be added to benzene through substitution reactions. This process is known as *nitration*. Although not

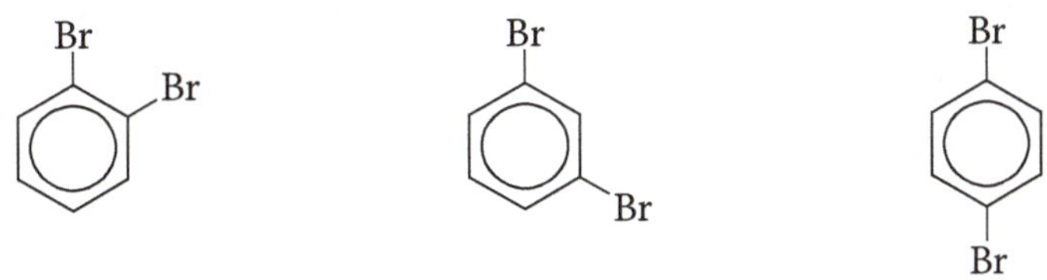

1,2-dibromobenzene	1,3-dibromobenzene	1,4-dibromobenzene
ortho-dibromobenzene	*meta*-dibromobenzene	*para*-dibromobenzene
o-dibromobenzene	*m*-dibromobenzene	*p*-dibromobenzene

Figure 13.17. Isomers of dibromobenzene. IUPAC names are shown in the first row. Common names and abbreviations for the common names are shown in the next two rows.

shown in Table 13.5, the nitro functional group consists of the nitronium ion, NO_2^+. The nitration reaction is as follows:

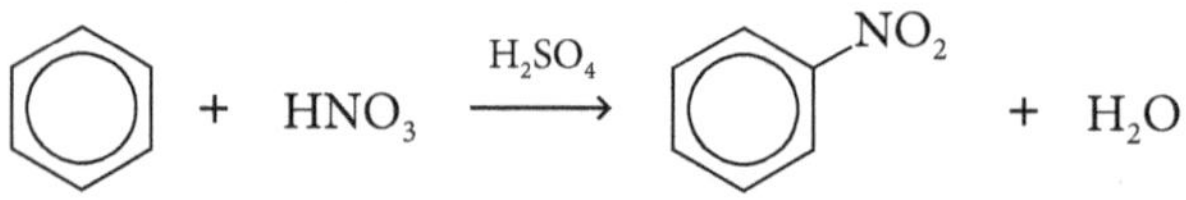

The compound formed is nitrobenzene. This reaction may be repeated to produce *ortho*-dinitrobenzene or the other dinitrobenzene isomers by the same reaction:

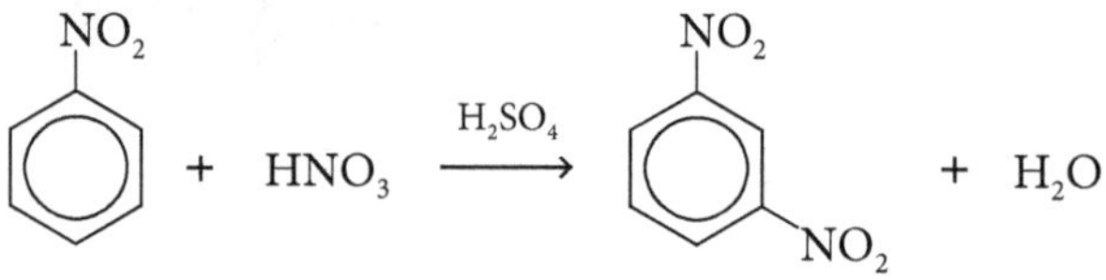

Alkyl groups may also be added to benzene rings by substitution reactions. Starting with an alkyl halide such as chloroethane and an appropriate catalyst, substitutions such as the following occur:

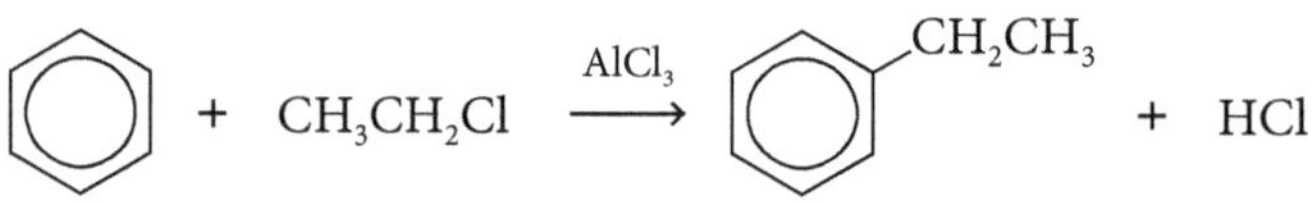

This compound, ethylbenzene, is an intermediate compound in the production of styrene, the precursor to polystyrene (commonly known by the trade name Styrofoam). The reaction above to produce ethylbenzene is an example of an important type of substitution reaction known as *alkylation*. In an alkylation reaction, an alkyl group is transferred from one molecule to another.

One exciting final example must not be overlooked. Toluene is an important aromatic hydrocarbon composed of a benzene ring with a methyl group attached:

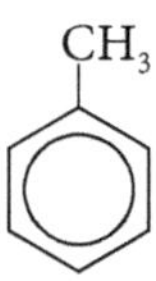

The IUPAC name for toluene is methylbenzene, but the common name is always used. Toluene is a common solvent and powerful fuel. It used to be the primary component of the fuel used by Formula One race cars, although nowadays F1 fuel is quite similar to ordinary gasoline composed of alkanes (with some spiffy additives).

Toluene may be produced by the alkylation reaction we just discussed. Using bromomethane and sodium metal, bromobenzene can be converted to toluene as follows:

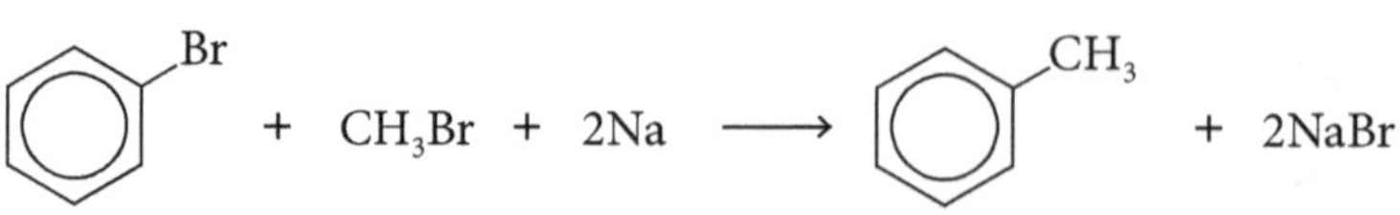

Now here comes the interesting part. Toluene can be nitrated with nitric acid and sulfuric acid to add nitro groups, just as we saw above in the preparation of nitrobenzene:

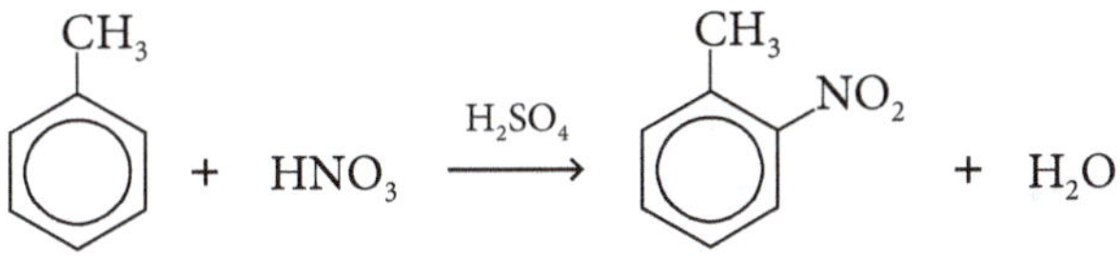

The compound produced by this reaction is 2-nitrotoluene. This compound is then nitrated again to produce 2,4-dinitrotoluene. Finally, a slightly different reaction is used to add a third nitro group to form 2,4,6-trinitrotoluene:

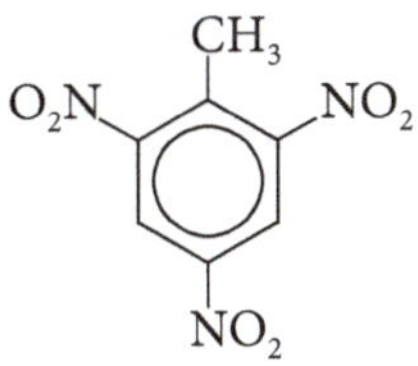

This is the interesting molecule known everywhere as *TNT*. If you are interested in a refresher on why TNT is used by road construction crews to blow holes through mountains, read again the box on page 135.

13.3 Other Organic Compounds

13.3.1 Alcohols and Ethers

Alcohols are hydrocarbon derivatives in which one or more hydrogen atoms are replaced by a hydroxyl group (—OH). The names of the alcohols derive from the parent hydrocarbon with the ending *-ol*. You can think of a hydroxyl group as a water molecule, with its bent geometry, with one hydrogen atom removed. The hydroxyl group always exhibits the same bent geometry at the carbon atom where it attaches to the hydrocarbon, like this:

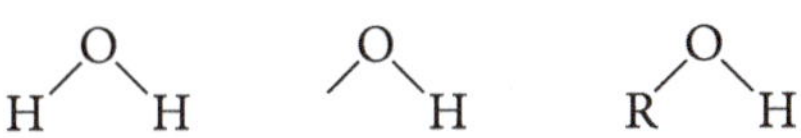

In Section 8.3.3, I discuss three of the simplest alcohols—methanol, ethanol, and isopropyl alcohol. There I note that because of the hydroxyl group, alcohols are water soluble and engage in hydrogen bonding. Because of the hydrogen bonding, these compounds have higher boiling points that their nonpolar parent hydrocarbons do.

Several of the common alcohols are shown in Figure 13.18. Note the use of *-diol* and *-triol* suffixes to indicate that two or three hydroxyl groups are attached to the parent hydrocarbon. Note also the use of numerals to indicate where the hydroxyl groups are attached.

Ethanol is the alcohol in found in alcoholic beverages, and is produced (commercially and in beverages) by yeast as they consume sugar. Ethylene glycol is the main component of automotive antifreeze and is quite toxic. Glycerol, also called glycerine, is used everywhere in the productions of foods, pharmaceuticals, cosmetics, and soaps. Unlike ethylene glycol, glycerine's toxicity is very low, enabling its use as a food additive. Isopropyl alcohol is in every home first-aid kit for disinfecting wounds. It is toxic. And phenol is used to make a host of compounds including phenolic plastics (such as Bakelite), aspirin and other drugs, detergents, and herbicides. Saying phenol is

Elimination Reactions

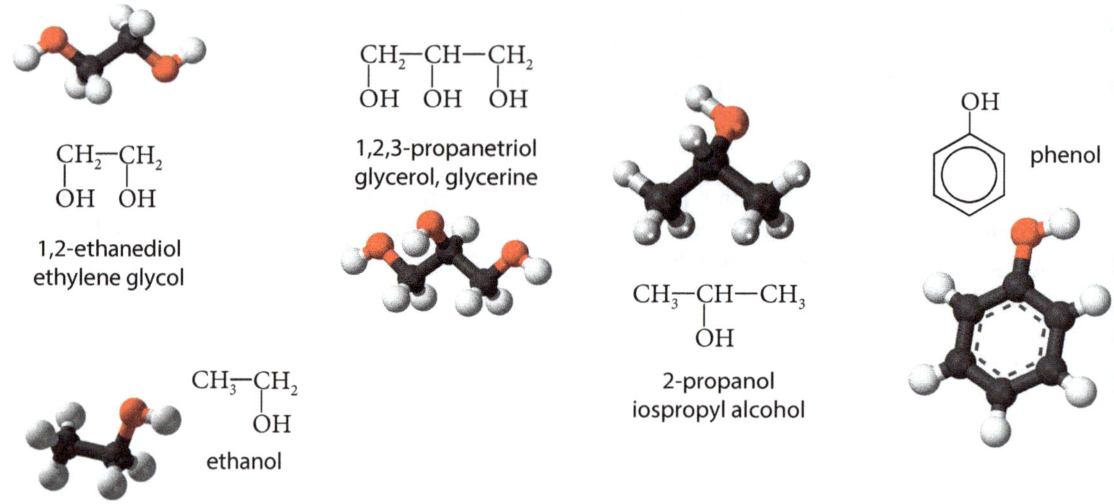

Figure 13.18. Various alcohols. IUPAC names are shown above common names.

toxic would be putting it mildly. Both phenol and its vapors are corrosive to human tissue and cause severe chemical burns.

Ethanol can be used as a precursor for producing ethylene. The reaction involved is a type called an *elimination reaction*. In an elimination reaction, a hydrogen atom and a hydroxyl group from adjacent carbon atoms of a larger molecule are removed as a molecule of water. The conversion of ethanol to ethylene uses heat and a sulfuric acid catalyst in the following elimination reaction. The hydrogen and hydroxyl group marked in red are removed to form the water molecule:

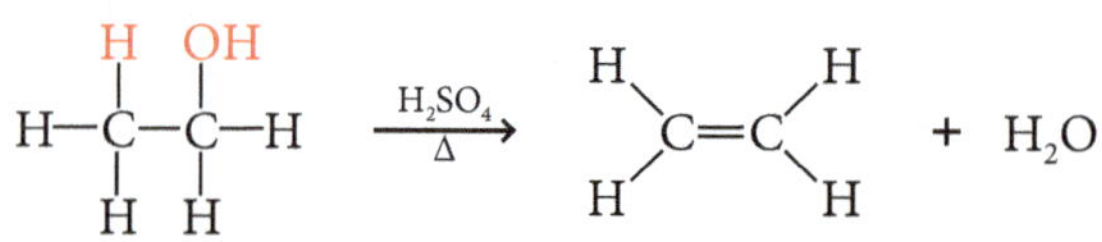

ethanol ethylene

The reason sulfuric acid acts as a catalyst in this reaction is that H_2SO_4 is strongly *hygroscopic*. This means it has a strong affinity for water and absorbs it.

An *ether* is formed when two hydrocarbon groups are joined to the same oxygen atom. The reason I have organized the ethers here in this section with the alcohols is this: just as alcohols may be thought of as a water molecule with one hydrogen removed and a hydrocarbon attached in its place, an ether may be thought of as a water molecule with both hydrocarbons removed and two hydrocarbons attached, like this:

Ethers were known well before the IUPAC rules for naming took effect, and as a result most ethers go by their common names. The most well-known ether is diethyl ether, which is simply called ether. The molecule is depicted in Figure 13.19.

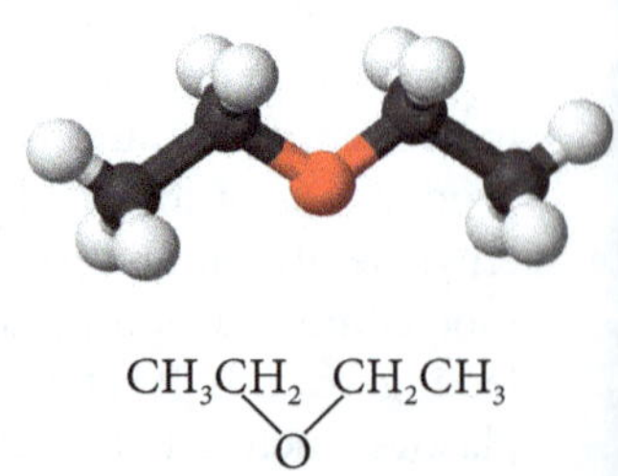

Figure 13.19. Diethyl ether.

Diethyl ether changed the world for those who underwent or performed surgeries. Its safe use as an anesthetic was demonstrated in the 1840s, after which patients could undergo surgery without having to endure the intense pain of surgery without anesthesia. A century ago, ether also had a reputation as an addictive narcotic used by some people for recreational purposes. This sometimes led to disaster because ether is highly flammable. Today, diethyl ether has been replaced as an anesthetic by other substances with fewer side effects. It is used today as a solvent and as a starter fluid for cars that have run out of gas.

13.3.2 Aldehydes and Ketones

The compound classes called *aldehydes* and *ketones* are characterized by the carbonyl functional group. When the carbon atom in the carbonyl group is attached to an alkyl group on one side and a hydrogen atom on the other side, an aldehyde results with this structure:

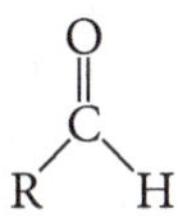

When alkyl groups are on both sides of the carbonyl group, the result is a ketone with the following structure:

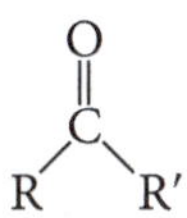

The simplest of these molecules are formaldehyde, acetaldehyde, and acetone. Their structures are shown in Figure 13.20. All three of the compounds are used in the manufacturing of countless other organic chemicals. However, each of them also has some shady episodes in their past. I can't resist passing on some of these details.

For over a century, formaldehyde has been synonymous with preserving laboratory specimens in jars. The liquid in those jars—which you will see in any biology laboratory and about a thousand old movies—is not pure formaldehyde; it is a solution. Formaldehyde is a gas, and the preserving solution used in labs—and for embalming—is a 37% solution by mass of formaldehyde in water. In solution, formaldehyde is unstable (it polymerizes, which we discuss soon), so to stabilize it, 10–15% of methanol may be added to the solution. The solution with the methanol in it is generally called formalin.

Formaldehyde is also used in the manufacturing of foam insulation and resins for construction materials. Its use in these materials became headline news during the 2005 Hurricane Katrina and 2008 Iowa flood disasters. Many disaster victims living in trailer homes supplied by the Federal Emergency Management Agency (FEMA) began developing health problems later traced to the formaldehyde used in the construction materials in the trailers. Since the late 1980s, formaldehyde has been classified as a carcinogen by several

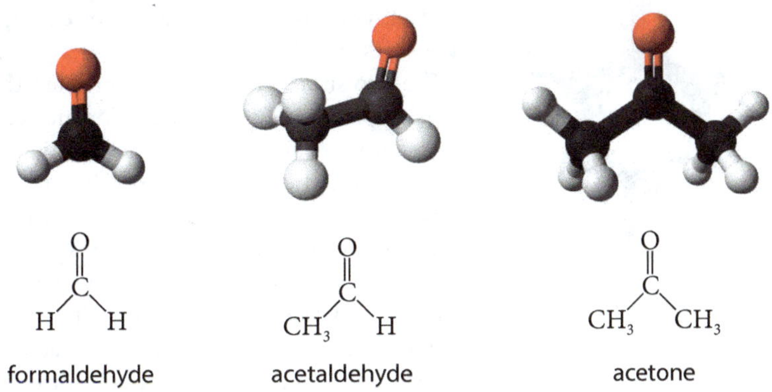

Figure 13.20. The three simplest aldehydes and ketones.

agencies. Notably, formaldehyde was the first polyatomic organic molecule discovered in interstellar space and as a result is now part of the lore of astronomical discovery.

Acetaldehyde is also a major industrial chemical, but is less well known to the public. It occurs naturally in bread, ripe fruit, plants, and coffee. It is used in the production process for polyvinyl acetate, a component widely used in different types of glue. It used to be the primary starting point for the manufacturing of acetic acid (the world's most famous weak acid, which I mention so many times in this text I have lost count). However, that has changed because techniques for producing acetic acid have shifted to a process using methanol, so worldwide demand for acetaldehyde is dropping. One common appearance of acetaldehyde does not, unfortunately, seem to be dropping: it is produced in the liver by the partial oxidation of ethanol, and is the cause of hangovers resulting from alcohol consumption.

Acetone is a very widely used solvent and the most important of all the ketones. The smell of acetone is easily recognizable because for a long time it has been the main ingredient in nail polish remover. An interesting bit of acetone history: it was used in the manufacturing of cordite, a smokeless replacement for gunpowder used by the British for propelling bullets and artillery shells from about 1890 to the end of World War I. As a result, the name cordite has found its way into many a wartime novel and film. They always refer to the recognizable smell of cordite.

13.3.3 Carboxylic Acids and Esters

The *carboxylic acids* and the *esters* both incorporate the carboxyl functional group, or a modification of it sometimes called the ester group. These structures are, respectively, as follows:

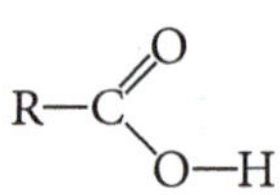

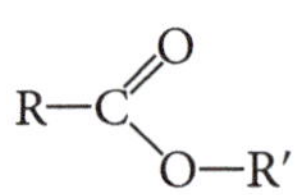

The carboxyl group is simply a hybrid between the carbonyl and hydroxyl groups (hence its name). The geometry of the carboxyl group leads to the formation of strongly polar molecules and thus strong intermolecular forces, including hydrogen bonding and dipole-dipole forces. Accordingly, the boiling points of carboxylic acids are high. The boiling point of formic acid, for example, is a bit higher than that of water—100.5°C.

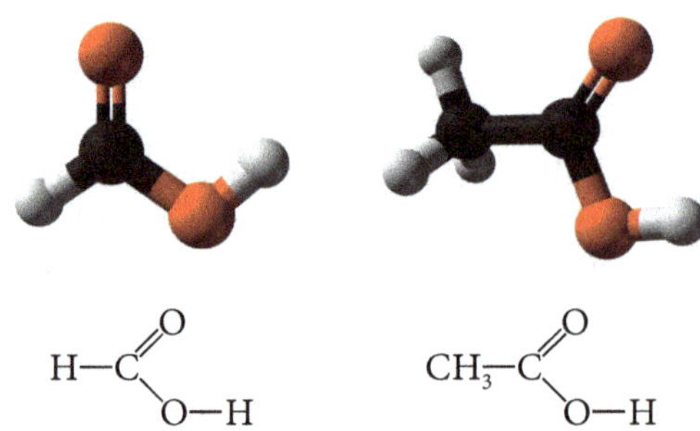

methanoic acid
formic acid

ethanoic acid
acetic acid

Figure 13.21. The two simplest carboxylic acids. IUPAC names are above and common names are below.

Figure 13.22. Benzoic acid.

The two simplest carboxylic acids are those with one and two carbon atoms, depicted in Figure 13.21. Formic acid, HCOOH, you may recall, is in the venom of ants. It is also the chemical that causes irritation from the scratches of nettles. As usual, formic acid is used in the manufacturing of many other chemicals. Industrially, acetic acid, CH_3COOH (here it is yet again), is the most important carboxylic acid. It is, of course, the main component of vinegar other than water. And like formaldehyde, acetic acid has been discovered in interstellar space.

IUPAC names for the carboxylic acids are formed from the alkane with the same number of carbon atoms. The *-ane* is dropped and the suffix *-oic* is added.

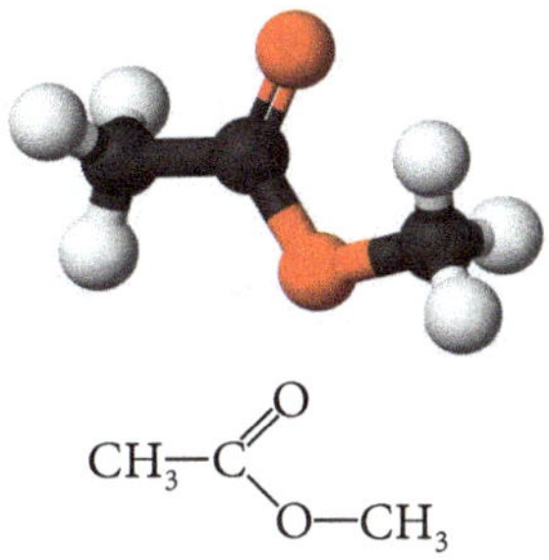

Figure 13.23. Methyl acetate.

Another important carboxylic acid is benzoic acid, shown in Figure 13.22. As with all the compounds we are encountering in this chapter, benzoic acid is used to produce innumerable other chemicals. Of particular note is the use of benzoic acid and its salt sodium benzoate as a food preservative. One often sees these listed as ingredients on food packaging.

Esters are common organic compounds and produce a pleasant, fruity odor. They are responsible for most of the aromas from fruit and flowers. As a result, they are used extensively in the production of flavorings and fragrances. A simple ester is methyl acetate, shown in Figure 13.23. In this molecule, as in all esters, the hydrogen atom from the hydroxyl portion of the carboxyl group has been replaced with a different group, the methyl group in this case.

Condensation Reactions

Esters can be produced by a type of reaction called a *condensation reaction*. Condensation reactions occur when two molecules (or two parts of the same molecule) combine, resulting in the removal of a small molecule such as water in the process. One type of condensation reaction is *esterification*. In this reaction, an acid reacts with an alcohol and the result is an ester. A representative case is this reaction between acetic acid and methanol to produce methyl acetate:

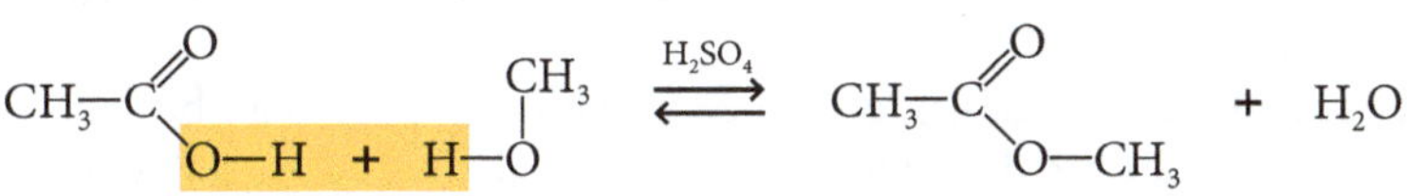

To form the water molecule in this reaction, the carboxylic acid contributes the hydroxyl group and the alcohol contributes the other hydrogen. Again, H_2SO_4 is used as a dehydrating agent.

To form the IUPAC names for esters, the alkyl group from the alcohol contributes the first part of the name, and the anion derived from the acid contributes the second part.

13.3.4 Nitrogen Compounds: Amines and Amides

It should come as no surprise to you that nitrogen is a significant element for living organisms. We have repeatedly encountered the Haber-Bosch process for making ammonia, from which is produced ammonium nitrate fertilizer. The last of the compound classes we review here are two classes involving nitrogen.

The first is the *amines*, which are basically derivatives of ammonia. In the discussion of alcohols above, I propose that one can think of an alcohol as a water molecule with one hydrogen replaced by an alkyl group. Likewise, an ether can be thought of as a water molecule with both hydrogens replaced by alkyl groups. Similarly, amines may be thought of as ammonia molecules with hydrogen atoms replaced by alkyl groups. Since there are three hydrogen atoms in an ammonia molecule, the number of hydrogens replaced with alkyl groups gives rise to three classes of amines—primary, secondary, and tertiary—de-

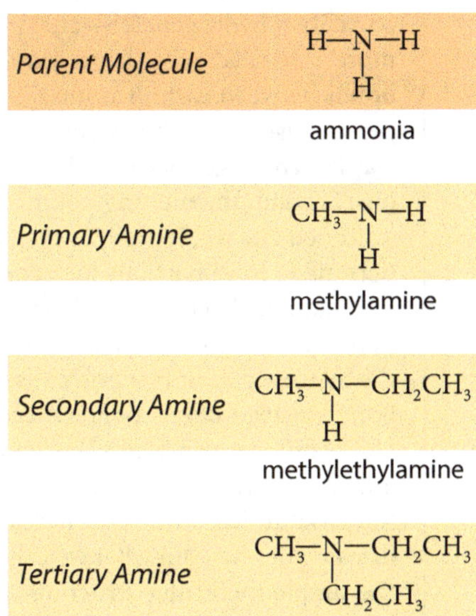

Figure 13.24. Representative amines.

Hmm... Interesting. Trans fats

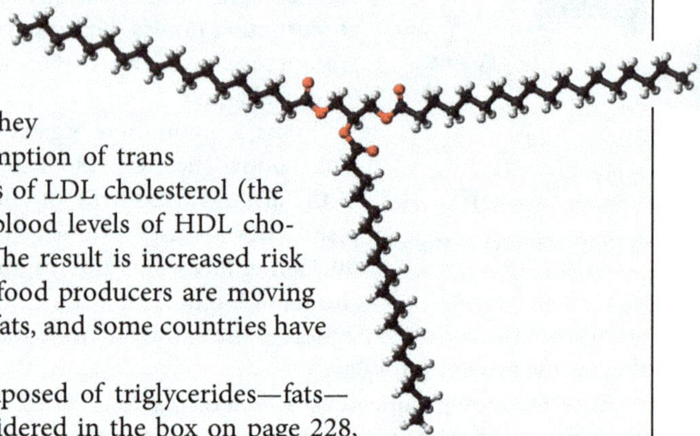

We have now covered all the necessary basics to support an explanation of trans fats in foods—what they are and how they arise. Research shows that consumption of trans fats in foods increases blood levels of LDL cholesterol (the "bad" cholesterol) and decreases blood levels of HDL cholesterol (the "good" cholesterol). The result is increased risk of coronary heart disease. Today, food producers are moving away from foods containing trans fats, and some countries have banned them altogether.

Natural vegetable oils are composed of triglycerides—fats—the same large molecules we considered in the box on page 228, shown again above. Triglycerides are esters composed of three long carboxylic acids called *fatty acids*, each connected to one of the oxygen atoms in a glycerol molecule. Glycerol is an alcohol; fatty acids are carboxylic acids. As described in the text, an alcohol connected to an acid forms an ester. The model below shows how one of the fatty acids connects to the glycerol molecule.

The fatty acids are aliphatic carbon chains, four to 28 carbons long, and can be saturated or unsaturated. The natural fatty acids found in the triglycerides of the vegetable oils used for cooking are unsaturated, and contain from one to six *cis-* double bonds.

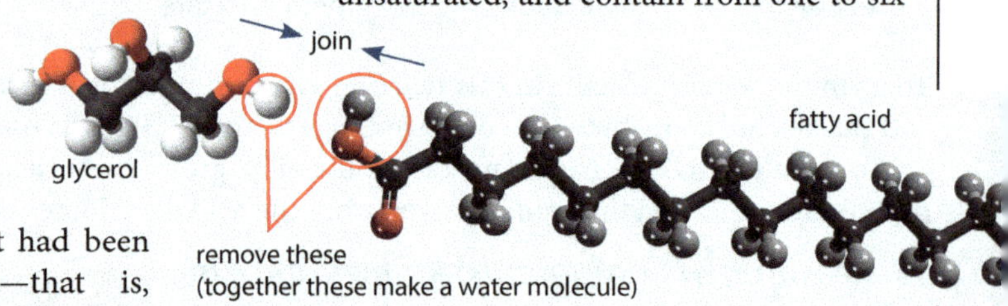

For most of the 20th century, many food processors and fast-food restaurants preferred to use oil that had been partially hydrogenated—that is, more saturated with fewer double bonds between carbon atoms. There are a couple of reasons why hydrogenation of vegetable oil became popular. First, hydrogenated oil stays fresh longer because the fatty acids don't oxidize the way they can in the unsaturated oils. Second, the melting points of saturated fatty acids are higher, so foods prepared with saturated fats feel less oily. Vegetable shortening and margarine are both examples of products made from partially hydrogenated vegetable oil.

That's the background. Now here's the problem. Partial hydrogenation eliminates many of the *cis-* double bonds in the fatty acids, but not all of them. And the heat associated with the hydrogenation reaction causes some of the double bonds that remain to become *trans-* bonds instead of *cis-* bonds.. Because of the geometry of carbon bonding, fatty acids with *cis-* bonds contain kinks, but fatty acids with *trans-* bonds—trans fats—are straight, just like the saturated fatty acids in the molecule shown at the top of the page. Explaining *why* the unsaturated *trans-* fats increase LDL cholesterol and increase the risk of coronary heart disease goes beyond what we can discuss here. But I hope you are intrigued by the fact that the simple molecular structures and reactions we have studied in this chapter allow us to understand the triglycerides in vegetable oil, what trans fats are, and where they come from.

pending on how many hydrogen atoms are replaced. These three classes are illustrated with representative examples in Figure 13.24.

Ammonia is a polar molecule, which is why it dissolves in water. Primary and secondary amines are polar as well and quite soluble. Tertiary amines are essentially nonpolar because the larger a molecule is, the less significant the nonbonding electron pair on the nitrogen atom is.

As with many of the compounds we have encountered, amines such as dimethylamine and trimethylamine are used in the manufacture of an abundance of industrial compounds. These two also both occur naturally. And here's a nice party fact to impress your friends with: trimethylamine, shown in Figure 13.25, has the distinction of being the compound that produces the odors of rotting fish and bad breath.

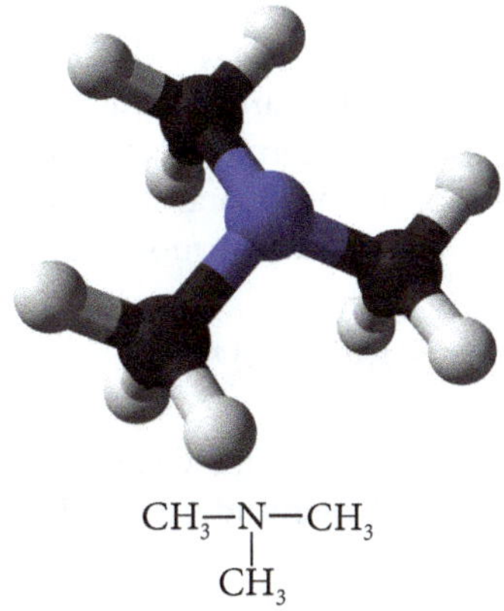

Figure 13.25. Trimethylamine.

Like ammonia, amines have an unshared electron pair on the nitrogen atom. Recall from Chapter 9 that a Lewis base is an electron pair donor. I mention in that chapter that the Lewis acid-base theory is particularly useful in organic chemistry. Here, finally, is an example of that. With the unshared electron pair on its nitrogen atom, an amine acts like a Lewis base. Amines also act like Brønsted-Lowry bases, since they can accept a proton. Acid-base neutralization reactions involving amines are typified by reactions such as that between dimethylamine and hydrochloric acid to produce a salt called dimethylammonium chloride:

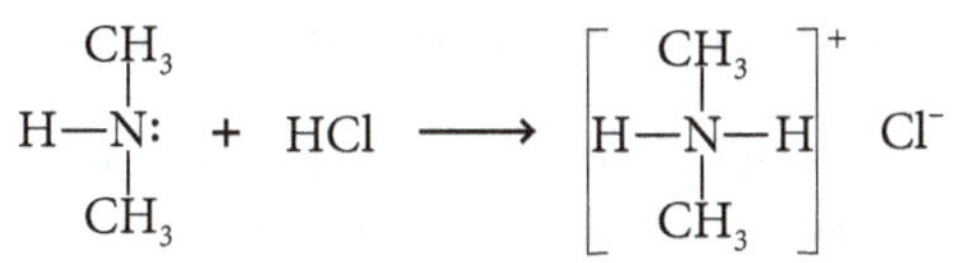

dimethylamine dimethylammonium chloride

Our final class of compounds is the *amides*—nitrogen-containing compounds incorporating the carboxamide functional group:

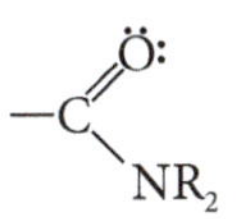

The R can represent various elements, but hydrogen is common and we will limit our discussion to such an example.

Let's get into this by describing how amides are formed in a two-step reaction process from other compounds you already know about. Consider the simple acid-base reaction between acetic acid and ammonia to produce the ammonium acetate salt. Prior to this chapter, we would have written the reaction this way:

$$CH_3COOH(aq) + NH_3(aq) \rightleftharpoons NH_4CH_3COO(aq)$$

This is a classic proton transfer reaction. Using structural formulas, we can write the reaction this way:

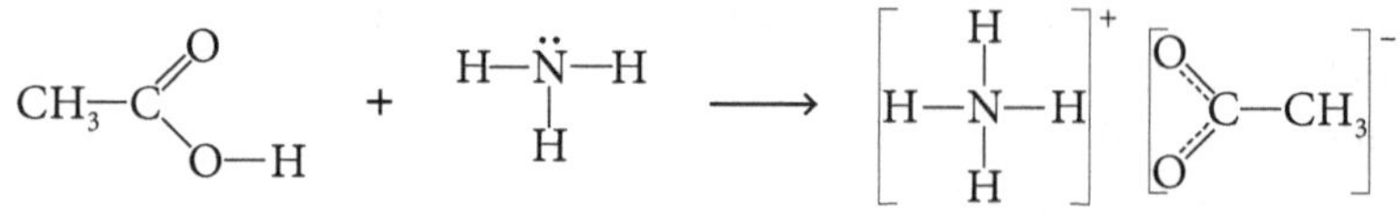

The structural formula reveals a lot about what happens, such as the formation of the resonance structure in the acetate ion.

In the second step, the ammonium acetate salt is heated to produce acetamide, CH_3CONH_2, and water. The molecular equation for this reaction is:

$$NH_4CH_3COO(s) \xrightarrow{\Delta} CH_3CONH_2(s) + H_2O(g)$$

The compound CH_3CONH_2 is an amide called acetamide. Written with structural formulas, we can better understand the molecules involved:

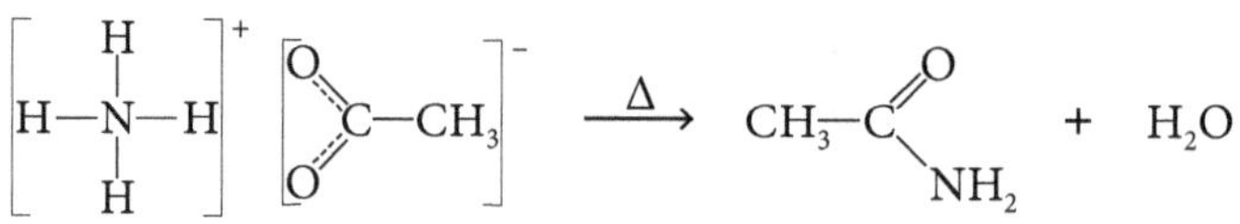

ammonium acetate acetamide

Now you can see how the structure of the acetamide molecule incorporates the carboxamide functional group. In this two-step example, we start with acetic acid, react it with a base to get a salt, and then heat the salt to get the corresponding amide. Other acids can be used to produce amides in the same way, and the name of the amide is formed by replacing the *-ic* suffix on the acid with *-amide*.

As with compounds formed from the carboxyl group, amides such as CH_3CONH_2 are strongly polar, which means, as usual, that amides have high boiling points and are soluble in water. These generalizations apply best to the smaller amides. Amides can get quite large—many drugs are amides—and with larger molecules, the significance of the polarity near the oxygen atom in the carboxamide functional groups decreases.

Finally, scientists have detected acetamide near the center of the Milky Way galaxy—the third organic compound we have come across that has been found in interstellar space.

13.4 Polymers

13.4.1 Types of Polymers

Polymers are fundamental to life. Cellulose is the polymer that provides structural strength in the cellular walls of green plants. Thus, cellulose is what forms the strands in cotton, celery, paper, and cardboard. Proteins, the huge macromolecules at the heart of cellular processes, are polymers. The plastics used to make plastic bottles, plastic bags, and polystyrene cups are polymers. And the synthetic materials used in synthetic fibers—nylon, polyester, acrylic, spandex, Kevlar, and others—are polymers. Two of these well-known polymers are depicted in Figure 13.26.

Polymers are long chains of shorter building-block molecules called *monomers* that chemically bind together in organic reactions. Many polymers, including those we consider in this section, are composed of identical monomers linked together one after the other. Polymers composed of two or more different monomers linked together in repeating units are call *copoly-*

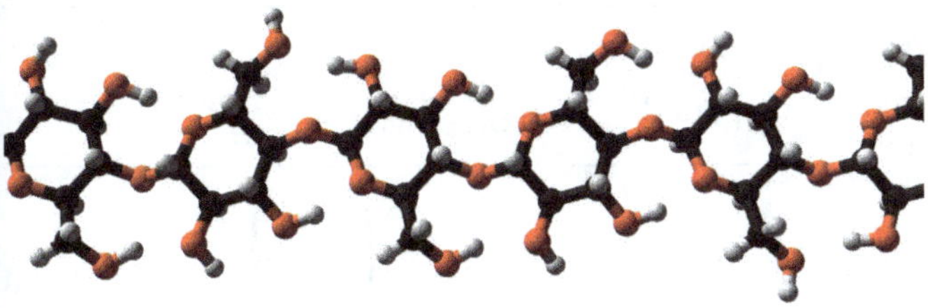

Figure 13.26. Two polymers, cellulose (left) and nylon (below).

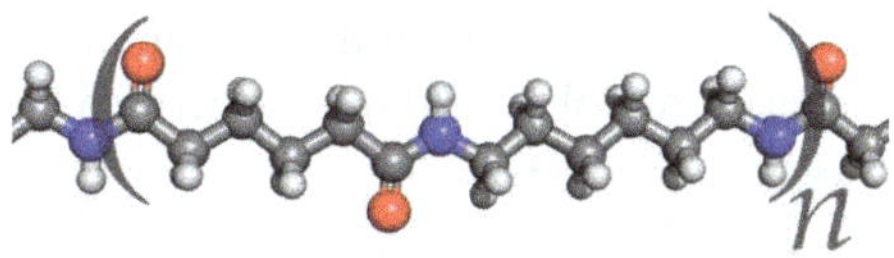

mers. A common example of a copolymer is ABS plastic, the material Lego bricks are made of. ABS is composed of three different monomers.

13.4.2 Polymerization

Polymerization is the term used for any reaction process that converts monomers into polymers. Polymers can be formed by two basic types of processes, known as *step-growth polymerization* and *chain-growth polymerization*.

Step-growth polymerization can be accomplished by either of two reactions, both of which we have discussed in this chapter—addition and condensation. Either way, polymers are built up from monomers one monomer at a time. In contrast, chain-growth polymerization occurs when monomers form short chains and the chains then combine together to form the polymer.

We first encountered addition reactions with the hydrogenation process in Section 13.2.5. An interesting example of an addition reaction in step-growth polymerization is the process used by the rubber plant to convert the monomer isoprene into the natural rubber polymer, polyisoprene. This reaction may be represented as follows:

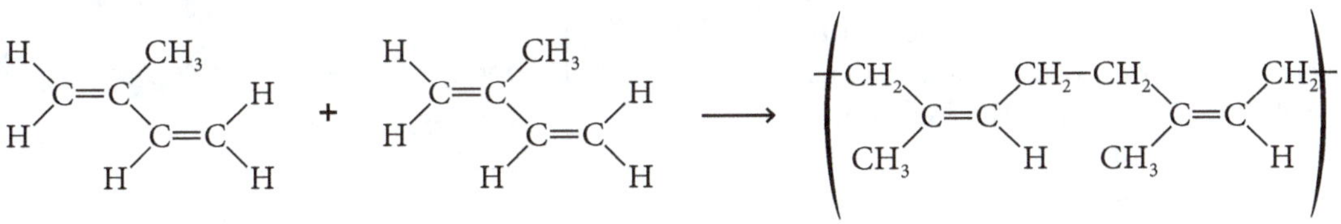

isoprene isoprene polyisoprene

As before, the large brackets around the polymer are there to indicate that the section of the molecule shown inside the brackets repeats many times.

We looked at condensation reactions in Section 13.3.3, where the reaction between acetic acid and methanol to produce methyl acetate was our example. In that reaction, a water molecule separates from the reactants. Nylon and polyester are both formed by condensation polymerization. We see an example of a condensation reaction being used for polymerization in the very short Section 13.5.

Chain-growth polymerization occurs with unsaturated monomers. The double and triple bonds in the monomers are able to convert to single bonds, enabling the carbon atoms in one monomer to bond to those on either side of it. One of the ways this can happen is by a free radical mechanism, similar to the free radical halogenation reaction we looked at in Section 13.2.1.

Examples of polymers formed by chain-growth polymerization include polyethylene, PVC, and polypropylene.

13.4.3 A Short Polymer Showcase

On the next two pages are images, models, and details of four common polymers. Notice that in each case, the double bond in the starting compound is replaced with a single bond in the monomer, enabling the monomers to chain together to form the polymer. Notice also the strong similarities between the four monomers.

Polyvinyl Chloride

You have probably heard of this material all your life—it's usually called PVC and it is a very popular piping material for plumbing and electrical wiring.

Figure 13.27. PVC tubing for use in plumbing.

Starting Compound	chloroethylene 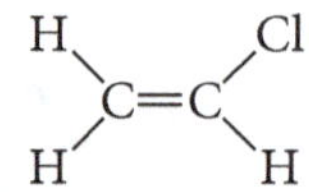
Monomer	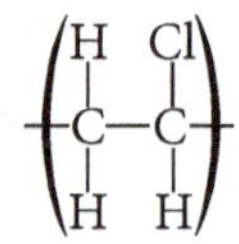
Model	

Figure 13.27 shows PVC pipe used for plumbing. PVC has found many other uses beside piping. PVC is even a popular material now for strange (shiny) sorts of clothing. (To each his own.)

Polystyrene

Polystyrene is one of the most widely used plastics in the world. The general-purpose form is used for the clear plastic covers of CD jewel cases and breakable plastic cups. The high-impact form is used for plastic cutlery, yogurt cups, and disposable razors. Dow Chemical invented the process to produce foam polystyrene, Styrofoam being the trademark name for that material. Polystyrene foam is used extensively for packing material, coffee cups, and flotation devices. A few of the thousands of polystyrene products are shown in Figure 13.28.

Starting Compound	phenylethene, aka styrene 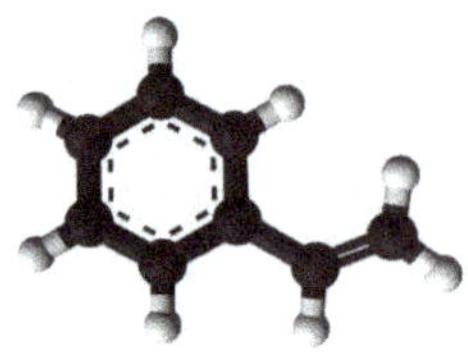
Monomer	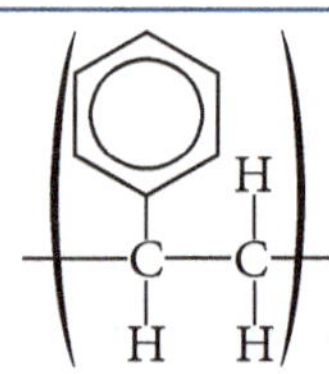
Model	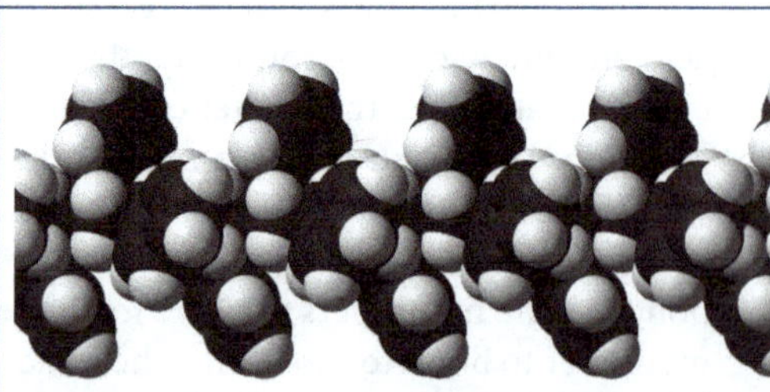

Figure 13.28. Polystyrene foam (left), and common polystyrene products (center and right).

Polyethylene Ethylene—the simplest alkene—is the starting compound for making polyethylene, the most heavily produced industrial chemical in the world. These chains are very long—typically around 100,000 monomers or 200,000 carbon atoms.

Starting Compound	ethylene 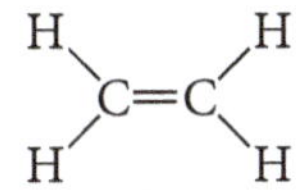
Monomer	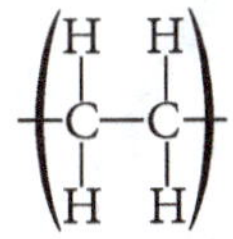
Model	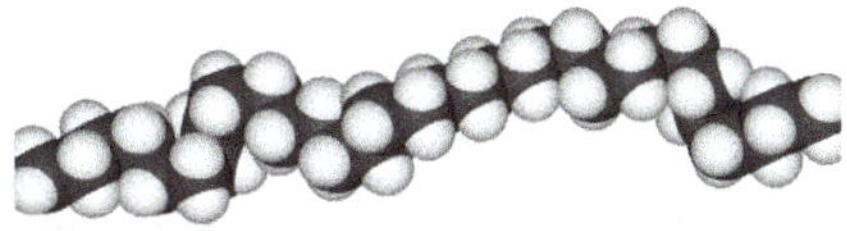

There are several forms of polyethylene, such as low-density polyethylene (LDPE), high-density polyethylene (HDPE), cross-linked (XLPE), and others. LDPE consists of strands like the one shown in the model. Plastic bags and sheeting are made of LDPE. HDPE polymers have branches extending from the sides all along the polymer. Intermolecular forces between all the branches make this material much stiffer, so it is used for all manner of plastic bottles and plastic toys. In XLPE, the polymer strands have connecting links between them throughout the material. These connecting links make the material very rigid and suitable for industrial piping, as illustrated in Figure 13.29.

Figure 13.29. A section of XLPE pipe.

PTFE PTFE stands for polytetrafluoroethylene, a polymer of tetrafluoroethylene. But most people know this polymer by its most popular brand name, Teflon, a brand owned by DuPont. You already know that PTFE (Teflon, that is) is a popular for coating for frying pans, waffle irons, and corn muffin pans. As shown in Figure 13.30, PTFE is also popular for plumbing tape (used for wrapping pipe threads to prevent leaks) and stopcocks in laboratory burets.

Starting Compound	tetrafluorloroethylene 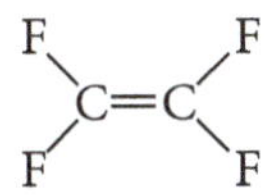
Monomer	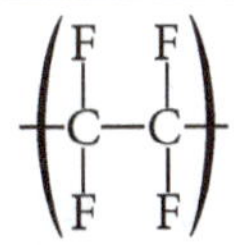
Model	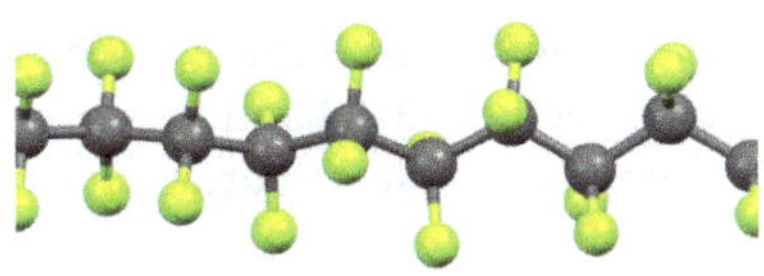

Valves used in industrial applications frequently have PTFE parts inside. Sometimes, valves used for especially corrosive fluids are completely lined with PTFE to protect the metal valve body from corrosion.

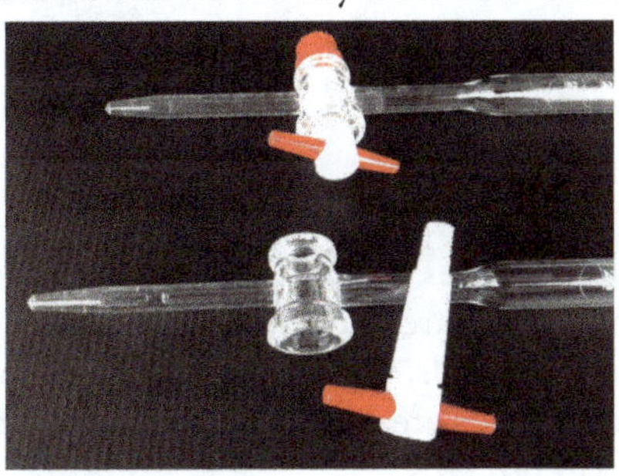

Figure 13.30. PTFE products.

Hmm... Interesting. Recycling codes

Everyone knows about the recycling codes on the bottom of plastic bottles, but now you can understand what some of them mean.

The codes go all the way from 1 to 99, but the plastics are represented by numbers 1–7 and 9. Codes 2, 3, and 4 are for materials discussed in this chapter. The codes for plastics are in the table below.

Code	Material	Uses
1	polyethylene terephthalate (PET or PETE), a type of polyester	soft drink bottles, water bottles, polyester fibers
2	high-density polyethylene (HDPE)	plastic bottles, toys, trash cans
3	polyvinyl chloride (PVC)	piping, conduit, window frames, bottles for chemicals
4	low-density polyethylene (LDPE)	plastic bags, plastic tubing, soap bottles
5	polypropylene	car bumpers, interior trim parts on cars
6	polystyrene	CD cases, food containers, foam packing, flower pots
7	all other plastics	
9	acrylonitrile butadiene styrene (ABS)	Lego bricks, TV and computer cabinets, coffee makers

13.5 A Wee Bit of Biochemistry

We do not have space to dive into the deep waters of biochemistry. But the carboxyl group from Section 13.3.3 and the amine group from Section 13.3.4 are all we need to illustrate a few things about one important biochemical topic—amino acids and proteins. We examine these briefly here.

If we connect an amine group and a carboxyl group to the same carbon atom, we get the following structure:

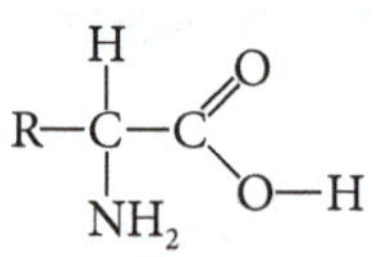

This is the general form of an *amino acid*. As usual, the R is a place holder and represents a multitude of different possible structures. Amino acids are the building blocks peptides and proteins are made of. Over 500 amino acids are known, 22 of which are the "standard" amino acids found in the proteins of plants and animals. Of the standard amino acids, 21 are found in the human body. In the body, the amino acids are either oxidized to provide the body with energy, or they are chained together to form proteins.

The construction of proteins from amino acids occurs by step-growth polymerization, and is another example of a condensation reaction, in which molecules join together and a water molecule is removed. The linking process to form proteins may be represented this way:

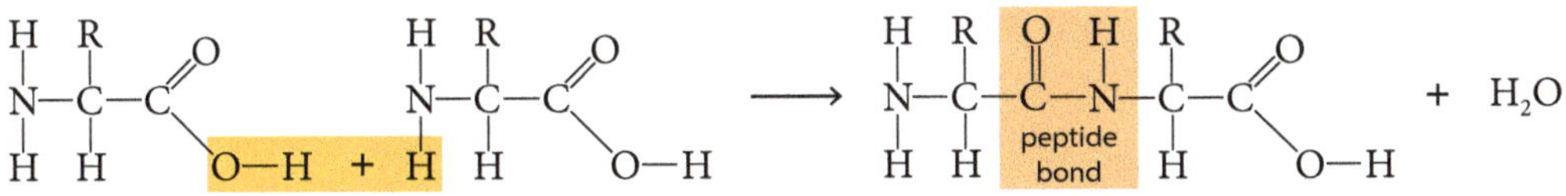

The hydroxyl group from one amino acid and one of the amine hydrogens in the second amino acid are removed to form a water molecule. The carbon in the carboxyl functional group where the hydroxyl group was formerly attached bonds to the nitrogen in the amine group that lost the hydrogen atom. The resulting bond is called a *peptide bond*. The four atoms shown in the colored region of the peptide bond are considered to be part of the peptide link. The partial amino acids left on either side of the peptide bond are called amino acid *residues*.

The compound formed from two amino acids is called a *dipeptide*. Peptides are short chains of amino acid residues linked by peptide bonds. Made from only two amino acids, the dipeptide shown above is the shortest possible peptide. Polypeptides are longer, unbranched chains of amino acids up to approximately 50 amino acids long. Proteins are even longer chains, consisting of one or many more peptides connected together, and possessing specific biological (and wonder-inducing) functionality.

Chapter 13 Exercises

SECTION 13.1

1. Distinguish between graphene and diamond and explain why one allotrope makes a good lubricant while the other makes a good edge for a cutting tool.
2. Is every compound containing carbon a part of organic chemistry? Explain.
3. Describe the property held in common by graphene and nanotubes.
4. What are fullerenes?

SECTION 13.2

5. Explain what isomers are.
6. For each of the pairs of molecules below, identify whether the two molecules shown are isomers. If they are, name each of them. If not, name the compound and explain why the two structures shown represent the same compound.

a.

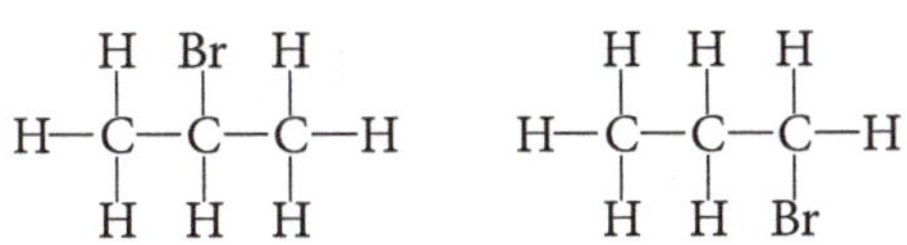

b.

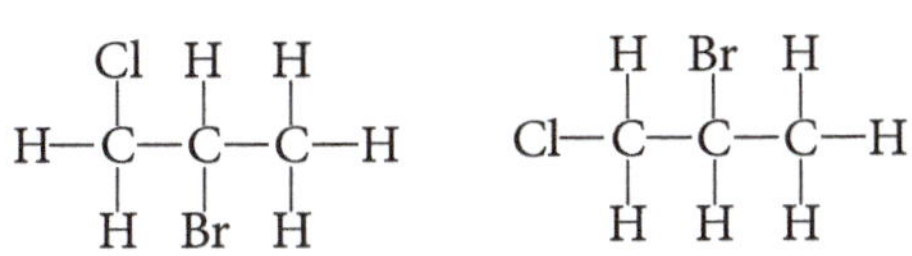

7. Why are alkenes regarded as unsaturated hydrocarbons?
8. Why are aromatic compounds regarded as unsaturated compounds?

9. Write the structural formulas and names for the first three alkenes, assuming a single double bond between the first two carbon atoms.

10. Write the IUPAC names for each of the following compounds.

a.
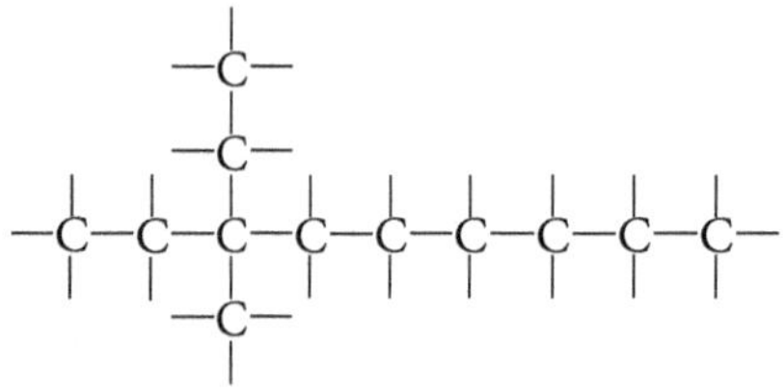

b.
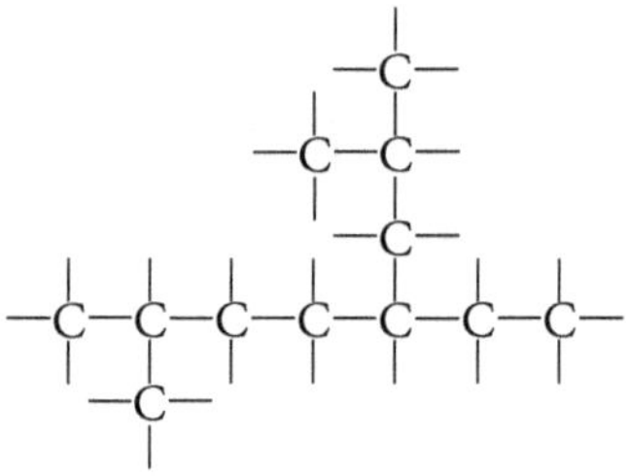

c.
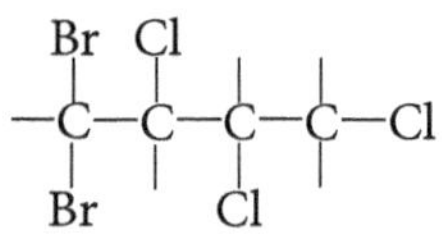

d.
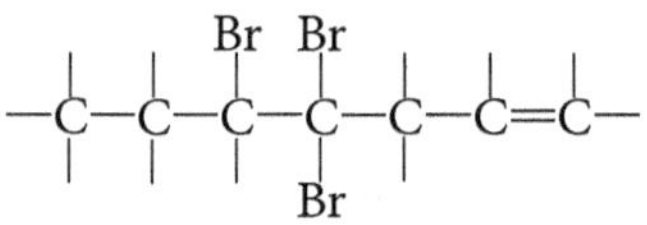

e.
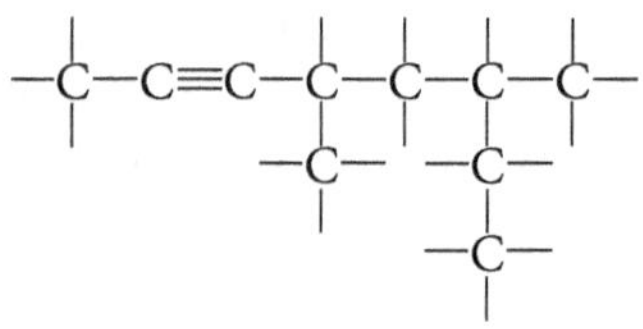

f.
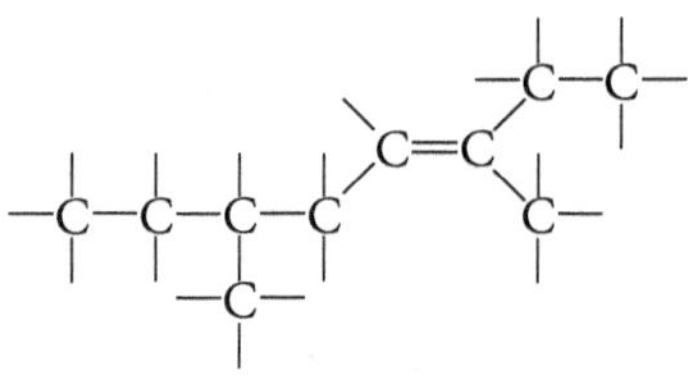

11. Write structural formulas for the following compounds.

a. 3,3-dimethylheptane
b. 4-ethyl-*trans*-2-hexene
c. 2,2,4,4-tetramethylpentane
d. 2,3-dimethyl-*cis*-3-ethylhexane
e. 5,6-dimethyl-3-octyne
f. 3-chloro-*trans*-2-butene

12. Write the IUPAC names for each of the following compounds.

a.
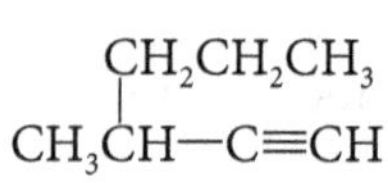

b.
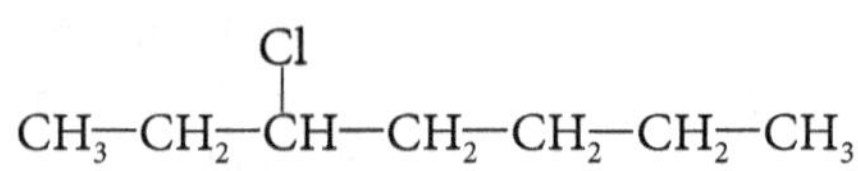

c.
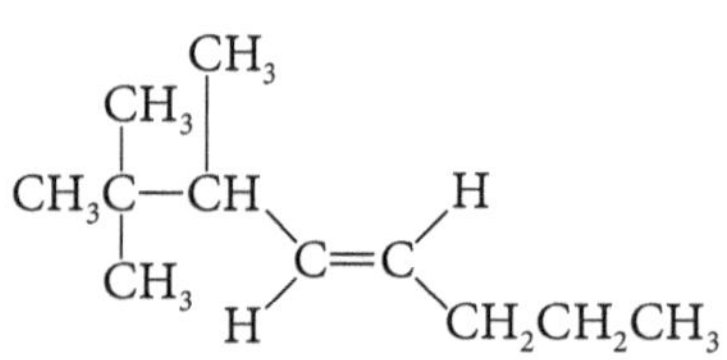

d.
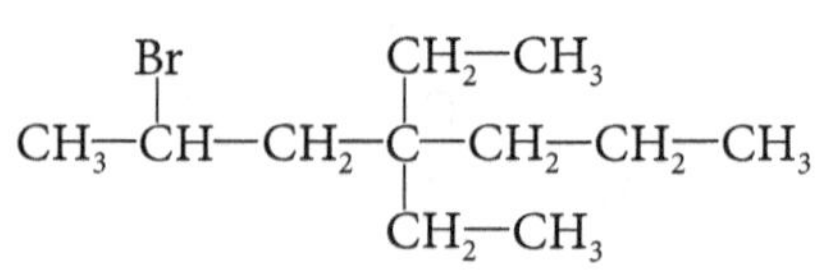

e.
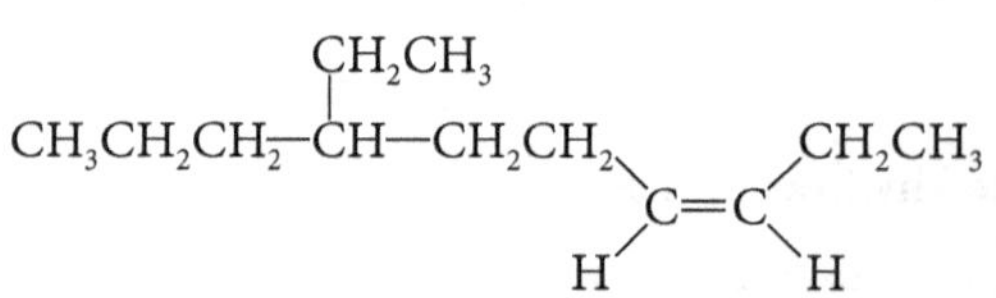

13. Write a structural equation describing the hydrogenation of propyne.
14. Explain the difference between halogenation and hydrogenation.
15. Write the IUPAC names for each of the following compounds.

a. 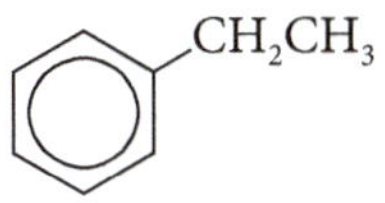

b. 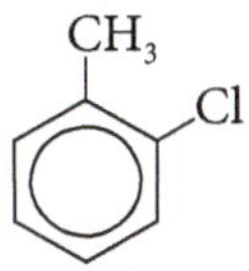

c. 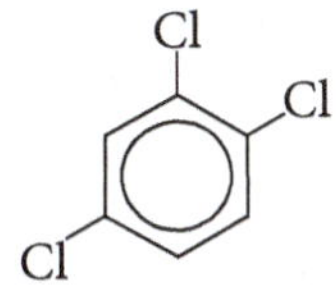

d. 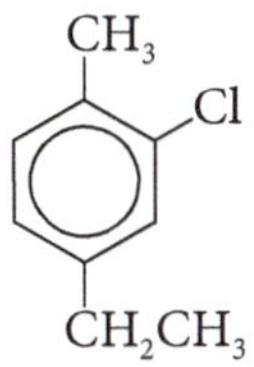

e. 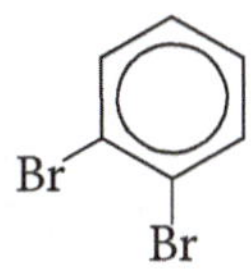

f. 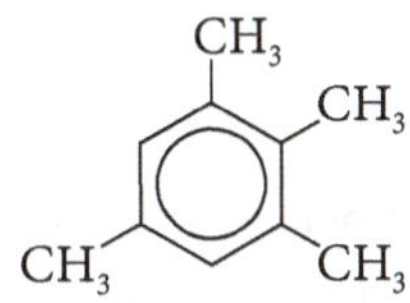

16. Write condensed structural formulas for the following compounds.

a. 1,2-dichlorobenzene
b. 1-methyl-3-ethylbenzene
c. 1,3,5-tribromobenzene
d. 4-methyltoluene

17. What structural feature must be present if a hydrocarbon can undergo a substitution reaction but not an addition reaction? What structural feature must be present for a hydrocarbon to undergo an addition reaction?
18. How are cycloalkanes different from aromatic hydrocarbons?
19. What is nitration?

SECTION 13.3

20. Alcohols are water soluble and have relatively high boiling points compared to alkanes of similar size, which are not water soluble and have lower boiling points. Account for these differences.
21. Distinguish between the roles of alkenes in addition and elimination reactions.
22. Describe the structural relationship between water, alcohol, and ethers.
23. Explain why acetaldehyde is an aldehyde rather than a ketone.
24. Compare and contrast acetone and formaldehyde.
25. Describe the structural relationship between ammonia and the amines.
26. Explain why amines act as bases in many chemical reactions.
27. Draw the structural formulas for the carbonyl, carboxyl, and carboxamide groups, and compare and contrast these three functional groups.
28. Explain how the structure of the carboxylic acids accounts for their high boiling points.
29. Use esterification as an example to explain condensation reactions and describe the structural features of the molecules involved that allow this type of reaction to occur.
30. Distinguish between ethers and esters, and provide one well-known fact about each.
31. Compare and contrast elimination and condensation reactions.

32. For each of the following molecules, identify the class of the compound and determine its name.

a. 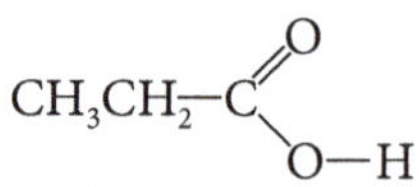

b. 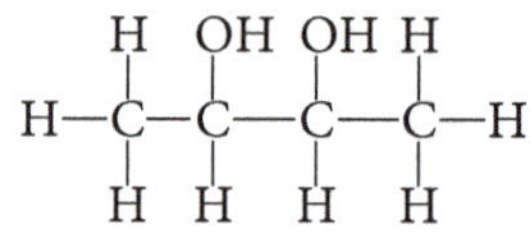

c. 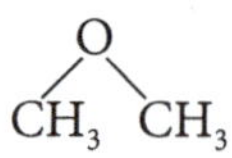

d. 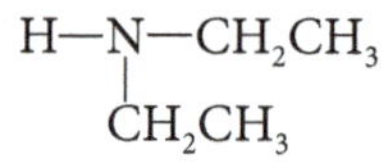

SECTION 13.4

33. Distinguish between monomers, polymers, and copolymers.

34. What is polymerization?

35. What is the difference between ethylene and the polyethylene monomer?

36. Describe step-growth and chain-growth polymerization and give examples of each.

37. Briefly describe the following four polymers: PVC, polystyrene, polyethylene, and PTFE. For each one, give the starting compound, the monomer, and one or two examples of how the polymer has been put to use.

SECTION 13.5

38. What is an amino acid? Answer this question in terms of the functional groups involved.

39. Write a structural equation for the formation of a dipeptide and explain why this reaction is a condensation reaction.

40. Distinguish between dipeptides, peptides, and proteins.

GENERAL REVIEW EXERCISES

41. Assume that the average composition of gasoline is octane and that air contains 20.95% O_2 by volume. Determine the following:

a. The molecular equation for the complete combustion of gasoline, assuming the reaction products are carbon dioxide and water.

b. The volume of air at STP required for complete combustion of 1.00 L of gasoline. The density of C_8H_{18} is 703 g/L.

c. The mass of water produced by the complete combustion of 1.00 L of gasoline.

42. For the gasoline combustion in the previous item, what volume of air is required at a temperature of 546°C?

43. Consider an aqueous acetic acid solution prepared with 625 g of acetic acid and enough water to bring the total volume to 1.00 L.

a. Determine the molarity of the solution.

b. Assuming that 1.3% of the mass of acetic acid ionizes, determine the following for this solution: $[H_3O^+]$, pH, $[OH^-]$, and pOH.

c. Predict the volume of 1.00 *M* NaOH required to neutralize the solution. (Only the ionized acid needs to be neutralized.)

44. Consider the combustion of liquid benzene to produce carbon dioxide and water.

a. Write the balanced chemical equation.

b. Determine the enthalpy of reaction for this reaction. Recall that enthalpy of formation data are in Appendix B, Table B.7.

45. In some states, ethanol is blended with gasoline up to 15% by volume to make a mixture called E15. Compute the change in the Gibbs free energy for the combustion of pure ethanol. See Appendix B for the necessary data.

46. A certain solution of 0.035 *M* formic acid has a pH of 2.60. Determine the acid dissociation constant for formic acid and compare your result to the entry in Table 11.1.

47. A chemist combines 645 mL of a 0.024 *M* solution of sodium hydroxide with 300.0 mL of a 0.111 M solution of calcium chloride. Determine if a precipitate will form. Solubility product data are found in Appendix B, Table B.10.

48. Balance the following redox reactions. Assume each reaction takes place in acidic conditions.

a. $CuS(s) + HNO_3(aq) \rightarrow Cu(NO_3)_2(aq) + NO(g) + S(s) + H_2O(l)$

b. $Fe(s) + SnCl_4(aq) \rightarrow FeCl_3(aq) + SnCl_2(aq)$

49. Balance the following redox reaction equations. Assume each reaction takes place in basic conditions. Note that the potassium compounds ionize into K^+ ions and tin-oxygen anions.

a. $Bi(OH)_3(s) + K_2SnO_2(aq) \rightarrow Bi(s) + K_2SnO_3(aq)$

b. $CO_2(g) + NH_2OH(aq) \rightarrow N_2(g) + CO(g) + H_2O(l)$

Glossary

A

absolute zero	The theoretical lowest temperature that can exist. At absolute zero all atomic and molecular motion ceases and the entropy of crystals is zero.
acid-base neutralization reaction	A chemical reaction in which an acid and base react to produce a salt, water, and sometimes carbon dioxide. In the process, the low-pH solution of the acid and the high-pH solution of the base are neutralized to form a solution with a neutral pH of approximately 7.
acid dissociation constant	The equilibrium constant for an acid in aqueous solution.
acid rain	Rain with a low pH due to the presence of sulfuric acid. The acid forms in the atmosphere from sulfur dioxide and moisture.
activated complex	A range of transitional molecular structures of partially formed bonds that results from an effective molecular collision with adequate energy. Once the activated complex is formed, a chemical reaction can commence.
activation energy	The amount of energy required to initiate a chemical reaction; the minimum energy necessary to transform the reactant molecules into the activated complex for the reaction.
activity series of metals	A list of metals in order of their chemical activity. More active metals replace less active metals in single replacement reactions.
actual yield	The amount of a given product in a chemical reaction that is produced in an actual experiment or chemical process.
addition reaction	A reaction in which atoms are added to a molecule. Hydrogenation is a typical example.
adsorption	The collection of molecules onto a solid surface, attracted there by electrostatic attraction or Van der Waals forces.
alcohol	Hydrocarbon derivatives in which one or more hydrogen atoms are replaced by a hydroxyl group.
aldehyde	A compound including the carbonyl functional group, the group having an alkyl group attached to one side and a hydrogen atom attached to the other side.
aliphatic hydrocarbon	Any alkane, alkene, or alkyne.
alkali metal	Any element in Group 1 of the Periodic Table of the Elements.
alkaline-earth metal	Any element in Group 2 of the Periodic Table of the Elements.
alkane	A hydrocarbon molecule containing only single bonds. Alkanes are saturated hydrocarbons.
alkene	A hydrocarbon molecule containing one or more double bonds between carbon atoms.
alkylation	A chemical reaction in which an alkyl group is transferred from one molecule to another.
alkyl group	A side branch, derivable from an alkane, attached to a hydrocarbon molecule.
alkyl halide	A haloalkane.
alkyne	A hydrocarbon molecule containing one or more triple bonds between carbon atoms.
allotrope	One of the atomic arrangements exhibited by a pure element whose atoms can be arranged in more than one way.
alloy	A homogeneous mixture of metals formed by mixing the molten metals together and allowing the solution to cool and solidify.

amine	An ammonia derivative in which one or more of the hydrogen atoms in the ammonia molecule is replaced with an alkyl group.
amino acid	The general term for the molecule produced when an amine group and a carboxyl group are joined to the same carbon atom.
amino acid residue	The parts of amino acids that remain linked together in a peptide after hydrogen and oxygen atoms are removed from the carboxyl group and a hydrogen atom is removed from the amine group to form the peptide bond.
amorphous carbon	An allotrope of carbon in which the carbon atoms have no regular geometry or shape.
amphiprotic species	Any species that can act either as a proton acceptor or a proton donor, depending on the conditions.
amphoteric species	Any species that can act either as an acid or as a base, depending on the conditions.
angular momentum quantum number	See azimuthal quantum number.
anion	A negatively charged ion.
anion hydrolysis	When the anion of a salt reacts with water to produce hydroxide ions and a basic solution.
anode	The electrode in an electrochemical cell where the oxidation half-reaction takes place.
aromatic hydrocarbon	A hydrocarbon containing one or more benzene rings.
atomic mass	The average mass of the atoms of a given element in unified atomic mass units, u.
atomic mass unit	An obsoleted unit used for relative atomic masses; replaced by the unified atomic mass unit, u.
atomic number	A number, denoted by Z, that designates the number of protons contained in the nucleus of atoms of a given element. The elements are arranged in the periodic table in order by their atomic numbers from 1 to 118.
atomic spectrum	The set of wavelengths of electromagnetic radiation emitted by the excited atoms of a particular element.
Aufbau principle	The principle that in the ground state electrons occupy orbitals in atoms starting with the lowest energy orbital and going sequentially up from there to fill orbitals associated with higher energy as required.
azimuthal quantum number	A number denoted by $l = 0, 1, 2, 3$ or 4 and used to designate specific subshells within the shells of an atom. The values of l are usually denoted by the letters *s*, *p*, *d*, *f*, and *g*.

B

barometer	A device that measures atmospheric pressure.
barometric pressure	Atmospheric pressure as measured by a barometer.
base dissociation constant	The equilibrium constant for a base in aqueous solution.
binary acid	An acid formed from hydrogen and one of the more electronegative nonmetals.
binary formula	A chemical formula for a compound consisting of two parts, such as the symbols for cation and anion.
birefringence	The property, exhibited by some crystals, of refracting light in two directions as the light passes through the crystal.
body-centered cubic lattice	A close-packed metallic lattice with layers in the sequence A–B–A–B, all A layers being aligned together, all B layers aligned together, and the atomic centers of A layers aligned with the interstitial centers of the B layers, and vice versa.
boiling	The phase transition that occurs when a liquid is at the boiling point temperature and additional heat is added, causing the liquid to transition to the vapor state; vaporization.

boiling point elevation	The increase in boiling point that occurs in a solution relative to the boiling point of the pure solvent; a colligative property.
bomb calorimeter	A calorimeter that uses a sealed canister in which a sample is placed in a pure oxygen environment. The canister is placed in a water chamber, and the contents of the canister are ignited with a spark. The change in water temperature is used to determine the enthalpy of combustion for the reaction.
bond energy	The energy holding two atoms together in a covalent bond; the energy that must be supplied to separate two atoms held together by a covalent bond.
bonding domain	An electron domain between two atoms covalently bonded together.
bond length	The distance between nuclei in the chemical bond between two atoms.
bond number	A measure of covalent bond strength between two atoms. A single, double, or triple bond between exactly two atoms has a bond number of 1, 2, or 3, respectively. If a bond is also included as part of a resonance structure, then the bond number is the sum of the single dedicated bond (1) and the bond's fractional share of the resonance structure bond.
Brownian motion	The jittery, random motion exhibited by particles such as pollen grains when the particles are suspended in water and observed through a microscope. The phenomenon was first observed by Robert Brown. In 1905, Albert Einstein demonstrated that the phenomenon is caused by collisions of water molecules with the suspended particles.
buckminsterfullerene	An allotrope of carbon (C_{60}) in which the atoms are arranged in a sphere in the pattern of the corners on the surface of a soccer ball.
buckyball	Buckminsterfullerene.
buffering	In acidic solution, adding a salt with the anion in common with the acid to the solution; in basic solution, adding a salt with the cation in common with the base to the solution. In either case, the effect is to make the solution resistant to changes in pH from the addition of a small amount of acid or base.
buret	A tall, thin article of glassware, fitted with a stopcock at the bottom, and marked along the side with a volume scale. Burets are used for dispensing precise amounts of liquid.

C

calorimeter	A device used for making temperature measurements on a sample, while keeping the contents of the calorimeter thermally isolated. Calorimeters are used for determining molar heat capacity, specific heat capacity, enthalpy of combustion, and other thermal parameters.
capillary action	The tendency of a liquid to be drawn upward against gravity into a thin tube or solid porous substance. Capillary action is a result of the forces involved in surface tension, and the polar character of the particles in the solid surface.
carbon nanotube	An allotrope of carbon in which sections of graphene are rolled into tubes.
carboxyl group	A functional group composed of a central carbon atom that bonds to a molecule, a hydroxyl group bonded to the carbon atom, and an oxygen atom bonded to the carbon atom with a double bond.
carboxylic acid	A compound incorporating the carboxyl functional group, the group having a hydrogen atom or alkyl group attached.
catalysis	The process of using a catalyst to expedite the rate of a reaction.
catalyst	A substance that expedites or facilitates a chemical reaction without itself being consumed in the reaction.
cathode	The electrode in an electrochemical cell where the reduction half-reaction takes place.
cathodic protection	The technique of protecting ship hulls, underground pipes and the like from corrosion due to redox reactions by attaching a sacrificial anode to metal parts.

cation	A positively charged ion.
cation hydrolysis	When the cation of a salt reacts with water to produce hydronium ions and an acidic solution.
chalcogen	Any element in Group 16 of the Periodic Table of the Elements. (The synthetic element 116 is not generally included with the chalcogens at this time, but is expected to exhibit chalcogen properties as well.)
charge	A fundamental property of subatomic particles. Each proton carries one unit of positive charge, each electron carries one unit of negative charge, and each neutron carries no charge. An electrical current is a moving stream of charged particles.
chemical activity	A reference to how readily or aggressively a substance engages in various chemical processes.
chemical change	A change involving a chemical reaction, in which new chemical species are formed, with chemical identities and chemical properties different from those of the original substances.
chemical equation	An equation written with the chemical formulas of the compounds involved in a chemical reaction that shows the reactants and the products, their ratios, and (optionally) the forms they are in.
chemical equilibrium	A state in which chemical species are transforming into each other at equal rates in an equilibrium system.
chemical formula	An expression of the elements in a compound and their ratios in which the elements are denoted by their chemical symbols.
chemical property	A property of pure substances related to a substance's tendency to engage in specific chemical reactions or processes; a property related to a substance's tendency to form or break chemical bonds under specific conditions. Examples of quantifiable chemical properties are number of valence electrons, oxidation state(s), electronegativity, ionization energy, and electron affinity. Examples of qualitative chemical properties include flammability, susceptibility to corrosion, and the types of bonds the element forms (ionic or covalent).
chemistry	The study of the elements, how they combine to form mixtures and compounds, the properties of these substances, and the processes involved.
colligative properties	Properties of solutions that depend only on the concentration of solute and not on the identity of the solute.
collision theory	The theory that chemical reactions occur when particles of reactants collide with adequate energy and correct orientation. Factors that increase the frequency of particle collisions increase the reaction rate.
colloid	A colloidal dispersion.
colloidal dispersion	A heterogeneous mixture formed when microscopic particles, ranging in size from 1 to 1,000 nm, are dispersed throughout a dispersing medium.
combustion reaction	A chemical reaction in which a substance combines with oxygen, releasing energy in the form of heat and light; burning.
common-ion effect	Reduced precipitation or reduced ionization that occurs when a soluble solute is added to a solution of a different solute, and the two solutes have an ion in common.
compound	A substance, either molecular or crystalline, in which the atoms of two or more different elements are chemically bonded together.
condensation	The phase transition that occurs when a vapor is at the boiling point temperature and heat is removed, causing the vapor to transition to the liquid state.
condensation reaction	A reaction in which two molecules combine, usually resulting in the removal of a small molecule such as water in the process.

condensed electron configuration	A shorter form of the electron configuration for an atom of a given element in which the noble gas of the preceding period in the periodic table is used to represent the atom's core electrons.
condensed structural formula	A short-hand version of the structural formulas used to represent organic compounds. Each carbon atom is listed, and after each one are listed the elements attached to it. Subscripts are used to indicate multiples in the attached elements.
conduction electron	The valence electrons from the metal atoms in a metallic lattice that are free to move about in the overlapping orbitals in the lattice under the influence of an electric potential. Collectively, the conduction electrons are referred to as the electron sea.
conjugate acid	According to Brønsted-Lowry acid-base theory, the species formed when a base accepts a proton.
conjugate base	According to Brønsted-Lowry acid-base theory, the species that remains when an acid donates a proton.
copolymer	A polymer composed of two or more different types of monomers.
core electron	As distinguished from valence electrons, any of the electrons in the inner shells of an atom that do not participate in bonding.
covalent bond	A chemical bond between two or more nonmetal atoms in which the atoms bond together in individual molecules through the sharing of electron pairs.
critical point	The point located on a pressure-temperature phase diagram by the critical temperature and critical pressure.
critical pressure	The lowest pressure at which a substance can exist as a liquid at the critical temperature.
critical temperature	The highest temperature at which a substance can exist as a liquid.
crude oil	Petroleum.
crystal lattice	A regular, three-dimensional, geometric array of positive and negative ions bound together by electrical attraction; one of the two fundamental atomic structures of matter (the other being molecules).
crystallogen	Any element in Group 14 of the Periodic Table of the Elements. (The synthetic element 114 is not generally included with the crystallogens at this time, but is expected to exhibit crystallogen properties as well.)
cyclo– hydrocarbon	Any alkane, alkene, or alkyne in which the carbon atoms form a closed loop and having two hydrogen atoms bonded to each carbon atom.

D

Debye force	The force of electrical attraction between molecules, one of which is a permanent dipole, and the other of which is a temporary, induced dipole due to the presence of the permanent dipole; one of the three Van der Waals forces.
decomposition reaction	A chemical reaction in which a compound is broken down into other compounds or elements.
delocalization	The condition of electrons that cannot be associated with the orbitals involved in a single pair of atoms, but are instead occupying an orbital that spreads across several atoms in a molecule of lattice. Examples are the electrons in bonds of resonance structures and the conduction electrons in a metallic lattice.
deposition	The phase transition in which a gas changes into a solid.
diamond	An allotrope of carbon with a strong crystal lattice.
diatomic molecule	A molecule consisting of two atoms, often referring to one of the elements that form diatomic molecules with themselves, such as hydrogen, fluorine, chlorine, bromine, and iodine.
diffusion	The tendency of molecules in fluids spontaneously to mix and mingle due to their random motions and collisions. Foreign atoms in a solid crystal lattice are also known diffuse or migrate through the lattice.

dipole — A molecule with one end or region more electrically positive than the other due to the bonding electrons being located more densely in one part of the molecule than another. Dipoles can be permanent dipoles, induced dipoles, or randomly occurring temporary dipoles. More generally, any object with one end more electrically positive than the other.

diprotic acid — An acid with two hydrogen atoms in the molecule, both capable of dissociating.

dissociation — The process of separating into independent ions in solution. This process is exhibited by soluble ionic compounds, and by acids and bases.

dissolution — The process of dissolving.

double bond — A covalent bond in which two atoms share two pairs of electrons.

double replacement reaction — A chemical reaction in which the cations and anions of two binary compounds switch places.

dry cell — An electrochemical cell that contains no liquid, such as cells used to power portable electronics.

ductility — The ability of a substance to deform under tension without snapping, thus permitting the substance to be drawn into a wire. This property is typical of many metals.

E

effective nuclear charge — A value representing the ability of an atomic nucleus to attract electrons, roughly calculated as the difference between the total number of electrons in an atom and the number of core electrons in the atom.

effusion — The process of a gas passing from one contained area to another through a hole so small that only one particle of gas can pass through at a time. Effusion occurs when the hole diameter is less than the mean free path of the gas particles. If the hole diameter is larger than the mean free path, multiple particles pass through simultaneously and diffusion occurs.

electrical conductivity — The ability of a substance to conduct electricity; the flow of electric charge (usually electrons).

electric potential — Voltage.

electrochemical cell — A container in which an electrochemical redox reaction can take place. The cell includes two half-cells, each of which contains an electrode in an electrolyte, and a means of ion transfer between the half-cells.

electrode — One of the two sites—anode or cathode—in an electrochemical cell where the redox half-reactions take place. The oxidation half-reaction occurs at the anode; the reduction half-reaction occurs at the cathode.

electrolyte — A substance that dissociates in water to produce an electrically conductive solution. In electrochemistry, the electrically conductive solution itself.

electrolytic cell — An electrochemical cell that consumes electricity. Electrochemical cells are used for industrial processes such as electroplating and smelting.

electromagnetic spectrum — The range of wavelengths of electromagnetic radiation found in nature, including radio waves, microwaves, infrared light, visible light, ultraviolet light, X-rays, and gamma rays.

electron — One of the three basic subatomic particles in atoms. Electrons carry the atom's negative charge and reside in the orbitals surrounding the atomic nucleus. The activity of electrons determines an atom's chemical behavior.

electron affinity — The amount of energy released when an electron is added to a ground-state, isolated, gaseous atom.

electron configuration — A shorthand notation that lists the orbitals containing electrons in the ground-state atoms of a given element.

electron domain — The region occupied by a shared pair of electrons in a covalent bond or around a central atom in a molecule.

electronegativity	A metric invented by Linus Pauling that indicates the relative attraction of atoms for bonding electron pairs in molecules. The electronegativity scale ranges from a low of 0.7 for the element francium to a high of 3.98 for the element fluorine.
electron sea	The vast number of conduction electrons in a metal.
element	A pure substance consisting entirely of atoms possessing the same number of protons, and listed in the Periodic Table of the Elements.
empirical formula	A formula that represents the ratios—rather than the actual numbers—of elements in a compound. The actual numbers of atoms of different elements in one molecule of a molecular compound are indicated by the molecular formula.
endpoint	The pH at which a pH indicator changes color, used in titration as an estimate for the equivalence point.
energy level	Any one of the energy states available to electrons according to the Bohr model of the atom.
enthalpy (change in)	ΔH, the heat that flows into (is absorbed by) or out of (is released from) a system during a process that occurs at constant pressure. If ΔH is positive, heat is absorbed during the process and the process is endothermic. If ΔH is negative, heat is released during the process and the process is exothermic.
enthalpy diagram	A diagram showing enthalpy levels at various steps in a reaction process.
enthalpy of combustion	The enthalpy of reaction for a combustion reaction.
enthalpy of reaction	The net amount of heat energy absorbed or released during a chemical reaction. *See* enthalpy (change in).
enthalpy of solution	The net amount of heat energy absorbed or released during the process of dissolution.
entropy	A thermodynamic variable associated with the degree of disorder present in a system. Entropy is also associated with the number of microstates available to particles in a system.
enzyme	A catalyst that expedites the chemical reactions in a biochemical system.
equilibrium	A state in which chemical or physical species are transforming into each other at equal rates in an equilibrium system.
equilibrium constant	Denoted by K, the ratio of the product of product concentrations, each raised to the power of their coefficients in the equilibrium equation, to the product of reactant concentrations, each raised to the power of their coefficients.
equivalence point	The point during a titration at which chemically equivalent amounts of acid and base are present in the titration solution.
ester	A compound incorporating the carboxyl functional group, formed when the group hydrogen atom is replaced by an alkyl group to create an ester functional group, and the resulting ester group is attached to an alkyl group. Esters are the result of esterification reactions.
esterification	Any reaction, typically between an acid and an alcohol, that produces an ester.
ether	A compound formed when two hydrocarbon groups are joined to the same oxygen atom.
evaporation	The spontaneous phase transition in which particles in a liquid transition to the vapor state, while the temperature of the liquid is below the boiling point.
evaporative cooling	The cooling that occurs during evaporation, when the most energetic liquid particles enter the vapor state, leaving behind particles in the liquid state with lower average energy and thus lower temperature.
exchange reaction	A double replacement reaction.
excitation	The increased energy state an atom enters upon absorbing a quantum of energy from a photon or through collision with another atomic or subatomic particle.
excited state	Any energy state of an atom that is higher than the ground state.
experiment	A test designed to confirm or disconfirm a particular hypothesis.

extensive property	Any property of a substance (such as mass and volume) that depends on the amount of the substance that is present.

F

face-centered cubic lattice	A close-packed metallic lattice with layers in the sequence A–B–C–A–B–C, all A layers being aligned together, all B layers aligned together, and all C layers aligned together. The atomic centers of B layers are displaced slightly from those of layer A, and the C layers are displaced slightly again. The fourth layer is displaced yet further and is again at the A alignment.
fatty acid	Any one of many carboxylic acids found in the triglycerides of oils.
formula mass	The mass, in unified atomic mass units (u), obtained by adding together the atomic masses of the elements in a compound, taking into account the number of atoms of each element represented in the chemical formula of the compound.
formula equation	An unbalanced chemical equation that lists reactants and products.
formula unit	Typically used in reference to crystalline substances, a formula unit is one set of the atoms represented by the chemical formula of the compound. In molecular substances, the formula unit is the molecule.
fractional distillation	A process used to separate mixtures of hydrocarbons according to molecular weight. The mixture is vaporized and then fed into a fractionating column in which the temperature gradually decreases from bottom to top. When molecules in the vapor phase arrive at the region in the column that is at their condensation temperature, they condense and are removed from the column.
fractionating column	A vessel used in the process of fractional distillation to separate mixtures of hydrocarbons according to molecular weight.
free radical	Any atom, molecule, or ion that has a lone, unpaired electron.
free radical halogenation	A three-step halogenation reaction that proceeds when free radicals formed from halogen molecules in turn form free radicals from hydrocarbon molecules. The hydrocarbon free radicals then attract a halogen free radical to the site formerly occupied by a hydrogen atom.
freezing	The phase transition that occurs when a liquid is at the freezing (melting) point temperature and heat is removed, causing the liquid to transition to the solid state.
freezing point depression	The reduction in freezing point that occurs in a solution relative to the freezing point of the pure solvent; a colligative property.
functional group	An atom or group of atoms that establishes a class of compounds with specific chemical properties.

G

galvanic cell	A voltaic cell.
gas	One of the four basic states of matter; the non-ionized state characterized by the highest internal energy, in which the particles of a substance are completely free from one another.
gauge pressure	A pressure measurement relative to atmospheric pressure; a pressure measurement that uses atmospheric pressure as the zero reference.
Gibbs free energy	Calculated as the enthalpy of a system, less the product of temperature and entropy. The algebraic sign of the change in Gibbs free energy for a given chemical reaction indicates whether the process occurs spontaneously. Also called *free energy*.
graphene	An allotrope of carbon in which atoms are arranged in a hexagonal pattern in a sheet one atom thick.
graphite	An allotrope of carbon composed of pieces of graphene.

ground state	The lowest possible energy state of an atom, when all electrons in the atom are in orbitals representing the lowest possible total energy.
group	A column in the Periodic Table of the Elements.

H

half-cell	One of the two compartments in an electrochemical cell. Each half-cell contains one of the electrodes immersed in an electrolyte.
half-reaction	An equation showing separately the oxidation or reduction process in a redox reaction.
half-reaction method	A method of balancing chemical equations in oxidation-reduction reactions that makes use of the oxidation and reduction half-reactions.
haloalkane	An alkane with one or more hydrogen atoms replaced by a halogen.
halogen	Any element in Group 17 of the Periodic Table of the Elements. (The synthetic element 117 is not generally included with the halogens at this time, but is expected to exhibit halogen properties as well.)
halogenation	A substitution reaction in which one or more hydrogen atoms in a hydrocarbon are replaced with halogen atoms.
heterogeneous catalyst	A catalyst in a different phase or state than the reactants in a chemical system.
heterogeneous mixture	A mixture of two or more pure substances in which the different substances appear in lumps containing many atoms or molecules. Typically the individual substances in a heterogeneous mixture may be discerned with the naked eye or with a microscope, because the individual particles in the mixture are large enough to reflect light.
heterogeneous reaction	A reaction in which more than one state of matter is present among the reactants.
homogeneous catalyst	A catalyst in the same phase or state as the reactants in a chemical system.
homogeneous mixture	A solution.
homogeneous reaction	A reaction in which all reactants are in the same phase or state.
homologous series	A series of compounds in which each compound differs from the one before it by the addition of the same structural unit.
Hund's rule	The principle that in an unfilled subshell, electrons occupy separate orbitals singly until each orbital contains one electron, after which the electrons begin doubling up in the orbitals.
hydrate	An ionic compound that contains water molecules trapped in the crystal lattice.
hydration	The process of polar water molecules being attracted to and surrounding ions in aqueous solution.
hydride	A hydrogen atom that has ionized by gaining an electron to acquire a charge of −1.
hydrocarbon	Any compound containing only carbon and hydrogen.
hydrogenation	An addition reaction in which hydrogen atoms are added to alkene or alkyne molecules, converting multiple bonds into single bonds and increasing the saturation of the molecule.
hydrogen bond	The attraction between nonbonding electron pairs on the central atom of a molecule and hydrogen atoms that are attached to oxygen, nitrogen, or fluorine atoms.
hydrogen halide	A compound composed of hydrogen and one of the halides, all acids in solution.
hydroxyl group	Symbolized as —OH, an oxygen atom with a hydrogen atom bonded to it, while the oxygen atom is in turn bonded to another atom in a molecule.
hygroscopic	Having a strong affinity for water.
hydrolysis	A chemical reaction between water molecules and the ions of a dissolved salt.

hypothesis — An informed prediction about what will happen in certain circumstances. Every hypothesis is based on a particular theory. It is hypotheses that are tested in scientific experiments.

I

icosagen — Any element in Group 13 of the Periodic Table of the Elements. (The synthetic element 113 is not generally included with the icosagens at this time, but is expected to exhibit icosagen properties as well.)

ideal gas — A gas modeled according to the assumptions of the kinetic-molecular theory of gases.

induced dipole — A molecule with temporary dipole character due to the proximity of a permanent dipole.

inner transition metal — Any one of the *f*-block elements, usually shown in two separate rows of elements under the main body of the Periodic Table of the Elements. Also known as rare-earth elements.

intensive property — Any bulk property of substances (such as electrical conductivity or density) that does not depend on the amount of the substance present.

intermediates — Chemical species produced at intermediate stages of a reaction that are not present at the end of the reaction.

intermolecular force — The forces of electrical attraction between molecules, relatively weak compared to the forces involved in ionic and covalent bonds. Intermolecular forces consist of hydrogen bonds and Van der Waals forces. The Van der Waals forces consist of Keesom forces between permanent dipoles, Debye forces between a dipole and an induced dipole, and London dispersion forces between instantaneously formed temporary dipoles. The presence and strength of intermolecular forces governs properties in molecular substances such as melting point, boiling point, and solubility.

internal energy — The total energy represented by the sum of the kinetic energies of each of the particles in a substance.

ion — (1) An atom that has gained or lost one or more electrons, thus acquiring a net electrical charge. (2) A molecule with a net electrical charge.

ionic bond — A chemical bond between atoms in which positive and negative ions are bound together by electrical attraction in a crystal lattice. Ions form when electrons are transferred from metal atoms to nonmetal atoms, so ionically bonded substances typically involve metal and nonmetal atoms. Ionic bonds also form between metals and polyatomic ions.

ionic equation — A chemical equation showing ionizable reactant and product compounds as individual ions rather than as compound formula units or molecular units.

ionization constant of water — K_w, the product of the hydronium ion and hydroxide ion concentrations in aqueous solution, a constant at any given temperature. At 25°C this product is equal to $K_w = 1.0 \times 10^{-14}\ M^2$.

ionization energy — The amount of energy required to remove a ground-state electron from an isolated, gaseous atom. The energy required to remove one electron is called the first ionization energy. The additional energy required to remove a second electron is the second ionization energy, and so on.

ionization equation — A chemical equation describing ionization and specifying the energy involved.

ionizing radiation — Particles of matter or electromagnetic radiation (photons) with sufficient energy to remove an electron from an atom, thus ionizing it. Types of ionizing radiation include high-speed subatomic particles and electromagnetic radiation at wavelengths equal to or shorter than those of the high-energy end of the ultraviolet region of the electromagnetic spectrum.

isoelectronic series — A list of atoms or ions in which each possesses the same number of electrons.

isomer — A molecule with the same numbers and types of atoms as another molecule but with a different structural arrangement.

isomerism	The occurrence of two or more isomers of a molecule.
isotope	Any of the atoms of a given element, distinguished from one another by the different numbers of neutrons the atoms can have in the nucleus.

K

Keesom force	The force of electrical attraction between molecules that are permanent dipoles; one of the three Van der Waals forces.
keytone	A compound including the carbonyl functional group, the group having alkyl groups attached to both sides.
kinetic energy	The energy associated with any object in motion. Kinetic energy is directly proportional to the object's mass and proportional to the square of the object's speed.

L

lattice energy	The total energy binding the ions together in the crystal lattice of one mole of an ionically bonded substance.
Lewis structure	A molecular representation in which atoms are represented by their chemical symbols and bonds between atoms are represented by line segments.
Lewis symbol	A representation of an atom and its valence electrons in which the atom is represented by its chemical symbol and the electrons are represented by dots placed around the element symbol.
limiting reactant	The reactant that is consumed first in a chemical reaction, thereby determining the quantities of products produced by the reaction.
limiting reagent	A limiting reactant.
liquid	One of the four basic states of matter; the state in which the particles of a substance have sufficient kinetic energy to remain separated from each other enough to move around, but not enough to break free entirely.
London dispersion forces	The forces of electrical attraction between molecules exhibiting instantaneous, temporary dipole character.
lone pair	A nonbonding electron domain; an unshared pair of electrons.
Lonsdaleite	An allotrope of carbon found in meteorites, also called "hexagonal diamond."
luster	Shininess.

M

Madelung rule	The mathematical rule that specifies the sequence in which an atom's orbitals fill with electrons in the ground state as they fill up from low-energy orbitals first and then to higher energy orbitals according to the Aufbau principle.
magnetic quantum number	A number denoted by m_l that denotes orbitals of specific shape and orientation within a given subshell in an atom.
main group element	Any element in Groups 1–2 or 13–18 of the Periodic Table of the Elements.
malleability	The ability of a substance to be shaped (rather than shattered) by pounding. This property is typical of many metals.
manometer	A thin, U-shaped tube filled with liquid and used to make measurements of small differences in pressure by measuring the height difference between the liquid columns in the two sides of the tube.
mass number	The total number of protons and neutrons in the nucleus of an atom.
melting	The phase transition that occurs when a solid is at the melting point temperature and heat is added, causing the solid to transition to the liquid state.
metal	An element exhibiting the physical and chemical properties of metals. The metals are located in Groups 1–12 (except hydrogen), in the lower parts of Groups 13–15, and in the region of the inner transition metals in the Periodic Table of the Elements.

metallic bond	The bonding arrangement that binds atoms together in the crystal lattice of a pure metal. In the lattice, the atomic shells and orbitals overlap and the valence electrons of the atoms occupy orbitals in more than one atom simultaneously. The energy that holds the electron in multiple orbitals at once has the effect of holding the atoms together in the crystal lattice.
metalloid	Any of the elements located in the region between the metals and nonmetals in the Periodic Table of the Elements, usually assumed to include elements 5, 14, 32, 33, 51, 51, and 84. The metalloids possess properties that are not clearly those of the metals nor those of the nonmetals.
metathesis	A double replacement reaction.
microstate	For an atom or molecule, one of the possible combinations of position and velocity.
miscible	Able to mix together in any proportions to form a solution.
mixture	A combination of two or more pure substances in which the individual substances retain their respective chemical identities and may be separated by physical means. The two main classes of mixtures are homogeneous mixtures, which are solutions, and heterogeneous mixtures.
molality	A measure of solution concentration, defined as the number of moles of solute per kilogram of solvent.
molar heat capacity	For a given substance, the amount of heat that must be added or removed to change the temperature of one mole of the substance by 1°C.
molar heat of fusion	For a given substance, the amount of heat per mole of substance that must be added to effect a phase transition from solid to liquid (melting) or removed to effect a phase transition from liquid to solid (freezing).
molar heat of vaporization	For a given substance, the amount of heat per mole of substance that must be added to effect a phase transition from liquid to vapor (vaporization or boiling) or removed to effect a phase transition from vapor to liquid (condensation).
molarity	A measure of solution concentration, defined as the number of moles of solute per liter of solution.
molar mass	The mass of one mole of a substance in units of grams per mole.
mole	The amount of a pure substance that contains Avogadro's number of particles of the substance.
molecular equation	A chemical equation showing the reactant and product compounds as formula units or molecular units (as opposed to ions).
molecular formula	A formula indicating the actual numbers of atoms of different elements in one molecule of a molecular compound.
molecular mass	For compounds that exist as molecules, the mass, in unified atomic mass units (u) obtained by adding together the atomic masses of the elements in the compound, taking into account the number of atoms of each element represented in the chemical formula of the compound. May also be expressed in grams (g) by adding together the gram masses of the atoms in the molecule, or by converting a molar mass to grams with the use of the Avogadro constant.
molecular orbital	A single orbital formed by the melding together of individual orbitals in individual atoms as the atoms bond together in a molecule.
molecule	A cluster of atoms chemically bound together by covalent bonds; one of the two fundamental atomic structures of matter (the other being the crystal lattice).
monomer	A molecule that can chemically bond with other identical or similar molecules to form long molecular chains called polymers.
monoprotic acid	An acid with one hydrogen atom in the molecule.
monovalence	The tendency exhibited by some elements, especially those in Groups 1–4, 16, 17, and parts of 15, always to acquire the same charge and oxidation state during ionization.

multivalence	The ability of many elements, particularly the transition metals, to ionize to any one of two or more different oxidation states, acquiring various ionic charges, depending on the circumstances.

N

net ionic equation	An ionic equation with the spectator ions removed.
neutral	Neither acidic nor basic, and thus having a pH of 7.0.
neutralization	An acid-base reaction, in which the solution after the reaction is neither acidic nor basic but neutral.
nitration	A substitution reaction in which a nitro group (NO_2^+) is added to a hydrocarbon molecule in place of a hydrogen atom that is removed from the molecule.
noble gas	Any element in Group 18 of the Periodic Table of the Elements. (The synthetic element 118 is not generally included with the noble gases at this time, but is expected to exhibit noble gas properties as well.)
nonbonding domain	An electron domain containing an unshared pair of electrons around a central atom in a molecule that does not bond the central atom to any other atom.
nonmetal	Elements on the extreme right end of the periodic table, to the right of the metalloids, exhibiting the physical and chemical properties of nonmetals.
nucleon	A proton or neutron, so called because these two particles are contained in the atomic nucleus.
neutralization	A chemical reaction in which an acid and base react to produce a salt, water, and sometimes carbon dioxide. In the process the low-pH solution of the acid and the high-pH solution of the base are neutralized to form a solution with a neutral pH of approximately 7.
neutron	One of the three basic subatomic particles in atoms. Neutrons carry no charge and have mass slightly higher than the mass of protons. Neutrons reside in the atomic nucleus.
nucleus	The tiny central region in an atom that contains the atom's protons and neutrons.
nuclide	Any isotope of any element.

O

octet	The eight electrons that occupy the *s* and *p* subshells in main group elements.
octet rule	The principle that when forming chemical bonds, main group elements seek to obtain a full octet by gaining, losing, or sharing electrons.
orbital	One of the various cloud-like regions surrounding the atomic nucleus where an atom's electrons are held. The orbitals are arranged in subshells and shells associated with different levels of energy. Each orbital can hold up to two electrons.
orbital diagram	A notation in which atomic orbitals are represented as boxes and electrons as arrows. The notation is used to provide a graphical representation of where the electrons are in the orbitals of an atom of a given element.
orbital hybridization theory	The theory that when forming covalent bonds, the orbitals in different subshells of a central atom in a molecule lose their distinct identities and merge together to form an equal number of hybrid orbitals identical in energy, and that the hybrid orbitals are oriented according to the geometries described by Valence Shell Electron Pair Repulsion theory.
olefin	An alkene.
open-chain hydrocarbon	Any alkane, alkene, or alkyne that is not a *cyclo–* hydrocarbon.
organic chemistry	The branch of chemistry devoted to the study of compounds containing carbon.
organic compound	A compound containing carbon. Carbonates, oxides, metal cyanides, and a few other compounds are not considered organic.

osmosis	The process in which particles of solvent pass through a semipermeable membrane in greater quantities in the direction toward the side of the membrane with the highest solute concentration.
osmotic pressure	The pressure required on the higher-concentration side of a semipermeable membrane to stop the process of osmosis.
oxidant	An oxidizing agent.
oxidation	A chemical change in which the oxidation state of an element increases.
oxidation number	The number (positive or negative) representing an element's oxidation state.
oxidation-reduction reaction	A chemical reaction in which elements are oxidized and reduced. Oxidation-reduction reactions are also called redox reactions.
oxidation state	For ions or elements in ionic compounds, the oxidation state is equal to the charge on a monatomic ion. In molecules, the oxidation state of an element in the molecule is a hypothetical or artificial number assigned to each element in the molecule as if bonding electrons were held completely by one atom or another. In molecules, the oxidation state is associated with the strength of the element's attraction to the bonding electrons, and thus is based on the element's electronegativity relative to the other elements in the molecule. The oxidation state in a molecule is not based on any actual charge.
oxidizer	An oxidizing agent.
oxidizing agent	An element that receives electrons to cause the oxidation of another element.
oxyacid	An acid containing oxygen and at least one other element, usually a nonmetal.
oxyanion	A polyatomic ion containing oxygen.

P

passivation	The phenomenon, exhibited by aluminum and some other metals, of forming a thin oxide coating on the surface of the metal which then protects the interior of the metal from further oxidation. Also called autopassivation.
Pauli exclusion principle	The principle that no two electrons in an atom can occupy the same quantum state.
peptide bond	The chemical bond between amino acid residues linking together the carboxyl group of one amino acid to the amine group of the next amino acid. The bond consists of the oxygen and carbon atoms in the carboxyl group of one amino acid, and the nitrogen and hydrogen atoms in the amine group of the adjacent amino acid.
percent composition	The percentages, by mass, of each element in a compound.
percent yield	In a chemical process, the ratio of actual yield to theoretical yield, multiplied by 100%.
periodicity	The cyclic variation of the physical and chemical properties of elements in the Periodic Table of the Elements.
periodic law	The fact that the elements, when arranged in order according to atomic number, exhibit periodicities in many physical and chemical properties. Notably, elements in the same group exhibit similar properties.
permanent dipole	A molecule with permanent dipole character due to differences in the electronegativities of the atoms in the molecule.
petroleum	A natural mixture of hydrocarbon molecules of many different sizes.
pH	A measure of acidity in a solution, calculated as the negative of the common logarithm of the hydronium ion concentration.
phase diagram	A diagram showing the phases or states of a particular substance with respect to pressure, temperature, and/or energy.
phase (of matter)	State (of matter).
phase transition	The change of state of a substance from one state or phase to another.

pH indicator	A weak acid or weak base that changes color as the pH of a solution increases or decreases through its transition interval. pH indicators are used in titrations to indicate when the equivalence point has been reached. The indicator in the solution changes color at its endpoint, and the endpoint is used as an estimate for the equivalence point. Some pH indicators are also used to measure the pH of a solution.
photon	A single packet of electromagnetic energy; a quantum of electromagnetic radiation; a particle of light.
physical change	A change of any kind in which the substances involved retain their chemical identities and chemical properties.
physical equilibrium	A state in an equilibrium system in which particles are transforming between two or more physical states at equal rates. Examples include different phases of a substance at the triple point, and gas molecules evaporating out of solution and dissolving into solution in a sealed container.
physical property	Properties of pure substances other than those that describe the substance's chemical behavior. Physical properties of substances include color, density, boiling point, melting point, electrical conductivity, malleability, vapor pressure, thermal conductivity, and many others.
plasma	A state of matter consisting of a gas with ionized particles.
pnictogen	Any element in Group 15 of the Periodic Table of the Elements. (The synthetic element 115 is not generally included with the pnictogens at this time, but is expected to exhibit pnictogen properties as well.)
pOH	A measure of basicity similar to pH for acidity. In an aqueous solution, pH + pOH = 14.0.
polarity	An imbalance in the distribution of electron electrical charge in a molecule due to the different electronegativities of the atoms in the molecule. Polarization causes one region of the molecule to be electrically positive relative to another region that is electrically negative.
polyatomic ion	A molecule formed by covalent bonds between atoms of nonmetals that possesses a positive or negative net electrical charge.
polycyclic aromatic hydrocarbon	An aromatic hydrocarbon formed from two or more benzene rings fused together.
polymer	A long molecule composed of chemically linked monomers.
polymerization	Any reaction process that converts monomers into polymers.
polyprotic acid	An acid with more than one hydrogen atom in the molecule, each capable of dissociating.
precipitate	(1) (noun) A solid that forms when reactants in solution combine to form an insoluble product. (2) (verb) The process of an insoluble solid coming out of solution.
precipitation	The process of forming a precipitate.
precursor	A compound used as the starting point for the manufacture of another compound.
pressure	The average amount of force exerted per unit area on a surface by collisions from the moving particles (atoms or molecules) in a substance.
principle quantum number	A number denoted by $n = 1, 2, 3$, etc., which designates one of the shells in an atom that contains subshells and orbitals in which electrons are contained.
probability distribution	A mathematical description resulting from a solution to the Schrödinger equation of where electrons are likely to be found in the orbitals of atoms.
protein	A large chain molecule formed typically from hundreds of amino acid residues joined together by peptide bonds, and possessing specific biological functionality.
proton	One of the three basic subatomic particles in atoms. Protons carry the atom's positive charge. The number of protons in an atom determines which element

	the atom represents. Protons have mass slightly lower than the mass of neutrons. Protons reside in the atomic nucleus.
protonation	The process of a molecule accepting a proton in aqueous solution (and thus acting like a Brønsted-Lowry base).
pure substance	Any element or compound.

Q

quantum	The smallest possible single unit of energy possessed by any subatomic particle or photon.
quantum model	A model of the atom based on the theory of energy quantization and the solutions to the Schrödinger equation.
quantum number	One of the four values that define the state of an electron in an atom. The four quantum numbers are the principle quantum number, the azimuthal (or angular momentum) quantum number, the magnetic quantum number, and the spin projection quantum number.
quantum state	The state of an electron in an atom defined by the four quantum numbers known as the principle quantum number, the azimuthal (or angular momentum) quantum number, the magnetic quantum number, and the spin projection quantum number.

R

rare-earth element	Any one of the *f*-block elements, usually shown in two separate rows of elements under the main body of the Periodic Table of the Elements. Also known as the inner transition metals.
rate constant	A constant that appears in a rate law. For a given reaction, the rate constant is constant at a given temperature.
rate-determining step	The slowest step in a reaction mechanism, and thus the step that determines the reaction rate.
rate law	A mathematical expression that characterizes a reaction rate in terms of the concentrations of the reactants.
reaction kinetics	The general subject of the physical mechanisms of chemical reactions and the factors that govern the rates of chemical reactions.
reaction mechanism	A model depicting a sequence of steps and intermediates involved in a chemical reaction.
reaction rate	The rate at which the molar concentrations of reactants and products are changing at any point during a chemical reaction.
real gas	A gas modeled according to the theory in the Van der Waals equation, a theoretical gas model based on the kinetic-molecular theory of gases, but with particle volume and intermolecular forces taken into account.
redox reaction	A chemical reaction in which elements are oxidized and reduced.
reducer	A reducing agent.
reducing agent	An element that loses electrons to cause the reduction of another element.
reductant	A reducing agent.
reduction	A chemical change in which the oxidation state of an element decreases.
representative element	A main group element.
resonance structure	A molecular structure in which one or more covalent bonds formed by electron pairs are shared or spread out over multiple bonding locations in a molecule. In such structures the electrons are said to be *delocalized*.

S

salt	One of the products of an acid-base reaction, formed from anion of the acid and the cation of the base.
salt bridge	A tube filled with salt ions embedded in a gel matrix that allows migration of charge (the ions) between the two half-cells of an electrochemical cell.
saturated hydrocarbon	Alkanes, so named because there are no double or triple carbon bonds, so each carbon atom has the maximum number of hydrogen atoms attached to it.
saturated solution	A solution containing the maximum amount of solute that will dissolve in the solvent at a particular temperature.
Schrödinger equation	A complex equation that provides through its solutions information about the probable locations of electrons in the orbitals of atoms.
scientific fact	A proposition based on a large amount of scientific data that is correct so far as we know.
screening constant	A constant used in the calculation of effective nuclear charge, roughly equal to the number of core electrons in an atom.
semipermeable membrane	A barrier such as the wall of a cell, human skin, or the skin of a vegetable that allows solvent (typically water) molecules to pass through, but not ions that may be in solution on either side of the membrane.
shield effect	The forming of an electrical screen by core electrons around an atomic nucleus, which has the effect of reducing the nuclear electrical attraction for valence electrons (also called atomic shielding).
shell	A term used to describe the major collections of subshells and orbitals in atoms where electrons are contained. The shells are denoted by the principle quantum number *n*, which takes on the integer values, 1, 2, 3, etc.
single replacement reaction	A chemical reaction in which an element in a compound is removed and replaced by a different element.
solid	One of the four basic states of matter; the state characterized by the lowest internal energy, in which the particles of a substance are bound together by electrical or bonding forces and do not have sufficient energy to break free. Particles in solids vibrate in place, but do not flow in bulk.
solubility equilibrium	An equilibrium system composed of a saturated solution with excess solute.
solubility product constant	The equilibrium constant for a saturated solution of a sparingly soluble compound.
soluble	Able to dissolve in a given solvent to an appreciable degree.
solute	The solid substance dissolved into a liquid solvent to form a solution.
solution	A mixture of two or more pure substances with uniform composition all the way down to the atomic or molecular level; a homogeneous mixture. Solutions may be solid (alloys), liquid, or gaseous. Individual particles in solutions are the size of molecules or atoms (typically < 1 nm in diameter), and thus are orders of magnitude too small to be seen with the eye, even under magnification.
solvation	The process of polar solvent molecules being attracted to and surrounding ions in solution. In aqueous solution, solvation is referred to as hydration.
solvent	The liquid into which a solid dissolves to form a solution.
sparingly soluble	Essentially insoluble except to a minute degree, typically far less than approximately 0.05 mol/L or 0.1 g/100 mL.
specific gravity	The ratio of the density of a substance to the density of water, symbolized as s.g. Substances with a specific gravity of s.g. < 1 will float on water.
spectator ion	An ion that appears on both the reactant side and the product side in an ionic equation, and thus takes no part in the chemical reaction represented by the equation.

spectroscopy	The science of identifying the presence of elements by means of the electromagnetic spectra they emit.
spin projection quantum number	The quantum number used to differentiate between the two possible spin states (spin up and spin down) electrons can possess in a single orbital in an atom. Each orbital can hold up to two electrons which must have opposite spin states to fulfill the Pauli exclusion principle.
standard hydrogen electrode	A specially designed electrode configuration made of platinum wire and hydrogen gas, established as the reference standard for measuring all standard reduction potentials.
standard reduction potential	The voltage a voltaic cell reads when a given reducing agent is oxidized in one half-cell while a standard hydrogen electrode reduces H^+ to H_2 in the other half-cell.
standard solution	A solution of precisely known molar concentration. Standard solutions are used in titration to determine the unknown concentration of an acid or base.
standard temperature and pressure	A set of conditions defined as 0°C and 100 kPa.
state (of matter)	Any of the solid, liquid, gas, or plasma states defined by the internal energy, thermal properties, and (in the case of plasmas) ionization of a substance.
Stock system	A system for naming ionic compounds in which the oxidation state of the cation is indicated by upper-case Roman numerals in parentheses after the name of the cation and before the name of the anion.
stoichiometry	The calculations associated with chemical equations, in which given quantities of reactants or products are used to determine the required amounts of other reactants or the amounts produced of other products.
stopcock	A valve in an article of laboratory glassware.
strong electrolyte	A substance that dissociates completely, or nearly so, in aqueous solution.
structural formula	Diagrams similar to Lewis structures used to represent organic compounds.
subatomic particle	A proton, neutron, or electron, or any of the smaller particles of which protons and neutrons are composed.
sublimation	The phase transition in which a solid changes into a gas.
subshell	The term used to denote the various ordered clusters of orbitals associated with a given principle quantum number (shell) and available in an atom for electrons to be contained in.
substitution reaction	A chemical reaction in which one or more atoms in a molecule are replaced by others. Halogenation is a typical example.
supercritical fluid	A substance at conditions above the critical temperature and critical pressure. At these conditions distinct liquid and vapor states do not exist. Instead, the fluid will be more vapor-like or more liquid-like depending on its pressure and temperature.
surface tension	A characteristic of liquids caused by intermolecular forces between particles at the liquid surface. Because of these forces, some liquids—notably water—form a meniscus, allow objects denser than the liquid to float (such as an insect walking on the surface), and exhibit capillary action.
suspension	A heterogeneous mixture formed when particles of size approximately 1 μm or larger are dispersed in a fluid medium. Particles in suspensions eventually settle out.
synthesis reaction	A chemical reaction in which two or more elements or compounds combine to form a single compound.

T

temporary dipole	A molecule with a temporary dipole character, either due to electrical induction from a nearby permanent dipole, or due to random fluctuations of electrons in the molecular orbitals.

temperature	A variable defined by thermodynamic theory that correlates directly with the internal energy of a substance.
theoretical yield	The amount of a given product produced in a chemical reaction, determined by stoichiometric calculation based on the available quantity of the limiting reactant and 100% reaction efficiency.
theory	A mental model that accounts for the data (facts) in a certain field of research, and attempts to relate them together, interpret them and explain them.
thermal conductivity	The ability of a substance to allow heat to flow by conduction.
thermochemical equation	A chemical equation that shows the physical states of all reactants and products, as well as the enthalpy of reaction.
titration	A procedure used to determine the unknown concentration of an acid or base, in which an acid or base of precisely known concentration, called a standard solution, is gradually added to a base or acid until chemically equivalent amounts of the acid and base are present in the solution, a condition called the equivalence point. The equivalence point is indicated by a pH indicator in the solution that changes color at its endpoint, and the endpoint is used as an estimate for the equivalence point.
TNT	2,4,6-trinitrotoluene.
transition interval	The range of pH values in which a pH indicator changes color, depending on whether the pH of the solution is higher or lower than the pH at the indicator's endpoint.
triple bond	A covalent bond in which two atoms share three pairs of electrons.
triple point	The particular combination of unique temperature and pressure that enables a substance to exist in equilibrium simultaneously as solid, liquid, and vapor.
triprotic acid	An acid with three hydrogen atoms in the molecule, each capable of dissociating.

U

unified atomic mass unit	A unit of relative atomic mass, defined as exactly 1/12 the mass of an atom of carbon-12.

V

valence-bond theory	The attempt to construct linear combinations of single-atom solutions from the Schrödinger equation in order to form multiple-atom solutions that predict the same orbital configurations as those envisioned by orbital hybridization theory.
valence electron	Any electron in the outer, unfilled shell of an atom. The valence electrons are the electrons that participate in ionization and chemical bonding.
valence shell	The outer, unfilled shell of an atom where the valence electrons are contained.
Valence Shell Electron Pair Repulsion Theory	The theory that bonding domains and nonbonding domains around central atoms in molecules repel each other, providing an explanation for three-dimensional molecular structure.
Van der Waals forces	Intermolecular forces, not including hydrogen bonding. Van der Waals forces include the Keesom force, the Debye force, and London dispersion forces.
vapor	Gas.
vaporization	The phase transition in which a liquid changes into a vapor.
vapor pressure	The pressure the vapor of a liquid assumes at a given temperature due to evaporation in a closed container; the pressure at which a liquid boils at a given temperature.
vapor pressure lowering	The reduction in vapor pressure that occurs in a solution relative to the vapor pressure of the pure solvent; a colligative property.
visible spectrum	The range of wavelengths in the electromagnetic spectrum that are visible to the human eye, approximately from 700 nanometers to 400 nanometers. Wave-

	lengths in the visible spectrum appear to humans as the colors in the rainbow, including red, orange, yellow, green, blue and violet.
volumetric flask	A flask marked with a specific volume and used to prepare solutions with a given molarity.
voltaic cell	An electrochemical cell that produces electricity, such as the dry cells used to power electronic devices.

W

weak electrolyte	A substance that dissociates only to a limited degree in aqueous solution.
wet cell	An electrochemical cell containing liquid, such as a car battery.
work	A form of energy, specifically, energy transferred from one object, machine, or person to another by means of applying a force to an object and moving the object through a distance. When the force applied and distance moved point in the same direction, the amount of work done, and thus energy transferred, is equal to the product of force and distance.

Answers to Selected Exercises

There are always errors in any technical writing project of this scope, and regrettable as it is, there are almost certainly errors in this list of answers to problems. If you find an answer you believe to be erroneous, first check the errata file on our website (centripetalpress.com/catalog/free-resources) to see if it is an error we know about and have posted the correction for. If you don't see it there, we would greatly appreciate it if you would send an email to us about it at info@centripetalpress.com.

Chapter 1

1. $E = 3.66 \times 10^{-19}$ J
2. $\lambda = 874.39$ nm, IR
3. $\lambda = 91.2$ nm, UV
4. 410 nm: 4.8×10^{-19} J
   434 nm: 4.58×10^{-19} J
   486 nm: 4.09×10^{-19} J
   656 nm: 3.03×10^{-19} J

12. a. Cl: $1s^2 2s^2 2p^6 3s^2 3p^5$

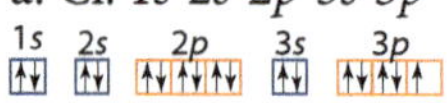

b. O: $1s^2 2s^2 2p^4$

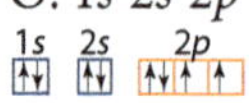

c. Ru: $1s^2 2s^2 2p^6 3s^2 3p^6 4s^2 3d^{10} 4p^6 5s^1 4d^7$

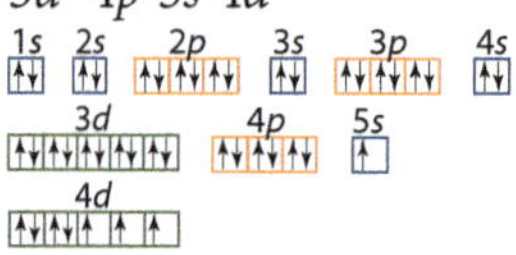

d. K: $1s^2 2s^2 2p^6 3s^2 3p^6 4s^1$

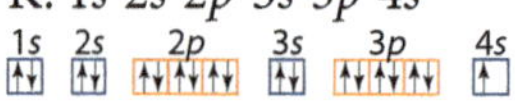

e. V: $1s^2 2s^2 2p^6 3s^2 3p^6 4s^2 3d^3$

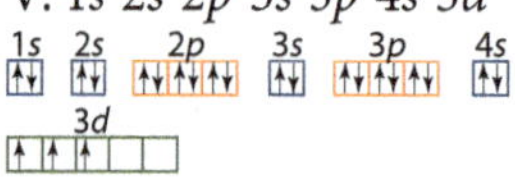

f. Br: $1s^2 2s^2 2p^6 3s^2 3p^6 4s^2 3d^{10} 4p^5$

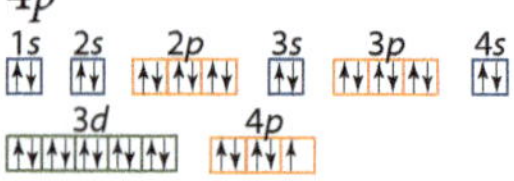

13. a. Cl: $[Ne]3s^2 3p^5$
b. N: $[He]2s^2 2p^3$
c. Al: $[Ne]3s^2 3p^1$
d. Y: $[Kr]5s^2 4d^1$
e. Sr: $[Kr]5s^2$
f. W: $[Xe]6s^2 4f^{14} 5d^4$
g. Cs: $[Xe]6s^1$
h. I: $[Kr]5s^2 4d^{10} 5p^5$
i: Nd: $[Xe]6s^2 4f^4$

16. Yb: $[Xe]6s^2 4f^{14}$
Es: $[Rn]7s^2 5f^{11}$
No: $[Rn]7s^2 5f^{14}$
18. 146
19. a. 6.94×10^{23}
b. 8.13×10^{23}
c. 4.9133×10^{24}
20. a. 39.10 g
b. 3×10^{-20} g
c. 0.131 g d. 14 g
e. 336.56 g f. 24.04 g
21. a. 0.34 mol
b. 0.71216 mol
c. 28.32 mol
d. 0.5255 mol
e. 0.313 mol
f. 3.40×10^{-6} mol
22. 1.9342 u, or 3.2118×10^{-24} g
23. 28.086 u
24. a. 17.0304 g/mol
b. 44.010 g/mol
c. 70.9054 g/mol
d. 159.610 g/mol
e. 132.089 g/mol
f. 342.30 g/mol
g. 46.069 g/mol
h. 44.096 g/mol
i. 60.0843 g/mol
25. a. 95.2104 u
b. 164.088 u
c. 96.064 u
d. 159.610 u
e. 67.806 u
f. 153.822 u
26. 382 g
27. a. 14.6 mol
b. 8.81×10^{24} atoms
c. 9.50×10^{22} atoms

28. a. 2.10×10^2 mol
b. 2.53×10^{26} atoms
c. 2.91×10^{22} atoms
29. CH_3O, $C_2H_6O_2$
30. a. 92.261% C, 7.739% H
b. CH c. C_6H_6
31. 8.913×10^{23} particles
32. a. 27.367% Na, 1.1998% H, 14.298% C, 57.136% O
b. 74.1858% Na, 25.8142% O
c. 69.943% Fe, 30.0567% O
d. 63.49928% Ag, 8.24539% N, 28.2553% O
e. 25.339% Ca, 30.375% C, 3.8234% H, 40.4620% O
f. 60.002% C, 4.4756% H, 35.5226% O
33. 43.854%
34. Na_2SO_4
35. $C_4H_8O_4$
36. a. $C_4H_5N_2O$, $C_8H_{10}N_4O_2$
b. $C_{13}H_{18}O_2$ c. C_3H_8
d. $C_{14}H_{18}N_2O_5$ e. CH, C_2H_2
37. 91.2% C, 8.75% H, C_7H_8

Chapter 2

6. a. 0.920 Å, 9.09%
b. 1.225 Å, 8.60%
c. 1.225 Å, 5.60%
d. 1.653 Å, 5.29%
9. Xe < Ag < Y < Rb
10. Mg < Na < Ba < Cs
11. Ca > Mg > Mg^{2+} > Be^{2+}
12. S^{2-} > Cl^- > Ar > K^+ > Ca^{2+}
14. 5, 2, 7, 15
18. Cu^{2+}: $[Ar]3d^9$
As^{5+}: $[Ar]3d^{10}$

Ag^+: [Kr]$4d^{10}$
Au^{3+}: [Xe]$4f^{14}5d^8$
20. Ba < Sr < Mg < P < Ar
22. Rb < S < Br
29. Cs < Ta < Ni < Se < Cl < F
33. E = 13 eV
34. λ = 390 nm (UV)
35. a. P: p, 3, 15
b. Tl: p, 6, 13 c. Mo: d, 5, 6
d. V: d, 4, 5
36. 16, 32
37. 1.53×10^{24}
38. 2.0553% H, 32.694% S, 65.251% O; H: 4.30×10^{23} atoms; S: 2.15×10^{23} atoms; O: 8.60×10^{24} atoms
40. 52%
41. 1.20×10^2 g
42. WCl_6

Chapter 3

6. a. $SrBr_2$, b. MgO, c. $CuCl_2$,
d. Li_2Se, e. Ni_3P_2, f. K_3N,
g. $BeCl_2$ h. NaI i. Cr_2O_3
j. Ag_2Se k. MnF_4 l. $SbBr_3$
m. Rb_2S n. Sc_2O_3 o. AlF_3
p. MgF_2 q. $ZnCl_2$ r. Li_2S
s. Co_2O_3 t. VI_5
7. a. strontium bromide
b. magnesium oxide
c. copper(II) chloride; cupric chloride
d. lithium selenide
e. nickel phosphide
f. potassium nitride
g. beryllium chloride
h. sodium iodide
i. chromium(III) oxide; chromous oxide
j. silver selenide
k. manganese(IV) fluoride; manganic fluoride
l. antimony(III) bromide; antimonous bromide
m. rubidium sulfide
n. scandium oxide
o. aluminum fluoride
p. magnesium fluoride
q. zinc chloride
r. lithium sulfide
s. cobalt(III) oxide; cobaltic oxide
t. vanadium(V) iodide; vanadic iodide

12.
a.
H—C≡C—Cl

b.
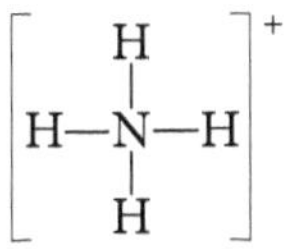

c.
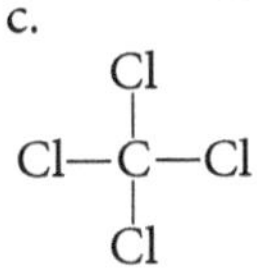

d.
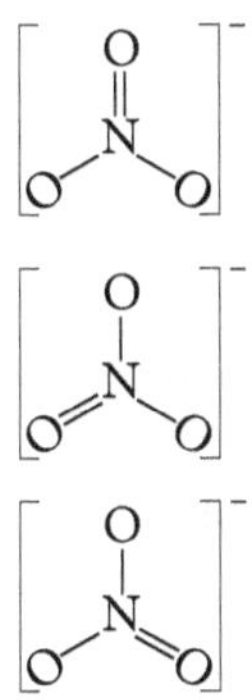

e.
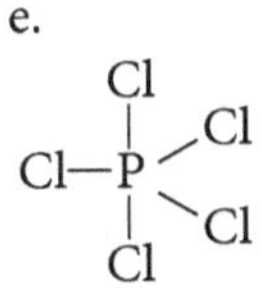

f.
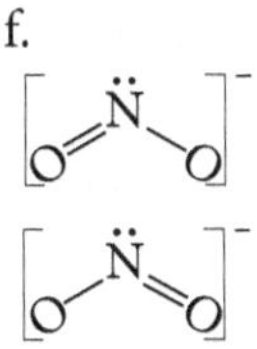

g.
C≡O

h.
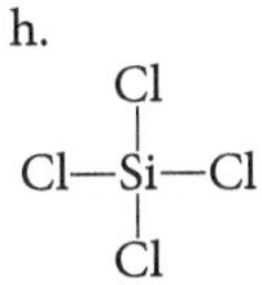

i.
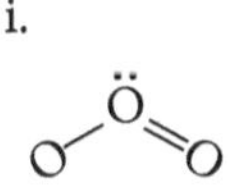

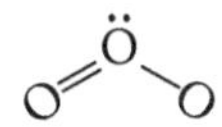

j.
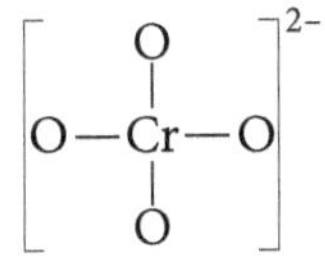

k.
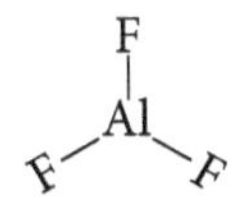

l.
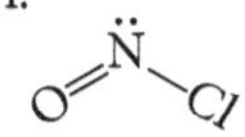

m.
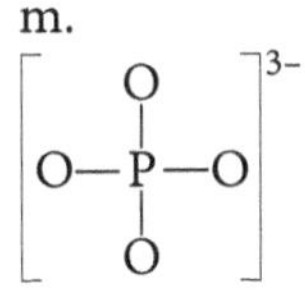

n.
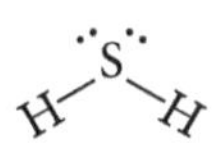

o.
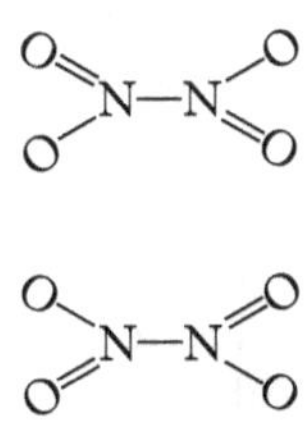

p.
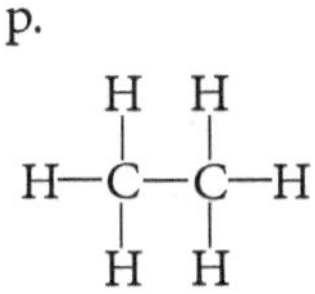

q.
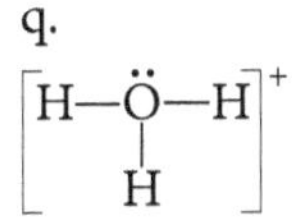

r.
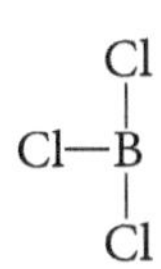

s.

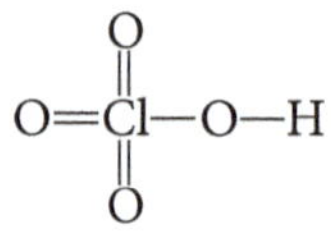

t.

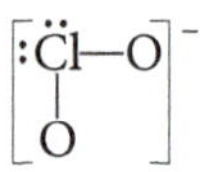

u.

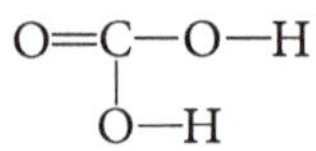

v.

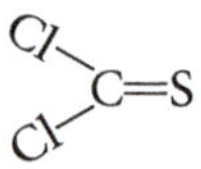

w.

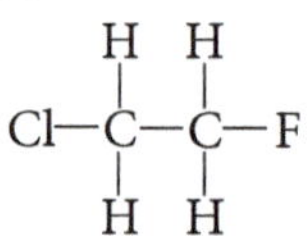

x.

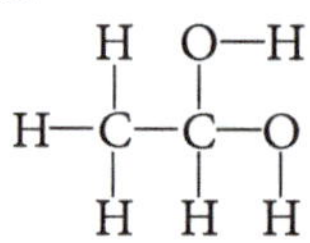

13. c. carbon tetrachloride
e. phosphorus pentachloride
g. carbon monoxide
h. silicon tetrachloride
k. aluminum trifluoride
n. dihydrogen sulfide (common name hydrogen sulfide)
o. dinitrogen tetroxide
r. boron trichloride

14. a. (exception)

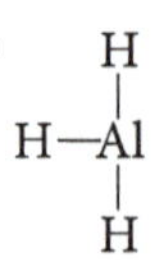

b. S=C=S

c.

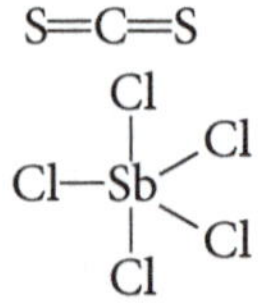

(exception)

d. 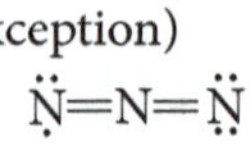 (free radical)

e.

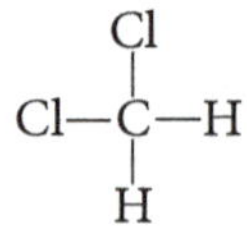

f.

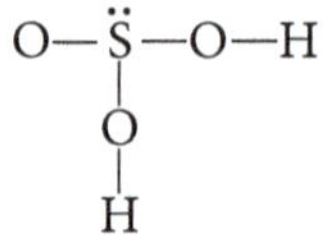

g. 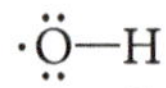 (free radical)

h. 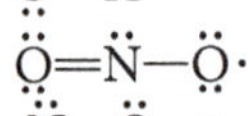

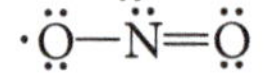 (free radical)

15. a. sulfuric acid
b. hydrofluoric acid
c. carbonic acid
d. perchloric acid
e. nitrous acid
f. bromous acid
g. hydrobromic acid
h. phosphoric acid

18. a. $MgCO_3$ magnesium carbonate
b. Na_2CrO_4 sodium chromate
c. $Ca_3(PO_4)_2$ calcium phosphate
d. $(NH_4)_3PO_4$ ammonium phosphate
e. $CuSO_4$ copper(II) sulfate
f. $NaHCO_3$ sodium bicarbonate
g. $Fe(ClO_3)_3$ iron(III) chlorate
h. $Al(OH)_3$ aluminum hydroxide
i. $Cr(SO_3)_3$ chromium(VI) sulfite
j. $Ca(CH_3COO)_2$ calcium acetate

20. bond numbers: a—2; b—1; c—1.5; b,c,a (increasing strength)
a,c,b (increasing length)

22. a. O—F: nonpolar covalent, least polar; C—F: polar covalent; Be—F: ionic, most polar
b. F—F: least polar, nonpolar covalent; S—O: polar covalent; B—F: ionic, most polar
c. O—Cl: nonpolar covalent, least polar; C—P: nonpolar covalent; S—Cl: polar covalent, most polar

23. red: a, yellow: b, orange: c

24. 1.8051×10^{24} atoms of O

25. S: 21.955%, F: 78.0457%

27. Cl: $1s^2 2s^2 2p^6 3s^2 3p^5$

Sc: $1s^2 2s^2 2p^6 3s^2 3p^6 4s^2 3d^1$

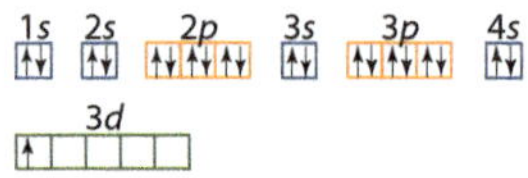

Ag: $1s^2 2s^2 2p^6 3s^2 3p^6 4s^2 3d^{10} 4p^6 5s^1 4d^{10}$

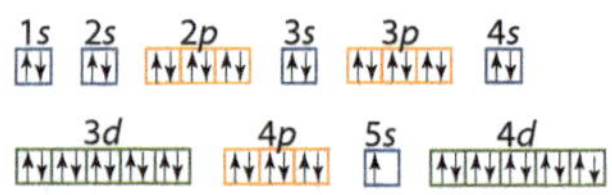

Gd: $1s^2 2s^2 2p^6 3s^2 3p^6 4s^2 3d^{10} 4p^6 5s^2 4d^{10} 5p^6 6s^2 4f^7 5d^1$

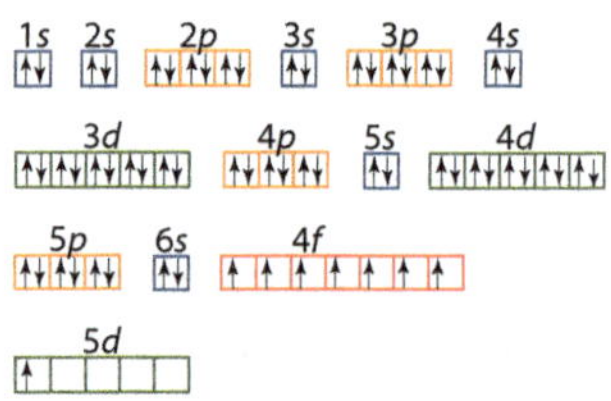

28. 12.01 u; 0%

29. 2.94×10^{22} Cl atoms

31. $Mg^{2+} < Ca^{2+} < K^+ < Rb^+ < Rb$

32. $CuSO_4 \cdot 5H_2O$; copper sulfate pentahydrate

33. 533 nm

Chapter 4

2.

a.

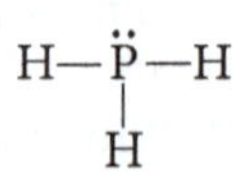

tetrahedral/pyramidal

b.

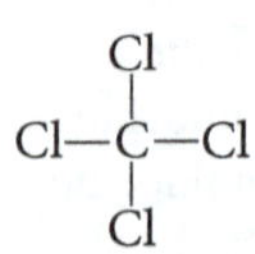

tetrahedral/tetrahedral

c. 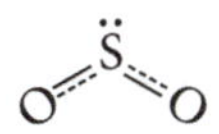

trigonal/bent

d. 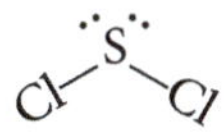

tetrahedral/bent

e. 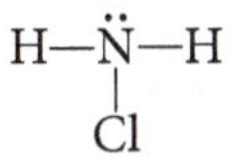

tetrahedral/pyramidal

f. 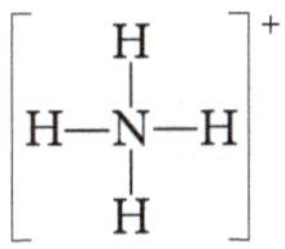

tetrahedral/tetrahedral

g. 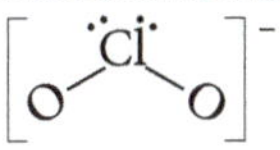

tetrahedral/bent

h. 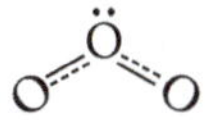

trigonal/bent

i. H—C≡C—H

linear/linear

j. S=C=S

linear/linear

k. 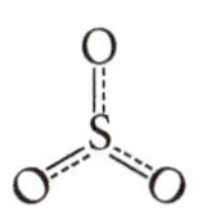

trigonal/trigonal

l. 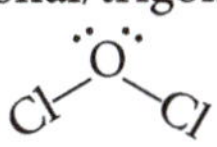

tetrahedral/bent

m. H—Cl:

linear/linear

n. 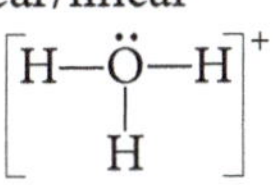

tetrahedral/pyramidal

o. H—C≡N

linear/linear

3. a. sp^3 b. sp^3 c. sp^3 d. sp
e. N: sp^2, O: sp^3 f. sp^3

5. a. 109.5° b. 104.5°
c. 109.5° d. 109.5° e. 120°
f. 104.5° g. 107°

7. a. tetrahedral/bent
b. tetrahedral
c. tetrahedral/pyramidal
d. trigonal bipyramidal
e. octahedral/square planar

10. a. tetrahedral
b. tetrahedral
c. tetrahedral/pyramidal
d. trigonal planar
e. tetrahedral/bent

23. 89.094 g/mol; 40.444% C, 15.721% N, 35.916% O, 7.9189% H

24. 2.06 eV

25. a. polar covalent b. ionic
c. nonpolar covalent
d. polar covalent e. nonpolar covalent f. polar covalent

26. b, c, d, f

27. $F < F^- < Cl^- < I^- < Te^{2-}$

28. Hg: $[Xe]6s^24f^{14}5d^{10}$
Ru: $[Kr]5s^14d^7$
Br: $[Ar]4s^23d^{10}4p^5$
Ba: $[Xe]6s^2$
Cr: $[Ar]4s^13d^5$
S: $[Ne]3s^23p^4$

31. Sr

33. a. 3.64 mol
b. 0.00718 mol c. 0.002 mol
d. 138 mol e. 10.5 mol
f. 47.635 mol

34. a. 1.4×10^{23} Hg atoms
b. 2.26×10^{25} Fe atoms
c. 6.0221×10^{23} Ca atoms
d. 1.5×10^{23} Sr atoms
e. 2.409×10^{24} Na atoms
f. 1.92×10^{25} Ca atoms

35.

Sc:

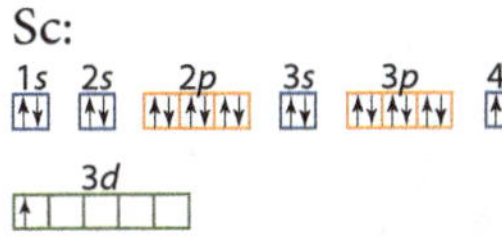

Fe:

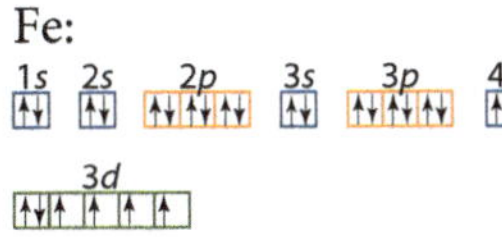

Ne:

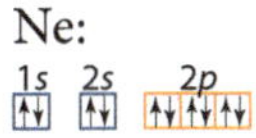

As:

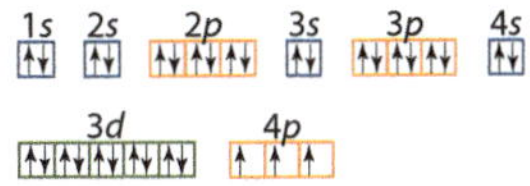

I:

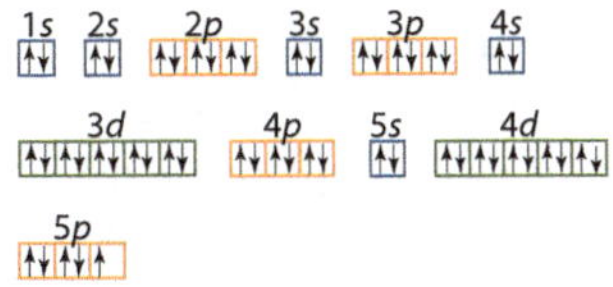

36. 55.855 u

40. 40.0% C, 6.72% H, 53.28% O; CH_2O; $C_6H_{12}O_6$

Chapter 5

1.
$2C_4H_{10} + 13O_2 \rightarrow$
$8CO_2 + 10H_2O$

2.
$FeS + 2HCl \rightarrow$
$H_2S + FeCl_2$

3.
$2C_2H_2 + 5O_2 \rightarrow$
$4CO_2 + 2H_2O$

4.
$CO + 2H_2 \rightarrow CH_3OH$

5.
$K_2CO_3 + H_2SO_4 \rightarrow$
$K_2SO_4 + CO_2 + H_2O$

6.
$WO_3 + 3H_2 \rightarrow W + 3H_2O$

7.
$2NaN_3 \rightarrow 2Na + 3N_2$

8. a. P: +3, Br: −1
b. O: −2, As: +5
c. O: −2, Cl: +7
d. F: 0 e. F: −1, U: +6
f. O: −2, C: +4
g. O: −2, H: +1, P: +5
h. O: −2, N: +5, Zn: +2
i. H: +1, C: −2.5
j: O: −2, Cr: +6
k. H: −1, K: +1 l. Fe: 0

10. butane: none

FeS: double replacement
C_2H_2: none
CO: synthesis
K_2CO_3: none
WO_3: single replacement
NaN_3: decomposition
11. a. Fe reduced, Al oxidized, single replacement
b. O reduced, S oxidized
c. N reduced, H oxidized, synthesis
d. Cl reduced, I oxidized, single replacement
13.
a.
$$Mg(OH)_2 \rightarrow MgO + H_2O$$
decomposition
b.
$$2NaCl + H_2SO_4 \rightarrow Na_2SO_4 + 2HCl$$
double replacement
c.
$$CaO + H_2O \rightarrow Ca(OH)_2$$
synthesis
d.
$$NH_4HCO_3 + NaCl \rightarrow NaHCO_3 + NH_4Cl$$
double replacement
14.
a.
$$2Ca(s) + O_2(g) \rightarrow 2CaO(s)$$
b. no reaction
c. no reaction
d.
$$2Al(s) + 3H_2SO_4(aq) \rightarrow Al_2(SO_4)_3(aq) + 3H_2(g)$$
e.
$$Ni(s) + CuCl_2(aq) \rightarrow NiCl_2(aq) + Cu(s)$$
f.
$$Ba(s) + 2H_2O(l) \rightarrow Ba(OH)_2(aq) + H_2(g)$$
g. no reaction
h.
$$2Al(s) + 3Pb(NO_3)_2(aq) \rightarrow 3Pb(s) + 2Al(NO_3)_3(aq)$$
i.
$$Li(s) + KI(aq) \rightarrow LiI(aq) + K(s)$$
j. no reaction
15.
a.
$$2LiClO_3(s) \rightarrow 2LiCl(s) + 3O_2(g)$$
b.
$$Cu(OH)_2(s) \rightarrow CuO(s) + H_2O(l)$$
c.
$$H_2CO_3(aq) \rightarrow CO_2(g) + H_2O(l)$$
16. 660 g
17. 787.5 mol O_2, 1575 mol H_2O
18. 2379 g
19. a. 0.029 mol HCl
b. 0.52 g H_2O
20. 1.75×10^4 kg
21. a. NaOH
b. 52.47 g Na_2SO_4
c. 13.31 g H_2O
22. a. 4375 mol O_2
b. 1.4 kg H_2O
c. 105 kg O_2
23. a. O_2
b. 1310 g NO
c. 96.9%
24. a. H_2SO_4
b. 8.5 g $Al(OH)_3$
c. 37.0 g $Al_2(SO_4)_3$, 11.7 g H_2O
25. a. N_2O_4
b. 2349 kg H_2O
c. 1.178×10^{29} N atoms
26. a. 54.1 mol HF
b. 102 g H_2SiF_6
c. 1300 mol HF
d. SiO_2
e. 2.89 kg H_2SiF_6
f. 91.9%
27. a. 90.5 g C_6H_5Br b. 69.9%
28. a. 5.70×10^{-6} mol NaI
b. 18.6 g NaI
c. NaI, 1232 g I_2
d. 5.846×10^{24} I atoms
29. a. 67.2% C, 6.94% H, 12.1% N, 13.8% O
b. $C_{13}H_{16}N_2O_2$
c. 1.0×10^{20} C atoms
30. 545.9 nm
31. Y: $[Kr]5s^2 4d^1$
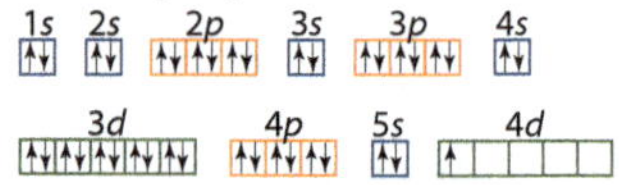

Sn: $[Kr]5s^2 4d^{10} 5p^2$
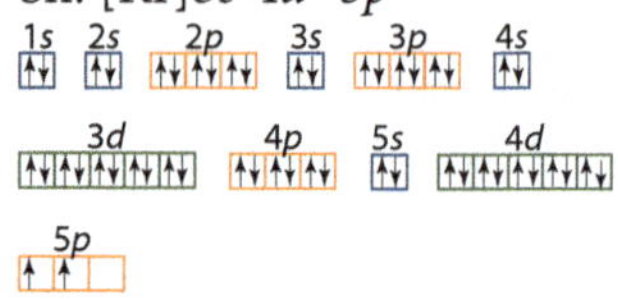

Cr: $[Ar]4s^1 3d^5$
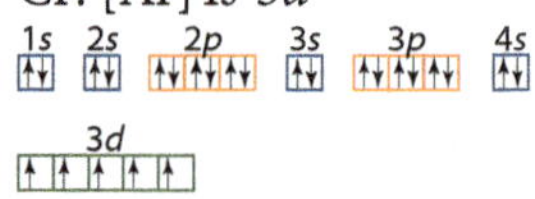

I: $[Kr]5s^2 4d^{10} 5p^5$
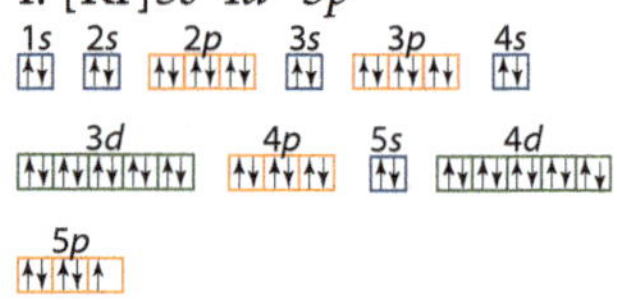

33. 28.085 u
34.
$Br^- > Kr > Rb^+ > Sr^{2+} > Y^{3+}$
35. Fr < Ba < Fe < Se < Cl < F
36.
a.
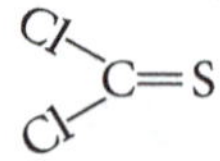

b.

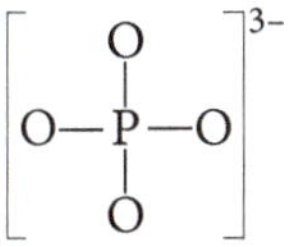

c.

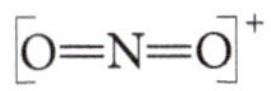

d.

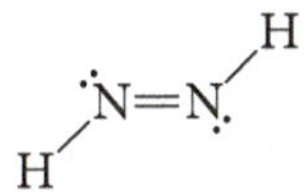

e.

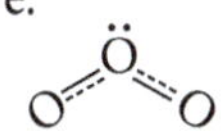

f.

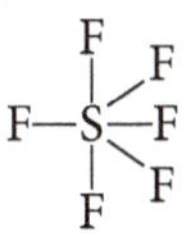

g.
:C≡O:

h.

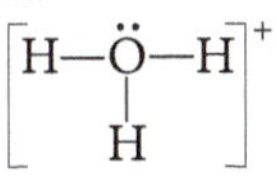

37. a. trigonal planar
b. tetrahedral
c. linear
d. trigonal planar/bent at each N atom
e. tetrahedral/bent
f. trigonal bipyramidal
g. linear
h. tetrahedral/pyramidal

Chapter 6

3. 0–900 m/s, 350 m/s
5. 10.4 m
6. 12,800,000 Pa, 12,800 kPa, 95,700 Torr, 126 atm, 128 bar
7. 53 atm
16. ethanol
19. a. 7.8 kJ b. 162 kg
c. 4.00×10^3 kJ d. 176 mL
e. 1920 kJ
20. a. −75°C b. −55°C, 218 K, 4 atm, 3000 Torr
c. −55°C, −35°C d. 40 atm
e. 15 atm, 1000 atm
f. 35 atm
22. 1500 Torr
25. 93°C
26.
Br: $1s^2 2s^2 2p^6 3s^2 3p^6 4s^2 3d^{10} 4p^5$

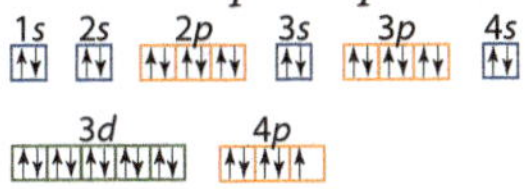

Y: $1s^2 2s^2 2p^6 3s^2 3p^6 4s^2$ $3d^{10} 4p^6 5s^2 4d^1$

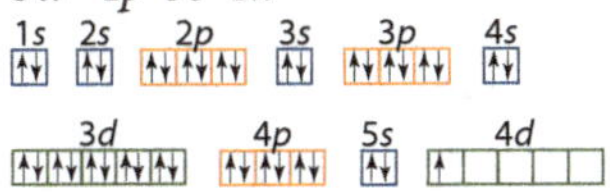

27. a. 62.5 mol
b. 9170 g
28.
$$2NH_4NO_3(s) \rightarrow 2N_2(g) + O_2(g) + 4H_2O(g)$$
29. a. $AgNO_3$ b. 1.67 g
c. 1.92 g
30. $C_{13}H_{18}O_2$ (empirical and molecular)
31. S: +6, C: +4
32. C—N: nonpolar covalent
I—I: nonpolar covalent
Al—Cl: polar covalent
34.
a.

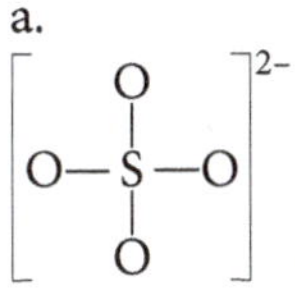

tetrahedral

b.

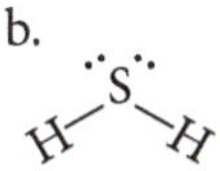

bent

c.

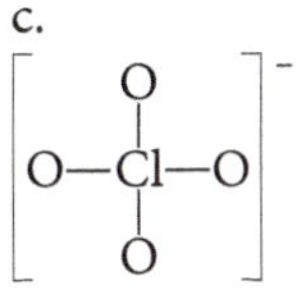

tetrahedral

Chapter 7

1. a. 2.52 L b. 463 K
2. 0.77 bar
3. 2010 cm^3
9. 2271 L
10. 4950°F
11. 15.5 L
12. 0.698 mol
13. a. 68,300 mol
b. 1.10×10^3 kg
c. 0.707 kg/m^3 d. 4.11×10^{28} carbon atoms
14. a. 834 kg b. 4.20×10^2 kg
15. a. 1.876 mol b. chlorine
16. a. 11.2 mol b. 821 L
17. 28 psig
18. a. 58.69 mol b. 1333 L
19. a. 20.8 g/mol
b. 6.86 g/mol
20. 21,600,000 K
21. CO_2: 1.938 g/L, O_2: 1.409 g/L, N_2: 1.234 g/L, CH_4: 0.7064 g/L
25. $P_{nitrogen} = 1.32$ atm
$P_{argon} = 0.193$ atm
$P_{methane} = 0.097$ atm
26. a. 602 Torr
b. 5.40 mol
27. $X_{propane} = 0.479$,
$X_{methane} = 0.521$,
$P_{propane} = 2.55$ bar,
$P_{methane} = 2.77$ bar,
$P_T = 5.33$ bar
28. a. $n_{oxygen} = 0.128$ mol
$n_{hydrogen} = 0.0432$ mol
b. $P_{oxygen} = 1.03$ atm
$P_{hydrogen} = 0.350$ atm
c. $P_T = 1.38$ atm
29. $n_{nitrogen} = 0.011$ mol
$P_{nitrogen} = 0.97$ atm
$X_{nitrogen} = 0.97$
$n_{water} = 0.000307$ mol
$P_{water} = 0.026$ atm
$X_{water} = 0.026$
30. 0.467 g
31. 6.01 g
32. 2.5 L O_2, 5.0 L CO_2
33. 150.0 m^3 O_2, 120.0 m^3 H_2O, 90.00 m^3 CO_2
34. 4.5 g $Fe(OH)_3$
3.3 g Fe_2O_3
35. 71.4 kg
36. 71.4 kg
37. 0.016 g
38. 0.351 L
39. a. 0.682 mol KI,

0.341 mol Cl_2,
0.682 mol KCl, 0.341 mol I_2
b. 113 g KI, 24.2 g Cl_2,
50.8 g KCl, 86.6 g I_2
40. a. Al
b. 34.80 L
41. a. CO b. 55.0 m^3
c. 605.0 m^3
42. a. 0.17 L b. 5.0 g
43. 0.033 mol/s
44. 0.0443 m/s
45. 2.0×10^2 g/mol
46. 1/1.8
47. 62.1 nm (UV)
48. 15.867% C, 2.2192% H, 18.5040% N, 63.4094% O
49. a. 84.0%
b. 4.85×10^{26} molecules
c. 121 kg
50. a. plutonium b. bromine
c. plutonium d. strontium
e. bromine f. bromine
51. 9180 kJ
52. gas, solid

Chapter 8

14. insoluble: a, e, f, soluble: b, c, d, g, h, i
20. a. 68%
b. 0.0015, 0.9985
d. 23% decrease (from 0.7 to 0.54)
21. b. 31.7
d.

	$Al_2(SO_4)_3$	H_2O
% of total mass	0.282	0.718
% of total moles	0.020	0.980

22. a. 7.20×10^2 g b. 901 g
23. 3.94 *M*
24. 636 g
25. 110 mL
26. 2.2 *M*
27. a.
$Na_2SO_4(aq) + Ba(NO_3)_2(aq)$
$\rightarrow BaSO_4(s) + 2NaNO_3(aq)$
b. 67 g
28. 164 mL
29. 45,000 L
30. a. 365 g
b. 367 g
31. 1.32 *m*
32. 720 mL
33. a. $CaCO_3$
b.
$Na_2CO_3(aq) + Ca(NO_3)_2(aq)$
$\rightarrow 2NaNO_3(aq) + CaCO_3(s)$
c.
$2Na^+(aq) + CO_3^{2-}(aq)$
$+ Ca^{2+}(aq) + 2NO_3^-(aq)$
$\rightarrow 2Na^+(aq) + 2NO_3^-(aq)$
$+ CaCO_3(s)$

$CO_3^{2-}(aq) + Ca^{2+}(aq) \rightarrow$
$CaCO_3(s)$
d. Na^+, NO_3^-
34. precipitate: AgCl
spectator ions: Li^+, NO_3^-
35.a.
$MgCl_2(aq) + Pb(NO_3)_2(aq)$
$\rightarrow Mg(NO_3)_2(aq) + PbCl_2(s)$

$Mg^{2+}(aq) + 2Cl^-(aq)$
$+ Pb^{2+}(aq) + 2NO_3^-(aq)$
$\rightarrow Mg^{2+}(aq) + 2NO_3^-(aq)$
$+ PbCl_2(s)$

$Pb^{2+}(aq) + 2Cl^-(aq)$
$\rightarrow PbCl_2(s)$
b. spectator ions: Mg^{2+}, NO_3^-, precipitate: $PbCl_2$
c. 5.6 g
39. 100.520°C
40. 3.46 *m*, 1.98 *m*, 6.23 *m*
41. 9.7 *m*
42. 73.3 g
43. 670 g
44. −23.7°C
45. −123.0°C
46. ΔT_b = +0.26°C, +0.39°C, +0.26°C (this last value is ideal, but is likely to be less)
47. T_f = −0.15°C, −0.07°C, −0.04°C, −0.11°C
48. −7.44°C
49. 11.7 g/mol
50. 17.5 L $CO_2(g)$, 35.0 L $H_2O(g)$
52. 8.50 atm
53. 55 kg
54. 22.52 kg
55. 60.871% C, 4.3783% H, 34.7503% O
56. +5

Chapter 9

4. a. hydrofluoric acid (monoprotic) b. hypochlorous acid (monoprotic) c. nitrous acid (monoprotic) d. sulfurous acid (diprotic) e. hydrobromic acid (monoprotic) f. phosphoric acid (triprotic) g. hydrosulfuric acid (diprotic) h. chromic acid (diprotic) i. acetic acid (monoprotic)
6. a. $HClO_2$ b. $HBrO_4$ c. HF d. HI
8. a.

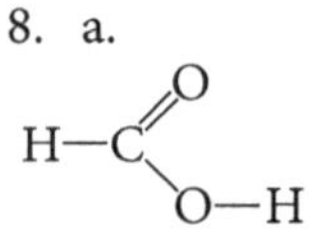

b.

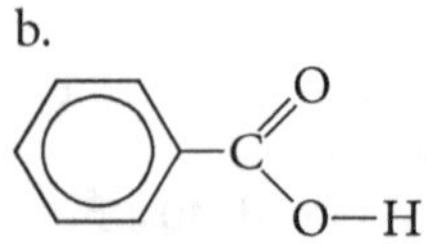

13.

	acid	base	conj. acid	conj. base
a	HNO_2	CN^-	HCN	NO_2^-
b	H_2O	HCO_3^-	H_2CO_3	OH^-
c	HSO_4^-	$C_2O_4^{2-}$	$HC_2O_4^-$	SO_4^{2-}
d	NH_4^+	HPO_4^{2-}	$H_2PO_4^-$	NH_3

19.

a.

$HCl+KOH \rightarrow KCl+H_2O$

$H^+ + Cl^- + K^+ + OH^- \rightarrow K^+ + Cl^- + H_2O$

$H^+ + OH^- \rightarrow H_2O$

b.

$HNO_3+NaOH \rightarrow NaNO_3+H_2O$

$H^+ + NO_3^- + Na^+ + OH^- \rightarrow Na^+ + NO_3^- + H_2O$

$H^+ + OH^- \rightarrow H_2O$

c.

$H_2SO_4+Ca(OH)_2 \rightarrow CaSO_4+2H_2O$

$2H^+ + SO_4^{2-} + Ca^{2+} + 2OH^- \rightarrow Ca^{2+} + SO_4^{2-} + 2H_2O$

$2H^+ + 2OH^- \rightarrow 2H_2O$

d.

$H_2SO_3 + 2KOH \rightarrow K_2SO_3 + 2H_2O$

$2H^+ + SO_3^{2-} + 2K^+ + 2OH^- \rightarrow 2K^+ + SO_3^{2-} + 2H_2O$

$2H^+ + 2OH^- \rightarrow 2H_2O$

20.

a.

$2Al+3H_2SO_4 \rightarrow Al_2(SO_4)_3+3H_2$

$2Al+6H^+ +3SO_4^{2-} \rightarrow 2Al^{3+} +3SO_4^{2-} +3H_2$

$2Al+6H^+ \rightarrow 2Al^{3+} +3H_2$

b.

$Zn+2HCl \rightarrow ZnCl_2+H_2$

$Zn+2H^+ +2Cl^- \rightarrow Zn^{2+} +2Cl^- +H_2$

$Zn+2H^+ \rightarrow Zn^{2+} +H_2$

21.

a.

$2HNO_3+CaCO_3 \rightarrow Ca(NO_3)_2 +H_2CO_3$

b.

$2HCl+BaCO_3 \rightarrow BaCl_2+H_2CO_3$

c.

$H_2SO_4+Na_2CO_3 \rightarrow Na_2SO_4+H_2CO_3$

d.

$2HNO_3+MgCO_3 \rightarrow Mg(NO_3)_2 +H_2CO_3$

Note: These answers all show the formation of carbonic acid, H_2CO_3. This acid is unstable and immediately breaks down to CO_2 and water. Thus, each equation could show the products as: ... + CO_2 + H_2O.

22.

a.

$H_3PO_4+F^- \rightarrow H_2PO_4^- +HF$

right

b.

$HSO_4^- +CO_3^{2-} \rightarrow SO_4^{2-} +HCO_3^-$

right

c.

$ClO_2^- +H_3O^+ \rightarrow HClO_2+H_2O$

right

d.

$HClO_3+HCO_3^- \rightarrow ClO_3^- +H_2CO_3$

right

e.

$H_3O^+ +HS^- \rightarrow H_2O+H_2S$

right

f.

$HPO_4^{2-} +NH_3 \rightarrow PO_4^{3-} +NH_4^+$

left

23. a. 81.4 g, 42.0 L b. 616 mL

24. a. 160 g b. 22.7 L

25. a. 110 g b. 1.1 L

28. basic: a, c, f, g
acidic: b, e
neutral: d, h

32.

a.
$[H_3O^+] = 1.56 \times 10^{-3}$ *M*,
$[OH^-] = 6.4 \times 10^{-12}$ *M*

b.
$[H_3O^+] = 4.5 \times 10^{-13}$ *M*,
$[OH^-] = 0.0220$ *M*

c.
$[H_3O^+] = 4.0 \times 10^{-15}$ *M*,
$[OH^-] = 2.50$ *M*

d.
$[H_3O^+] = 8.911 \times 10^{-5}$ *M*,
$[OH^-] = 1.1 \times 10^{-10}$ *M*

33. a. pOH = 1.602,
pH = 12.4

b. pH = −0.78, pOH = 14.8

c. pH = −1.08, pOH = 15.1

d. pH = 12.6, pOH = 1.391

34.

a.
$[H_3O^+] = 0.10$ *M*,
$[OH^-] = 1.0 \times 10^{-13}$ *M*

b.
$[H_3O^+] = 3 \times 10^{-4}$ *M*,
$[OH^-] = 3 \times 10^{-11}$ *M*

c.
$[H_3O^+] = 1 \times 10^{-4}$ *M*,
$[OH^-] = 1 \times 10^{-10}$ *M*

d.
$[H_3O^+] = 1.3 \times 10^{-12}$ *M*,
$[OH^-] = 7.6 \times 10^{-3}$ *M*

e.
$[H_3O^+] = 1 \times 10^{-12}\ M$,
$[OH^-] = 8 \times 10^{-3}\ M$
f.
$[H_3O^+] = 1 \times 10^{-7}\ M$,
$[OH^-] = 1 \times 10^{-7}\ M$
g.
$[H_3O^+] = 1.0 \times 10^{-10}\ M$,
$[OH^-] = 1.0 \times 10^{-4}\ M$
h.
$[H_3O^+] = 1 \times 10^{-14}\ M$,
$[OH^-] = 1\ M$

35.

	pH	pOH	$[H_3O^+]$	$[OH^-]$
a	5.6	8.4	$3 \times 10^{-6}\ M$	$4 \times 10^{-9}\ M$
b	4.7	9.26	$1.8 \times 10^{-5}\ M$	$5.5 \times 10^{-10}\ M$
c	9.00	5.00	$1.0 \times 10^{-9}\ M$	$1.0 \times 10^{-5}\ M$
d	2.5	11.5	$3 \times 10^{-3}\ M$	$3 \times 10^{-12}\ M$
e	2.35	11.66	$4.5 \times 10^{-3}\ M$	$2.2 \times 10^{-12}\ M$
f	13.0	1.00	$1.0 \times 10^{-13}\ M$	$0.10\ M$
g	6.8	7.22	$1.7 \times 10^{-7}\ M$	$6.0 \times 10^{-8}\ M$
h	13.5	0.5	$3.2 \times 10^{-14}\ M$	$0.31\ M$
i	2.0	12.0	$1 \times 10^{-2}\ M$	$1 \times 10^{-12}\ M$
j	12.51	1.49	$3.1 \times 10^{-13}\ M$	$3.2 \times 10^{-2}\ M$

36. a. $2.14 \times 10^{-3}\ M$
b. $1.325\ M$
c. $0.01855\ M$
d. $2.780\ M$
37. $5.82 \times 10^{-3}\ M$
38. $3.77 \times 10^{-2}\ M$
39. 7.3%
40. 0.146 *M* max
41. 0.4552 *M*
42. 16.8 kJ
43. a. HNO_3 b. 0.7500 mol
c. 0.7500 mol d. 0.1388 mol
44.
a.
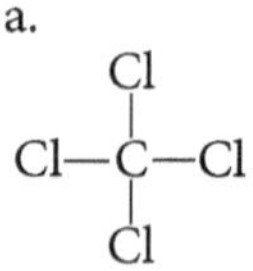
b.
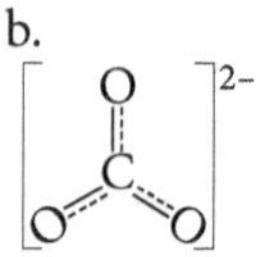
c.
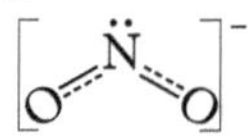
d.
H—C≡C—H
e.
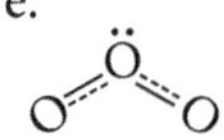
f.
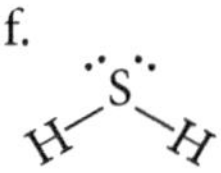
g.
O=C=O
h.
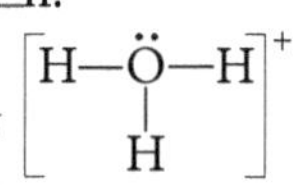

45.
a.
$$Na_2CO_3 + Ba(NO_3)_2 \rightarrow 2NaNO_3 + BaCO_3$$
$$2Na^+ + CO_3^{2-} + Ba^{2+} + 2NO_3^- \rightarrow 2Na^+ + 2NO_3^- + BaCO_3$$
$$CO_3^{2-} + Ba^{2+} \rightarrow BaCO_3$$
b. spectator ions: Na^+, NO_3^-
precipitate: $BaCO_3$
c. 0.25 g
46. 2.3×10^{-3} mol/s
47. 0.276 g
48. a. 1.353 mol b. 43.29 g
49. a. $CuSO_4$ b. 34.1 g
c. 12.5 g
50. 105 g/mol
51. 5600 eV
52. 58.5 g/mol

Chapter 10

4. b. −890 kJ
c.
$$CH_4(g) + 2O_2(g) \rightarrow CO_2(g) + 2H_2O(l)$$
$$\Delta H = -890\ \text{kJ}$$
$$CH_4(g) + 2O_2(g) \rightarrow CO(g) + 2H_2O(l) + \frac{1}{2}O_2(g)$$
$$\Delta H = -607\ \text{kJ}$$
$$CO(g) + 2H_2O(l) + \frac{1}{2}O_2(g) \rightarrow CO_2(g) + 2H_2O(l)$$
$$\Delta H = -283\ \text{kJ}$$
9.
a.
$$2C(s) + H_2(g) \rightarrow C_2H_2(g)$$
$$\Delta H_f^\circ = 227.4\ \text{kJ/mol}$$
b.
$$2Li(s) + S(s) + 2O_2(g) \rightarrow Li_2SO_4(s)$$
$$\Delta H_f^\circ = -1436.5\ \text{kJ/mol}$$
c.
$$2S(s) + 3O_2(g) \rightarrow 2SO_3(g)$$
$$\Delta H_f^\circ = -395.7\ \text{kJ/mol}$$
d.
$$N_2(g) + 3H_2(g) \rightarrow 2NH_3(g)$$
$$\Delta H_f^\circ = -45.9\ \text{kJ/mol}$$
10. a. −58.1 kJ/mol
b. −24.8 kJ/mol
c. 108.5 kJ/mol
d. 179.2 kJ/mol
e. 37.1 kJ/mol
12. a. 52.6 kJ/mol
b. −424.7 kJ/mol
c. 49.2 kJ/mol
13. a. −725.9 kJ/mol
b. −1366.8 kJ/mol
14. a. neg, b. neg, c. pos
19. a. 175.2 kJ
b. −394.3 kJ
20. a. 97 kJ/mol (not spontaneous) b. −135.9 kJ/mol (spontaneous) c. −569 kJ/mol (spontaneous)
21. 25°C: −32.7 kJ/mol
500°C: 61.4 kJ/mol

22.

a.

$\Delta G^\circ = \Sigma G_p^\circ - \Sigma G_r^\circ =$ 148.4 kJ/mol

$\Delta G^\circ = \Delta H^\circ - T\Delta S =$ 148.4 kJ/mol

b.

$\Delta G^\circ = \Sigma G_p^\circ - \Sigma G_r^\circ =$ −550.8 kJ/mol

$\Delta G^\circ = \Delta H^\circ - T\Delta S =$ −550.8 kJ/mol

c.

$\Delta G^\circ = \Sigma G_p^\circ - \Sigma G_r^\circ =$ 107.8 kJ/mol

$\Delta G^\circ = \Delta H^\circ - T\Delta S =$ 107.7 kJ/mol

d.

$\Delta G^\circ = \Sigma G_p^\circ - \Sigma G_r^\circ =$ −157.1 kJ/mol

$\Delta G^\circ = \Delta H^\circ - T\Delta S =$ −157.0 kJ/mol

e.

$\Delta G^\circ = \Sigma G_p^\circ - \Sigma G_r^\circ =$ 69.0 kJ/mol

$\Delta G^\circ = \Delta H^\circ - T\Delta S =$ 69.0 kJ/mol

27. $R = k[A][B]^2$, tripling [B] increases R by factor of 9

28. $A_2 + B \rightarrow C + E_2$, step 1, $R = k[A_2]^2$

29.

a.

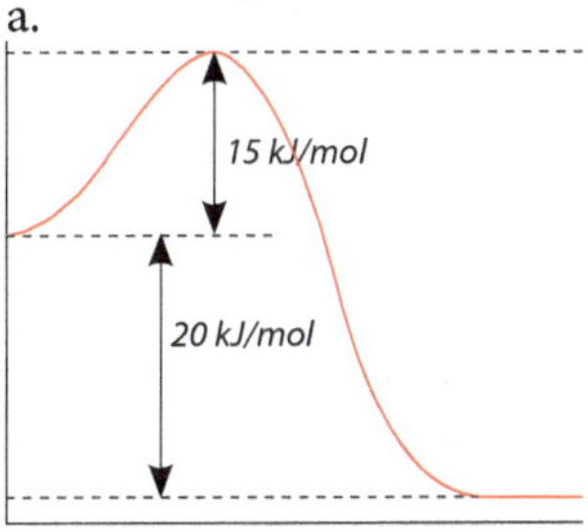

b.

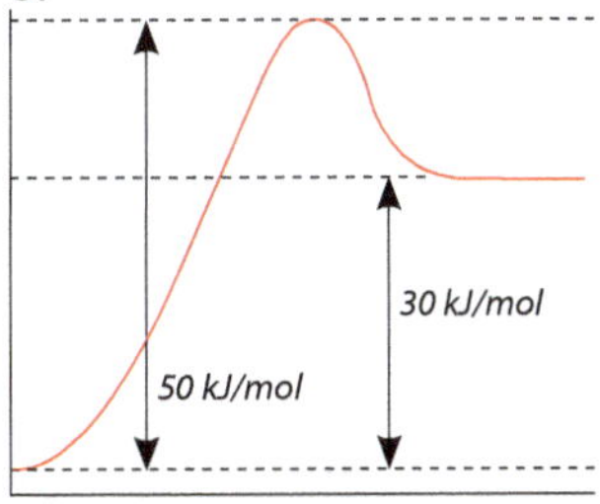

c.

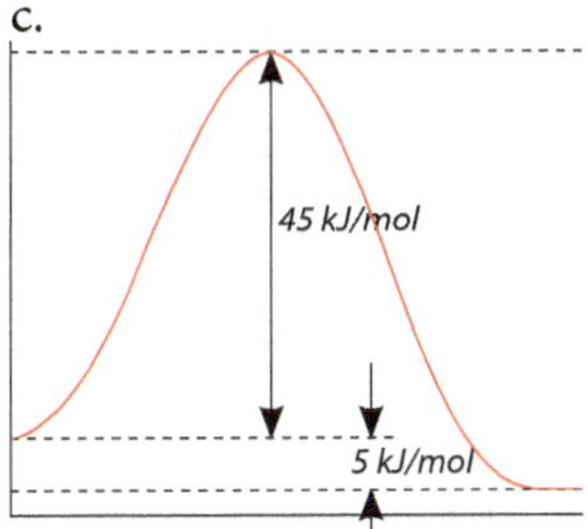

31. a. $R = k[A]^2[B]$

b. $k = 0.058\ s^{-1}M^{-2}$

c. 0.00091 M/s

32. 1.01 L

33. 1.9 g

34. 11.00 L HF, 1.833 L O_3

35. 0.2763 M

36. $[H_3O^+] = 1.5 \times 10^{-10}\ M$, pOH = 4.2, $[OH^-] = 6.7 \times 10^{-5}\ M$

37. $T_f = -12.2$°C, $T_b = 103.32$°C

38. 0.597 M

Chapter 11

1. a. reverse b. forward c. neither

2.

a.

$$K = \frac{[CH_4][H_2S]^2}{[H_2]^4[CS_2]}$$

b.

$$K = [Cd^{2+}][OH^-]^2$$

c.

$$K = \frac{[CO_2][H_2]}{[CO][H_2O]}$$

d.

$$K = \frac{[H_3O^+][NO_2^-]}{[HNO_2]}$$

e.

$$K = \frac{[SO_3]^2}{[SO_2]^2[O_2]}$$

f.

$$K = \frac{[Ag^+][CN^-]^2}{[Ag(CN)_2^-]}$$

g.

$$K = \frac{[H^+][NO_2^-]}{[HNO_2]}$$

h.

$$K = \frac{[CO][H_2]}{[H_2O]}$$

3. a. beginning b. after equilibrium is reached

7. 8.2×10^{13}

8. 117 M

9. a. left b. right c. at equilibrium

10. $4.9 \times 10^4\ M$

11. a. most: N_2, O_2, least: NO

b. most: HBr, least: H_2, Br_2

12. b. 0.0250

17. 1.3×10^{-4}

18. 3.9×10^{-3}

19. pH = 1.879, $K_a = 0.00201$

20. $[H_3O^+] = 1.10 \times 10^{-2}\ M$
$[ClCH_2COO^-] = 1.10 \times 10^{-2}\ M$
$[ClCH_2COOH] = 8.90 \times 10^{-2}\ M$
$K_a = 1.36 \times 10^{-3}$

21. $8.5 \times 10^{-2}\ M$

22. 0.0064%, 0.0088%, 0.00018%

23. $1.0 \times 10^{-7}\ M$

25. a. $[C_{10}H_{15}ON] = 0.035\ M$
$[C_{10}H_{15}ONH^+] = [OH^-] = 2.1 \times 10^{-3}\ M$

b. 1.3×10^{-4}

26. a. NH_3 b. NH_3OH^+

c. NH_4^+: $K_a = 5.6 \times 10^{-10}$
NH_3OH^+: $K_a = 1.1 \times 10^{-6}$

30. 7.63×10^{-9}
31. 4.93×10^{-5}
32. 1.72×10^{-7}
33. 1.7×10^{-5}
34. 4.33×10^{-9}
35. 4.23×10^{-6} *M*
36. a. 2.14×10^{-5} g/100 mL
b. 4.82×10^{-3} g/100 mL
c. 2.24×10^{-10} g/100 mL
d. 8.99×10^{-6} g/100 mL
e. 1.8×10^{-4} g/100 mL
f. 3.53×10^{-6} g/100 mL
g. 3.15×10^{-3} g/100 mL
37. $PbCO_3$, $CdCO_3$, Ag_2CO_3, $SrCO_3$, $BaCO_3$, $MgCO_3$
38. a.
$1.39 \times 10^{-8} < 5.02 \times 10^{-6}$; no precipitate
b. $7.51 \times 10^{-6} < 1.20 \times 10^{-5}$; no precipitate
c. $0.0109 > 1.77 \times 10^{-10}$; precipitate will form
d.
$1.5 \times 10^{-21} < 4.87 \times 10^{-17}$; no precipitate
e. $4.1 \times 10^{-8} > 2.58 \times 10^{-9}$; precipitate will form
39. empirical: C_3H_4O
molecular: $C_9H_{12}O_3$
40. a. 1.79 mol
b. 2.08 kg c. 3,003 g
41. a. 68.1% b. 79.3%
42. 10.9 L
43. a. 201.37 kPa
b. 0.00123 mol c. 0.0419 g
44. 10.51 L
45. -0.94°C
46. NO: +2, NO_2: +4, N_2O: +1, N_2: 0, NO_2^-: +3, HCN: -3, HNO_3: +5, NH_3 -3
47. 44.4 kJ/mol

Chapter 12

2.

a.

+3−1 +1 −2 +1+3−2 +1−1

$PBr_3 + 3H_2O \rightarrow H_3PO_3 + 3HBr$

Not a redox reaction.

b.

+1+5−2 +1−1 0

$2KClO_3 \rightarrow 2KCl + 3O_2$

Cl is reduced; it is the oxidizing agent. O is oxidized; it is the reducing agent.

c.

+1+6 −2 +1 −1

$2H_2SO_4 + 2NaBr \rightarrow$

0 +4 −2 +1+6−2 +1 −2

$Br_2 + SO_2 + Na_2SO_4 + 2H_2O$

S is reduced; it is the oxidizing agent. Br is oxidized; it is the reducing agent.

d.

+1 −1 +1+5−2

$NaCl + AgNO_3 \rightarrow$

+1−1 +1+5−2

$AgCl + NaNO_3$

Not a redox reaction.

e.

+1−1 +1 +1−2 +1−1 +1+5−2

$NaI + 3HClO \rightarrow 3HCl + NaIO_3$

Cl is reduced; it is the oxidizing agent. I is oxidized; it is the reducing agent.

f.

+4−2 +1+5−2 +1−2

$3SO_2 + 2HNO_3 + 2H_2O$

+1+6−2 +2−2

$\rightarrow 3H_2SO_4 + 2NO$

N is reduced; it is the oxidizing agent. S is oxidized; it is the reducing agent.

3.

	strongest	weakest
a.	Ca	NO
b.	Li	Br^-
c.	Mg	Cr^{3+}
d.	Sn	F^-

4. a. yes b. yes c. no d. no

6.

a.

$SO_3^{2-} + 2OH^- \rightarrow$

$SO_4^{2-} + H_2O + 2e^-$

b.

$N_2 + 6H_2O + 6e^- \rightarrow$

$2NH_3 + 6OH^-$

c.

$N_2 + 8H^+ + 6e^- \rightarrow 2NH_4^+$

d.

$Sn^{2+} \rightarrow Sn^{4+} + 2e^-$

e.

$ClO_3^- + 6H^+ + 6e^- \rightarrow$

$Cl^- + 3H_2O$

f.

$4OH^- \rightarrow O_2 + 2H_2O + 4e^-$

g.

$H_2SO_3 + H_2O \rightarrow$

$SO_4^{2-} + 4H^+ + 2e^-$

h.

$O_2 + 2H_2O + 4e^- \rightarrow 4OH^-$

7.

a.

$2FeCl_3 + H_2S \rightarrow$

$2FeCl_2 + 2HCl + S$

oxidizing agent: Fe
reducing agent: S

b.

$I_2 + 5HClO + H_2O \rightarrow$

$2HIO_3 + 5HCl$

oxidizing agent: Cl
reducing agent: I

c.

$4KMnO_4 + 5CH_3OH$

$+ H_2O \rightarrow 4Mn(OH)_2$

$+ 5HCOOH + 4KOH$

oxidizing agent: Mn
reducing agent: C

d.

$H_2O_2 + 2ClO_2 \rightarrow$

$2HClO_2 + O_2$

oxidizing agent: Cl
reducing agent: O

e.

$2Al + LiCl + LiNO_2 + 2H_2O \rightarrow 2LiAlO_2 + NH_4Cl$

oxidizing agent: N
reducing agent: Al

f.

$14HCl + 2HMnO_4 \rightarrow 5Cl_2 + 2MnCl_2 + 8H_2O$

oxidizing agent: Mn
reducing agent: Cl

g.

$2S + 2HNO_3 + H_2O \rightarrow 2H_2SO_3 + N_2O$

oxidizing agent: N
reducing agent: S

h.

$2HMnO_4 + HBr \rightarrow HBrO_3 + 2MnO_2 + H_2O$

oxidizing agent: Mn
reducing agent: Br

9. b. oxidizing agent: Fe, reducing agent: Al
d. 22 g
e. 45 g

19.

	cathode	anode	E°_{cell}
a.	Ag	Sn	0.9371 V
b.	Fe	Co	1.05 V
c.	Br	Na	3.78 V
d.	Br	I	0.530 V

24. a. 80.5% b. 97 g c. NO_2
25. 65.2 atm
26. 0.0115 mol
27. 1.96×10^{24} carbon atoms
28. 1.08 L
29. –1468.1 kJ/mol

Chapter 13

6. a. 2-bromopropane, 1-bromopropane
b. 1-chloro-2-bromopropane

9.

a.

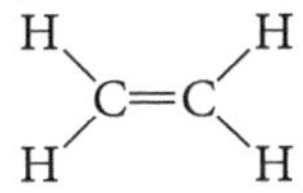

ethene, aka, ethylene

b.

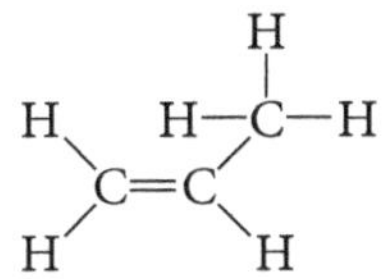

propene

c.

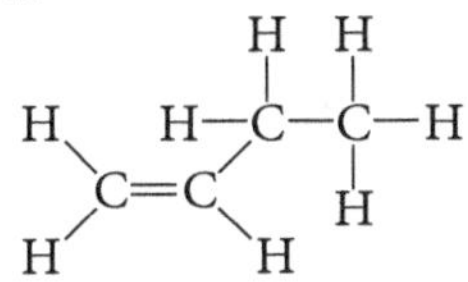

1-butene

10.

a. 3-methyl-3-ethylnonane
b. 2,7-methyl-4-ethyloctane
c. 1,1-dibromo-2,3,4-trichlorobutane
d. 4,4,5-tribromo-1-heptene
e. 4,6-methyl-2-octyne
f. 3,6-methyl-*trans*-3-octene

11.

a.

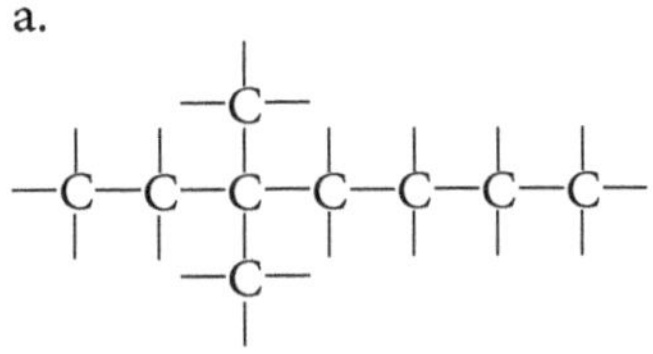

b.

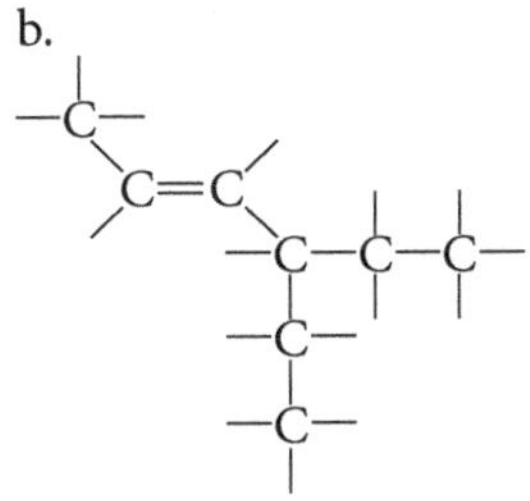

c.

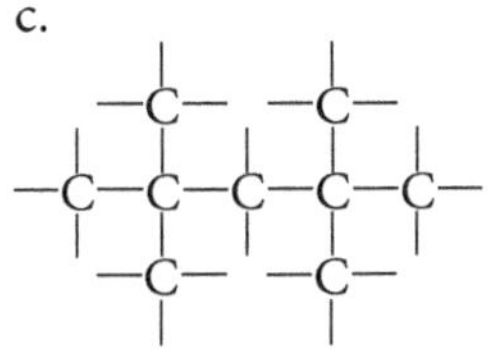

d.

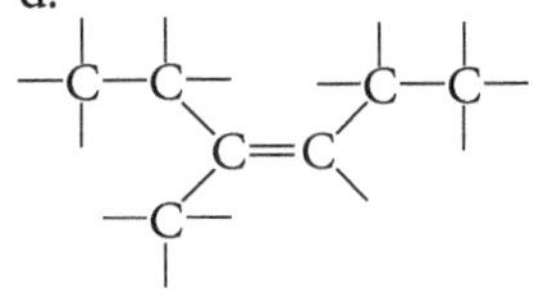

e.

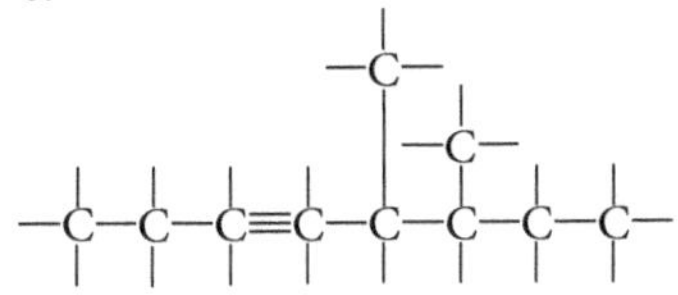

f.

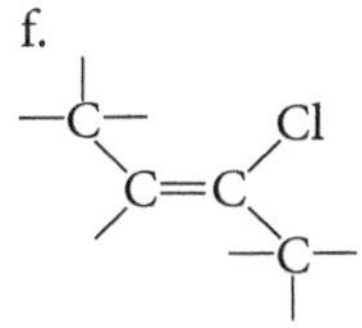

12.

a. 4-methyl-1-hexyne
b. 3-chloroheptane
c. 2,2,3-trimethyl-*trans*-4-octene
d. 2-bromo-4,4-diethylheptane
e. 7-ethyl-*cis*-3-decene

13. example:

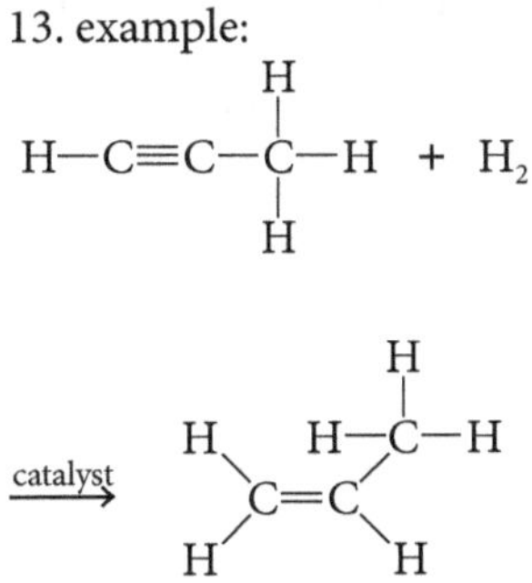

15.

a. ethylbenzene
b. 2-chlorotoluene or 1-chloro-2-methylbenzene
c. 1,2,5-trichlorobenzene
d. 2-chloro-4-ethyltoluene or 1-chloro-2-methyl-5-ethylbenzene
e. 1,2-dibromobenzene
f. 1,2,3,5-tetramethylbenzene or 2,3,5-trimethyltoluene

16.

a.

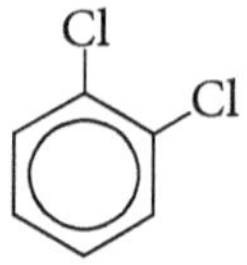

b.

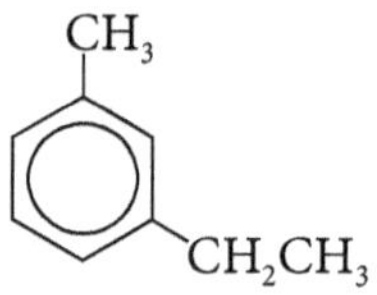

c.

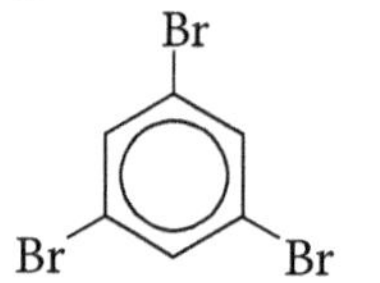

d.

32. a. carboxylic acid; propanoic acid
b. alcohol; 2,3-dibutanediol
c. ether; dimethyl ether
d. secondary amine; diethylamine

41.

a.

$$2C_8H_{18} + 25O_2 \rightarrow$$

$$16CO_2 + 18H_2O$$

b. 8340 L c. 998 g

42. 2.50×10^4 L

43. a. 10.4 *M*
b. $[H_3O^+] = 0.135$ *M*,
pH = 0.870, pOH = 13.1,
$[OH^-] = 7.4 \times 10^{-14}$ *M*
c. 0.135 L

44.

a.

$$2C_6H_6 + 15O_2 \rightarrow$$

$$12CO_2 + 6H_2O$$

b. −3267.5 kJ/mol

45. −1325.3 kJ/mol

46. 1.8×10^{-4}

47. $9.01 \times 10^{-6} > 5.02 \times 10^{-6}$; precipitate will form

48.

a.

$$3CuS + 8HNO_3 \rightarrow 3Cu(NO_3)_2$$

$$+2NO + 3S + 4H_2O$$

b.

$$2Fe + 3SnCl_4 \rightarrow$$

$$2FeCl_3 + 3SnCl_2$$

49.

a.

$$2Bi(OH)_3 + 3K_2SnO_2 \rightarrow$$

$$2Bi + 3K_2SnO_3 + 3H_2O$$

b.

$$CO_2 + 2NH_2OH \rightarrow$$

$$N_2 + CO + 3H_2O$$

Appendix A

A dash indicates that it would be either silly or incorrect to write the answer that way, so I didn't. Silly would be because there are simply too many zeros, or no zeros at all. Incorrect would be because we are unable to express the result this way and still show the correct number of significant digits.

	Standard Notation	Scientific Notation
1	5,740 ft	5.74×10^{3} ft
2	0.00354 kg	3.54×10^{-3} kg
3	0.04111 L	4.111×10^{-2} L
4	400,000 mi	4×10^{5} mi
5	–	1.5499×10^{-15} m
6	0.00075 m^3	7.5×10^{-4} m^3
7	983,600,000 ft/s	9.836×10^{8} ft/s
8	605,000 s	6.05×10^{5} s
9	5.57 g/cm^3	–
10	–	1.0×10^{1} m^3/min
11	60,000,000,000 L/hr	6×10^{10} L/hr
12	0.0052 m^3	5.2×10^{-3} m^3
13	0.876 in^2	8.76×10^{-1} in^2
14	9.802 m/s^2	–
15	0.0003001 kg/min	3.001×10^{-4} kg/min
16	4,771 kg/m^3	4.771×10^{3} kg/m^3
17	13,600,000,000 mg/m^3	1.36×10^{10} mg/m^3
18	–	1.5×10^{13} cm
19	29 m/s	2.9×10^{1} m/s
20	0.0000249 in	2.49×10^{-5} in
21	33,700,000 mph	3.37×10^{7} mph

Appendix A

Units, Unit Conversions, Significant Digits, and Scientific Notation

Note: At the end of this appendix is a problem set students may use to practice correct use of metric prefixes, unit conversions, significant digits, and scientific notation.

A.1 Units of Measure

Units of measure are crucial in science. Science is about making measurements, and a measurement without its units of measure is a meaningless number. For this reason, your answers to computations in scientific calculations must always show the units of measure.

The two major systems of units students should know about are the *International System of Units*, known as the *SI system* or the metric system, and the *U.S. Customary System*, or USCS. You have probably studied these systems before and should be already familiar with some of the SI units and prefixes. In this course, we do not make much use of the measurement system you are most familiar with—the USCS. For scientific work, the entire international scientific community uses the SI system. But here I address the USCS briefly before moving on.

A.1.1 The US Customary System

Americans are generally comfortable with measurements in feet, miles, gallons, inches, and degrees Fahrenheit because they grow up using this system and are very familiar with it. But in fact, the USCS is rather cumbersome. One problem is that there are many different units of measure for every kind of physical quantity. Just for measuring length or distance, for example, we have the inch, foot, yard, and mile. The USCS is also full of random numbers such as 3, 12, and 5,280. A third problem is that there is no inherent connection between units for different types of quantities. Gallons have nothing whatsoever to do with feet, and quarts have nothing to do with miles.

The USCS may be familiar ground, and it may even feel patriotic to prefer it, but it is not the system of measurement scientists use. Scientists everywhere use the SI, and it is to that system we now turn.

Unit	Symbol	Quantity
meter	m	length
kilogram	kg	mass
second	s	time
ampere	A	electric current
kelvin	K	temperature
candela	Cd	luminous intensity
mole	mol	amount of substance

Table A.1. Base units in the SI unit system.

A.1.2 The SI Unit System

In contrast to the USCS, the SI system is simple and has many advantages. There is usually only one basic unit for each kind of quantity, such as the meter for measuring length. Instead of having many unrelated units of measure for measuring quantities of different sizes, prefixes based on powers of ten are used on all the units to accommodate various sizes of quantities. And

Unit	Symbol	Quantity
joule	J	energy
newton	N	force
cubic meter	m^3	volume
watt	W	power
pascal	Pa	pressure

Table A.2. Some SI System derived units.

units for different types of quantities relate to one another in some way. Unlike the gallon and the foot, which have nothing to do with each other, the cubic meter is 1,000,000 cubic centimeters. For all these reasons, the USCS is not used much at all in scientific work. The SI system is the international standard.

There are seven *base units* in the SI System, listed in Table A.1. All other SI units of measure, such as the joule (J) for measuring quantities of energy and the newton (N) for measuring amounts of force, are based on these seven base units. Units based on combinations of the seven base units are called *derived units*. A few common derived units are listed in Table A.2.

You are already familiar with the SI unit for time: the second. You may or may not be familiar with some of the other base units, so here are some facts and photos to help familiarize you with these. A meter is just a few inches longer than a yard (3 feet). Figure A.1 shows a wooden measuring rule one meter long, commonly called a meter stick, along with a metal yardstick for comparison.

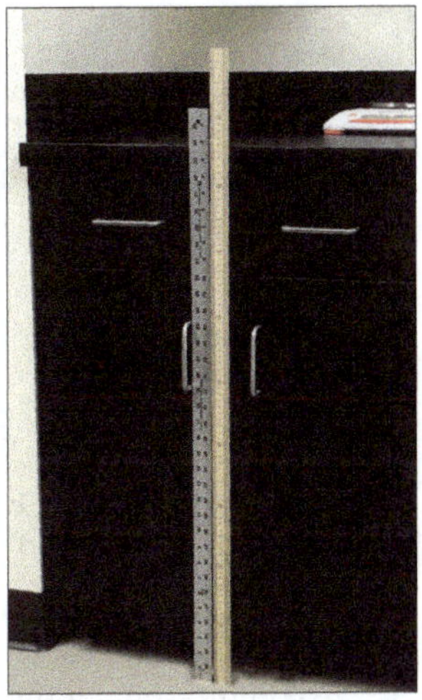

Figure A.1. A meter stick (right), with a yardstick for comparison.

On earth, a mass of one kilogram weighs about 2.2 pounds. The six-volt lantern battery shown in Figure A.2 weighs just under 2.2 pounds, so the mass of the battery is just about one kilogram.

Figure A.2. The mass of this battery is about one kilogram.

I can't show you a picture of one ampere of electric current, but it may be helpful to know that a standard electrical receptacle (or "outlet") such as the one shown in Figure A.3 is rated to carry 15 amperes of current. (However, the largest continuous current that the receptacle is allowed to supply is 80% of its rating, or 12 amperes. This is why vacuum cleaners are often advertised as having 12-amp motors. That's the upper limit of the current available to run them.)

Figure A.3. A standard receptacle in American homes is rated for a current of 15 amperes.

Regarding temperature units, the Celsius scale is generally used for making scientific temperature measurements, but the Kelvin scale must be used for nearly all scientific calculations involving temperature. You must become familiar with both scales.

On the Celsius scale, water freezes at 0°C and boils at 100°C. Since Celsius temperature measurements can take on negative values, the Celsius scale (like the Fahrenheit scale) is not an absolute temperature scale. The Kelvin scale is an absolute scale, with 0 K (0 kelvins) being equal to absolute zero, theoretically the lower limit of possible temperatures. Note that the term "degrees" is not used when stating or writing values in kelvins.

A temperature change of one kelvin is the same as a temperature change of one degree Celsius, and both are almost double the change that a change of one degree Fahrenheit is. For reference, room temperature on the three scales is 72°F, 22.2°C, and 295.4 K.

All seven of the SI base units are defined in terms of physical constants, each one of which is defined with its own exact value. For example, the speed of light in a vacuum is defined to be 299,792,458 meters per second. The meter is defined from this: the distance light travels in 1/299,792,458 seconds. (And the second has its own definition.) Formerly, units such as the kilogram and the meter were defined by man-made physical objects (artifacts), such as the one

Figure A.4. Standard meter bar number 27, owned by the U.S. and used as the standard meter from 1893 to 1960.

shown in Figure 1.5. The bar shown in the figure was the standard in the U.S. for the meter from 1893 to 1960. But this method of definition was not at all convenient.

In 1960, the definition of the meter was changed so that the meter was equal to a certain number of wavelengths of a certain color of light emitted by a certain isotope of the element krypton. In 1983, the definition of the meter was changed again to its present definition in terms of the speed of light and the second. The kilogram has a similar history, and its definition in terms of an artifact only ended in 2019.

The SI system includes not only the base and derived units, but all the other units that can be formed by adding metric prefixes to these units. We address the prefixes in the next section. But first I will mention a particular subset of the SI system known as the MKS system. MKS stands for meter-kilogram-second. The MKS system uses only the base and derived units without the prefixes (except for the kilogram, the only base unit with a prefix). The nice thing about the MKS system is that any calculation performed with MKS units produces a result in MKS units. For this reason, the MKS system is used almost exclusively in physics. However, in chemistry, it is common to use SI units that are not MKS units. Some commonly used non-MKS units are the gram (g), the centimeter (cm), the cubic centimeter (cm^3), the liter (L), and the milliliter (mL). The liter is not actually an official SI unit, but it is used all the time in chemistry anyway.

A.1.3 Metric Prefixes

In the system of units commonly used in the U.S., different units are used for different sizes of objects. For example, for short distances we might use the inch or the foot, whereas for longer distances we switch to the mile. For the small volumes used in cooking, we use the fluid ounce (or pint, quart, teaspoon, tablespoon, etc.), but for larger volumes like the gasoline in the gas tank of a car, we switch to the gallon. (That's six different volume units I just listed!)

The SI System is much simpler. Each type of quantity—such as length or volume—has one main unit of measure. Instead of using several different units for different sizes of quantities, the SI System uses multipliers on the units to multiply them for large quantities, or to scale them down for smaller quantities. We call these multipliers the *metric prefixes*. The complete list of the 20 metric prefixes is in Table A.3. You do not need to memorize all these; some are rarely used. But you do need to memorize some of them. I recommend that all science students commit to memory the prefixes listed in Table A.4.

Multiples	Prefix	deca–	hecto–	kilo–	mega–	giga–	tera–	peta–	exa–	zetta–	yotta–
	Symbol	da	h	k	M	G	T	P	E	Z	Y
	Factor	10	10^2	10^3	10^6	10^9	10^{12}	10^{15}	10^{18}	10^{21}	10^{24}
Fractions	Prefix	deci–	centi–	milli–	micro–	nano–	pico–	femto–	atto–	zetto–	yocto–
	Symbol	d	c	m	μ	n	p	f	a	z	y
	Factor	1/10	$1/10^2$	$1/10^3$	$1/10^6$	$1/10^9$	$1/10^{12}$	$1/10^{15}$	$1/10^{18}$	$1/10^{21}$	$1/10^{24}$

Table A.3. The SI System prefixes.

Fractions			Multiples		
Prefix	Symbol	Factor	Prefix	Symbol	Factor
centi–	c	$1/10^2$	kilo–	k	10^3
milli–	m	$1/10^3$	mega–	M	10^6
micro–	μ	$1/10^6$	giga–	G	10^9
nano–	n	$1/10^9$	tera–	T	10^{12}
pico–	p	$1/10^{12}$			

Table A.4. Metric prefixes to memorize.

Table A.5 shows a few representative examples of how to use the prefixes to represent multiples (quantities larger than the SI base unit) and fractions (quantities smaller than the SI base unit). As illustrations, let's look at a couple of these more closely.

The prefix *kilo–* is a *multiple*, and it means 1,000. One kilogram is 1,000 grams, one kilometer is

	Prefix	Symbol	Meaning	Examples of usage
Multiples	kilo–	k	1,000	One kilojoule is 1,000 joules. There are 1,000 joules in one kilojoule, so 1,000 J = 1 kJ.
	mega–	M	1,000,000	One megawatt is 1,000,000 watts. There are 1,000,000 watts in one megawatt, so 1,000,000 W = 1 MW.
Fractions	centi–	c	1/100	One centimeter is 1/100 of a meter. There are 100 centimeters in one meter, so 100 cm = 1 m.
	milli–	m	1/1,000	One milligram is 1/1,000 of a gram. There are 1,000 milligrams in one gram, so 1,000 mg = 1 g.
	micro–	μ	1/1,000,000	One microliter is 1/1,000,000 of a liter. There are 1,000,000 microliters in one liter, so 1,000,000 μL = 1 L.

Table A.5. Examples of correct usage of metric prefixes.

1 g

1,000 g

Figure A.5. The bowl in the photo above contains one gram (1 g) of chocolate chip cookie dough ice cream. Each of the two bowls shown to the right contains 500 g of chocolate chip cookie dough ice cream, so together they contain 1,000 g, or one kilogram of ice cream.

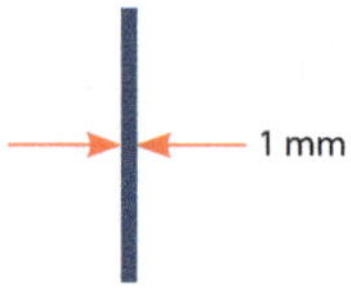

Figure A.6. One millimeter, which is one thousandth of a meter.

1,000 meters, and so on. Figure A.5 illustrates with my favorite dairy product—chocolate chip cookie dough ice cream. One gram is only a fraction of a taste (and contains only two chocolate chips and no cookie dough). A kilogram of ice cream is 1,000 grams of ice cream, equivalent to two large bowls of ice cream.

The prefix *milli–* is a *fraction*, and it means one thousandth. One millimeter is one thousandth of a meter, and so on. The wooden rule in Figure A.1 is one meter in length. A millimeter is one thousandth of this length, equal to the width of the line in Figure A.6.

We conclude this introduction to metric prefixes with a few brief notes. First, when using the prefixes for quantities of mass, prefixes are never added to the kilogram. Prefixes are only added to the gram, even though the kilogram is the base unit in the SI system, not the gram. Second, note that when writing the symbols for metric prefixes, the case of the letter matters: *kilo–* always takes a lower-case k, *mega–* always takes an upper-case M, and so on. Third, one of the prefix symbols is not an English letter. The prefix μ for *micro–* is the lower-case Greek letter *mu*, the m in the Greek alphabet. Finally, pay close attention to the difference between multiplier prefixes and fraction prefixes. Learning to use the fraction prefixes properly is the most challenging part of mastering the SI System of units, and using them incorrectly in unit conversion factors (our next topic) is a common student error.

A.2 Converting Units of Measure

A.2.1 Basic Principles of Unit Conversion Factors

For scientists and engineers, one of the most commonly used skills is re-expressing quantities into equivalent quantities with different units of measure. These calculations are called *unit conversions*. Mastery of this skill is essential for all students studying science. In this section, I describe what unit conversion factors are and how they are used.

Let's begin with the basic principles of how unit conversions work. First, we all know that multiplying any value by *unity* (one) leaves its value unchanged. Second, we also know that in any fraction, if the numerator and denominator are equivalent, the value of the fraction is unity (one). For example, the expression "12 bricks over 12 bricks" is equal to one:

$$\frac{12 \text{ bricks}}{12 \text{ bricks}} = 1$$

This is because the numerator and denominator are equivalent, and any time this is the case, the value of the fraction is one, or unity.

A *unit conversion factor* is simply a fractional expression in which the numerator and denominator are equivalent ways of writing the same physical quantity with different units of measure. This means a conversion factor is just a special way of writing unity (one).

The third basic principle is that when multiplying fractions, factors that appear in both the numerator and denominator may be "cancelled out." So when performing ordinary unit conversions, what we are doing is repeatedly multiplying a given quantity by unity so that cancellations alter the units of measure until they are expressed the way we wish. Since all we are doing is multiplying by one, the value of our original quantity is unchanged; it simply looks different because it is expressed with different units of measure.

There are many different units of measure and there are many different conversion factors used for performing unit conversions. Table B.3 in Appendix B lists a number of important ones.

Let me elaborate a bit more on the idea of unity I mentioned above, using one common conversion factor as an example. American school kids all learn that there are 5,280 feet in one mile, which means 5,280 ft = 1 mi. One mile and 5,280 feet are equivalent ways of writing the same length. If we place these two expressions into a fraction, the numerator and denominator are equivalent, so the value of the fraction is unity, regardless of the way we write it. The equation 5,280 ft = 1 mi can be written as a conversion factor two different ways, and the fraction equals unity either way:

$$\frac{5280 \text{ ft}}{1 \text{ mi}} = \frac{1 \text{ mi}}{5280 \text{ ft}} = 1 \qquad \text{(A.1)}$$

Now, I need to make an important clarification about the use of the equal sign in the expressions I just wrote. In mathematics, the equal sign means *identity*. An expression such as 3 miles = 3 miles is a mathematical *identity*; the first expression, "3 miles" is *identical* to the second expression, also "3 miles." But when we are dealing with converting units of measure from one set of units to a different set of units, we don't use the equal sign to mean identity. Obviously, the expression

$$\frac{5280 \text{ ft}}{1 \text{ mi}}$$

is not *identical* to the expression

$$\frac{1 \text{ mi}}{5280 \text{ ft}}$$

One of these has units of ft/mi and the other has units of mi/ft. But even though these expressions are not identical, they are *equivalent*. When we are dealing with converting units of measure, this is the sense in which we interpret the equal sign. We are using it to mean *equivalent*. This is why I can write Equation (A.1) using equal signs. The three terms in Equation (A.1) are not identical, but they are equivalent.

Suppose you have a measurement such as 43,000 feet that you wish to re-express in miles. To convert the units from feet to miles, first write down the quantity you are given, with its units of measure:

43,000 ft

Next, select a unit conversion factor containing the units you presently have and the ones you want to convert to. (This is not always possible. Sometimes more than one conversion factor is required, as Example 1.1 below illustrates.) In this case, those units are feet and miles, and the conversion factors containing these units are the two written in Equation (A.1). As I explain below, the one we need for converting 43,000 ft into miles is the second one. So to perform the conversion, you multiply your given quantify by the conversion factor. Then you cancel any units that appear both in the numerator and the denominator, as follows:

$$43{,}000 \text{ ft} \cdot \frac{1 \text{ mi}}{5280 \text{ ft}} = 8.1 \text{ mi}$$

There are two important comments to make here. First, since any conversion factor can be written two ways (depending on which quantity is placed in the numerator), how do we know which way to write the conversion factor? Well, we know from algebra that when we have quantities in the numerator of a fraction that are multiplied (factors), and quantities in the denominator of the fraction that are multiplied (factors), any quantities that appear in both the numerator and denominator can be cancelled out. In the example above, we want to cancel out the "feet" in the given quantity (which is in the numerator), so the conversion factor needs to be written with feet in the denominator and miles in the numerator.

Second, if you perform the calculation above (43,000 ÷ 5,280), the result that appears on your calculator screen is 8.143939394. So why didn't I write down all those digits in my result? Why did I round my answer off to simply 8.1 miles? The answer to that question has to do with the significant digits in the value 43,000 ft that we started with. We address the issue of significant digits later in this appendix, but in the examples that follow I always write the results with the correct number of significant digits for the values involved in the problem.

The following example illustrates the use of conversion factors based on metric prefixes. This example also illustrates how to perform a conversion when more than one conversion factor is required.

▼ Example A.1

Convert the value 2,953,000 μg into kilograms.

Referring to Tables A.4 and A.5, you see that the symbol μ means *micro–*, which means one millionth. Thus, μg means millionths of a gram. We use this information to make conversion factors. Since it takes 1,000,000 millionths of a gram to make one gram, 1,000,000 μg = 1 g, and thus

$$\frac{1{,}000{,}000\ \mu g}{1\ g} = \frac{1\ g}{1{,}000{,}000\ \mu g} = 1$$

Note that converting from μg to g only gets us part of the way toward the solution. We need another conversion factor to get from g to kg. Looking again at Tables A.4 and A.5, we see that there are 1,000 grams in one kilogram, or 1,000 g = 1 kg. From this we can make additional conversion factors:

$$\frac{1000\ g}{1\ kg} = \frac{1\ kg}{1000\ g} = 1$$

Now to perform the conversion, first convert micrograms into grams. Then convert grams into kilograms. You can do this two-step conversion at the same time by simply multiplying both conversion factors at the same time as follows:

$$2{,}953{,}000\ \cancel{\mu g} \cdot \frac{1\ \cancel{g}}{1{,}000{,}000\ \cancel{\mu g}} \cdot \frac{1\ kg}{1000\ \cancel{g}} = 0.002953\ kg$$

The μg in the given quantity cancels with the μg in the denominator of the first conversion factor. The g in the first conversion factor cancels with the g in the denominator of the second conversion factor. The units we are left with are the kg in the numerator of the second conversion factor. The kg did not cancel out with anything, so these are the units of our result.

▲

A.2.2 Tips for Converting Units of Measure

There are several important points you must remember in order to perform unit conversions correctly. I illustrate them below with examples. You should rework each of the examples on your own paper as practice to make sure you can do them correctly. The conversion factors used in the examples below are all listed in Table B.3 in Appendix B.

Point 1 *Never use slant bars in your unit fractions. Use only horizontal bars.*

In printed materials, we often sees values written with a slant fraction bar in the units, as in the value 35 m/s. Although writing the units this way is fine for a printed document, you should not write values this way when you are performing unit conversions. This is because it is easy to get confused and not notice that one of the units is in the denominator in such an expression (s, or seconds, in my example), and the conversion factors used must take this into account.

▼ Example A.2

Convert 57.66 mi/hr into m/s.

Writing the given quantity with a horizontal bar makes it clear that the "hours" is in the denominator. This helps you write the hours-to-seconds factor correctly.

To perform this conversion, we must convert the miles in the given quantity into meters, and we must convert the hours into seconds. From Table B.3, we select the two conversion factors we need. Then we multiply the given quantity by them to convert the mi/hr into m/s. When doing the multiplying, we write all the unit fractions with horizontal bars.

$$57.66\ \frac{\text{mi}}{\text{hr}} \cdot \frac{1609\ \text{m}}{1\ \text{mi}} \cdot \frac{1\ \text{hr}}{3600\ \text{s}} = 25.77\ \frac{\text{m}}{\text{s}}$$

As you see, the miles cancel and the hours cancel, leaving meters in the numerator and seconds in the denominator. Now that you have your result, you may write it as 25.77 m/s if you wish, but do not use slant fraction bars in the units when you are working out the unit conversion.

▲

Point 2 *The term "per," abbreviated* p, *implies a fraction.*

Some units of measure are commonly written with a "p" for "per," such as mph for miles per hour, or gps for gallons per second. Change these expressions to fractions with horizontal bars when you work out the unit conversion.

▼ Example A.3

Convert 472.2 gps to L/hr.

When you write down the given quantity, change the gps to gal/s and write these units with a horizontal bar:

$$472.2\ \frac{\text{gal}}{\text{s}} \cdot \frac{3.785\ \text{L}}{1\ \text{gal}} \cdot \frac{3600\ \text{s}}{\text{hr}} = 6{,}434{,}000\ \frac{\text{L}}{\text{hr}}$$

▲

Point 3 *Use the* [×] *and* [÷] *keys correctly when entering values into your calculator.*

When dealing with several numerator terms and several denominator terms, multiply all the numerator terms together first, hitting the [×] key between each, then hit the [÷] key and enter all the denominator terms, hitting the [÷] key between each. This way you do not need to write down intermediate results, and you do not need to use any parentheses.

▼ Example A.4

Convert 43.2 mm/hr into km/yr.

The setup with all the conversion factors is as follows:

$$43.2\ \frac{\text{mm}}{\text{hr}}\cdot\frac{1\ \text{m}}{1000\ \text{mm}}\cdot\frac{1\ \text{km}}{1000\ \text{m}}\cdot\frac{24\ \text{hr}}{1\ \text{dy}}\cdot\frac{365\ \text{dy}}{1\ \text{yr}}=0.378\ \frac{\text{km}}{\text{yr}}$$

To execute this calculation in your calculator, you enter the values and operations in this sequence:

$$43.2\times24\times365\div1000\div1000=$$

If you do so, you get 0.37843200. (Again, significant digits rules require us to round to 0.378.)

▲

Point 4 *When converting units for area and volume such as* cm^2 *or* m^3, *you must use the appropriate length conversion factor twice for areas and three times for volumes.*

The units "cm^2" for an area mean the same thing as "cm × cm." Likewise, "m^3" means "m × m × m." So when you use a length conversion factor such as 100 cm = 1 m or 1 in = 2.54 cm, you must use it twice to get squared units (areas) or three times to get cubed units (volumes).

▼ Example A.5

Convert 3,550 cm^3 to m^3.

$$3550\ \text{cm}^3\cdot\frac{1\ \text{m}}{100\ \text{cm}}\cdot\frac{1\ \text{m}}{100\ \text{cm}}\cdot\frac{1\ \text{m}}{100\ \text{cm}}\cdot=0.00355\ \text{m}^3$$

Notice in this example that the unit cm occurs three times in the denominator, giving us cm^3 when they are all multiplied together. This cm^3 term in the denominator cancels with the cm^3 term in the numerator. And since the m unit occurs three times in the numerator, they multiply together to give us m^3 for the units in our result.

▲

The issue of needing to repeat conversion factors only arises when you are using a unit raised to a power, such as a when a length unit is used to represent an area or a volume. When using a conversion factor such as 3.785 L = 1 gal, the units of measure are written using units that are strictly volumetric (liters and gallons), and are not obtained from lengths the way they are with in^2, ft^2, cm^3, and m^3. Another common unit that uses an exponent is acceleration, which has units of m/s^2 in the MKS unit system.

▼ Example A.6

Convert 5.85 mi/hr^2 into MKS units.

The MKS unit for length is meters (m), so we must convert miles to meters. The MKS units for time is seconds (s), so we must convert hr^2 into s^2.

$$5.85\ \frac{\text{mi}}{\text{hr}^2}\cdot\frac{1609\ \text{m}}{1\ \text{mi}}\cdot\frac{1\ \text{hr}}{3600\ \text{s}}\cdot\frac{1\ \text{hr}}{3600\ \text{s}}=0.000726\ \frac{\text{m}}{\text{s}^2}$$

With this example, you see that since the "hours" unit is squared in the given quantity, the conversion factor converting the hours to seconds must appear twice in the conversion calculation. The "miles" unit in the given quantity has no exponent, so the conversion factor used to convert miles to meters only appears once in the calculation.

▲

A.2.3 Converting Temperature Units

Converting temperature values from one scale to another requires the use of equations rather than conversion factors. This is due to the fact that the Fahrenheit and Celsius scales are not absolute temperature scales. If all temperature scales were absolute scales like the Kelvin scale is, temperature conversions could be performed with conversion factors just as other conversions are. To convert a temperature in degrees Fahrenheit (T_F) into degrees Celsius (T_C), use this equation:

$$T_C=\frac{5}{9}\left(T_F-32°\right)$$

Using a bit of algebra, we can work this around to give us an equation that can be used to convert Celsius temperatures to Fahrenheit values:

$$T_F=\frac{9}{5}T_C+32°$$

To convert a temperature in degrees Celsius into kelvins (T_K), use this equation:

$$T_K=T_C+273.15$$

Again, some algebra gives us the equation the other way around.

$$T_C=T_K-273.15$$

All four of the temperature conversion equations above are exact. This is important to know later when we discuss significant digits. These equations are listed in Table B.3 in Appendix B.

▼ Example A.7

The normal temperature of the human body is 98.6°F. Express this value in degrees Celsius and kelvins.

Since the given value is in degrees Fahrenheit, write down the equation that converts values from °F to °C.

$$T_C=\frac{5}{9}\left(T_F-32°\right)$$

Now insert the Fahrenheit value and calculate the Celsius value.

$$T_C = \frac{5}{9}(98.6° - 32°) = 37.0°\text{C}$$

Now we are able to use the Celsius value to compute the Kelvin value.

$$T_K = T_C + 273.15 = 37.0 + 273.15 = 310.2\text{ K}$$

The reason the answer is 310.2 K instead of 310.15 K is due to the significant digits rule for addition. Again, the topic of significant digits is coming up next.

▲

▼ Example A.8

The melting point of aluminum is 933.5 K. Express this temperature in degrees Celsius and degrees Fahrenheit.

Write down the equation that converts Kelvin values to Celsius values.

$$T_C = T_K - 273.15$$

From this we calculate the Celsius value as

$$T_C = T_K - 273.15 = 933.5 - 273.15 = 660.3°\text{C}$$

Next, write down the equation for converting a Celsius value to a Fahrenheit value.

$$T_F = \frac{9}{5}T_C + 32°$$

Insert the Celsius value into this equation and calculate the Fahrenheit value.

$$T_F = \frac{9}{5}T_C + 32° = \frac{9}{5} \cdot 660.3°\text{C} + 32° = 1220.5°\text{F}$$

▲

A.3 Accuracy and Precision

A.3.1 Distinguishing Between Accuracy and Precision

The terms *accuracy* and *precision* refer to the practical limitations inherent in making measurements. Science is all about investigating nature, and to do that we must make measurements.

Accuracy relates to error—that is, to the lack of it. Error is the difference between a measured value and the true value. The lower the error is in a measurement, the better the accuracy. Error arises from many different sources, including human mistakes, malfunctioning equipment, incorrectly calibrated instruments, vibrations, changes in temperature or humidity, or unknown causes that are influencing a measurement without the knowledge of the experimenter. All measurements contain error, because (alas!) perfection is simply not a thing we have access to in our world.

Precision refers to the resolution or degree of "fine-ness" in a measurement. The limit to the precision that can be obtained in a measurement is ultimately dependent on the instrument being used to make the measurement. If you want greater precision, you must use a more precise instrument. The degree of precision in every measurement is signified by the measurement value itself because the precision is a built-in part of the measurement. *The precision of a measurement is indicated by the number of significant digits (or significant figures) included in the measurement value when the measurement is written down* (see below).

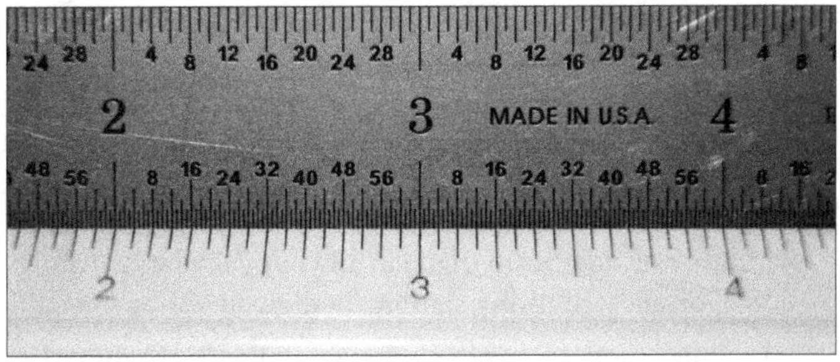

Figure A.7. The accuracy of these two scales is the same, but the machinist's rule (above) is more precise than the architect's scale (below).

Here is an example that illustrates the idea of precision and also helps distinguish between precision and accuracy. The photograph in Figure A.7 shows a machinist's rule and an architect's scale set one above the other. Since the marks on the two scales line up consistently, *these two scales are equally accurate.* But the machinist's rule (on top) is more precise. The architect's scale is marked in 1/16-inch increments, but the machinist's rule is marked in 1/64-inch increments. Thus, *the machinist's rule is more precise.*

It is important that you are able to distinguish between accuracy and precision. Here is another example to help illustrate the difference. Let's say Shana and Marius each buy digital thermometers for their homes. The thermometer Shana buys costs $10 and measures to the nearest 1°F. Marius pays $40 and gets one that reads to the nearest 0.1°F. Shana reads the directions and properly installs the sensor for her new thermometer in the shade. Marius doesn't read the directions and mounts his sensor in the direct sunlight, which causes a significant error in the thermometer reading when the sun is shining on it; thus Marius' measurements are not accurate. The result is that Shana has lower-precision, higher-accuracy measurements!

A.3.2 Significant Digits

The precision in any measurement is indicated by the number of *significant digits* it contains. Thus, the number of digits we write in any measurement we deal with in science is very important. The number of digits is meaningful because it shows the precision inherent in the instrument used to make the measurement.

Let's say you are working a computational exercise in a science book. The problem tells you that a person drives a distance of 110 miles at an average speed of 55 miles per hour and wants you to calculate how long the trip takes. The correct answer to this problem *is different* from the correct answer to a similar problem with given values of 110.0 miles and 55.0 miles per hour. And if the given values are 110.0 miles and 55.00 miles per hour, the correct answer is different yet again. Mathematically, of course, all three answers are the same. If you drive 110 miles at 55 miles per hour, the trip takes two hours. But scientifically, the correct answers to these three problems are different: 2.0 hours, 2.00 hours, and 2.000 hours, respectively. The difference between these cases is in the precision indicated by the given data, which are *measurements.* (Even though this is just a made-up problem in a book and not an actual measurement someone made in an experiment, the given data are still measurements. There is no way to talk about distances or speeds without talking about measurements, even if the measurements are only imaginary or hypothetical.)

So when you perform a calculation with physical quantities (measurements), you can't simply write down all the digits shown by your calculator. The precision inherent in the measurements used in a computation governs the precision in any result you might calculate from those measurements. And since the precision in a measurement is indicated by the number of significant digits, data and calculations must be written with the correct numbers of significant digits. To do this, you need to know how to count significant digits, and you must use the correct number of significant digits in all your calculations and experimental data.

Correctly counting significant digits involves four different cases:

1. Rules for determining how many significant digits there are in a given measurement.
2. Rules for writing down the correct number of significant digits in a measurement you are making and recording.
3. Rules for computations you perform with measurements—multiplication and division.
4. Rules for computations you perform with measurements—addition and subtraction.

We address each of these cases below, in order.

Case 1 We begin with the rule for determining how many significant digits there are in a given measurement value. The rule is as follows:

- *The number of significant digits in a number is found by counting all the digits from left to right beginning with the first nonzero digit on the left. When no decimal is present, trailing zeros are not considered significant.*

Let's apply this rule to the following values to see how it works.

15,679 This value has five significant digits.

21.0005 This value has six significant digits.

37,000 This value has only two significant digits because when there is no decimal, trailing zeros are not significant. Notice that the word *significant* here is a reference to the precision of the measurement, which in this case is rounded to the nearest thousand. The zeros in this value are certainly *important*, but they are not *significant* in the context of precision.

0.0105 This value has three significant digits because we start counting digits with the first nonzero digit on the left.

0.001350 This value has four significant digits. Trailing zeros count when there is a decimal.

The significant digit rules enable us to tell the difference between two measurements such as 13.05 m and 13.0500 m. Again, these values are obviously equivalent *mathematically*. But they are different in what they tell us about the process of how the measurements were made—and science deals in measurements. The first measurement has four significant digits. The second measurement is more precise—it has six significant digits and was made with a more precise instrument.

Now, just in case you are bothered by the zeros at the end of 37,000 that are not significant, here is one more way to think about significant digits that may help. The precision in a measurement depends on the instrument used to make the measurement. If we express the measurement in different units, this cannot change the precision of the value. A measurement of 37,000 grams is equivalent to 37 kilograms, as shown in the following calculation:

$$37{,}000\ \text{g} \cdot \frac{1\ \text{kg}}{1000\ \text{g}} = 37\ \text{kg}$$

Whether we express this value in grams or kilograms, it still has two significant digits.

Case 2 The second case addresses the rules that apply when you are recording a measurement yourself, rather than reading a measurement someone else has made. When you make measurements yourself, as when conducting the laboratory experiments in this course, you must know the rules for which digits are significant in the reading you are making on the measurement instrument. The rule for making measurements depends on whether the instrument you are using is a digital instrument or an analog instrument. Here are the rules for these two possibilities:

- *Rule 1 for digital instruments* *For the digital instruments commonly found in introductory science labs, assume all the digits in the reading are significant except leading zeros.*
- *Rule 2 for analog instruments* *The significant digits in a measurement include all the digits known with certainty, plus one digit at the end that is estimated between the finest marks on the scale of your instrument.*

The first of these rules is illustrated in Figure A.8. The reading on the left has leading zeros, which do not count as significant. Thus, the first reading has three significant digits. The second reading also has three significant digits. The third reading has five significant digits.

The fourth reading also has five significant digits because with a digital display the only zeros that don't count are the leading zeros. Trailing zeros are significant with a digital instrument. *However*, when you write this measurement down, you must write it in a way that shows those zeros to be significant. The way to do this is by using scientific notation. When a value is written in scientific notation, *the digits that are written down in front of the power of 10* (the stem, also called the mantissa) *are the significant digits*. Thus, the right-hand value in Figure A.8 must be written as 4.2000×10^4. We address scientific notation in more detail in the next section.

Figure A.8. With digital instruments, all digits are significant except leading zeros. Thus, the numbers of significant digits in these readings are, from left to right, three, three, five, and five.

Dealing with digital instruments is actually more involved than the simple rule above implies, but the issues involved go way beyond what we can deal with in introductory science classes. So, simply make your readings and assume that all the digits in the reading except leading zeros are significant.

Now let's look at some examples illustrating the rule for analog instruments. Figure A.9 shows a machinist's rule being used to measure the length in millimeters (mm) of a brass block. We know the first two digits of the length with certainty; the block is clearly between 31 mm and 32 mm long. We have to estimate the third significant digit. The scale on the rule is marked in increments of 0.5 mm. Comparing the edge of the block with these marks, I estimate the next digit to be a 6, giving a measurement of 31.6 mm. Others might estimate the last digit to be a 5 or a 7; these small differences in the last digit are unavoidable because the last digit is estimated. Whatever you estimate the last digit to be, two digits of this measurement are known with certainty, the third digit is estimated, and the measurement has three significant digits.

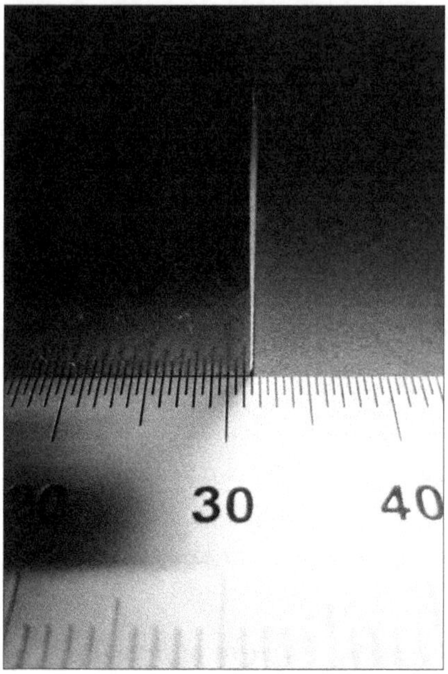

Figure A.9. Reading the significant digits with a machinist's rule.

The photograph in Figure A.10 shows a liquid volume measurement in milliliters (mL) being made with an article of apparatus called a *buret*. Notice in this figure that when measuring liquid volume the surface of the liquid curls up at the edge of the cylinder, forming a bowl-shaped surface on the liquid. This curved surface is called a *meniscus. For most liquids, liquid measurement readings are taken at the bottom of the meniscus.* Liquid mercury is the major exception, because the meniscus in liquid mercury is inverted—liquid mercury curves down at the edges. In that case, the measurement is read at the top of the meniscus. But that is an unusual case. For most liquids, the reading is made at the bottom of the meniscus.

For the buret in the figure, you can see that the scale is marked in increments of 0.1 milliliters (mL). This means we are to estimate to the nearest 0.01 mL. To one person, it may look like the bottom of the meniscus (where the black curve touches the bottom of the silver bowl) is just above 2.2 mL, so that person would call this measurement 2.19 mL. To someone else, it may seem that the bottom of the meniscus is right on 2.2, in which case that person would call the reading 2.20 mL. Either way, the reading has three significant digits and the last digit is estimated to be either 9 or 0.

The third example involves a liquid volume measurement with an article of apparatus called a *graduated cylinder*. The scales on small graduated cylinders like this one are marked in increments of 1 mL. In the photo of Figure A.11, the entire meniscus appears silvery in color with a black curve at the bottom. For the liquid shown in the figure, we know the first two digits of the volume measurement with certainty because the reading at the bottom of the meniscus is clearly between 82 mL and 83 mL. We have to estimate the third digit, and I estimate the edge of the meniscus to be at 60% of the distance between 82 and 83, giving a reading of 82.6 mL. Others may prefer a different value for that third digit.

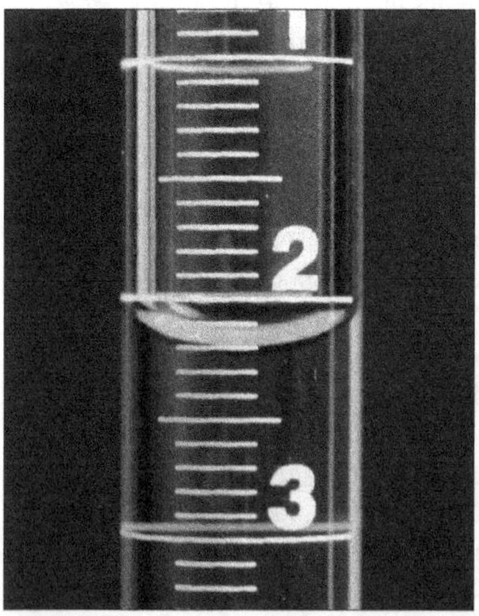

Figure A.10. Reading the significant digits on a buret.

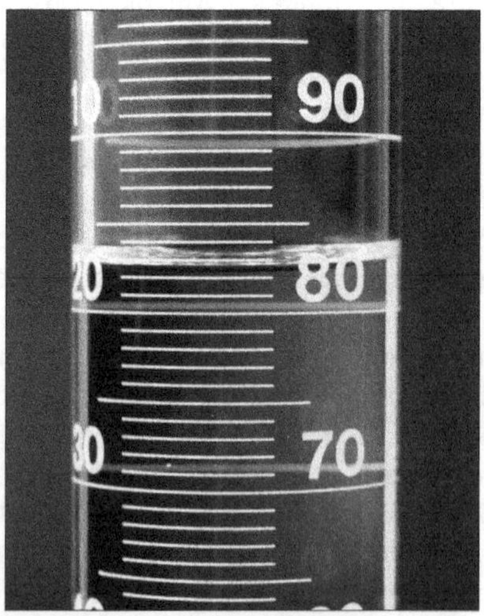

Figure A.11. Reading the significant digits on a graduated cylinder.

It is important for you to keep the significant digits rules in mind when you are making measurements and entering data for your lab reports. The data in your lab journal and the values you use in your calculations and report should correctly reflect the use of the significant digits rules as they apply to the actual instruments you use to make your measurements.

Case 3 The third and fourth cases of rules for significant digits apply to the calculations you perform with measurements. In Case 3, we deal with multiplication and division. The main idea behind the rule for multiplying and dividing is that the precision you report in your result cannot be higher than the precision that is in the measurements you start with. The precision in a measurement depends on the instrument used to make the measurement, nothing else. Multiplying and dividing things cannot increase that precision, and thus your results can be no more precise than the measurements used in the calculations. In fact, your result can be no more precise than the *least precise value* used in the calculation. The least precise value is, so to speak, the "weak link" in the chain, and a chain is no stronger than its weakest link.

Here are the two rules for using significant digits in calculations involving multiplication and division:

- *Rule 1 When multiplying or dividing, count the significant digits in each of the values you using in the calculation, including any conversion factors involved. (However, note: Conversion factors that are exact are not considered.) Determine how many significant digits there are in the least precise of these values. The result of your calculation must have this same number of significant digits.*

- *Rule 2 When performing a multi-step calculation, keep at least one extra digit during intermediate calculations, and round off to the final number of significant digits you need at the very end. This practice ensures that small round-off errors don't accumulate during a multi-step calculation. This extra digit rule also applies to unit conversions performed as part of the computation.*

I illustrate the two rules above, along with some more unit conversions, in the following example problem and calculation.

Example A.9

At a chemical research lab, a stream of a reactant solution is flowing into a reaction vessel at a rate of 56.75 μL per second. A volume of 1.0 ft^3 of this solution is required in the vessel for the reaction. Determine the amount of time needed for the required volume to be collected. State your result in hours.

First note that the value of the flow rate has four significant digits, and the required volume has two significant digits. The two-digit value is the least precise of these, so our result must be rounded to two significant digits. But to avoid rounding error, we must work with values having at least three significant digits (one more than we need) until the very end.

One of the volumes in this problem is in μL and the other is in ft^3. I begin by converting the required volume from ft^3 to μL so our volumes all have the same units. We have no conversion factor that goes directly from ft^3 to μL, so we must use a chain of conversion factors that we know or have available. Since we are dealing with relatively small volumes based on length units, the main conversion from USCS units to SI units is the inch to centimeter factor of 1 in = 2.54 cm. This factor is exact and should be committed to memory. If we were starting with gallons instead of ft^3, we might use the conversion 1 gal = 3.785 L, although this factor is not exact.

We first convert ft^3 to in^3, then from in^3 to cm^3, then from cm^3 to L, and finally from L to μL.

$$1.0\ \text{ft}^3 \cdot \frac{12\ \text{in}}{1\ \text{ft}} \cdot \frac{12\ \text{in}}{1\ \text{ft}} \cdot \frac{12\ \text{in}}{1\ \text{ft}} \cdot \frac{2.54\ \text{cm}}{1\ \text{in}} \cdot \frac{2.54\ \text{cm}}{1\ \text{in}} \cdot \frac{2.54\ \text{cm}}{1\ \text{in}} \cdot \frac{1\ \text{L}}{1000\ \text{cm}^3} \cdot \frac{1 \times 10^6\ \mu\text{L}}{1\ \text{L}} = 28{,}300{,}000\ \mu\text{L}$$

This result has three significant digits. We are keeping one extra digit during the intermediate calculations. It is not incorrect to write down all the digits your calculator shows. But it *is* pointless. A person who writes all the digits regardless of whether they are needed simply shows that he or she doesn't understand significant digits. Those extra digits are meaningless.

Notice that all the conversion factors used in the calculation above are exact; none of them are approximations. Since they are all exact, they play no role in limiting the significant digits in our result. Note that if you must use a conversion factor that is approximate, you should make sure the precision of the value in the conversion factor is at least as high as the precision in your data (two significant digits, in this case). That way your conversion factor does not limit the precision of your result. If, for some reason, you do not have a conversion factor with as many significant digits as your least precise measurement, then the precision of your result must match the precision of the conversion factor. The least precise value in the entire calculation always governs the precision in your result.

Now we compute the time required by dividing the required volume by the flow rate.

$$t = \frac{28{,}300{,}000 \ \mu\text{L}}{56.75 \ \frac{\mu\text{L}}{\text{s}}} = 499{,}000 \text{ s}$$

This value also has three significant digits—one more than we need. We now convert this value from seconds to hours as the problem statement requires.

$$499{,}000 \text{ s} \cdot \frac{1 \text{ hr}}{3600 \text{ s}} = 139 \text{ hr}$$

Finally, we round the result to the required two significant digits. The second non-zero digit from the left (3) is in the tens place, so we round to the nearest ten.

t = 140 hr

▲

Case 4 The fourth case of rules for significant digits also applies to the calculations you perform with measurements. In Case 4, we deal with addition and subtraction.

The rule for addition and subtraction is completely different from the rule for multiplication and division. When performing addition, it is not the number of significant digits that governs the precision of the result. Instead, it is the *place value of the last digit that is farthest to the left in the numbers being added* that governs the precision of the result. This rule is quite wordy and is best illustrated by an example. Consider the following addition problem:

$$\begin{array}{r} 13.65 \\ 1.9017 \\ \underline{+\ 1{,}387.069} \\ 1{,}402.62 \end{array}$$

Of the three values being added, 13.65 has digits out to the hundredths place, 1.9017 has digits out to the ten thousandths place, and 1,387.069 has digits out to the thousandths place. Looking at the final digits of these three, you can see that the final digit farthest to the left is the 5 in 13.65, which is in the hundredths place. This is the digit that governs the final digit of the result.

There can be no digits to the right of the hundredths place in the result. The justification for this rule is that one of our measurements is precise only to the nearest hundredth, even though the other two are precise to the nearest thousandth or ten thousandth. We are going to add these values together, and one of them is precise only to the nearest hundredth. It makes no sense to have a result that is precise to a place more precise than that, so hundredths are the limit of the precision in the result.

Correctly performing addition problems in science (where nearly everything is a measurement) requires that you determine the place value governing the precision of your result, perform the addition, then round the result. In the above example, the sum is 1,4602.6207. Rounding this value to the hundredths place gives 1,4602.62.

Going back to Example 1.7, we saw the following equation for converting a temperature from kelvins to degrees Celsius:

$$T_K = T_C + 273.15 = 37.0 + 273.15 = 310.2 \text{ K}$$

The two values we are adding are 37.0, which has digits out to the tenths place, and 273.15, which has digits out to the hundredths place. The final digit in 37.0 is the one farther to the left, so it governs the final digit we can have in the sum. The final digit in 37.0 is in the tenths place, so the final digit in the sum must also be in the tenths place. Adding the values gives 310.15 K. Rounding this to the tenths place gives us 310.2 K for our result.

A.4 Other Important Math Skills

A.4.1 Scientific Notation

No doubt you have studied *scientific notation* in your math classes. However, beginning in high school, scientific notation is used all the time in scientific study. Knowing how to use scientific notation correctly—including the use of the special key found on scientific calculators for working with values in scientific notation—is very important.

Mathematical Principles

Scientific notation is a way of expressing very large or very small numbers without all the zeros, unless the zeros are *significant*. This is of enormous benefit when one is dealing with a value such as 0.0000000000001 cm (the approximate diameter of an atomic nucleus). The basic idea will be clear from a few examples.

Let's say we have the value 3,750,000. This number is the same as 3.75 million, which can be written as $3.75 \times 1{,}000{,}000$. Now, 1,000,000 itself can be written as 10^6 (which means one followed by six zeros), so our original number can be expressed equivalently as 3.75×10^6. This expression is in scientific notation. The number in front, the stem, is always written with one digit followed by the decimal and the other digits. The multiplied 10 raised to a power has the effect of moving the decimal over as many places as necessary to recreate our original number.

As a second example, the current population of earth is about 7,200,000,000, or 7.2 billion. One billion has nine zeros, so it can be written as 10^9. So we can express the population of earth in scientific notation as 7.2×10^9.

When dealing with extremely small numbers such as 0.000000016, the process is the same, except the power on the 10 is negative. The easiest way to think of it is to visually count how many places the decimal in the value has to be moved over to get 1.6. To get 1.6, the decimal has to be moved to the right 8 places, so we write our original value in scientific notation as 1.6×10^{-8}.

Using Scientific Notation with a Scientific Calculator

All scientific calculators have a key for entering values in scientific notation. This key is labeled [EE] or [EXP] on most calculators, but others use a different label.[1] It is very common for those new to scientific calculators to use this key incorrectly and obtain incorrect results. So read carefully as I outline the procedure.

The whole point of using the [EE] key is to make keying in the value as quick and error free as possible. *When using the scientific notation key to enter a value, you do not press the* [×] *key, nor do you enter the 10.* The scientific calculator is designed to reduce all this key entry, and the potential for error, by use of the scientific notation key. You only enter the stem of the value and the power on the ten and let the calculator do the rest.

Here's how. To enter a value, simply enter the digits and decimal in the stem of the number. Then hit the [EE] key, and then enter the power on the ten. The value is then in the calculator and you may do with it whatever you need to. As an example, to multiply the value 7.2×10^9 by 25 using a standard scientific calculator, the sequence of key strokes is as follows:

7.2 [EE] 9 [×] 25 [=]

Notice that between the stem and the power, the only key pushed is the [EE] key.

When entering values in scientific notation with negative powers on the 10, the [+/−] key is used before the power to make the power negative. Thus, to divide 1.6×10^{-8} by 36.17, the sequence of key strokes is:

1.6 [EE] [+/−] 8 [÷] 36.17 [=]

Again, neither the 10 nor the × sign that comes before it are keyed in. The [EE] key has these built in.

Students sometimes wonder why it is *incorrect* to use the [10^x] key for scientific notation. To calculate 7.2×10^9 times 25, they are tempted to enter the following:

7.2 [×] [10^x] 9 [×] 25 [=]

The problem with this approach is that sometimes it works and sometimes it doesn't, and calculator users need to use key entries that *always* work. The scientific notation key ([EE]) keeps all the parts of a value in scientific notation together as one number. That is, when the [EE] key is used, a value such as 7.2×10^9 is not two separate numbers to the calculator; it is a single numerical value. But when the [×] key is manually inserted, the calculator treats the numbers separated by the [×] key as two separate values, and this can caused the calculator to use a different order of operations than you intend. For example, using the [10^x] key causes the calculator to render an incorrect answer for a calculation such as this:

$$\frac{3.0 \times 10^6}{1.5 \times 10^6}$$

The denominator of this expression is exactly half the numerator, so the value of this fraction is obviously 2.0. But when using the [10^x] key, the 1.5 and the 10^6 in the denominator are separated and treated as separate values. The calculator then performs the following calculation:

---

1 One infuriating model uses the extremely unfortunate label [$\times 10^x$] which looks a *lot* like [10^x], a different key with a completely different function.

$$\frac{3.0\times10^{6}}{1.5}\times10^{6}$$

This comes out to 2,000,000,000,000 (2×10^{12}), which is not the same as 2.0!

The bottom line is that the EE key, however it may be labeled, is the correct key to use for scientific notation.

Finally, when writing a result in scientific notation, it is not acceptable to write it using the EE notation your calculator uses. For example, your calculator might display a result as 3.14EE8 or 3.14E8, but you must write this as 3.14×10^{8}.

A.4.2 Calculating Percent Difference

One of the conventional calculations in science experiments is the so-called "experimental error." Experimental error is typically defined as the difference between a predicted value and an experimental value, expressed as a percentage of the predicted value, or

$$\text{experimental error}=\frac{|\text{predicted or accepted value}-\text{experimental value}|}{\text{predicted or accepted value}}\times100\%$$

Although the term "experimental error" is widely used, it is a poor choice of words. When there is a mismatch between theory and experiment, the experiment may not be the source of the error. Often, it is the theory that is found wanting—this is how science advances.

I now prefer to use the phrase *percent difference* to describe the value computed by the above equation. When quantitative results are compared to quantitative predictions or accepted values, students should compute the percent difference as

$$\text{percent difference}=\frac{|\text{predicted or accepted value}-\text{experimental value}|}{\text{predicted or accepted value}}\times100\%$$

Appendix A Exercises

Perform each of the following unit conversions. Express each result using the correct number of significant digits. Where possible and appropriate, express your result in both standard notation and scientific notation. Answers are at the end of the Answers to Selected Exercises section just before this appendix.

	Convert This Quantity	Into These Units
1.	1,750 meters (m)	feet (ft)
2.	3.54 grams (g)	kilograms (kg)
3.	41.11 milliliters (mL)	liters (L)
4.	7×10^8 m (radius of the sun)	miles (mi)
5.	1.5499×10^{-12} millimeters (mm)	m
6.	750 cubic centimeters (cm^3 or cc) (size of the engine in my old motorcycle)	m^3
7.	2.9979×10^8 meters/second (m/s) (speed of light in a vacuum)	ft/s
8.	168 hours (hr) (one week)	s
9.	5,570 kilograms/cubic meter (kg/m^3) (average density of the earth)	g/cm^3
10.	45 gallons per second (gps) (flow rate of Mississippi River at the source)	m^3/minute (m^3/min)
11.	600,000 cubic feet/second (ft^3/s) (flow rate of Mississippi River at New Orleans)	liters/hour (L/hr)
12.	5,200 mL (volume of blood in a typical man's body)	m^3
13.	5.65×10^2 mm^2 (area of a postage stamp)	square inches (in^2)
14.	32.16 ft/s^2 (acceleration of gravity, or one "g")	m/s^2
15.	5,001 µg/s	kg/min
16.	4.771 g/mL	kg/m^3

	Convert This Quantity	Into These Units
17.	13.6 g/cm^3 (density of mercury)	mg/m^3
18.	93,000,000 mi (distance from earth to the sun)	cm
19.	65 miles per hour (mph) (typical highway speed limit)	m/s
20.	633 nanometers (nm) (wavelength of light from a red laser)	in
21.	5.015% of the speed of light	mph

Appendix B

Reference Data

On this page and the next, items marked in yellow are approximate. All others are exact.

Table B.1. SI prefixes.

Multiples	Prefix	*deca–*	*hecto–*	*kilo–*	*mega–*	*giga–*	*tera–*	*peta–*	*exa–*	*zetta–*	*yotta–*
	Symbol	da	h	k	M	G	T	P	E	Z	Y
	Factor	10	10^2	10^3	10^6	10^9	10^{12}	10^{15}	10^{18}	10^{21}	10^{24}
Fractions	Prefix	*deci–*	*centi–*	*milli–*	*micro–*	*nano–*	*pico–*	*femto–*	*atto–*	*zetto–*	*yocto–*
	Symbol	d	c	m	μ	n	p	f	a	z	y
	Factor	1/10	$1/10^2$	$1/10^3$	$1/10^6$	$1/10^9$	$1/10^{12}$	$1/10^{15}$	$1/10^{18}$	$1/10^{21}$	$1/10^{24}$

Table B.2. Physical constants.

Quantity	Symbol	Value
speed of light in a vacuum	c	$299{,}792{,}458\ \text{m}\cdot\text{s}^{-1}$
acceleration due to gravity, sea level	g	$9.80\ \text{m}\cdot\text{s}^{-2}$
atmospheric pressure	P_{atm}	101,325 Pa
		1 atm
		14.7 psi
proton mass	m_p	1.672622×10^{-27} kg
		1.007276 u
neutron mass	m_n	1.674927×10^{-27} kg
		1.008665 u
electron mass	m_e	9.109382×10^{-31} kg
		0.0005486 u
Avogadro constant	N_A	$6.02214076 \times 10^{23}\ \text{mol}^{-1}$
Planck constant	h	$6.62607015 \times 10^{-34}\ \text{J}\cdot\text{s}$
Boltzmann constant	k_B	$1.380649 \times 10^{-23}\ \text{J}\cdot\text{K}^{-1}$
Rydberg constant	R	$1.097373 \times 10^{7}\ \text{m}^{-1}$
unified atomic mass unit	u	1.660539×10^{-27} kg
electron charge	e	$1.602176634 \times 10^{-19}$ C
ideal gas constant	R	$8.31446261815324\ \text{J}\cdot\text{mol}^{-1}\cdot\text{K}^{-1}$
		$0.082057\ \text{L}\cdot\text{atm}\cdot\text{mol}^{-1}\cdot\text{K}^{-1}$
		$8.314462\ \text{L}\cdot\text{kPa}\cdot\text{mol}^{-1}\cdot\text{K}^{-1}$
		$62.36337\ \text{L}\cdot\text{Torr}\cdot\text{mol}^{-1}\cdot\text{K}^{-1}$

Table B.3. Unit conversion factors.

Pressure

Unit Name	Symbol	Definition
pascal	Pa	1 Pa = 1 N/m^2
kilopascal	kPa	1 kPa = 1000 Pa
atmosphere	atm	1 atm = 101,325 Pa
bar	bar	1 bar = 100,000 Pa
pounds per square inch	psi	1 psi = 6894.757 Pa
torr	Torr	1 Torr = 1/760 atm

Length

1 mi = 5280 ft	1 ft = 12 in	1 in = 2.54 cm	1 cm = 10 mm
1 yd = 3 ft	1 mi = 1609 m	1 ft = 0.3048 m	1 Å = 0.1 nm

Temperature

$T_C = \frac{5}{9}(T_F - 32°)$	$T_F = \frac{9}{5}T_C + 32°$	$T_K = T_C + 273.15$	$T_C = T_K - 273.15$

Volume

1 gal = 3.78541 L	1 L = 1000 cm^3	1 mL = 1 cm^3	1 m^3 = 1000 L
1 tsp = 4.92892 mL	1 Tbsp = 14.7868 mL	1 cp = 236.588 mL	1 pt = 473.176 mL
1 qt = 0.946353 L	1 gal = 4 qt	1 ft^3 = 7.480523 gal	

Force

1 lb = 4.448 N

Energy and Mass

1 eV = 1.60218 × 10^{-19} J	1 u = 1.660539 × 10^{-27} kg

Time

1 min = 60 s	1 hr = 60 min	1 hr = 3600 s	1 dy = 24 hr
1 yr = 365 dy			

Table B.4. Vapor pressure of water.

T (°C)	P (kPa)	P (Torr)	T (°C)	P (kPa)	P (Torr)
0.01	0.6117	4.588	52	13.631	100.2
2	0.7060	5.295	54	15.022	112.7
4	0.8136	6.103	56	16.533	124.0
6	0.9354	7.016	58	18.171	136.3
8	1.0730	8.048	60	19.946	149.6
10	1.2282	9.212	62	21.867	164.0
12	1.4028	10.52	64	23.943	179.6
14	1.5990	11.99	66	26.183	196.4
16	1.8188	13.64	68	28.599	214.5
18	2.0647	15.49	70	31.201	234.0
20	2.3393	17.55	72	34.000	255.0
22	2.6453	19.84	74	37.009	277.6
24	2.9858	22.40	76	40.239	301.8
25	3.1699	23.78	78	43.703	327.8
26	3.3639	25.23	80	47.414	355.6
28	3.7831	28.38	82	51.387	385.4
30	4.2470	31.86	84	55.635	417.3
32	4.7596	35.70	86	60.173	451.3
34	5.3251	39.94	88	65.017	487.7
36	5.9479	44.61	90	70.182	526.4
38	6.6328	49.75	92	75.684	567.7
40	7.3849	55.39	94	81.541	611.6
42	8.2096	61.58	96	87.771	658.3
44	9.1124	68.35	98	94.390	708.0
46	10.099	75.75	100	101.42	760.7
48	11.177	83.83	102	108.87	816.6
50	12.352	92.45	104	116.78	875.9

Table B.5. Constants for water.

Constant	Symbol	Value
molar heat capacity (ice, −10.0°C)	C	$0.0364\ \frac{\text{kJ}}{\text{mol}\cdot\text{K}}$
molar heat capacity (water)	C	$0.0752\ \frac{\text{kJ}}{\text{mol}\cdot\text{K}}$
molar heat capacity (steam, 110°C)	C	$0.0369\ \frac{\text{kJ}}{\text{mol}\cdot\text{K}}$
molar heat of fusion	H_f	$6.01\ \frac{\text{kJ}}{\text{mol}}$
molar heat of vaporization	H_v	$40.7\ \frac{\text{kJ}}{\text{mol}}$
density (4.0°C)	ρ	$0.9999749\ \frac{\text{g}}{\text{cm}^3}$
density (22.0°C)	ρ	$0.9978\ \frac{\text{g}}{\text{cm}^3}$
density (25.0°C)	ρ	$0.9970\ \frac{\text{g}}{\text{cm}^3}$

Table B.6. Molal freezing point and boiling point constants for representative solvents at P = 1 atm. (p) = polar solvent; (np) = nonpolar solvent

Solvent	Freezing Point (°C)	Molal f.p. Constant, K_f (°C/m)	Boiling Point (°C)	Molal b.p. Constant, K_b (°C/m)
acetic acid (p)	16.6	3.63	117.9	3.22
acetone (p)	−94.7	2.67	56.1	1.80
benzene (np)	5.5	5.07	80.1	2.64
ethanol (p)	−114.4	1.959	78.3	1.23
methanol (p)	−97.5	2.56	64.6	0.86
phenol (p)	40.9	6.84	181.9	3.54
toluene (np)	−95.0	3.55	110.6	3.40
water (p)	0.00	1.86	99.974	0.513

Table B.7. Standard molar enthalpy of formation values for representative substances at T = 25°C and P = 1 bar.

Substance	Formula	ΔH_f° (kJ/mol)	Substance	Formula	ΔH_f° (kJ/mol)
acetic acid (*l*)	CH_3COOH	–484.3	lead(II) sulfate (*s*)	$PbSO_4$	–920.0
acetylene (*g*)	C_2H_2	227.4	lithium chloride (*s*)	$LiCl$	–408.6
ammonia (*g*)	NH_3	–45.9	lithium sulfate (*s*)	Li_2SO_4	–1436.5
ammonium chloride (*s*)	NH_4Cl	–314.4	magnesium chloride (*s*)	$MgCl_2$	–641.3
barium chloride (*s*)	$BaCl_2$	–855.0	magnesium hydroxide (*s*)	$Mg(OH)_2$	–924.5
barium nitrate (*s*)	$Ba(NO_3)_2$	–988.0	magnesium oxide (*s*)	MgO	–601.6
barium sulfate (*s*)	$BaSO_4$	–1473.2	magnesium sulfate (*s*)	$MgSO_4$	–1284.9
benzene (*g*)	C_6H_6	82.9	methane (*g*)	CH_4	–74.6
benzene (*l*)	C_6H_6	49.1	methanol (*l*)	CH_3OH	–239.2
calcium carbonate (*s*)	$CaCO_3$	–1207.6	nitric oxide (*g*)	NO	91.3
calcium chloride (*s*)	$CaCl_2$	–795.4	nitrogen dioxide (*g*)	NO_2	33.2
calcium hydroxide (*s*)	$Ca(OH)_2$	–985.2	nitrous oxide (*g*)	N_2O	81.6
calcium oxide (*s*)	CaO	–634.9	ozone (*g*)	O_3	142.7
calcium sulfate (*s*)	$CaSO_4$	–1434.5	potassium chloride (*g*)	KCl	–436.5
carbon monoxide (*g*)	CO	–110.5	potassium hydroxide (*s*)	KOH	–424.6
carbon dioxide (*g*)	CO_2	–393.5	potassium oxide (*s*)	K_2O	–361.5
carbon tetrachloride (*l*)	CCl_4	–128.2	potassium sulfate (*s*)	K_2SO_4	–1437.8
copper(II) chloride (*s*)	$CuCl_2$	–222.1	propane (*g*)	C_3H_8	–103.8
copper(II) nitrate (*s*)	$Cu(NO_3)_2$	–302.9	silicon dioxide (*s*)	SiO_2	–910.7
copper(II) sulfate (*s*)	$CuSO_4$	–771.4	silicon tetrachloride (*l*)	$SiCl_4$	–687.0
ethane (*g*)	C_2H_6	–84.0	silver chloride (*s*)	$AgCl$	–127.0
ethanol (*l*)	C_2H_5OH	–277.6	silver nitrate (*s*)	$AgNO_3$	–124.4
ethylene (*g*)	C_2H_4	52.4	silver sulfide (*s*)	Ag_2S	–32.6
formic acid (*l*)	CH_2O_2	–425.0	sodium chloride (*s*)	$NaCl$	–411.2
glucose (*s*)	$C_6H_{12}O_6$	–1273.3	sodium hydroxide (*s*)	$NaOH$	–425.8
hydrogen bromide (*g*)	HBr	–36.3	sodium nitrate (*s*)	$NaNO_3$	–467.9
hydrogen chloride (*g*)	HCl	–92.3	sodium oxide (*s*)	Na_2O	–414.2
hydrogen fluoride (*g*)	HF	–273.3	sodium sulfate (*s*)	Na_2SO_4	–1387.1
hydrogen iodide (*g*)	HI	26.5	sucrose (*s*)	$C_{12}H_{22}O_{11}$	–2226.1
hydrogen peroxide (*l*)	H_2O_2	–187.8	sulfur dioxide (*g*)	SO_2	–296.8
iron(III) chloride (*s*)	$FeCl_3$	–399.5	sulfur trioxide (*g*)	SO_3	–395.7
iron(II) oxide (*s*)	FeO	–272.0	water (*g*)	H_2O	–241.8
iron(III) oxide (*s*)	Fe_2O_3	–824.2	water (*l*)	H_2O	–285.8
iron(II) sulfate (*s*)	$FeSO_4$	–928.4	zinc nitrate (*s*)	$Zn(NO_3)_2$	–483.7
lead(II) oxide (*s*)	PbO	–217.3	zinc oxide (*s*)	ZnO	–350.5
lead(II) nitrate (*s*)	$Pb(NO_3)_2$	–451.9	zinc sulfate (*s*)	$ZnSO_4$	–982.8

Table B.8. Standard molar entropy values for representative substances at $T = 25°C$ and $P = 1$ bar.

Substance	Formula	$S°$ (J/mol·K)	Substance	Formula	$S°$ (J/mol·K)
acetylene (*g*)	C_2H_2	200.9	lithium sulfate (*s*)	Li_2SO_4	115.1
ammonia (*g*)	NH_3	192.8	magnesium chloride (*s*)	$MgCl_2$	89.6
ammonium chloride (*s*)	NH_4Cl	94.6	magnesium hydroxide (*s*)	$Mg(OH)_2$	63.2
barium chloride (*s*)	$BaCl_2$	123.7	magnesium oxide (*s*)	MgO	27.0
barium nitrate (*s*)	$Ba(NO_3)_2$	214.0	magnesium sulfate (*s*)	$MgSO_4$	91.6
barium sulfate (*s*)	$BaSO_4$	132.2	methane (*g*)	CH_4	186.3
benzene (*g*)	C_6H_6	269.2	methanol (*l*)	CH_3OH	126.8
benzene (*l*)	C_6H_6	173.4	nitric oxide (*g*)	NO	210.8
calcium carbonate (*s*)	$CaCO_3$	91.7	nitrogen (*g*)	N_2	191.6
calcium chloride (*s*)	$CaCl_2$	108.4	nitrogen dioxide (*g*)	NO_2	240.1
calcium hydroxide (*s*)	$Ca(OH)_2$	83.4	nitrous oxide (*g*)	N_2O	220.0
calcium oxide (*s*)	CaO	38.1	oxygen (*g*)	O_2	205.2
calcium sulfate (*s*)	$CaSO_4$	106.5	ozone (*g*)	O_3	238.9
carbon monoxide (*g*)	CO	197.7	potassium chloride (*g*)	KCl	82.6
carbon dioxide (*g*)	CO_2	213.8	potassium hydroxide (*s*)	KOH	81.2
chlorine (*g*)	Cl_2	223.1	potassium sulfate (*s*)	K_2SO_4	175.6
copper(II) chloride (*s*)	$CuCl_2$	108.1	propane (*g*)	C_3H_8	270.3
copper(II) sulfate (*s*)	$CuSO_4$	109.2	silicon dioxide (*s*)	SiO_2	41.5
ethane (*g*)	C_2H_6	229.2	silicon tetrachloride (*l*)	$SiCl_4$	239.7
ethanol (*l*)	C_2H_5OH	160.7	silver chloride (*s*)	AgCl	96.3
ethylene (*g*)	C_2H_4	219.3	silver nitrate (*s*)	$AgNO_3$	140.9
fluorine (*g*)	F_2	202.8	silver sulfide (*s*)	Ag_2S	144.0
hydrogen (*g*)	H_2	130.7	sodium chloride (*s*)	NaCl	72.1
hydrogen bromide (*g*)	HBr	198.7	sodium hydroxide (*s*)	NaOH	64.4
hydrogen chloride (*g*)	HCl	186.9	sodium nitrate (*s*)	$NaNO_3$	116.5
hydrogen fluoride (*g*)	HF	173.8	sodium oxide (*s*)	Na_2O	75.1
hydrogen iodide (*g*)	HI	206.6	sodium sulfate (*s*)	Na_2SO_4	149.6
hydrogen peroxide (*l*)	H_2O_2	109.6	sulfur (*s*)	S	22.6
iron(III) chloride (*s*)	$FeCl_3$	142.3	sulfur dioxide (*g*)	SO_2	248.2
iron(III) oxide (*s*)	Fe_2O_3	87.4	sulfur trioxide (*g*)	SO_3	256.8
iron(II) sulfate (*s*)	$FeSO_4$	107.5	water (*g*)	H_2O	188.8
lead(II) oxide (*s*)	PbO	68.7	water (*l*)	H_2O	70.0
lead(II) sulfate (*s*)	$PbSO_4$	148.5	zinc oxide (*s*)	ZnO	43.7
lithium chloride (*s*)	LiCl	59.3	zinc sulfate (*s*)	$ZnSO_4$	110.5

Table B.9. Standard molar Gibbs free energy of formation values for representative substances at T = 25°C and P = 1 bar.

Substance	Formula	ΔG_f° (kJ/mol)	Substance	Formula	ΔG_f° (kJ/mol)
acetylene (*g*)	C_2H_2	209.9	lithium sulfate (*s*)	Li_2SO_4	-1321.7
ammonia (*g*)	NH_3	-16.4	magnesium chloride (*s*)	$MgCl_2$	-591.8
ammonium chloride (*s*)	NH_4Cl	-202.9	magnesium hydroxide (*s*)	$Mg(OH)_2$	-833.5
barium chloride (*s*)	$BaCl_2$	-806.7	magnesium oxide (*s*)	MgO	-569.3
barium nitrate (*s*)	$Ba(NO_3)_2$	-792.6	magnesium sulfate (*s*)	$MgSO_4$	-1170.6
barium sulfate (*s*)	$BaSO_4$	-1362.2	methane (*g*)	CH_4	-50.5
benzene (*g*)	C_6H_6	129.7	methanol (*l*)	CH_3OH	-166.6
benzene (*l*)	C_6H_6	124.5	nitric oxide (*g*)	NO	87.6
calcium carbonate (*s*)	$CaCO_3$	-1129.1	nitrogen dioxide (*g*)	NO_2	51.3
calcium chloride (*s*)	$CaCl_2$	-748.8	nitrous oxide (*g*)	N_2O	103.7
calcium hydroxide (*s*)	$Ca(OH)_2$	-897.5	ozone (*g*)	O_3	163.2
calcium oxide (*s*)	CaO	-603.3	potassium chloride (*g*)	KCl	-408.5
calcium sulfate (*s*)	$CaSO_4$	-1322.0	potassium hydroxide (*s*)	KOH	-379.4
carbon monoxide (*g*)	CO	-137.2	potassium sulfate (*s*)	K_2SO_4	-1321.4
carbon dioxide (*g*)	CO_2	-394.4	propane (*g*)	C_3H_8	-23.4
copper(II) chloride (*s*)	$CuCl_2$	-175.7	silicon dioxide (*s*)	SiO_2	-856.3
copper(II) sulfate (*s*)	$CuSO_4$	-662.2	silicon tetrachloride (*l*)	$SiCl_4$	-619.8
ethane (*g*)	C_2H_6	-32.0	silver chloride (*s*)	$AgCl$	-109.8
ethanol (*l*)	C_2H_5OH	-174.8	silver nitrate (*s*)	$AgNO_3$	-33.4
ethylene (*g*)	C_2H_4	68.4	silver sulfide (*s*)	Ag_2S	-40.7
hydrogen bromide (*g*)	HBr	-53.4	sodium chloride (*s*)	$NaCl$	-384.1
hydrogen chloride (*g*)	HCl	-95.3	sodium hydroxide (*s*)	$NaOH$	-397.7
hydrogen fluoride (*g*)	HF	-275.4	sodium nitrate (*s*)	$NaNO_3$	-367.0
hydrogen iodide (*g*)	HI	1.7	sodium oxide (*s*)	Na_2O	-375.5
hydrogen peroxide (*l*)	H_2O_2	-120.4	sodium sulfate (*s*)	Na_2SO_4	-1270.2
iron(III) chloride (*s*)	$FeCl_3$	-334.0	sulfur dioxide (*g*)	SO_2	-300.1
iron(III) oxide (*s*)	Fe_2O_3	-742.2	sulfur trioxide (*g*)	SO_3	-374.2
iron(II) sulfate (*s*)	$FeSO_4$	-820.8	water (*g*)	H_2O	-228.6
lead(II) oxide (*s*)	PbO	-187.9	water (*l*)	H_2O	-237.1
lead(II) sulfate (*s*)	$PbSO_4$	-813.0	zinc oxide (*s*)	ZnO	-320.5
lithium chloride (*s*)	$LiCl$	-384.4	zinc sulfate (*s*)	$ZnSO_4$	-871.5

Table B.10. Solubility product constants at 25°C.

Substance	Formula	K_{sp}	Substance	Formula	K_{sp}
barium carbonate	$BaCO_3$	2.58×10^{-9}	mercury(I) carbonate	Hg_2CO_3	3.6×10^{-17}
barium chromate	$BaCrO_4$	1.17×10^{-10}	mercury(I) chloride	Hg_2Cl_2	1.43×10^{-18}
barium fluoride	BaF_2	1.84×10^{-7}	mercury(I) fluoride	Hg_2F_2	3.10×10^{-6}
barium sulfate	$BaSO_4$	1.08×10^{-10}	mercury(I) sulfate	Hg_2SO_4	6.5×10^{-7}
barium sulfite	$BaSO_3$	5.0×10^{-10}	mercury(II) bromide	$HgBr_2$	6.2×10^{-20}
cadmium carbonate	$CdCO_3$	1.0×10^{-12}	mercury(II) iodide	HgI_2	2.9×10^{-29}
cadmium hydroxide	$Cd(OH)_2$	7.2×10^{-15}	nickel(II) carbonate	$NiCO_3$	1.42×10^{-7}
cadmium fluoride	CdF_2	6.44×10^{-3}	nickel(II) hydroxide	$Ni(OH)_2$	5.48×10^{-16}
calcium carbonate	$CaCO_3$	3.36×10^{-9}	nickel(II) phosphate	$Ni_3(PO_4)_2$	4.72×10^{-32}
calcium fluoride	CaF_2	3.45×10^{-11}	silver bromate	$AgBrO_3$	5.38×10^{-5}
calcium hydroxide	$Ca(OH)_2$	5.02×10^{-6}	silver bromide	$AgBr$	5.35×10^{-13}
calcium phosphate	$Ca_3(PO_4)_2$	2.07×10^{-33}	silver carbonate	Ag_2CO_3	8.46×10^{-12}
calcium sulfate	$CaSO_4$	4.93×10^{-5}	silver chloride	$AgCl$	1.77×10^{-10}
cobalt(II) hydroxide	$Co(OH)_2$	5.92×10^{-15}	silver chromate	Ag_2CrO_4	1.12×10^{-12}
cobalt(II) phosphate	$Co_3(PO_4)_2$	2.05×10^{-35}	silver iodide	AgI	8.52×10^{-17}
copper(I) bromide	$CuBr$	6.27×10^{-9}	silver sulfate	Ag_2SO_4	1.20×10^{-5}
copper(I) chloride	$CuCl$	1.72×10^{-7}	silver sulfite	Ag_2SO_3	1.50×10^{-14}
copper(I) iodide	CuI	1.27×10^{-12}	strontium carbonate	$SrCO_3$	5.60×10^{-10}
iron(II) carbonate	$FeCO_3$	3.13×10^{-11}	strontium fluoride	SrF_2	4.33×10^{-9}
iron(II) fluoride	FeF_2	2.36×10^{-6}	strontium sulfate	$SrSO_4$	3.44×10^{-7}
iron(II) hydroxide	$Fe(OH)_2$	4.87×10^{-17}	thallium(I) bromate	$TlBrO_3$	1.10×10^{-4}
lead(II) bromide	$PbBr_2$	6.60×10^{-6}	thallium(I) bromide	$TlBr$	3.71×10^{-6}
lead(II) carbonate	$PbCO_3$	7.40×10^{-14}	thallium(I) chloride	$TlCl$	1.86×10^{-4}
lead(II) chloride	$PbCl_2$	1.70×10^{-5}	thallium(I) chromate	Tl_2CrO_4	8.67×10^{-13}
lead(II) fluoride	PbF_2	3.3×10^{-8}	tin(II) hydroxide	$Sn(OH)_2$	5.45×10^{-27}
lead(II) hydroxide	$Pb(OH)_2$	1.43×10^{-20}	yttrium carbonate	$Y_2(CO_3)_3$	1.03×10^{-31}
lead(II) sulfate	$PbSO_4$	2.53×10^{-8}	yttrium fluoride	YF_3	8.62×10^{-21}
magnesium carbonate	$MgCO_3$	6.82×10^{-6}	yttrium hydroxide	$Y(OH)_3$	1.00×10^{-22}
magnesium fluoride	MgF_2	5.16×10^{-11}	zinc carbonate	$ZnCO_3$	1.46×10^{-10}
magnesium hydroxide	$Mg(OH)_2$	5.61×10^{-12}	zinc fluoride	ZnF_2	3.04×10^{-2}
magnesium phosphate	$Mg_3(PO_4)_2$	1.04×10^{-24}	zinc hydroxide	$Zn(OH)_2$	3×10^{-17}
manganese(II) carbonate	$MnCO_3$	2.24×10^{-11}	zinc selenide	$ZnSe$	3.6×10^{-26}
mercury(I) bromide	$HgBr$	6.40×10^{-23}			

Appendix C

Scientists to Know About

Wolfgang von Goethe once wrote that "the history of science is science." I agree, and this is why knowing about the contributions of significant scientists is an important component in any science class. A number of important scientists make an appearance in this text. Instead of including learning objectives about the scientists on the chapter Objectives lists, the scientists are listed here. This way, teachers can decide what knowledge about the scientists they will require, and students have an outline of names all in one place to help with organizing their studies.

Section	Scientist	Life Dates, Birth Country	Contribution
1.1.1	Max Planck	1858–1947, Germany	First proposed the quantization of energy as a mathematical trick.
1.1.1	Albert Einstein	1879–1955, Germany	First theorized that energy and light are actually, physically quantized.
1.1.3	Johann Balmer	1825–1898, Switzerland	Discovered the formula predicting the lines in the hydrogen atom visible spectrum.
1.1.3	Johannes Rydberg	1854–1919, Sweden	Worked out the general formula for wavelengths in the hydrogen spectrum.
1.1.3	Theodore Lyman	1875–1954, U.S.A.	First observed the ultraviolet series in the hydrogen spectrum.
1.1.3	Friedrich Paschen	1865–1947, Germany	First observed the infrared series in the hydrogen spectrum.
1.2	Niels Bohr	1885–1962, Denmark	Proposed the Bohr model of the atom, which includes electrons in fixed energy levels.
1.3.1	Erwin Schrödinger	1887–1961, Austria	Developed the Schrödinger equation describing particles in quantum systems.
1.3.1	Wolfgang Pauli	1900–1958, Austria	Proposed the Pauli exclusion principle.
1.5.4	Jean Perrin	1870–1942, France	Determined the value of the Avogadro constant.
1.5.4	Amedeo Avogadro	1776–1856, Italy	Proposed that at a given pressure and temperature, the volume of a gas is proportional to the number of particles of the gas. (See also 3.3.1 and 7.1.3.)
2.1.1	Dmitri Mendeleev	1834–1907, Russia	Developed the first fully-formed periodic table.
2.4.5	Linus Pauling	1901–1994, U.S.A.	Developed the electronegativity scale. (See also 4.1.4.)
3.3.1	Stanislao Cannizzaro	1826–1910, Italy	Demonstrated that Amedeo Avogadro's molecular theory is correct.
3.3.1	Amedeo Avogadro	1776–1856, Italy	First introduced the theory that atoms in gases bond together to form molecules. (See also 1.5.4 and 7.1.3.)
3.3.6	Gilbert N. Lewis	1875–1946, U.S.A.	Developed the theory of covalent bonding. (See also 7.2.2 and 9.2.3.)
3.3.8	Michael Faraday	1791–1867, England	First isolated and identified benzene.
3.3.8	August Kekulé	1829–1896, Germany	Discovered the structure of benzene ring. (See also 13.2.3.)

4.1.4	Linus Pauling	1901–1994, U.S.A.	Developed the theory of orbital hybridization. (See also 2.4.5.)
5.1.2	Antoine Lavoisier	1743–1794, France	Discovered the law of conservation of mass in chemical reactions.
5.1.2	Joseph Priestley	1733–1804, England	First isolated oxygen.
6.1.2	James Clerk Maxwell	1831–1879, Scotland	Developed the velocity distribution for particles in a gas.
6.1.2	Ludwig Boltzmann	1844–1906, Austria	Developed the explanatory theory behind the Maxwell-Boltzmann velocity distribution.
6.1.4	Blaise Pascal	1623–1662, France	Developed many of the general principles pertaining to pressure in liquids and gases.
6.1.4	Evangelista Torricelli	1608–1647, Italy	Invented the mercury barometer.
7.1.1	Robert Boyle	1627–1691, Ireland	Discovered Boyle's law.
7.1.2	Jacques Charles	1746–1823, France	Discovered Charles' law.
7.1.2	Joseph Gay-Lussac	1778–1850, France	Published Charles' law and credited it to Jacques Charles.
7.1.3	Amedeo Avogadro	1776–1856, Italy	Proposed that at a given pressure and temperature, equal volumes of gases contain equal numbers of particles. (See also 1.5.4 and 3.3.1.)
7.2.2	Johannes Van der Waals	1837–1923 The Netherlands	Developed the first equations to model the behavior of real gases.
7.2.2	Gilbert N. Lewis	1875–1946, U.S.A.	Statements on the distinction between science and truth claims. (See also 3.3.6 and 9.2.3.)
7.3.1	John Dalton	1766–1844, England	Developed the first scientific atomic model and the law of partial pressures.
7.4.2	Thomas Graham	1805–1869, Scotland	Developed Graham's law of effusion.
8.1.2	Robert Brown	1773–1858, Scotland	First observed Brownian motion.
8.2.1	Svante Arrhenius	1859–1927, Sweden	Published the first theory of ionic dissociation.
8.3.6	Henri Louis Le Châtelier	1850–1936, France	Formulated Le Châtelier's principle.
8.3.6	William Henry	1774–1836, England	Formulated Henry's Law.
9.2.2	Johannes Brønsted	1879–1947, Denmark	Conceptualized acid-base reactions in terms of proton transfer.
9.2.2	Martin Lowry	1874–1936, England	Conceptualized acid-base reactions in terms of proton transfer.
9.2.3	Gilbert N. Lewis	1875–1946, U.S.A.	Proposed the Lewis acid-base theory. (See also 3.3.6 and 7.2.2.)
10.3.3	Josiah Gibbs	1839–1903, U.S.A.	Developed the concept of the free energy in a thermodynamic system.
13.1.3	Kathleen Lonsdale	1903–1971, Ireland	Carbon allotrope lonsdaleite named in honor of her. (See also 13.2.7.)
13.2.3	August Kekulé	1829–1896, Germany	Helped organize the IUPAC. (See also 3.3.8.)
13.2.7	Kathleen Lonsdale	1903–1971, Ireland	Proved the benzene ring is flat. (See also 13.1.3.)

Appendix D

Memory Requirements

We list here the various items students must know from memory for courses using this text. Students must also be mindful of the chapter Objectives Lists, where all the tasks students are responsible for are listed. However, the prefixes, constants, and conversion factors/equations below are listed here for convenience, since most of them are not listed in the Objectives Lists. Values shaded in yellow are not exact and thus must be taken into account when dealing with precision in computations (significant digits).

1. The following metric prefixes (including their symbols and meanings):

Fractions			Multiples		
Prefix	Symbol	Factor	Prefix	Symbol	Factor
centi–	c	$1/10^2$	kilo–	k	10^3
milli–	m	$1/10^3$	mega–	M	10^6
micro–	μ	$1/10^6$	giga–	G	10^9
nano–	n	$1/10^9$	tera–	T	10^{12}
pico–	p	$1/10^{12}$			

2. The following physical constants:

Avogadro constant	$6.02214076 \times 10^{23}$ mol^{-1}	number of particles in one mole of substance
ρ_{water}	998 kg/m^3 = 0.998 g/cm^3	density of water at room temperature
g	9.80 m/s^2	acceleration at sea level due to gravity

3. The following conversion factors and equations:

2.54 cm = 1 in

1000 cm^3 = 1 L

1000 L = 1 m^3

1 mL = 1 cm^3

$T_C = \frac{5}{9}(T_F - 32°)$ $\quad$ $T_F = \frac{9}{5}T_C + 32°$ $\quad$ $T_K = T_C + 273.15$ $\quad$ $T_C = T_K - 273.15$

References and Citations

The following works were consulted in the writing of this text:

Brown, Theodore L., LeMay, H. Eugene, Jr., Bursten, Bruce E., Murphy, Catherine J. *Chemistry: The Central Science*, 11th ed., Pearson, 2009.

Cobb, Cathy and Fetterrolf, Monty L. *The Joy of Chemistry*, Prometheus, 2010.

Coffee, Patrick. *Cathedrals of Science: The Personalities and Rivalries that Made Modern Chemistry*, Oxford, 2008.

CRC Handbook of Physics and Chemistry, 91st ed., CRC Press, 2010.

Kean, Sam. *The Disappearing Spoon*, Back Bay Books, 2010.

King, G. Brooks and Caldwell, William E., *The Fundamentals of College Chemistry*, American Book, 1954.

Le Couteur, Penny and Burreson, Jay. *Napoleon's Buttons: 17 Molecules that Changed History*, Tarcher/Penguin, 2004.

McKeague, Charles, and Turner, Mark, *Trigonometry*, 6th ed., Cengage Learning, 2008.

Modern Chemistry, Holt, Rinehart and Winston, 2005.

Pauling, Linus. *College Chemistry*, W. H. Freeman, 1964.

Pauling, Linus. *The Nature of the Chemical Bond*, Cornell UP, 1960.

Peters, Edward I. *Introduction to Chemical Principles*, Saunders, 1986.

Scerri, Eric R. *The Periodic Table: A Very Short Introduction*, Oxford, 2011.

Stwertka, Albert. *A Guide to the Elements*, Oxford, 2012.

Unless otherwise noted, all tabulated data are from the *CRC Handbook of Physics and Chemistry*, 91st ed.

In addition to the sources above, the excellent science pages at en.wikipedia.com were used as a source of general information and for fact checking.

Specific credit is due for the following:

Section 1.4.1	The metaphor attributed to Wolfgang Pauli is from Kean, *The Disappearing Spoon*, p. 360.
Section 2.1	Atomic mass values for the Periodic Table of the Elements are from Stwertka, *A Guide to the Elements*.
Section 2.1	The arrangement for elements 57, 89, 71, and 103 is based on Scerri, *The Periodic Table: A Very Short Introduction*, p. 134–136.
Section 2.4.2	The effective nuclear charge values for Li, Na, and K are from Brown, et al., *Chemistry: The Central Science*, p. 259.

Section 3.4.2	The color sequence for % ionic character is from Pauling, *The Nature of the Chemical Bond.*
Example 5.10	Some details about the Haber-Bosch process are from Coffee, *Cathedrals of Science.*
Chapter 5 Exercises	Problem 25 is adapted from Holt, *Modern Chemistry.*
Section 7.2.2	The quote by G. N. Lewis is from Coffee, *Cathedrals of Science*, p. 185.
Example 7.8	Adapted from Brown, et al., *Chemistry: The Central Science.*
Chapter 7 Exercises	Problems 25, 26, and 35 are adapted from Brown, et al., *Chemistry: The Central Science.*
Section 8.2.1	The quote, "the wild army of the Ionists" is from Coffee, *Cathedrals of Science*, p. 23.
Section 8.3.2	The solubility guidelines in Table 8.3 are adapted from Brown, et al., *Chemistry: The Central Science.*
Section 8.3	The boxed explanation for how soap works is adapted from Le Couteur and Burreson, *Napoleon's Buttons: 17 Molecules that Changed History*, p. 285–289.
Section 8.6.2	Working temperature data for salting roads is from chemistry.about.com.
Section 9.2.4	Ionizing percentage for HSO_4^- is from Peters, *Introduction to Chemical Principles.*
Example 10.3	Adapted from Holt, *Modern Chemistry.*
Example 10.4	Adapted from Holt, *Modern Chemistry.*
Example 10.6	Data in Table 10.4 are adapted from Brown, et al., *Chemistry: The Central Science.*
Section 11.1.3	Concentration data for HI equilibrium from Peters, *Introduction to Chemical Principles.*
Chapter 11 Exercises	In problem 8, the value of the equilibrium constant was obtained from Holt, *Modern Chemistry.*
Chapter 11 Exercises	Problem 9 is adapted from Brown, et al., *Chemistry: The Central Science.*
Section 12.3.5	The half-reactions for alkaline dry cells are from Brown, et al., *Chemistry: The Central Science.*
Section A.3.2	The definition for significant digits is quoted from Charles McKeague and Mark Turner, *Trigonometry*, 6th ed.

Image Credits

Cover. Nada Orlic, http://nadaorlic.info/ 2. Dhatfield, public domain. 4. Ben Mills, public domain. 5. water: Ben Mills, public domain. ice: Solid State, licensed under CC-BY-SA-3.0. 6. drop on leaf: tanakawho, licensed under CC-BY-SA-2.0. drops on wood: neekoh.fi, licensed under CC-BY-SA-2.0. 8. flame: John D. Mays. photosynthesis: John D. Mays based on At09kg, licensed under CC-BY-SA-3.0. 10. crystal model: Benjah-bmm27, public domain. fluorite: CarlesMilan, licensed under CC-BY-SA-3.0. 11. John D. Mays. 14. John D. Mays. 16. Gringer, public domain. 17. Planck: Public domain. Einstein: Nobel Prize photograph, public domain. 19. atoms: John D. Mays. EAT: Ctankcycles, licensed under CC-BY-SA-3.0. phonons: John D. Mays based on FlorianMarquardt at the English Language Wikipedia, licensed under CC-BY-SA-3.0. 20. John D. Mays 21. spectrum: Gringer, public domain. 22. John D. Mays. 23. Schrodinger: Public domain. Pauli: Bettina Katzenstein/ETH Zürich, licensed under CC-BY-SA-3.0. 26. Dhatfield, public domain. 27. John D. Mays. 28. Lithium_flame_test.JPG via https://commons.wikimedia.org/wiki/File:Lithium_flame_test.JPG. Author: Chemicalinterest, public domain. Copper_flame_test.JPG via https://commons.wikimedia.org/wiki/File:Copper_flame_test.JPG. Author: Chemicalinterest, public domain. 37. Public domain. 40. John D. Mays. 50. top: Dmitri Mendeleev, public domain. middle left: Marco Piazzalunga, licensed under CC-BY-SA-3.0. middle right: Mardeg at en.wikipedia, licensed under CC-BY-SA-3.0. lower left: Bastianow/Lidia/Fourdraine/J. Scholten, licensed under CC-BY-SA-3.0. lower right: Richard Powell (Singinglemon), public domain. 52. Public domain. 53–64. John D. Mays. 66. Pauling: Public domain. other: John D. Mays. 67. John D. Mays. 68. NASA, Hui Yang University of Illinois ODNursery of New Stars, public domain. 72. Dover: Immanuel Giel, licensed under CC-BY-SA-3.0. model: Benjah-bmm27, public domain. 74. Swiss Banker, public domain. 76–77. John D. Mays. 78. photo: Licensed under CC-BY-SA-3.0. model: Benjah-bmm27, public domain. 79. photo: Walkerma, public domain. models: Benjah-bmm27, public domain. 82. photo: Public domain. models: Benjah-bmm27, public domain. 83. John D. Mays. 84. Arpingstone, public domain. 95. Kekulé: Public domain. models: Benjah-bmm27, public domain. 96. John D. Mays. 98. chlorine: Greenhorn1, public domain. bromine: Jurii, licensed under CC-BY-SA-3.0. 99. green models: John D. Mays. water model: Benjah-bmm27, public domain. 101. glass model: John D. Mays based on Jdrewitt, public domain. beta quartz model: Benjah-bmm27, public domain. Crystal photo: Rob Lavinsky, iRocks.com, licensed under CC-BY-SA-3.0. 104: Benjah-bmm27, public domain. 107–114 John D. Mays. 115. model: John D. Mays. tin photo: Alchemist-hp, licensed under CC-BY-SA-3.0. 116. John D. Mays. 117. Jurii, licensed under CC-BY-SA-3.0. 118. water model: John D. Mays. 3-D models: Benjah-bmm27, public domain. 119. shoes: Yarnalgo, licensed under CC-BY-SA-2.0. models: John D. Mays. 120. w:User:Lpm, licensed under C-BY-SA-3.0. 122. Benjah-bmm27, public domain. 124. Tubifex, licensed under CC-BY-SA-3.0. 126. John D. Mays 127. Lavoisier: Public domain. Priestley: Public domain. mercury oxide: Materialscientist, licensed under CC-BY-SA-3.0. 128. Materialscientist, CC-BY-SA-3.0. 129. sulfur sample: Benjah-bmm27, public domain. model: Benjah-bmm27, public domain. burning sulfur: Johannes 'volty' Hemmerlein, licensed under CC-BY-SA-3.0. 130. Laitr Keiows, licensed under CC-BY-SA-3.0. 131. Martin Walker, public domain. 132. Tubifex, licensed under CC-BY-SA-3.0. 135. Tomas er, licensed under CC-BY-SA-3.0. 136. calcium hydroxide: Walkerma, public domain. trees: Nipik, public domain. 137. Toby Hudson, licensed under CC-BY-SA-3.0. 138. John D. Mays. 139. magnesium: John D. Mays. aqua regia: Thejohnnler, public domain. 140. Sculptor and engraver Erik Lindberg, public domain. 141. Tubifex, licensed under CC-BY-SA-3.0. 145. Mondalor, licensed under CC-BY-SA-3.0. 149. Drahkrub, licensed under CC-BY-SA-3.0. 153. Třinecké železárny, free use license. 154. 3-D model: Benjah-bmm27, public domain. 156. NASA/GRC/Martin Brown and Quentin Schwinn, public domain. 159. Maxwell: G. J. Stodart, public domain. Boltzmann: unbekannt, public domain. graph: John D. Mays based on Pdbailey, public domain. 161. Pascal: Anonymous, public domain. pressure diagram: John D. Mays. 162. Torricelli: S. L. Pelaco, public domain. barometer: John D. Mays. 163. John D. Mays. 165–166. John D. Mays. 167. bricks: Hankwang, licensed under CC-BY-SA-3.0. other: John D. Mays. 168. John D. Mays. 169. sushi: Justinc, licensed under CC-BY-SA 2.0. nitrogen: Jurii, licensed under CC-BY-SA 3.0. lightning: Unknown, public domain. aurora: NASA, public domain. 170. John D. Mays. 171. Schnobby, licensed under CC-BY-SA-3.0. 172–176. John D. Mays. 177. Henry Gray, Gray's Anatomy, public domain. 178. John D. Mays. 180. John D. Mays. 182. Alchemist-hp, licensed under CC-BY-SA-3.0. 184. Boyle: Johann Kerseboom, public domain. other: John D. Mays. 185. balloon: Antoine Sergent dit Sergent-Marceau (?), public domain. Gay-Lussac:

Francois Seraphin Delpech, public domain. 186–187. John D. Mays. 188. C. Sentier, public domain. 190. Unknown, public domain. 193. John D. Mays. 199. Thomas Phillips, public domain. 201, 205, 211. John D. Mays. 214. Zephyris, licensed under CC-BY-SA-3.0. 216. John D. Mays. 217. Brown: Henry William Pickersgill, public domain. blue glass: Public domain. red glass: Schtone, licensed under CC-BY-SA-3.0. 218. solutions: John D. Mays. Arrhenius: Public domain. 219–224. John D. Mays. 225. acid model: Benjah-bmm27, public domain. other: John D. Mays. 227. ethanol, methanol, acetone: Benjah-bmm27, public domain. isopropyl: Jynto, public domain. 228. top: Jynto, public domain. middle: Benjah-bmm27, public domain. bottom: Matt18224, public domain. 229. micelle: John D. Mays, based on Smokefoot, public domain. other: John D. Mays. 230. Alfred Löhr, public domain. 231–234. John D. Mays. 237. Der Kreole, licensed under CC-BY-SA-3.0. 239–240. John D. Mays. 241. Jacqueline Arneson, used by permission. 242. Bengt Oberger, licensed under CC-BY-SA-3.0. 246, 251. John D. Mays. 252. left: Riley Huntley (Huntley Photography), licensed under CC-BY-SA-3.0. right: Joanne Bergenwall Aw, licensed under CC-BY-SA-3.0. 254. John D. Mays. 255. Supermartl, altered by Haltopub, licensed under CC-BY-SA-3.0. 258. Benjah-bmm27, public domain. 259. John D. Mays. 260. Brønsted: Peter Elfelt, public domain. Lowry: Public domain. 261. Photographersnature, licensed under CC-BY-SA-3.0. 263, 265. John D. Mays. 267. Benjah-bmm27, public domain. 271. John D. Mays. 273. Martina Steiner, public domain. 274. Chemicalinterest, public domain. 275–279. John D. Mays. 286. Taavi Adamberg, licensed under CC-BY-SA-3.0. 288. NASA, public domain. 289–293. John D. Mays. 295. Harbor1, licensed under CC-BY-SA-3.0. 296. Benjah-bmm27, public domain. 299–304. John D. Mays. 305. Public domain. 309. John D. Mays. 310. Petty Officer 3rd Class Benjamin Crossley, U.S. Navy, public domain. 311. John D. Mays. 312. John D. Mays based on public domain. 313–319. John D. Mays. 322. PRHaney, licensed under CC-BY-SA-3.0. 324–327. John D. Mays. 333. John D. Mays, based on Francis E. Williams, licensed under CC-BY-SA-3.0. 334–340. John D. Mays. 341. Rogeriopfm, licensed under CC-BY-SA-3.0. 342, 343. John D. Mays. 347. top: Walkerma, public domain. bottom: CarlesMilan, licensed under CC-BY-SA-3.0. 354. Alfred T. Palmer, public domain. 356. Apollo: NASA, public domain. verdigris: Beyond My Ken, licensed under CC-BY-SA-3.0. batteries: Mathieu BOIS, licensed under CC-BY-SA-3.0. 357–365. John D. Mays. 369. Jynto, public domain. 373–377. John D. Mays. 378. Sabisteb, licensed under CC-BY-SA-1.0. 379–382. John D. Mays. 383. top: John D. Mays. bottom: John D. Mays based on Tympanus, public domain. 384. Top: Sunny.solanki, licensed under CC-BY-SA-3.0. bottom: Ronald Saunders, licensed under CC-BY-SA-2.0. 385. top: Swoolverton, licensed under CC-BY-SA-3.0. middle: FiveRings, public domain. bottom: Polyparadigm, public domain. 386. top: Rémi Kaupp, licensed under CC-BY-SA-3.0. middle: Zwergelstern, licensed under CC-BY-SA-3.0. bottom: Knotnic, public domain. 387. John D. Mays. 389. Schuyler S., licensed under CC-BY-SA-2.5. 392. Dwight Burdette, licensed under CC-BY-SA-3.0. 394–395. Benjah-bmm27, public domain. 396. top: Michael Ströck, licensed under CC-BY-SA-3.0. bottom: Cédric THÉVENET, licensed under CC-BY-SA-3.0. 398. methane Ball-and-stick: Benjah-bmm27, public domain. methane space-filling: Dbc334 and Jynto, public domain. ethane and propane models: Benjah-bmm27, public domain. butane ball-and-stick: Ben Mills and Jynto, public domain. butane space-filling: Benjah-bmm27, public domain. 399. Ben Mills and Jynto, public domain. 400. top: Benjah-bmm27, public domain. middle: NASA, public domain. bottom: NASA, public domain. 408. Benjah-bmm27, public domain. 409. top: Luigi Chiesa, licensed under CC-BY-SA-3.0. bottom: John D. Mays, based on H Padleckas, licensed under CC-BY-SA-2.5. 412. 3-D models: Benjah-bmm27, public domain. 413. photo: Benjah-bmm27, public domain. 416. isopropyl: Jynto, public domain. other 3-D models: Benjah-bmm27, public domain. 417. 3-D models: Benjah-bmm27, public domain. 418. formic and acetic: Benjah-bmm27, public domain. benzoic: Ben Mills and Jynto, public domain. 419. 3-D model: Ben Mills and Jynto, public domain. 420. top: Jynto, public domain. glycerol: Benjah-bmm27, public domain. fatty acid: Matt18224, public domain. 421. 3-D model: Benjah-bmm27, public domain. 423. cellulose: Benjah-bmm27, public domain. nylon: YassineMrabet, created with PyMOL, licensed under CC-BY-SA-3.0. 424. PVC parts: Nerjip, licensed under CC-BY-SA-3.0. 3-D models: Benjah-bmm27, public domain. polystyrene foam: Phrexian, licensed under CC-BY-SA-3.0. yogurt cup: Hispalois, licensed under CC-BY-SA-3.0. razor: Hispalois, licensed under CC-BY-SA-3.0. 425. red pipe: Arnulf zu Linden, licensed under CC-BY-SA-3.0. 3-D models: Benjah-bmm27, public domain. frying pan: Andrevan, licensed under CC-BY-SA-2.5. burets: John D. Mays. PTFE tape: Dr. Reiner Düren aka (RedPiranha), licensed under CC-BY-SA-3.0. 426. recycle code: Tomia, licensed under CC-BY-SA-3.0. cup: Licensed under CC-BY-SA-3.0. 467. John D. Mays. 468. Unknown, public domain. 469–480. John D. Mays.

Index

26 **Fe** 1.83 +2 , +3	atomic number chemical symbol electronegativity common oxidation states

	1	2	3	4	5	6	7	8	9	10	11	12	13	14	15	16	17	18
1	1 **H** 2.20 +1																	2 **He**
2	3 **Li** 0.98 +1	4 **Be** 1.57 +2											5 **B** 2.04 +3	6 **C** 2.55 +4, −4	7 **N** 3.04 −3	8 **O** 3.44 −2	9 **F** 3.98 −1	10 **Ne**
3	11 **Na** 0.93 +1	12 **Mg** 1.31 +2											13 **Al** 1.61 +3	14 **Si** 1.90 +4	15 **P** 2.19 −3	16 **S** 2.58 −2	17 **Cl** 3.16 −1	18 **Ar**
4	19 **K** 0.82 +1	20 **Ca** 1.00 +2	21 **Sc** 1.36 +3	22 **Ti** 1.54 +4	23 **V** 1.63 +4, +5	24 **Cr** 1.66 +3. +6	25 **Mn** 1.55 +2, +4	26 **Fe** 1.83 +2 , +3	27 **Co** 1.88 +2, +3	28 **Ni** 1.91 +2	29 **Cu** 1.90 +1, +2	30 **Zn** 1.65 +2	31 **Ga** 1.81 +3	32 **Ge** 2.01 +2, +4	33 **As** 2.18 +3, +5	34 **Se** 2.55 −2	35 **Br** 2.96 −1	36 **Kr**
5	37 **Rb** 0.82 +1	38 **Sr** 0.95 +2	39 **Y** 1.22 +3	40 **Zr** 1.33 +4	41 **Nb** 1.6 +5	42 **Mo** 2.16 +4, +6	43 **Tc** 2.10 +4, +7	44 **Ru** 2.2 +3, +4	45 **Rh** 2.28 +3, +4	46 **Pd** 2.20 +2, +4	47 **Ag** 1.93 +1	48 **Cd** 1.69 +2	49 **In** 1.78 +3	50 **Sn** 1.96 +2, +4	51 **Sb** 2.05 +3, +5	52 **Te** 2.1 −2	53 **I** 2.66 −1	54 **Xe**
6	55 **Cs** 0.79 +1	56 **Ba** 0.89 +2	71 **Lu** 1.0	72 **Hf** 1.3 +4	73 **Ta** 1.5 +5	74 **W** 1.7 +6	75 **Re** 1.9 +6, +7	76 **Os** 2.2 +4	77 **Ir** 2.2 +3, +4	78 **Pt** 2.2 +2, +4	79 **Au** 2.4 +1, +3	80 **Hg** 1.9 +2, +4	81 **Tl** 1.8 +1	82 **Pb** 1.8 +2, +4	83 **Bi** 1.9 +3, +5	84 **Po** 2.0 +4	85 **At** 2.2	86 **Rn**
7	87 **Fr** 0.7 +1	88 **Ra** 0.9 +2	103 **Lr**	104 **Rf**	105 **Db**	106 **Sg**	107 **Bh**	108 **Hs**	109 **Mt**	110 **Ds**	111 **Rg**	112 **Cn**	113 **Nh**	114 **Fl**	115 **Mc**	116 **Lv**	117 **Ts**	118 **Og**

liquid at room temperature

radioactive

Period	1	2	3	4	5	6	7	8	9	10	11	12	13	14	15	16	17	18
1	1 **H** Hydrogen 1.0079																	2 **He** Helium 4.0026
2	3 **Li** Lithium 6.941	4 **Be** Beryllium 9.0122											5 **B** Boron 10.811	6 **C** Carbon 12.011	7 **N** Nitrogen 14.0067	8 **O** Oxygen 15.9994	9 **F** Fluorine 18.9984	10 **Ne** Neon 20.1797
3	11 **Na** Sodium 22.9898	12 **Mg** Magnesium 24.3050											13 **Al** Aluminum 26.9815	14 **Si** Silicon 28.0855	15 **P** Phosphorus 30.9738	16 **S** Sulfur 32.066	17 **Cl** Chlorine 35.4527	18 **Ar** Argon 39.948
4	19 **K** Potassium 39.098	20 **Ca** Calcium 40.078	21 **Sc** Scandium 44.9559	22 **Ti** Titanium 47.88	23 **V** Vanadium 50.9415	24 **Cr** Chromium 51.9961	25 **Mn** Manganese 54.9380	26 **Fe** Iron 55.847	27 **Co** Cobalt 58.9332	28 **Ni** Nickel 58.6934	29 **Cu** Copper 63.546	30 **Zn** Zinc 65.39	31 **Ga** Gallium 69.723	32 **Ge** Germanium 72.61	33 **As** Arsenic 74.9216	34 **Se** Selenium 78.96	35 **Br** Bromine 79.904	36 **Kr** Krypton 83.80
5	37 **Rb** Rubidium 85.468	38 **Sr** Strontium 87.62	39 **Y** Yttrium 88.9059	40 **Zr** Zirconium 91.224	41 **Nb** Niobium 92.9064	42 **Mo** Molybdenum 95.94	43 **Tc** Technetium 98.9072	44 **Ru** Ruthenium 101.07	45 **Rh** Rhodium 102.9055	46 **Pd** Palladium 106.42	47 **Ag** Silver 107.8682	48 **Cd** Cadmium 112.411	49 **In** Indium 114.82	50 **Sn** Tin 118.710	51 **Sb** Antimony 121.76	52 **Te** Tellurium 127.60	53 **I** Iodine 126.9045	54 **Xe** Xenon 131.29
6	55 **Cs** Cesium 132.905	56 **Ba** Barium 137.327	71 **Lu** Lutetium 174.967	72 **Hf** Hafnium 178.49	73 **Ta** Tantalum 180.9479	74 **W** Tungsten 183.85	75 **Re** Rhenium 186.207	76 **Os** Osmium 190.2	77 **Ir** Iridium 192.22	78 **Pt** Platinum 195.08	79 **Au** Gold 196.9665	80 **Hg** Mercury 200.59	81 **Tl** Thallium 204.3833	82 **Pb** Lead 207.2	83 **Bi** Bismuth 208.9804	84 **Po** Polonium 208.9824	85 **At** Astatine 209.9871	86 **Rn** Radon 222.0176
7	87 **Fr** Francium 223.0197	88 **Ra** Radium 226.0254	103 **Lr** Lawrencium 262.11	104 **Rf** Rutherfordium 261.11	105 **Db** Dubnium 262.114	106 **Sg** Seaborgium 263.118	107 **Bh** Bohrium 262.12	108 **Hs** Hassium (265)	109 **Mt** Meitnerium (266)	110 **Ds** Darmstadtium (281)	111 **Rg** Roentgenium (281)	112 **Cn** Copernicium (285)	113 **Nh** Nihonium (284)	114 **Fl** Flerovium (289)	115 **Mc** Moscovium (288)	116 **Lv** Livermorium (293)	117 **Ts** Tennessine (294)	118 **Og** Oganesson (294)

57 **La** Lanthanum 138.9055	58 **Ce** Cerium 140.115	59 **Pr** Praseodymium 140.9077	60 **Nd** Neodymium 144.24	61 **Pm** Promethium 144.9127	62 **Sm** Samarium 150.36	63 **Eu** Europium 151.965	64 **Gd** Gadolinium 157.25	65 **Tb** Terbium 158.9253	66 **Dy** Dysprosium 162.50	67 **Ho** Holmium 164.9303	68 **Er** Erbium 167.26	69 **Tm** Thulium 168.9342	70 **Yb** Ytterbium 173.04
89 **Ac** Actinium 227.0278	90 **Th** Thorium 232.0381	91 **Pa** Protactinium 231.0359	92 **U** Uranium 238.0289	93 **Np** Neptunium 237.0482	94 **Pu** Plutonium 244.0642	95 **Am** Americium 243.0614	96 **Cm** Curium 247.0703	97 **Bk** Berkelium 247.0703	98 **Cf** Californium 251.0796	99 **Es** Einsteinium 252.083	100 **Fm** Fermium 257.0951	101 **Md** Mendelevium 258.10	102 **No** Nobelium 259.1009